VOLUME FORTY SEVEN

Annual Reports in MEDICINAL CHEMISTRY

VOLUME FORTY SEVEN

ANNUAL REPORTS IN MEDICINAL CHEMISTRY

Sponsored by the Division of Medicinal Chemistry of the American Chemical Society

Editor-in-Chief

MANOJ C. DESAI

Gilead Sciences, Inc.
Foster City, CA, USA

Section Editors

ROBICHAUD • STAMFORD • WEINSTEIN • McALPINE • PRIMEAU • LOWE • BERNSTEIN • BRONSON

ELSEVIER

AMSTERDAM • BOSTON • HEIDELBERG • LONDON
NEW YORK • OXFORD • PARIS • SAN DIEGO
SAN FRANCISCO • SINGAPORE • SYDNEY • TOKYO
Academic Press is an imprint of Elsevier

Academic Press is an imprint of Elsevier
525 B Street, Suite 1900, San Diego, CA 92101-4495, USA
225 Wyman Street, Waltham, MA 02451, USA
The Boulevard, Langford Lane, Kidlington, Oxford, OX51GB, UK
32, Jamestown Road, London NW1 7BY, UK
Radarweg 29, PO Box 211, 1000 AE Amsterdam, The Netherlands

First edition 2012

ISBN: 978-0-12-396492-2
ISSN: 0065-7743

For information on all Academic Press publications
visit our website at www.store.elsevier.com

Printed and bound in USA
12 13 14 15 11 10 9 8 7 6 5 4 3 2 1

CONTENTS

Part 5
Infectious Diseases

Editor: John Primeau, Westford, Massachusetts

CONTRIBUTORS

PREFACE

The *Annual Report in Medicinal Chemistry* is dedicated to furthering legitimate interest in learning, chronicling, and sharing information on the discovery of compounds and the methods that lead to new therapeutic advances. *ARMC*'s tradition is to provide disease-based reviews and highlight emerging technologies of interest to medicinal chemists.

A distinguishing feature of *ARMC* is its knowledgeable section editors in the field who evaluate invited reviews for scientific rigor. This volume contains 31 chapters in eight sections. The first five sections have a therapeutic focus. Their topics include CNS diseases (edited by Albert J. Robichaud); cardiovascular and metabolic diseases (edited by Andrew Stamford); inflammatory, pulmonary, and gastrointestinal diseases (edited by David S. Weinstein); oncology (edited by Shelli R. McAlpine); and infectious diseases (edited by John Primeau). Sections VI and VII review important topics in biology (edited by John A. Lowe) and new technologies for drug optimization (edited by Peter R. Bernstein). The last section deals with drugs approved by the FDA in the previous year (edited by JoAnne Bronson).

While committed to maintaining the *ARMC* tradition, the current volume experiments with new ways to keep it relevant to medicinal chemists. The format for Volume 47 follows previous issues but has two new features. Added for the first time are personal essays, written by MEDI hall of famers, which express the personal stories and scientific careers of "drug hunters." The volume opens with essays by Paul Anderson, Peter Bernstein, and Ross Johnson. Additionally, we have included for the first time a chapter in the last section listing new chemical entities that have entered Phase III in 2011.

This would not have been possible without our panel of section editors, to whom I am indeed very grateful. I would like to thank the authors of this volume for their hard work, patience, dedication, and scholarship in the lengthy process of writing, editing, and making last-minute edits and revisions to their contributions. Further, I extend my sincere thanks to the following reviewers who have provided independent edits to the chapters: Myra Beaudoin Bertrand, George Chang, Bert Chenard, Michael Clarke, Andrew Combs, Chris Cox, Kevin Currie, Mike DeNinno, Robert Dow, Carolyn Dzierba, Gary Flynn, Randall Halcomb, David Kimball, Michael Kort, Lori Krim-Gavrin, Katerina Leftheris, Mary Mader, James R. Merritt, Ryan Moslin, Gregory Notte, Chris O'Donnell, Anandan

Palani, Paul Renhowe, Ralph Robinson, Paul Roethle, David Rotella, Greg Roth, Joachim Rudolf, David Sperandio, Vicky Steadman, James Taylor, Chandrasekar Venkataramani, and Will Watkins.

Finally, I would like to thank John E. Macor, who was the editor-in-chief for Volumes 42 through 46, for his outstanding stewardship. I am immensely grateful to John in the transition to this volume and for handing over the reins on a solid path.

I hope the material provided in this volume will serve as a precious resource on important aspects of medicinal chemistry and, in this way, maintain the tradition of excellence that *ARMC* has brought to us for more than four decades. I am excited about our new initiatives; I look forward to hearing from you about them and welcome your suggestions for future content.

Manoj C. Desai, Ph.D.
Gilead Sciences, Inc.
Foster City, CA

Personal Essays

CHAPTER ONE

Reflections on Medicinal Chemistry at Merck, West Point

Paul S. Anderson
Lansdale, PA, USA

Much has changed since I arrived at Merck's West Point laboratories in 1964 as an organic chemist interested in learning how to do drug discovery. It quickly became apparent that this endeavor would require a prolonged period of study and learning. Later, I realized that it would, in fact, require life-long learning. The learning experience began immediately as others taught me the process by which therapeutic targets were selected and pursued at that time. Medicinal chemistry was make a one compound at a time endeavor that mainly relied on data from experiments in whole animals for SAR information. Safety pharmacology and drug metabolism studies were reserved mostly for development candidates. Despite these limitations, the pharmaceutical industry had been remarkably successful in the discovery of new medicines. One such example was the thiazide diuretics.

The West Point laboratories had successfully discovered and developed the thiazide diuretic chlorothiazide **1**.[1] At the time, whole animal pharmacology was central to this major accomplishment. The biological assays in place tended to measure pharmacological activity without necessarily defining the mechanism by which the activity was achieved. It was clear, however, that physiological understanding of kidney function had played a critical role in the course of events that led to the discovery of the thiazide diuretics. The discovery road began with an earlier observation that the antibacterial agent sulfanilamide produced alkaline diuresis in patients. This effect was tracked to inhibition of carbonic anhydrase in the kidney. Carbonic anhydrase inhibitors induced excretion of sodium bicarbonate by the kidney. Inhibition of sodium reabsorption by exchange of sodium for hydrogen ion in the distal part of the kidney therefore was thought to account for the alkaline diuresis induced by these enzyme inhibitors. Karl Beyer set forth the idea that a drug that worked in the proximal portion of the kidney would be better tolerated by excreting sodium chloride instead of sodium bicarbonate.[2] His efforts to find such a molecule in collaboration with medicinal chemists James Sprague and Frederick Novello led to the

Annual Reports in Medicinal Chemistry, Volume 47
ISSN 0065-7743
http://dx.doi.org/10.1016/B978-0-12-396492-2.00001-1

discovery of dichlorfenamide **2**, a potent carbonic anhydrase inhibitor with increased chloride secretion relative to acetazolamide. Further SAR work revealed that addition of an amino group to the aromatic ring of bis-sulfonamides decreased carbonic anhydrase inhibition without loss of chloride secretion. A subsequent molecule in which the amino group and a sulphonamido group were incorporated into a ring became known as chlorothiazide **1**, a potent diuretic that did not inhibit carbonic anhydrase. It was the first of many thiazide diuretics that found broad clinical use in treating edema and hypertension.

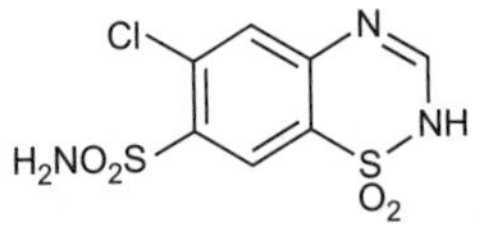

1

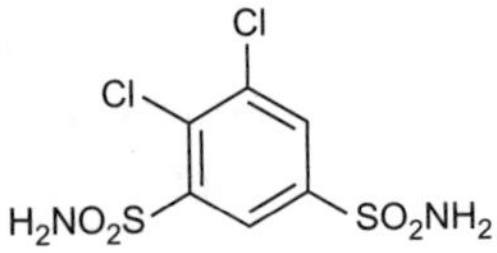

2

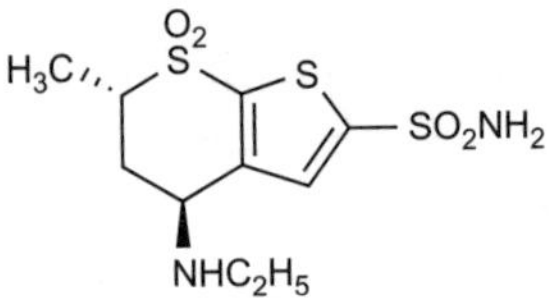

3

Interestingly, carbonic anhydrase reappeared as a target of interest at Merck in the 1980s as Robert Smith and his colleagues sought to design an inhibitor for topical use in the eye to treat glaucoma. Although many carbonic anhydrase inhibitors were known, those previously approved for clinical use as oral medications failed to lower intraocular pressure on topical administration. We believed that a topically administered drug would avoid the side effects associated with the use of the oral medications. Eventually, we were successful in finding such an inhibitor. Success required gaining knowledge about ocular transport issues from *in vitro* models and finding the right animal model for testing these insights *in vivo*. It also required the identification of a potent carbonic anhydrase inhibitor that was soluble in water and had good transport properties as well as a long residence time in the ciliary process tissue where the enzyme resides in the eye. At West Point, Jack Baldwin accomplished this difficult task with the discovery of dorzolamide **3**, the first topically effective carbonic anhydrase inhibitor for the treatment of glaucoma.[3] In the course of this work, we had learned that the challenges of getting acceptable physical, metabolic, and transport properties in a drug candidate could be even greater than those associated with optimization for potency and selectivity. One of the ancillary benefits of this program was an early opportunity to explore the use of X-ray crystal structures of an inhibitor bound to an enzyme in drug discovery.[4,5] This early experience in the manipulation of physical properties and the use of protein–ligand crystal structures in drug design would pay a dividend in later work on the design of inhibitors for HIV protease.

In the 1970s, Merck had initiated a program to expand access to new compounds being synthesized in academic chemistry departments by offering to screen these materials for biological activity under a collaborative agreement. One compound that proved to be interesting was 5,6,7,8-tetrafluoro-1,4-dihydronaphthalene-1,4-imine, which unexpectedly was found to have benzodiazepine-like behavioral activity as well as potent antiseizure activity in rodents without the usual sedative side effects of this well-known class of compounds. Pursuit of this interesting observation led to a more drug-like molecule dizocilpine **4**, better known as MK-801.[6] Studies focused on the mechanism of action of MK-801 used (^{3}H)MK-801 to identify specific high-affinity binding sites in rat brain membranes that proved to be associated with NMDA receptors. Neurophysiological studies were then used to establish that MK-801 was a noncompetitive antagonist of the NMDA receptor.[7] While MK-801 was not developed as a drug, it has served as a valuable, commonly used tool in neuropharmacology. The MK-801 work taught us a great deal about the importance of mechanism of drug action studies in drug discovery work.

While the MK-801 work was in progress, Dan Veber's group at West Point was pursuing several other opportunities for drug discovery based on hypotheses that antagonists of certain peptide hormones might be useful therapeutic agents. The peptide hormone CCK was selected as the initial target based on the current knowledge of its physiology and pharmacology. CCK-A receptors were believed to mediate pancreatic and biliary secretion as well as gut motility, while CCK-B receptors were implicated in gastric acid secretion and panic–anxiety attacks. The plan was to screen fermentation broths for molecules with affinity for CCK receptors using a radio receptor assay in the hope of finding nonpeptidal compounds that would prove to be antagonists. The screening effort identified asperlicin as a modestly potent nonpeptidal CCK antagonist. Subsequent fragmentation and reassembly analysis of the natural product sought to identify and optimize the pharmacophore responsible for this activity. As the project proceeded, Ben Evans and Mark Bock were successful in designing **5** and **6** that were orally bioavailable, potent, and selective antagonists for CCK-A and -B receptors, respectively.[8,9] While these molecules demonstrated the expected pharmacological consequences of selective receptor blockage, neither became a drug because of side effect and efficacy issues. A similar approach to the study of the peptide hormone oxytocin also gave rise to new nonpeptidal antagonists that were not subsequently developed as drugs. It became apparent that there are many targets for which a preclinical hypothesis for a mechanism of drug action can be developed, but few of these hypotheses actually translate to new medicines for a variety of reasons. One that did achieve the desired objective was the West Point effort to discover and develop an antagonist for the αIIbβ3 integrin receptor, which recognizes the RGD sequence present in many integrins. The Merck effort led by Ruth Nutt and George Hartman produced both peptide and nonpeptide antagonists. The nonpeptide antagonist was developed as tirofiban **7**,[10] which was approved by the FDA in 1998 for treatment of unstable angina. These experiences with MK-801 and the search for peptide antagonists as new medicines taught us that target selection is a critical component of the drug discovery process that requires careful analysis before committing to a course of action. A clear understanding of what will be required to translate the preclinical hypothesis into a successful clinical outcome needs to be a key part of this analysis.

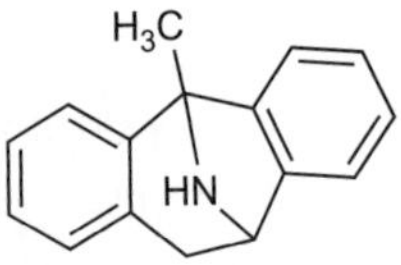

4

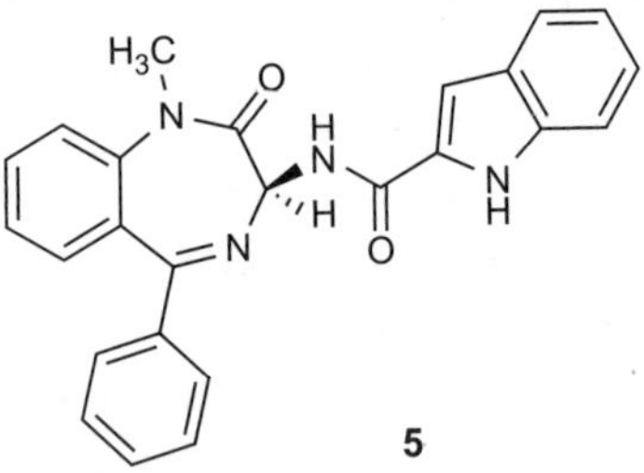

5

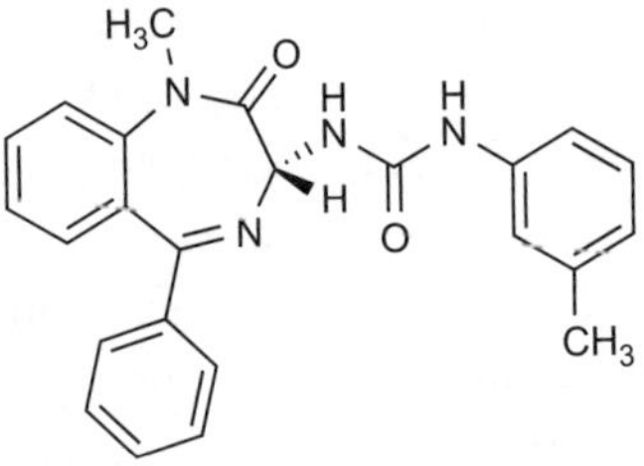

6

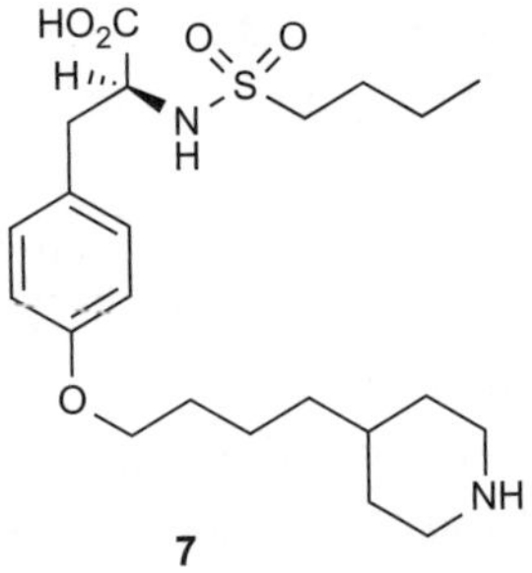

7

Discovery of drugs to manage HIV infection was a major goal of the pharmaceutical industry during the closing years of the twentieth century. Once human immunodeficiency virus was identified as the etiological agent of AIDS, understanding of molecular events in viral replication revealed the virus-specific enzymes HIV protease, reverse transcriptase, and integrase to be excellent targets for therapeutic intervention. Both established and early stage pharmaceutical companies joined the effort to find drugs that worked by these mechanisms. At Merck, our entry into the fray was spearheaded by the synthesis of HIV protease as an initial source of the enzyme for inhibitor assay development and protein crystallographic studies.[11] Recognition of HIV protease as an aspartic acid protease provided insight for inhibitor design gleaned from earlier study on other aspartic acid proteases such as renin. The Merck sample collection contained a number of peptides and peptide-like molecules that had been designed to be renin inhibitors. These molecules all had a core secondary alcohol component believed to be a transition-state mimic for the cleavage site in substrates. Several of these renin inhibitors were found to also be HIV protease inhibitors. While they provided guidance for the SAR optimization work, they lacked potency in cells and were not orally bioavailable. X-ray crystal structures of HIV protease with and without early inhibitors bound to it facilitated the design process, as did molecular modeling. These structures confirmed the importance of proper positioning of the secondary alcohol in the enzyme active site and revealed other interesting interactions between inhibitors and the flap region of the active site. With these insights in hand, the march toward molecules with high potency in cell culture was fairly rapid, however, finding molecules with acceptable ADMET properties proved to be challenging. Very few molecules satisfied the ADMET criteria for development as an orally active drug. Eventually, the West Point team led by Joel Huff and Joe Vacca overcame these challenges with the discovery of indinavir **8**.[12]

Nonnucleoside reverse transcriptase inhibitors also have become important in the treatment of HIV infection. Screening of small molecule sample collections for leads to this class of compounds was an important step toward their discovery. Several companies including Johnson & Johnson, Boehringer-Ingelheim, Merck, and Pharmacia Upjohn obtained leads from their sample collections that evolved into clinical candidates. The Merck lead evolved into the widely used drug efavirenz **9**, which was discovered by Steve Young.[13] The Boehringer-Ingelheim lead progressed to nevirapine and the Pharmacia Upjohn compound became delavirdine. These inhibitors bind to a flexible, allosteric site in the enzyme, which may account for the structural diversity permitted by this target. Crystal structures of inhibitors bound to this site were subsequently obtained. Although this information did not contribute to the early work on nonnucleoside inhibitors, it has enhanced our understanding of this flexible binding site.

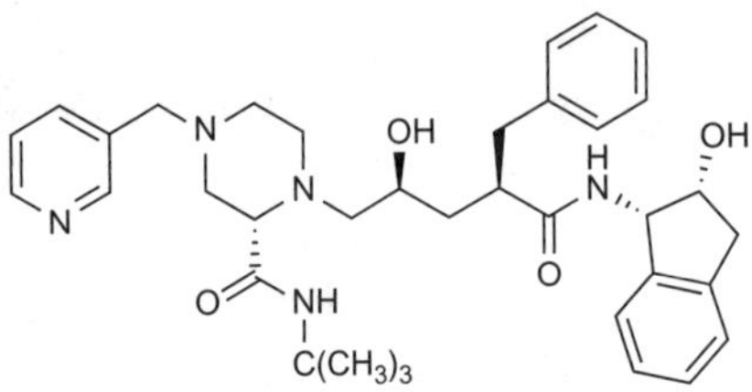

8

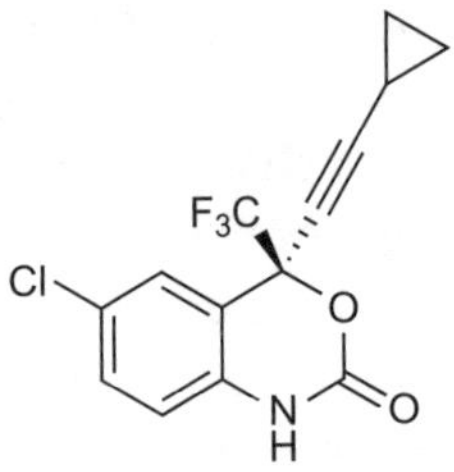

9

One of the most important contributions to medicine in the second half of the twentieth century was the development of the statins. The conversion of 3-hydroxy-3-methylglutaryl coenzyme A (HMG-CoA) to mevalonic acid by the enzyme HMG-CoA reductase is the rate-limiting step in cholesterol biosynthesis. Thus this enzyme was an obvious target for the discovery of drugs with utility in controlling cholesterol in the blood. Subsequent understanding of the regulatory role that HMG-CoA reductase plays in LDL receptor expression further enhanced the desire to find inhibitors with drug-like properties. An enzyme assay suitable for use in screening fermentation broths for inhibitory activity was developed and used to discover natural product inhibitors such as compactin and lovastatin **10**. Merck developed and launched lovastatin as a product for controlling blood cholesterol level. This successful product served as a template for further work to improve its drug-like properties. It was noted that metabolic hydrolysis of the ester in lovastatin gave rise to a metabolite with much less enzyme inhibitory activity. Increasing bulk by adding a methyl group alpha to the ester carbonyl in lovastatin enhanced metabolic stability and more than doubled *in vivo* potency. The new molecule called simvastatin **11** became a very successful treatment for hypercholesterolemia.[14] Both of these inhibitors are prodrugs because the active species is the beta hydroxy acid formed on metabolic hydrolysis of the lactone. It was assumed that this part of the inhibitor resembled an intermediate in the reaction pathway of HMG-CoA reductase and therefore was key to how the enzyme recognizes and binds to this class of inhibitor. This assumption suggested that it might be possible to replace the stereochemically complex decalin part of simvastatin with a less complex hydrocarbon structure. Exploration of this hypothesis at Merck led to the discovery of a family of biphenyl replacements for the decalin substructure such as **12**.[15] A common feature of these and latter HMG-CoA reductase inhibitors was the *p*-fluorophenyl entity thought at the time to be a replacement for the hydrophobic ester present in simvastatin. Subsequent X-ray crystal structures of enzyme-bound inhibitors validated this concept.[16]

The new biphenyl inhibitors were very potent but did not appear to have additional preclinical advantages that merited clinical development. However, other companies pursued these structures to arrive at fluvastatin and atorvastatin using a strategy that often is referred to as variation of a known active compound. Over time, this has been one of the most productive approaches to the discovery of new medicines, as one can readily see by reviewing the discovery of beta blockers, histamine antagonists, ACE inhibitors, and many other majors classes of drugs.

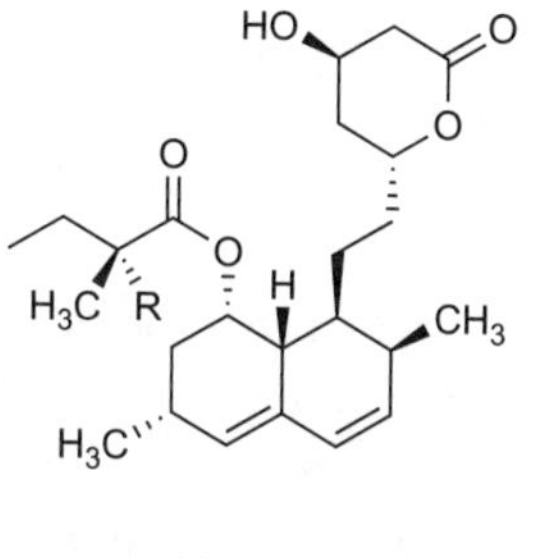

10 R = H
11 R = CH_3

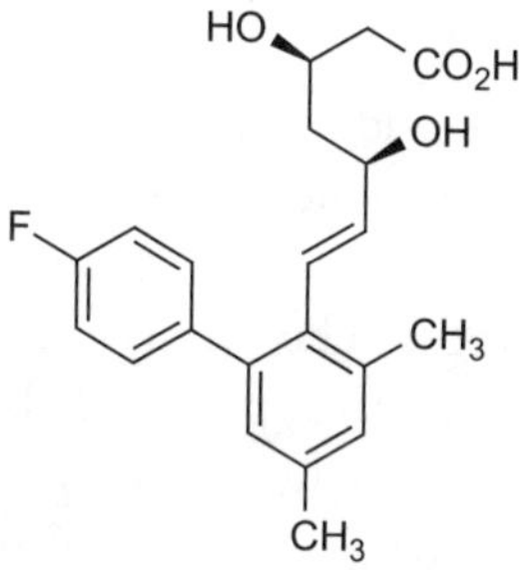

12

The above advances in health care offered useful teaching about the nature of drug discovery. Studying the physiology of organ function can put one on a productive track for drug discovery, as was the case with the thiazide diuretics. Understanding mechanism of drug action also has great value, as was the case with the integrin antagonists, statins, and HIV infection treatments. It was clear that the hurdles for drug discovery are high. Discovering a molecule that is selective for the desired mechanism of drug action and has an acceptable ADMET profile for the intended route of administration is a formidable challenge even with the tools that are available today. In most projects, very few molecules satisfy all of the necessary criteria.

New tools such as genomics, metabolomics, bioinformatics, high-throughput screening and chemistry, fragment-based drug design, and structure-based drug design have been incorporated into the drug discovery process without fundamentally changing the paradigm or increasing productivity. This can be put into perspective by realizing that technology supports drug discovery, but people discover drugs. Process makes you functional, but process does not make you creative. Creativity is still a key factor in the transformation of a drug discovery concept into a new medicine. To this end, target selection is key, because most new target opportunities begin with a preclinical hypothesis that lacks formal clinical validation. History tells us that most of these hypotheses fail to achieve clinical validation. However, it is likely that some of the new tools may eventually reduce failure rate by helping drug discovery scientists to pick the best targets for the discovery of new medicines. To this end, Sir David Jack[17] has reminded us that successful organizations engage all of their people in the selection of the best ideas for drug discovery with recognition that these are likely to be the ones that are "simple, practicable with available resources and novel enough to yield medicines that are likely to be better than probable competitors in ways that will be obvious to both doctors and their patients." In a related statement, George Merck addressed the challenge of balancing business interests and the interest of patients by saying in a 1950 speech at the Medical College of Virginia, "We try to remember that medicine is for the patient. . .it is not for the profits. The profits follow, and if we have remembered that, they have never failed to appear. The better we have remembered it, the larger they have been." It would be good for us to remember this as the drug discovery enterprise moves forward in the twenty-first century. The pharmaceutical industry has undergone changes in recent years such that building shareholder value has come to dominate creation of value for patients and doctors. Better balance is needed between business interests and the interests of patients and their physicians in order to have a productive industry that can better serve the healthcare needs of society in the twenty-first century. A return to better balance will enable creative medicinal chemists to continue to discover important new medicines.

REFERENCES

(1) Novello, F.C.; Sprague, J.M. *J. Am. Chem. Soc.* **1957**, *79*, 2028.
(2) Beyer, K.H. *Trends Pharmacol. Sci.* **1980**, *1*, 114.
(3) Sugrue, M.F.; Harris, A.; Adamsons, I. *Drugs Today* **1997**, *33*, 283.
(4) Baldwin, J.J.; Ponticello, G.S.; Anderson, P.S.; Christy, M.E.; Murcko, M.A.; Randall, W.C.; Schwam, H.; Sugrue, M.F.; Springer, J.P.; Gautheron, P.; Grove, J.; Mallorga, P.; Viader, M.P.; McKeever, B.M.; Navia, M.A. *J. Med. Chem.* **1989**, *32*, 2510.
(5) Smith, G.M.; Alexander, R.S.; Christianson, D.V.; McKeever, B.M.; Ponticello, G.S.; Springer, J.P.; Randall, W.C.; Baldwin, J.J.; Habecker, C.N. *Protein Sci.* **1994**, *3*, 118.
(6) Thompson, W.J.; Anderson, P.S.; Britcher, S.F.; Lyle, T.A.; Thies, J.E.; Magill, C.A.; Varga, S.L.; Schwering, J.E.Z.; Lyle, P.A.; Christy, M.E.; Evans, B.E.Z.; Colton, C.D.; Holloway, M.K.; Springer, J.P.; Hirshfield, J.M.; Ball, R.G.; Amato, J.S.; Larsen, R.D.; Wong, E.H.F.; Kemp, J.A.; Tricklebank, M.D.; Singh, L.; Oles, R.; Priestly, T.; Marshall, G.M.; Knight, A.R.; Middlemiss, D.N.; Woodruff, G.N.; Iversen, L.L. *J. Med. Chem.* **1990**, *33*, 789.

(7) Wong, E.H.F.; Kemp, J.A.; Priestley, T.; Knight, A.R.; Woodruff, G.N.; Iversen, L.L. *Proc. Natl. Acad. Sci. U.S.A.* **1986**, *83*, 7104.
(8) Evans, B.E.; Bock, M.G.; Rittle, K.E.; DiPardo, R.M.; Whitter, W.L.; Veber, D.F.; Anderson, P.S.; Freidinger, R.M. *Proc. Natl. Acad. Sci. U.S.A.* **1986**, *83*, 4918.
(9) Bock, M.G.; DiParto, R.M.; Evans, B.E.; Rittle, K.E.; Whitter, W.L.; Veber, D.F.; Anderson, P.S.; Freidinger, R.M. *J. Med. Chem.* **1989**, *32*, 13.
(10) Lynch, J.F.; Cook, J.J.; Sitko, G.R.; Holahan, M.A.; Ramjit, D.R.; Mellott, M.J.; Stranieri, M.T.; Stabilito, I.I.; Zhang, G.; lynch, R.J.; Manno, P.D.; Cjang, C.T.-C.; Egbertson, M.S.; Halczenko, W.; Duggan, M.E.; Laswell, W.L.; Vassallo, L.M.; Shafer, J.A.; Anderson, P.S.; Friedman, P.A.; Hartman, G.D.; Gould, R.J. *JPET* **1995**, *272*, 20.
(11) Nutt, R.F.; Brady, S.F.; Darke, P.L.; Ciccarone, T.M.; Colton, C.D.; Nutt, E.M.; Rodkey, J.A.; Bennett, C.D.; Waxman, L.H.; Sigal, I.S.; Anderson, P.S.; Veber, D.F. *Proc. Natl. Acad. Sci. U.S.A.* **1988**, *85*, 7129.
(12) Vacca, J.P.; Dorsey, B.D.; Schleif, W.A.; Levin, R.B.; McDaniel, S.L.; Darke, P.L.; Zugay, J.; Quintero, J.C.; Blahy, O.M.; Roth, E.; Sardana, V.V.; Schlabach, A.J.; Graham, P.I.; Condra, J.H.; Gotlib, L.; Holloway, M.K.; Lin, J.; Chen, I.-W.; Vastag, K.; Ostovic, D.; Anderson, P.S.; Emini, E.A.; Huff, J.R. *Proc. Natl. Acad. Sci. U.S.A.* **1994**, *91*, 4096.
(13) Young, S.D.; Britcher, S.F.; Tran, L.O.; Payne, L.S.; Lumma, W.C.; Lyle, T.A.; Huff, J.R.; Anderson, P.S.; Olsen, D.B.; Carrol, S.S.; Pettibone, D.J.; O'Brien, J.A.; Ball, R.G.; Balani, S.K.; Lin, J.A.; Chen, I.-W.; Schleif, W.A.; Sardana, V.V.; Long, W.J.; Barnes, V.W. *Antimicrob. Agents Chemother.* **1995**, *39*, 2602.
(14) Hoffman, W.F.; Alberts, A.W.; Anderson, P.S.; Chen, J.S.; Smith, R.L.; Willard, A.K. *J. Med. Chem.* **1986**, *29*, 849.
(15) Stokker, G.E.; Alberts, A.W.; Anderson, P.S.; Cragoe, E.J.; Deana, A.A.; Gilfillan, J.L.; Hirshfield, J.; Holtz, W.J.; Hoffman, W.F.; Huff, J.W.; Lee, T.J.; Novello, F.C.; Prugh, J.D.; Rooney, C.S.; Smith, R.L.; Willard, A.K. *J. Med. Chem.* **1986**, *29*, 170.
(16) Istvan, E.S.; Deisenhofer, J. *Science* **2001**, *292*, 1160.
(17) Jack, D. *Creating the Right Environment for Drug Discovery*. Quay Publishing: Lancaster, **1991**.

CHAPTER TWO

My Path in Seeking New Medicines

Peter R. Bernstein
PharmaB LLC, Rose Valley, PA, United States

Contents

I was born in New York City, the middle of Moses and Helen Bernstein's three sons. My father was a pharmacist, my mother a bookkeeper. Our parents were very supportive of us, asking only that we do our very best in whatever field we chose.

From an early age, science fascinated me; I got my first Gilbert chemistry set when I was still very young. In seventh grade, I learned that teachers were not infallible. My science teacher was introducing us to the concept of radioactive decay and half-lives. He drew a bar graph showing exactly half the original radioactive material remaining at the end of each half-life period, but it also seemed to indicate that, until that period expired, 100% of the original material was still there. When I raised my hand to ask whether the decay, in fact, occurred constantly over time, we ended up in an argument, followed quickly by a note home to my parents. Their advice was simple: "Yes you are right, but you shouldn't embarrass the teacher by correcting him!"

A key moment for me was during my sophomore year at the University of Rochester when I took Organic Chemistry, taught by Professor Jack Kampmeier. His passion for organic chemistry was infectious, and I caught the bug. The following summer, I won an NSF undergraduate research participation grant and chose to work with Professor Richard Schlessinger. Working with him was hard as he was very demanding. I soon learned from the local NSF program Director that he was not happy with the number of hours I was working. After investigating, the Director learned I was being productive and working about 65 h a week. He suggested that I try to work out an understanding with Professor Schlessinger, and if that did not pan out,

Annual Reports in Medicinal Chemistry, Volume 47
ISSN 0065-7743
http://dx.doi.org/10.1016/B978-0-12-396492-2.00002-3

he offered me a place in his lab or help in finding another spot. I succeeded with Professor Schlessinger and by the end of the summer, I was obsessed with synthesis. Over the next 2 years, I took many graduate chemistry courses, spent a summer working at Kodak in their color synthesis group, and did my senior research in the labs of Professor Andrew Kende.

By then, I knew my goal was a Ph.D. in Organic Synthesis and that I wanted to prepare biologically active molecules as potential medications. Somewhere along the line, I had read a copy of *Burger's Medicinal Chemistry* (in those days, only a single volume), and this had narrowed my focus. I was accepted at several graduate schools and, lured by the West Coast, visited both Stanford and Berkeley in the spring of my senior year. But I found that discussions with several distinguished professors, including W.S. Johnson, often ended with commentary like, "With your set of interests and having been accepted to Columbia, why aren't you going to work with Professor Stork?" Though I no longer relished living in New York City, I accepted that my immediate future might lay on the East Coast after all. Columbia it was.

My time there rapidly flew by, and working in the labs of Professor Gilbert Stork turned out to be a dream. There were many brilliant scientists in his group who had an impact on me—Paul Wender, William Greenlee, and Minoru Isobe, to name just a few. Columbia was an open environment, with students encouraged to discuss their science broadly across the chemistry department. I was impressed by this culture of openness, which I have tried to emulate ever since. Upon graduating, I did a postdoctoral fellowship with Professor Barry Trost at the University of Wisconsin. I felt that working with Professor Trost, who was then focused on developing new synthetic methodology, would help balance my training, which had thus far centered on synthetic route design and delivery of complex molecules. Little did I know that my time with Professor Trost, which resulted in a formal total synthesis of Vitamin D, would be so influenced by my prior experience.[1]

At this point, industry came calling, and I accepted an offer from ICI Pharmaceuticals in Wilmington, Delaware. It was a great opportunity in the U.S. pharmaceuticals division of one of the world's biggest chemical conglomerates. My hope was that working at ICI in a smaller group would provide some of the flexibility and entrepreneurship of a start-up or small company, with the stability of a major firm.

My first assignment was to a selective antimuscarinic program targeting asthma. Within about 6 months, we disproved some of the project's foundations, and I was asked to propose a new antiasthma target. This was a challenging assignment that required me to read the literature broadly. I offered two options. The first was blocking the leukotrienes (LT), either with an antagonist or with a biosynthesis inhibitor. The second involved antagonism of platelet-activating factor. Management chose LT antagonists and asked me to start before we actually had biologists on the team. My first task was to develop an in-house synthesis that would allow us to generate LTs in sufficient quantity and quality to use as standard agonists. As I worked on this, the management team used the promise of such a supply to lure key bioscientists to join us.

The synthesis was exceedingly difficult, time-consuming, inspiring, frustrating, and rewarding. It would have not been successful without the efforts of my colleagues Jim Maloney, Ed Vacek, and Tom Maduskuie. At times, our findings were groundbreaking, and this was a bit heady for someone so new to drug discovery. To achieve our synthetic goal, we needed to introduce prep-HPLC and spend lots of time on the department's new high-field ^{1}H NMR. Working on the leukotrienes led to interactions with leading pulmonary clinicians as we established various collaborations to study the effects of our synthetic leukotrienes in humans. After writing the chemistry portion of a Drug Master File that was submitted in support of those studies, I also had direct interaction with reviewing chemists at the FDA.

Three years into my new job, I realized that, although I was working hard, sometimes till 11:00 p.m., my expectations of recognition in a large organization were not being met. I responded by ramping up my professional activities outside work, including becoming involved with the local section of the ACS—chairing symposia, becoming an alternate councilor and joining committees. I also decided to add a hobby. After trying several, I settled on whitewater kayaking. Kayaking taught me that one has to make decisions even with imperfect data. Planning, acting, and having a back-up plan are critical when one is about to enter a set of rapids or go over a waterfall. Worse than making a "wrong" decision is to delay making one, only to be carried along out of control.

Back on the leukotriene project, I was joined in my efforts by Alvin Willard, Ying Yee, and Fred Brown. Together, we made analogs of the leukotrienes and of FPL-55712,[2] the only known antagonist. Hybridization of our output led us to a novel indole series that we focused upon as a shared effort. This was a key decision, underpinning the eventual success of our program, which we made under the guidance of Barrie Hesp. Our team did not have the experience to understand how valuable a lead the indoles were, although we all knew that there were many competitors working on analogs of the LTs and FPL-55712. Victor Matassa joined our team to replace Alvin Willard and, shortly after that, ICI-204219 was submitted as our development candidate, zafirlukast.[3]

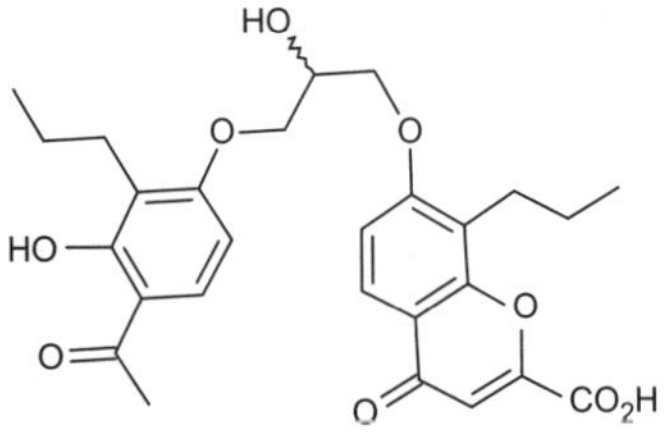

FPL-55712

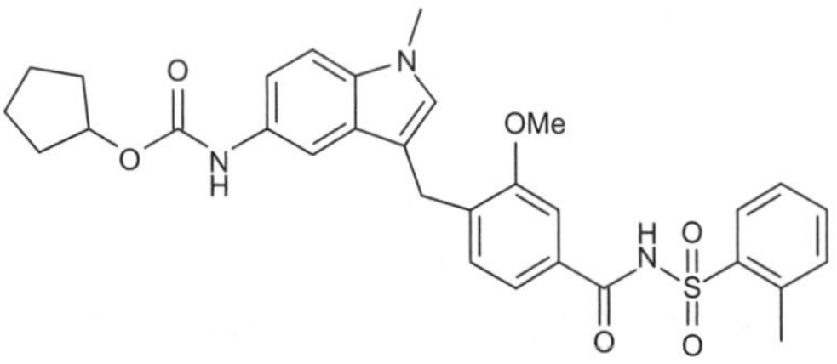

ICI-20419 [zafirlukast]

It was about this time that another major change occurred in my life when I fell in love with Ala Hamilton-Day, a complex and beautiful trial attorney. We married after a whirlwind courtship and soon had two children, Sarah and Matthew. At work, I remained on the leukotriene project, and, in addition to supporting the development of zafirlukast, I also strove to deliver a back-up. The goal was driven by our desire for a compound that did not have an aryl amine substructure. We succeeded with ZD3523,[4] which was out-licensed to Mitsubishi when, ultimately, we did not need it as a replacement for zafirlukast. It had been so long since pharmaceuticals had out-licensed a compound that the finance department was unsure which group should be credited with the licensing fees, a fact I learned on receiving a letter from the head of R&D thanking me for my role in delivering an unexpected windfall to his budget. This was Zeneca R&D because ICI had just "demerged" and the pharmaceuticals group had gone to the newly created Zeneca Pharmaceuticals.

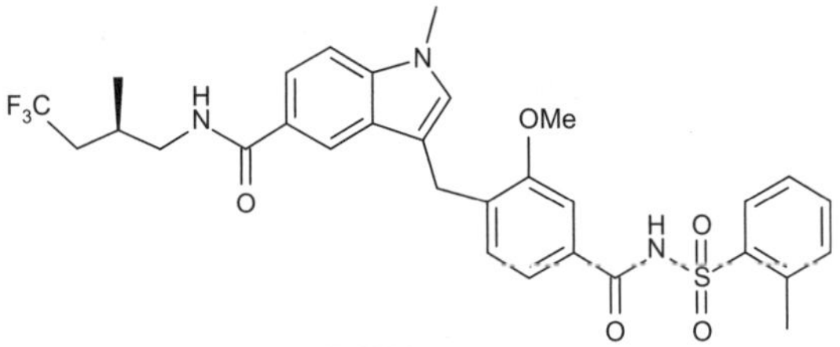

ZD-3523

With the closure of the discovery effort on leukotrienes, I chose to move to the human leukocyte elastase (HLE) inhibition project. Although it was a very different type of biological target, a circulating enzyme rather than a membrane-bound G protein-coupled receptor, the project itself remained on familiar territory, focusing as it did on chronic obstructive pulmonary disease. While this was not asthma, it was another pulmonary disease, and I felt working on it would allow me to continue to develop my knowledge of respiratory diseases. The impetus behind my decision was that I had come to believe that one of the keys to successfully driving a small molecule drug discovery project forward was having at least one chemist on the team who understood enough about the bioassays to adequately discuss them with the lead bioscientists. Many times over my career, we have considered assays for a project that seem great in isolation but, upon detailed inspection, fall short of our needs. For example, the team might want to use qSAR to drive prediction of future results, but the test output is too removed from the quantitative interaction (e.g., binding constant or enzyme inhibition value) to be used for this purpose. Or perhaps the precision of the assay after it was "optimally" configured (to give higher output/lower cost) was not high enough to understand the pharmacology of the drug candidate.

In the HLE inhibitor project, Peter Warner had just made the breakthrough discovery of a nonpeptide backbone for such inhibitors[5] and I joined it when he returned to England. Our goal was to increase potency and deliver an easy-to-synthesize analog that was orally efficacious. During that effort, we learned important lessons about needing to understand the physical properties of drug candidates: solubility, lipophilicity, and non-specific protein binding that previously had been examined only in retrospect after compounds had become fairly advanced. Our chemistry department created its own section devoted to generating this data.

While working on this project, I had a lesson of my own in effective communication. I learned that a computational scientist was frustrated by my refusal to honor his request to prepare a phosphinamide derivative that he predicted would be an excellent compound. While researching routes to comply with his request, however, I had learned that such aryl phosphinamides were inherently unstable to disproportionation and could not be isolated. Unfortunately, he interpreted my report that I *could not* make the derivative as I *would not* make it, a fact I discovered only later. Had I offered a fuller explanation of the roadblock I encountered, I could have avoided the ensuing confusion.

In the end, in spite of significant progress on the nonpeptidic series,[6] the two development candidates that resulted from these efforts, ZD0892 and ZD8321, were both peptidic.[7] My gamble to stay in respiratory diseases had paid off. At this point in my career, I had produced many papers, presentations, and reviews and had helped deliver development candidates opposite two targets. Internally, I received a promotion to ICI's scientific ladder; externally, I was invited to meetings of the *Collegium Internationale Allergologicum*, to join the *American Thoracic Society* and to participate in discussions defining similarities and differences between asthma and rhinitis.

My next project targeted Neurokinin antagonists for treating asthma and was well underway when I came on board. The timing of my return to an asthma project was auspicious because zafirlukast had been approved a short time before and was facing some postlaunch challenges. On a personal level, I also learned the joy that you feel when people reap the benefits of a drug you have helped develop. A kayaking friend who had asthma went on zafirlukast and sent me a personal note on how it had changed her life. She had been doing an exercise routine that respected the limits her condition placed on her. If she started her aerobic exercise without taking a β-agonist, she would need to stop after a certain time because she was out of breath. Alternately, if she took the β-agonist, she would have to stop after a different amount of time because her heart was pounding. However, after starting zafirlukast, she discovered that she did not know when to stop. Her routine had actually been disrupted because she was no longer limited by shortness of breath or a pounding heart.

On the neurokinin project, the team had already delivered ZD7944, a selective NK2 antagonist, when I joined it. Project pharmacologists had just discovered that simultaneous administration of NK2 and NK1 antagonists resulted in synergistic increases in efficacy in several disease models. Therefore, I was asked to explore the development of dual NK1/NK2 antagonists.

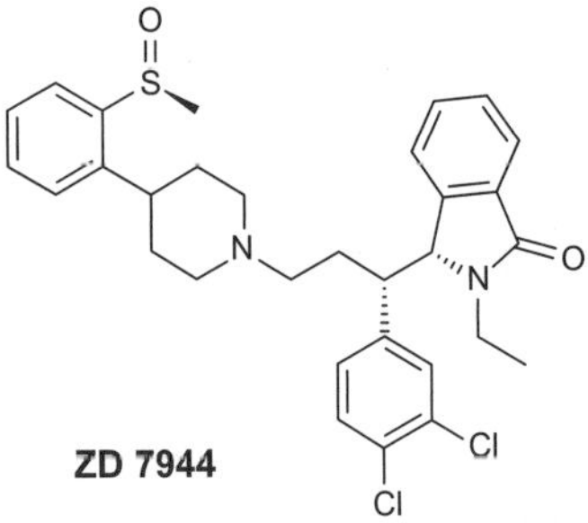

ZD 7944

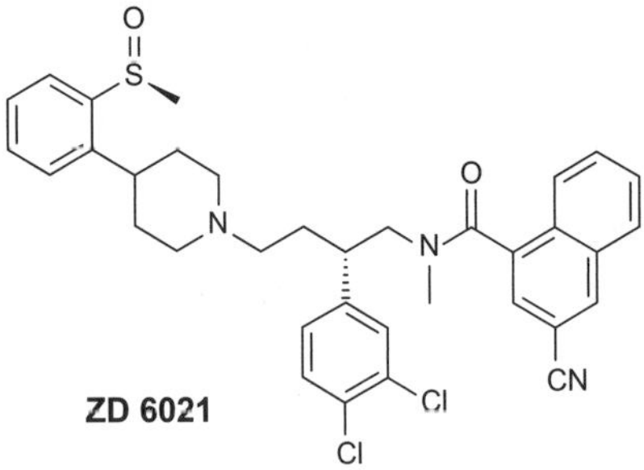

ZD 6021

My approach was to use the newly developed technology of robotic-assisted synthesis to rapidly explore the SAR space around ZD7944 and see if we could find lead-like dual agents related to this selective NK2 antagonist. Our robotic technology group was located in the United Kingdom, and I collaborated with them to produce hundreds of analogs for us on the >100 mg/compound scale. The U.K. team, however, did not then have automated purification and analytical instrumentation that could handle this sort of output. Luckily, we did in the United States, and by combining the strengths of both teams, we delivered several libraries. Most compounds were >95% pure; a few were lower, but none with <85% purity was submitted. In a couple of iterations, we learned that not only were dual agents possible, but we had found a compound that almost fit our criteria for a development candidate. (It fell short, however, as it was a nitro naphthalene derivative that we were sure would be Ames positive.) Upon considering replacements for NO_2 that would yield the desired profile, we decided to make the corresponding cyano-compound.

My experiences at this time reinforced my respect for the power of new technology both to deliver and to corrupt. It delivered, as we would never have progressed so far and fast without the robotic-aided synthesis. It also corrupted, since some of my colleagues could not accept that my associate and I had gone from making so many compounds to so few. We were now under "quota," and that was bad. That we had good theoretical reasons to target one difficult-to-make compound only worked in our favor after it was made and profiled well enough to be a development candidate, ZD6021.[8] How difficult efficient synthesis of the cyano-naphthoic acid proved is best exemplified by a series of papers from my process chemistry colleagues.[9]

We knew ZD6021 had two weaknesses: relatively rapid metabolism and moderate permeability. Working closely with a DMPK team led by Karin Kirkland, we chose to determine where metabolism was occurring and then prepare analogs that would specifically block that site. I believe that this was the first time in Wilmington that a medicinal chemist and a DMPK scientist had collaborated in such a way. Together, we learned that oxidative metabolism was occurring in both the naphthalene and piperidine regions of the molecule. Adding blocking groups to the aryl piperidine substituent led to improvements in clearance and afforded a dual-acting backup ZD2249.[10] Concurrently, to confirm the site of metabolism in the naphthalene, we prepared two isomeric hydroxy-substituted analogs and chose to make them via the intermediate methoxy analogs. Although the methoxy compounds had been made as intermediates, we submitted them for testing and made a surprising discovery. The 2-methoxy substituent not only decreased clearance and increased permeability but also had a dramatic impact on receptor selectivity, converting our dual agent to a highly selective, brain-permeable NK1 antagonist.

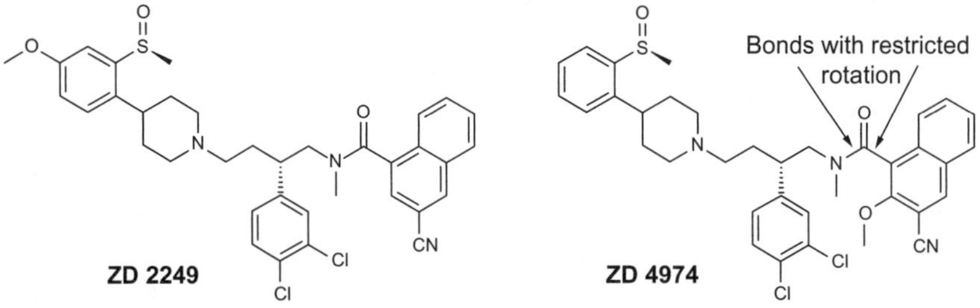

The timing of this discovery was fortuitous, since it occurred about when Astra and Zeneca merged and at a time when Pfizer and Merck had reported results with NK1 antagonists as potential central nervous system (CNS) drugs. A result of the AstraZeneca merger was that respiratory disease research stopped in Wilmington, with future research limited to CNS and pain. Our NK1-selective and improved compound was nominated as ZD4974 for development as an antidepressant.[11] Our experience with it led to our explaining NK1/NK2 selectivity via the existence of specific orientations of the aryl group and the amide linker and to discussions on the impact of atropisomers in drug development.[12]

In some ways, this was the most frustrating set of projects I had worked on because the NK1, NK1/NK2, and NK2 teams nominated six compounds for development, and none of them made it to humans. All were stopped for toxicological reasons that varied between compounds. Our analysis of their structures and the toxicological data showed that the only feature all of these compounds had in common was a 3,4-dichlorophenyl-substituted methine.[13] Looking more broadly within the AstraZeneca and MDDR databases, we realized that this feature was very common in advanced candidates but at that time was found in only one marketed drug. Candidates opposite many disease targets had failed to deliver for a variety of reasons. This was our first inkling that substructures might exist that led to potent receptor interactions such that they were found in a broad array of candidates but that they were generally problematic, with rare exceptions.

Leaving neurokinins, my team and I worked for short periods on β-estrogen antagonists, γ-secretase inhibitors, NMDA antagonists, and H_3-antagonists before we settled into 5-HT$_{1B}$ antagonists for another significant effort. The5-HT$_{1B}$ team had already nominated two compounds for clinical development, ARA-2 and AZD1134. These had been stopped due to toxicology findings, and our goal was discovering compounds with improved profiles. I joined the project just before Bob Jacobs, the project chemistry leader, left AstraZeneca. This was fortunate for me because I ended up taking over a very well-thought-out and advanced chemistry program. The goal was to find a compound with reduced phospholipogenic potential that also had improved DMPK (permeability and clearance) and physical (solubility) properties. Using an *in vitro* PLD screen, newly developed pharmacophore models and computer-based predictive tools to guide us, we delivered AZD3783.[14] This compound had an improved profile and progressed into clinical development. We also worked to deliver a backup to AZD3783. Unfortunately, long-term toxicology studies on AZD3783 led to the termination of its development and also caused us to reject any backup that had even a theoretical possibility of inducing phospholipidosis. By this point, I was the project leader and my chemistry team, led by David Nugiel, succeeded in preparing a proof-of-concept compound **1**.[15] This compound was the first "nonbasic" 5-HT_{1B} antagonist and, therefore, had no potential to be phospholipogenic. Regrettably, AstraZeneca was backing away from such targets, and the project closed without the opportunity to deliver a compound suitable for development.

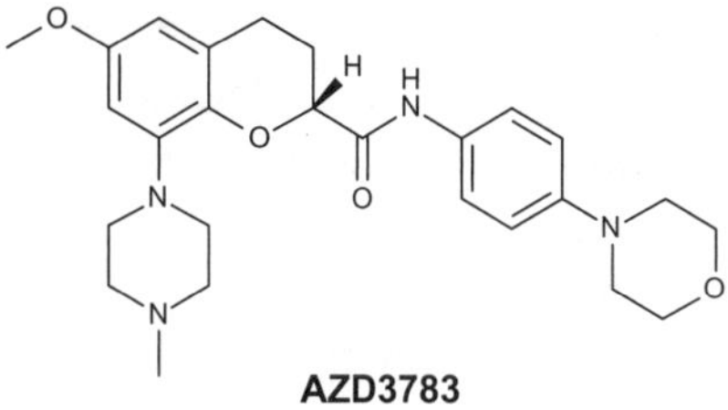

AZD3783

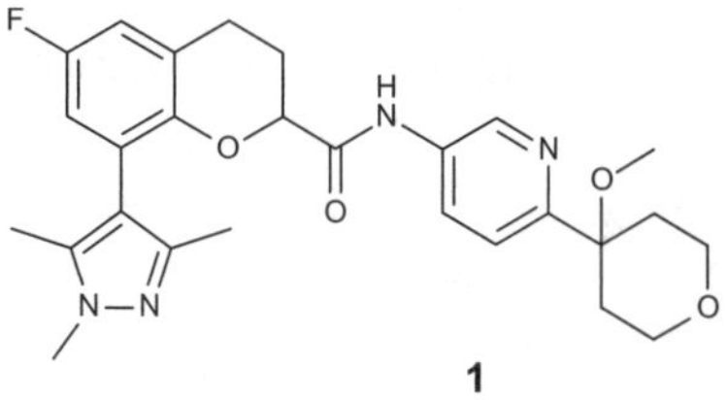

1

My final AstraZeneca project was developing a dual NET/DAT inhibitor as a follow-up to nomifensine, an effective antidepressant that had been pulled off the market due to idiosyncratic toxicology. Like both the 5-HT_{1B} and the H_3 programs, an increasingly important component of this drug discovery project was based on information technology. Computational chemistry played critical roles in compound design, information tracking, and data analysis validation. The last is easily overlooked due to the desire to set specific biochemical parameters as criteria for succession to candidate. The theory is that rigid criteria make it "easier" to decide whether a compound should advance. Unfortunately, the assays are not always accurate enough to give significant answers to the specific questions. In this case, the key to progress was recognizing that the assays, as originally configured, did not provide a precise enough number for the ratio of NET and DAT activities to deliver a compound meeting the candidate drug target profile. Only after modification were they adequate for project progression. Although significant progress was made opposite this target, my site and all psychiatric research were terminated before we could deliver a clinical candidate.

After retiring, I established Pha*R*maB LLC as a platform from which to remain professionally active—as a scientific and/or editorial advisory board member, lecturer, editor, consultant, symposium organizer, and author—while also seeking to find a new balance point in life.

REFLECTIONS

I have had an extremely fortunate career as a medicinal chemist. I joined the field just as an explosion of scientific knowledge was opening up vast opportunities for drug discovery. This led to what I consider a Golden Age of drug discovery in the late 1980s and early 1990s, during which success via utilization of new technology seemed assured. Unfortunately, all good things come to an end, and I also experienced the crashing of the old Pharma model. In an attempt to meet unrealistic financial expectations, many companies focused on potential "blockbusters," merged, cut back operations, and engaged experts from outside the R&D environment to increase R&D efficiency.

As this happened, some companies forgot that key scientists are not fungible and misapplied otherwise effective tools for increasing efficiency. For example, many processes in discovery (e.g., running an assay or reporting results) may be greatly improved through the implementation of efficiency analyses; however, invention requires going outside the known. Because of this, and as reported in a *Business Week* article on 3M, some ways to improve the production process (e.g., lean-six-sigma analyses) can squelch the innovation required to invent something new.[16]

More recently, I have seen several major pharmaceutical companies take successful steps to reclaim productivity in both discovery and development. Moreover, vitality and passion essential to early drug discovery efforts are often now found in smaller organizations. For those companies, the primary focus is less on getting a compound that fits rigid criteria and a specific development model and more on determining how to successfully deliver their compound to patients.

Early in my career, there seemed to be greater openness and civility in drug discovery. Although there was competition, there was also a belief that patients would be better served with multiple drugs acting at the same target and that more than one drug could be successful. That viewpoint led to greater sharing of precompetitive information, which then helped the overall drug discovery development process. In the intervening years, an often-expressed view (in the popular press, by politicians and reimbursement agencies) is that only the first drug in a class should be reimbursed. The trope is that other compounds are just "me-too" drugs being pushed by greedy pharmaceutical companies. That there may only be one acceptable drug has led to heightened competition and less sharing of information.

It is my aspiration that the growing acceptance of personalized medicine will help to reverse this mentality. Personalized medicine recognizes that not all people with a given condition will respond identically to the same drug. As this concept gains greater acceptance, I hope that more open sharing will take place because I believe it will lead to greater overall success rates, making a genuine difference in the lives of others.

REFERENCES

(1) Trost, B.M.; Bernstein, P.R.; Funfschilling, P.C. *J. Am. Chem. Soc.* **1979**, *101*, 4378.
(2) Augstein, J.; Farmer, J.B.; Lee, T.B.; Sheard, P.; Tattersall, M.L. *Nat. New Biol.* **1973**, *245*, 215.
(3) Krell, R.D.; Buckner, C.K.; Keith, R.A.; Snyder, D.W.; Brown, F.J.; Bernstein, P.R.; Matassa, V.; Yee, Y.K.; Hesp, B.; Giles, R.E.J. *J. Allergy Clin. Immunol.* **1988**, *81*, 276.
(4) Jacobs, R.T.; Bernstein, P.R. *Drugs Fut.* **1995**, *20*, 1233.
(5) Brown, F.J.; Andisik, D.A.; Bernstein, P.R.; Bryant, C.; Ceccarelli, C.; Damewood, J.R.; Earley, R.; Edwards, P.D.; Feeney, S.; Green, R.; Gomes, B.C.; Kosmider, B.J.; Krell, R.D.; Shaw, A.; Steelman, G.B.; Thomas, R.M.; Vacek, E.P.; Veale, C.A.; Warner, P.; Williams, J.C.; Wolanin, D.J.; Woolson, S.A. *J. Med. Chem.* **1994**, *37*, 1259.
(6) Bernstein, P.R.; Gomes, B.C.; Kosmider, B.J.; Vacek, E.P.; Williams, J.C. *J. Med. Chem.* **1995**, *38*, 212.
(7) Veale, C.A.; Bernstein, P.R.; Bohnert, C.M.; Brown, F.; Bryant, C.; Damewood, J.; Earley, R.; Edwards, P.; Feeney, S.; Gomes, B.; Hulsizer, J.; Kosmider, B.J.; Krell, R.D.; Moore, G.; Salcedo, T.; Shaw, A.; Silberstein, D.S.; Steelman, G.B.; Stein, M.; Strimpler, A.; Thomas, R.M.; Vacek, E.; Williams, J.C.; Wolanin, D.J.; Woolson, S. *J. Med. Chem.* **1997**, *40*, 3173.
(8) Bernstein, P.R.; Aharony, D.; Albert, J.S.; Andisik, D.; Barthlow, H.G.; Bialecki, R.; Davenport, T.; Dedinas, R.F.; Dembofsky, B.T.; Koether, G.; Kosmider, B.J.; Kirkland, K.; Ohnmacht, C.J.; Potts, W.; Rumsey, W.L.; Shen, L.; Shenvi, A.; Sherwood, S.; Stollman, D.; Russell, K. *Bioorg. Med. Chem. Lett.* **2001**, *11*, 2769.
(9) Ashworth, I.W.; Bowden, M.C.; Dembofsky, B.; Levin, D.; Moss, W.; Robinson, E.; Szczur, N.; Virica, J. *Org. Proc. Res. Dev.* **2003**, 7, 74.
(10) Shenvi, A.; Bernstein, P.; Ohnmacht, C.; Albert, J.; Hill, D.; Green, R.; Shen, L.; Stollman, D.; Dembofsky, B.; Andisik, D.; Dedinas, R.; Sherwood, S.; Koether, G.;

Kosmider, B.; Davenport, T.; Kirkland, K.; Potts, W.; Bialecki, R.; Aharony, D.; Rumsey, W.; Russell, K. *The discovery of novel orally active dual NK1/NK2 antagonist ZD2249. Tachykinins* **2000**, La Grande Motte, France, Oct. 17–20, 2000.

(11) Rumsey, W.; Aharony, D.; Bialecki, R.; Abbott, B.; Barthlow, H.; Caccese, R.; Ghanekar, S.; Lengel, D.; McCarthy, M.; Wenrich, B.; Undem, B.; Kirkland, K.; Potts, W.; Albert, J.; Ohnmacht, C.; Shenvi, A.; Bernstein, P.; Russell, K. *Pharmacological characterization of a novel, orally active antagonist of the NK1 receptor, ZD4974. Tachykinins* **2000**, La Grande Motte, France, Oct. 17-20, 2000.

(12) Albert, J.S.; Ohnmacht, C.; Bernstein, P.R.; Rumsey, W.L.; Aharony, D.; Alelyunas, Y.; Russell, D.J.; Potts, W.; Sherwood, S.A.; Shen, L.; Dedinas, R.F.; Palmer, W.E.; Russell, K. *J. Med. Chem.* **2004**, *47*, 519.

(13) Bernstein, P. Abstracts of papers 241st ACS national meeting, # 06-BMGT, Anaheim, CA, **2011**.

(14) Zhang, M.; Zhou, D.; Wang, Y.; Maier, D.L.; Widzowski, D.V.; Sobotka-Briner, C.D.; Brockel, B.J.; Potts, W.M.; Shenvi, A.B.; Bernstein, P.R.; Pierson, M.E. *J. Pharmacol. Exp. Ther.* **2011**, *319*, 1.

(15) Nugiel, D.A.; Krumrine, J.R.; Hill, D.; Damewood, J.R.; Bernstein, P.R.; Sobotka-Briner, C.D.; Liu, J.; Zacco, A.; Pierson, M.E. *J. Med. Chem.* **1876**, *2010*, 53.

(16) Hindo, B. At 3M, A Struggle Between Efficiency And Creativity. Business Week, June 11, **2007**.

CHAPTER THREE

Tales of Drug Discovery

M. Ross Johnson
Parion Sciences, Inc., 2525, Meridian Parkway, Suite 260, Durham, NC 27713

Contents

I'm a Med Chem Junkie. I have a lifelong "addiction" to drug discovery using the tools and concepts of medicinal and organic chemistry. Besides my family, it is the first thing I think about when I wake up and the last thing I remember going to sleep. Looking back, I was excited each and every day because every day carried within it a chance to find a cure for a disease, help ease someone's medical condition, or even save someone's life. There were many research highs and lows along the way, but it helped to believe in my dreams and work hard.

I started off the research phase of my scientific career working for Joe Casanova at LA State University as an NSF undergraduate fellow studying lead tetraacetate oxidations. When I then went on to Berkeley for my BS degree, I worked with the internationally renowned organic chemist Henry Rapoport in natural products synthesis and George Payne (of the "Payne Rearrangement") at Shell Development Company in Emeryville doing sulfur ylid chemistry. After receiving my BS in Chemistry from Berkeley, I went to UCSB for my Ph.D. to work with Bruce Rickborn in physical organic chemistry. Bruce was my mentor and lifelong friend who trained me, not as a medicinal chemist, but as a problem solver. I think this philosophy served me well in the ensuing years in drug discovery. After a short post doc back at Cal with Fritz Jensen in organometallic chemistry, I was hired by

Annual Reports in Medicinal Chemistry, Volume 47
ISSN 0065-7743
http://dx.doi.org/10.1016/B978-0-12-396492-2.00003-5

Pfizer in Groton Connecticut. It was an exciting time for me and my family as we moved East but filled with a little trepidation, since I did not have the slightest idea what I was getting into.

Despite the vast array of tools at one's disposal in the medicinal chemist's armamentarium, it is the way in which one approaches a target and chooses the appropriate tools which will ultimately determine a successful outcome. First of all, I have found it useful to always have a hypothesis—even if it turns out wrong. For example, in the prostaglandin program, we predicted that if we could block major metabolic routes, we would improve duration of action and then influence tissue selectivity by introducing bioisosteres. In the cannabinoids, we postulated a structural similarity between HHC and PGE2 that allowed us to focus on three key interaction points, which led to highly novel cannabinoid structures and beyond. Those first two programs led me to discover just how important natural products can be as prototype leads. In the pursuit of better "glitazones," we postulated that a conformationally rigid structural approach would improve potency. Having a hypothesis also insures you think about the compounds you are about to create. There should be a reason for each and every compound you make!

Once a hypothesis is in place, one can take three basic approaches in designing drugs: (1) The "Soaring Eagle"; (2) The Giant Leap; and (3) Methyl, Ethyl, Butyl, Futile. In practice, successful projects use all three. The Soaring Eagle Approach is akin to an eagle hunting its prey. It starts at the top of the canyon and, with sharp eyes, circles down ever lower to capture its prey. This approach is usually employed in areas where there is already a lot of SAR and mechanism available, and you have many competitors, like we did in the diabetes and opiate analgesic programs. The second approach, The Giant Leap, is used when you have homed in on a group of structures and find key activity parameters in specific regions of the molecule. Rather than defaulting to the third approach, change things radically. In the cannabinoids, we removed rings; in the prostaglandins, we used aryl groups to inhibit oxidation; and in the epithelial sodium channel (ENaC) program, we added additional electrostatic binding regions. Now, as much as every medicinal chemist hates it, there is a role for the third approach. In the short-acting opiate program, we were able to titrate the duration of action simply by varying the alkyl group, thus influencing metabolic inactivation by nonspecific esterases. This latter approach is an end-stage tactic, as there is not much more you can do after this. Once you have successfully generated your target molecule, you will now start to experience the real thrill of discovery: you will get biological feedback from your biological counterpart. Thus begins the exciting and up-and-down iterative process that will ultimately produce a drug. The following are a few of my own tales of drug discovery.

1. PROSTAGLANDINS

My medicinal chemistry career began in 1971 at Pfizer, where I worked under the direction of Dr. Hans Hess on modifying prostaglandins to improve selectivity for a variety of therapeutic uses. Along with Tom Schaaf, Jasjit Bindra, and Jim Eggler, we focused our research primarily on two areas to increase the potency and duration of action of the natural prostaglandins: the metabolic stabilization of the upper carboxylic side chain to prevent beta-oxidation and modification of the lower side chain to prevent oxidation (15-dehydrogenase). We made extensive contributions to this area that resulted in novel 16-aryl and 16-aryloxy prostaglandin congeners which, when modified in the upper side chain with a bioisostere of carboxylic acids, led to the discovery and ultimate commercialization of sulprostone (Nalador®).[1–4] Up to this point, the widespread use of prostaglandins for a variety of gynecological and obstetric uses was limited by the lack of tissue selectivity and metabolic stability of the natural prostaglandins and simple analogs thereof.

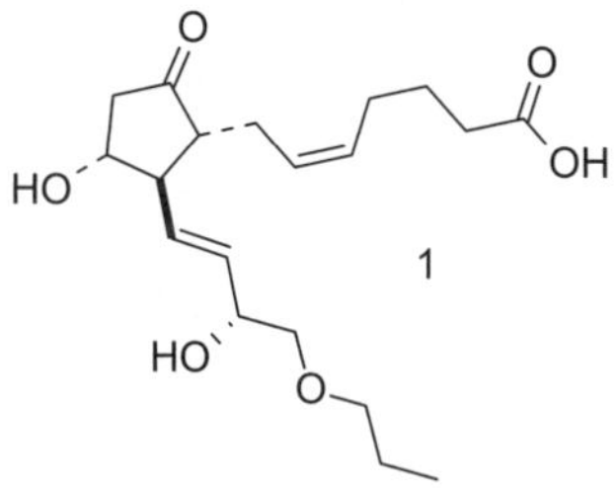

1

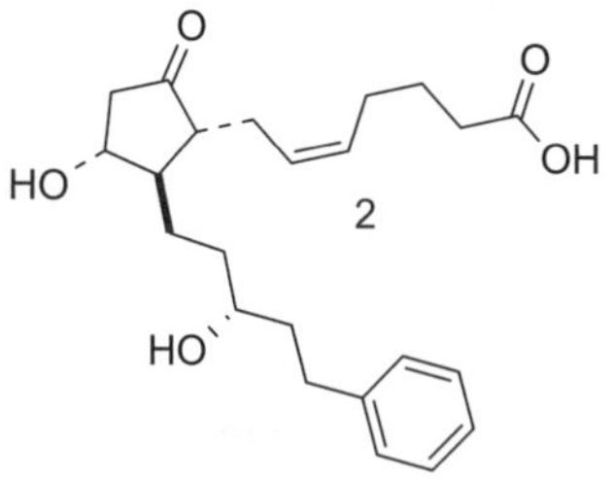

2

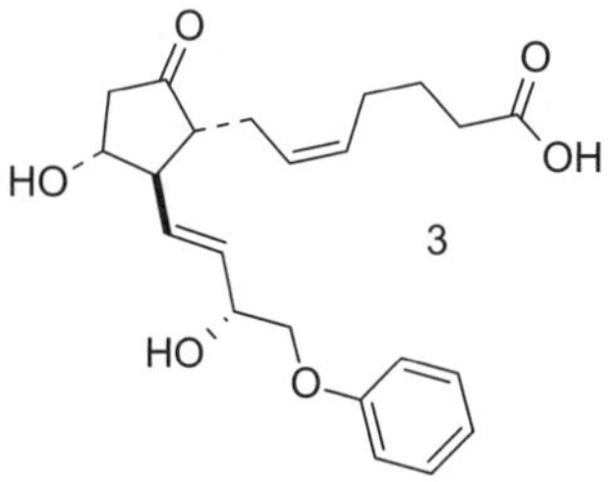

3

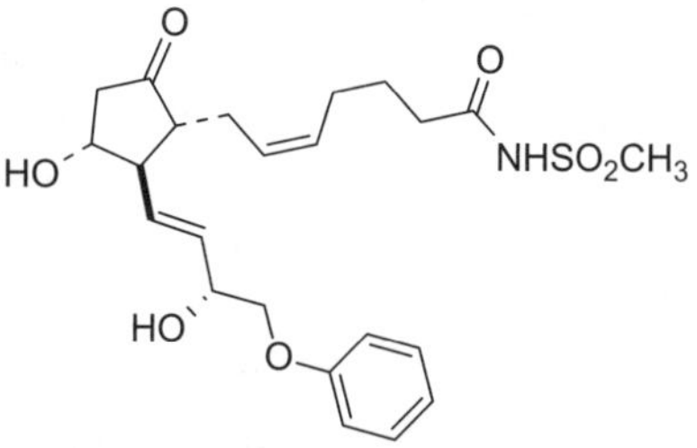

sulprostone/Nalador®

We had increased potency and selectivity by first showing that the 17-oxa and 17-phenyl moieties could be combined to produce 16-phenoxy PGE_2, which enhanced potency *in vivo* due to an increased stability to C_{15}-hydroxyprostaglandin dehydrogenase. Further work showed that tissue selectivity could be enhanced by replacing the carboxylic acid with methane sulfonimide and simultaneously eliminating beta-oxidation. The resultant new chemical entity (sulprostone) incorporated both of these features to produce a drug which was 30 times more selective than PGE_2 and is still used worldwide today for a number of gynecological and obstetric purposes. The 16-phenoxy moiety was subsequently used in a number of marketed prostaglandin and prostanoid drugs. Our work in this field showed the power of using metabolic information and bioisosterism to overcome shortcomings in natural products.

2. CANNABINOID ANALGESICS

In 1976, I entered CNS research at Pfizer and was assigned the daunting task of producing a nonnarcotic analgesic as potent as morphine. This problem had been studied intensively for nearly 50 years prior to my work by L.F. Small and E.L. May, two giants in the field of opioid analgesics. I was intrigued by the potent, specific analgesic activity of 9-beta-hydroxyhexahydocannabinol (HHC) that was discovered by May.[5] Working with Dr. George Milne at Pfizer, we showed that HHC did not act through opioid receptors. Based on these observations and the molecular differences between morphine and HHC, we proposed a structural hypothesis that highlighted the differences between the planar prostaglandins and T-shaped structure of morphine (the "Prostaglandin Overlap" model).[6] Thus began a program of synthesis and SAR with Dr. Larry Melvin at Pfizer that would have a lasting impact on the cannabinoid field that continues to this day. Not only did this hypothesis provide novel classic cannabinoid structures but also the totally ingenious "non-classical", cannabinoids possessing potent nonnarcotic analgesic activity substantially greater than morphine. These compounds also possessed potent antiemetic activity.

By 1983, our group had amassed a large number of diverse structures exhibiting similar SAR.[7] The exquisite potency, stereospecificity, and lack of a unifying link to existing neurotransmitter system led me to propose that compounds like levonantradol and CP-55,244 bind to a specific (at the time unknown) CNS receptor system. Based on this insight and armed with our structurally diverse novel compounds, including the tritium-labeled drug

$[^3H]$CP-55,940, the elucidation and biochemical identification of the cannabinoid receptor was accomplished in collaboration with A.C. Howlett, then at St. Louis University.[8] Following this landmark work, Miles Herkenham at the NIH, in collaboration with myself, Kenner Rice, and Brian DeCosta,[9] conducted the definitive autoradiographic cannabinoid receptor localization study using $[^3H]$ CP-55,940 in several mammalian species, including humans. These anatomical findings explained many of the properties of the cannabinoids, including their almost complete lack of acute toxicity. The studies were central to Lisa Matsuda's subsequent discovery of the cloned human cannabinoid receptor at the NIH. A sequential, quantitative autoradiographic study in 40 regions of the brain and spinal cord with $[^3H]$CP-55,940 provided powerful support for the physiologic and pharmacological relevance of the receptor.[10]

The medicinal chemistry we were able to generate and the excellent work utilizing these tools by my group of collaborators in the cannabinoid area have been a major driving force in the advancement of the field to its present state, emphasizing once again the indispensable role of medicinal chemistry in solving multidisciplinary scientific problems. The compounds and concepts that we advanced over 25 years ago continue to be valuable tools in present-day cannabinoid research. It is well worth noting that these three papers,[8–10] all of which have over 1000 citations each, are all primary research publications. So while I was disappointed that we never got a marketed drug out of this research, we did make an impact on the basic science of the field, and I hope current work based on our earlier findings will yet prove beneficial to mankind.

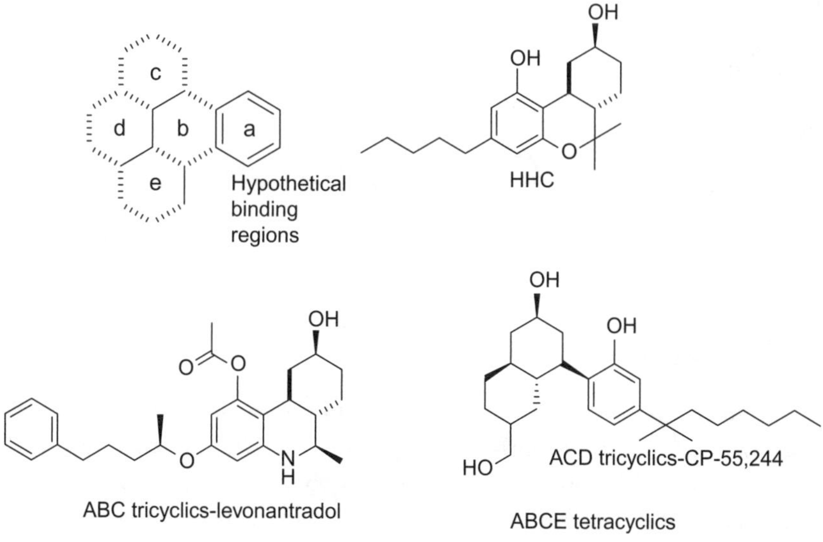

AC bicyclics

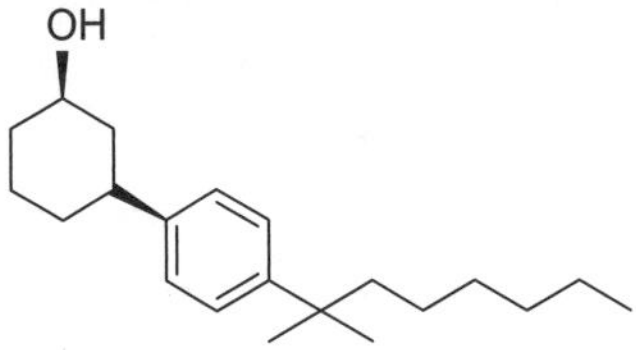

AC bicyclics-CP-47,497

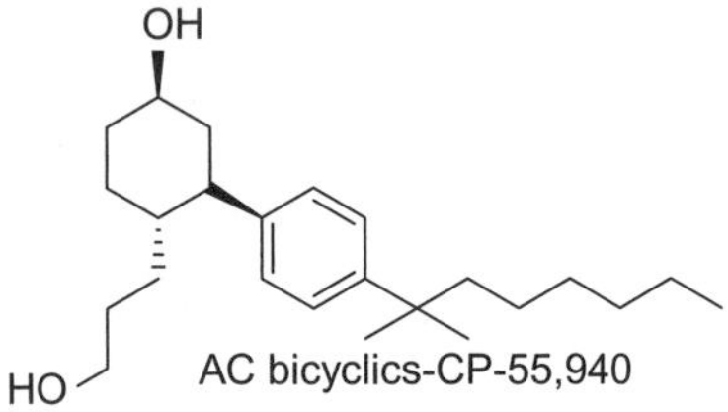

AC bicyclics-CP-55,940

3. HYPOGLYCEMIC AGENTS

In 1985, I became the manager of metabolic diseases at Pfizer. Along with Bob Volkman, Dave Clark, and Jim Eggler, we became involved with the design of second-generation hypoglycemics based on ciglitazone. We proposed and synthesized a series of conformationally restricted dihydrobenzofurans and dihydrobenzopyrans. On the basis of a dramatic increase of *in vivo* potency as a hypoglycemic agent in the *ob/ob* mouse, one of the resolved, conformationally restricted enantiomers was moved into human trials under the name of englitazone.[11] It is worth noting that this class of antidiabetic agents, known collectively as thiazolidinediones (TZDs), or "glitazones," is an important addition to the physician's ability to keep Type II diabetes under control. More recently, these molecules show promise in delaying the onset of Alzheimer's disease. TZDs lower blood glucose levels by stimulating the growth of mitochondria. The theory is that by stimulating mitochondrial growth in the brain, these drugs can delay the ability of Alzheimer's to kill brain neurons. This is an age-old story in medicinal chemistry: Yesterday's discoveries in one therapeutic area often provide the basis for clinical discoveries in other areas.

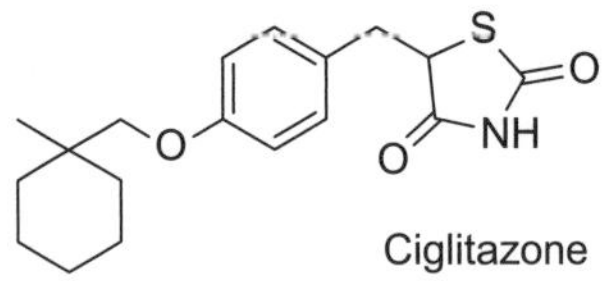

Ciglitazone

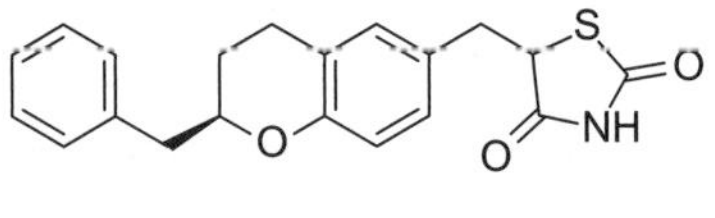

Englitazone

4. OPIOID ANALGESICS

In early 1988, I moved to Glaxo, Inc. in Research Triangle Park, NC as one of the founding scientists of research in the United States. Working out of temporary laboratories at the University of North Carolina, my young group of chemists sought to produce an ultra short-acting opioid analgesic.

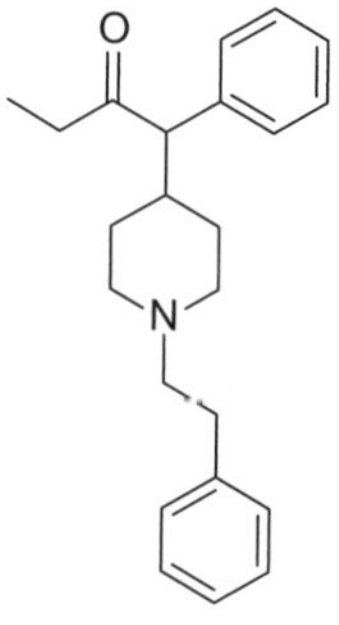

Fentanyl

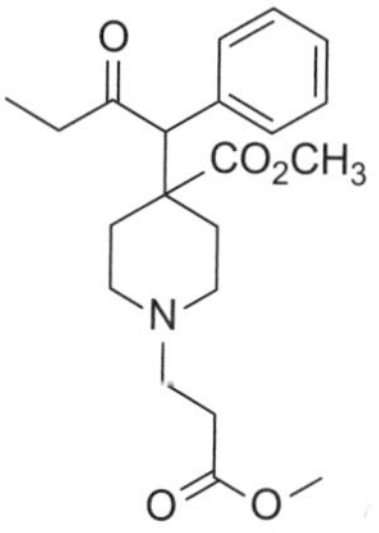

Remifentanyl/Ultiva

We proposed that the ideal analgesic would have a maximum biological half-life of 10–30 min and be biotransformed into inactive metabolites minimizing accumulation and redistribution with prolonged administration, thereby avoiding the respiratory depression and muscle rigidity commonly associated with opiates. Having thus hypothesized the ideal short-acting narcotic analgesic, my protégé, Paul Feldman, and I focused on the synthesis of analogs based on several classes of 4-anilidopiperidines. We circled in on molecules that would be rapidly inactivated enzymatically in plasma and thus be independent of hepatic metabolism, which was of further benefit to renally compromised patients. In one series, we found a compound that possessed potent mu opioid activity with a high degree of analgesic efficacy and an extremely short duration of action.[12] The resultant drug, remifentanyl, showed a profile that could be essentially titrated when used and that completely returns the patient to baseline values in minutes rather than hours when discontinued. Remifentanyl was the first commercial drug marketed by Glaxo Inc. and was sold under the trade name Ultiva®. Since its introduction, this drug has proven to be a major adjunct in surgical anesthesia and is the drug of choice in certain situations. Ultiva® has recently been dubbed "the ultimate opioid."

5. HIV FUSION INHIBITORS

In 1995, I decided to enter the biotech industry. I joined Trimeris as its Chief Scientific Officer and subsequently became CEO. Researchers Tom Mathews and Dani Bolognese at Duke University had found that a 36-amino acid peptide (T-20) blocked HIV cell–cell fusion *in vitro*. I was convinced it should work in humans, so we assembled a team at Trimeris to advance this compound into human trials. The group demonstrated very early on that the drug reaches the target lymph and blood compartments *in vivo*, but a huge problem in its development remained. Early versions of the drug required

106 steps by traditional solid-state methods, so I assembled some of my old development chemistry team at Glaxo to solve this hurdle. This team of chemists, led by M.C. Kang and inspired by Brian Bray, achieved a vastly more efficient method of manufacturing T-20, which now stands as the most complex synthetic peptide ever manufactured on such a massive scale (>1 metric ton per year). Having solved this important issue, we moved the program into full development. Our first clinical trial, which showed T-20 to lower HIV viral load below detectable levels, is still a landmark study in the field.[13] This program led to the approval of enfuritide, a life-saving drug, sold as Fuzeon®. It is the first and only fusion inhibitor on the market to treat HIV in combination with mechanistically distinct inhibitors. Fuzeon is particularly effective in treating patients who are failing on other drug classes and truly makes the difference between life and death for this patient population. There was no medicinal chemistry magic here—just good old-fashioned organic chemistry and drug development, which carried the day.

6. EPITHELIAL SODIUM CHANNEL BLOCKERS

In 1999, I teamed with the head of the Cystic Fibrosis Center at the University of North Carolina (Dr. Ric Boucher) to design and test drugs to increase mucociliary clearance (MCC) in humans. Early SAR studies had given me unique insights that led to the medicinal chemical conceptualization of the ENaC and the subsequent discovery of auxiliary binding sites resulting in the design of several classes of novel and potent compounds.[14] Compound PS-552 was selected for clinical studies based on the increased potency (100 ×), reduced reversibility (5 ×), and increased selectivity for nonrenal elimination (9 ×) when compared to the prototype ENaC blocker amiloride.[15] PS-552 has shown that it produces MCC in humans and was advanced to Phase II clinical trials for the treatment of cystic fibrosis and xerostomia caused by Sjogren's disease. Following the advancement of PS-552, we continued to improve these potential therapeutics by designing nearly perfect antedrugs based on our earlier hypothesis[14] in the ENaC field. Compared to the well-characterized ENaC

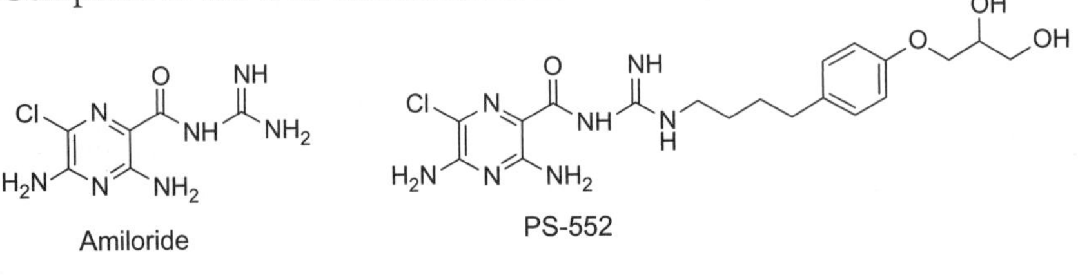

blocker amiloride, these new drugs were designed to provide (1) enhanced blocking activity on ENaCs, (2) increased duration of blocking ENaC, (3) decreased rate of absorption across the airway epithelium, (4) enhanced solubility in hypertonic saline, (5) improved retention of ASL (airway surface liquid) volume on airway surfaces, and (6) improved drug safety by incorporating antedrug properties that make it susceptible to epithelial biotransformation which, in turn, produces less active systemic metabolites. Drugs from this class were superior in all these attributes, and clinical trials were initiated as reported in the *C&E News* article, "Breathing Easier" (September 1, 2008).

7. CONCLUSIONS AND RAMBLINGS

To be successful in any life endeavor, I believe one needs to set priorities. I think the three most important things are family, your particular belief system, and your profession. With regard to the latter, drug discovery is a team game, and your mentors, colleagues, and friends are an important part of the success paradigm. As I look back on the breadth and depth of our body of work and the contributions made to medicinal chemistry over a nearly 40-year career, my coworkers' efforts acknowledged below are truly remarkable. My special thanks to Bruce Rickborn, Hans Hess, and Kenner Rice for their lifelong support and mentoring.

ACKNOWLEDGMENTS

- Prostaglandins: Hans Hess, Jasjit Bindra, Tom Schaaf, Jim Eggler, Derek Tichner, George Monsam, Eddie Kleinman
- Cannabinoids: Larry Melvin, George Milne, Jim Eggler, Charles Harbert, Tom Althuis, Werner Kappler, Hans Wiederman, Charles Norris, Allyn Howlett, Miles Herkenham, Kenner Rice, Brian de Costa, Scott Richardson, Bill Devane, Billy Martin, Everette May, Ray Wilson
- Englitazone: Dave Clark, Steve Goldstein, Bob Volkmann, Jim Eggler, Jerry Holland, Brett Hulin, Nancy Hudson
- Opiate Analgesics (Remifentanil/Ultiva): Paul Feldman, Marcus Brackeen, Jeff Leighton
- Antitumor (Campothecin Analogs): Mike Luzzio, Mike Evans, Milana Dezube, Sal Profeta, Jeff Besterman
- Campothecin GI147211C: Frank Fang, M.C. Kang, Craig LeHoulier, George Lewis and John Partridge
- Cobalt Porphyrins (Satiety Agents): Steven Frye
- HIV (Carbovir): Mike Peel, Dan Sternbach
- Insulin Second Messenger: Jeff Cobb, Pedro Cuatrecasas

- HIV (Fuzeon): M.C. Kang, Brian Bray, Bill Lackey, Tom Matthews, Peter Jeffs, Katherine Mader, Dani Bolognesi, Dennis Lambert, Sam Hopkins
- Sodium Channel Blockers: Ric Boucher, Andrew Hirsh, Bill Thelin, Karl Donn, Ann Stevens, John Ansede, Ben Yerxa, Gary Phillips, Bill Baker, Eric Dowdy, Bruce Molino, Bruce Sargent, Jim Zhang, Ron Aungst, Guihui Chen, Don Kuhla

REFERENCES

(1) Hess, H.-J.; Schaaf, T.K.; Bindra, J.S.; Johnson, J.R.; Constantine, J.W. In *International Sulprostone Symposium: Structure-Activity Considerations Leading to Sulprostone*; Friebel, K., Schneider, A., Wurtel, H., Eds.; Medical Scientific Dept., Schering, A.G.: Berlin and Bergkamen, 1979; pp 29–37.

(2) Johnson, M.R.; Schaaf, T.K.; Constantine, J.W.; Hess, H.J. *Prostaglandins* **1980**, *20*, 515.

(3) Schaaf, T.K.; Bindra, J.S.; Eggler, J.F.; Plattner, J.J.; Nelson, A.J.; Johnson, M.R.; Constantine, J.W.; Hess, H.J.; Elger, W. *J. Med. Chem.* **1981**, *24*, 1353.

(4) Schaaf, T.K.; Johnson, M.R.; Constantine, J.W.; Bindra, J.S.; Elger, W. *J. Med. Chem.* **1983**, *26*, 328–334.

(5) Wilson, R.S.; May, E.L. *J. Med. Chem.* **1974**, *17*, 475.

(6) Johnson, M.R.; Melvin, L.S.; Milne, G.M. *Life Sci.* **1982**, *31*, 1703.

(7) Johnson, M.R.; Melvin, L.S. In *Cannabinoids as Therapeutic Agents; Chapter 7: The Discovery of Nonclassical Cannabinoid Analgetics*; Mechoulam, R., Ed.; CRC Press, Inc.: Boca Raton, FL, 1986; pp 121–145.

(8) Devane, W.A.; Dysarz, F.A., III; Johnson, M.R.; Melvin, L.S.; Howlett, A.C. *Mol. Pharm.* **1988**, *34*, 605.

(9) Herkenham, M.; Lynn, A.B.; Johnson, M.R.; Melvin, L.S.; de Costa, B.R.; Rice, K.C. *Proc. Natl. Acad. Sci. U.S.A.* **1932**, *1990*, 87.

(10) Herkenham, M.; Lynn, A.B.; Johnson, M.R.; Melvin, L.S.; de Costa, B.R.; Rice, K.C. *J. Neurosci.* **1991**, *11*, 563–583.

(11) Clark, D.A.; Goldstein, S.W.; Volkmann, R.A.; Eggler, J.F.; Holland, G.F.; Hulin, B.; Stevenson, R.W.; Kreutter, D.K.; Gibbs, E.M.; Krupp, M.N.; Merrigan, P.; Kelbaugh, P.L.; Andrews, E.G.; Tickner, D.L.; Suleske, R.T.; Lamphere, C.H.; Rajeckas, F.J.; Kappeler, W.H.; McDermott, R.E.; Hutson, N.J.; Johnson, M.R. *J. Med. Chem.* **1991**, *34*, 319–325.

(12) Feldman, P.L.; James, M.K.; Brackeen, M.F.; Bilotta, J.M.; Schuster, S.V.; Lahey, A.P.; Lutz, M.W.; Johnson, M.R.; Leighton, H.J. *J. Med. Chem.* **1991**, *34*, 2202.

(13) Kilby, J.M.; Hopkins, S.; Venetta, T.M.; DiMassimo, B.; Cloud, G.A.; Lee, J.Y.; Alldredge, L.; Hunter, E.; Lambert, D.; Bolognesi, D.; Matthews, T.; Johnson, M.R.; Nowak, M.A.; Shaw, G.M.; Saag, M. *Nat. Med.* **1998**, *11*, 1302.

(14) Hirsch, A.J.; Molino, B.F.; Zhang, J.; Astakhova, N.; Sargent, B.D.; Swenson, B.D.; Usyatinsky, A.; Wyle, M.J.; Boucher, R.C.; Smith, R.T.; Zamurs, A.; Johnson, M.R. *J. Med. Chem.* **2006**, *49*, 4098.

(15) Hirsch, A.J.; Zhang, J.; Zamurs, A.; Fleegle, J.; Thelin, W.R.; Caldwell, R.A.; Sabater, J.R.; Abraham, W.M.; Donowitz, M.; Cha, B.; Johnson, K.B.; St. George, J.A.; Johnson, M.R.; Boucher, R.C. *JPET* **2008**, *325*(1), 77.

PART 1

Central Nervous System Diseases

Editor: Albert J. Robichaud
Sage Therapeutics, Inc. Cambridge, Massachusetts

CHAPTER FOUR

Recent Developments in Targeting Neuroinflammation in Disease

Allen T. Hopper*, Brian M. Campbell†, Henry Kao†, Sean A. Pintchovski†, Roland G.W. Staal†

*Discovery Chemistry and DMPK, Lundbeck Research, Paramus, New Jersey, USA
†Neuroinflammation Disease Biology Unit, Lundbeck Research, Paramus, New Jersey, USA

Contents

1. INTRODUCTION

Neuroinflammation (NI) is the process by which an organism attempts to remove an injurious stimulus in the central nervous system (CNS) and initiate the healing process to protect the cells and overall function of the brain. NI may be accompanied by increased vascular permeability, invasion of peripheral immune cells, release of inflammatory mediators (cytokines, reactive oxygen species, etc.), and tissue dysfunction.[1] The primary mediators of NI are microglia, the only immune cells residing within the CNS. However, neuroinflammatory responses can also be driven by astrocytes, oligodendrocytes, and neurons, as well as cells of the peripheral immune system such as monocytes/macrophages and T cells.[2] These cell

Annual Reports in Medicinal Chemistry, Volume 47
ISSN 0065-7743
http://dx.doi.org/10.1016/B978-0-12-396492-2.00004-7

types are all capable of releasing various pro- and anti-inflammatory mediators. Typically, inflammation is an acute process whose ultimate goal is resolution and repair of injured cells/tissue. However, when inflammation becomes excessive or prolonged, it becomes pathological and is associated with a variety of diseases.

The case can be made that most CNS disorders, including Alzheimer's disease (AD), amyotrophic lateral sclerosis (ALS), autism, epilepsy, HIV-associated neurocognitive disorder, Huntington's disease (HD), multiple sclerosis (MS), neuropathic pain, Parkinson's disease (PD), schizophrenia, stroke, and spinal cord injury (SCI), exhibit NI pathology.[3–12] For some indications, such as MS, the peripheral immune system is the primary cause and driver of disease progression. In other cases, such as SCI, the peripheral immune system is not a cause, but certainly a well-characterized contributor to inflammation. For neurodegenerative disease, NI is typically chronic and may begin a decade or more before the onset of symptoms (e.g., HD[13]). Furthermore, it is well established that modulating NI genetically and pharmacologically can alter the course of disease in animal models of these diseases. There is also convincing evidence, mostly from preclinical models, that microglia play an integral role in the development of neuropathic pain.[14] Since few if any CNS disorders are completely or adequately treated by any single drug, finding an effective therapy that slows the progression of a disease would be an important advancement. Reducing NI may not address the cause of the disease, but it may mitigate or resolve the inflammatory response that drives disease progression. As few drugs on the market specifically target NI, this approach presents a promising alternative, or cooperative, drug discovery strategy for the treatment of many CNS disorders.

Modulating the interaction and communication between microglia, neurons, and cells of the immune system, by targeting ion channels, G-protein-coupled receptors, enzymes, and kinases, for example, may ameliorate NI associated with CNS diseases. Metabotropic and ionotropic neurotransmitter receptors, enzymes such as cyclooxygenase and inducible nitric oxide synthase, antioxidants, and biologics targeting adhesion molecules or anti-inflammatory peptides are well reviewed elsewhere.[15] Kinases such as MAPK, MEK, CDK5, GSK, JNK, and IRAKs are not very specific to a particular pathway, cell type, or biology and are also reviewed elsewhere.[16,17] Therefore, in this review, we focus on examples from several major drug target classes where recent medicinal chemistry advancements have been made, in order to provide a broad overview of the diverse strategies to target NI.

2. ION CHANNELS

2.1. Purinergic receptors and neuroinflammation

Purinergic receptors are ion channels expressed on various cells in the CNS and immune system and can be classified into two subfamilies: G-protein-coupled metabotropic (P2Y) and ligand-gated ionotropic (P2X) receptors. Although metabotropic purinergic signaling has been implicated in neuroinflammation,[18] the focus here centers on recent discoveries of the P2X receptors, notably $P2X_4$ and $P2X_7$, which are the predominant ligand-gated purinergic receptors expressed on microglia.

2.1.1 $P2X_4$ receptor ($P2X_4R$)

Activation of $P2X_4R$ results in increased intracellular Ca^{2+}, K^+, and Na^+ and subsequent activation of p38-MAPK signaling. One factor released in response to activation of this receptor is BDNF which acts on the TRK-B receptor on lamina I neurons in the dorsal horn of the spinal cord, resulting in their hyperexcitability.[19,20] Blockade of $P2X_4R$ function significantly ameliorates neuropathic pain in preclinical models, a finding that is supported by a variety of genetic and pharmacological studies. This was first demonstrated by using the antagonist of $P2X_{1-4}Rs$ (TNP-ATP) to reverse tactile allodynia in a spinal nerve injury model.[21] In contrast, pyridoxalphosphate-6-azophenyl-2′4′-disulphonic acid, an antagonist of $P2X_{1-3,5,7}Rs$, but not $P2X_4R$ was ineffective at reversing allodynia in rats with nerve injury. Genetic approaches such as knockdown of the $P2X_4R$ using antisense oligonucleotides[21] and genetic ablation of the $P2X_4R$ gene[22] also resulted in a significant reduction in pain behavior. While most data support a role for $P2X_4R$ in modulating neuropathic pain, $P2X_4$ receptors also regulate microglial responses, making them attractive targets for other CNS indications.

There is a scarcity of published $P2X_4R$ chemical matter suggesting potential issues of chemical tractability around this target.[23] However, several chemotypes have been identified with modest potency, including benzofuro-1,4-diazepin-2-one analog **1** ($P2X_4R$ IC_{50} 0.5 μM) and 1-aryl-2-phenoxymethyl-piperazine analogs, exemplified by the SSRI paroxetine ($P2X_4R$ $IC_{50} \sim 3$ μM).[23,24] More recently, diazapinedione derivatives,[25] exemplified by **2** and 4-aryl-2-quinazolinone compounds[26] exemplified by **3** were reported with $P2X_4R$ IC_{50} values from 0.16 μM. The utility of these analogs as tools to probe $P2X_4R$ function *in vivo* remains unclear since their receptor selectivity, brain penetration, and PK data have not yet been reported.

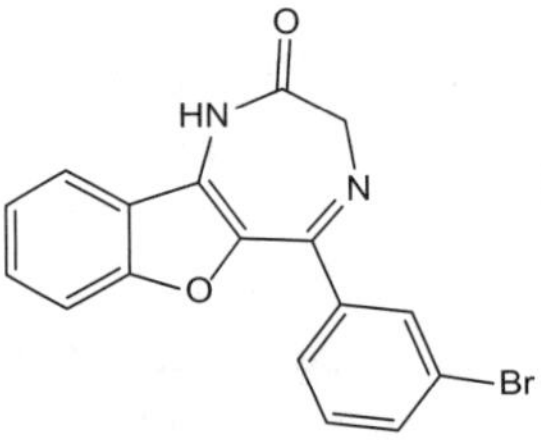

1

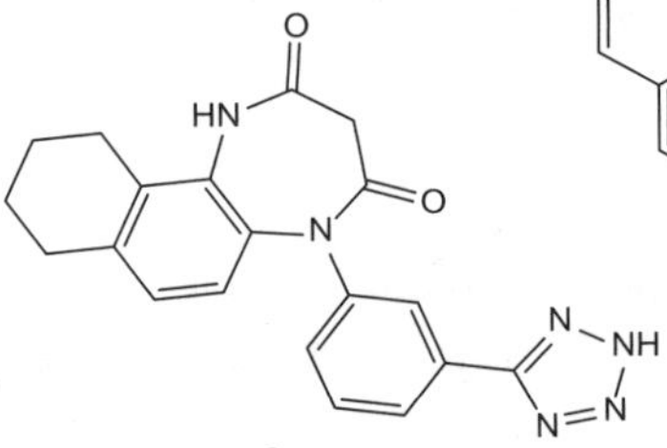

2

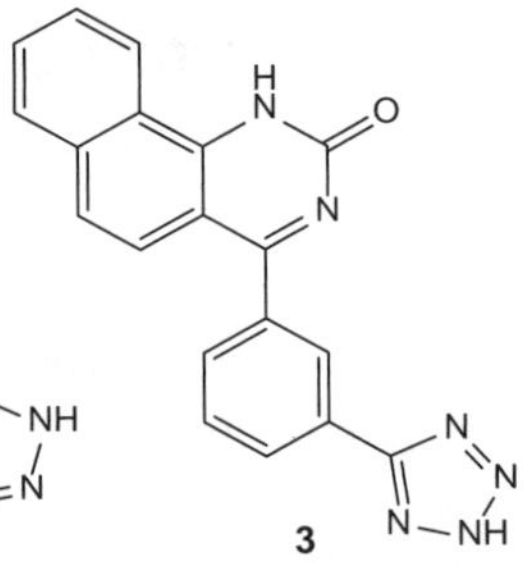

3

2.1.2 P2X$_7$ receptor (P2X$_7$R)

The P2X$_7$R is activated by high extracellular concentrations of ATP (> 100 μM) that leads to Ca^{2+} influx/K^+ efflux and initiation of a cascade leading to maturation and release of the proinflammatory cytokine IL-1β. In the CNS, P2X$_7$R is primarily expressed on microglia, oligodendrocytes, and activated astrocytes, though the data supporting neuronal expression are controversial. Elevated receptor expression and function have been associated with various CNS diseases, including depression, AD, MS, epilepsy, ALS, and neuropathic pain.[18] Several compounds have entered clinical trials for peripheral indications, but to date, this target has not been probed clinically for a CNS indication. 1-Hydroxycycloheptyl analog **4** (CE-224,535) is a potent human P2X$_7$R antagonist that inhibits IL-1β release from lipopolysaccharide (LPS)- and ATP-stimulated human whole blood with IC_{50} and IC_{90} values of 1.0 and 4.7 nM, respectively.[27] This candidate was evaluated in a clinical trial for rheumatoid arthritis, but failed to show significant efficacy, even with trough free plasma concentrations significantly above the IC_{90} in human whole blood. The CNS penetrability of this compound is not reported, and it lacks sufficient rodent P2X$_7$R potency for testing in preclinical models, a common issue with P2X$_7$R antagonists due to wide differences in receptor homology across species.

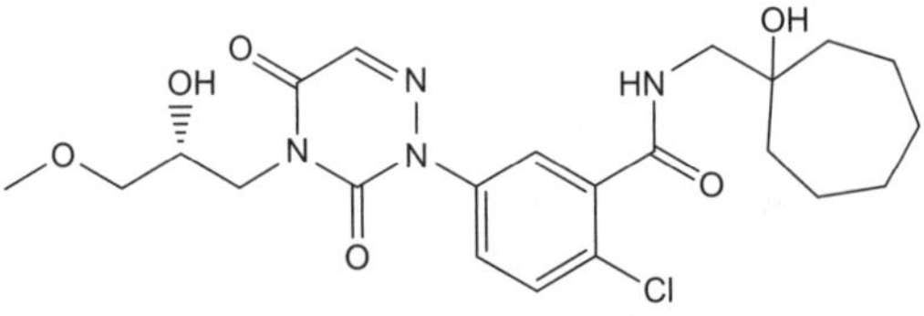

4

Many of the early $P2X_7R$ antagonists either lack sufficient CNS penetration or rodent $P2X_7R$ antagonist activity to support testing in preclinical *in vivo* NI models. However, there have been recent advances in overcoming these challenges. For instance, 2-cyano-guanidine analog **5** (A-804598) is one of the few selective $P2X_7R$ antagonists with good CNS penetration and similar activity in human, rat, and mouse ($P2X_7R$ IC_{50}: 11, 10, and 9 nM, respectively), thus making it a potential tool compound to probe preclinical rodent models where NI is a component. In addition, tritiation of **5** at the 4-position of the phenyl ring (**6**) provides a radioligand with high specific activity.[28]

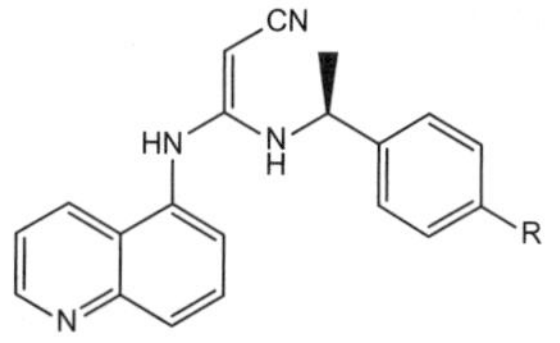

5 R = H
6 R = T

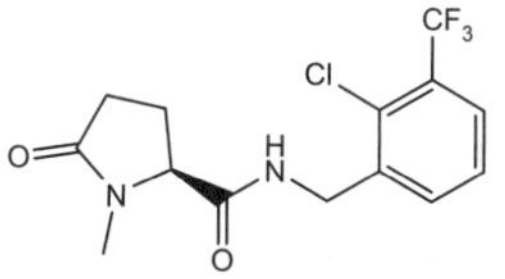

7

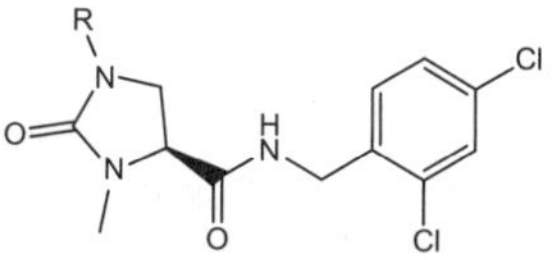

8 R = H
9 R = 3-(2-methyl)pyridine

A series of pyroglutamic acid amides, represented by **7**, with a human pIC_{50} of 8.5 and a rat pIC_{50} of 6.5 were reported recently. While **7** is 100-fold less active in rat, efficacy was still observed in a rat model of Freund's complete adjuvant (FCA)-induced centralized inflammatory knee joint pain and in a chronic constriction injury model of neuropathic pain. Exposure at the minimal effective doses provided free drug fractions in excess of the rat pIC_{50} in both the periphery and the CNS.[29] The importance of CNS activity in the FCA knee joint pain model was supported by a comparative study of 2-oxo-4-imidazolidinecarboxamide analogs **8** and **9** differing in their brain penetration capabilities. Compound **9** which was inactive in the FCA model had superior free plasma exposure and $P2X_7R$ potency, but poor brain penetration. On the other hand, the less potent, brain penetrant $P2X_7R$ antagonist **8** showed efficacy in this knee joint pain model.[30]

In summary, significant progress has been made in identifying tools that offer both sufficient CNS penetration and rodent potency to probe the utility of $P2X_7R$ antagonism for the treatment of disease where NI is present.

3. ENZYMES

3.1. Kynurenine pathway in neuroinflammation

The kynurenine pathway of tryptophan metabolism is an area of growing interest for NI and has received much attention in recent years.[13,31–33] Throughout the body, including the CNS, the primary fate of

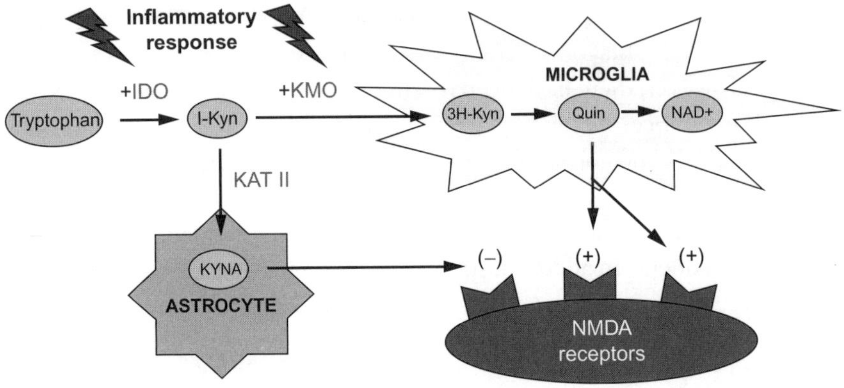

Figure 4.1 Kynurenine pathway. Enzymes regulating kynurenine metabolism in the CNS are reportedly upregulated in response to inflammation. Selective inhibition at points within the kynurenine pathway may be beneficial in treating NI-related conditions. Indolamine 2,3-dioxygenase (IDO), kynurenine 3-monooxygenase, kynurenine aminotransferase II (KAT II). (See Color Plate 4.1 in Color Plate Section.)

tryptophan is conversion to l-kynurenine. In turn, l-kynurenine is processed into several metabolites including kynurenic acid (KYNA) and quinolinic acid (QUIN) that subsequently activate or inhibit NMDA neurotransmission. KYNA and QUIN are produced in distinct cell types in the brain based on the distribution of the catabolic enzymes (Fig. 4.1).[34] The functional roles of the kynurenine pathway enzymes have been previously reviewed[35] and are currently the target of several drug discovery efforts. Recent evidence suggests that kynurenine metabolic enzymes are activated by inflammatory responses and their products may also impact NI and immune function.[36,37]

3.1.1 Indole 2,3-dioxygenase

IDO (IDO1; INDO) is the first and rate-limiting step in the kynurenine metabolic pathway of tryptophan in the CNS. TNFα and INFγ, elevated in inflamed tissue, stimulate IDO expression which regulates T-helper and T-regulatory cell populations[38] making IDO an attractive target for immune-related therapies. Within the CNS, inflammation-mediated IDO expression increases tryptophan metabolism, thereby depleting its availability for serotonin production.[39] Therefore, IDO inhibitors are speculated to have antidepressant activity for those suffering from chronic CNS inflammatory conditions.

Progress has been achieved in identifying new chemical matter for IDO. Historically, medicinal chemistry efforts have focused on tryptophan analogs, which act in a competitive fashion. More recently, noncompetitive β-carboline inhibitors have been identified.[40] In general, these compounds have IDO activity in the 10–100 μM range. The cocrystal structure of human IDO with the noncompetitive inhibitor phenyl-imidazole was solved recently,[41] and this key achievement has led to the rational design of new chemical motifs, some with improved potency. A recent patent application expands upon the phenyl-imidazoles with a number of compounds, such as **10** in the submicromolar range.[42,43] Several new noncompetitive chemotypes have been reported recently, which broadens the chemical space for additional discovery efforts.[44,45] A series of competitive inhibitors identified from a high-throughput screen, with submicromolar activity, was recently reported.[46] Specifically, lead compound **11** inhibits IDO with an IC_{50} of 67 nM. In a mouse *in vivo* model, **11** reduces plasma kynurenine levels by 60% at free plasma exposures 2.5-fold above the IC_{50}.

Identification of selective analogs has also become more challenging with the discovery of a second enzyme that metabolizes tryptophan into l-kynurenine, IDO2 (INDOL1), discovered in 2007 with 43% homology to IDO1.[47] Recently, a series of thiazolopyrazolo analogs with excellent IDO1 selectivity were described. For example, compound **12** has an IDO1 IC_{50} of 3 μM and is >80-fold selectivity versus IDO2.[48] This selectivity over IDO2 is corroborated by docking studies based on the crystal structure of IDO1 and a homology model of IDO2.

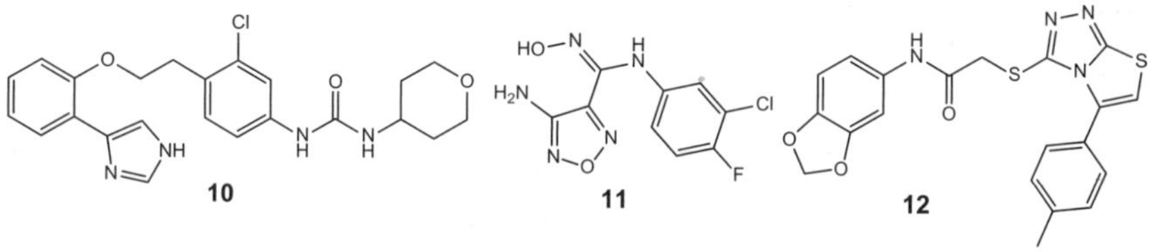

10 **11** **12**

3.1.2 Kynurenine aminotransferase II

The KYNA branch of the kynurenine pathway is regulated by kynurenine aminotransferases (KATs I–IV), and specifically KAT II in the brain.[49] Though often referred to as the "neuroprotective" branch of the kynurenine pathway due to an inhibitory effect on glutamate neurotransmission, KYNA has recently been shown to also act as an endogenous agonist of the aryl hydrocarbon receptor (AhR)[50] suggesting a potential role in inflammation and immune responses. IL-1β-mediated release of IL-6 from astrocytes was synergistically augmented by KYNA through the activation of AhR. While this field is still in its infancy, these data are intriguing in the context of neuroscience where elevated IL-6 levels are the most commonly associated change in IL expression in inflammation-mediated mood disturbances.[47]

Historically, there has been a paucity of selective and potent tools available to elucidate the role of KYNA in the brain. The discovery of agents such as **13**, *S*-ESBA, was an important advancement in KYNA biology. However, due to its poor CNS penetration, direct injection into the brain is required for *in vivo* studies. While this may be appropriate for measuring effects on glutamate neurotransmission, intrathecal injection compromises NI evaluation since local drug application compromises the blood–brain barrier and is likely to elicit an acute inflammatory reaction. However, with the discovery of the KAT II crystal structure,[51] progress is being made in the development of more potent and brain penetrable agents. In 2010, a second generation KAT II inhibitor, BFF-122 (**14**), was disclosed which covalently binds to the enzyme cofactor pyridoxal phosphate (PLP) in the catalytic pocket creating an irreversible adduct.[52] Though still possessing poor brain penetration, **14** was the first reported compound with submicromolar affinity for the human KAT II enzyme representing an important improvement in potency. More recently, PF-4589989 (**15**) was disclosed as the first low nanomolar affinity brain penetrable KAT II inhibitor.[53,54] Like BFF-122, PF-4859989 reportedly forms a covalent adduct with PLP in the binding pocket of KAT II resulting in inactivation of the enzyme. While irreversible inhibitors create challenges in defining the biological off-rate of their effects, these agents will be critical in understanding both the central and peripheral roles of KYNA production in inflammation and immune biology.

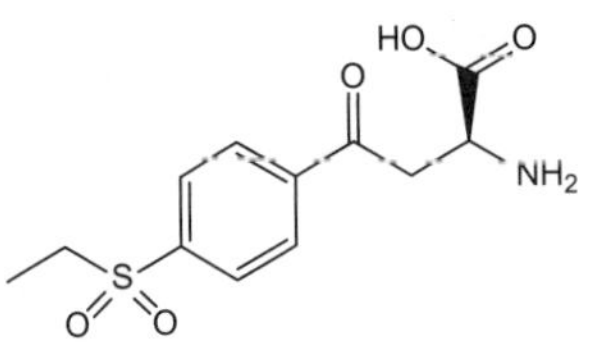

13

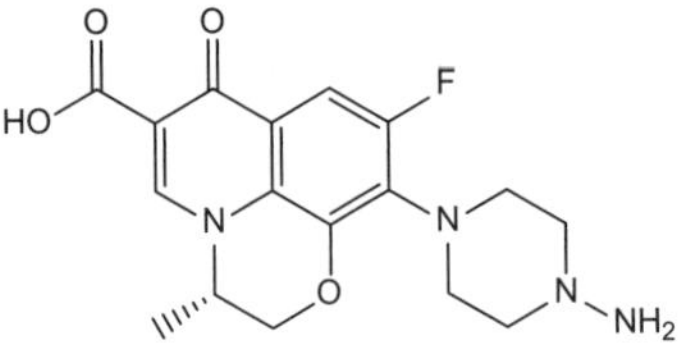

14

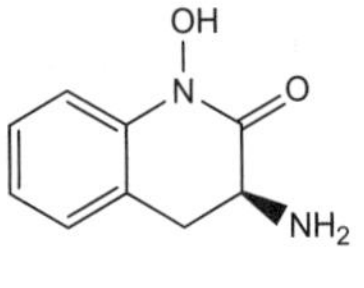

15

4. RECEPTORS

4.1. Toll-like receptors

Toll-like receptors (TLRs) play a critical role in the proper coordination of innate and adaptive immune responses to foreign pathogens and injury.[55,56] TLRs are type 1 transmembrane glycoproteins, expressed either on plasma membranes or on intracellular membranes, which form homo- or heterodimers that undergo conformational shifts upon ligand binding. This shift leads to the recruitment of adaptor and signaling molecules which, in turn, induce specific patterns of cytokine and chemokine expression. There are 10 known TLRs in humans, and despite a high degree of structural similarity, each receptor complex recognizes a distinct pathogen (PAMP)- or danger-associated molecular pattern (DAMP). TLRs are traditionally thought of as peripheral immune cell (leukocyte) targets[57–59]; however, a growing body of evidence indicates that TLRs (in particular, TLRs 2, 3, 4, 8, and 9) also play a crucial role within the CNS.[17,60] For example, increased TLR expression and NI have been observed in the brains of AD patients as well as in animal models of AD.[61] Furthermore, TLR-induced leukocyte activity in the periphery can regulate the pathology of certain neurodegenerative CNS disorders in animal models.[62–64] Such findings hint at the possibility of treating specific CNS disorders by peripheral modulation of TLRs,[65] perhaps bypassing the need for CNS penetrant compounds. This section will focus on two representative TLRs of high biological interest, TLR4 (plasma membrane bound) and TLR9 (endolysosomal membrane bound).

4.1.1 TLR4

TLR4 has several known agonist ligands, both pathogen derived as well as endogenous (HSP60, HSP90, beta-amyloid, α-synuclein, fibrinogen, and opioids).[61,66–69] Binding of LPS to MD2 triggers a structural rearrangement that is transduced to TLR4 and its cytosolic toll and interleukin 1 receptor (TIR) like domain resulting in clustering of TLR4 receptors enabling interactions with other signaling proteins.[70] In the CNS, TLR4 is primarily expressed on microglia although expression on astrocytes and neurons has been reported.[71] Clinical and preclinical data implicate TLR4 in a variety of CNS diseases including AD, ALS, epilepsy, MS, neuropathic pain, PD, stroke, and ischemia.[17,60,61,71–76] For stroke prevention and plaque clearance in AD, an agonist would be desirable based on preclinical models, although extreme TLR4 activation could result in sepsis. For treatment of stroke, PD, neuropathic pain, and ALS, an antagonist would be required; however, immunosuppression could be a drawback. While safety issues are a potential concern as a result of modulating the immune system, there is compelling biological rational suggesting that TLR4 modulation could have substantial clinical impact on a variety of diseases.

Several recent reviews highlight the chemical matter for TLR4.[57,77–79] TLR4 antagonists block LPS signaling by a variety of modes including interfering with TLR4 directly or interfering with the TLR4 and LPS binding proteins interaction.[80] The small molecule irreversible inhibitor **16**, TAK-242, is one of only a few examples of small molecules that interact directly with TLR4 through binding to cys-747 in its intracellular domain. TAK-242 was in phase III clinical development for the treatment of sepsis but was recently discontinued due to lack of efficacy. Eritoran tetrasodium is a lipid A analog in phase 2 clinical trials for sepsis and is administered by IV injection. This compound binds to MD2, thus blocking the interaction of LPS with the TLR4–MD2 complex. Eritoran tetrasodium inhibits LPS-induced TNF-α release in human whole blood with an IC_{50} of 10 nM. In a similar mode of action, small molecule benzylpyrazoles, exemplified by **17** ($EC_{50}=18.7\ \mu M$), were reported[81] to inhibit LPS-induced nitric oxide production by binding to the MD2 region that interacts with LPS. These analogs may offer a new starting point for medicinal chemistry efforts to provide orally available TLR4 antagonists.

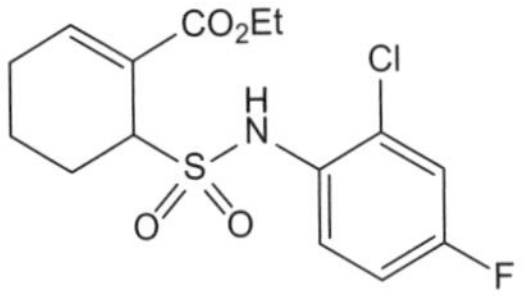

16
TAK-242

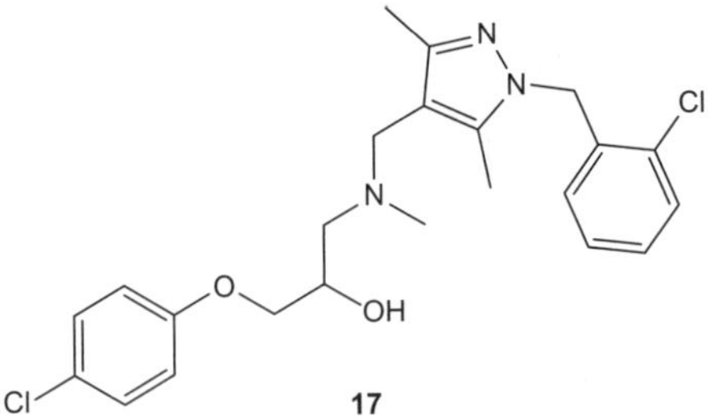

17

4.1.2 TLR9

While most TLRs are expressed on the cell surface of leukocytes, TLR9 primarily functions within subcellular endolysosomes (digestive organelles)[82–84] where it is ideally situated to detect PAMP ligands from internalized and digested DNA material. TLR9 biology has been implicated in a range of CNS infection, injury, and disease settings including meningitis and herpes,[85] AD,[62,86] MS,[87,88] Guillain–Barré syndrome,[89] and ischemic stroke.[63] Depending on disease context, either TLR9 agonists or antagonists could serve as therapeutic agents.

TLR9 agonists: Short fragments of single-stranded DNA (ssDNA) containing unmethylated cytosine-phospho-guanine (CpG) motifs, which are overrepresented in bacterial and viral, but not mammalian DNA, serve as naturally occurring PAMP ligands for TLR9.[90] Agonist SARs for synthetic

CpG ssDNA sequences and their sugar backbone modifications have been extensively reviewed.[59,90–92] However, recent work has identified novel sequence and structural modifications, as well as novel ssDNA carriers, which improve the pharmacokinetic properties of CpG ssDNA. While most CpG ssDNA has limited secondary or tertiary structure, three-dimensional "origami" CpG ssDNA structures have been described that elicit more robust immunological responses than equivalent amounts of standard CpG ssDNA.[93] In addition, the immunostimulatory effects of CpG ssDNA are enhanced when formulated with novel liposome carriers[94] or carbon nanotubes[95] or boron nitride nanospheres,[96] perhaps due to increased delivery and exposure of CpG ssDNA within the TLR9-expressing endolysosomes.[82] To date, only CpG ssDNA-based TLR9 agonists have been disclosed.

TLR9 antagonists: Both nucleotide and small molecule TLR9 antagonists have been described. For example, novel 4,5-fused pyrimidine derivatives **18** and **19** have been shown to inhibit CpG-induced cytokine expression *in vitro* and *in vivo*.[97] Gold nanoparticles, ideal for medical applications due to their bio-inert and noncytotoxic properties, have also recently been shown to inhibit TLR9-specific signaling in a particle size- and concentration-dependent manner.[98] In addition, ssDNA inhibitors of TLR9 have been previously described.[99]

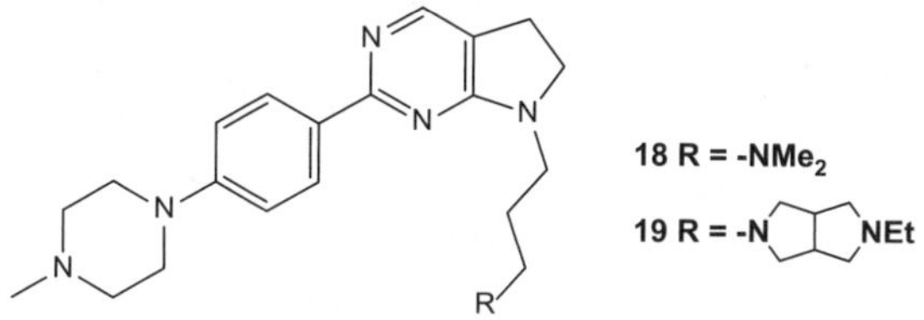

4.2. Chemokine receptors

Chemokines have previously been reviewed in Annual Report in Medicinal Chemistry. (ARMC see 30, 209; 35, 191; 39, 117). Chemokines are defined and classified by conserved cysteine residues that form intramolecular disulfide bonds. The number of cysteine pairs as well as the number of amino acids separating the two internal residues determines their class and name. For example, the chemokine **C**X3**C**L1 has two internal *cysteine residues* separated by 3 amino acids (X3) where L denotes ligand and R denotes receptor CX3C***R***1. In all, there are at least 46 chemokine ligands and at least 18 functional receptors. All chemokines are secreted in relatively large quantities except CXCL6 and

CX3CL1, as they are produced as membrane-bound ligands. Furthermore, there is tremendous promiscuity in the receptor/ligand interactions with most chemokine receptors binding multiple ligands and most chemokine ligands binding to multiple receptors. This biology has been reviewed in detail.[100]

4.2.1 CX3CR1

CX3C*R*1 (*R*eceptor for CX3CL1) is an attractive target because it has a 1:1 binding specificity with its ligand, CX3CL1 (also known as fractalkine). CX3CL1 is the sole member of the class of chemokines whose cysteines are separated by 3 amino acids. While CX3CL1 is produced as a membrane-bound ligand by neurons, constitutive cleavage liberates soluble chemokine that acts as a chemoattractant. Under stress, chemokine cleavage is increased via a different set of inducible proteases.[101,102] In the periphery, CX3CR1 is expressed on T cells, dendritic cells, and a small subpopulation of monocytes,[102] while in the CNS, the receptor is exclusively expressed on microglia. Peripherally, CX3CL1 is expressed on vascular endothelial tissue where it may be involved in leukocyte extravasation into tissues, and in the CNS, the ligand is predominantly expressed on neurons. There is considerable evidence that in models of neuropathic pain CX3CL1 signals neuronal stress to microglia, resulting in microglia-induced enhancement of pain sensation. Additionally, several *in vivo* studies have been published that suggest that altering (increasing or decreasing) CX3CL1 signaling can significantly modulate neurodegeneration and pathology in animal models of ALS, AD, PD, and SCI,[99,103–107] making this a promising target for therapeutic intervention in diseases with NI.

New small molecule-based chemical matter for CX3CR1 has been relatively limited with the exception of a group of published patents on a series of thiazolopyrimidines and thiazolopyrimidones.[108–111] These compounds were identified based on cross-reactivity profiling of CXCR2 antagonists. The reference compound benzylthio analog **20** has a K_i of 54 nM for inhibiting ^{125}I-CX3CL1 binding in membrane preparations from cells expressing CX3CR1 but is nonselective over CXCR2. Addition of an alpha-methyl to the benzyl group, as in **21**, improves the CX3CR1 K_i to 7.8 nM and reduces the K_i for CXCR2 to 1359 nM. The chirality of the alpha-methyl group is important for driving CX3CR1 selectivity as the *R*-diastereomer **22** has similar potency as **21** at CX3CR1, but has a K_i of only 240 nM at CXCR2.

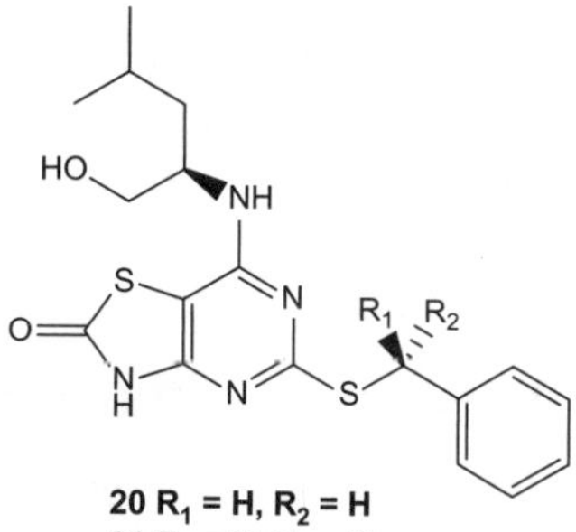

20 R_1 = H, R_2 = H
21 R_1 = H, R_2 = Me
22 R_1 = Me, R_2 = H

5. CONCLUDING REMARKS

In summary, NI is an important factor of many CNS diseases, and alleviating NI is anticipated to reduce disease severity and improve patient outcome in a majority of cases. There are a variety of traditionally druggable targets for NI such as enzymes, receptors, and ion channels. In this review, targets were highlighted where advances in medicinal chemistry had been achieved in the recent past. In terms of future perspectives, a better understanding of inflammation in the brain and the relationship between peripheral and central inflammation is required. Since most currently approved anti-inflammatory drugs target the periphery, it is important to understand the extent to which these compounds affect NI. Even though increased risk of infection is a potential issue for NI targets owing to immunomodulatory effects,[112,113] there is significant opportunity to discover new molecules for the alleviation of NI in CNS diseases.

REFERENCES

(1) Broussard, G.J.; Mytar, J.; Li, R.-C.; Klapstein, G.J. *Inflammopharmacology* **2012**, *20*, 109–126.
(2) Kim, S.U.; de Vellis, J. *J. Neurosci. Res.* **2005**, *81*, 302–313.
(3) Papadimitriou, D.; Le Verche, V.; Jacquier, A.; Ikiz, B.; Przedborski, S.; Re, D.B. *Neurobiol. Dis.* **2010**, *37*, 493–502.
(4) Przedborski, S. *Mov. Disord.* **2010**, *25*(Suppl. 1), S55–S57.
(5) Schwab, C.; McGeer, P.L. *J. Alzheimers Dis.* **2008**, *13*, 359–369.
(6) Grovit-Ferbas, K.; Harris-White, M.E. *Immunol. Res.* **2010**, *48*, 40–58.
(7) Najjar, S.; Pearlman, D.; Miller, D.C.; Devinsky, O. *Neurologist* **2011**, *17*, 249–254.
(8) Schnieder, T.P.; Dwork, A.J. *Biol. Psychiatry* **2011**, *69*, 134–139.
(9) Vargas, D.L.; Nascimbene, C.; Krishnan, C.; Zimmerman, A.W.; Pardo, C.A. *Ann. Neurol.* **2005**, *57*, 67–81.
(10) Thiel, A.; Heiss, W.D. *Stroke* **2011**, *42*, 507–512.
(11) Alexander, J.K.; Popovich, P.G. *Prog. Brain Res.* **2009**, *175*, 125–137.

(12) Vallejo, R.; Tilley, D.M.; Vogel, L.; Benyamin, R. *Pain Pract.* **2010**, *10*, 167–184.
(13) Moller, T. *J. Neural Transm.* **2010**, *117*, 1001–1008.
(14) Wieseler-Frank, J.; Maier, S.F.; Watkins, L.R. *Neurochem. Int.* **2004**, *45*, 389–395.
(15) Nimmo, A.J.; Vink, R. *Recent Pat. CNS Drug Discov.* **2009**, *4*, 86–95.
(16) English, J.M.; Cobb, M.H. *Trends Pharmacol. Sci.* **2002**, *23*, 40–45.
(17) Okun, E.; Griffioen, K.J.; Mattson, M.P. *Trends Neurosci.* **2011**, *34*, 269–281.
(18) Di Virgilio, F.; Ceruti, S.; Bramanti, P.; Abbracchio, M.P. *Trends Neurosci.* **2009**, *32*, 79–87.
(19) Coull, J.A.; Beggs, S.; Boudreau, D.; Boivin, D.; Tsuda, M.; Inoue, K.; Gravel, C.; Salter, M.W.; De Koninck, Y. *Nature* **2005**, *438*, 1017–1021.
(20) Trang, T.; Beggs, S.; Salter, M.W. *Exp. Neurol.* **2012**, *234*, 354–361.
(21) Tsuda, M.; Shigemoto-Mogami, Y.; Koizumi, S.; Mizokoshi, A.; Kohsaka, S.; Salter, M.W.; Inoue, K. *Nature* **2003**, *424*, 778–783.
(22) Ulmann, L.; Hatcher, J.P.; Hughes, J.P.; Chaumont, S.; Green, P.J.; Conquet, F.; Buell, G.N.; Reeve, A.J.; Chessell, I.P.; Rassendren, F. *J. Neurosci.* **2008**, *28*, 11263–11268.
(23) Gum, R.J.; Wakefield, B.; Jarvis, M.F. *Purinergic Signal.* **2012**, *8*, 41–56.
(24) Gunosewoyo, H.; Kassiou, M. *Expert Opin. Ther. Pat.* **2010**, *20*, 625–646.
(25) Sakuma, S.; Arai, M.; Kobayashi, K.; Watanabe, Y.; Imai, T.; Inoue, K. Patent Application WO 2012008478, **2012**.
(26) Ushioda, M.; Sakuma, S.; Imai, T.; Inoue, K. Patent Application WO 2012017876, **2012**.
(27) Duplantier, A.J.; Dombroski, M.A.; Subramanyam, C.; Beaulieu, A.M.; Chang, S.-P.; Gabel, C.A.; Jordan, C.; Kalgutkar, A.S.; Kraus, K.G.; Labasi, J.M.; Mussari, C.; Perregaux, D.G.; Shepard, R.; Taylor, T.J.; Trevena, K.A.; Whitney-Pickett, C.; Yoon, K. *Bioorg. Med. Chem. Lett.* **2011**, *21*, 3708–3711.
(28) Donnelly-Roberts, D.L.; Namovic, M.T.; Surber, B.; Vaidyanathan, S.X.; Perez-Medrano, A.; Wang, Y.; Carroll, W.A.; Jarvis, M.F. *Neuropharmacology* **2008**, *56*, 223–229.
(29) Abdi, M.H.; Beswick, P.J.; Billinton, A.; Chambers, L.J.; Charlton, A.; Collins, S.D.; Collis, K.L.; Dean, D.K.; Fonfria, E.; Gleave, R.J.; Lejeune, C.L.; Livermore, D.G.; Medhurst, S.J.; Michel, A.D.; Moses, A.P.; Page, L.; Patel, S.; Roman, S.A.; Senger, S.; Slingsby, B.; Steadman, J.G.A.; Stevens, A.J.; Walter, D.S. *Bioorg. Med. Chem. Lett.* **2010**, *20*, 5080–5084.
(30) Abberley, L.; Bebius, A.; Beswick, P.J.; Billinton, A.; Collis, K.L.; Dean, D.K.; Fonfria, E.; Gleave, R.J.; Medhurst, S.J.; Michel, A.D.; Moses, A.P.; Patel, S.; Roman, S.A.; Scoccitti, T.; Smith, B.; Steadman, J.G.A.; Walter, D.S. *Bioorg. Med. Chem. Lett.* **2010**, *20*, 6370–6374.
(31) Dobos, N.; de Vries Erik, F.J.; Kema Ido, P.; Patas, K.; Prins, M.; Nijholt Ingrid, M.; Dierckx Rudi, A.; Korf, J.; den Boer Johan, A.; Luiten Paul, G.M.; Eisel Ulrich, L.M. *J. Alzheimers Dis.* **2012**, *28*, 905–915.
(32) Zinger, A.; Barcia, C.; Herrero Maria, T.; Guillemin Gilles, J. *Parkinsons Dis.* **2011**, *2011*, 716859.
(33) Kincses, Z.T.; Toldi, J.; Vecsei, L. *J. Cell. Mol. Med.* **2010**, *14*, 2045–2054.
(34) Amori, L.; Guidetti, P.; Pellicciari, R.; Kajii, Y.; Schwarcz, R. *J. Neurochem.* **2009**, *109*, 316–325.
(35) Schwarcz, R. *Curr. Opin. Pharmacol.* **2004**, *4*, 12–17.
(36) Chen, Y.; Guillemin, G.J. *Int. J. Tryptophan Res.* **2009**, *2*, 1–19.
(37) Kolodziej, L.R.; Paleolog, E.M.; Williams, R.O. *Amino Acids* **2011**, *41*, 1173–1183.
(38) Xu, H.; Zhang, G.-X.; Ciric, B.; Rostami, A. *Immunol. Lett.* **2008**, *121*, 1–6.
(39) Oxenkrug Gregory, F. *Isr. J. Psychiatry Relat. Sci.* **2010**, *47*, 56–63.

(40) Huang, Q.; Zheng, M.; Yang, S.; Kuang, C.; Yu, C.; Yang, Q. *Eur. J. Med. Chem.* **2011**, *46*, 5680–5687.
(41) Sugimoto, H.; Oda, S.-i.; Otsuki, T.; Hino, T.; Yoshida, T.; Shiro, Y. *Proc. Natl. Acad. Sci. U.S.A.* **2006**, *103*, 2611–2616.
(42) Mautino, M.R.; Kumar, S.; Jaipuri, F.; Waldo, J.; Kesharwani, T.; Zhang, X. Patent Application WO 2011056652, **2011**.
(43) Di Pucchio, T.; Danese, S.; De Cristofaro, R.; Rutella, S. *Expert Opin. Ther. Pat.* **2010**, *20*, 229–250.
(44) Dolusic, E.; Larrieu, P.; Blanc, S.; Sapunaric, F.; Pouyez, J.; Moineaux, L.; Colette, D.; Stroobant, V.; Pilotte, L.; Colau, D.; Ferain, T.; Fraser, G.; Galleni, M.; Frere, J.-M.; Masereel, B.; Van den Eynde, B.; Wouters, J.; Frederick, R. *Eur. J. Med. Chem.* **2011**, *46*, 3058–3065.
(45) Smith, J.R.; Evans, K.J.; Wright, A.; Willows, R.D.; Jamie, J.F.; Griffith, R. *Bioorg. Med. Chem.* **2012**, *20*, 1354–1363.
(46) Yue, E.W.; Douty, B.; Wayland, B.; Bower, M.; Liu, X.; Leffet, L.; Wang, Q.; Bowman, K.J.; Hansbury, M.J.; Liu, C.; Wei, M.; Li, Y.; Wynn, R.; Burn, T.C.; Koblish, H.K.; Fridman, J.S.; Metcalf, B.; Scherle, P.A.; Combs, A.P. *J. Med. Chem.* **2009**, *52*, 7364–7367.
(47) Metz, R.; DuHadaway, J.B.; Kamasani, U.; Laury-Kleintop, L.; Muller, A.J.; Prendergast, G.C. *Cancer Res.* **2007**, *67*, 7082–7087.
(48) Meininger, D.; Zalameda, L.; Liu, Y.; Stepan, L.P.; Borges, L.; McCarter, J.D.; Sutherland, C.L. *Biochim. Biophys. Acta, Proteins Proteomics* **2011**, *1814*, 1947–1954.
(49) Han, Q.; Cai, T.; Tagle, D.A.; Li, J. *Cell. Mol. Life Sci.* **2010**, *67*, 353–368.
(50) Di Natale, B.C.; Murray, I.A.; Schroeder, J.C.; Flaveny, C.A.; Lahoti, T.S.; Laurenzana, E.M.; Omiecinski, C.J.; Perdew, G.H. *Toxicol. Sci.* **2010**, *115*, 89–97.
(51) Rossi, F.; Garavaglia, S.; Montalbano, V.; Walsh, M.A.; Rizzi, M. *J. Biol. Chem.* **2008**, *283*, 3559–3566.
(52) Rossi, F.; Valentina, C.; Garavaglia, S.; Sathyasaikumar, K.V.; Schwarcz, R.; Kojima, S.-i.; Okuwaki, K.; Ono, S.-i.; Kajii, Y.; Rizzi, M. *J. Med. Chem.* **2010**, *53*, 5684–5689.
(53) Dounay, A.B.; Anderson, M.; Bechle, B.M.; Campbell, B.M.; Claffey, M.M.; Evdokimov, A.; Evrard, E.; Fonseca, K.R.; Gan, X.; Ghosh, S.; Hayward, M.M.; Horner, W.; Kim, J.-Y.; McAllister, L.A.; Pandit, J.; Paradis, V.; Parikh, V.D.; Reese, M.R.; Rong, S.; Salafia, M.A.; Schuyten, K.; Strick, C.A.; Tuttle, J.B.; Valentine, J.; Wang, H.; Zawadzke, L.E.; Verhoest, P.R. *ACS Med. Chem. Lett.* **2012**, *3*, 187–192.
(54) Claffey, M.M.; Dounay, A.B.; Gan, X.; Hayward, M.M.; Rong, S.; Tuttle, J.B.; Verhoest, P.R. Patent Application WO 2010146488, **2010**.
(55) Medzhitov, R. *Nat. Rev. Immunol.* **2001**, *1*, 135–145.
(56) Iwasaki, A.; Medzhitov, R. *Nat. Immunol.* **2004**, *5*, 987–995.
(57) Hennessy, E.J.; Parker, A.E.; O'Neill, L.A.J. *Nat. Rev. Drug Discov.* **2010**, *9*, 293–307.
(58) Keogh, B.; Parker, A.E. *Trends Pharmacol. Sci.* **2011**, *32*, 435–442.
(59) Spyvee, M.; Hawkins, L.D.; Ishizaka, S.T. *Annu. Rep. Med. Chem.* **2010**, *45*, 191–207.
(60) Carty, M.; Bowie, A.G. *Biochem. Pharmacol.* **2011**, *81*, 825–837.
(61) Lee, C.Y.; Landreth, G.E. *J. Neural Transm.* **2010**, *117*, 949–960.
(62) Scholtzova, H.; Kascsak, R.J.; Bates, K.A.; Boutajangout, A.; Kerr, D.J.; Meeker, H.C.; Mehta, P.D.; Spinner, D.S.; Wisniewski, T. *J. Neurosci.* **2009**, *29*, 1846–1854.
(63) Stevens, S.L.; Ciesielski, T.M.P.; Marsh, B.J.; Yang, T.; Homen, D.S.; Boule, J.-L.; Lessov, N.S.; Simon, R.P.; Stenzel-Poore, M.P. *J. Cereb. Blood Flow Metab.* **2008**, *28*, 1040–1047.

(64) Rosenzweig, H.L.; Lessov, N.S.; Henshall, D.C.; Minami, M.; Simon, R.P.; Stenzel-Poore, M.P. *Stroke* **2004**, *35*, 2576–2581.
(65) McAllister, A.K.; Water, J.v.d. *Neuron* **2009**, *64*, 9–12.
(66) Hutchinson, M.R.; Zhang, Y.; Shridhar, M.; Evans, J.H.; Buchanan, M.M.; Zhao, T.X.; Slivka, P.F.; Coats, B.D.; Rezvani, N.; Wieseler, J.; Hughes, T.S.; Landgraf, K.E.; Chan, S.; Fong, S.; Phipps, S.; Falke, J.J.; Leinwand, L.A.; Maier, S.F.; Yin, H.; Rice, K.C.; Watkins, L.R. *Brain Behav. Immun.* **2010**, *24*, 83–95.
(67) Stefanova, N.; Fellner, L.; Reindl, M.; Masliah, E.; Poewe, W.; Wenning, G.K. *Am. J. Pathol.* **2011**, *179*, 954–963.
(68) Fassbender, K.; Walter, S.; Kuhl, S.; Landmann, R.; Ishii, K.; Bertsch, T.; Stalder, A.K.; Muehlhauser, F.; Liu, Y.; Ulmer, A.J.; Rivest, S.; Lentschat, A.; Gulbins, E.; Jucker, M.; Staufenbiel, M.; Brechtel, K.; Walter, J.; Multhaup, G.; Penke, B.; Adachi, Y.; Hartmann, T.; Beyreuther, K. *FASEB J.* **2004**, *18*, 203–205.
(69) Gay, N.J.; Gangloff, M. *Annu. Rev. Biochem.* **2007**, *76*, 141–165.
(70) Palsson-McDermott, E.M.; O'Neill, L.A. *Immunology* **2004**, *113*, 153–162.
(71) Buchanan, M.M.; Hutchinson, M.; Watkins, L.R.; Yin, H. *J. Neurochem.* **2010**, *114*, 13–27.
(72) Walter, S.; Letiembre, M.; Liu, Y.; Heine, H.; Penke, B.; Hao, W.; Bode, B.; Manietta, N.; Walter, J.; Schulz-Schuffer, W.; Fassbender, K. *Cell. Physiol. Biochem.* **2007**, *20*, 947–956.
(73) Lehnardt, S. *Glia* **2010**, *58*, 253–263.
(74) Maroso, M.; Balosso, S.; Ravizza, T.; Liu, J.; Aronica, E.; Iyer, A.M.; Rossetti, C.; Molteni, M.; Casalgrandi, M.; Manfredi, A.A.; Bianchi, M.E.; Vezzani, A. *Nat. Med.* **2010**, *16*, 413–419.
(75) Casula, M.; Iyer, A.M.; Spliet, W.G.; Anink, J.J.; Steentjes, K.; Sta, M.; Troost, D.; Aronica, E. *Neuroscience* **2011**, *179*, 233–243.
(76) Zhang, R.; Hadlock, K.G.; Do, H.; Yu, S.; Honrada, R.; Champion, S.; Forshew, D.; Madison, C.; Katz, J.; Miller, R.G.; McGrath, M.S. *J. Neuroimmunol.* **2011**, *230*, 114–123.
(77) Hu, Y.; Xie, G.H.; Chen, Q.X.; Fang, X.M. *Curr. Drug Targets* **2011**, *12*, 256–262.
(78) Czarniecki, M. *J. Med. Chem.* **2008**, *51*, 6621–6626.
(79) Wittebole, X.; Castanares-Zapatero, D.; Laterre, P.F. *Mediators Inflamm.* **2010**, No pp given.
(80) Peri, F.; Piazza, M. *Biotechnol. Adv.* **2012**, *30*, 251–260.
(81) Bevan, D.E.; Martinko, A.J.; Loram, L.C.; Stahl, J.A.; Taylor, F.R.; Joshee, S.; Watkins, L.R.; Yin, H. *ACS Med. Chem. Lett.* **2010**, *1*, 194–198.
(82) Brinkmann, M.M.; Spooner, E.; Hoebe, K.; Beutler, B.; Ploegh, H.L.; Kim, Y.M. *J. Cell. Biol.* **2007**, *177*, 265–275.
(83) Kim, T.S.; Lim, H.K.; Lee, J.Y.; Kim, D.J.; Park, S.; Lee, C.; Lee, C.U. *Neurosci. Lett.* **2008**, *436*, 196–200.
(84) Ewald, S.E.; Engel, A.; Lee, J.; Wang, M.; Bogyo, M.; Barton, G.M. *J. Exp. Med.* **2011**, *208*, 643–651.
(85) Sorensen, L.N.; Reinert, L.S.; Malmgaard, L.; Bartholdy, C.; Thomsen, A.R.; Paludan, S.R. *J. Immunol.* **2008**, *181*, 8604–8612.
(86) Doi, Y.; Mizuno, T.; Maki, Y.; Jin, S.; Mizoguchi, H.; Ikeyama, M.; Doi, M.; Michikawa, M.; Takeuchi, H.; Suzumura, A. *Am. J. Pathol.* **2009**, *175*, 2121–2132.
(87) Prinz, M.; Garbe, F.; Schmidt, H.; Mildner, A.; Gutcher, I.; Wolter, K.; Piesche, M.; Schroers, R.; Weiss, E.; Kirschning, C.J.; Rochford, C.D.; Bruck, W.; Becher, B. *J. Clin. Invest.* **2006**, *116*, 456–464.
(88) Marta, M.; Meier, U.C.; Lobell, A. *Autoimmun. Rev.* **2009**, *8*, 506–509.

(89) Wang, Y.-Z.; Liang, Q.-H.; Ramkalawan, H.; Zhang, W.; Zhou, W.-B.; Xiao, B.; Tian, F.-F.; Yang, H.; Li, J.; Zhang, Y.; Xu, N.-A. *Immunol. Invest.* **2012**, *41*, 171–182.
(90) Krieg, A.M. *Nat. Rev. Drug Discov.* **2006**, *5*, 471–484.
(91) Narayanan, S.; Dalpke, A.H.; Siegmund, K.; Heeg, K.; Richert, C. *J. Med. Chem.* **2003**, *46*, 5031–5044.
(92) Meng, W.; Yamazaki, T.; Nishida, Y.; Hanagata, N. *BioMed Cent. Biotechnol.* **2011**, *11*.
(93) Schueller, V.J.; Heidegger, S.; Sandholzer, N.; Nickels, P.C.; Suhartha, N.A.; Endres, S.; Bourquin, C.; Liedl, T. *Am. Chem. Soc. Nano* **2011**, *5*, 9696–9702.
(94) Kim, D.; Kwon, S.; Ahn, C.-S.; Lee, Y.; Choi, S.-Y.; Park, J.; Kwon, H.-Y.; Kwon, H.-J. *Biochem. Mol. Biol. Rep.* **2011**, *44*, 758–763.
(95) Zhao, D.; Alizadeh, D.; Zhang, L.; Liu, W.; Farrukh, O.; Manuel, E.; Diamond, D.J.; Badie, B. *Clin. Cancer Res.* **2010**, *17*, 771–782.
(96) Zhi, C.; Meng, W.; Yamazaki, T.; Bando, Y.; Golberg, D.; Tang, C.; Hanagata, N. *J. Mater. Chem.* **2011**, *21*, 5219–5222.
(97) Asano, S.; Kamimoto, K.; Isobe, Y. Patent Application WO 2011152485, **2011**.
(98) Tsai, C.-Y.; Lu, S.-L.; Hu, C.-W.; Yeh, C.-S.; Lee, G.-B.; Lei, H.-Y. *J. Immunol.* **2012**, *188*, 68–76.
(99) Fuhrmann, M.; Bittner, T.; Jung, C.K.; Burgold, S.; Page, R.M.; Mitteregger, G.; Haass, C.; LaFerla, F.M.; Kretzschmar, H.; Herms, J. *Nat. Neurosci.* **2010**, *13*, 411–413.
(100) Zlotnik, A.; Yoshie, O.; Nomiyama, H. *Genome Biol.* **2006**, 7, 243.
(101) D'Haese, J.G.; Demir, I.E.; Friess, H.; Ceyhan, G.O. *Expert Opin. Ther. Targets* **2010**, *14*, 207–219.
(102) Ludwig, A.; Mentlein, R. *J. Neuroimmunol.* **2008**, *198*, 92–97.
(103) Cardona, A.E.; Pioro, E.P.; Sasse, M.E.; Kostenko, V.; Cardona, S.M.; Dijkstra, I.M.; Huang, D.; Kidd, G.; Dombrowski, S.; Dutta, R.; Lee, J.C.; Cook, D.N.; Jung, S.; Lira, S.A.; Littman, D.R.; Ransohoff, R.M. *Nat. Neurosci.* **2006**, *9*, 917–924.
(104) Bhaskar, K.; Konerth, M.; Kokiko-Cochran, O.N.; Cardona, A.; Ransohoff, R.M.; Lamb, B.T. *Neuron* **2010**, *68*, 19–31.
(105) Denes, A.; Ferenczi, S.; Halasz, J.; Kornyei, Z.; Kovacs, K.J. *J. Cereb. Blood Flow Metab.* **2008**, *28*, 1707–1721.
(106) Lee, S.; Varvel, N.H.; Konerth, M.E.; Xu, G.; Cardona, A.E.; Ransohoff, R.M.; Lamb, B.T. *Am. J. Pathol.* **2010**, *177*, 2549–2562.
(107) Donnelly, D.J.; Longbrake, E.E.; Shawler, T.M.; Kigerl, K.A.; Lai, W.; Tovar, C.A.; Ransohoff, R.M.; Popovich, P.G. *J. Neurosci.* **2011**, *31*, 9910–9922.
(108) Nordvall, G.; Rein, T.; Sohn, D.; Zemribo, R. Patent Application WO 2005033115, **2005**.
(109) Johansson, R.; Karlstroem, S.; Kers, A.; Nordvall, G.; Rein, T.; Slivo, C. Patent Application WO 2008039138, **2008**.
(110) Johansson, R.; Karlstroem, S.; Kers, A.; Nordvall, G.; Rein, T.; Slivo, C. Patent Application WO 2008039139, **2008**.
(111) Dahlstroem, M.; Nordvall, G.; Rein, T.; Starke, I. Patent Application WO 2009120140, **2009**.
(112) Piccotti, J.R.; Lebrec, H.N.; Evans, E.; Herzyk, D.J.; Hastings, K.L.; Burns-Naas, L.A.; Gourley, I.S.; Wierda, D.; Kawabata, T.T. *J. Immunotoxicol.* **2009**, *6*, 1–10.
(113) Culver, E.L.; Travis, S.P.L. *Curr. Drug Targets* **2010**, *11*, 198–218.

CHAPTER FIVE

Secretase Inhibitors and Modulators as a Disease-Modifying Approach Against Alzheimer's Disease

Harrie J.M. Gijsen, François P. Bischoff
Neuroscience Medicinal Chemistry, Janssen Research & Development, Pharmaceutical Companies of Johnson & Johnson, Beerse, Belgium

Contents

1. INTRODUCTION

With an aging population across the world, the prevalence of Alzheimer's disease (AD) and the consequent burden to society are rapidly rising. The currently approved medications for AD only offer symptomatic treatment of limited duration without affecting the progression of the

Annual Reports in Medicinal Chemistry, Volume 47
ISSN 0065-7743
http://dx.doi.org/10.1016/B978-0-12-396492-2.00005-9

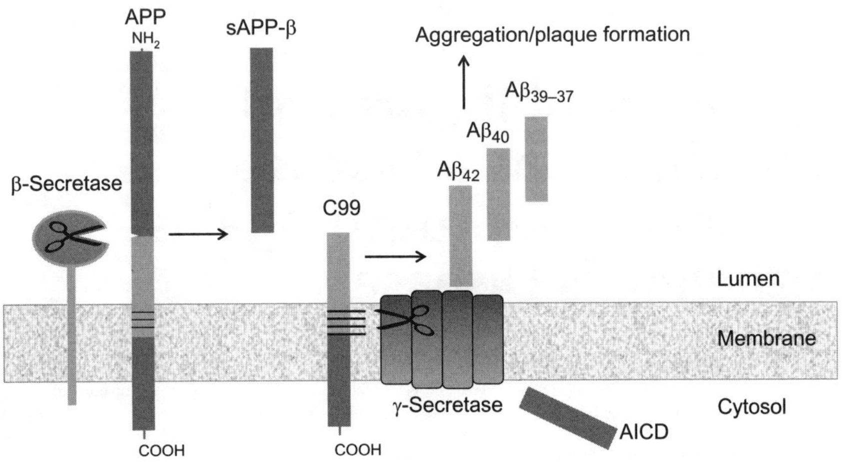

Figure 5.1 Production of amyloid peptides by the sequential cleavage of BACE1 and GS.

disease. Therefore, disease-modifying approaches are urgently needed. A hallmark pathology of AD is the presence of amyloid plaques in the brain, which are mainly aggregates of amyloid beta (Aβ) peptides, among which Aβ42 is the most neurotoxic. These peptides are formed via proteolytic processing of the amyloid precursor protein (APP) by two aspartyl proteases: beta-site APP-cleaving enzyme 1 (BACE1 or β-secretase) and γ-secretase (GS) (Fig. 5.1). Consequently, BACE1 and GS are attractive targets to prevent both the build-up of the amyloid plaques and the formation of toxic amyloid dimers and oligomers.

This report highlights key progress and findings toward γ-secretase inhibitors (GSIs)/modulators and β-secretase inhibitors, since the last time these topics were reviewed in this journal.[1]

2. INHIBITORS OF GS

2.1. Function of GS

GS is a member of the intramembrane-cleaving aspartyl protease family consisting of four integral membrane proteins: presenilin (PS) 1 or 2, nicastrin (Nct), anterior pharynx-defective 1 (Aph-1), and presenilin enhancer protein 2 (Pen-2).[2] PSs constitute the aspartic protease catalytic subunit of the GS complex which cleaves C99, the 99-residue membrane stub generated after cleavage of APP by BACE1, in a progressive manner at the ε, ζ, and γ sites, resulting in Aβ species of varying lengths.[3] Initial proteolysis releases the APP intracellular domain (AICD) in the cytoplasm, leaving membrane-embedded Aβ49 or Aβ48. GS then successively cleaves Aβ49 and Aβ48 by releasing tripeptides to give rise to Aβ40 and Aβ42, respectively.[4] These are either released extracellularly or further cleaved to generate shorter isoforms Aβ39–Aβ37, of which Aβ38 results from the product line containing Aβ42. The GS complex also functions in many other cellular processes, and more than 50 different substrates have been identified to date, including Notch,[5] creating the potential for adverse effects upon inhibition of GS.

2.2. Current status of GSIs

Several GSIs have moved to the clinic, and their progress has been reviewed recently.[6,7] The most advanced compound, semagacestat **1** (LY450,139) entered two large Phase III trials in mild-to-moderate AD patients which were prematurely interrupted in August 2010.[7] Instead of slowing disease progression, **1** was associated with a statistically significant decline in cognition. In addition, an increased risk of skin cancer was reported. Multiple reasons have been proposed to explain the lack of efficacy and observed side effects,[7] including the rebound effect on Aβ42 at low levels of **1**, poor selectivity versus Notch cleavage (threefold *in vitro*), and accumulation of neurotoxic C99.[8]

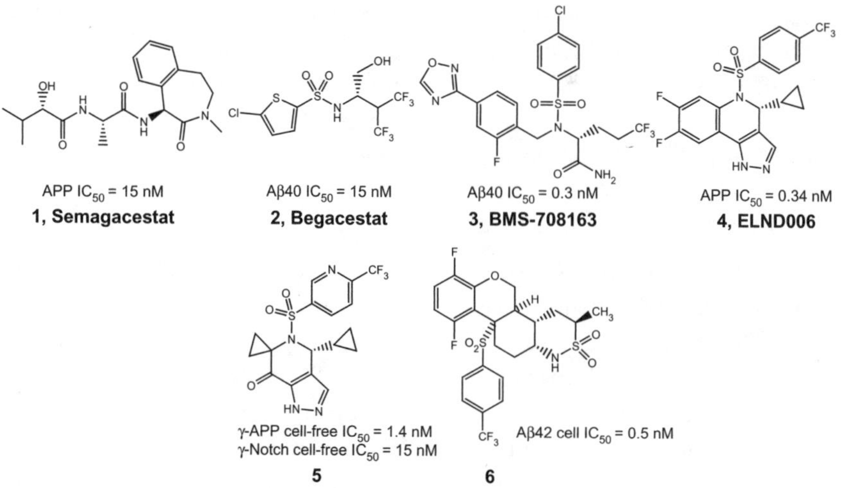

APP IC_{50} = 15 nM **1, Semagacestat**

Aβ40 IC_{50} = 15 nM **2, Begacestat**

Aβ40 IC_{50} = 0.3 nM **3, BMS-708163**

APP IC_{50} = 0.34 nM **4, ELND006**

γ-APP cell-free IC_{50} = 1.4 nM
γ-Notch cell-free IC_{50} = 15 nM
5

Aβ42 cell IC_{50} = 0.5 nM
6

Two other GSIs studied in the clinic, begacestat **2** (GSI-953) and ELND006 **4**, have 17- and 16-fold Notch sparing selectivity, respectively, in cellular assays.[9,10] Despite achieving Aβ lowering in human, development of both compounds has been halted, with **4** showing liver toxicity.[7,11] Avagacestat **3** (BMS-708163) is another potent GSI[12] undergoing clinical testing. Dosing **3** at 200 and 400 mg reduced Aβ42 levels in CSF by 32% and 34%, respectively. In a Phase II study with **3**, a trend for cognitive decline and potentially Notch-related side effects were noted at the high dose, despite a reported 193-fold selectivity against Notch.[12,13] Further development of **3** is targeting prodromal AD.[7]

Medicinal chemistry efforts around GSIs have been reviewed recently.[6] Since then, new arylsulfonamide-containing GSIs have been reported. For instance, pyrazole sulfonamide-based dihydroquinoline derivative **5**, with optimized PK properties, reduced mouse brain Aβ40 levels by 27% (1 mg/kg, p.o.).[14] Tetracyclic sulfones have also been reported as potent GSIs.[15] For instance, **6** (SCH 1500022) demonstrated subnanomolar cellular activity and reduced mouse brain Aβ40 levels by 57% (10 mg/kg, p.o.). Moreover, **6** was identified as a selective PS1 inhibitor, which may provide insights into developing new GSIs with improved side-effect profiles.[16]

3. MODULATORS OF GS

In comparison with GSIs, modulators (GSMs) cause a product shift from the longer Aβ peptide isoforms to shorter, more soluble, and less amyloidogenic isoforms, without inhibiting APP or NOTCH proteolytic processing. As such, modulating GS may avoid some of the adverse effects observed with GSIs. Since the late stage clinical failure of the NSAID-derived GSM tarenflurbil in 2008, considerable progress has been made toward discovering more potent and better brain penetrable compounds derived from both NSAID- and non-NSAID scaffolds.[17]

3.1. Mechanism of modulation

The molecular mechanism of GSMs is still a subject of debate.[18] Recent photoaffinity labeling studies with carboxylic acid-[19] and imidazole-derived GSMs[20] suggest that both bind to the N-terminal fragment of PS, although their binding sites only partially overlap.[20] All GSMs reduce Aβ42 levels without a major affect on total Aβ levels, but reduction of Aβ40 tends to be more pronounced for non-NSAID-derived GSMs.[21] The enhanced processing of the longer Aβ peptides toward the shorter isoforms Aβ39–Aβ37 supports the hypothesis that GSMs actually activate GS and thus counter the loss of GS function linked to many familial AD causing mutations.[22] In a study to further discriminate between modulation and inhibition, sustained cognitive improvement was achieved with a GSM, but not with GSIs, the latter leading to accumulation of C99.[8]

3.2. NSAID-derived modulators

The failure of tarenflurbil **7** to improve conditions in mild AD patients has been attributed to insufficient potency (Aβ42 IC_{50} ~ 300 μM) and a plasma-to-CSF ratio of only 0.5–1%.[23] A slight improvement in these parameters has been achieved with CHF5074 (**8**) (Aβ42 IC_{50} = 40 μM, brain penetration 3–5%), which is currently undergoing clinical testing.[24] Medicinal chemistry programs around new NSAID-derived GSMs have led to increasingly potent and brain-penetrant compounds. The orally active NSAID-derived GSM **9** (JNJ-40418677) was reported to inhibit Aβ42 with an IC_{50} of 200 nM and have excellent brain penetration with a B/P ratio around 1 in mice.[25] In a preventive study, chronic treatment of Tg2576 mice with **11** dose dependently reduced both plaque number and area.[25] Brain penetration of NSAID-derived GSMs is improved by introduction of a basic nitrogen atom as in **10** (BIIB042). A brain concentration of 4.6 μM of **10** was achieved in mice (10 mg/kg, p.o.), resulting in a 40% reduction in brain Aβ42.[26] Replacement of the central phenyl ring with piperidine, as in **11** and **12**, has been reported by several groups. Heterocyclic aromatic groups were introduced in order to lower the lipophilicity of the compounds, with **11** demonstrating high brain levels (4 μM) in mice (5 mg/kg, p.o.).[27] 4,4-Difluoropiperidine **12** resulted in improved pharmacokinetic properties and dose-dependently lowered Aβ42 in rats (ED_{50} 5 mg/kg, with brain/plasma levels at 1/3.7 μM).[28]

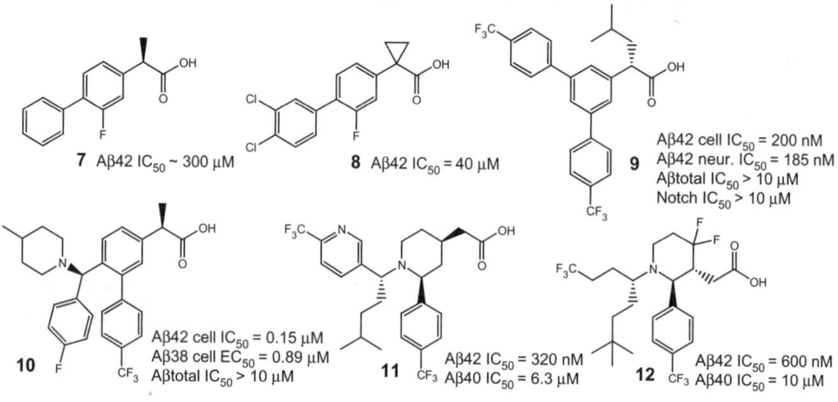

3.3. Imidazole-derived modulators

A class of noncarboxylic acid, imidazole-containing GSMs has been developed, which contains an anilinothiazole core represented by **13**. In line with NSAID-derived **9**, chronic treatment with **13** resulted in a significant inhibition of plaque deposition.[21] Related analog **14** (E-2012) had progressed toward a Phase I clinical trial and reduced plasma Aβ42 levels dose dependently, with a maximum reduction of ~50% after a 400-mg dose.[29] Further development of **14** was suspended as lenticular opacity was observed in a preclinical study in rats.[17]

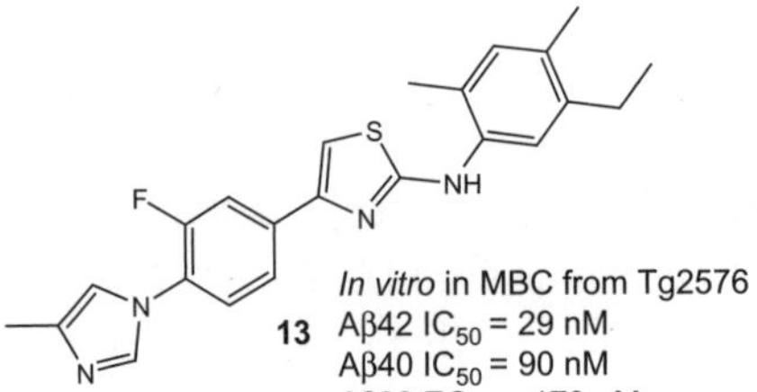

13 *In vitro* in MBC from Tg2576
Aβ42 IC_{50} = 29 nM
Aβ40 IC_{50} = 90 nM
Aβ38 EC_{50} = 170 nM
Aβtotal unchanged

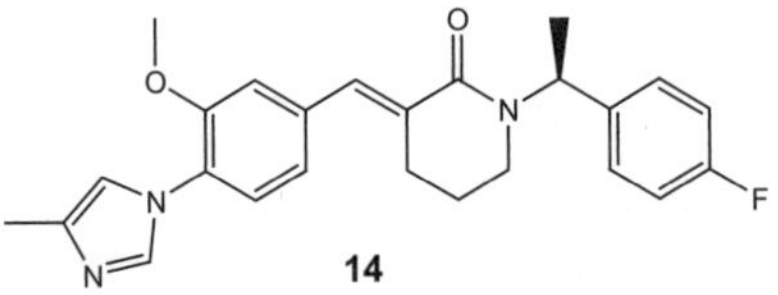

14

These two series have served as a starting point for a substantial amount of work that has been recently reviewed.[17] For example, a novel series of potent pyridazine- and pyridine-derived GSMs have been described, exemplified by **15**, which reduced rat brain Aβ42 by 28% with a corresponding brain concentration of 8.9 μM.[30] Several chemical subclasses with additional conformational restrictions led to a further increase in *in vitro* and *in vivo* potency. Among these, benzimidazole **16** induced a 63% lowering of Aβ42 levels and a 91% increase of Aβ38 levels in mouse brain (30 mg/kg, p.o.).[31] Potent pyrazolopyridine **17** produced a 45% reduction of rat CSF Aβ42 levels at a corresponding brain concentration of 1.22 μM.[32] Application of amide isosteres to **14** resulted in cyclic hydroxylamidines, as exemplified by **18**.[33] In rat, treatment with **18** (3 mg/kg, p.o.) reduced cortical Aβ42 by 37%. Replacement of the characteristic imidazole moiety in these series with other heterocycles, as exemplified with **19**, has also been reported.[34]

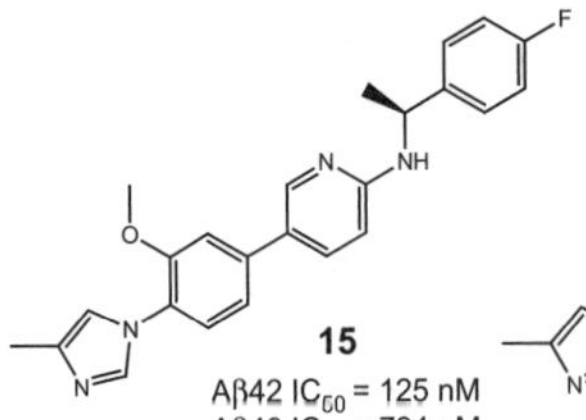

15
Aβ42 IC_{50} = 125 nM
Aβ40 IC_{50} = 794 nM

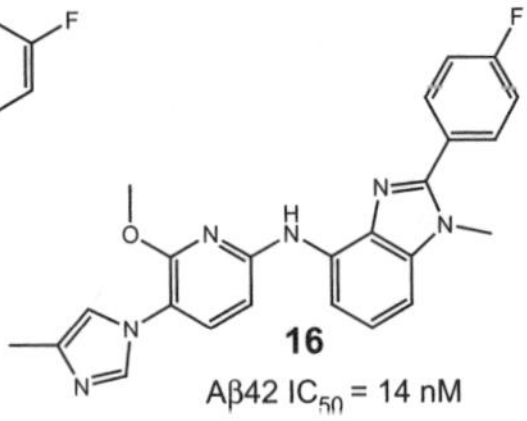

16
Aβ42 IC_{50} = 14 nM

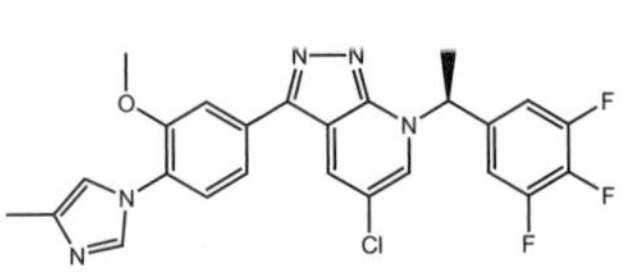

17
Aβ42 IC_{50} = 107 nM

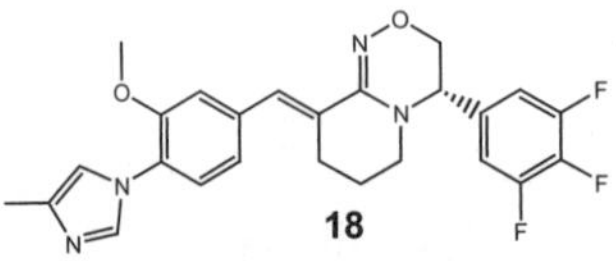

18
Aβ42 IC_{50} = 33 nM
Aβ40 IC_{50} = 123 nM
Aβtotal IC_{50} = 20 μM

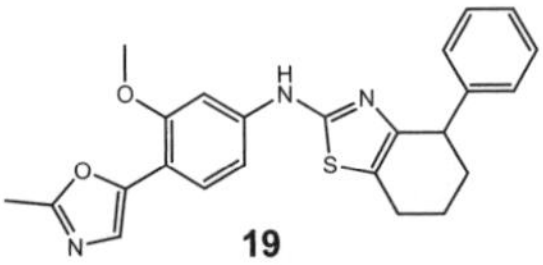

19
Aβ42 IC_{50} = 44 nM

3.4. Triterpene-derived modulators

Structurally distinct compounds such as **20**, semi-synthetically derived from triterpene glycosides extracted from ginkgo or black cohosh, were reported as GSMs with a unique profile as they selectively lowered both Aβ42 as well as Aβ38 while sparing Aβ40.[35] In mice, **20** had a B/P ratio of 1.6 and an oral bioavailability of 37% (30 mg/kg, p.o).[36]

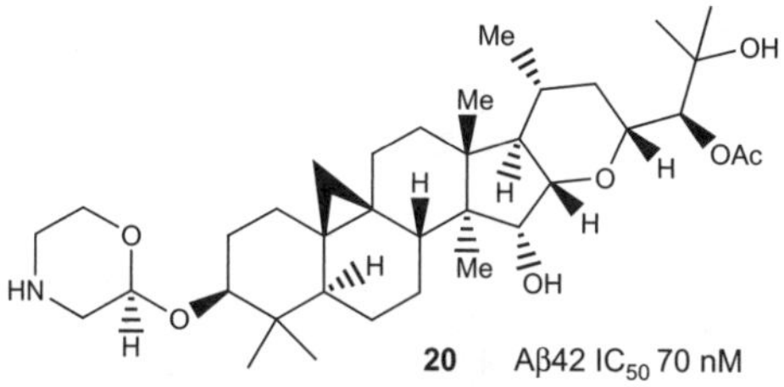

20 Aβ42 IC_{50} 70 nM

4. INHIBITORS OF β-SECRETASE

4.1. BACE as a druggable target

Membrane-bound aspartyl protease BACE1 catalyzes the initial cleavage step in the formation of Aβ peptides and thus has been a prime target for interference in Aβ production since its discovery in 1999. Although potent inhibitors have been developed over the past decade, it has been extremely difficult to combine *in vitro* potency with sufficiently high drug concentrations in brain to achieve optimal *in vivo* efficacy. Since the last report in this series in 2007, considerable progress has been made in this regard, which will be highlighted in the following sections. Several reviews on the biology[37] and medicinal chemistry challenges[38,39] of BACE1 have been published in recent years, including a book entirely devoted to this target.[40] Although key questions related to the desired level of inhibition and on- and off-target selectivity remain unanswered, over the past 2 years, a growing number of companies moved into human clinical trials with BACE inhibitors.

4.2. Main classes of BACE inhibitors

BACE1 inhibitors have been designed starting from various approaches, ranging from substrate-derived inhibitors to fragment-based screening.[41] Structure-based design has been routinely applied, with over 170 crystal structures of the soluble, catalytic domain of both apo and complexed BACE1 deposited in the PDB. Most BACE1 inhibitors contain a "warhead" which interacts via hydrogen bonds with the catalytic aspartyl dyad. Historically, a large group of BACE1 inhibitors are substrate transition-state analogs (TSAs) mimicking the tetrahedral intermediate formed during catalysis. They include (nor)statine, aminostatine, hydroxyethyleneamine, hydroxyethylene, and aminoethylene TSAs. An overview of these structures can be found in several recent reviews.[39,41,42] TSA-derived inhibitors are often highly potent, as exemplified by hydroxyethylene TSA **21** (BACE1 IC_{50} = 0.3 nM),[43] but often lack sufficient brain exposure for *in vivo* efficacy. However, compound **22** with a modest BACE1 IC_{50} of 230 nM, but good permeability and reduced PgP liability showed about 30% reduction in guinea pig brain and CSF Aβ levels (30 mg/kg).[44]

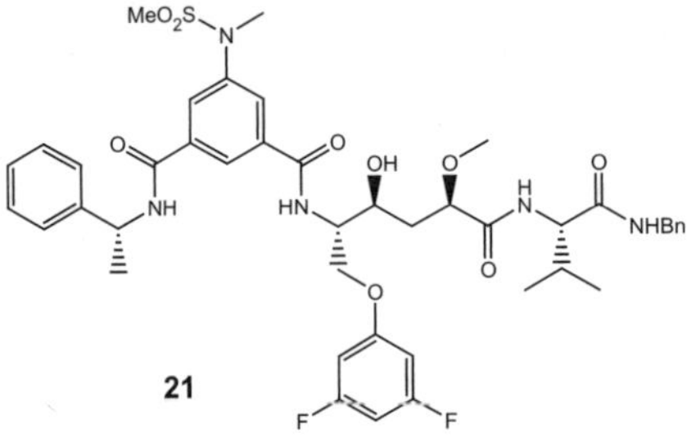

21

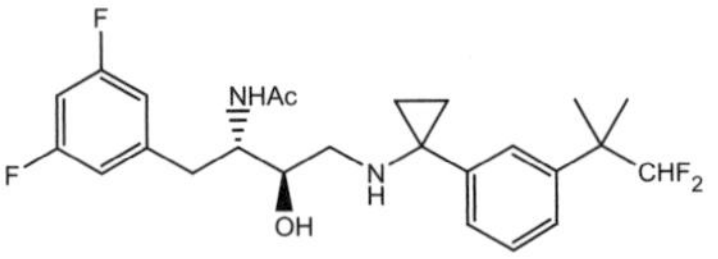

22

The first BACE compound to enter the clinic, CTS-21166, has been described as a TSA, although the structure remains undisclosed.[45] Animal studies indicated good PK/PD properties, including a brain/plasma ratio of 0.44, and oral bioavailability. Dosing CTS-21166 for 6 weeks in rats (4 mg/kg, i.p.) reduced brain Aβ levels by 35–38% and plaque load by 40%. In a human proof-of-concept study, CTS-21166 reduced plasma Aβ levels by up to 80% at the highest dose (225 mg, i.v.), but no CSF data were available. Despite the apparent success of CTS-21166, in general, TSAs display poor drug-like properties due to their bulkiness and often still peptidic nature. In recent years, this has translated in a significant decrease in newly reported TSAs, especially in the patent literature. Compounds **23–26** represent alternative heterocyclic classes of inhibitors.

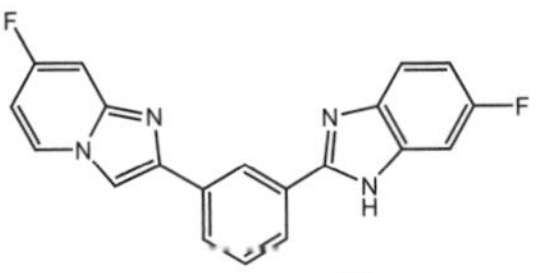

23

BACE1 IC_{50} 18 nM

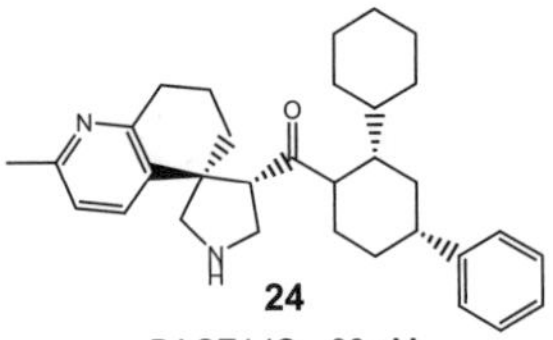

24

BACE1 IC_{50} 36 nM

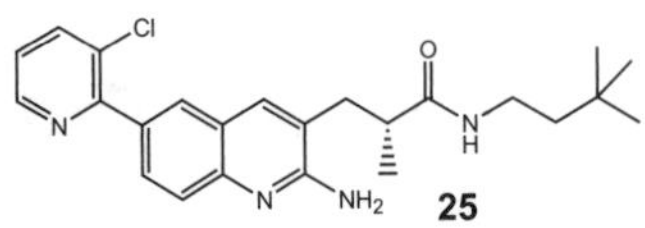

25

BACE1 IC_{50} 11 nM

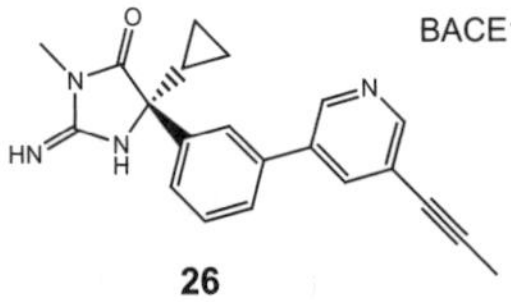

26

BACE1 IC_{50} 5.4 nM

Extensive rigidification of hydroxyethylene TSAs, to remove all peptidic nature, led to **23**.[46] In **24**, the pyrrolidine nitrogen forms a hydrogen bond network with the catalytic aspartyl residues.[47] Introduction of the weakly basic pyridyl nitrogen in **24** resulted in an acceptable PK profile and oral bioavailability. Fragment-based screening led to heterocyclic scaffolds with amidine or guanidine motifs such as in **25** and **26**.[48,49] These motifs form an optimal hydrogen-bonding network with the catalytic aspartates, as apparent from cocrystal structures with BACE1. The latter represent a majority of the recently reported structures and will be discussed further in the next section.

4.3. Amidine- and guanidine-derived inhibitors

The surge in patents around amidine- and guanidine-containing scaffolds is undoubtedly related to the combination of improved potency and blood–brain barrier penetration resulting in good to excellent *in vivo* reduction of Aβ levels in brain and CSF. A representative of this class is aminothiazine **27** (LY2811376), which was shown to be highly brain penetrant in PDAPP mice with a B/P ratio of ~2 and reduced CSF Aβ levels in dog with ~70% (9 h after 5 mg/kg, p.o.). In a Phase I clinical trial, **27** reduced CSF Aβ dose dependently, with an average reduction of CSF Aβ40 over 24 h of ~56% after a 90-mg oral dose.[50] Further development of **27** was stopped due to adverse eye effects in rats, which importantly was shown to be unrelated to BACE1 inhibition. Elongation of the biaryl moiety in **27** to amides as in **28** has led to an increase in potency while maintaining excellent *in vivo* efficacy (85% reduction in mouse brain Aβ40 levels, at 100 mg/kg, s.c.).[51] In addition, **28** also inhibits BACE2 with an IC_{50} of 6 nM, thus may find application in the treatment of diabetes.[52]

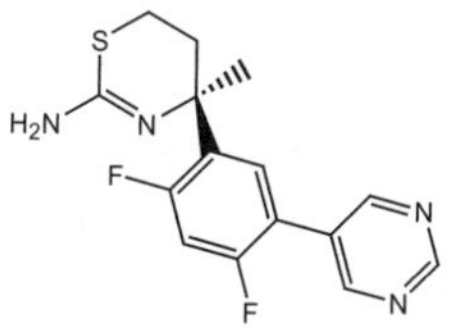

27
BACE1 IC_{50} 239 nM
BACE2 IC_{50} 2880 nM

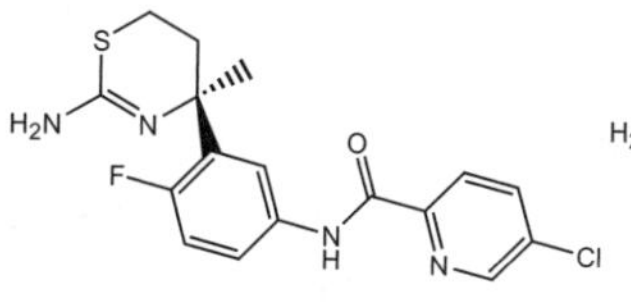

28
BACE1 IC_{50} 18 nM
BACE2 IC_{50} 6 nM

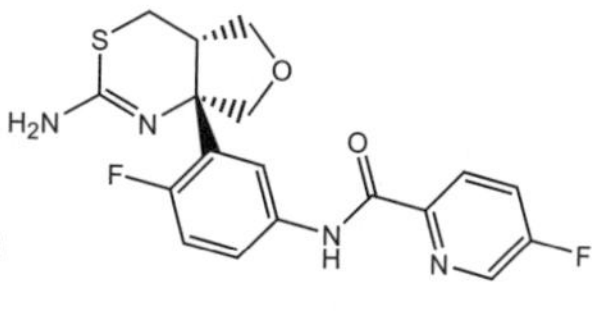

29
BACE1 IC_{50} 20 nM

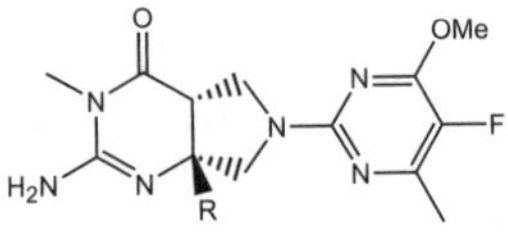

30

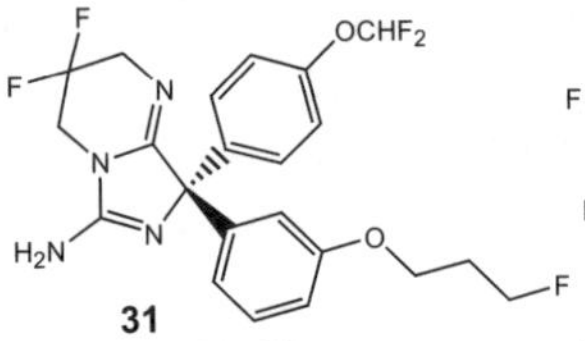

31
BACE1 IC_{50} 31 nM

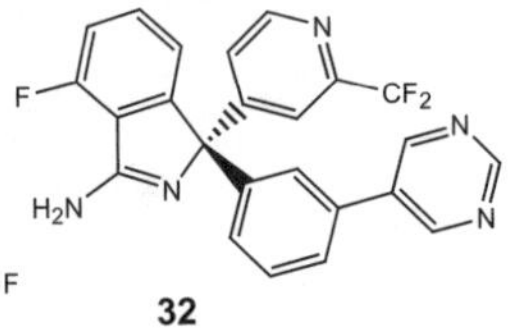

32
BACE1 IC_{50} 63 nM

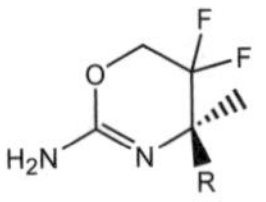

33

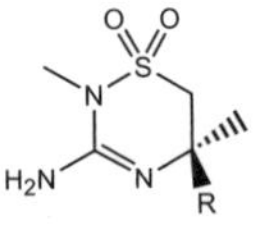

34

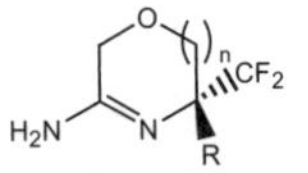

n = 1,2
35

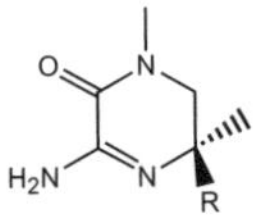

36

Ring-fused analogs have also been reported, with **29** reducing CSF Aβ levels by 62% (10 mg/kg, p.o.).[53] As further exemplified by **30** and **31**, many variations of substitution pattern, ring size, and ring fusion have appeared.[54,55] Alternative amidine-containing scaffolds, as exemplified by **32–36**, have also been described as potent inhibitors.[56–60] As for **28**, some of these compounds have limited to no selectivity over BACE2.[57] Lowering brain Aβ levels by all these variations is greatly dependent upon the level of brain exposure. Despite a reduced efflux ratio *in vitro*, **31** only achieved high brain concentrations and Aβ reduction *in vivo* by concomitant dosing with a PgP inhibitor.[55] In contrast, a 60-mg/kg dose of a brain-penetrant derivative of **36** resulted in a 73% reduction in mouse brain Aβ.[60]

4.4. Alternative methods for inhibiting BACE activity

All of the inhibitor classes mentioned above target the catalytic site by occupying the APP substrate pockets. Recently, highly selective BACE1 antibodies have been reported which target an exosite and reduce Aβ levels *in vitro* and *in vivo*.[61,62] Brain uptake can be elegantly increased by engineering a dual specific antibody with high affinity for BACE1 and low affinity for the transferrin receptor.[63] The antibodies were shown to bind noncompetitively and were highly selective over BACE2 and Cathepsin D. TAK-070 (**37**) is a small molecule which was also shown to bind noncompetitively to BACE1.[64] It was only binding to full-length BACE1, but not to truncated BACE1 lacking the transmembrane domain. Chronic treatment of Tg2576 mice reduced cerebral Aβ deposition and normalized behavioral impairments in cognitive tests.

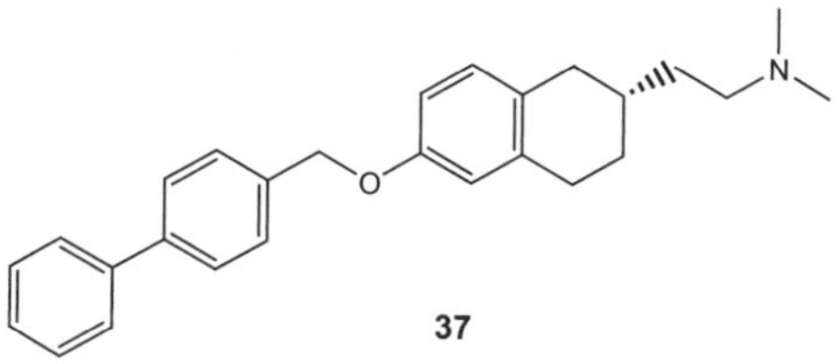

37

5. CONCLUDING REMARKS

Consensus is growing that for a successful disease-modifying therapy of AD with amyloid-targeting drugs, treatment has to take place before the disease has progressed too far and most likely before disease symptoms become apparent (prodromal AD).[65] The need for early and chronic treatment will require high safety margins. Inhibition of either GS or BACE has a potential for on-target-related side effects, as they both process multiple proteins. While this has become manifest in clinical trials with GSIs, BACE1 knockout animals display a fairly normal phenotype. GSMs may circumvent the Notch-related toxicity related to GSIs, but the relatively high micromolar concentrations required for reduction of Aβ levels increases the likelihood for off-target side effects.

Although the validity of the amyloid hypothesis is supported by genetics, so far clinical trial results with GSM tarenflurbil and GSI semagacestat have been disappointing. Currently ongoing Phase III trials with the monoclonal antibodies bapineuzumab and solanezumab, for which the first data are expected in 2012, may provide renewed support for the amyloid hypothesis.[66] Multitargeted anti-Alzheimer agents, for example, displaying dual BACE/acetyl cholinesterase inhibition or GSM/PPARγ activity, may provide additional value.[67] With the more potent GSMs, PS1-selective GSIs, and brain-penetrant BACE inhibitors described in this report, new tools have become available to clinically test the therapeutic potential of intervening in the amyloid peptide production.

REFERENCES

(1) Olson, R.E.; Marcin, L.R. *Annu. Rep. Med. Chem.* **2007**, *42*, 27.
(2) Li, H.; Wolfe, M.S.; Selkoe, D.J. *Structure* **2009**, *17*, 326.
(3) Tolia, A.; De Strooper, B. *Semin. Cell Dev. Biol.* **2009**, *20*, 211.
(4) Takami, M.; Nagashima, Y.; Sano, Y.; Ishihara, S.; Morishima-Kawashima, M.; Funamoto, S.; Ihara, Y. *J. Neurosci.* **2009**, *29*, 13042.
(5) Lleo, A. *Curr. Top. Med. Chem.* **2008**, *8*, 9.
(6) Kreft, A.F.; Martone, R.; Porte, A. *J. Med. Chem.* **2009**, *52*, 6169.

(7) D'Onofrio, G.; Panza, F.; Frisardi, V.; Solfrizzi, V.; Imbimbo, B.P.; Paroni, G.; Cascavilla, L.; Seripa, D.; Pilotto, A. *Expert Opin. Drug Discovery* **2012**, 7, 19.
(8) Mitani, Y.; Yarimizu, J.; Saita, K.; Uchino, H.; Akashiba, H.; Shitaka, Y.; Ni, K.; Matsuoka, N. *J. Neurosci.* **2012**, *32*, 2037.
(9) Mayer, S.C.; Kreft, A.F.; Harrison, B.; Abou-Gharbia, M.; Antane, M.; Aschmies, S.; Atchison, K.; Chlenov, M.; Cole, D.C.; Comery, T.; Diamantidis, G.; Ellingboe, J.; Fan, K.; Galante, R.; Gonzales, C.; Ho, D.M.; Hoke, M.E.; Hu, Y.; Huryn, D.; Jain, U.; Jin, M.; Kremer, K.; Kubrak, D.; Lin, M.; Lu, P.; Magolda, R.; Martone, R.; Moore, W.; Oganesian, A.; Pangalos, M.N.; Porte, A.; Reinhart, P.; Resnick, L.; Riddell, D.R.; Sonnenberg-Reines, J.; Stock, J.R.; Sun, S.-C.; Wagner, E.; Wang, T.; Woller, K.; Xu, Z.; Zaleska, M.M.; Zeldis, J.; Zhang, M.; Zhou, H.; Jacobsen, J.S. *J. Med. Chem.* **2008**, *51*, 7348.
(10) Hopkins, C.R. *ACS Chem. Neurosci.* **2011**, *2*, 279.
(11) Hopkins, C.R. *ACS Chem. Neurosci.* **2012**, *3*, 3.
(12) Gillman, K.W.; Starrett, J.E.; Parker, M.F.; Xie, K.; Bronson, J.J.; Marcin, L.R.; McElhone, K.E.; Bergstrom, C.P.; Mate, R.A.; Williams, R.; Meredith, J.E.; Burton, C.R.; Barten, D.M.; Toyn, J.H.; Roberts, S.B.; Lentz, K.A.; Houston, J.G.; Zaczek, R.; Albright, C.F.; Decicco, C.P.; Macor, J.E.; Olson, R.E. *ACS Med. Chem. Lett.* **2010**, *1*, 120.
(13) Salloway, S.; Coric, V.; Brody, M.; Andreasen, N.; van Dyck, C.; Soininen, H.; Thein, S.; Shiovitz, T.; Kumar, S.; Pilcher, G.; Colby, S.; Rollin, L.; Feldman, H.; Berman, R. AAIC, Paris, France, **2011**; Abstract O4-06-08.
(14) Ye, X.M.; Konradi, A.W.; Smith, J.; Aubele, D.L.; Garofalo, A.W.; Marugg, J.; Neitzel, M.L.; Semko, C.M.; Sham, H.L.; Sun, M.; Truong, A.P.; Wu, J.; Zhang, H.; Goldbach, E.; Sauer, J.-M.; Brigham, E.F.; Bova, M.; Basi, G.S. *Bioorg. Med. Chem. Lett.* **2010**, *20*, 3502.
(15) Sasikumar, T.K.; Burnett, D.A.; Asberom, T.; Wu, W.-L.; Bennett, C.; Cole, D.; Xu, R.; Greenlee, W.J.; Clader, J.; Zhang, L.; Hyde, L. *Bioorg. Med. Chem. Lett.* **2010**, *20*, 3645.
(16) Lee, J.; Song, L.; Terracina, G.; Bara, T.; Josien, H.; Asberom, T.; Theodros, S.; Thavalakulamgar, K.; Burnett, D.A.; Clader, J.; Parker, E.M.; Zhang, L. *Biochemistry* **2011**, *50*, 4973.
(17) Oehlrich, D.; Berthelot, D.J.-C.; Gijsen, H.J.M. *J. Med. Chem.* **2011**, *54*, 669.
(18) Crump, C.J.; Johnson, D.S.; Li, Y.-M. *EMBO J.* **2011**, *30*, 4696.
(19) Ohki, Y.; Higo, T.; Uemura, K.; Shimada, N.; Osawa, S.; Berezovska, O.; Yokoshima, S.; Fukuyama, T.; Tomita, T.; Iwatsubo, T. *EMBO J.* **2011**, *30*, 4815.
(20) Ebke, A.; Lübbers, T.; Fukumori, A.; Shirotani, K.; Haass, C.; Baumann, K.; Steiner, H. *J. Biol. Chem.* **2011**, *286*, 37181.
(21) Kounnas, M.Z.; Danks, A.M.; Cheng, S.; Tyree, C.; Ackerman, E.; Zhang, X.; Ahn, K.; Nguyen, P.; Comer, D.; Mao, L.; Yu, C.; Pleynet, D.; Digregorio, P.J.; Velicelebi, G.; Stauderman, K.A.; Comer, W.T.; Mobley, W.C.; Li, Y.-M.; Sisodia, S.S.; Tanzi, R.E.; Wagner, S.L. *Neuron* **2010**, *67*, 769.
(22) Chávez-Gutiérrez, L.; Bammens, L.; Benilova, I.; Vandersteen, A.; Benurwar, M.; Borgers, M.; Lismont, S.; Zhou, L.; Van Cleynenbreugel, S.; Esselmann, H.; Wiltfang, J.; Serneels, L.; Karran, E.; Gijsen, H.; Schymkowitz, J.; Rousseau, F.; Broersen, K.; De Strooper, B. *EMBO J.* **2012**, *31*, 2261.
(23) Green, R.C.; Schneider, L.S.; Amato, D.A.; Beelen, A.P.; Wilcock, G.; Swabb, E.A.; Zavitz, K.H. *J. Am. Med. Assoc.* **2009**, *302*, 2557.
(24) Imbimbo, B.; Mackintosh, D.; Aponte, T.; Frigerio, E.; Breda, M.; Fernandez, M.; Giardino, L.; Calzà, L.; Norris, D.; Wagner, K.; Shenouda, M. AAIC, Paris, France, **2011**; Abstract P2-466.

(25) Van Broeck, B.; Chen, J.-M.; Treton, G.; Desmidt, M.; Hopf, C.; Ramsden, N.; Karran, E.; Mercken, M.; Rowley, A. *Br. J. Pharmacol.* **2011**, *163*, 375.
(26) Peng, H.; Talreja, T.; Xin, Z.; Cuervo, J.H.; Kumaravel, G.; Humora, M.J.; Xu, L.; Rohde, E.; Gan, L.; Jung, M.; Shackett, M.N.; Chollate, S.; Dunah, A.W.; Snodgrass-Belt, P.A.; Arnold, H.M.; Taveras, A.G.; Rhodes, K.J.; Scannevin, R.H. *ACS Med. Chem. Lett.* **2011**, *2*, 786.
(27) Hall, A.; Elliott, R.L.; Giblin, G.M.P.; Hussain, I.; Musgrave, J.; Naylor, A.; Sasse, R.; Smith, B. *Bioorg. Med. Chem. Lett.* **2010**, *20*, 1306.
(28) Stanton, M.G.; Hubbs, J.; Sloman, D.; Hamblett, C.; Andrade, P.; Angagaw, M.; Bi, G.; Black, R.M.; Crispino, J.; Cruz, J.C.; Fan, E.; Farris, G.; Hughes, B.L.; Kenific, C.M.; Middleton, R.E.; Nikov, G.; Sajonz, P.; Shah, S.; Shomer, N.; Szewczak, A.A.; Tanga, F.; Tudge, M.T.; Shearman, M.; Munoz, B. *Bioorg. Med. Chem. Lett.* **2010**, *20*, 755.
(29) Nagy, C.; Schuck, E.; Ishibashi, A.; Nakatani, Y.; Rege, B.; Logovinsky, V. AAIC, Honolulu, Hawaii, **2010**; Abstract P3-415.
(30) Wan, Z.; Hall, A.; Sang, Y.; Xiang, J.-N.; Yang, E.; Smith, B.; Harrison, D.C.; Yang, G.; Yu, H.; Price, H.S.; Wang, J.; Hawkins, J.; Lau, L.-F.; Johnson, M.R.; Li, T.; Zhao, W.; Mitchell, W.L.; Su, X.; Zhang, X.; Zhou, Y.; Jin, Y.; Tong, Z.; Cheng, Z.; Hussain, I.; Elliott, J.D.; Matsuoka, Y. *Bioorg. Med. Chem. Lett.* **2011**, *21*, 4832.
(31) Bischoff, F.; Berthelot, D.; De Cleyn, M.; Macdonald, G.; Minne, G.; Oehlrich, D.; Pieters, S.; Surkyn, M.; Trabanco, A.A.; Tresadern, G.; Van Brandt, S.; Velter, I.; Zaja, M.; Borghys, H.; Masungi, C.; Mercken, M.; Gijsen, H.J.M. *J. Med. Chem.* [Online early access]. http://dx.doi.org/10.1021/jm201710f. Published Online: May 31, 2012.
(32) Qin, J.; Zhou, W.; Huang, X.; Dhondi, P.; Palani, A.; Aslanian, R.; Zhu, Z.; Greenlee, W.; Cohen-Williams, M.; Jones, N.; Hyde, L.; Zhang, L. *ACS Med. Chem. Lett.* **2011**, *2*, 471.
(33) Sun, Z.-Y.; Asberom, T.; Bara, T.; Bennett, C.; Burnett, D.; Chu, I.; Clader, J.; Cohen-Williams, M.; Cole, D.; Czarniecki, M.; Durkin, J.; Gallo, G.; Greenlee, W.; Josien, H.; Huang, X.; Hyde, L.; Jones, N.; Kazakevich, I.; Li, H.; Liu, X.; Lee, J.; MacCoss, M.; Mandal, M.B.; McCracken, T.; Nomeir, A.; Mazzola, R.; Palani, A.; Parker, E.M.; Pissarnitski, D.A.; Qin, J.; Song, L.; Terracina, G.; Vicarel, M.; Voigt, J.; Xu, R.; Zhang, L.; Zhang, Q.; Zhao, Z.; Zhu, X.; Zhu, Z. *J. Med. Chem.* **2012**, *55*, 489.
(34) Lübbers, T.; Flohr, A.; Jolidon, S.; David-Pierson, P.; Jacobsen, H.; Ozmen, L.; Baumann, K. *Bioorg. Med. Chem. Lett.* **2011**, *21*, 6554.
(35) McKee, T. D.; Loureiro, R. B.; Xia, W.; Austin, W. F.; Bronk, B. S.; Creaser, S. P.; Fuller, N. O.; Hubbs, J. L.; Tate, B. AAIC, Paris, France, 2011; Abstract P1-028.
(36) Bronk, B. S.; Austin, W. F.; Creaser, S. P.; Fuller, N. O.; Hubbs, J. L.; Ma, J.; Shen, R.; Tate, B.; Loureiro, R. B.; McKee, T. D.; Ives, J. L.; Xia, W. 243rd ACS National Meeting, San Diego, CA, **2012**; Abstract MEDI-224.
(37) Kandalepas, P.C.; Vassar, R. *J. Neurochem.* **2012**, *120*(S1), 55.
(38) Stachel, S.J. *Drug Dev. Res.* **2009**, *70*, 101.
(39) Ghosh, A.K.; Brindisi, M.; Tang, J. *J. Neurochem.* **2012**, *120*(S1), 71.
(40) Varghese, J., Ed.; In *BACE: Lead target for orchestrated therapy of Alzheimer's disease*; John Wiley & Sons Inc: Hoboken, NJ, 2010.
(41) Holloway, M.K.; Hunt, P.; McGaughey, G.B. *Drug Dev. Res.* **2009**, *70*, 70.
(42) Iserloh, U.; Cumming, J.N. In *Aspartic acid proteases as therapeutic targets*; Ghosh, A.K., Ed.; Wiley-VCH: Weinheim, 2010; p 441.
(43) Björklund, C.; Oscarson, S.; Benkestock, K.; Borkakoti, N.; Jansson, K.; Lindberg, J.; Vrang, L.; Hallberg, A.; Rosenquist, A.; Samuelsson, B. *J. Med. Chem.* **2010**, *53*, 1458.

(44) Truong, A.P.; Toth, G.; Probst, G.D.; Sealy, J.M.; Bowers, S.; Wone, D.W.G.; Dressen, D.; Hom, R.K.; Konradi, A.W.; Sham, H.L.; Wu, J.; Peterson, B.T.; Ruslim, L.; Bova, M.P.; Kholodenko, D.; Motter, R.N.; Bard, F.; Santiago, P.; Ni, H.; Chian, D.; Soriano, F.; Cole, T.; Brigham, E.F.; Wong, K.; Zmolek, W.; Goldbach, E.; Samant, B.; Chen, L.; Zhang, H.; Nakamura, D.F.; Quinn, K.P.; Yednock, T.A.; Sauer, J.-M. *Bioorg. Med. Chem. Lett.* **2010**, *20*, 6231.

(45) Koelsch, G. Alzheimer's disease. Keystone Symposium; Keystone, CO, **2008**. http://www.alzforum.org/new/detail.asp?id=1790.

(46) Al-Tel, T.H.; Semreen, M.H.; Al-Qawasmeh, R.A.; Schmidt, M.F.; El-Awadi, R.; Ardah, M.; Zaarour, R.; Rao, S.N.; El-Agnaf, O. *J. Med. Chem.* **2011**, *54*, 8373.

(47) Stachel, S.J.; Steele, T.G.; Petrocchi, A.; Haugabook, S.J.; McGaughey, G.; Holloway, K.M.; Allison, T.; Munshi, S.; Zuck, P.; Colussi, D.; Tugasheva, K.; Wolfe, A.; Graham, S.L.; Vacca, J.P. *Bioorg. Med. Chem. Lett.* **2012**, *22*, 240.

(48) Cheng, Y.; Judd, T.C.; Bartberger, M.D.; Brown, J.; Chen, K.; Fremeau, R.T.; Hickman, D.; Hitchcock, S.A.; Jordan, B.; Li, V.; Lopez, P.; Louie, S.W.; Luo, Y.; Michelsen, K.; Nixey, T.; Powers, T.S.; Rattan, C.; Sickmier, E.A.; St. Jean, D.J.; Wahl, R.C.; Wen, P.H.; Wood, S. *J. Med. Chem.* **2011**, *54*, 5836.

(49) Cumming, J.N.; Smith, E.M.; Wang, L.; Misiaszek, J.; Durkin, J.; Pan, J.; Iserloh, U.; Wu, Y.; Zhu, Z.; Strickland, C.; Voigt, J.; Chen, X.; Kennedy, M.E.; Kuvelkar, R.; Hyde, L.A.; Cox, K.; Favreau, L.; Czarniecki, M.F.; Greenlee, W.J.; McKittrick, B.A.; Parker, E.M.; Stamford, A.W. *Bioorg. Med. Chem. Lett.* **2012**, *22*, 2444.

(50) Ma, P.C.; Dean, R.A.; Lowe, S.L.; Martenyi, F.; Sheehan, S.M.; Boggs, L.N.; Monk, S.A.; Mathes, B.M.; Mergott, D.J.; Watson, B.M.; Stout, S.L.; Timm, D.E.; LaBell, E.S.; Gonzales, C.R.; Nakano, M.; Jhee, S.S.; Yen, M.; Ereshefsky, L.; Lindstrom, T.D.; Calligaro, D.O.; Cocke, P.J.; Hall, G.; Friedrich, S.; Citron, M.; Audia, J.E. *J. Neurosci.* **2011**, *31*, 16507.

(51) Kobayashi, N.; Ueda, K.; Itoh, N.; Suzuki, S.; Sakaguchi, G.; Kato, A.; Yukimasa, A.; Hori, A.; Kooriyama, Y.; Haraguchi, H.; Yasui, K.; Kanda, Y. U.S. Patent Application 0160290-A12010, **2010**.

(52) Esterházy, D.; Stützer, I.; Wang, H.; Rechsteiner, M.P.; Beauchamp, J.; Döbeli, H.; Hilpert, H.; Matile, H.; Prummer, M.; Schmidt, A.; Lieske, N.; Boehm, B.; Marselli, L.; Bosco, D.; Kerr-Conte, J.; Aebersold, R.; Spinas, G.A.; Moch, H.; Migliorini, C.; Stoffel, M. *Cell Metab.* **2011**, *14*, 365.

(53) Audia, J. E.; Mergott, D. J.; Shi, C. E.; Vaught, G. M.; Watson, B. M.; Winneroski, L. L. Patent Application WO 005738-A1, **2011**.

(54) Stamford, A. W.; Zhaoning, Z.; Mandal, M.; Wu, Y.; Cumming, J. N.; Liu, X.; Li, G.; Iserloh, U. Patent Application WO 131975-A1, **2009**.

(55) Swahn, B.-M.; Holenz, J.; Kihlström, J.; Kolmodin, K.; Lindström, J.; Plobeck, N.; Rotticci, D.; Sehgelmeble, F.; Sundström, M.; Von Berg, S.; Fälting, J.; Georgievska, B.; Gustavsson, S.; Neelissen, J.; Ek, M.; Olsson, L.-L.; Berg, S. *Bioorg. Med. Chem. Lett.* **2012**, *22*, 1854.

(56) Bergström, P. -O.; Minidis, A.; Stranne, R. U. J.; Wernersson, M. Patent Application WO 142716-A1, **2011**.

(57) Banner, D.; Guba, W.; Hilpert, H.; Mauser, H.; Mayweg, A. V.; Narquizian, R. P. E.; Power, E.; Rogers-Evans, M.; Woltering, T.; Wostl, W. Patent Application WO 069934-A1, **2011**.

(58) Scott, J. D.; Stamford, A. W.; Gilbert, E. J.; Cumming, J. N.; Iserloh, U.; Misiaszek, J. A.; Li, G. Patent Application WO 044181-A1, **2011**.

(59) Badiger, S.; Chebrolu, M.; Frederiksen, M.; Holzer, P.; Hurth, K.; Lueoend, R. M.; Machauer, R.; Moebitz, H.; Neumann, U.; Ramos, R.; Rueeger, H.; Tintelnot-Blomley, M.; Veenstra, S. J.; Voegtle, M. Patent Application WO 009943-A1, **2011**.

(60) Tresadern, G.; Delgado, F.; Delgado, O.; Gijsen, H.; Macdonald, G.J.; Moechars, D.; Rombouts, F.; Alexander, R.; Spurlino, J.; Van Gool, M.; Vega, J.A.; Trabanco, A.A. *Bioorg. Med. Chem. Lett.* **2011**, *21*, 7255.
(61) Atwal, J.K.; Chen, Y.; Chiu, C.; Mortensen, D.L.; Meilandt, W.J.; Liu, Y.; Heise, C.E.; Hoyte, K.; Luk, W.; Lu, Y.; Peng, K.; Wu, P.; Rouge, L.; Zhang, Y.; Lazarus, R.A.; Scearce-Levie, K.; Wang, W.; Wu, Y.; Tessier-Lavigne, M.; Watts, R.J. *Sci. Transl. Med.* **2011**, *3*, 84ra43.
(62) Zhou, L.; Chavez-Gutierrez, L.; Bockstael, K.; Sannerud, R.; Annaert, W.; May, P.C.; Karran, E.; De Strooper, B. *J. Biol. Chem.* **2011**, *286*, 8677.
(63) Yu, Y.J.; Zhang, Y.; Kenrick, M.; Hoyte, K.; Luk, W.; Lu, Y.; Atwal, J.; Elliott, J.M.; Prabhu, S.; Watts, R.J.; Dennis, M.S. *Sci. Transl. Med.* **2011**, *3*, 84ra44.
(64) Fukumoto, H.; Takahashi, H.; Tarui, N.; Matsui, J.; Tomita, T.; Hirode, M.; Sagayama, M.; Maeda, R.; Kawamoto, M.; Hirai, K.; Terauchi, J.; Sakura, Y.; Kakihana, M.; Kato, K.; Iwatsubo, T.; Miyamoto, M. *J. Neurosci.* **2010**, *30*, 11157.
(65) Karran, E.; Mercken, M.; De Strooper, B. *Nat. Rev. Drug Discov.* **2011**, *10*, 698.
(66) Panza, F.; Frisardi, V.; Imbimbo, B.P.; Seripa, D.; Solfrizzi, V.; Pilotto, A. *Expert Opin. Biol. Ther.* **2011**, *11*, 679.
(67) Bajda, M.; Guzior, N.; Ignasik, M.; Malawska, B. *Curr. Med. Chem.* **2011**, *18*, 4949.

CHAPTER SIX

mGluR2 Activators and mGluR5 Blockers Advancing in the Clinic for Major CNS Disorders

Sylvain Célanire, Guillaume Duvey, Sonia Poli, Jean-Philippe Rocher
Addex Therapeutics, Geneva, Switzerland

Contents

1. INTRODUCTION

Metabotropic glutamate receptors (mGluRs) are membrane-type receptors for glutamate, which is the major excitatory neurotransmitter in the brain. They belong to the Class C G-protein coupled receptor and consist of a family of eight receptor subtypes, as summarized in Table 6.1. Each receptor subtype has a unique pharmacological profile and their therapeutic potential as drug targets is predicted mainly based on their respective localization in the central and peripheral nervous systems.[1,2] Beyond the classical approach targeting the orthosteric binding site of these receptors, the development of allosteric modulators represents an emerging class of

Annual Reports in Medicinal Chemistry, Volume 47
ISSN 0065-7743
http://dx.doi.org/10.1016/B978-0-12-396492-2.00006-0

Table 6.1 mGluR subtypes: function and signaling pathway

Receptor class	Subtype	Function	Signaling
Group I	mGluR1 mGluR5	Excitatory	Gq/G_{11}; ↑PLC
Group II	mGluR2 mGluR3	Inhibitory	Gq/G_{11}; ↑PLC
Group III	mGluR4 mGluR6 mGluR7 mGluR8	Inhibitory	Gi/G_0; ↓AC

PLC, phospholipase C; AC, adenylyl cyclase.

orally available small molecules offering greater selectivity and better modulatory control at disease mediating receptors.[3,4] This approach has been successively applied to mGluRs and extensively reviewed.[5,6] The mGluR2 activators and mGluR5 blockers are the first drug targets that have demonstrated clinical proof-of-concept in patients suffering from major CNS disorders, such as schizophrenia and anxiety, Parkinson's disease levodopa-induced dyskinesia (PD-LID), fragile X syndrome (FXS), and major depressive disorders. This review focuses on the drug discovery effort and clinical status of mGluR2 activators acting at orthosteric or allosteric binding sites, and negative allosteric modulators (NAMs) of mGluR5.

2. DISCOVERY AND DEVELOPMENT OF MGLUR2 ACTIVATORS

2.1. Localization and functions of mGluR2

Both mGluR2 and mGluR3 receptors are coupled to Gi/Go proteins, and their activation inhibits cAMP formation (Table 6.1).[7] Both receptors share ~70% sequence homology, highly conserved for the ligand binding site and most of the mGluR2 agonists identified so far were dual mGluR2/mGluR3 (mGluR2/3) activators. mGluR2 is exquisitely localized in neurons, specifically in the preterminal region of axons, while mGluR3 is located both pre- and postsynaptically, and in glial cells.[8] Presynaptic mGluR2/3 receptors are either activated by an excess of glutamate or by its release from astrocytes through the cystine–glutamate membrane antiporter.[9] mGluR2 is highly expressed in limbic-related regions, making its ligands promising therapeutic agents for neuropsychiatric disorders.[10] Recently, an mGluR2-5-HT_{2A} receptor complex has been characterized,[11] postulated to be the target of antipsychotic drugs acting as 5-HT_2R blocker and/or mGluR2 activators.[12] Site-directed mutagenesis studies have clarified the molecular interaction of mGluR2 orthosteric agonists (and antagonists) with the active site[13] as well as mGluR2 PAM (and NAMs) with the seven-transmembrane domain.[14,15]

2.2. mGluR2 orthosteric agonists

Since the identification of the conformationally constrained bicyclo[3.1.0] alkane-based glutamate mGluR2/3 agonist LY354740 (**1**, eglumetad) in 1997, Lilly scientists have dedicated intensive efforts to delineate this unique chemotype into numerous nanomolar dual mGluR2/3 agonists, represented by LY379268 (**2**), LY389795 (**3**), and LY459477 (**4**),[16] as well as novel mGluR2 agonist/mGluR3 antagonist compounds, such as LY541850 (**5**).[17,18] All these compounds display 100- to >600-fold selectivity versus the other mGluR subtypes.

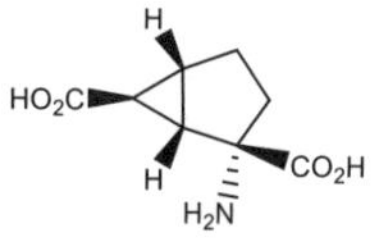

1, LY354740

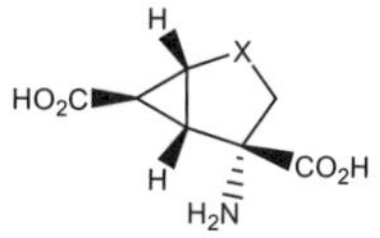

2, LY379268 (X=O)
3, LY389795 (X=S)

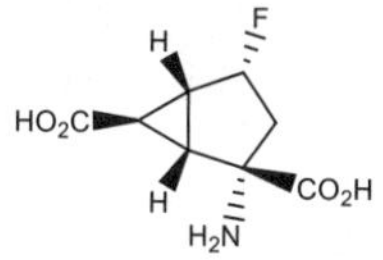

4, LY459477

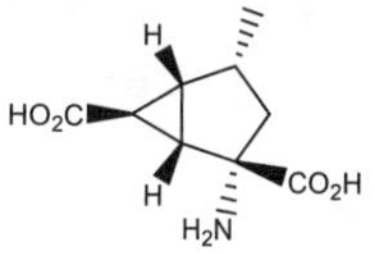

5, LY541850

Peptidyl prodrugs have been developed in order to improve oral absorption and brain penetration, providing the first development candidates LY544344 (**6**, talaglumetad)[19] and LY2140023 (**8**, pomaglumetad), prodrug of LY404039 (**7**, hmGluR2/mGluR3 $EC_{50}=23/48$ nM).[20] Lilly scientists have pursued their pioneering discovery efforts leading to the first selective mGluR2 agonists **9** (LY2812223; hmGluR2 $EC_{50}=23$ nM, hmGluR3 $EC_{50}>30$ μM, calcium mobilization assay). LY2812223 has demonstrated *in vivo* efficacy in rodent models of schizophrenia, bipolar disorders, anxiety, and depression.[21] Its L-alanine prodrug LY2979165 (**10**) has been reported to be orally absorbed via the intestinal absorption transporter PepT1 and is effectively hydrolyzed in the human intestine and liver.[21,22]

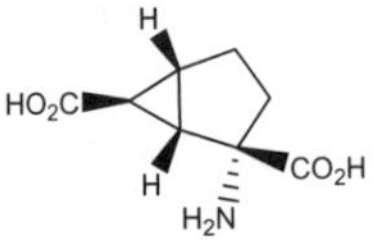

1, LY354740

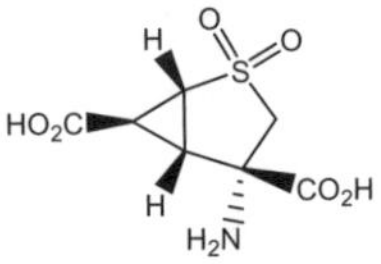

7, LY404039

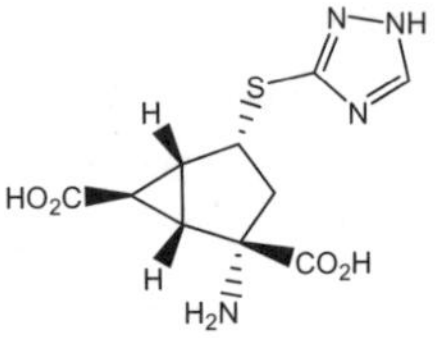

9, LY2812223

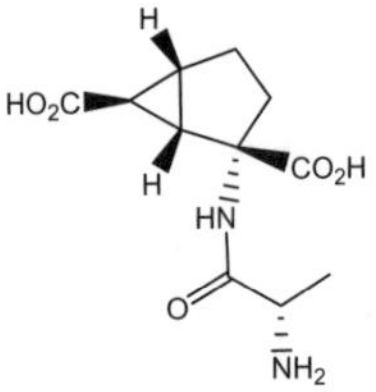

6, LY544344

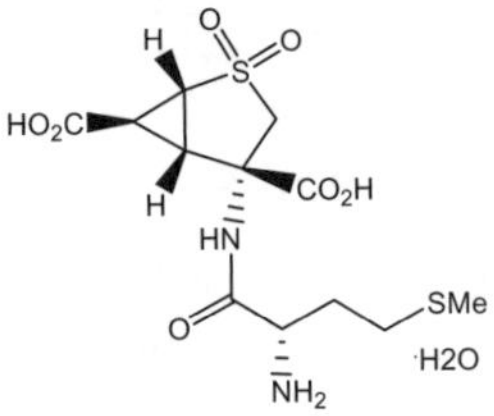

8, LY2140023

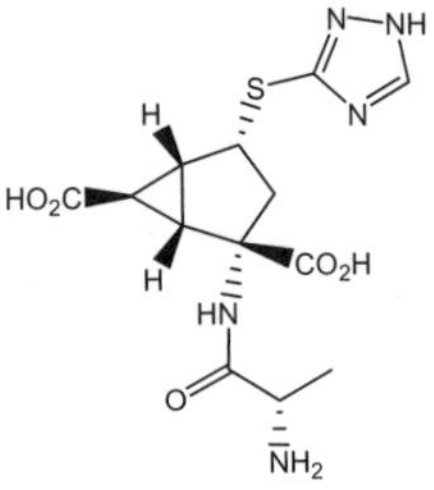

10, LY2979165

2.3. mGluR2 positive allosteric modulators

Two major reviews have reported the large chemical diversity of mGluR2 positive allosteric modulators (PAMs) based on the initial discovery of the first modulators in the early 2000s.[23,24] New structural enrichment has been recently disclosed and described hereafter.

2.3.1 *Benzimidazoles*

Researchers from Pfizer have described the optimization of benzimidazole derivatives, such as compound **11** (rmGluR2 EC_{50} = 4790 nM), identified from an HTS campaign. Exhibiting D_2R-antagonism off-target activities, a subsequent optimization led to the selective azabenzimidazole **12** (EC_{50} = 35 nM), which demonstrated oral bioavailability in rats (F = 50%; 3 mg/kg, p.o.) and efficacy in psychosis mice models (minimum effective dose (MED) ranging from 10 to 17.8 mg/kg, s.c.).[25] A similar strategy was independently reported leading to piperazine **13** (GSK1331258; EC_{50} = 79 nM).[26]

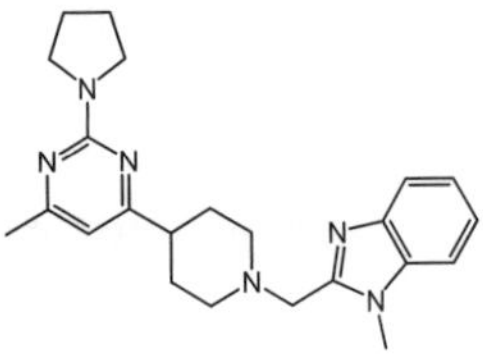

11

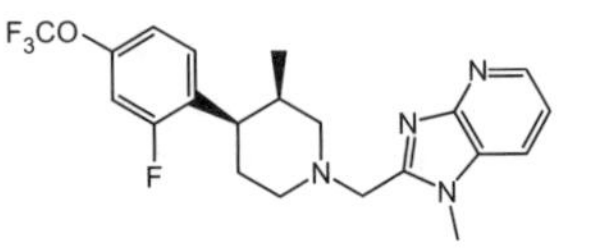

12

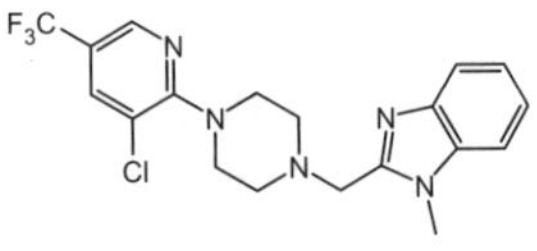

13 GSK1331258

2.3.2 Indanone and isosteres

The improved potency and PK properties of the historical mGluR2 PAM biphenylindanone A (**14**, BINA) via modification of the indanone core has been recently disclosed, as shown in compounds **15** and **16** (rmGluR2, $EC_{50}=50$ and 170 nM, respectively; thallium flux assay).[27,28] In particular, **16** exhibited oral bioavailability ($F=86\%$) and high brain/plasma ratio of 4.8 (10 mg/kg, p.o.) and demonstrated efficacy in a rat cocaine self-administration model (MED = 20 mg/kg, p.o.).[29]

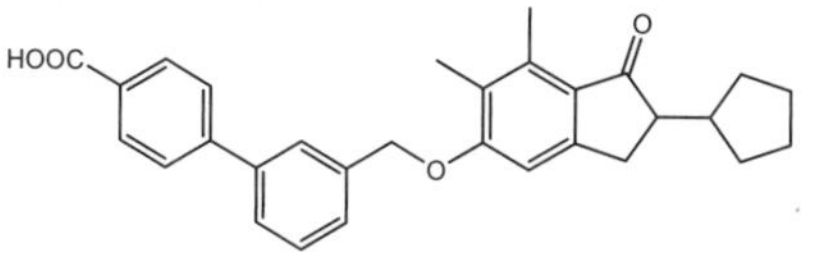

14 BINA

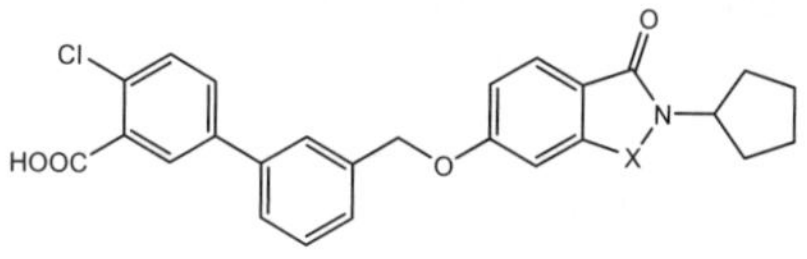

15 X= CH_2
16 X=S

2.3.3 Benzotriazoles, benzimidazolones, and benzothiazolones

N-cyclopropylmethylene-halobenzotriazole series has been broadly exemplified[30,31] yielding potent derivatives, like the direct-branched aryl **17** ($EC_{50}=8$ nM) or pyridyloxy derivative **18**, being the most active compound within this series ($EC_{50}=3$ nM).

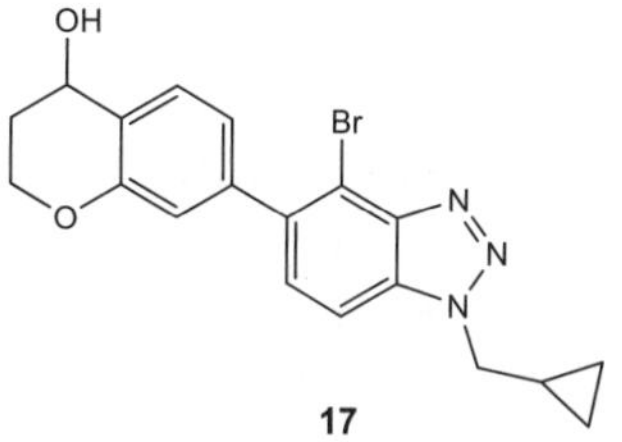

17

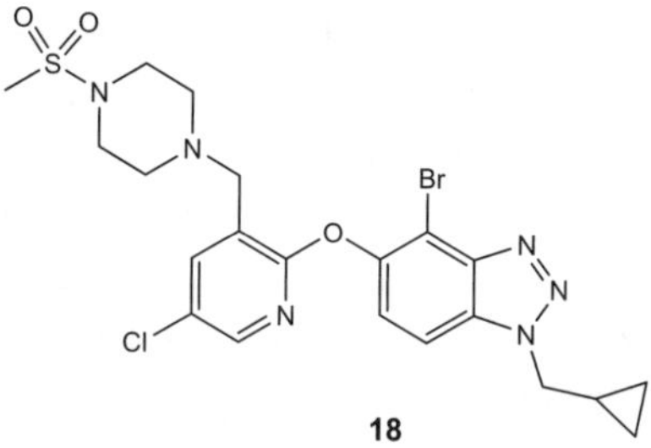

18

A potent *N*-alkylated azabenzimidazolone series has been reported, in which the methyl-*tert*-butyl group was recognized as a privileged substituent, represented by compound **19** ($EC_{50}=11$ nM).[32] An optimization of this template led to additional cores, such as diazasulfone **20**, being the most potent mGluR2 PAM reported to date in a calcium mobilization assay ($EC_{50}=0.5$ nM).[33]

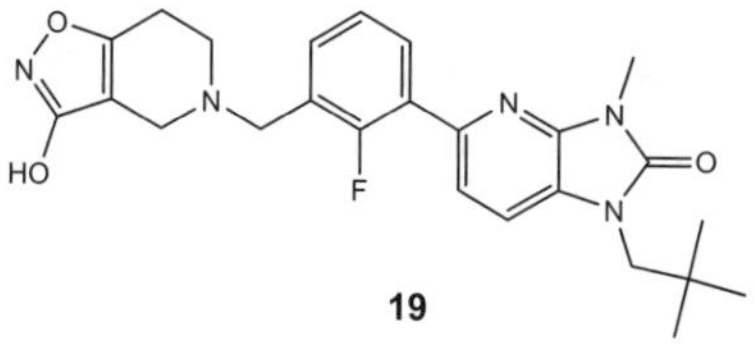

19

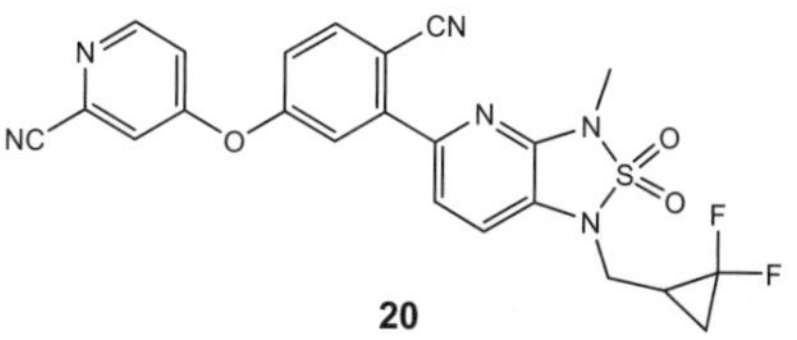

20

2.3.4 Pyridones and azolopyridines

Since the first disclosure of novel pyridone-based mGluR2 PAM chemical series in 2006,[34] scientists at Addex Therapeutics pursued their efforts around the 1,4-disubstituted pyridine HTS hit **21** (EC_{50} = 8 μM, [^{35}S] GTPγS assay). This compound showed an adequate overlap with the reported mGluR2 PAM 3D-pharmacophore,[35] and therefore allowed subsequent identification of more potent compounds, such as the compound **22** (EC_{50} = 42 nM). The more soluble pyridyl representative **23** (EC_{50} = 316 nM) displayed a full selectivity over other mGluR subtypes, an improved hERG channel interaction (21% inhibition at 3 μM) and brain exposure. Compound **23** demonstrated inhibition of REM sleep in a rat sleep-wave EEG model at 10 mg/kg (s.c.).

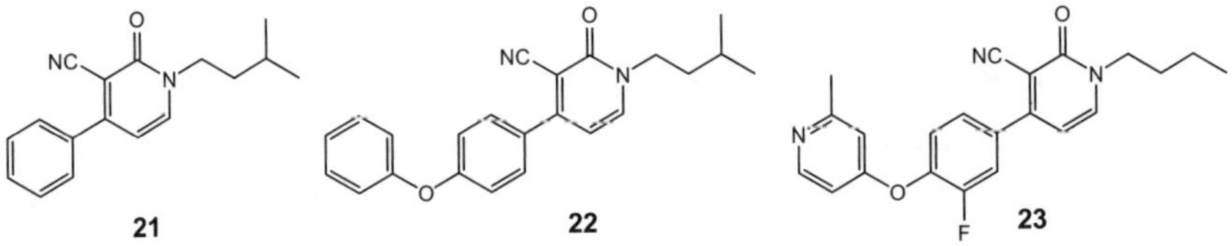

21 **22** **23**

The recently disclosed imidazopyridine series is the result of a scaffold hopping exercise on the 1,4-pyridone series (chemotype **B**).[36] Since compound **24** (EC_{50} = 186 nM) had poor oral bioavailability, it was further optimized into indole derivative **25** (EC_{50} = 85 nM), exhibiting an improved oral pharmacokinetic profile and brain penetration.[37] Compound **25** showed modulation of REM sleep in a rat sleep–wake EEG model at 10 mg/kg orally, consistent with mGluR2 receptor activation.[38]

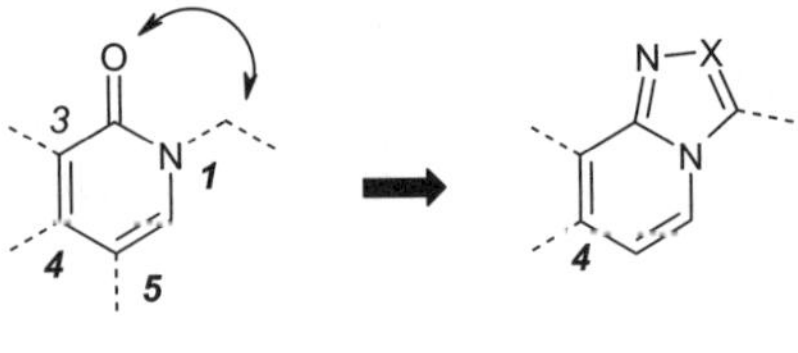

A **B (X=CH)** **C (X=N)**

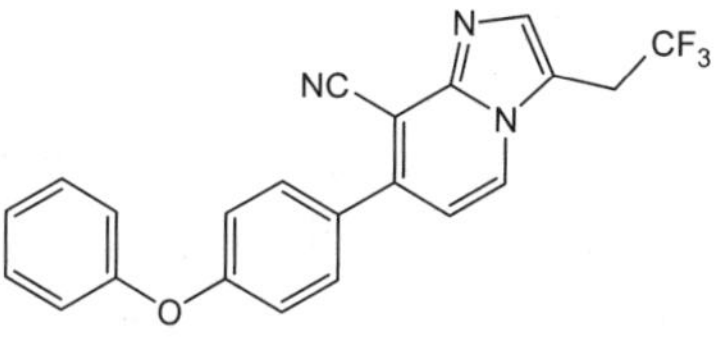

24

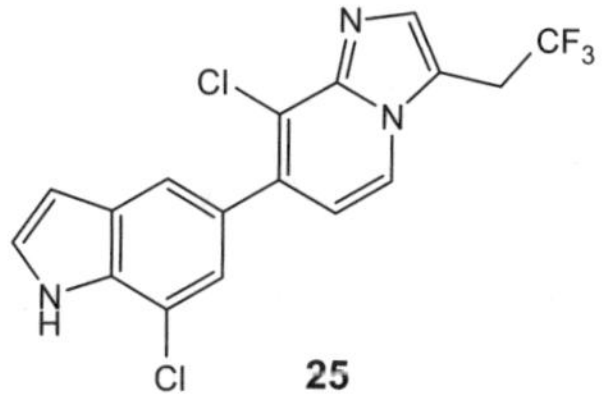

25

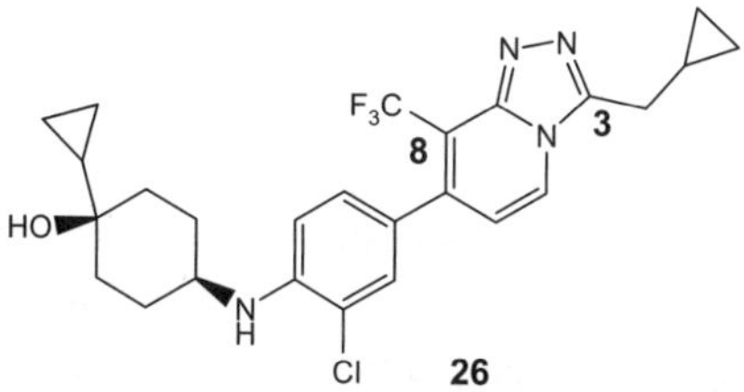

26

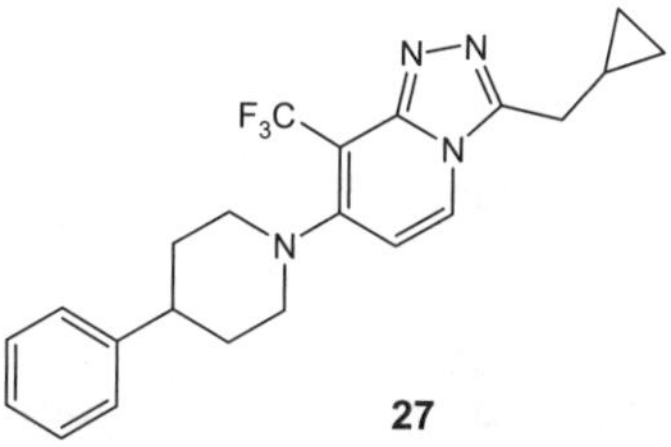

27

The triazolo[4,3:a]pyridine series (chemotype **C**) was exemplified in several patent applications with compounds demonstrated efficacy in different models of psychosis,[39,40] illustrated by compounds **26** and **27** (EC_{50} = 1.6 and 16.2 nM, respectively). Phenyl-piperidine **27** exhibited efficacy in PCP-induced hyperlocomotion in mice (ED_{50} = 5.4 mg/kg) and conditioned avoidance response in rats (ED_{50} = 16.3 mg/kg, p.o.).[41]

2.3.5 Oxazolidinones

The oxazolidinone derivative **28** (EC_{50} = 450 nM) was identified in an HTS calcium mobilization assay on hmGluR2 cell lines. Further optimization led to the identification of the benzonitrile **29** (EC_{50} = 82 nM) exhibiting moderate-to-high dog clearance (17.7 ml/min/kg), brain penetration in rat (CSF/plasma unbound ratio of 1), and a clean selectivity profile over other mGluR subtypes, D_2 and 5-HT_{2A} receptors. Compound **29** demonstrated efficacy in the ketamine-induced locomotor activity at 100 mg/kg after i.p. administration.[42]

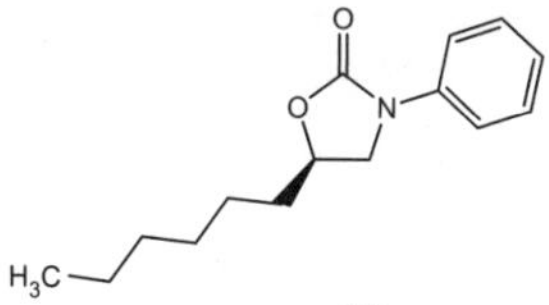

28

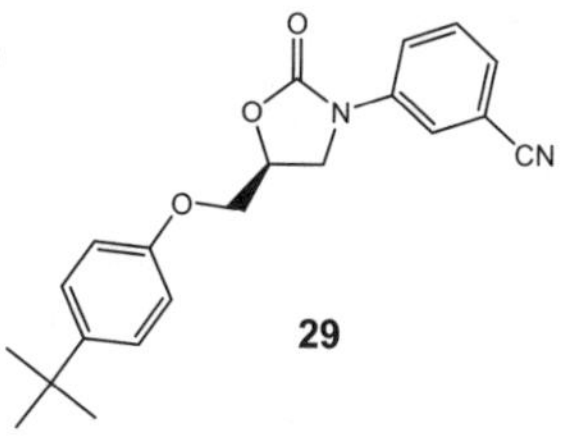

29

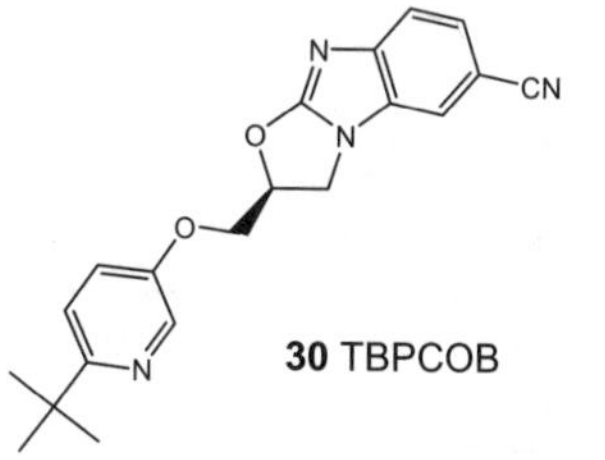

30 TBPCOB

The fusion of the oxazolidinone and the cyano-phenyl rings led to the discovery of the oxazolobenzimidazole series. The resulting *ter*-butyl-pyridyloxymethyl-cyano-oxazolobenzimidazole **30** combined high potency (hmGluR2 $EC_{50}=29$ nM, rmGluR2 $EC_{50}=42$ nM) and improved solubility and selectivity over mGluR, D_2, and 5-HT_{2A} receptors. Compound **30** demonstrated brain exposure and robust efficacy in a PCP-induced hyperlocomotion model in rats (MED = 30 mg/kg, p.o.).[43]

2.4. Clinical trials

Today, several mGluR2 activators have successfully reached clinical development phases for the treatment of severe CNS disorders, including schizophrenia and anxiety (see Table 6.2).[10,50,51] LY2140023 (**8**)[20] is the first mGluR2/3 agonist demonstrating its tolerability and significant improvement of positive and negative symptoms of schizophrenia in patients in a Phase IIa trial.[44] However, a second Phase II study proved inconclusive due to lack of separation from placebo of both **8** and olanzapine.[52] LY2140023 has now reached Phase III clinical stage while being further investigated in additional Phase II studies.[53] Preclinical[54,55] and clinical[56] evidence also supports a significant benefit of mGluR2 activators on working memory processes and cognitive symptoms in schizophrenia. More recently, two additional specific mGluR2 activators LY2979165 (**10**)[45] and ADX71149/JNJ40411813 (undisclosed structure) have reached Phases I and IIa, respectively, for psychiatric disorders.[46]

Table 6.2 mGluR2 activators in Phase I and Phase II trials

Compound	MoA	Indications	Clinical status
LY2140023 (**8**)	mGluR2/3 agonist	Schizophrenia	Phase II/III[44]
LY2979165 (**10**)	mGluR2 agonist	Bipolar disorders	Phase I[45]
ADX71149/JNJ40411813	mGluR2 PAM	Schizophrenia	Phase II[46]
LY544344 (**6**)	mGluR2/3 agonist	Generalized anxiety disorders	Phase II[47,48]
LY2300559 (**31**)	mGluR2 agonist/CysLT1 antagonist	Migraine	Phase II[49]

LY544344 **6** has demonstrated clinical improvement in patients with generalized anxiety disorder.[47,48] LY354740 **1** has failed to reach clinical significance compared to placebo in a double-blind clinical trial in patients with panic disorder.[57] In a separate study, LY354740 has, however, demonstrated anxiolytic-like effect in the fear-potentiated startle paradigm.[58]

LY2300559 (**31**), a dual mGluR2 PAM ($EC_{50}=50$ nM) and cysteinyl-leukotriene 1 (CysLT1) antagonist ($K_b=22$ nM),[59,60] has entered a Phase II placebo-controlled proof-of-concept study in patients with migraine. Despite its low brain-to-plasma ratio of 0.01, **31** has demonstrated efficacy in a preclinical rodent model of migraine and was reported to be well tolerated in rat and dog toxicological studies.[49]

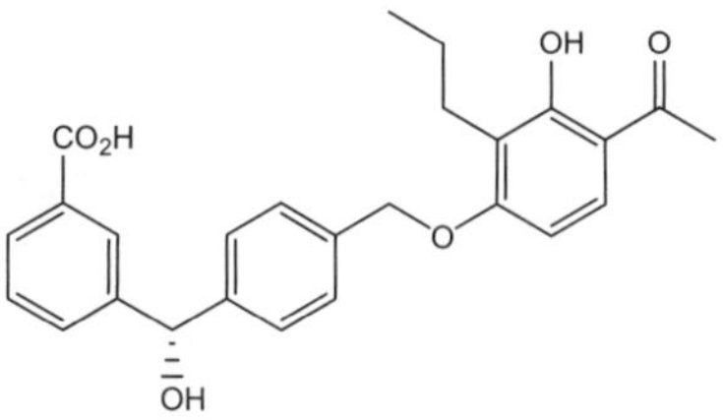

31

3. DRUG DISCOVERY AND DEVELOPMENT OF MGLUR5 NAMS

3.1. Localization and functions of mGluR5

mGluR5 is coupled to Gq/G_{11} protein (Table 6.1), is localized postsynaptically, and is highly expressed in the hippocampus, corpus striatum, cerebral cortex, and in basal ganglia,[5,61] a key region for motor control.[62] In addition, cross talk between mGluR5, dopamine D_2 receptors, and adenosine A_{2A} receptors, which are all highly expressed in the striatum, is hypothesized to play a role in the motor dysfunction observed in Parkinson's disease.[63]

3.2. Alkyne-based mGluR5 negative allosteric modulators

Many examples of mGluR5 NAMs have been reported in recent reviews[64,65] since the discovery of the diarylalkynes MPEP **32** and MTEP **33**. Potent mGluR5 modulators have been discovered through extensive exploration of the SAR in these narrow series.[66] Unless otherwise noted, biological activities reported in this section have been demonstrated using a calcium mobilization assay (FLIPR) using a human mGluR5 clone.

Dihydroquinolinone MRZ-8676 (**34**, $IC_{50}=20$ nM) has been reported to demonstrate activity in a rat model of L-dopa-induced dyskinesia (MED = 8.33 mg/kg, p.o.), and no tolerance was observed after repeat dosing (6 days) at 75 mg/kg.[67] The imidazolyl derivative CTEP (**35**) is another potent and selective mGluR5 NAM ($IC_{50}=11.4$ nM) and demonstrated efficacy in the Vogel conflict drinking rat model (MED = 0.3 mg/kg, p.o.).[68]

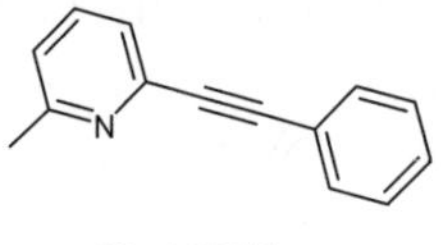

32, MPEP

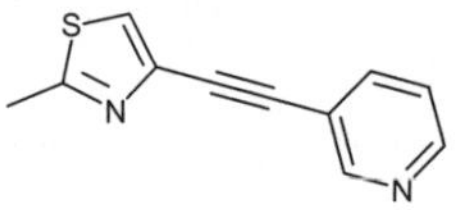

33, MTEP

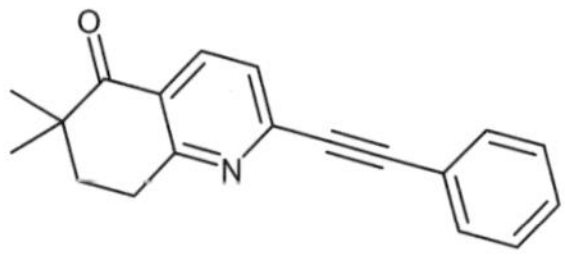

34, MRZ8676

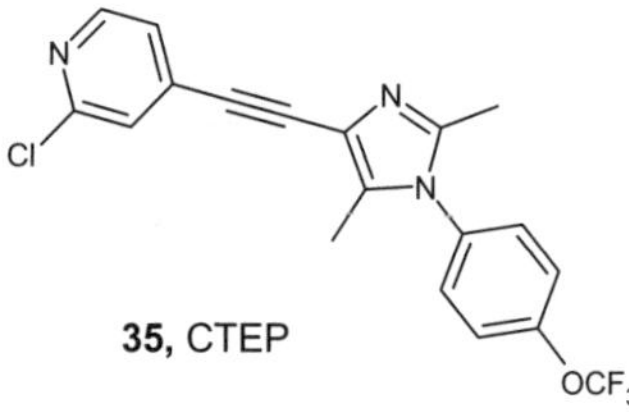

35, CTEP

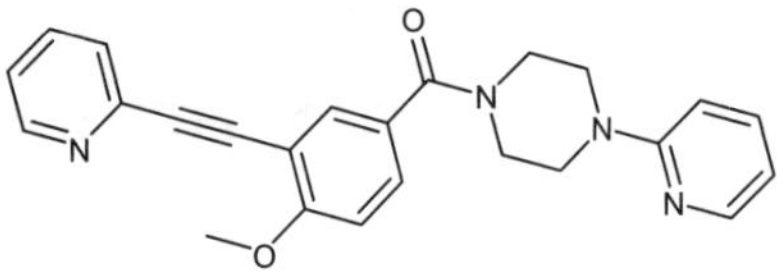

36

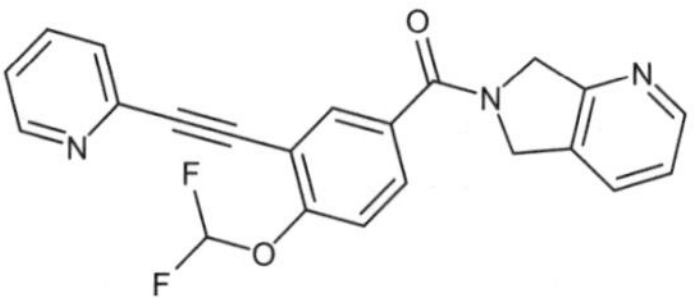

37, WYE304529

Further exploration around ethynylbenzamide **36** ($IC_{50}=8.0$ nM)[69] led to pyrido-isoindoline **37** (WYE304529/GRN529). Compound **37** displayed improved potency ($IC_{50}=3.1$ nM), above 1000-fold selectivity versus mGluR1, and dose-dependent efficacy in a wide range of rodent models of anxiety, depression, and pain.[70,71] Since the initial discovery of the Addex clinical candidate ADX10059 (**38**, raseglurant), intensive efforts led to the identification of the ethynylpyridine development candidate ADX48621 (**39**, dipraglurant; mGluR5, $IC_{50}=21$ nM).[64] ADX48621 has demonstrated antiparkinsonian activity in two preclinical models: the haloperidol-induced catalepsy in rats (ED50 = 2.8 mg/kg, p.o.) and the MPTP-treated macaque model (MED = 10 mg/kg, p.o.).[72] Another mGluR5 NAM clinical candidate, mavoglurant **40** (AFQ056, Novartis Pharmaceuticals), was able to restore the prepulse inhibition of startle response in Fmr1 KO mice, confirming the interest of this molecule in FXS.[73]

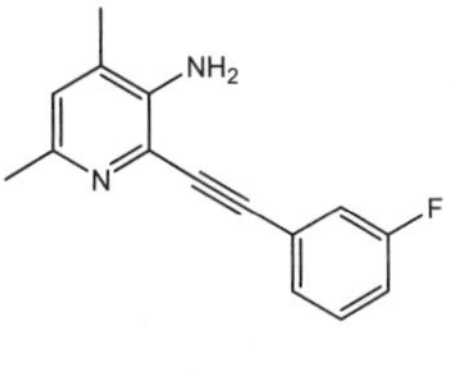

38, ADX10059

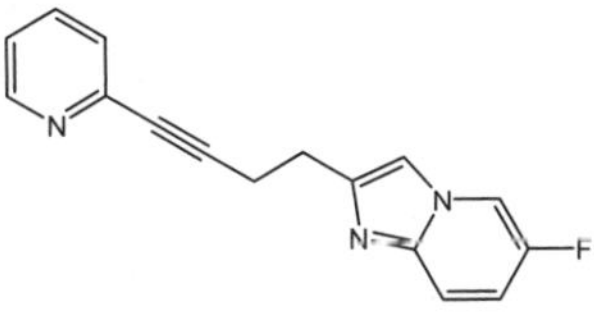

39, ADX48621

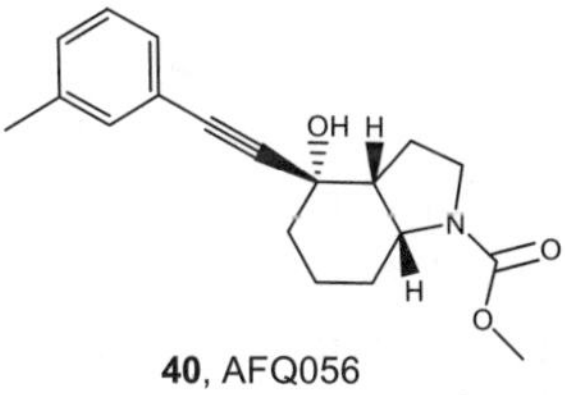

40, AFQ056

3.3. Non alkyne-based mGluR5 chemotypes

Bioisosteric replacement of the ethynyl bond, which keeps the molecular topology and the established SAR within the alkyne series, is exhibited by compounds such as 1,2,4-oxadiazole **41** (VU0285683) and the carboxamide **44** derivatives. As illustrated, many combinations preserved the 3-cyano-5-fluorophenyl moiety which can be considered as a privileged fragment.

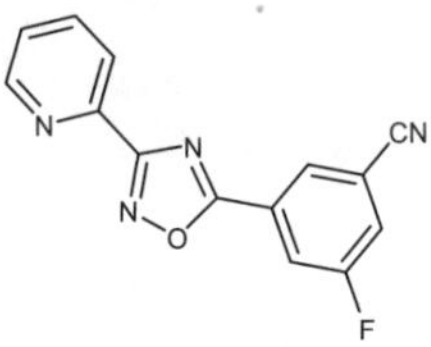

41, VU0285683

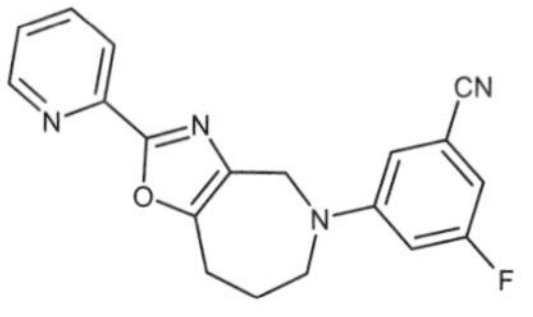

42

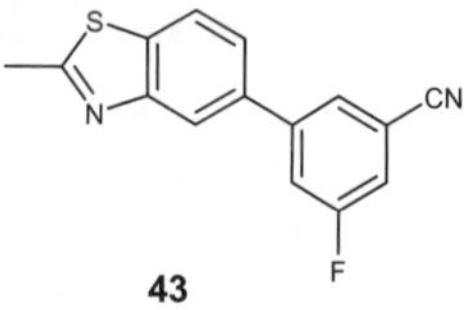

43

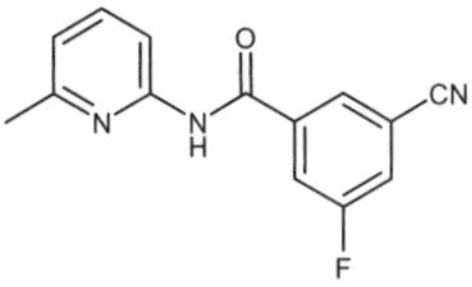

44

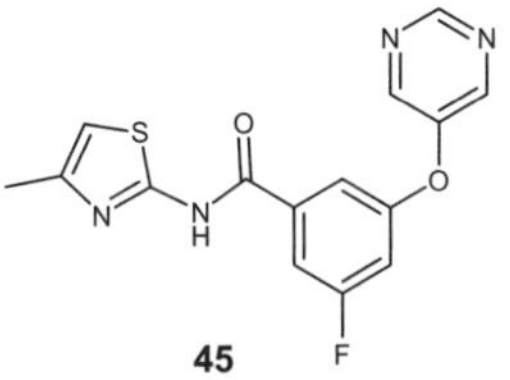

45

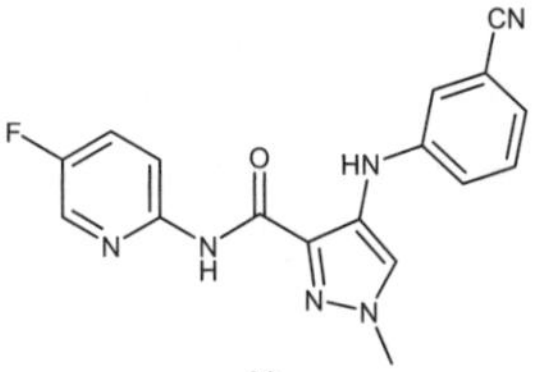

46

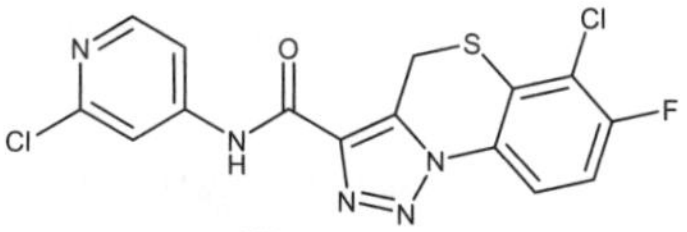

47

A series of tetrahydro-oxazolopyridines have been optimized leading to the identification of the azepine **42** (IC_{50} = 16 nM). Unfortunately, its short half-life in rat and monkey precluded further development.[74] Alternative heteroaryl ring systems were identified, such as the 2-methylbenzothiazole **43** (IC_{50} = 61 nM).[75] Carboxamide **44** (rmGluR5, IC_{50} = 24 nM) demonstrated efficacy in anxiety models similar to MTEP.[76] Compound **45** ($IC_{50}=25$ nM)[77] or pyrazole **46** ($IC_{50}=6.9$ nM)[78] showed efficacy in the marble burying mouse model. A novel constrained tricyclic series has been reported, illustrated by compound **47**, with subnanomolar activity ($pIC_{50}=9.1$).[79]

Another interesting set of potent mGluR5 NAM has been recently disclosed. Benzimidazol-3-yl pyridine **48** ($IC_{50}=24$ nM) showed oral efficacy in the anxiety fear-potentiated startle rat model (MED = 1 mg/kg).[80] Pyridinyl derivative **49** ($IC_{50}=17$ nM) exhibited moderate exposure after oral dosing due to extensive hepatic metabolism.[81] The piperidino-tetrazole **50** ($pIC_{50}=6.7$) resulted from a hit to lead campaign starting from a 3,5-oxadiazole hit.[82] Interestingly, *S*-enantiomers generally exhibited better potency.[83]

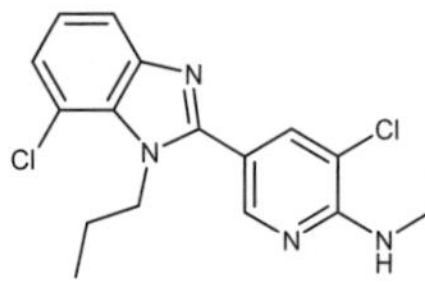

48

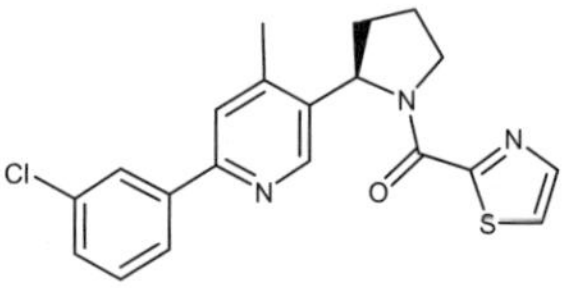

49

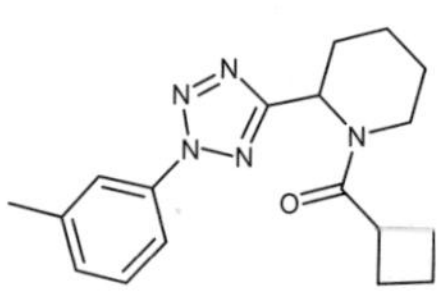

50

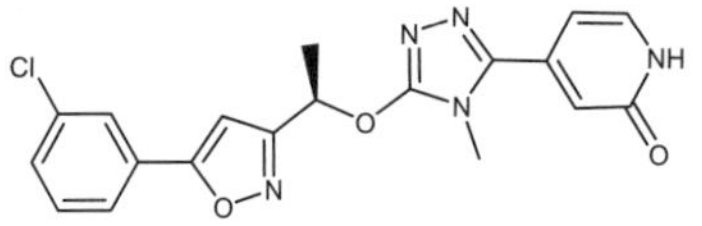

51

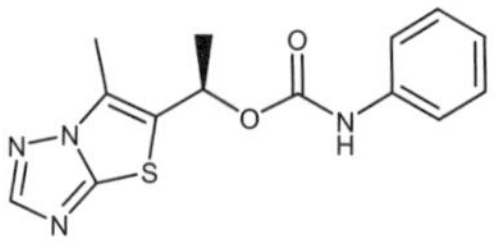

52, GSK2210875

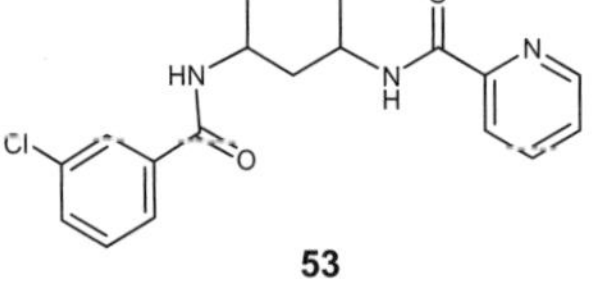

53

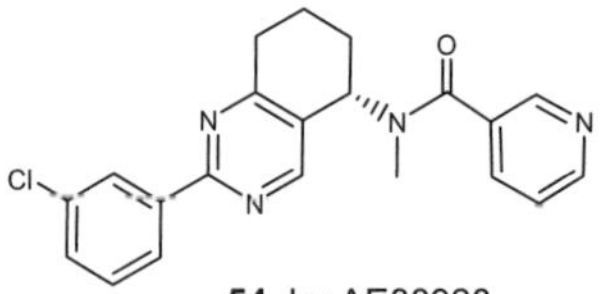

54, Lu AE88928

A number of related series where the heterocyclic spacer can be replaced by linear linkers have been extensively investigated, as illustrated methyloxy-triazole **51** ($IC_{50}=10$ nM)[84] or carbamate **52** (GSK2210875, $pIC_{50}=7.4$).[85] The tetrahydroquinazolinyl derivative **54** (LuAE88928, $IC_{50}=13$ nM), inspired from the cyclohexyldiamide **53** ($IC_{50}=10$ nM), has shown favorable PK profile and oral efficacy at 30 mg/kg in the Geller-Seifter conflict test.[86]

The mGluR5 NAM field of research is becoming mature, and the mining of a large collection of mGluR5 ligands and the recognition of their pharmacophoric patterns have been successfully applied to the discovery of novel chemotypes with several progressing into development.

3.4. Clinical trials

3.4.1 Parkinson's disease

To date, two compounds have been reported to be in clinical development for Parkinson's disease levodopa-induced dyskinesia. AFQ056 **40** has shown activity in two randomized controlled clinical trials in PD patients with moderate-to-severe LID and severe LID on stable dopaminergic therapy.[87] Patients received 25–150 mg AFQ056 or placebo twice daily for 16 days. AFQ056-treated patients showed significant improvements in dyskinesia on day 16 versus placebo, but no significant changes were seen from baseline to day 16 in the Unified Parkinson's Disease Rating Scale (UPDRS Part III), suggesting that AFQ056 does not affect the antiparkinsonian activity of levodopa and does not improve or worsen motor signs in these patients.

Dipraglurant (Addex Therapeutics) has entered a double-blind, placebo-controlled randomized Phase IIa study in patients with moderate or severe PD-LID, and positive top line data have been reported.[88] Dipraglurant demonstrated a statistically significant reduction in LID severity with both 50- and 100-mg doses, without interfering with levodopa efficacy. During week 4, patients reported a reduction in daily off-time of 50 min, suggesting an effect on parkinsonian motor symptoms in addition to the observed reductions in LID. In a subset of patients, dipraglurant appears to have reduced dystonia severity in addition to chorea, the two major LID components.

The antidyskinetic effect observed with AFQ056 and ADX48621 have demonstrated that mGluR5 inhibition is one of the most promising potential treatments for LID and a clinically relevant therapeutic target.

3.4.2 Other CNS disorders

FXS is an X-linked condition associated with intellectual disability and behavioral problems. AFQ056 has recently demonstrated a beneficial effect on the behavioral symptoms of fully methylated FXS patients versus partially methylated patients in a crossover study of 30 male FXS patients.[89] RO4917523 (undisclosed structure; Roche) entered a Phase II clinical trial in early 2012.[90] STX107 (undisclosed structure; Seaside Therapeutics) has completed Phase I trials.[91]

RO4917523 is currently being studied in patients with major depressive disorders, who show inadequate response to ongoing antidepressant therapies. Phase II studies have been completed, but no data have yet been released.[92]

ADX10059 (**38**) has shown the first clinical evidence for analgesic effects of mGluR5 NAMs. Data from a proof-of-concept study in episodic migraine patients demonstrated a significant improvement following acute treatment of ADX10059.[93]

4. CONCLUSION

Tremendous progress has been made recently toward the development of mGluR2 activators and mGluR5 NAM drugs. These are the first two subtypes from the whole class of mGluR family to reach mid-stage clinical testing and demonstrate positive outcomes in patients suffering from severe disorders. The first group II mGluR2/3 orthosteric agonists have demonstrated proof-of concept in humans for psychiatric and anxiety disorders. In addition, for the first time, pure mGluR2 PAMs have entered Phase II clinical trials in schizophrenia patients. Within group I mGluRs, three mGluR5 NAMs have achieved Phase II human proof-of-concept in PD-LID, FXS, or major depressive disorders.[94] In light of the wide range of chemotypes reported during the past 2–3 years, it is likely that novel potent and selective mGluR2 PAMs and mGluR5 NAMs will advance into the clinic in the coming years.

REFERENCES

(1) Nicoletti, F.; Bockaert, J.; Collingridge, G.L.; Conn, P.J.; Ferraguti, F.; Schoepp, D.D.; Wroblewski, J.T.; Pin, J.P. *Neuropharmacology* **2011**, *60*, 1017.
(2) Niswender, C.M.; Conn, P.J. *Annu. Rev. Pharmacol. Toxicol.* **2010**, *50*, 295.
(3) Kenakin, T. *Trends Pharmacol. Sci.* **2007**, *28*, 407.
(4) Christopoulos, A.; Kenakin, T. *Pharmacol. Rev.* **2002**, *54*, 323.
(5) Urwyler, S. *Pharmacol. Rev.* **2011**, *63*, 59.
(6) Gregory, K.J.; Dong, E.N.; Meiler, J.; Jeffrey, C.P. *Neuropharmacology* **2011**, *60*, 66.
(7) Pin, J.P.; Duvoisin, R. *Neuropharmacology* **1995**, *34*, 1.
(8) Tamaru, Y.; Nomura, S.; Mizuno, N.; Shigemoto, R. *Neuroscience* **2001**, *106*, 481.
(9) Kalivas, P.W. *Nat. Rev. Neurosci.* **2009**, *10*, 561.
(10) Fell, M.J.; McKinzie, D.L.; Monn, J.A.; Svensson, K.A. *Neuropharmacology* **2012**, *62*, 1473.
(11) Prezeau, L.; Rives, M.L.; Comps-Agrar, L.; Maurel, D.; Kniazeff, J.; Pin, J.P. *Curr. Opin. Pharmacol.* **2010**, *10*, 6.
(12) Fribourg, M.; Moreno, J.L.; Holloway, T.; Provasi, D.; Baki, L.; Mahajan, R.; Park, G.; Adney, S.K.; Hatcher, C.; Eltit, J.M.; Ruta, J.D.; Albizu, L.; Li, Z.; Umali, A.; Shim, J.;

Fabiato, A.; MacKerell, A.D.; Brezina, V.; Sealfon, S.C.; Filizola, M.; Gonzalez-Maeso, J.; Logothetis, D.E. *Cell* **2011**, *147*, 1011.
(13) Lundstrom, L.; Kuhn, B.; Beck, J.; Borroni, E.; Wettstein, J.G.; Woltering, T.J.; Gatti, S. *ChemMedChem* **2009**, *4*, 1086.
(14) Peeters, L.; Lavreysen, H.; Masure, S.; Gabriëls, L.; Tresadern, G.; Cid, J.; Atack, J.R. *Curr. Neuropharmacol.* **2011**, *9*, 1.
(15) Lundstrom, L.; Bissantz, C.; Beck, J.; Wettstein, J.G.; Woltering, T.J.; Wichmann, J.; Gatti, S. *Br. J. Pharmacol.* **2011**, *164*(2b), 521.
(16) Wright, R.A.; Johnson, B.G.; Zhang, C.; Salhoff, C.; Kingston, A.E.; Calligaro, D.O.; Monn, J.A.; Schoepp, D.D.; Marek, G.J. *Neuropharmacology* **2012**, http://dx.doi.org/10.1016/j.neuropharm.2012.01.019.
(17) Ceolin, L.; Kantamneni, S.; Barker, G.R.; Hanna, L.; Murray, L.; Warburton, E.C.; Robinson, E.S.; Monn, J.A.; Fitzjohn, S.M.; Collingridge, G.L.; Bortolotto, Z.A.; Lodge, D. *J. Neurosci.* **2011**, *31*, 6721.
(18) Hanna, L.; Ceolin, L.; Lucas, S.; Monn, J.; Johnson, B.; Collingridge, G.; Bortolotto, Z.; Lodge, D. *Neuropharmacology* **2012**, http://dx.doi.org/10.1016/j.neuropharm.2012.02.023.
(19) Monn, J.A.; Massey, S.M.; Valli, M.J.; Henry, S.S.; Stephenson, G.A.; Bures, M.; Herin, M.; Catlow, J.; Giera, D.; Wright, R.A.; Johnson, B.G.; Andis, S.L.; Kingston, A.; Schoepp, D.D. *J. Med. Chem.* **2007**, *50*, 233.
(20) Mezler, M.; Geneste, H.; Gault, L.; Marek, G.J. *Curr. Opin. Investig. Drugs* **2010**, *11*, 833.
(21) Monn, J.; Prietto, L.; Tabaoda Martinez, L.; Montero Salgado, C.; Shaw, B. *Patent Application WO2011084437*. **2011.**
(22) Swanson, S. *AAPS Prodrug Symposium*. Oct2011 http://www.aaps.org/Meetings_and_Professional_Development/Past_Meetings/2011_AAPS_Annual_Meeting_and_Exposition/Thu_0930_147AB_Swanson.
(23) Trabanco, A.A.; Cid, J.M.; Lavreysen, H.; Macdonald, G.J.; Tresadern, G. *Curr. Med. Chem.* **2011**, *18*, 47.
(24) Fraley, M.E. *Expert Opin. Ther. Pat.* **2009**, *19*, 1259.
(25) Zhang, L.; Brodney, M.A.; Candler, J.; Doran, A.C.; Duplantier, A.J.; Efremov, I.V.; Evrard, E.; Kraus, K.; Ganong, A.H.; Haas, J.A.; Hanks, A.N.; Jenza, K.; Lazzaro, J.T.; Maklad, N.; McCarthy, S.A.; Qian, W.; Rogers, B.N.; Rottas, M.D.; Schmidt, C.J.; Siuciak, J.A.; Tingley, F.D., III; Zhang, A.Q. *J. Med. Chem.* **2011**, *54*, 1724.
(26) D'Alessandro, P.L.; Corti, C.; Roth, A.; Ugolini, A.; Sava, A.; Montanari, D.; Bianchi, F.; Garland, S.L.; Powney, B.; Koppe, E.L.; Rocheville, M.; Osborne, G.; Perez, P.; Delafuente, J.; DeLosFrailes, M.; Smith, P.W.; Branch, C.; Nash, D.; Watson, S.P. *Bioorg. Med. Chem. Lett.* **2010**, *20*, 759.
(27) Dhanya, R.P.; Sidique, S.; Sheffler, D.J.; Nickols, H.H.; Herath, A.; Yang, L.; Dahl, R.; Ardecky, R.; Semenova, S.; Markou, A.; Conn, P.J.; Cosford, N.D. *J. Med. Chem.* **2011**, *54*, 342.
(28) Cosford, N.D.; Panickar, D.R.; Sidique, S. *Patent Application WO 2011116356*. 2011.
(29) Jin, X.; Semenova, S.; Yang, L.; Ardecky, A.; Sheffler, D.J.; Dahl, R.; Conn, J.P.; Cosford, N.D.P.; Markou, M. *Neuropsychopharmacology* **2010**, *35*, **2012.**
(30) Beshore, D.C.; Dudkin, V.; Kuduk, S.D.; Skudlarek, J.W.; Wang, C. *Patent Application WO 2010141360*. **2010.**
(31) Dudkin, V.; Fraley, M.E.; Wang, C.; Garbaccio, R.M.; Beshore, D.C.; Kuduk, S.D.; Skudlarek, J.W. *Patent Application WO 2011022312*. **2011.**
(32) Arrington, K.L.; Dudkin, V.; Laytov, M.E.; Pero, J.E.; Reif, A.J. *Patent Application WO 2011156245*. **2011.**
(33) Layton, M.E.; Kelly, M.J.; Hartingh, T.J. *Patent Application WO 2011109277*. **2011.**

(34) Imogai, H.; Cid-Nuñez, J.M.; Duvey, G.; Boléa, C.M.; Nhem, V.; Finn, T.P.; Le Poul, E.C.; Rocher, J-.P.; Lutjens, R.J. *Patent Application WO 2006030032.* **2006.**

(35) Cid, J.M.; Duvey, G.; Tresadern, G.; Nhem, V.; Furnari, R.; Cluzeau, P.; Vega, J.A.; de Lucas, A.I.; Matesanz, E.; Alonso, J.M.; Linares, M.L.; Andres, J.I.; Poli, S.M.; Lutjens, R.; Himogai, H.; Rocher, J.P.; Macdonald, G.J.; Oehlrich, D.; Lavreysen, H.; Ahnaou, A.; Drinkenburg, W.; Mackie, C.; Trabanco, A.A. *J. Med. Chem.* **2012**, *55*, 2388.

(36) Tresadern, G.; Cid, J.M.; Macdonald, G.J.; Vega, J.A.; De Lucas, A.I.; García, A.; Matesanz, E.; Linares, M.L.; Oehlrich, D.; Lavreysen, H.; Biesmans, I.; Trabanco, A.A. *Bioorg. Med. Chem. Lett.* **2010**, *20*, 175.

(37) Trabanco, A.A.; Tresadern, G.; Macdonald, G.J.; Vega, J.A.; de Lucas, A.I.; Matesanz, E.; Garcia, A.; Linares, M.L.; Alonso de Diego, S.A.; Alonso, J.M.; Oehlrich, D.; Ahnaou, A.; Drinkenburg, W.; Mackie, C.; Andres, J.I.; Lavreysen, H.; Cid, J.M. *J. Med. Chem.* **2012**, *55*, 2688.

(38) Ahnaou, A.; Dautzenberg, F.M.; Geys, H.; Imogai, H.; Gibelin, A.; Moechars, D.; Steckler, T.; Drinkenburg, W.H. *Eur. J. Pharmacol.* **2009**, *603*, 62.

(39) Cid-Nuñez, J.M.; De Lucas Olivares, A.I.; Trabanco-Suárez, A.A.; Macdonald, G.J. *Patent Application WO 2010130423.* **2010.**

(40) Cid-Nuñez, J.M.; Oehlrich, D.; Trabanco-Suárez, A.A.; Tresadern, G.J.; Vega Ramiro, J.A.; Macdonald, G.J. *Patent Application WO 2010130424.* **2010.**

(41) Trabanco, A.A.; Duvey, G.; Cid, J.M.; Macdonald, G.J.; Cluzeau, P.; Nhem, V.; Furnari, R.; Behaj, N.; Poulain, G.; Finn, T.; Lavreysen, H.; Poli, S.; Raux, A.; Thollon, Y.; Poirier, N.; D'Addona, D.; Andres, J.I.; Lutjens, R.; Le Poul, E.; Imogai, H.; Rocher, J.P. *Bioorg. Med. Chem. Lett.* **2011**, *21*, 971.

(42) Brnardic, E.J.; Fraley, M.E.; Garbaccio, R.M.; Layton, M.E.; Sanders, J.M.; Culberson, C.; Jacobson, M.A.; Magliaro, B.C.; Hutson, P.H.; O'Brien, J.A.; Huszar, S.L.; Uslaner, J.M.; Fillgrove, K.L.; Tang, C.; Kuo, Y.; Sur, S.M.; Hartman, G.D. *Bioorg. Med. Chem. Lett.* **2010**, *20*, 3129.

(43) Garbaccio, R.M.; Brnardic, E.J.; Fraley, M.E.; Hartman, G.D.; Hutson, P.H.; O'Brien, J.A.; Magliaro, B.C.; Uslaner, J.M.; Huszar, S.L.; Fillgrove, K.L.; Small, J.H.; Tang, C.; Kuo, Y.; Jacobson, M.A. *ACS Med. Chem. Lett.* **2010**, *1*, 406.

(44) Patil, S.T.; Zhang, L.; Martenyi, F.; Lowe, S.L.; Jackson, K.A.; Andreev, B.V.; Avedisova, A.S.; Bardenstein, L.M.; Gurovich, I.Y.; Morozova, M.A.; Mosolov, S.N.; Neznanov, N.G.; Reznik, A.M.; Smulevich, A.B.; Tochilov, V.A.; Johnson, B.G.; Monn, J.A.; Schoepp, D.D. *Nat. Med.* **2007**, *13*, 1102.

(45) http://clinicaltrials.gov/ct2/results?term=JNJ-40411813. NCT01248052.

(46) http://clinicaltrials.gov/ct2/results?term=JNJ-40411813. NCT01101659, NCT0135-8006, NCT01323205.

(47) Dunayevich, E.; Erickson, J.; Levine, L.; Landbloom, R.; Schoepp, D.D.; Tollefson, G.D. *Neuropsychopharmacology* **2008**, *33*, 1603.

(48) Michelson, D.; Levine, L.R.; Dellva, M.A.; Mesters, P.; Schoepp, D.D.; Dunayevich, E.; Tollefson, G.D. *Neuropharmacology* **2005**, *49*, 257.

(49) Johnson, M.P.; Schkeryantz, J.; Swanson, S.; Perkins, A.N.; Johnson, K.W. *Curr. Neuropharmacol.* **2011**, *9*, 1.

(50) Moghaddam, B.; Javitt, D. *Neuropsychopharmacology* **2012**, *37*, 4.

(51) Swanson, J.; Bures, M.; Johnson, M.P.; Linden, A.M.; Monn, J.A.; Schoepp, D.D. *Nat. Rev. Drug Discov.* **2005**, *4*, 131.

(52) Kinon, B.J.; Zhang, L.; Millen, B.A.; Osuntokun, O.O.; Williams, J.E.; Kollack-Walker, S.; Jackson, K.; Kryzhanovskaya, L.; Jarkova, N. *J. Clin. Psychopharmacol.* **2011**, *31*, 349.

(53) http://clinicaltrials.gov/ct2/results?term=LY2140023. NCT00845026, NCT01086748, NCT01052103, NCT01129674, NCT01125358, NCT01129674, NCT01487083, NCT01307800, NCT01328093, NCT01452919.

(54) Horiguchi, M.; Huang, M.; Meltzer, H.Y. *Psychopharmacology* **2011**, *217*, 13.

(55) Helton, D.R.; Tizzano, J.P.; Monn, J.A.; Schoepp, D.D.; Kallman, M.J. *J. Pharmacol. Exp. Ther.* **1998**, *284*, 651.

(56) Nikiforuk, A.; Popik, P.; Drescher, K.U.; Van Gaalen, M.; Relo, A.L.; Mezler, M.; Marek, G.; Schoemaker, H.; Gross, G.; Bespalov, A. *J. Pharmacol. Exp. Ther.* **2010**, *335*, 665.

(57) Bergink, V.; Westenberg, H.G. *Int. Clin. Psychopharmacol.* **2005**, *20*, 291.

(58) Grillon, C.; Cordova, J.; Levine, L.R.; Morgan, C.A., III *Psychopharmacology* **2003**, *168*, 446.

(59) http://clinical-trials.healia.com/doc/NCT01184508.

(60) Aicher, T.; Cortez, G.; Groendyke, T.; Khilevich, A.; Knolbelsdorf, J.; Magnus, N.; Marmsater, F.; Schkeryantz, J.; Tang, T. *Patent Application WO2006057870.* **2006.**

(61) Romano, C.; Sesma, M.A.; McDonald, C.T.; O'Malley, K.; van den Pol, A.N.; Olney, J.W. *J. Comp. Neurol.* **1995**, *355*, 455.

(62) Conn, P.J.; Battaglia, G.; Marino, M.J.; Nicoletti, F. *Nat. Rev. Neurosci.* **2005**, *6*, 787.

(63) Kniazeff, J.; Prezeau, L.; Rondard, P.; Pin, J.P.; Goudet, C. *Pharmacol. Ther.* **2011**, *130*, 9.

(64) Rocher, J.P.; Bonnet, B.; Bolea, C.; Lutjens, R.; Le Poul, E.; Poli, S.; Epping-Jordan, M.; Bessis, A.S.; Ludwig, B.; Mutel, V. *Curr. Top. Med. Chem.* **2011**, *11*, 680.

(65) Emmitte, K.A. *ACS Chem. Neurosci.* **2011**, *2*, 411.

(66) Alagille, D.; Dacosta, H.; Chen, Y.; Hemstapat, K.; Rodriguez, A.; Baldwin, R.M.; Conn, J.P.; Tamagnan, G. *Bioorg. Med. Chem. Lett.* **2011**, *21*, 3243.

(67) Dekundy, A.; Gravius, A.; Hechenberger, M.; Pietraszek, M.; Nagel, J.; Tober, C.; van der Elst, M.; Mela, F.; Parsons, C.G.; Danysz, W. *J. Neural Transm.* **2011**, *118*, 1703.

(68) Lindemann, L.; Jaeschke, G.; Michalon, A.; Vieira, E.; Honer, M.; Spooren, W.; Porter, R.; Hartung, T.; Kolczewski, S.; Buttelmann, B.; Flament, C.; Diener, C.; Fischer, C.; Gatti, S.; Prinssen, E.P.; Parrott, N.; Hoffmann, G.; Wettstein, J.G. *J. Pharmacol. Exp. Ther.* **2011**, *339*, 474.

(69) Gilbert, A.M.; Bursavich, M.G.; Lombardi, S.; Adedoyin, A.; Dwyer, J.M.; Hughes, Z.; Kern, J.C.; Khawaja, X.; Rosenzweig-Lipson, S.; Moore, W.J.; Neal, S.J.; Olsen, M.; Rizzo, S.J.; Springer, D. *Bioorg. Med. Chem. Lett.* **2011**, *21*, 195.

(70) Sperry, J.B.; Levent, M.; Varsolona, R.J.; Farr, R.M.; Ghosh, M.; Hoagland, S.M.; Rapisardi, V.; Fung, P.; Sarveiya, V.; Sutherland, K. ORGN-168, *242nd ACS National Meeting, Denver, CO.* Aug **2011.**

(71) Hugues, Z.A.; Neal, S.J.; Smith, D.L.; Sukoff Rizzo, S.J.; Pulicicchio, C.M.; Lotarski, S.; Bryce, D.; Lu, S.; Dwyer, J.; Olsen, M.; Bender, C.N.; Kouranova, E.; Springer, D.; Li, D.; O'Neil, S.V.; Andree, T.H.; Whiteside, G.T.; Dunlop, J.; Brandonm, N.J.; Schechter, L.E.; Leonard, S.K.; Rosenzweig-Lipson, S.; Ring, R.H. *Curr. Neuropharmacol.* **2011**, *9*, 1.

(72) Hill, M.P.; Girard, F.; Keywood, C.; Poli, S.-M.; Le Poul, E.; Crossman, A.R.; Ravenscroft, P.; Li, Q.; Bezard, E.; Mutel, V. *Society for Neuroscience Meeting, San Diego.* Nov 2010 Poster 557.12/M18.

(73) Levenga, J.; Hayashi, S.; de Vrij, F.M.; Koekkoek, S.K.; van der Linde, H.C.; Nieuwenhuizen, I.; Song, C.; Buijsen, R.A.; Pop, A.S.; Gomezmancilla, B.; Nelson, D.L.; Willemsen, R.; Gasparini, F.; Oostra, B.A. *Neurobiol. Dis.* **2011**, *42*, 311.

(74) Burdi, D.F.; Hunt, R.; Fan, L.; Hu, T.; Wang, J.; Guo, Z.; Huang, Z.; Wu, C.; Hardy, L.; Detheux, M.; Orsini, M.A.; Quinton, M.S.; Lew, R.; Spear, K.Y. *J. Med. Chem.* **2010**, *53*, 7107.

(75) Lindsley, C.W.; Bates, B.S.; Menon, U.N.; Jadhav, S.B.; Kane, A.S.; Jones, C.K.; Rodriguez, A.L.; Conn, P.J.; Olsen, C.M.; Winder, D.G.; Emmitte, K.A. *ACS Chem. Neurosci.* **2011**, *2*, 471.
(76) Rodriguez, A.L.; Grier, M.D.; Jones, C.K.; Herman, E.J.; Kane, A.S.; Smith, R.L.; Williams, R.; Zhou, Y.; Marlo, J.E.; Days, E.L.; Blatt, T.N.; Jadhav, S.; Menon, U.N.; Vinson, P.N.; Rook, J.M.; Stauffer, S.R.; Niswender, C.M.; Lindsley, C.W.; Weaver, C.D.; Conn, P.J. *Mol. Pharm.* **2010**, *78*, 1105.
(77) Conn, P.J.; Lindsley, C.W.; Emmitte, K.A.; Weaver, C.D.; Rodriguez, A.L.; Felts, A.; Jones, C.K.; Bates, B.S. *Patent Application WO 2011035209.* **2011.**
(78) Kanuma, K.; Tamita, T.; Imura, K. *Patent Application WO 2009078432.* 2009.
(79) Bertani, B.; Cremonesi, S.; Garzya, V.; Micheli, F.; Rupcic, R.; Sabbatini, F.M. *Patent Application WO 2011151361.* **2011.**
(80) Carcache, D.; Vranesic, I.; Blanz, J.; Desrayaud, S.; Fendt, M.; Glatthar, R. *ACS Med. Chem. Lett.* **2011**, *2*, 58.
(81) Weiss, J.M.; Jimenez, H.N.; Li, G.; April, M.; Uberti, M.A.; Bacolod, M.D.; Brodbeck, R.M.; Doller, D. *Bioorg. Med. Chem. Lett.* **2011**, *21*, 4891.
(82) Wágner, G.; Wéber, C.; Nyéki, O.; Nógrádi, K.; Bielik, A.; Molnár, L.; Bobok, A.; Horváth, A.; Kiss, B.; Kolok, S.; Nagy, J.; Kurkó, D.; Gál, K.; Greiner, I.; Szombathelyi, Z.; Keseru, G.M.; Domány, G. *Bioorg. Med. Chem. Lett.* **2010**, *20*, 3737.
(83) Stauffer, S.R. *ACS Chem. Neurosci.* **2011**, *2*, 450.
(84) Isaac, M.; Slassi, A.; Edwards, L.; Dove, P.; Xin, T.; Stefanac, T. *Patent Application WO 2008/041075.* **2008.**
(85) Pilla, M.; Andreoli, M.; Tessari, M.; Delle-Fratte, S.; Roth, A.; Butler, S.; Brown, F.; Shah, P.; Betini, E.; Cavallini, P.; Benedetti, R.; Minick, D.; Smith, P.; Tehan, B.; D'Alessandro, P.; Lorthioir, O.; Ball, C.; Garzya, V.; Goodacre, C.; Watson, S. *Bioorg. Med. Chem. Lett.* **2010**, *20*, 7521.
(86) Li, G.; April, M.; Austin, J.; Zhou, H.; Pu, X.; Boyle, N.J.; Uberti, M.A.; Brodbeck, R.M.; Bacolod, M.D.; Smith, D.G.; Doller, D. *Curr. Neuropharmacol.* **2011**, *9*, 1.
(87) Berg, D.; Godau, J.; Trenkwalder, C.; Eggert, K.; Csoti, I.; Storch, A.; Huber, H.; Morelli-Canelo, M.; Stamelou, M.; Ries, V.; Wolz, M.; Schneider, C.; Di, P.T.; Gasparini, F.; Hariry, S.; Vandemeulebroecke, M.; Abi-Saab, W.; Cooke, K.; Johns, D.; Gomez-Mancilla, B. *Mov. Disord.* **2011**, *26*, 1243.
(88) http://www.addextherapeutics.com/rd/pipeline/dipra-ir/, Mar 21, **2012**.
(89) Jacquemont, S.; Curie, A.; des Portes, V.; Torrioli, M.G.; Berry-Kravis, E.; Hagerman, R.J.; Ramos, F.J.; Cornish, K.; He, Y.; Paulding, C.; Neri, G.; Chen, F.; Hadjikhani, N.; Martinet, D.; Meyer, J.; Beckmann, J.S.; Delange, K.; Brun, A.; Bussy, G.; Gasparini, F.; Hilse, T.; Floesser, A.; Branson, J.; Bilbe, G.; Johns, D.; Gomez-Mancilla, B. *Sci. Transl. Med.* **2011**, *3 64ra1.*
(90) http://www.roche-trials.com/trialDetailsGet.action?studyNumber=NP27936.
(91) http://clinicaltrials.gov/ct2/show/NCT01325740.
(92) http://clinicaltrials.gov/ct2/show/NCT01437657.
(93) Marin, J.C.; Goadsby, P.J. *Expert Opin. Investig. Drugs* **2010**, *19*, 555.
(94) Duncan, J.R.; Lawrence, A.J. *Pharmacol. Biochem. Behav.* **2012**, *100*, 811.

CHAPTER SEVEN

NMDA Antagonists of GluN2B Subtype and Modulators of GluN2A, GluN2C, and GluN2D Subtypes—Recent Results and Developments

Kamalesh B. Ruppa*, Dalton King†, Richard E. Olson†
*NeurOp, Inc., 58 Edgewood Ave NE, Atlanta, GA 30303, USA
†Bristol-Myers Squibb, Richard L. Gelb Center for Pharmaceutical Research and Development, Neuroscience Discovery Chemistry, 5 Research Parkway, Wallingford, CT 06492-7660, USA

Contents

1. INTRODUCTION

N-methyl-D-aspartate (NMDA) receptors are members of a family of ionotropic glutamate receptors (iGluR).[1–10] They are widely expressed in the brain with specific anatomical localization and populations for different subunit types and thought to be involved in numerous physiological and pathological processes.[5] A common feature of iGluRs is

Annual Reports in Medicinal Chemistry, Volume 47
ISSN 0065-7743
http://dx.doi.org/10.1016/B978-0-12-396492-2.00007-2

their construction as tetrameric assemblies of subunits. NMDA tetramers are assembled from two subunits of glycine-binding GluN1 and two subunits of glutamate-binding GluN2 (GluN2A, GluN2B, GluN2C, or GluN2D), or in some cases GluN3 subunits (GluN3A or GluN3B). NMDA receptor ligand-gated channels are normally blocked by extracellular Mg^{2+} at resting membrane potential. They are activated upon binding to two coagonists, glycine or D-serine, at the GluN1 subunit ligand-binding domain (LBD), and glutamate at the GluN2 subunit LBD, resulting in the opening of the ion channel and permitting passage of Na^{+}, K^{+}, and Ca^{2+} ions in a nonselective manner. Overactivation of NMDA receptors generates an uncontrolled influx of calcium ions, which then activates a cascade of excitotoxic processes leading to neurodegeneration. Accordingly, NMDA antagonism is thought to be useful in the treatment of many CNS disorders including ischemic stroke, Parkinson's disease, Alzheimer's disease, neuropathic pain, and depression, among many others.[11–13]

Among the NMDA receptor subtypes, it is known that the GluN2B subunit can be allosterically regulated by the binding of ligands at multiple sites. These allosteric binding sites recognize ions (H^{+}, Mg^{2+}, Zn^{2+}), small molecules (ifenprodil-like ligands), or polyamines. Thus, the function of NMDA receptors could be altered by blocking or interacting with one of the following binding sites, (i) glycine agonist site, (ii) glutamate agonist site, (iii) ion channel pore, and (iv) allosteric sites on the amino terminal domain (ATD). The ATD, being distinct for subunit types, currently offers the greatest advantage for designing highly subunit selective antagonists.

The major focus of this review is to summarize recent medicinal chemistry advances in the area of GluN2B receptors, and emerging antagonists and modulators of other NMDA receptor subtypes.

2. X-RAY STRUCTURAL STUDIES AND HOMOLOGY MODELING OF NMDA RECEPTORS

Many reports have appeared describing the structural features and function of NMDA receptors.[14–23] Based on mutagenesis studies and molecular modeling, it has generally been accepted that the mechanism of action of phenylethanolamine ligands such as ifenprodil (**1**), traxoprodil (CP-101,606, **2**), and Ro 25-6981 (**3**) is exerted *via* allosteric binding to the ATD of GluN2B.[14–16]

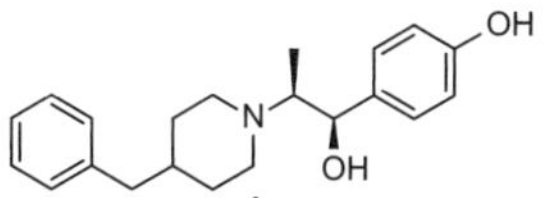
1

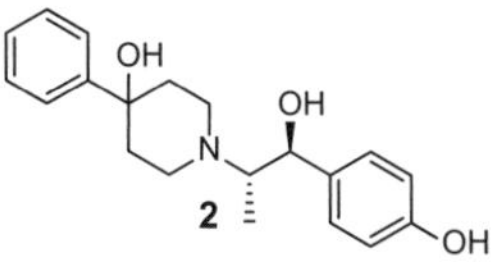
2

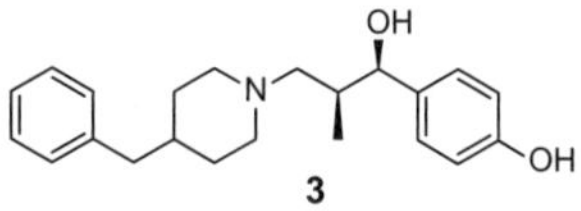
3

These studies, though, have only provided limited direct evidence regarding the detailed structure of the phenylethanolamine binding site.[15,16] A major step forward was recently made by Karakas *et al.*[17] with the publication of cocrystal structures of ifenprodil and Ro 25-6981 bound to a GluN1b–GluN2B ATD complex. The ATDs of GluN1b from *Xenopus laevis* and GluN2B from *Rattus norvegicus* form a unique heterodimer whose secondary structure is characterized by an asymmetrical set of interactions between the R1 and R2 domains of the GluN1b and GluN2B ATDs. The GluN1b R2 domain is not directly involved in the GluN1b–GluN2B interaction, leaving sufficient room for conformational flexibility in the opposing GluN2B R2 domain. Movement of the GluN2B R2 domain is thought to be important in mediating the allosteric regulation observed for this receptor.[18,19]

Ifenprodil and Ro 25-6981 were shown to bind at the GluN1b–GluN2B subunit interface through an induced-fit mechanism. The binding pocket is defined by residues from GluN1b R1, GluN2B R1, and GluN2B R2, with no overlap of the Zn^{2+} binding site located in the GluN2B ATD cleft. Furthermore, binding is characterized by three distinct types of interactions in the complex (Fig. 7.1): (1) hydrophobic interactions between the benzylpiperidine group common to both molecules and a cluster of hydrophobic residues from the GluN1b $\alpha2$ and $\alpha3$ helices and the GluN2B $\alpha1'$ and $\alpha2'$ helices, (2) hydrophobic interactions between the phenol groups and GluN1b Leu135, GluN2B Phe176, and GluN2B Pro177, and (3) direct polar interactions with Ser132 of GluN1b, Gln110 of GluN2B, and Asp236 of GluN2B. Although the benzylpiperidine groups of ifenprodil and Ro 25-6981 orient in a similar fashion, the methyl and hydroxyl groups in the propanol moiety face opposite directions, which may account for the difference in their binding affinities (Fig. 7.1C). In support of the physiological relevance of the identified binding site, the authors used a set of mutagenesis studies to show that critical residues in the site not only impart crucial binding interactions but also are important in producing the functional effect. The information contained in these cocrystal structures will undoubtedly serve as a molecular

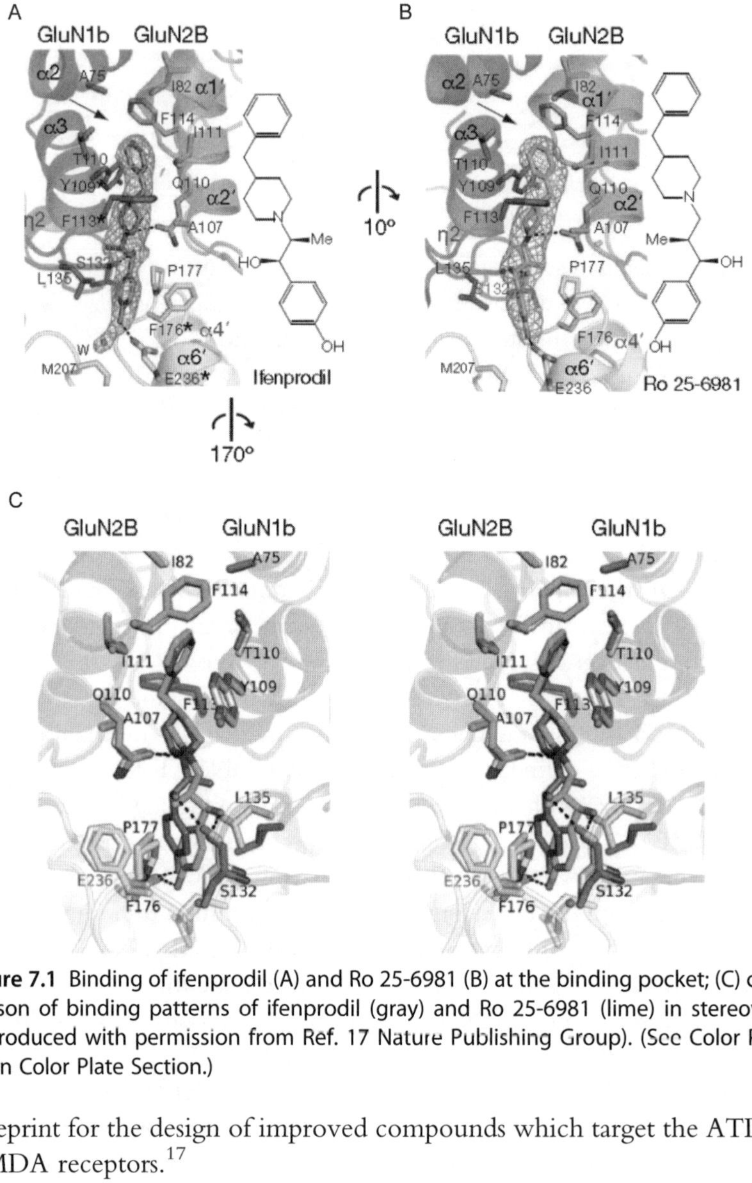

Figure 7.1 Binding of ifenprodil (A) and Ro 25-6981 (B) at the binding pocket; (C) comparison of binding patterns of ifenprodil (gray) and Ro 25-6981 (lime) in stereoview (reproduced with permission from Ref. 17 Nature Publishing Group). (See Color Plate 7.1 in Color Plate Section.)

blueprint for the design of improved compounds which target the ATD of NMDA receptors.[17]

Vance *et al.* reported the crystal structures of isolated GluN2D LBD in complex with four different agonist ligands (L-glutamate, D-glutamate, L-aspartate, and NMDA).[23] This study reveals that the binding of L-glutamate induces a unique conformation at the backside of the ligand-binding site. These data suggest that the activity of the GluN1/GluN2D NMDA receptor is controlled distinctively by the endogenous neurotransmitter L-glutamate.

3. CHANNEL BLOCKERS

NMDA channel blockers are noncompetitive antagonists and generally lack subunit selectivity. In addition, known channel blockers have a wide range of off-target activities (e.g., D_2, 5-HT, GABA, μ-, κ-opioid, σ, mAChR) and block channels other than NMDA (Na, nAChR, HCN1, and K_{ATP}). Examples of high-affinity blockers are ketamine (**4**), phenylcyclidine (PCP, **5**), and MK-801 (**6**), while memantine (**7**) and amantadine (**8**) are low-affinity blockers ($K_i > 1$ μM).

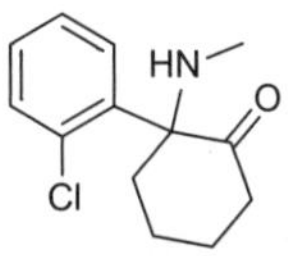

4

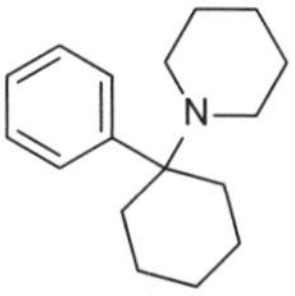

5

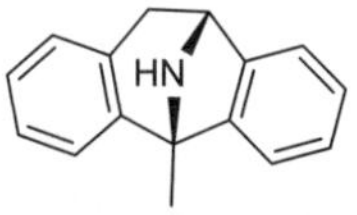

6

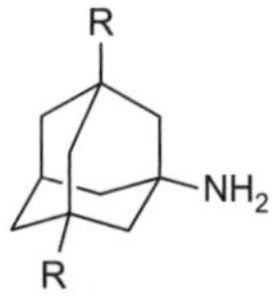

7 R = H
8 R = CH_3

Both ketamine and PCP produce psychotomimetic effects and are abused as recreational drugs. Subanaesthetic doses of ketamine are reported to cause a rapid and sustained antidepressant effect.[24] Related to this clinical finding, Duman et al., reported that the mTOR cellular signaling pathway in the prefrontal cortex (PFC) of the rat brain is activated as a response to ketamine. This leads to increased synaptic signaling proteins and a reversal of atrophy of spine synapses in the PFC.[24] The authors suggest that the functional reconnection of neurons resulting from this process underlies the rapid behavioral responses which are observed.[24] While channel blockers represent an interesting pharmacological class for the treatment of depression, ample opportunities remain for the development of drugs with a better tolerability profile.

A recent study described the identification of analogs of ketamine and PCP with reduced lipophilicity.[25] A heteroatom was introduced into the phenyl ring of PCP as in **9** and a novel lactam scaffold was employed as in ketamine analog **10**, and functional activity was tested in a Ca^{2+} flux FLIPR assay. Both ketamine and PCP showed moderate potencies at both GluN2A and GluN2B receptors, whereas **9** and **10** were 2- to 10-fold less potent. Dexoxadrol (**11**) is a high-affinity channel blocker ($K_i = 11$ nM), which like MK-801 ($K_i = 3$ nM) has PCP-like dissociative properties.[26] Since the severe side effects are attributed to high NMDA receptor affinity, analogs with moderate affinities between those of **11** and memantine ($K_i = 1.2$ μM) may display desired pharmacological activity with an improved side effect profile. Fluorinated analogs **12** and **13** were indeed two- and sevenfold less potent than **11** as evaluated in a [^{3}H]-MK-801 binding assay. Additionally, **12** and **13** showed reduced affinities at σ_1 and σ_2 receptors relative to **11**.

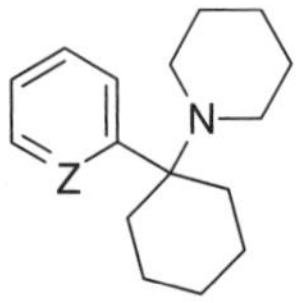

5 Z = CH
9 Z = N

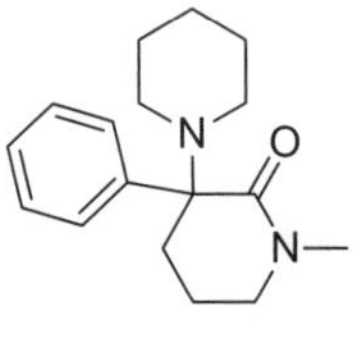

10

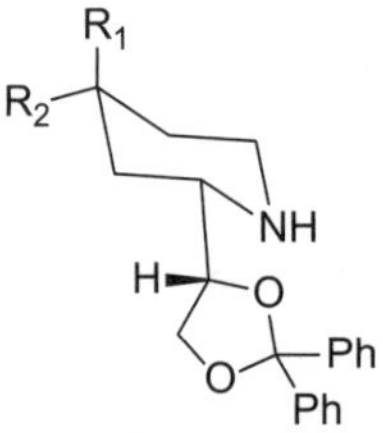

11 $R_1=R_2=H$
12 $R_1=F, R_2=H$
13 $R_1=R_2=F$

4. INHIBITORS OF GLUN2B SUBTYPE RECEPTOR

A number of GluN2B antagonists, such as ifenprodil (**1**), CP-101,606 (**2**), Ro 25-6981 (**3**), RGH-896 (radiprodil, **14**), EVT-101 (**15**), MK-0657 (**16**), have progressed as far as Phase II clinical trials; however, progress beyond this point has been hindered by the presence of undesirable CNS and other off-target side effects. Despite early setbacks, there still remains a high interest in developing orally administered, subtype selective NMDA antagonists with an acceptable side effect profile. Allosteric inhibitors of GluN2B have been hypothesized to have the highest potential to offer an improved therapeutic index. With the recent identification of the allosteric binding site of ifenprodil and Ro 25-6981,[17] a detailed template for the design of highly refined allosteric inhibitors of GluN2B now exists.

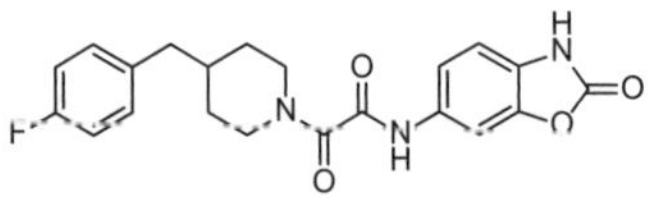

14

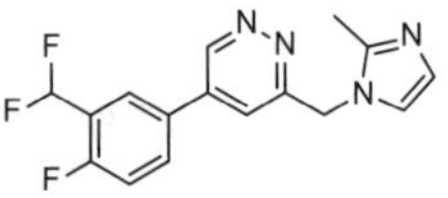

15

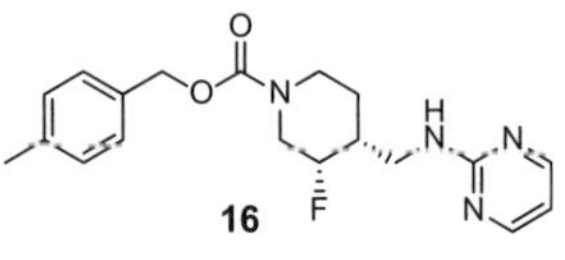

16

Early allosteric inhibitors such as the phenylethanolamines, ifenprodil and CP-101,606, required a terminal phenol group. In later compounds, phenol bioisosteres were employed. Additional studies focused on removing the metabolically labile OH group in the chain, while others transformed the basic tertiary amine center to a less basic ionizable form or to an amide center to minimize prominent off-target side effects at hERG and α_1-adrenergic receptors.

4.1. Phenylethanolamines and related compounds

The themes continue to be explored in recent examples. Wee and coworkers characterized the high-affinity protein binding domain of the 5-substituted benzimidazoles **17** and **18**.[27] Compounds **17** and **18**, subnanomolar inhibitors of GluN2B, were shown to bind directly to the ATD of the GluN2B subunit. In a series bearing some resemblance to **17**, an indole scaffold was identified by a molecular modeling strategy.[28] Optimum members of the series (**19–21**) showed potent GluN2B binding (K_i=17–25 nM). Compound **19** reduced NMDA receptor-mediated current in CA1 pyramidal neurons from rat hippocampus, suggesting that it might be useful as a neuroprotective agent.

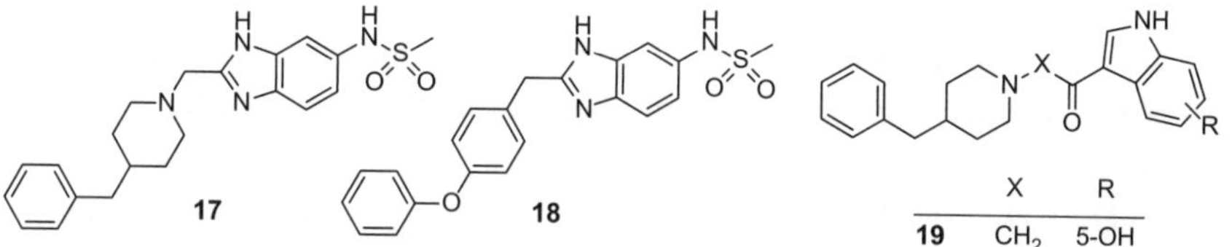

17 **18**

	X	R
19	CH_2	5-OH
20	CH_2	6-OH
21	CO	6-OH

Cyclic analogs of ifenprodil, **22** and **23**, based on a tetrahydro-3-benzazepine scaffold, had high GluN2B affinities in a [^{3}H]-ifenprodil-binding assay (K_i=14 and 5.5 nM, respectively).[29,30] Benzyl ether **24** had a loss of GluN2B affinity (K_i=187 nM). Despite its higher GluN2B binding affinity, the methyl ether **23** (IC_{50}=360 nM) was substantially less potent in an excitotoxicity assay than the phenol **22** (IC_{50}=18.4 nM), suggesting that the phenol group is necessary for efficient functional inhibition.

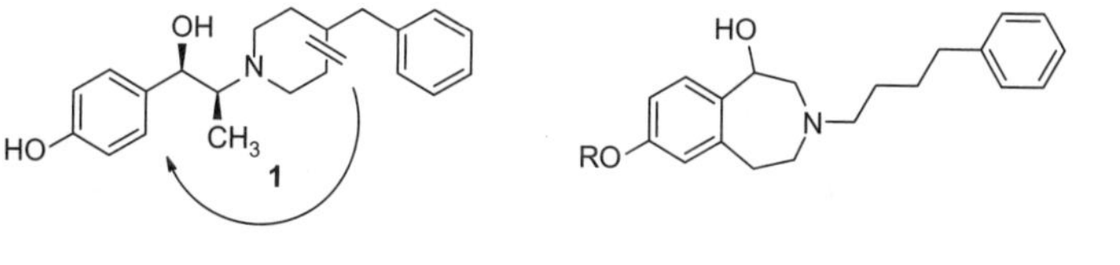

	R
22	**H**
23	**CH_3**
24	**CH_2Ph**

A series of isomeric *N*-methyl isoindolines was unique in lacking a phenol or bioisostere typically found in ifenprodil-like molecules.[31] Compounds **25** and **26**, both [1*S*, 1′*S*] configuration, were only moderately potent in a competitive binding assay with [^{3}H]-ifenprodil (430 and 270 nM, respectively) and also displayed functional inhibition of [^{3}H] MK-801 binding at the channel pore (210 and 500 nM, respectively).

A commercial database was evaluated in a virtual screening approach for GluN2B selective antagonists.[32] The primary hit compound **27** (IC_{50} 2.7 μM) was only 10-fold less potent than ifenprodil. Further, refinement of this chemotype will be necessary in order to decrease the potential for formation of covalent intermediates *in vivo*.

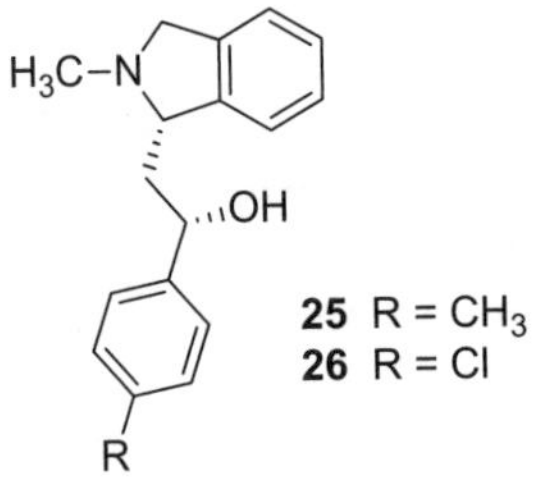

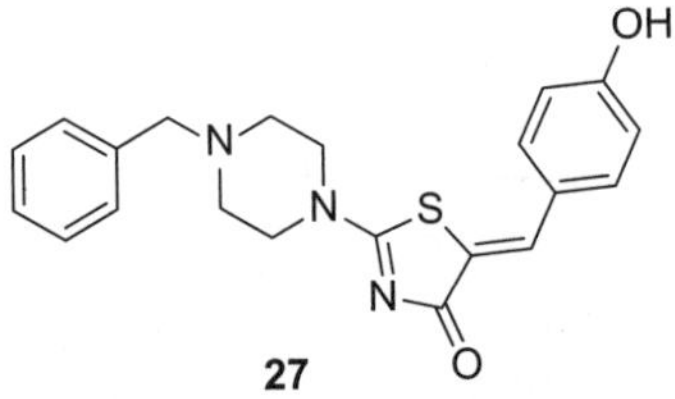

Members of a series of novel 3-substituted aminocyclopentanes were identified as orally bioavailable, highly potent and selective GluN2B receptor antagonists.[33] Starting from the initially identified carbamate **28** (similar to benchmark **29**), **30** was conceived by successively introducing ring constraints and isosteric groups for the benzamide and phenol. Two diastereomers of **30** (*SR*- and *SS*-) were as potent (K_i = 2.8 and 3.6 nM, respectively) against GluN2B as compound **29** and had low hERG potency (> 1000-fold vs. GluN2B) but were susceptible to human P_{gp} efflux *in vitro*. Replacement of the carbamate with an isosteric 1,2,4-oxadiazole and optimization of the phenyl group substitution pattern then yielded **31** (K_i = 0.88 nM) with low hERG potency (> 20,000-fold over GluN2B) and an acceptable P_{gp} liability. Compound **31** was active in a spinal nerve ligation model of neuropathic pain in rats (Chung model) and was efficacious in the haloperidol-induced catalepsy model as an acute rodent model of Parkinson's disease. In addition, **31** showed no measurable effect on motor coordination when dosed orally at 100 mg/kg in the rotarod assay, suggestive of a significant therapeutic margin.

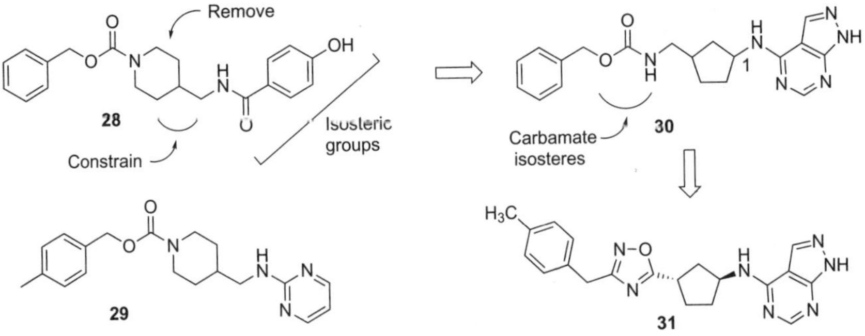

In addition to the established efficacy of GluN2B antagonists in models of depression and other CNS disorders, Smith *et al.* have reported that Ro 25-6981 and CP-101,606 reverse cognitive deficits induced by PCP and MK-801 in rats, using standard models of visual attention and working memory. Altogether, these data suggest that cognitive benefits might be achieved by selective blockade of GluN2B receptors.[34] Moreover, Ro 25-6981 preserved the beneficial effects of general NMDA antagonists on the expression of conditioned fear in comparison to fluoxetine.[35]

4.2. Atypical selective inhibitors of GluN2B

EVT-101 is a selective antagonist of GluN2B, but because its structure is more compact and it lacks a classic phenol bioisostere, it falls outside the typical motif of the phenylethanolamines and related isosteric compounds. Likewise, the 2,6-disubstituted pyrazines and corresponding matched pyridines **32–36** reported by Brown *et al.* are atypical GluN2B antagonists.[36] Although several compounds with different substituents on the phenyl group were identified, 4-F-2-OMePh was the optimum group for GluN2B potency. A cyclopentylaminomethyl group was optimal *meta*- to the phenyl substituent. To optimize GluN2B and decrease hERG potencies, pyridine (**33**, **35**, and **36**) and phenyl (**34**) moieties were substituted in place of the pyrazine (**32**). Pyridine isomer **33** was found to have the highest binding affinity for GluN2B, with an adequate safety index relative to hERG. Evaluated in the mouse forced swim test model of depression, **33** was active at 60 mg/kg. The brain/plasma ratio at 30 min post-dose was >10 with a dose-proportional exposure.

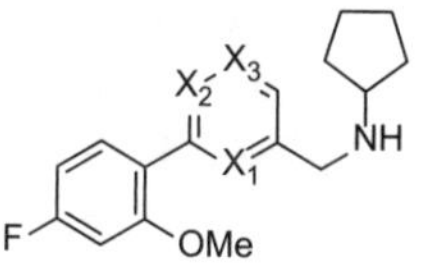

	X_1	X_2	X_3	GluN2B[a] K_i, nM	hERG[b] IC_{50}, μM
32	N	C	N	54	16.0
33	C	C	N	12	1.89
34	C	C	C	48	2.15
35	N	C	C	66	6.23
36	C	N	C	8900	3.85

[a] Inhibition of binding of [^{3}H]CP-101,606 to rat brain membranes
[b] IC_{50} values determined in CHOK1-hERG cells using EP measurements

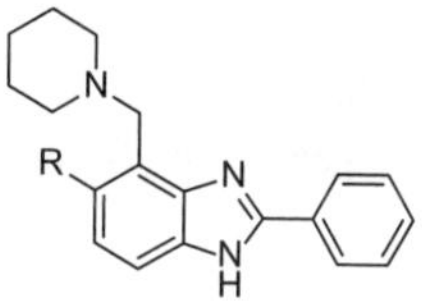

	R	IC_{50}[c], nM
37	H	398.1
38	OH	7.9
39	F	79.4

[c] [^{3}H]-Ro 25-6981 binding assay

Another compact scaffold identified by HTS as GluN2B-selective NMDA receptor antagonists had a benzimidazole core.[37] Although the potent molecule **38** had a phenolic OH group, the structure does not resemble the typical framework of phenylethanolamines. The replacement of the OH group in **38** with fluorine as in compound **39** resulted in retention of GluN2B potency (IC_{50} < 100 nM). In contrast, the H-analog **37** lost 50-fold potency.

4.3. Radioligands and imaging tool compounds

Several radioligands for the GluN2B subunit have been developed (e.g., [^{11}C]-CP-101,606, [^{11}C]-benzylamidine derivative, [^{11}C]-EMD-95885), but these ligands have only proven useful *in vitro* probably due to nonspecific binding.[38] A new study reports the development of high-affinity benzimidazole derivatives **40** and **41** as new SPECT ligands for GluN2B subunits. *In vitro* autoradiography experiments demonstrated that **40** and **41** bind selectively to GluN2B in rat brain slices.

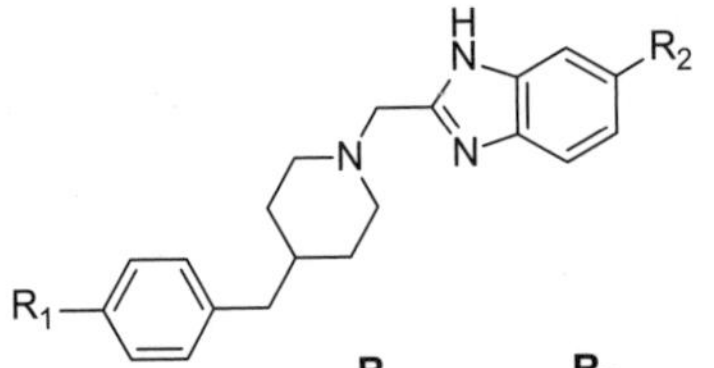

	R_1	R_2
40	^{125}I	OH
41	^{125}I	$NHSO_2Me$
42	^{18}F	OH

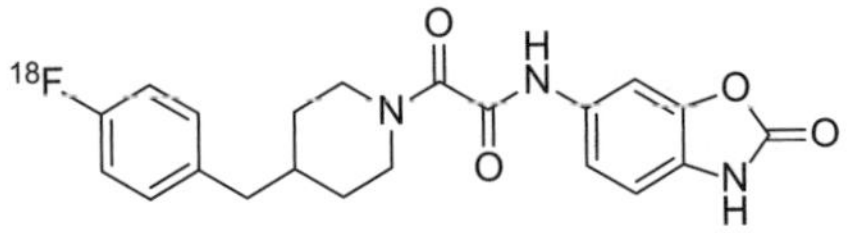

[^{18}F]-14

Recently, Labas *et al.* have reported the radiotracers [^{18}F]-**42** and [^{18}F]-RGH-896 ([^{18}F]-**14**) for PET imaging.[39] Both have demonstrated identical *in vivo* properties in rats. However, lower brain uptake and high accumulation of radioactivity in bone and cartilage were noticed.

Ifenprodil scaffold conjugates **43** and **44** were reported as fluorescent probes for confocal microscopy imaging of the GluN2B receptor.[40] Both **43** and **44** showed moderate GluN2B affinity (IC_{50} 1.4 and 1.8 μM, respectively).

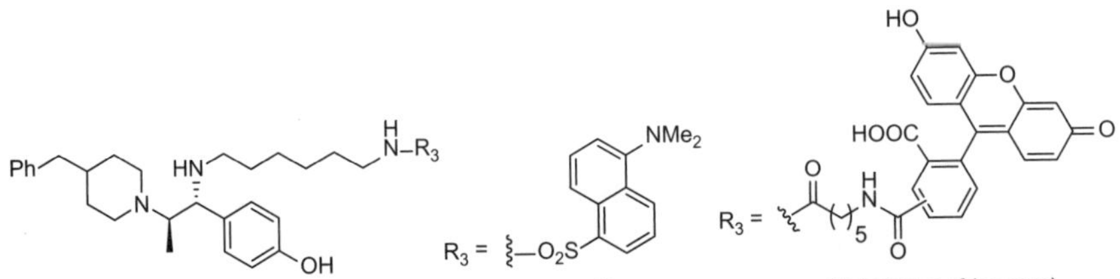

43 **44** (mixture of isomers)

5. INHIBITORS AND MODULATORS OF GLUN2A, GLUN2C, AND GLUN2D SUBTYPE RECEPTORS

In contrast to the abundance of molecules reported to modulate GluN2B subtype selective NMDA receptors, selective ligands for GluN2A, GluN2C, and GluN2D have only recently begun to emerge. The inhibition of both GluN2C and GluN2D by quinazolin-4-ones **45** and **46**,[41,42] and quinolone **47** have been reported recently.[43] A homology modeling study of chimeric NMDA receptors identified two key residues Gln701 and Leu705 in the lower lobe (membrane-proximal portion) of the GluN2 agonist binding domain that control the selectivity of **47**.[43]

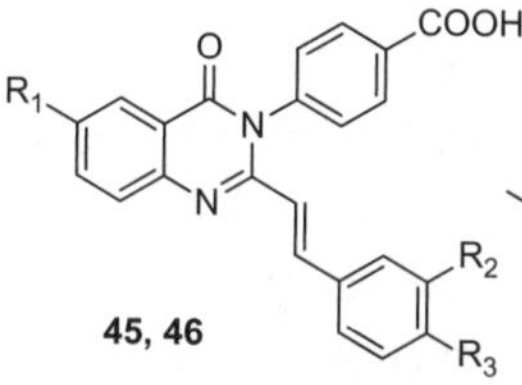

45, 46

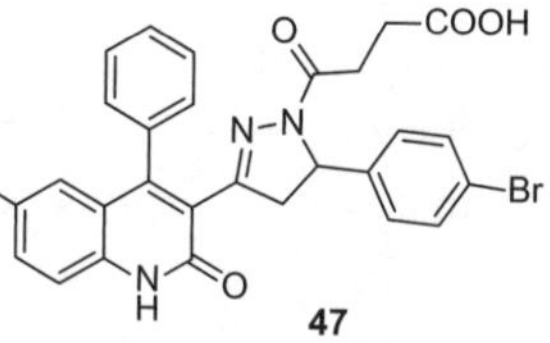

47

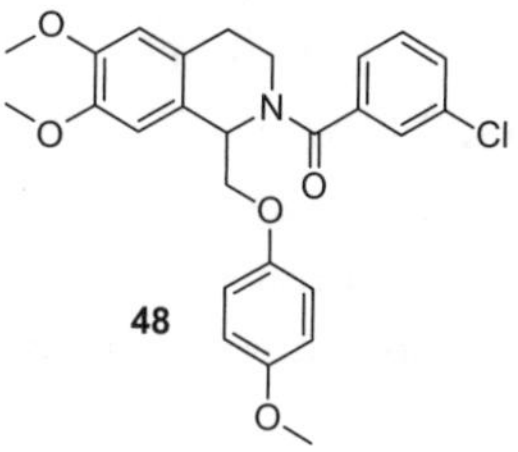

48

	R_1	R_2	R_3	GluN2C IC_{50} (μM)	GluN2D IC_{50} (μM)
45	I	H	NO_2	2.0	1.0
46	OMe	NO_2	H	7.1	3.9
47	-	-	-	7.0	2.7

Mullasseril *et al.* reported a novel subunit-selective potentiator of GluN2C and GluN2D of tetrahydroisoquinoline derivatives.[44] Compound **48** enhances by ~twofold the NMDA receptor responses of GluN2C (233% potentiation, $EC_{50}=2.7$ μM) and GluN2D (205% potentiation, $EC_{50}=2.8$ μM) activated by the partial agonists NMDA and glycine.

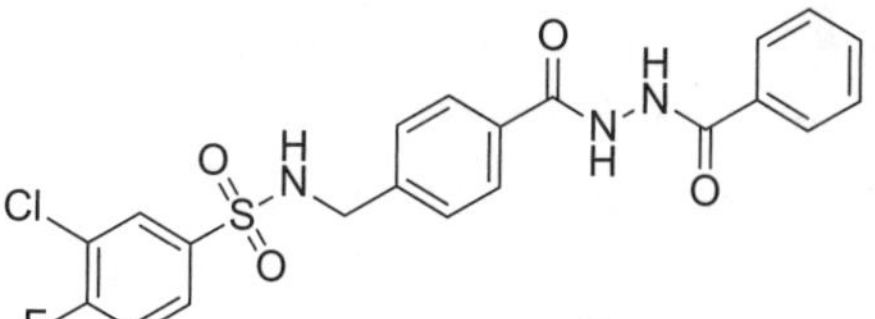

49

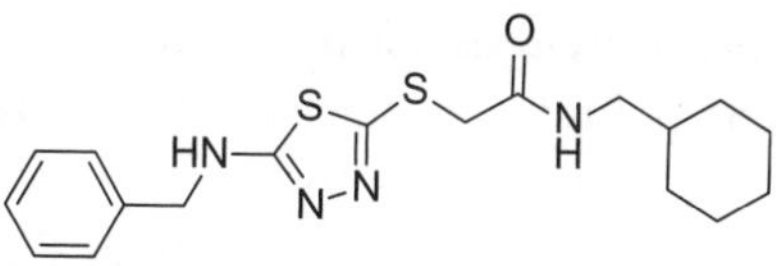

50

	49	**50**
IC_{50} GluN2A* (μM)	0.158	3.98

* FLIPR/Ca^{2+} assays. Corresponding IC_{50} for GluN2B were >50 μM

Sulfonamide **49** and thiadiazole **50** represent a new class of GluN2A antagonists identified through a HTS campaign.[45] Compound **49** was the most potent among the reported list of analogues. The replacement of the right side phenyl group with furanyl, pyridinyl, or cyclohexyl groups resulted in loss of potency. However, the displacement of glutamate site antagonist [^{3}H]CGP 39653 by **49** implies partial interaction with an agonist binding site.

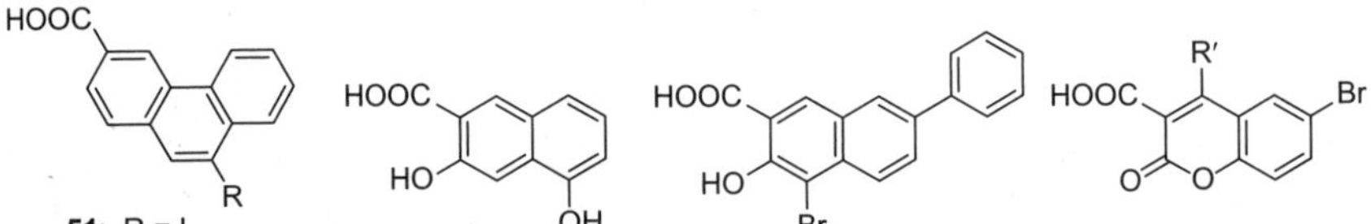

51: R = I,
52: R = *c*-Pr,
53: R = 4-Me-pentyl

54

55

56 R′ = H
57 R′ = CH_3

Costa *et al.* and Irvine *et al.*, reported the identification of both negative and positive allosteric modulators and inhibitors of subtype specific GluN2A/2C/2D NMDA receptors of polycyclic aryl carboxylic acid derivatives **51–57** and their related analogues.[46,47] The IC_{50}s were in the range of $\geq$2 μM but the observed effects in *X. laevis* oocyte two-electrode voltage clamp studies were significant with minor variations of substituent groups.

6. CLINICAL TRIALS

Among subtype selective NMDA antagonists, CP-101,606 alone has achieved proof of concept in human clinical trials of depression.[11] The non-selective antagonist ketamine has also been reported to demonstrate rapid clinical efficacy in treatment-resistant depression (TRD).[24] Together, this body of work serves to highlight the potential for developing GluN2B antagonists with an improved CNS side effect profile.

AZD6765, which is described as a low-trapping, mixed GluN2A/2B antagonist, is currently in multiple Phase IIb trials as an i.v. formulation for treatment-resistant major depressive disorder. Neither the structure of this compound nor the study results have been made public.[48]

Another GluN2B subtype selective antagonist, radiprodil (RGH-896), has been evaluated by Gedeon Richter and Forest Laboratories in a phase IIb study for neuropathic pain associated with diabetic peripheral neuropathy. In 2010, radiprodil failed to meet the primary endpoint of reducing mean daily pain scores compared with placebo, and no further development has been reported.[49] The Merck GluN2B antagonist, MK-0657, has been evaluated in Phase I trials for the treatment of both Parkinson's disease[50] and depression. It was reported that blood pressure effects were observed in some patients and that efficacy had not been achieved. Neither trial is ongoing.[51] The Evotec drug EVT-101, also a GluN2B-specific antagonist, has recently been evaluated as an oral agent in a Phase II proof-of-concept study for TRD. Following difficulties in the enrollment of the patients under the study protocol, clinical development was terminated during 2011. The company announcement also cited the need to improve "sharpen" the toxicology profile of EVT-101 and the potential need for an adjustment to the dosage scheme.[52] A follow-up compound EVT-103, also reported to be a GluN2B-specific antagonist, completed a Phase I study for TRD and was found to be safe and well tolerated.[53] The structure of EVT-103 has not been released.

7. CONCLUSIONS

Improvements to existing treatments for CNS diseases represent an unmet medical need, and NMDA receptors have become an important target for new therapeutic approaches. Complicating the development of new drugs has been the requirement that therapies target aberrant NMDA function while preserving normal CNS function across other glutamate or excitatory receptors. Progress in this area has been largely driven by the discovery of subtype selective GluN2B NMDA antagonists which act at an allosteric site of the ATD without direct block of the ion channel pore. With the determination of the structure and the molecular contacts of selective antagonists bound within that allosteric site, a template now exists for the design of molecules with improved potency and selectivity. Recent advances have not been limited to subtype selective GluN2B receptors. For the first time, subunit selectivity has been achieved with the development of novel antagonists and modulators of GluN2A/2C/2D receptors. It is generally accepted that these ligands do not bind within the ATD, LBD, or ion channel pore, but instead at a novel site on the NMDA receptor complex. Though improvements in potency, selectivity, and drug-like properties are still needed, this class presents new opportunities to improve our understanding of NMDA receptors and their pharmacology. With these new tools in hand, the potential for medicinal chemists to address the limitations of current therapies in a number of CNS diseases is greater than ever.

ACKNOWLEDGMENTS

The authors thank BMS for generous support to cover figure reproduction charges. KR thanks the National Institute of Mental Health of the National Institutes of Health (R43MH096363), and National Institute of Neurological Disorders and Stroke (1U44N507165701A1) for their support.

REFERENCES

(1) Dingledine, R.; Borges, K.; Browie, D.; Traynelis, S.F. *Pharmacol. Rev.* **1999**, *51*, 7.
(2) Monaghan, D.T.; Irvine, M.W.; Costa, B.M.; Fang, G.; Jane, D.E. *Neurochem. Int* **2012**. (http://dx.doi.org/10.1016/j.neuint.2012.01.004).
(3) Traynelis, S.F.; Wollmuth, L.P.; McBain, C.J.; Menniti, F.S.; Vance, K.M.; Ogden, K.K.; Hansen, K.B.; Yuan, H.; Myers, S.J.; Dingledine, R. *Pharmacol. Rev.* **2010**, *62*, 405.
(4) Mayer, M.L. *Curr. Opin. Neurobiol.* **2011**, *21*, 283.
(5) Paoletti, P. *Eur. J. Neurosci.* **2011**, *33*, 1351.
(6) Ogden, K.K.; Traynelis, S.F. *Trends Pharmacol. Sci.* **2011**, *32*, 726.
(7) Mony, L.; Kew, J.N.C.; Gunthorpe, M.J.; Paoletti, P. *Br. J. Pharmacol.* **2009**, *157*, 1301.
(8) Mattes, H.; Carcache, D.; Kalkman, H.O.; Koller, M. *J. Med. Chem.* **2010**, *53*, 5367.
(9) Koller, M.; Urwyler, S. *Expert Opin. Ther. Patents* **2010**, *20*, 1683.
(10) Layton, M.E.; Kelly, M.J., III; Rodzinak, K.J. *Curr. Top. Med. Chem.* **2006**, *6*, 697.
(11) Preskorn, S.H.; Baker, B.; Kolluri, S.; Menniti, F.S.; Krams, M.; Landen, J.W. *J. Clin. Psychopharmacol.* **2008**, *28*, 631.
(12) Ji, G.; Horváth, C.; Neugebauer, V. *Mol. Pain* **2009**, *5*(21), 1.
(13) Sanacora, G.; Zarate, C.A.; Krystal, J.H.; Manji, H.K. *Nat. Rev. Drug Discov.* **2008**, *7*, 426.
(14) Perin-Dureau, F.; Rachline, J.; Neyton, J.; Paoletti, P. *J. Neurosci.* **2002**, *22*, 5955.
(15) Karakas, E.; Simorowski, N.; Furukawa, H. *EMBO J.* **2009**, *28*, 3910.
(16) Mony, L.; Krzaczkowski, L.; Leonetti, M.; Le Goff, A.; Alarcon, K.; Neyton, J.; Bertrand, H.-O.; Acher, F.; Paoletti, P. *Mol. Pharmacol.* **2009**, *75*, 60.
(17) Karakas, E.; Simorowski, N.; Furukawa, H. *Nature* **2011**, *475*, 249.
(18) Kumar, J.; Schuck, P.; Jin, R.; Mayer, M.L. *Nat. Struct. Mol. Biol.* **2009**, *16*, 631.
(19) Jin, R.; Singh, S.K.; Gu, S.; Furukawa, H.; Sobolevsky, A.I.; Zhou, J.; Jin, Y.; Gouaux, E. *EMBO J.* **2009**, *28*, 1812.
(20) Mony, L.; Zhu, S.; Carvalho, S.; Paoletti, P. *EMBO J.* **2011**, *30*, 3134.
(21) Salussolia, C.L.; Prodromou, M.L.; Borker, P.; Wollmuth, L.P. *J. Neurosci.* **2011**, *31*, 11295.
(22) Hansen, K.B.; Furukawa, H.; Traynelis, S.F. *Mol. Pharmacol.* **2010**, *78*, 535.
(23) Vance, K.M.; Simorowski, M.; Traynelis, S.F.; Furukawa, H. *Nat. Commun.* **2011**, *2*, 294.
(24) Duman, R.S.; Li, N.; Liu, R.-J.; Duric, V.; Aghajanian, G. *Neuropharmacology* **2012**, *62*, 35.
(25) Zarantonello, P.; Bettini, E.; Paio, A.; Simoncelli, C.; Terreni, S.; Cardullo, F. *Bioorg. Med. Chem. Lett.* **2011**, *21*, 2059.
(26) Banerjee, A.; Schepmann, D.; Wünsch, B. *Bioorg. Med. Chem.* **2010**, *18*, 4095.
(27) Wee, X.-K.; Ng, K.-S.; Leung, H.-W.; Cheong, Y.-P.; Kong, K.-H.; Ng, F.-M.; Soh, W.; Lam, Y.; Low, C.-M. *Br. J. Pharmacol.* **2010**, *159*, 449.
(28) Gitto, R.; De Luca, L.; Ferro, S.; Buemi, M.R.; Russo, E.; De Sarro, G.; Costa, L.; Ciranna, L.; Prezzavento, O.; Arena, E.; Ronsisvalle, S.; Bruno, G.; Chimirri, A. *J. Med. Chem.* **2011**, *54*, 8702.
(29) Tewes, B.; Frehland, B.; Schepmann, D.; Schmidtke, K.-U.; Winckler, T.; Wünsch, B. *Bioorg. Med. Chem.* **2010**, *18*, 8005.
(30) Tewes, B.; Frehland, B.; Schepmann, D.; Schmidtke, K.-U.; Winckler, T.; Wünsch, B. *ChemMedChem* **2010**, *5*, 687.
(31) Müller, A.; Höfner, G.; Renukappa-Gutke, T.; Parsons, C.G.; Wanner, K.T. *Bioorg. Med. Chem. Lett.* **2011**, *21*, 5795.
(32) Mony, L.; Triballeau, N.; Paoletti, P.; Acher, F.C.; Bertrand, H.-O. *Bioorg. Med. Chem. Lett.* **2010**, *20*, 5552.
(33) Layton, M.E.; Kelly, M.J., III; Rodzinak, K.J.; Sanderson, P.E.; Young, S.D.; Bednar, R.A.; DiLella, A.G.; Mcdonald, T.P.; Wang, H.; Mosser, S.D.; Fay, J.F.; Cunningham, M.E.; Reiss, D.R.; Fandozzi, C.; Trainor, N.; Liang, A.; Lis, E.V.; Seabrook, G.R.; Urban, M.O.; Yergey, J.; Koblan, K.S. *ACS Chem. Neurosci.* **2011**, *2*, 352.

(34) Smith, J.W.; Gastambide, F.; ilmour, G.; Dix, S.; Foss, J.; Lloyd, K.; Malik, N.; Tricklebank, M. *Psychopharmacology* **2011**, *217*, 255.
(35) Haller, J.; Nagy, R.; Toth, M.; Pelczer, K.G.; Mikics, E. *Behav. Pharmacol.* **2011**, *22*, 113.
(36) Brown, D.G.; Maier, D.L.; Sylvester, M.A.; Hoerter, T.N.; Menhaji-Klotz, E.; Lasota, C.C.; Hirata, L.T.; Wilkins, D.E.; Scott, C.W.; Trivedi, S.; Chen, T.; McCarthy, D.J.; Maciag, C.M.; Sutton, E.J.; Cumberledge, J.; Mathisen, D.; Roberts, J.; Gupta, A.; Liu, F.; Elmore, C.S.; Alhambra, C.; Krumrine, J.R.; Wang, X.; Ciaccio, P.J.; Wood, M.W.; Cambell, J.B.; Johannson, M.J.; Xia, J.; Wen, X.; Jiang, J.; Wang, X.; Peng, Z.; Hu, T.; Wang, J. *Bioorg. Med. Chem. Lett.* **2011**, *21*, 3399.
(37) Davies, D.J.; Crowe, M.; Lucas, N.; Quinn, J.; Miller, D.D.; Pritchard, S.; Grose, D.; Bettini, E.; Calcinaghi, N.; Virginio, C.; Abberley, L.; Goldsmith, P.; Michel, A.D.; Chessell, L.P.; Kew, J.N.C.; Miller, N.D.; Gunthorpe, M.J. *Bioorg. Med. Chem. Lett.* **2012**, *22*, 2620.
(38) Fuchigami, T.; Yamaguchi, H.; Ogawa, M.; Biao, L.; Nakayama, M.; Haratake, M.; Magata, Y. *Bioorg. Med. Chem.* **2010**, *18*, 7497.
(39) Labas, R.; Gilbert, G.; Nicole, O.; Dhilly, M.; Abbas, A.; Tirel, O.; Buisson, A.; Henry, J.; Barré, L.; Debruyne, D.; Sobrio, F. *Eur. J. Med. Chem.* **2011**, *46*, 2295.
(40) Marchand, P.; Becerril-Ortega, J.; Mony, L.; Bouteiller, C.; Paoletti, P.; Nicole, O.; Barré, L.; Buisson, A.; Perrio, C. *Bioconjugate Chem.* **2012**, *23*, 21.
(41) Mosley, C.A.; Acker, T.M.; Hansen, K.B.; Mullasseril, P.; Andersen, K.T.; Le, P.; Vellano, K.M.; Bräuner-Osborne, H.; Liotta, D.C.; Traynelis, S.F. *J. Med. Chem.* **2010**, *53*, 5476.
(42) Hansen, K.B.; Traynelis, S.F. *J. Neurosci.* **2011**, *31*, 3650.
(43) Acker, T.M.; Yuan, H.; Hansen, K.B.; Vance, K.M.; Ogden, K.K.; Jensen, H.S.; Burger, P.B.; Mullasseril, P.; Snyder, J.P.; Liotta, D.C.; Traynelis, S.F. *Mol. Pharmacol.* **2011**, *80*, 782.
(44) Mullasseril, P.; Hansen, K.B.; Vance, K.M.; Ogden, K.K.; Yuan, H.; Kurtkaya, N.L.; Santangelo, R.; Orr, A.G.; Le, P.; Vellano, K.M.; Liotta, D.C.; Traynelis, S.F. *Nat. Commun.* **2010**, *1*, 1.
(45) Bettini, E.; Sava, A.; Griffante, C.; Carignani, C.; Buson, A.; Capelli, A.M.; Negri, M.; Andreetta, F.; Senar-Sancho, S.A.; Guiral, L.; Cardullo, F. *J. Pharmacol. Exp. Ther.* **2010**, *335*, 636.
(46) Costa, B.M.; Irvine, M.W.; Fang, G.; Eaves, R.J.; Mayo-Martin, M.B.; Laube, B.; Jane, D.E.; Monaghan, D.T. *Neuropharmacology* **2012**, *62*, 1730.
(47) Irvine, M.W.; Costa, B.M.; Volianskis, A.; Fang, G.; Ceolin, L.; Collingridge, G.L.; Monaghan, D.T.; Jane, D.E. *Neurochem. Int.* **2012**. http://dx.doi.org/10.1016/j.neuint.2011.12.020.
(48) http://clinicaltrials.gov/ct/show/NCT00491686. *Note added in proof*—A report of an early clinical study has recently appeared in which AZD6765 showed evidence of antidepressant effects: William Deakin JF, et al. CINP World Congress of Neuropsychopharmacology, 3–7 June 2012, Stockholm, Sweden, and related posters.
(49) http://www.frx.com/, Forest Laboratories press release, June 28, **2010**.
(50) Addy, C.; Assaid, C.; Hreniuk, D.; Stroh, M.; Xu, Y.; Herring, W.J.; Ellenbogen, A.; Jinnah, H.A.; Kirby, L.; Leibowitz, M.T.; Stewart, R.M.; Tarsy, D.; Tetrud, J.; Stoch, S.A.; Gottesdiener, K.; Wagner, J. *J. Clin. Pharmacol.* **2009**, *49*, 856.
(51) Gater, D. *Thomson Reuters Pharma, overnight report, 32nd national medicinal chemistry symposiumMinnesota, MN, USA*. Minnesota, MN, USA, June 6, 2010.
(52) http://www.evotec.com, Evotec press release, May 18, **2011**.
(53) http://www.evotec.com, Evotec press release, March 10, **2010**.

CHAPTER EIGHT

Recent Advances in the Development of PET and SPECT Tracers for Brain Imaging

Lei Zhang, Anabella Villalobos
Pfizer Worldwide Research and Development, Neuroscience Medicinal Chemistry, Cambridge, Massachusetts, USA

Contents

1. INTRODUCTION

Positron emission tomography (PET) and single photon emission computed tomography (SPECT) are radiotracer-based noninvasive imaging techniques that provide quantitative binding information on specific target areas of interest.[1,2] PET and SPECT have been proven to be particularly valuable for imaging targets in the central nervous system (CNS) due to inaccessibility of the human brain, providing impact at various stages of the drug discovery process, from informing basic pharmacology to enabling critical decision-making in clinical evaluations of novel pharmaceuticals. To enable PET and SPECT brain imaging, suitable

Annual Reports in Medicinal Chemistry, Volume 47
ISSN 0065-7743
http://dx.doi.org/10.1016/B978-0-12-396492-2.00008-4

radiotracers need to be developed. Such processes, however, can often be arduous and lengthy due to the challenge of meeting a demanding set of prerequisite attributes for a tracer.[3] In this review, we will compare these two imaging modalities, highlight recent advances in CNS PET tracer development, and review recent data in the field of SPECT in brain imaging.

2. PET AND SPECT

While PET and SPECT bear similarities in their functions in brain imaging, they have their own strengths and limitations. In terms of radionuclides (Table 8.1), PET requires positron-emitting radionuclides such as [^{11}C] and [^{18}F], while SPECT requires gamma-emitting radionuclides such as [^{123}I] and [^{99m}Tc]. Routinely used PET radionuclides such as [^{11}C] and [^{18}F] are isotopes of elements commonly found in drug-like molecules that can be incorporated into these molecules with minimal impact on their physicochemical and pharmacological properties. Thereby, these radionuclides offer greater flexibility in tracer design in comparison to SPECT radionuclides which may result in modification of physicochemical and pharmacological properties upon incorporation into molecules. On the other hand, the shorter half-lives of PET radionuclides, [^{11}C] ($T_{1/2}$=20 min) and [^{18}F] ($T_{1/2}$=110 min), constitute a disadvantage in that rapid synthesis/purification as well as proximity to a cyclotron facility is required for their successful preparation. In comparison, SPECT radionuclides, such as [^{123}I] ($T_{1/2}$=13 h) and [^{99m}Tc] ($T_{1/2}$=6 h), have relatively longer half-lives and can be bought commercially or synthesized onsite with

Table 8.1 PET and SPECT radionuclides

PET radionuclides	$T_{1/2}$	SPECT radionuclides	$T_{1/2}$ (h)
^{11}C	20 min	^{99m}Tc	6
^{18}F	110 min	^{123}I	13
^{13}N	10 min	^{201}Tl	73
^{15}O	2 min	^{67}Ga	78
^{64}Cu	12.8 h	^{111}In	68
^{68}Ga	68 min	^{133}Xe	127
^{82}Rb	1.3 min	^{131}I	192

low-cost generators, thus offering more flexibility in tracer synthesis, lower cost, and better access to the technique.[4] In terms of brain imaging, currently PET offers superior spatial resolution (as low as 1.2 mm) and sensitivity to allow quantification of tracer concentration in brain regions of interest with higher degree of accuracy. In contrast, SPECT has lower spatial resolution (~10–14 mm) and sensitivity (2–3 orders of magnitude lower than PET). This drawback in sensitivity mainly stems from the limited geometry efficiencies of SPECT collimators which can only detect photons within a small angular range, thus recording a smaller percentage of emission events. In addition, both spatial resolution and sensitivity of SPECT are position dependent, with increased attenuation and scattering of the signal as it travels through the body and dense tissues. Therefore, SPECT, in general, requires a larger number of attenuation corrections and longer scan times, often yielding noisier images that are more challenging to quantify.[5]

3. CNS PET TRACERS: GENERAL REQUIREMENTS AND KEY HURDLES

There are many criteria that must be met by successful PET ligands. From a chemical structure point of view, a PET tracer should have functional groups that allow for rapid radiolabeling to accommodate the short half-lives of PET radionuclides. In terms of pharmacology, a PET tracer must be highly potent and selective toward its intended pharmacological target, which often requires subnanomolar potency and greater than 100-fold selectivity over other targets. In terms of metabolism, tracers that form permeable radioactive metabolites should be avoided to prevent confounding quantification results as the PET detector measures total radioactivity without distinction of origin. In addition to the aforementioned criteria, a suitable PET tracer also needs to be brain permeable and must have low nonspecific binding (NSB) to achieve requisite signal-to-noise ratio. In fact, poor brain permeability and high NSB are the most frequent causes for failure in CNS PET ligand development.[6] Historically, lipophilicity parameters such as $\log P$ and $\log D$ have been used as criteria to predict brain permeability and NSB. A general notion within the PET field is that lower lipophilicity typically leads to lower NSB, yet lipophilicity needs to reside in a certain range (e.g., $\log P$ 1.5–2.5 or $\log D$ 1–3) in order to retain brain permeability.[7,8] This over simplified approach, however, has not been sufficient to predict PET tracers performance *in vivo*. Furthermore, the rather narrowly defined ranges of such parameters significantly restrict the scope of potential substrates for PET tracer development. Therefore, deeper understanding of molecular properties that are required for optimum brain permeability and low NSB as well as new research tools and assays that predict these two important parameters *in vivo* will likely provide significant impact in improving the overall efficiency of the CNS PET development process.

4. RECENT ADVANCES IN CNS PET TRACER DEVELOPMENT

4.1. Approaches to predict *in vivo* NSB

A biomathematical modeling approach to predict *in vivo* performance of PET radiotracers has been reported.[9] This method estimates the coefficients of variation of binding potential (%COV[BP]$_{ND}$), through Monte Carlo simulations based on *in silico* and *in vitro* physicochemical and pharmacological properties of candidate PET molecules including lipophilicity, molecular volume, free fractions in plasma and tissue, target density, affinity, perfusion, capillary surface area, and apparent aqueous volume in plasma and tissue. The performance of this approach was evaluated against a dataset containing 28 candidate PET tracers with preclinical *in vivo* endpoints available for comparison. In general, ranking of the tracers was consistent with their performance *in vivo*, with lower %COV[BP]$_{ND}$ values generated for clinically proven tracers such as [^{11}C]flumazenil and [^{11}C]raclopride and higher values observed for poor performing tracers such as [^{11}C]PK11195 and [^{11}C]GR205171.

Several methods evaluating the affinity of ligands for lipid membranes have also been reported as approaches to predict NSB. A high-throughput electrokinetic chromatography (EKC) method was developed based on a concept that the NSB of a given radiotracer primarily arises from its interaction with lipid cell membranes.[10] Three liposome and one surfactant vesicle systems were explored as *pseudo*-stationary phases to estimate membrane affinity of candidate molecules. Results indicated that the EKC analysis associated with an AOT (docusate sodium salt) vesicle system was shown to be the best approach, offering short run times, low cost, high reproducibility, and, importantly, a statistically significant correlation of retention times with *in vitro* or *in vivo* measured NSB parameters. Based on a similar tracer-lipid membrane interaction concept, an *ab initio* method showed a good correlation of the calculated drug–lipid interaction energy E_{int} with the *in vivo* measured NSB values of 10 clinically validated PET tracers.[11] A recent publication from the same group further validated this hypothesis by expanding the validation dataset to a set of 22 new candidate CNS PET tracers.[12] Full *ab initio* quantum mechanical calculations appear to be necessary for a successful correlation as further attempts to speed up the calculation using several semiempirical methods failed to reproduce the correlation. This might limit the use of this method as a virtual screen tool due to slow computational processes. Recently, a similar correlation between *in vivo* specific binding (BP_{ND}) and compound–membrane interactions (K_m) measured by immobilized artificial membrane chromatography was reported using a validation set consisting of 10 known tracers. Higher K_m values were shown to correlate with lower BP_{ND} values, thus higher *in vivo* NSB, and $K_m \leq 250$ was proposed as a preferred range to prioritize compounds for tracer development.[13] Furthermore, correlation of *in vivo* brain permeability with additional *in vitro* HPLC-measured properties, such as lipophilicity ($\log P$), permeability (P_m), and plasma protein binding (PPB), was also explored using the same validation set. The results demonstrated that $\log P$ was not a reliable predictor for brain permeability, whereas a stronger correlation was observed with high P_m and a specific range of PPB (45–85%).

4.2. Design and selection parameters for PET tracer development

In another approach to increase the ability to predict brain permeability and NSB and enable the selection of successful PET tracers in an accelerated way, a systematic analysis of a dataset containing 62 clinically validated PET tracers and 15 failed tracers due to high *in vivo* NSB was carried out. The intent of this analysis was to define a desired property space for CNS PET tracers based on key differences between successful and failed tracers with respect to physicochemical properties and *in vitro* ADME endpoints.[14] For analysis of the physicochemical properties, a simple multiparameter optimization design tool (CNS MPO Desirability Score) was used to assess drug-like and brain penetration properties of tracers in the dataset.[15] The CNS MPO helps assess the alignment of six fundamental physicochemical properties that lead to drug-like, brain penetrant molecules and offers greater flexibility in CNS compound design beyond the use of single parameters or hard cutoffs. The *in vitro* ADME endpoints examined included human and rat liver microsomal clearance, RRCK Papp permeability, MDR1 (P-gp) efflux, and fraction unbound in brain (Fu_b) and plasma (Fu_p). Based on this analysis, in addition to the existing knowledge of *in vitro* pharmacology and selectivity criteria ($B_{max}/K_d > 10$ and selectivity $> 30\times - 100\times$), a set of preferred design and selection parameters for a successful PET tracer were proposed. In terms of physicochemical properties, CNS MPO desirability scores >5 (CNS MPO2 >3)[16] and/or $\log D \leq 3$ should be targeted in the design and selection of PET ligands. In terms of PK parameters, RRCK Papp AB $>5 \times 10^{-6}$ cm/s for moderate to good passive permeability, MDR1 BA/AB <2.5 for low P-gp liability, and cFu_b (calculated fraction unbound in brain) >0.05 and cFu_p (calculated fraction unbound in brain) >0.15 for low risk of *in vivo* NSB should be targeted. This set of defined parameters was further evaluated against the profiles of 10 top performing CNS PET tracers selected based on their frequency of use, acceptance in the field, robust test–retest reliability, and good properties for quantification. This set of top performing tracers showed good alignment with the criteria defined above, suggesting that the above design and selection parameters could have the potential to steer novel CNS PET ligand development efforts toward higher performing ligands rather than marginal performers.

5. NEW CNS PET TRACERS

In the past year, a number of novel PET tracers for several hotly pursued targets in the pharmaceutical industry have been disclosed, driven by the increased use of PET imaging in clinical development to define receptor or target occupancy, exposures associated with efficacy, and dose selection. A better understanding of design principles; availability of chemical matter with suitable attributes such as better potency, selectivity, and brain permeability; and more efficient radiolabeling methodologies have also led to an increase in the development of novel PET ligands. New PET tracers developed for three targets, PDE10a, ORL-1, and FAAH, were selected to illustrate the important advances in the PET field.

5.1. PDE10a PET tracers

Phosphodiesterase 10a (PDE10a) is a unique dual specificity enzyme that hydrolyzes both cAMP and cGMP and plays a key role in regulating cyclic nucleotide signaling cascade. A suitable PET tracer would allow molecular imaging of the PDE10a enzyme in living subjects and facilitate the assessment of PDE10a as a drug target for various CNS disorders. The first reported PDE10a PET effort was the radiosynthesis and *in vivo* microPET evaluation of [^{11}C]-labeled papaverine (**1**), an early PDE10a inhibitor with moderate potency and selectivity.[17] While initial higher accumulation was

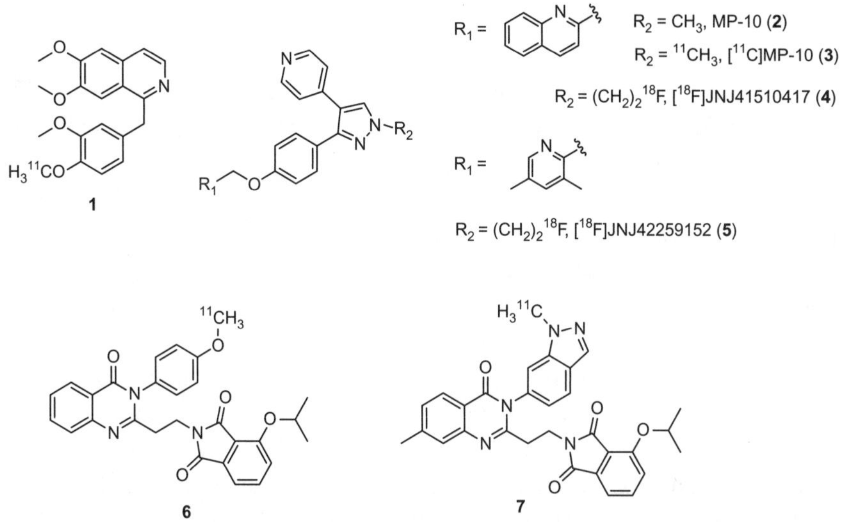

Figure 8.1 Structures of PDE10a PET tracers.

observed in striatum, a PDE10a-enriched region, the rapid washout in both rats and rhesus monkeys indicated that **1** was not a suitable PET tracer. Following the publication of a highly potent and selective PDE10a inhibitor MP-10 (**2**), several reports of novel PET tracers based on this chemotype emerged recently (Fig. 8.1). Recently, two research groups reported the development of [^{11}C]MP-10 (**3**) radiolabeled at the *N*-methyl of the central pyrazole ring.[18,19] Rapid brain uptake and preferential binding to striatum were observed with maximum striatum-to-cerebellum ratio of 6.55 and 1.5–2 in rats and rhesus monkeys, respectively. Similar brain uptake and preferential striatal binding were observed in porcine and baboons as well. Interestingly, while **3** achieved equilibrium in baboons roughly 40–60 min post dose, in rhesus monkeys, **3** showed continuous accumulation in striatum and cerebellum and did not reach equilibrium even at 120 min post dose. Metabolite analysis revealed the extensive formation of a brain permeable lipophilic radioactive metabolite in monkey, which may limit the clinical utility of **3**. Meanwhile, two [^{18}F]-labeled PET tracers derived from analogs of **2** were reported. A *N*-[^{18}F] fluoroethyl pyrazole derivative [^{18}F]JNJ4150417 (**4**) was found to have high PDE10a affinity (pIC$_{50}$ = 9.3) and showed specific and reversible binding to PDE10a in rats (peak striatum-to-cerebellum ratio of 4.6). However, the high lipophilicity ($C\log P$ = 4.2) and exceptionally high protein binding in plasma (99.5%) of **4**, coupled with its high affinity, led to undesirable slow kinetics.[20] Subsequent efforts aimed at the optimization of lipophilicity and affinity by replacing the 2-quinoline with more polar monocyclic pyridines yielded a superior radiotracer [^{18}F] JNJ42259152 (**5**) with lower lipophilicity (*C*log*P*=3.66) and sufficient affinity (*p*IC$_{50}$=8.8).[21] In rats and monkeys, **5** showed not only higher striatum-to-cerebellum ratios but also faster kinetics. The favorable overall profile positioned **5** to enter a first-in-human study for clinical PET imaging of PDE10a. However, it is worth noting that similar to **3**, brain permeable radioactive metabolites were also detected with **5**. Therefore, there is still a need for additional PDE10a PET tracers that are free of such metabolite issues. In addition to the MP-10-based PET tracers described above, two [^{11}C]-labeled PDE10a PET tracers **6** and **7**, from an alternative 4-oxo-3,4-dihydroquinazoline chemotype, were recently disclosed in the patent literature.[22] Cold reference compounds of **6** and **7** are highly potent PDE10a inhibitors with K_i values of 0.15 and 0.024 nM, respectively. The *in vivo* imaging results of these tracers were not discussed in the patent and have yet to be disclosed. It would be interesting to see whether tracers based on this alternative chemotype would offer differentiation in their metabolic profile, thereby minimizing the brain permeable radioactive metabolite issues associated with MP-10-derived PET tracers.

5.2. ORL1 PET tracers

Opioid receptor-like 1 (ORL1) receptors, also known as nociceptin/orphanin FQ peptide receptors, are widely expressed in CNS with potential links to many neuropsychiatric disorders. Development of a suitable PET tracer for the ORL1 receptor has been challenging due to its low expression level (B_{max}) and the high lipophilicity of literature leads. The first successful ORL1 PET tracer, [^{11}C]-(*S*)-3-(2′-fluoro-4′,5-dihydrospiro[piperidine-4,7′-thieno-[2,3-*c*]pyran-1-yl]-2(2-fluorobenzyl)-*N*-methylpropanamide (**8**), was reported in 2011 (Fig. 8.2).[23] Through structure–activity relationship (SAR) investigations around the amide alkyl chains and benzyl substitutions and subsequent *ex vivo* LC–MS/MS measurement in rats, **8** was identified as the best lead with potent ORL1 affinity (K_i=0.15 nM) and preferential uptake in ORL1-rich brain regions. Baseline imaging in monkeys revealed good brain uptake and a brain biodistribution consistent with that expected of the ORL1 receptor. Importantly, the radioactivity in ORL1-rich brain regions could be blocked by predosing with a high dose of naloxone, an opiate receptor antagonist, confirming the specificity of **8** to

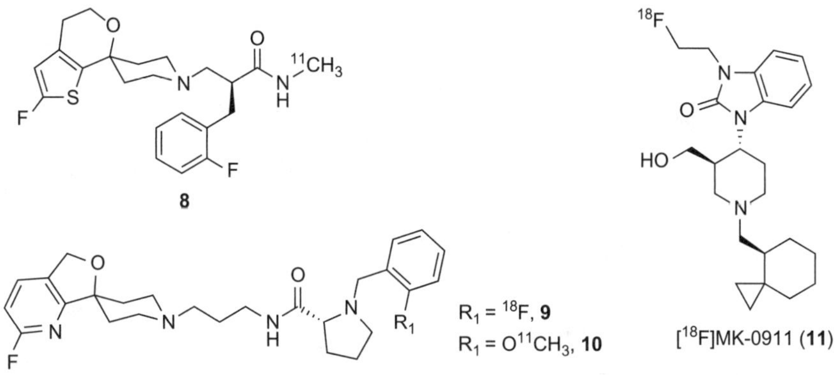

Figure 8.2 Structures of ORL PET tracers.

ORL1 receptor *in vivo* with an estimated specific to nonspecific ratio of 1.28 [30–90 min brain average area under the curve]. More recently, the characterization of several radiolabeled ORL1 antagonists (Fig. 8.2) were reported.[24] Analogs in a structurally similar spiropiperidine series, **9** ($IC_{50}=0.54$ nM) and **10** ($IC_{50}=0.71$ nM), though potent *in vitro*, showed poor brain permeability potentially due to high P-gp efflux (ratio > 3). In comparison, a structurally distinct benzo[*d*]imidazolone [^{18}F]MK-0911(**11**) showed low P-gp efflux (P-gp ratio = 1.4) while retaining favorable potency ($IC_{50}=0.56$ nM), which nicely translated into high brain uptake and preferential binding in ORL1 enriched brain regions in monkey microPET imaging studies. The brain binding of **11** was blocked by a selective ORL1 antagonist MK-0584 in a dose-responsive manner, with a specific to nonspecific ratio of ~2, achieved at the high dose of MK-0584. The favorable profiles of **8** and **11** suggest that they may be suitable PET tracers for imaging ORL1 receptor in humans.

5.3. FAAH PET tracers

The fatty acid amide hydrolase (FAAH) enzyme is a serine hydrolase responsible for degrading the fatty acid amide family of signaling lipids, including the endocannabinoid anandamide. The involvement of FAAH in pain and nervous system disorders has made it an attractive target for molecular imaging. Three [^{11}C]-labeled FAAH inhibitors have been recently reported as potential PET tracers for FAAH brain imaging (Fig. 8.3). Two were based on the irreversible covalent carbamate inhibitor URB597 (**12**) and one was based on a reversible FAAH inhibitor (MK-3168, **15**). **12** has been shown to inhibit FAAH irreversibly via carbamylation of FAAH's catalytic Ser^{241}, which acts as a nucleophile. A moderately active URB597 analog **13**

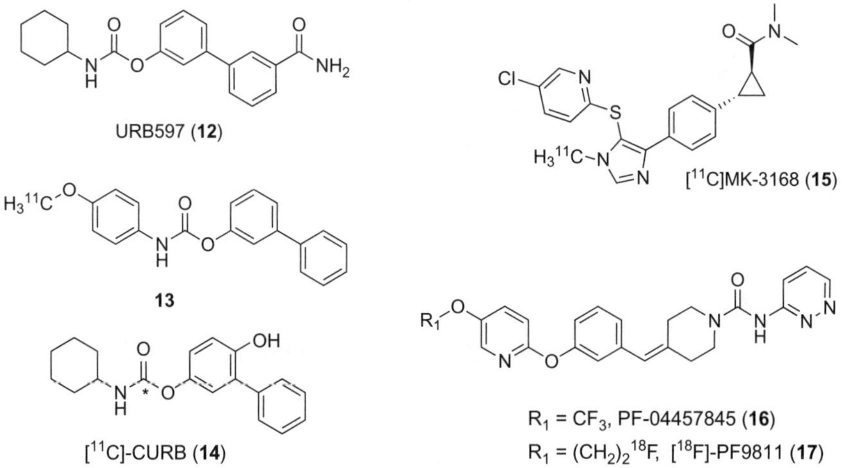

Figure 8.3 Structure of FAAH PET tracers.

($IC_{50}=436$ nM) was targeted as a PET lead, and a [^{11}C]-methoxy group was introduced into the aniline moiety which was expected to remain attached to the enzyme upon carbamylation. Subsequent biodistribution studies showed no retention of radioactivity in brain, substantial peripheral metabolism, and minimal differences in the biodistribution patterns of wild-type and FAAH knock-out mice.[25] As a follow-up to this effort, the same group reported PET imaging results of an improved URB597 analog, [^{11}C]CURB (**14**), which was labeled at the carbonyl position using a novel [^{11}C]CO_2 fixation methodology. In contrast to the previous tracer, **14** showed heterogeneous brain binding with little washout over time which was consistent with irreversible binding. The specific binding of **14** to FAAH was demonstrated by blocking experiments with a high dose of URB597, with the highest specific binding ratio observed in cortex and the lowest in hypothalamus.[26] More recently, another novel FAAH PET tracer, [^{11}C]MK-3168 (**15**), was disclosed, which unlike previously described tracers binds to FAAH in a reversible noncovalent manner with a K_d of 0.8 nM (human cortex tissue binding). PET imaging in rhesus monkeys demonstrated heterogeneous, specific brain uptake, consistent with known regional FAAH distribution.[27] Clinical imaging studies have been carried out with **15** to provide receptor occupancy information supporting the development of a clinical candidate. Good brain uptake and test/retest variability have been reported.

Recently, our group disclosed SAR efforts in a novel urea series leading to the identification of PF-04457845 (**16**) as a clinical candidate.[28] The excellent potency, selectivity, and pharmacokinetic properties of **16** make it an attractive scaffold for PET tracer development. Toward this end, we developed [^{18}F]PF-9811 (**17**), based on a close-in analog of **16**, wherein the trifluoromethyl moiety was replaced with a fluoroethoxy group, without impact in *in vitro* FAAH potency (IC_{50} = 16 nM) and *in vivo* FAAH inhibition activity (complete inhibition of FAAH *in vivo* at 10 mg/kg p.o. in C57B1/6 mice). Biodistribution experiments of **17** in rats showed good uptake in all regions of the brain, with preferential binding in the cortex, hippocampus, and cerebellum, and a statistically significant radiosignal increase from the 10 to 90-min time points, consistent with the characteristics of an irreversible inhibitor. Specificity of **17** for FAAH was demonstrated by pretreatment with **16**, which reduced uptake across all brain regions (37–73% at 90 min). In addition, **17** was evaluated in rat microPET studies and the results largely mirrored those of the biodistribution study, with high brain uptake and specific FAAH binding. The favorable outcome suggests that **17** represents a promising PET tracer for FAAH imaging with the potential advantage of a [^{18}F] radionuclide for higher resolution and flexibility in scan times.[29]

6. SPECT IN BRAIN IMAGING

SPECT has been successfully used in brain imaging, and the structures of representative clinically validated CNS SPECT tracers are shown in Fig. 8.4. The prerequisite attributes for a suitable CNS SPECT radiotracer

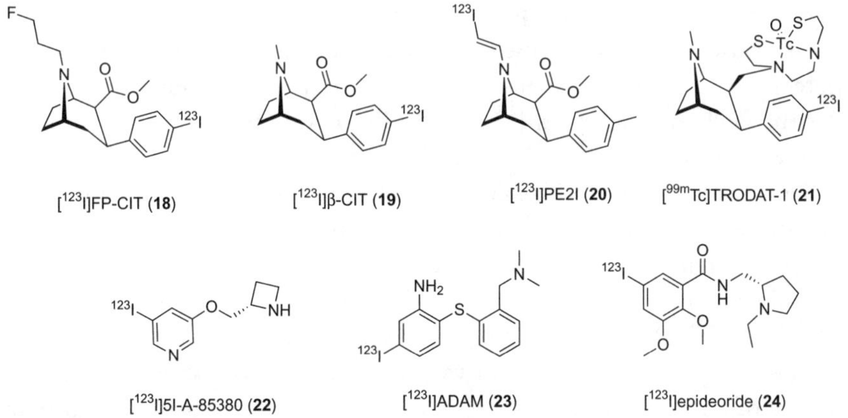

Figure 8.4 Structures of representative CNS SPECT tracers.

are similar to those of CNS PET tracers. However, there are added challenges for a SPECT tracer to meet the same set of criteria due to the nature and intrinsic properties of its radionuclides. SPECT radionuclides, such as [^{123}I] and [^{99m}Tc], are uncommon to CNS drug-like molecules, and incorporation of such radionuclides may impact physicochemical and pharmacological properties. For example, the incorporation of an [^{123}I] atom could add significant lipophilicity to the parent molecule, therefore increasing the risk of NSB. In the same way, introduction of [^{99m}Tc] would require tethering a metal complex to a small molecule, which may have significant impact on affinity for the target. Therefore, it is not surprising that most currently available CNS SPECT tracers are centered around certain type of targets, those with polar ligands which can tolerate an increase in lipophilicity (e.g., serotonin transporter, nicotinic acetylcholine receptors, and dopamine receptors) or those with high expression levels which can tolerate some loss in potency upon introduction of the SPECT radionuclide. As an example, [^{99m}Tc]-labeled CNS SPECT tracers have been successfully developed for the dopamine transporter due to the exceptionally high brain expression level of this transporter. One such example is [^{99m}Tc]TRODAT-1 (**21**), which is shown in Fig. 8.4.

Despite the limited number of clinically proven SPECT tracers available to date,[30] SPECT brain imaging studies on various neurological disorders have yielded highly impactful information and answers to important clinical neuropharmacological questions. Dopamine transporter SPECT tracers, such as [^{123}I]FP-CIT (**18**) and [^{123}I]β-CIT (**19**), are sensitive markers of dopamine neurodegeneration and have been successfully used in clinical studies to assess Parkinson's disease (PD) based on the correlation between reduction in nigrostriatal binding and disease severity.[31,2] In a recent report, SPECT imaging using **18** was also used to generate a striatal asymmetry index as a potential predictor of responsiveness to L-DOPA in patients with PD.[31] A structurally similar tracer, [^{123}I]PE2I (**20**), has shown promise as a highly sensitive and specific diagnostic tool to detect striatal neurodegeneration in patients with minor Parkinsonian symptoms.[32] SPECT tracers for other neuroreceptors have also shown promise in detecting subtle changes in receptor binding or expression levels as biomarkers for tracking various neurological disorders. For example, the nicotinic α4β2 SPECT tracer, [^{123}I]5I-A-85380 (**22**), was recently used to detect alterations in α4β2 receptor binding in patients with vascular dementia who showed decreased uptake of tracer in subcortical regions such as the dorsal thalamus and right caudate compared to age controlled healthy volunteers.[33] In similar studies,

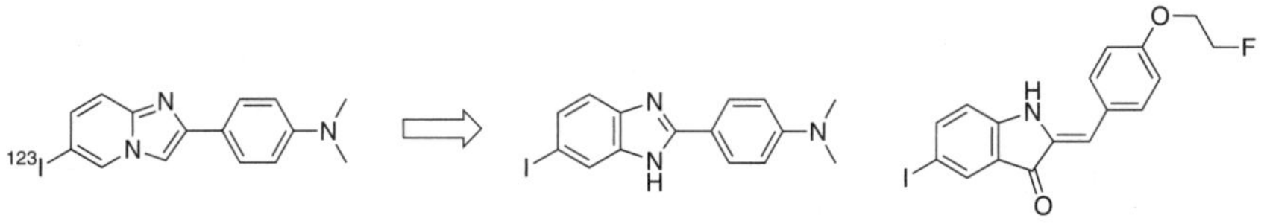

Figure 8.5 Recent disclosed SPECT tracer leads for β-amyloid plaques detection.

a serotonin transporter tracer, [^{123}I]ADAM (**23**), was used to assess the potential involvement of the mesopontine serotonergic system in the pathophysiology of migraine, revealing an increase in serotonin transporter binding in the brainstem of patients who suffer from migraines.[34] Recent publications on the development of SPECT tracers for *in vivo* detection of β-amyloid plaques are also encouraging. Two iodine-containing compounds, benzoimidazole **26** and aurone derivative FIAR **27**, were identified as improved SPECT tracer leads over [^{123}I]IMPY (**25**), the most advanced SPECT tracer for β-amyloid plaque imaging which suffers from a low signal-to-noise ratio in humans. Both compounds were reported to have high affinity for Aβ aggregates and demonstrated good signal-to-noise ratios and washout rate in mice biodistribution studies (Fig. 8.5).[35,36]

Technical advancements in the field have also dramatically improved the spatial resolution associated with SPECT measurements. For example, recently developed NanoSPECT/computed tomography (CT) can achieve resolution as low as 0.4–0.6 mm, thus allowing efficient imaging in preclinical rodent models which had been unattainable in the past.[2] This higher resolution SPECT technique could in principle be used in combination with microPET to determine the precise structural and functional anatomy of CNS disorders. The distinct advantages of SPECT in terms of cost and availability should not be overlooked. If suitable chemotypes are available and acceptable affinity can be achieved by incorporating SPECT radionuclides, development of a SPECT radiotracer should be considered as it could serve as a more economical alternative tool to PET for translational or diagnostic purposes.

7. CONCLUSIONS

In light of the considerable value of PET and SPECT brain imaging, it is important to continue to focus on the development of suitable CNS PET and SPECT tracers. A high performing PET or SPECT tracer will serve as a powerful tool to enable clinical evaluation of candidate compounds in terms of receptor occupancy, efficacious exposures, and doses, ensuring appropriate testing of mechanisms. To enable effective decision-making in clinical studies with tracers, it will be optimal to start the tracer development early in the discovery process. Recent advances in *in vitro* screening and *in silico* tools for NSB prediction as well as tractable selection parameters for medicinal chemistry design should facilitate and accelerate the discovery process of novel CNS PET and SPECT tracers with higher success rates and fewer resources. The exciting group of new PET tracers in PDE10, ORL1, and FAAH will undoubtedly bring additional proof of the value of PET imaging in clinical validation of these hotly pursued targets. In addition, new technical advancements, particularly in SPECT, have led to better resolution/sensitivity and an expanded role in small animal preclinical imaging. Both imaging modalities, together with other imaging technologies such as CT and magnetic resonance imaging, provide an exciting translational research platform in which each of these technologies brings their unique strengths to enable high-quality preclinical and clinical studies. PET and SPECT will remain impactful technologies in the discovery of future pharmaceuticals for the treatment of neurological disorders of high unmet medical need.

REFERENCES

(1) Ametamey, S.M.; Honer, M.; Schubiger, P.A. *Chem. Rev.* **2008**, *108*, 1501.
(2) Sharma, S.; Ebadi, M. *Neurochem. Int.* **2008**, *52*, 352.
(3) Magnus, S.; Pike, V.W.; Halldin, C. *Curr. Top. Med. Chem.* **2007**, 7, 1806.
(4) Halldin, C.; Gulyas, B.; Langer, O.; Farde, L. *Q. J. Nucl. Med.* **2001**, *45*, 139.
(5) Rahmim, A. *Iran. J. Nucl. Med.* **2006**, *14*, 1.
(6) McCarthy, D.J.; Halldin, C.; Andersson, J.D.; Pierson, M.E. *Annu. Rep. Med. Chem.* **2009**, *44*, 501.
(7) Cunningham, V.J.; Park, C.A.; Rabiner, E.A.; Gee, A.D.; Gunn, R.N. *Drug Discov. Today* **2005**, *2*, 311.
(8) Van de Waterbeemd, H.; Camenisch, G.; Folkers, G.; Chretien, J.; Raevsky, O. *J. Drug Target* **1998**, *6*, 151.
(9) Guo, Q.; Brady, M.; Gunn, R.N. *J. Nucl. Chem.* **2009**, *50*, 1715.
(10) Jiang, Z.J.; Reilly, J.; Everatt, B.; Briard, E. *J. Pharm. Biomed. Anal.* **2011**, *54*, 722.
(11) Rosso, L.; Gee, A.D.; Gould, I.R. *J. Comput. Chem.* **2008**, *29*, 2397.
(12) Dickson, C.J.; Gee, A.D.; Bennacef, I.; Gould, I.R.; Rosso, L. *Phys. Chem. Chem. Phys.* **2011**, *13*, 21552.
(13) Tavares, A.A.S.; Lewsey, J.; Dewar, D.; Pimlott, A.L. *Nucl. Biol. Med.* **2012**, *39*, 127.
(14) Zhang, L.; Villalobos, A.; Anderson, D.; Beck, E.; Blumberg, L.; Bocan, T.; Bronk, B.; Chen, L.; Brown-Proctor, C.; Grimwood, A.; Heck, S.; Skaddan, M.; McCarthy, T.; Zasadny, K. *J. Labelled Comp. Radiopharm.* **2011**, *54*, S292.
(15) Wager, T.; Hou, X.; Verhoest, P.; Villalobos, A. *ACS Chem. Neurosci.* **2010**, *1*, 435.
(16) Wager, T.; Chandrasekaran, R.Y.; Hou, X.; Troutman, M.D.; Verhoest, P.R.; Villalobos, A.; Will, Y. *ACS Chem. Neurosci.* **2010**, *1*, 420.
(17) Tu, Z.; Xu, J.; Jones, L.A.; Li, S.; Mach, R.H. *Nucl. Med. Biol.* **2010**, *37*, 509.

(18) Tu, Z.; Fan, J.; Li, S.; Jones, L.A.; Cui, J.; Padakanti, P.K.; Xu, J.; Zeng, D.; Shoghi, K.I.; Perlmutter, J.S.; Mach, R.H. *Bioorg. Med. Chem.* **2011**, *19*, 1666.
(19) Plisson, C.; Salinas, C.; Weinzimmer, D.; Labaree, D.; Lin, S.-F.; Ding, Y.-S.; Jakobsen, S.; Smith, P.W.; Eiji, K.; Carlson, R.E.; Gunn, R.N.; Rabiner, E.A. *Nucl. Med. Biol.* **2011**, *38*, 875.
(20) Celen, S.; Koole, M.; Angelis, M.D.; Sannen, I.; Chitneni, S.K.; Alcazar, J.; Dedeurwaerdere, S.; Moechars, D.; Schmidt, M.; Verbruggen, A.; Langlois, X.; Laere, K.V.; Andres, J.I.; Bormans, G. *J. Nucl. Med.* **2010**, *51*, 1584.
(21) Andres, J.-I.; Angelis, M.D.; Alcazar, J.; Iturrino, L.; Langlois, X.; Dedeurwaerdere, S.; Lenaerts, I.; Vanhoof, G.; Celen, S.; Bormans, G. *J. Med. Chem.* **2011**, *54*, 5820.
(22) Hostetler, E.; Cox, C.D.; Fan, H. WO 2010138577, **2010**.
(23) Pike, V.W.; Rash, K.S.; Chen, Z.; Pedregal, C.; Statnick, M.A.; Kimura, Y.; Hong, J.; Zoghbi, S.S.; Fujita, M.; Toledo, M.A.; Diaz, N.; Gackenheimer, S.L.; Tauscher, J.T.; Barth, V.N.; Innis, R.B. *J. Med. Chem.* **2011**, *54*, 2687.
(24) Hostetler, E.; Sanabria-Bohorquez, S.; Eng, W.S.; Joshi, A.; Gibson, R.; Patel, S.; O'Malley, S.; Krause, S.; Ryan, C.; Riffel, K.; Okamoto, O.; Ozaki, S.; Ohta, H.; Cook, J.; Burns, H.D.; Hargreaves, R. *J. Labelled Comp. Radiopharm.* **2011**, *54*, S78.
(25) Wyffels, L.; Muccioli, G.G.; Kapanda, C.N.; Labar, G.; Bruyne, S.D.; Vos, F.D.; Lambert, D.M. *Nucl. Med. Biol.* **2010**, *37*, 665.
(26) Wilson, A.A.; Garcia, A.; Parkes, J.; Houle, S.; Tong, J.; Vasdev, N. *Nucl. Med. Biol.* **2011**, *38*, 247.
(27) Li, W.; Sanabria-Bohórquez, S.; Joshi, A.; Cook, J.; Holahan, M.; Posavec, D.; Purcell, M.; DeVita, R.; Chobanian, H.; Liu, P.; Chioda, M.; Nargund, R.; Lin, L.; Zeng, Z.; Miller, P.; Chen, T.; O'Malley, S.; Riffel, K.; Williams, M.; Bormans, G.; Van Laere, K.; De Groot, T.; Evens, N.; Serdons, K.; Depre, M.; de Hoon, J.; Sullivan, K.; Hajdu, R.; Shiao, L.; Alexander, J.; Blanchard, R.; DeLepeleire, I.; Declercq, R.; Hargreaves, R.; Hamill, T. *J. Labelled Comp. Radiopharm.* **2011**, *54*, S38.
(28) Johnson, D.S.; Stiff, C.; Lazerwith, S.E.; Kesten, S.R.; Fay, L.K.; Morris, M.; Beidler, D.; Liimatta, M.; Smith, S.E.; Dudley, D.T.; Sadagopan, N.; Bhattachar, S.N.; Kesten, S.J.; Nomanbhoy, T.K.; Cravatt, B.F.; Ahn, K. *ACS Med. Chem. Lett.* **2011**, *2*, 91.
(29) Skaddan, M.B.; Zhang, L.; Johnson, D.S.; Zhu, A.; Zasadny, K.; Coelho, R.V.; Kuszpit, K.; Currier, G.; Fan, K.-H.; Beck, E.; Chen, L.; Drozda, S.E.; Balan, G.; Niphakis, M.; Cravatt, B.F.; Ahn, K.; Bocan, T.; Villalobos, A. *Nucl. Med. Biol.*, **2012**, *in press*.
(30) Gomes, C.M.; Abrunhosa, A.J.; Ramos, P.; Pauwels, E.K.J. *Adv. Drug Delivery Rev.* **2011**, *63*, 547.
(31) Contrafatto, D.; Mostile, G.; Nicoletti, A.; Raciti, L.; Luca, A.; Dibilio, V.; Lanzafame, A.; Distefano, A.; Drago, F.; Zappia, M. *Clin. Neuropharmacol.* **2011**, *34*, 71.
(32) Ziebell, M.; Andersen, B.B.; Thomsen, G.; Pinborg, L.H.; Karlsborg, M.; Hasselbalch, S.G.; Knudsen, G.M. *Eur. J. Nucl. Med. Mol. Imaging* **2012**, *39*, 242.
(33) Colloby, S.J.; Firbank, M.J.; Pakrasi, S.; Perry, E.K.; Pimlott, S.L.; Wyper, D.J.; McKeith, I.G.; Williams, E.D.; O'Brien, J.T. *Neurobiol. Aging* **2011**, *32*, 293.
(34) Schuh-Hofer, S.; Richter, M.; Geworski, L.; Villringer, A.; Israel, H.; Wenzel, R.; Munz, D.L.; Arnold, G. *J. Neurol.* **2007**, *254*, 789.
(35) Chu, M.; Ono, M.; Kimura, H.; Kawashima, H.; Liu, B.L.; Saji, H. *Nucl. Med. Biol.* **2011**, *38*, 313.
(36) Watanabe, H.; Ono, M.; Kimura, H.; Kagawa, S.; Nishii, R.; Fuchigami, T.; Haratake, A.; Nakayama, M.; Saji, H. *Bioorg. Med. Chem. Lett.* **2011**, *21*, 6519.

PART 2

Cardiovascular and Metabolic Diseases

Editor: Andy Stamford
Merck Research Laboratories, Rahway, New Jersey

CHAPTER NINE

Case History: Eliquis™ (Apixaban), a Potent and Selective Inhibitor of Coagulation Factor Xa for the Prevention and Treatment of Thrombotic Diseases

Donald J.P. Pinto, Pancras C. Wong, Robert M. Knabb, Ruth R. Wexler
Bristol-Myers Squibb, Princeton, New Jersey, USA

Contents

1. INTRODUCTION

Despite substantial advances in the prevention and treatment of cardiovascular diseases, they continue to be the leading cause of death in developed countries and are increasingly a major cause of morbidity and mortality in developing countries as well. Although the underlying causes of cardiovascular

ISSN 0065-7743
http://dx.doi.org/10.1016/B978-0-12-396492-2.00009-6

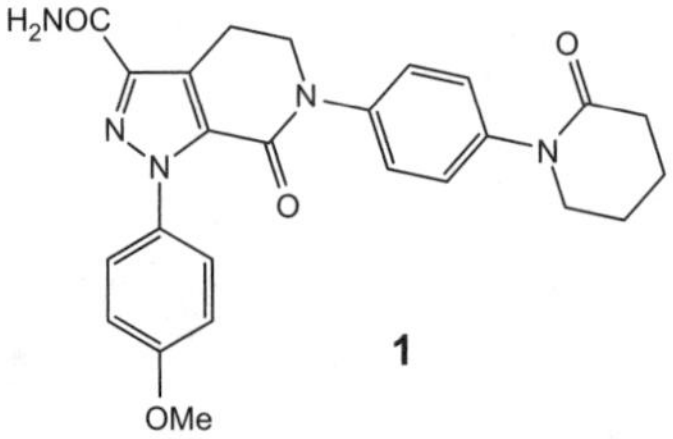

1

Figure 9.1 Eliquis™ (Apixaban).

diseases involve multiple, complex mechanisms, a common component of the end stage of these diseases is thrombosis. Safe and effective antithrombotic drugs are therefore critical to effective treatment of cardiovascular diseases. Paradoxically, many patients who are at the highest risk for thromboembolic diseases, including the very elderly, are often less likely to be taking highly effective antithrombotic drugs due to risk of bleeding. Due to the large remaining unmet medical need, there has been intense activity to develop new antithrombotic therapies with improved efficacy and safety. In recent years, the strategy of targeting coagulation factor Xa (FXa) has received substantial clinical validation. Eliquis™ (Apixaban **1**, Fig. 9.1) is one of the first compounds acting by this mechanism to complete late-stage clinical studies and enter clinical practice. Along with other novel oral anticoagulants, it is poised to usher in a new era of antithrombotic therapy.

2. RATIONALE FOR TARGETING FXa

Vitamin K antagonists (VKAs), such as warfarin, are no longer the only available oral anticoagulants. Direct thrombin inhibitors, such as dabigatran etexilate, and FXa inhibitors, such as rivaroxaban and apixaban, have been developed and shown to be effective oral anticoagulants.[1–4] To date, there is no direct clinical evidence favoring one target over the other. However, there is some theoretical and preclinical evidence to support that the FXa mechanism may positively differentiate from thrombin as a preferred antithrombotic target.

First, as blood coagulation involves sequential steps of activation and amplification of coagulation proteins, generation of one molecule of FXa results in the production of hundreds of thrombin molecules.[2] In theory, therefore, inhibition of FXa may be more efficient in reducing fibrin formation than direct inhibition of thrombin activity. This principle is consistent with an *in vitro* observation that inhibition of FXa produced a more effective sustained reduction of thrombus-associated procoagulant activity than inhibition of thrombin activity.[3] Second, inhibition of FXa is not thought to affect existing levels of thrombin and its activity. In addition, reversible FXa inhibitors might not completely suppress the production of thrombin. These small amounts of thrombin might be enough to activate high-affinity platelet thrombin receptors to preserve hemostasis. Early work from several laboratories provided experimental evidence from animal studies suggesting that the antithrombotic efficacy of FXa inhibitors is accompanied by a lower risk of bleeding when compared with thrombin inhibitors.[4–8] In summary, inhibition of FXa may represent an attractive approach compared with thrombin inhibition for effective and safe antithrombotic therapy. However, head-to-head clinical studies to validate this hypothesis have not been performed.

3. MEDICINAL CHEMISTRY EFFORTS CULMINATING IN APIXABAN

3.1. Factor Xa program objectives

As we believed that Factor Xa had the hallmarks of a target that would positively differentiate from anticoagulant standard of care, our discovery objective was to continuously deliver compounds until an optimal compound was in full development. This was driven by our strong belief that a high-quality factor Xa inhibitor would be transformational. The objective was to optimize for the right balance of efficacy and safety. The goal of the program was to identify potent, highly selective noncovalent FXa inhibitors with good oral bioavailability ($>20\%$) and a half-life suitable for either twice daily (BID) or once daily (QD) dosing with low peak/trough to minimize the potential for bleeding liabilities. *In vivo*, the compounds were required to demonstrate efficacy in preclinical thrombosis models. The ideal candidate would also not have drug–drug or food interactions, particularly given that these are issues with warfarin.

3.2. Early preclinical leads

When the medicinal chemistry program began in the mid-1990s, the only published FXa inhibitors were dibasic compounds which were not orally bioavailable (Fig. 9.2) such as **2** (DABE), **3** (BABCH), and **4** (DX-9065a) (K_i=570, 13, and 41 nM, respectively).[9–11] The potency of these compounds resulted from a strong interaction between the amidine of the inhibitor with the S1 Asp189 residue, and a π-cation interaction between the hydrophobic residues in the S4 subsite and the remaining basic functionality.[10]

At the genesis of our program, homology models and the X-ray coordinates for the FXa dimer were used extensively to design a number of dibasic FXa inhibitors.[12,13] An initially designed compound, ketone **5**, though a weak inhibitor of FXa ($K_i = 5100$ nM) was rapidly improved to **6** ($K_i = 34$ nM), by the introduction of an ester group.[12] With the aid of molecular modeling, these compounds evolved into amidine-based benzimidazoles such as compound **7** ($K_i = 140$ nM, Fig. 9.3). The SAR was extended to include several additional scaffolds such as indole, indoline, and phenylpyrrolidine, all conferring improved FXa activity, albeit with poor selectivity over trypsin and no oral bioavailability.[13]

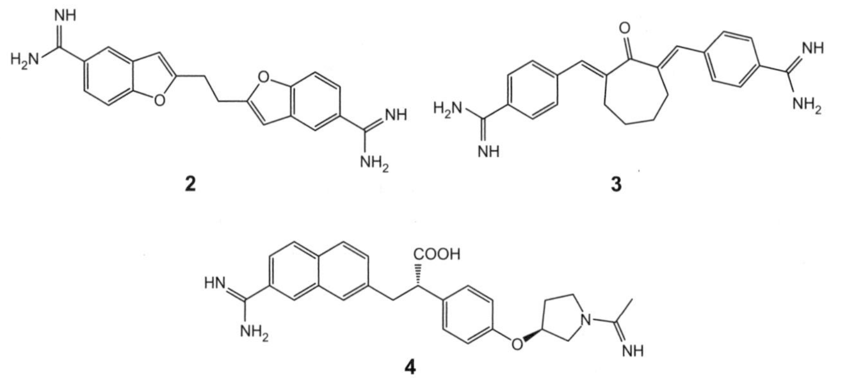

Figure 9.2 Published dibasic benzamidine FXa inhibitors.

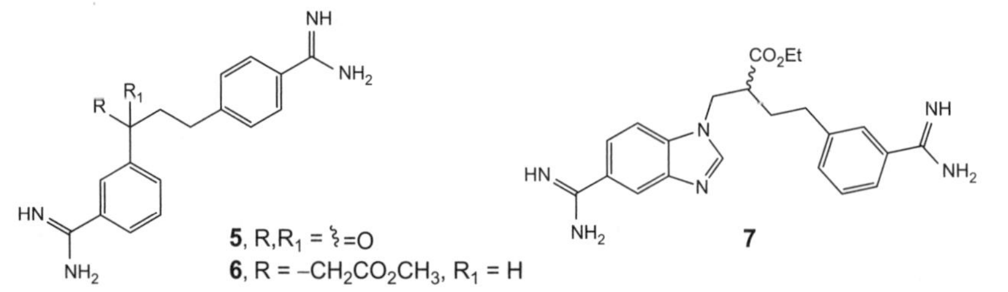

Figure 9.3 Early bis-benzamidine leads.

3.3. Screening library hits—The discovery of the isoxazoline scaffold

The second and more innovative approach resulted in a "focused screening" strategy which was coupled early with structure-based design to drive affinity. We recognized that the peptide sequence of ligands for the GPIIb/IIIa receptor Arg–Gly–Asp (RGD) and the two prothrombin cleavage sequences for FXa, namely, Glu–Gly–Arg (EGR), though reversed, shared some similarity. Based on these observations, and because known GPIIb/IIIa receptor antagonists contain a benzamidine group which is also found in FXa inhibitors such as **5–7**, our internal proprietary collection of small molecule GPIIb/IIIa antagonists was screened against FXa.[14]

This effort led to the identification of a weak isoxazoline inhibitor **8** (K_i=38.5 μM; Fig. 9.4). Not discouraged by the weak affinity of **8**, lead optimization was jump-started by expeditiously improving affinity to subnanomolar levels by enhancing hydrophobic interactions in the S1 and S4 pockets.

Replacement of the aspartate residue in **8** with a second benzamidine afforded compound **9** (K_i=1.4 μM), providing ~30-fold improvement in FXa affinity. Direct substitution of the carboxamide group onto the isoxazoline core, followed by substitution with a geminal carbonyl group designed to hydrogen bond to

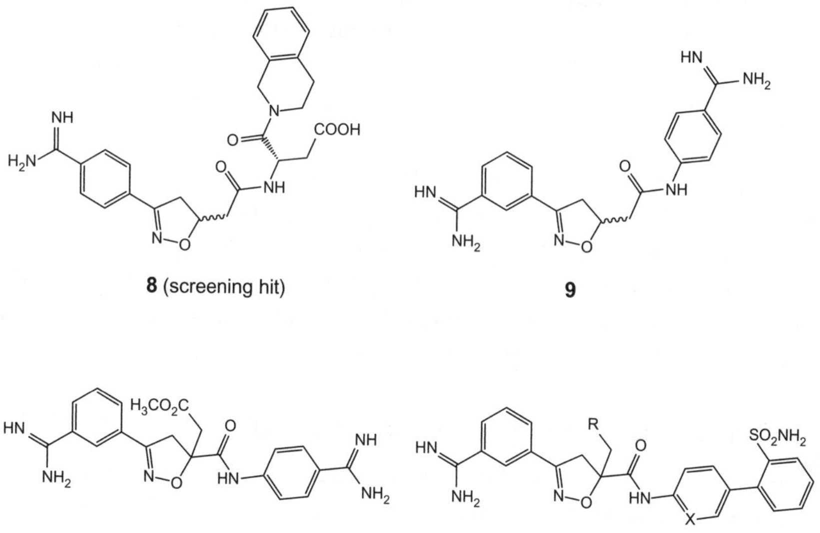

8 (screening hit)

9

10

11 R = CO_2CH_3, X = CH (–) enant.

12 R = N-tetrazole, X = N (–) enant.

Figure 9.4 Optimizing isoxazoline analogs.

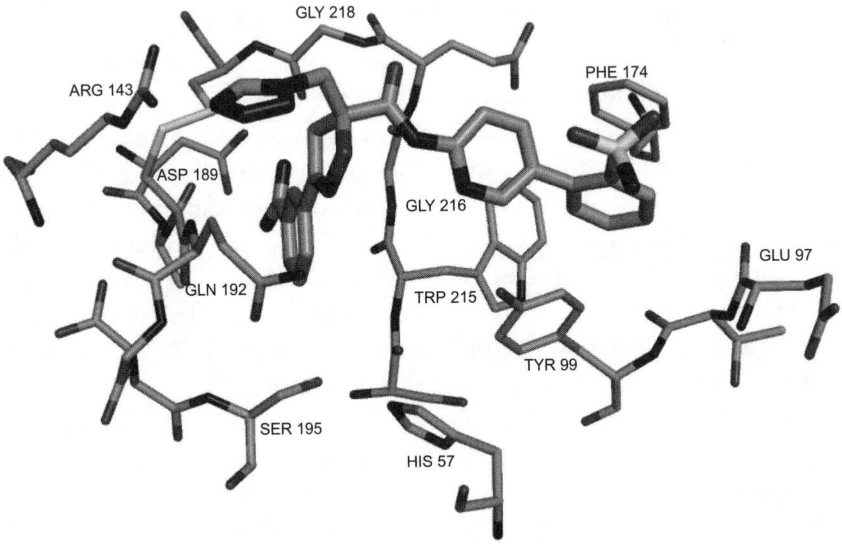

Figure 9.5 Model of isoxazoline **12** in the active site of FXa. (See Color Plate 9.5 in Color Plate Section.)

Tyr99 and the backbone of Gln192, led to 5,5-disubstituted isoxazoline **10** (K_i=94 nM). Finally, replacement of the basic P4 amidine with a biaryl moiety resulted in high-affinity inhibitors such as **11** (K_i=6.3 nM) and **12** (K_i=0.52 nM).[15,16] Based on modeling in the active site of FXa, the excellent affinity exhibited by these compounds was rationalized to be the result of bidentate interactions of the benzamidine with Asp189 in the S1 specificity pocket, favorable π stacking of the pendent phenyl ring with Tyr99 and Phe174, and an edge-to-face interaction with Tyr215 in the S4 pocket (Fig. 9.5). The biarylsulfonamide P4 motif represented a major milestone, as for the first time high affinity was achieved with a compound containing a neutral P4 moiety, which was used extensively during lead optimization.[17,18] Interestingly, inhibitors from both the pharmacophore approach and the "focused" screening approach converged on a similar binding motif, where inhibitors bound in the active site in an L-shaped configuration.

3.4. The discovery of pyrazole-based inhibitors

Efforts continued to focus on driving FXa-binding affinity to picomolar levels, recognizing that the ultimate goal of replacing the benzamidine with a less basic/neutral P1 functionality was expected to achieve permeability at the expense of affinity. Scaffold optimization (Fig. 9.6) initially led to vicinally substituted isoxazoline compounds such as **13** (diastereomeric mixture,

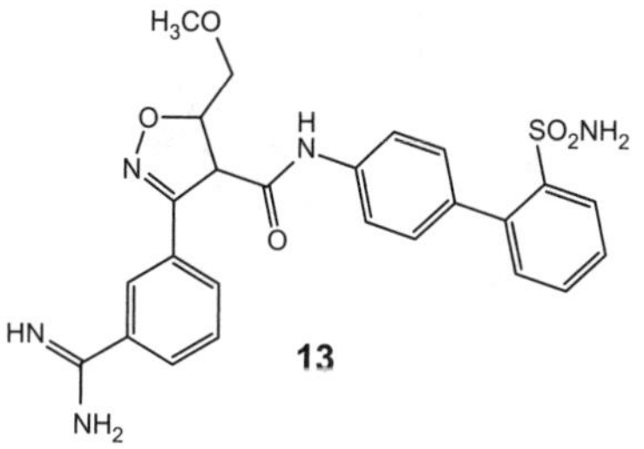

13

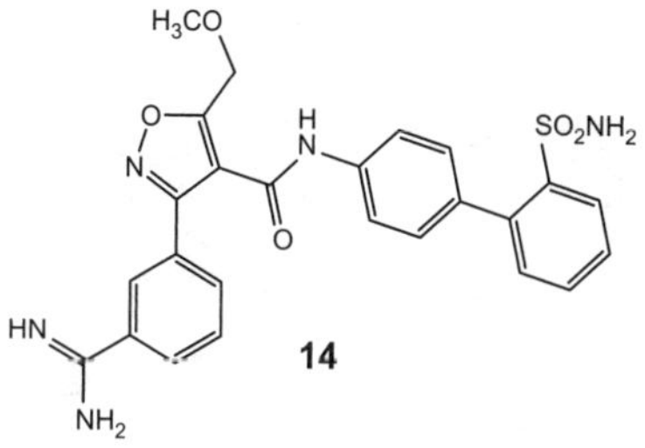

14

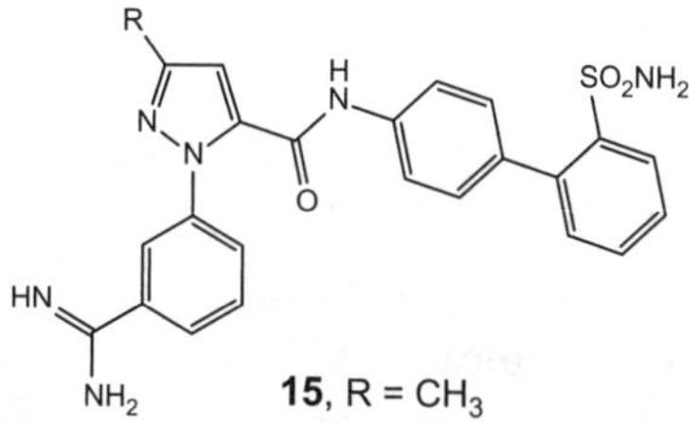

15, R = CH_3
16, R = CF_3

Figure 9.6 The discovery of pyrazole **15** (SN429).

K_i=0.5 nM) with more optimal complementarity with the S1 and S4 pockets.[17] The isoxazole analog **14** (K_i=0.15 nM) lacking the stereogenic centers showed similar affinity. Pyrazole **15** (SN429, K_i=0.013 nM), the result of an independent rational design effort, was a major program milestone based on its picomolar affinity.[18] An X-ray structure of **15** bound to trypsin confirmed the interaction of the carboxamide carbonyl with the Gly216 NH and the biarylsulfonamide P4 group optimally stacked in the hydrophobic S4 region.[18] While extremely potent, **15** had a short half-life, poor oral bioavailability, and lacked selectivity over trypsin-like serine proteases. Evaluation of numerous heteroaryl scaffolds affirmed the pyrazoles to be superior in terms of affinity.[17–19] In fact, the most potent compound synthesized in the program was achieved with the 3-trifluoromethylpyrazole analog **16** (K_i < 5 pM). This was a significant development for the program, as we achieved one of our main goals of driving the affinity to a high level thus enabling us to focus on addressing permeability and oral bioavailability for the series.

3.5. Benzamidine mimics—SAR leading to Razaxaban and preclinical properties

Successful replacement (Fig. 9.7) of the benzamidine with the less basic P1 benzylamine ($pK_a \sim 8.8$)[18] and insertion of a fluoro substituent on the inner phenyl ring of the P4 group led to **17** (K_i=2.7 nM, activated partial

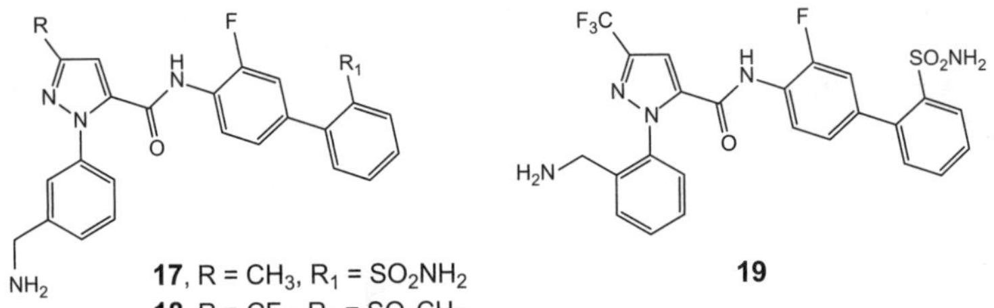

17, R = CH_3, R_1 = SO_2NH_2
18, R = CF_3, R_1 = SO_2CH_3

19

Figure 9.7 Lead optimization of pyrazoles to early clinical candidate **18** (DPC423).

thromboplastin time (APTT) IC_{2x}=2.3 μM). Replacement of the methylsulfonamide P4 aryl substituent with a methylsulfone and the pyrazole methyl substituent with a trifluoromethyl group afforded **18** (K_i=0.15 nM), which exhibited good Caco-2 permeability (P_{app}=4.86 × 10^{-6}cm/s) and high selectivity over other serine proteases, with the exception of trypsin and kallikrein (both K_i=60 nM).[20] The improved pharmacokinetic (PK) profile (*F*%=57, $T_{1/2}$=7.5 h), and efficacy in rabbit thrombosis models (A–V shunt, ID_{50}=1.1 μmol/kg/h), enabled advancement of **18** (DPC423) to clinical trials as the first clinical candidate from the program.

In Phase I, **18** showed desirable exposure at the doses studied and had a half-life of ~30 h.[20] However, further advancement was curtailed due to preclinical toxicity. With the advancement of **18**, a deep backup strategy for advancing compounds was adopted, which focused on optimizing the selectivity profile to minimize the potential for off-target safety issues. To our delight, the *o*-benzylamine analog, **19** (DPC602) was potent, highly selective (FXa K_i=0.9 nM, trypsin K_i=3500 nM), and demonstrated improved oral bioavailability.[21] Under basic conditions, chemical instability of **19** was observed; the amine cyclized on to the carbonyl of the pyrazole carboxamide, thereby liberating the biarylamine, and development of **19** was discontinued.

At the time this work was done, there was little in the serine protease inhibitor literature describing less basic benzamidine replacements. Therefore, the medicinal chemistry team embarked on a pioneering and comprehensive evaluation of less basic and neutral P1 groups to build in selectivity and achieve oral bioavailability.[22] This work is summarized schematically in Fig. 9.8, along with additional SAR trends. Emphasis was placed on large P1 moieties which could be accommodated in the S1 specificity pocket of FXa that is larger and more lipophilic than in trypsin-like serine proteases owing to the presence of Ala190 in the former, but that might be expected

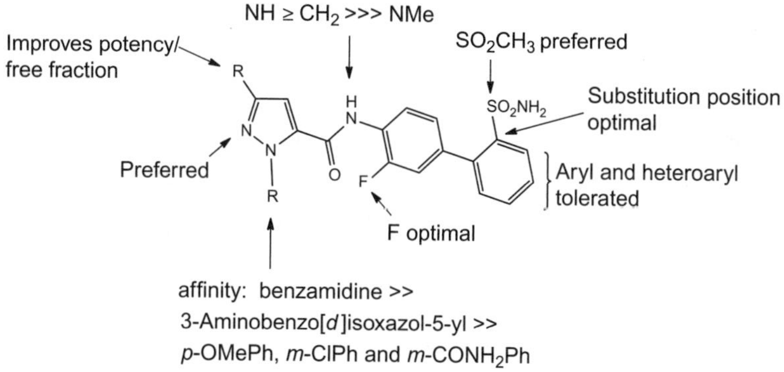

Figure 9.8 SAR trends for the pyrazole series.

to clash with the side chain of Ser190 present in trypsin-like serine proteases. Substitution at the 1-position of the pyrazole with the 3-aminobenzo[*d*]isoxazol-5-yl P1 group was shown to confer desirable affinity and selectivity.

Through a parallel synthesis effort, neutral P1 moieties including 4-methoxyphenyl, 3-carbamoylphenyl, and 3-chlorophenyl were also identified, and although 10- to 20-fold less potent, the compounds were more permeable than similar benzamidines and showed improved selectivity.[22] Success in exploiting the larger S1 pocket of FXa to afford selectivity and oral bioavailability was another highly significant milestone for our program. In addition, a novel and mild cross-coupling methodology (Chan-Lam) was developed in part to facilitate synthesis and scale-up of compounds containing these azole P1 moieties.[23]

Based on its greater potency and improved selectivity, the aminobenzisoxazole P1 series was selected for further optimization.[22,24] To this end, **20** (Fig. 9.9) showed greater affinity and selectivity (FXa $K_i = 0.16$ nM, trypsin $K_i > 3000$ nM) but demonstrated poor clotting activity and permeability. To overcome this, the biaryl P4 group was replaced with a phenylimidazole moiety (**21** $K_i = 0.70$ nM) to increase polarity, solubility, and permeability.[24] Further improvement in affinity and high selectivity (>40,000-fold) was realized with 2-aminomethylimidazole P4 analogs such as **22** ($K_i = 0.17$ nM, Caco-2 $P_{app} = 0.2 \times 10^{-6}$ cm/s) and **23** ($K_i = 0.19$ nM, Caco-2 $P_{app} = 5.56 \times 10^{-6}$ cm/s), with **23** showing improved permeability in the Caco-2 assay. When orally administered to dogs, **23** demonstrated high oral bioavailability, a high volume of distribution, and a moderate half-life ($F\% = 84\%$, Cl = 1.1 L/kg/h, $V_{dss} = 3.5$ L/kg, $T_{1/2} = 5.3$ h). A FXa-bound X-ray crystal structure of

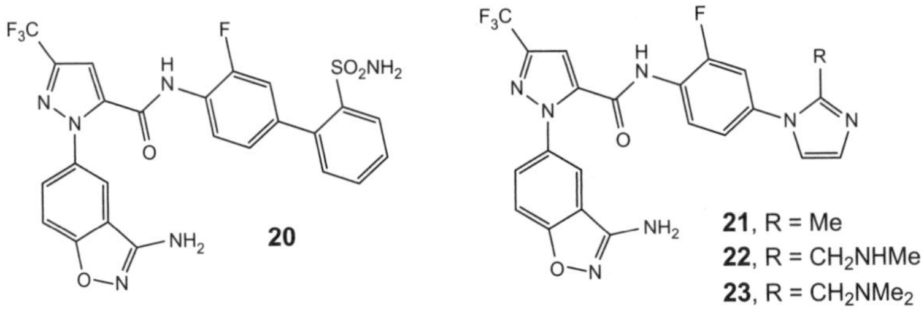

Figure 9.9 Lead optimization of pyrazoles to clinical candidate razaxaban **23**.

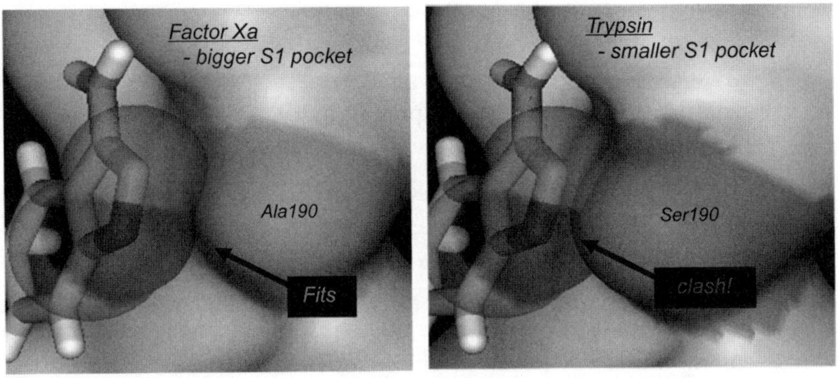

Figure 9.10 Aminobenzisoxazole **23** in the S1 specificity pocket of FXa and trypsin. (See Color Plate 9.10 in Color Plate Section.)

23 showed a binding orientation similar to that observed with previous pyrazole candidate **18**. The larger P1 aminobenzisoxazole successfully exploits the differences in the S1 specificity pockets as predicted, resulting in its favorable selectivity profile (>2000-fold, Fig. 9.10). The pendant P4 (dimethylaminomethyl)imidazole nitrogen interacts directly with Glu97 through a network of water molecules in the S4 pocket. Compound **23** was highly efficacious in rabbit models of thrombosis and, given its overall very good profile, was subsequently advanced to clinical development as razaxaban, DPC906 (BMS-562389). Razaxaban was the first small molecule, direct FXa inhibitor to complete a pivotal Phase II proof-of-principle study in deep vein thrombosis (DVT), demonstrating strong efficacy and a favorable bleeding profile.[25] The clinical profile of razaxaban was a major advance for the field and was critical for our program in establishing the dose projections for later clinical studies with apixaban.

3.6. Strategies leading to the dihydropyrazolopyridinone scaffold

Immediately after **23** was advanced, the goal was to identify a structurally diverse backup compound in the event of other unexpected issues. A key backup strategy was to design rigidified pyrazole scaffolds in order to address the potential for metabolic cleavage of the carboxamide moiety given the potential for generation of a mutagenic aniline fragment. Although this was never an issue for the P4 fragment of **23** or its predecessor compound **18**, Ames testing of aniline fragments for all new compounds was implemented, which was resource intensive and rate limiting. A two-pronged approach which leveraged new scaffolds was adopted to eliminate this concern (Fig. 9.11). Strategy 1 focused on tying back the amide NH onto the P4 inner phenyl ring affording compounds with bicyclic scaffolds such as indoline amides. Although these compounds were potent and selective, they had weak clotting activity and poor oral bioavailability.[26,27] Strategy 2 focused on cyclizing the amide moiety onto the pyrazole based on the crucial observation that the 5-pyrazole carboxamide is in a planar orientation with the pyrazole ring in analogs such as razaxaban. Accordingly, several 5,6- and 5,7-bicyclic pyrazole scaffolds were synthesized.[28–30]

From this series, only the pyrazolopyrimidinone **24** and the dihydropyrazolopyridinone **25** maintained a combination of a high level of

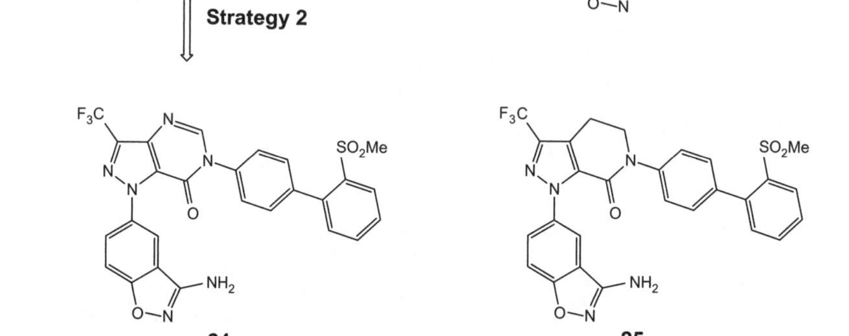

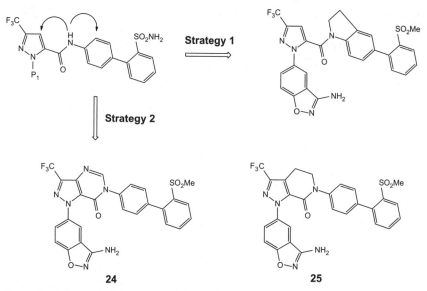

Figure 9.11 Strategies for structural diversification.

FXa affinity, potency in the clotting assay, and high oral bioavailability. Ultimately, instability of the pyrimidinone analogs under acidic conditions precluded them from further consideration. The dihydropyrazolopyridinone scaffold as in **25** showed good stability and versatility in tolerating a broader range of functional groups[29] and, hence, was selected as the scaffold of choice for further optimization.

3.7. SAR in the dihydropyrazolopyridinone series: Optimizing for ideal PK

Initially, the aminobenzisoxazole P1 group was maintained and extensive variation of the P4 moiety was carried out.[30] This led to potent compounds such as **26** and **27** (K_i=0.04 nM, and K_i=0.03 nM, respectively; Fig. 9.12). The overall efficacy and PK profile favored **27**, which was considered for development.

The high FXa affinity exhibited by the bicyclic pyrazole scaffold prompted a reevaluation of the less potent neutral P1 moieties. As part of this effort, the *p*-methoxyphenyl P1 group (Fig. 9.8) which had previously demonstrated high *in vivo* exposure was reintroduced. The 3-position of the pyrazole ring was targeted to optimize for affinity, polarity, and free fraction. This strategy provided compounds such as **28** (K_i=0.14 nM, prolongation of prothrombin time (PT) IC_{2x}=1.2 μM; Fig. 9.12), which demonstrated high affinity and a reasonable free fraction.[31] When administered to dogs, **28** had high oral bioavailability and a long half-life; however, the latter was the result of a high volume of distribution and moderate clearance (Cl=1.3 L/kg/h, V_{dss}=7.4 L/kg, $T_{1/2}$=7.3 h, $F\%$=56). While identifying compounds with a high volume of distribution was a strategy we had

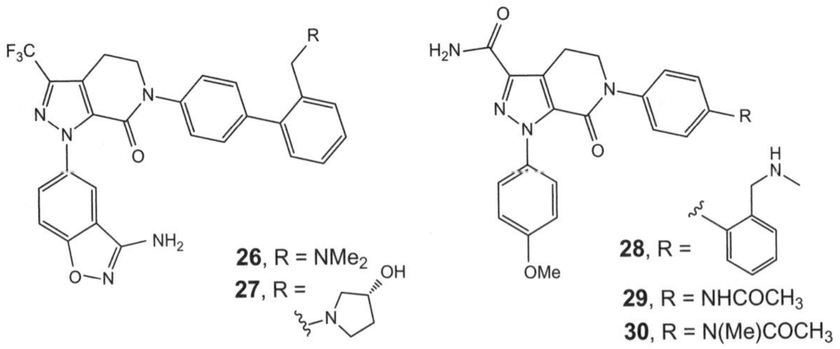

Figure 9.12 Optimization strategies leading to apixaban **1**.

used to achieve long half-life, many compounds with this property were discontinued due to safety issues during preclinical development. An innovative strategy ensued to identify compounds with low volume of distribution (to maintain high drug levels in the bloodstream) and low clearance (to provide acceptable half-lives) while maintaining high affinity, selectivity, and oral bioavailability. Evaluation of metabolic stability in human liver microsomes was included in the early assessment of new compounds to identify compounds likely to have low clearance.

To minimize volume of distribution, we returned to the P4 position, specifically examining polar, nonbasic functionality. Initial analogs such as acetamide **29** showed weak FXa activity (K_i = 180 nM). However, the activity was recovered with **30** (K_i = 0.61 nM, PT IC_{2x} = 3.1 μM). Rigidification of the acetamide group leads to the lactam **1** (K_i = 0.08 nM, PT IC_{2x} = 3.8 μM).[31] An X-ray structure of **1** bound to FXa (Fig. 9.13) showed an orientation similar to that observed for earlier compounds. The *p*-methoxyphenyl group was deep in the S1 pocket, and the lactam makes the requisite hydrophobic interactions in the S4 pocket.

The overall PK profile of **1** in dogs was favorable, demonstrating high oral bioavailability, extremely low clearance, low volume of distribution, and a moderate half-life (Cl = 0.02 L/kg/h, V_{dss} = 0.2 L/kg, $T_{1/2}$ = 5.8 h,

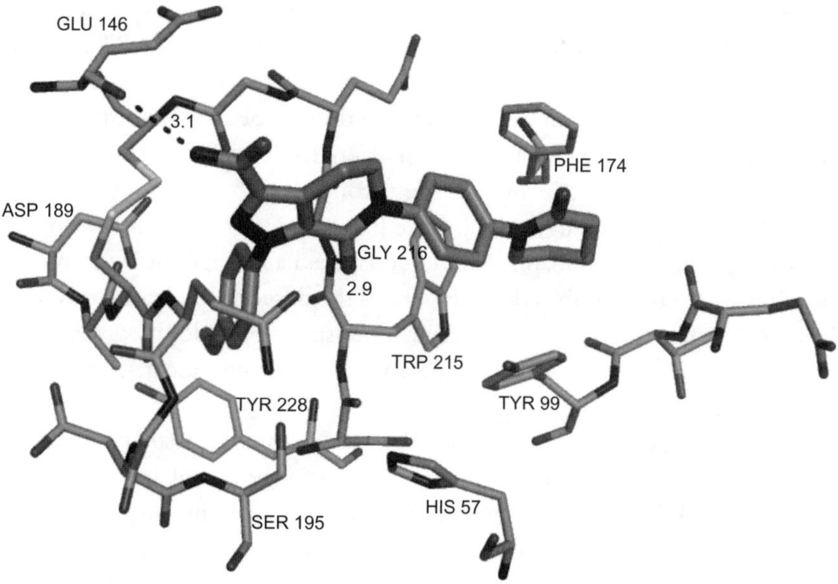

Figure 9.13 FXa-bound X-ray structure of apixaban **1**. (See Color Plate 9.13 in Color Plate Section.)

F% = 58).[32] In addition, compound **1** exhibited outstanding selectivity relative to other serine proteases,[33] was highly efficacious in various antithrombotic models and did not demonstrate any liabilities in safety studies.[34] This compound clearly met our predefined criteria for an ideal oral anticoagulant and was selected for clinical development as BMS-562247, apixaban.

4. PRECLINICAL PROPERTIES OF APIXABAN

A review of apixaban describing its preclinical pharmacology as well as its preclinical drug metabolism and PK profile has been published.[34] Apixaban is a potent, reversible, direct, active site inhibitor of FXa, with a K_i of 0.08 nM for human FXa and with greater than 30,000-fold selectivity over other human coagulation proteases. Unlike the indirect FXa inhibitor fondaparinux, apixaban does not require antithrombin III to inhibit FXa. It inhibits free, prothrombinase-bound as well as clot-bound FXa activity and reduces thrombin generation *in vitro*. Apixaban also inhibits FXa from rabbits, rats, and dogs, with K_i values of 0.16, 1.4, and 1.8 nM, respectively, which parallels its antithrombotic potency in these species. Although apixaban has no direct effects on platelet aggregation, it indirectly inhibits platelet aggregation induced by thrombin derived from the upstream proteases in the blood coagulation cascade. In standard clotting assays, apixaban is more potent in the prolongation of prothrombin time than APTT *in vitro* in rats, rabbits, dogs, and humans.

Apixaban given prophylactically caused dose-dependent antithrombotic activity in rats and rabbits, in models of arterial and venous thrombosis and prevented the growth of a preexisting thrombus. Effective concentrations tended to be higher in rats for which the FXa affinity of apixaban was lower. Dose–response studies of apixaban demonstrated a therapeutic window between the dose that inhibits thrombosis and the dose that increases provoked bleeding. When added on top of aspirin or aspirin plus clopidogrel at their clinically relevant doses, apixaban improved antithrombotic activity, without excessive increases in bleeding times.

Apixaban has good oral bioavailability, low clearance, a small volume of distribution in animals and humans, and a low potential for drug–drug interactions. Elimination pathways for apixaban include renal excretion, metabolism, and biliary/intestinal excretion. Although a sulfate conjugate of *O*-demethyl apixaban (*O*-demethyl apixaban sulfate) has been identified as the major circulating metabolite of apixaban in humans, it is biologically inert and inactive against human FXa. Apixaban was without toxicity in multiple preclinical toxicology studies, including chronic safety studies in rats and dogs, carcinogenic studies in mice and rats, reproductive toxicology studies in rats and rabbits, and mutagenic studies.[35]

Together, these preclinical findings have established the favorable pharmacological and safety profile of apixaban and support the potential use of apixaban in the clinic for the prevention and treatment of various thromboembolic diseases.

5. CLINICAL STUDIES OF APIXABAN

Apixaban has been evaluated in a series of late-stage clinical trials in multiple indications, including prevention and treatment of venous thromboembolism (VTE), secondary prevention of acute coronary syndromes (ACS), and stroke prevention in patients with atrial fibrillation (AF).

After initial studies in healthy human subjects showed a terminal half-life of 8–15 h,[36] it was considered for either QD or BID administration. A dose-ranging study for prevention of VTE in patients undergoing total knee replacement demonstrated that both antithrombotic efficacy and bleeding were dose dependent.[37] Whereas bleeding was similar when the same daily dose was administered QD versus BID, efficacy was consistently better with BID dosing, suggesting that a strategy that minimizes peak to trough fluctuations in concentration provides a more optimal benefit-to-risk profile. Phase 3 studies of apixaban have all utilized a BID regimen.

For prevention of VTE after knee or hip replacement, apixaban 2.5 mg BID demonstrated superior efficacy to enoxaparin 40 mg QD without increasing bleeding.[38,39] In a study versus a 30-mg BID regimen of enoxaparin used in North America, apixaban failed to prove noninferiority, despite similar rates of the primary endpoint, but resulted in less bleeding.[38]

A Phase 2 study for treatment of DVT for 3 months compared apixaban 5 or 10 mg BID, and 20 mg QD with standard treatment with low-molecular-weight heparin (LMWH) followed by warfarin.[40] Results of this study led to the selection of 5 mg BID after 1 week of 10 mg BID for evaluation versus LMWH/warfarin in the initial treatment of VTE in the ongoing AMPLIFY study (ClinicalTrials.gov: NCT00643201), and doses of 2.5 or 5 mg BID versus placebo for extended treatment in the AMPLIFY-EXT study (ClinicalTrials.gov: NCT00633893).

Treatment with apixaban on top of conventional mono or dual antiplatelet therapy was evaluated in patients who had recent ACS. Although promising results were observed in the Phase 2 APPRAISE-1 study,[41] the Phase 3 APPRAISE-2 study was terminated after an observation of a significantly increased risk of major bleeding, especially in patients who were receiving dual antiplatelet therapy.[42]

For prevention of stroke in patients with AF, apixaban 5 mg BID, or 2.5 mg BID in selected patients, was evaluated in two large Phase 3 studies. In AVERROES, apixaban resulted in a 55% reduction in stroke or systemic embolism versus aspirin in patients who were not suitable for treatment with a VKA.[43] Major bleeding was higher in patients treated with apixaban, but the difference was not statistically significant and there was no increase in intracranial or fatal bleeding. When compared with blinded warfarin in the ARISTOTLE study, apixaban demonstrated statistically significant reductions of 21% in stroke and systemic embolism, 31% in major bleeding, and 11% in all-cause death.[2]

6. CONCLUSION

Apixaban was the culmination of a succession of novel and innovative medicinal chemistry discoveries resulting from a structure-based design approach and extensive PK optimization to strike the optimal balance of efficacy and safety. During the lead optimization phase, computer-aided drug design and X-ray crystallography were highly leveraged to drive affinity, selectivity and facilitate oral bioavailability. Potency was achieved by optimizing to extremely high drug affinity (picomolar levels) and oral bioavailability was ultimately achieved by replacing the positively charged group with a neutral group. Selectivity versus other serine proteases resulted from capitalizing on the larger S1 pocket in FXa. The PK profile and favorable safety profile exhibited by apixaban is ideal for an anticoagulant and is due to the low volume of distribution and low clearance. Apixaban was well tolerated in preclinical studies and human clinical trials. Results of the clinical trial program of apixaban have led to its approval in Europe and other countries outside the USA for prevention of VTE after knee or hip replacement. It is currently under review in the USA, Europe, and other countries for stroke prevention in patients with AF.

ACKNOWLEDGMENTS

Declaration of interest. The authors are employees of Bristol-Myers Squibb Company and were previously involved in factor Xa inhibitor research.

REFERENCES

(1) Soff, G.A. *Arterioscler. Thromb. Vasc. Biol.* **2012**, *32*, 569.
(2) Mann, K.G.; Butenas, S.; Brummel, K. *Arterioscler. Thromb. Vasc. Biol.* **2003**, *23*, 17.
(3) McKenzie, C.R.; Abendschein, D.R.; Eisenberg, P.R. *Arterioscler. Thromb. Vasc. Biol.* **1996**, *16*, 1285.
(4) Harker, L.A.; Hanson, S.R.; Kelly, A.B. *Thromb. Haemost.* **1995**, *74*, 464.
(5) Morishima, Y.; Tanabe, K.; Terada, Y.; Hara, T.; Kunitada, S. *Thromb. Haemost.* **1997**, *78*, 1366.
(6) Sato, K.; Kawasaki, T.; Taniuchi, Y.; Hirayama, F.; Koshio, H.; Matsumoto, Y. *Eur. J. Pharmacol.* **1997**, *339*, 141.
(7) Wong, P.C.; Crain, E.J.; Watson, C.A.; Zaspel, A.M.; Wright, M.R.; Lam, P.Y.; Pinto, D.J.; Wexler, R.R.; Knabb, R.M. *J. Pharmacol. Exp. Ther.* **2002**, *303*, 993.
(8) Wong, P.C.; Crain, E.J.; Watson, C.A.; Xin, B. *J. Thromb. Haemost.* **2009**, *7*, 1313.
(9) Tidwell, R.R.; Webster, W.P.; Shaver, S.R.; Geratz, J.D. *Thromb. Res.* **1980**, *19*, 339.
(10) Sturzebecher, J.; Markwardt, F.; Walsmann, P. *Thromb. Res.* **1976**, *9*, 637.
(11) Nagahara, T.; Yokoyama, Y.; Inamura, K.; Katakura, S.; Komoriya, S.; Yamaguchi, H.; Hara, T.; Iwamoto, M. *J. Med. Chem.* **1994**, *37*, 1200.
(12) Maduskuie, T.P.; McNamara, K.J.; Ru, Y.; Knabb, R.M.; Stouten, P.F. *J. Med. Chem.* **1998**, *41*, 53.
(13) Quan, M.L.; Wexler, R.R. *Curr. Top. Med. Chem.* **2001**, *1*, 137.
(14) Quan, M.L.; Pruitt, J.R.; Ellis, C.D.; Liauw, A.Y.; Galemmo, R.A.; Stouten, P.F.W.; Wityak, J.; Knabb, R.M.; Thoolen, M.J.; Wong, P.C.; Wexler, R.R. *Bioorg. Med. Chem. Lett.* **1997**, *7*, 2813.
(15) Quan, M.L.; Liauw, A.Y.; Ellis, C.D.; Pruitt, J.R.; Carini, D.J.; Bostrom, L.L.; Huang, P.P.; Harrison, K.; Knabb, R.M.; Thoolen, M.J.; Wong, P.C.; Wexler, R.R. *J. Med. Chem.* **1999**, *42*, 2752.
(16) Quan, M.L.; Ellis, C.D.; Liauw, A.Y.; Alexander, R.S.; Knabb, R.M.; Lam, G.; Wright, M.R.; Wong, P.C.; Wexler, R.R. *J. Med. Chem.* **1999**, *42*, 2760.
(17) Pruitt, J.R.; Pinto, D.J.; Estrella, M.J.; Bostrom, L.L.; Knabb, R.M.; Wong, P.C.; Wright, M.R.; Wexler, R.R. *Bioorg. Med. Chem. Lett.* **2000**, *10*, 685.
(18) Pinto, D.J.P.; Orwat, M.J.; Wang, S.; Fevig, J.M.; Quan, M.L.; Amparo, E.; Cacciola, J.; Rossi, K.A.; Alexander, R.S.; Smallwood, A.M.; Luettgen, J.M.; Liang, L.; Aungst, B.J.; Wright, M.R.; Knabb, R.M.; Wong, P.C.; Wexler, R.R.; Lam, P.Y.S. *J. Med. Chem.* **2001**, *44*, 566.
(19) Fevig, J.M.; Pinto, D.J.; Han, Q.; Quan, M.L.; Pruitt, J.R.; Jacobson, I.C.; Galemmo, R.A.; Wang, S.; Orwat, M.J.; Bostrom, L.L.; Knabb, R.M.; Wong, P.C.; Lam, P.Y.S.; Wexler, R.R. *Bioorg. Med. Chem. Lett.* **2001**, *11*, 641.
(20) Pinto, D.J.P.; Smallheer, J.M.; Cheney, D.L.; Knabb, R.M.; Wexler, R.R. *J. Med. Chem.* **2010**, *53*, 6243.
(21) Pruitt, J.R.; Pinto, D.J.P.; Galemmo, R.A.; Alexander, R.S.; Rossi, K.A.; Wells, B.L.; Drummond, S.; Bostrom, L.L.; Burdick, D.; Bruckner, R.; Chen, H.; Smallwood, A.; Wong, P.C.; Wright, M.R.; Bai, S.; Luettgen, J.M.; Knabb, R.M.; Lam, P.Y.S.; Wexler, R.R. *J. Med. Chem.* **2003**, *46*, 5298.
(22) Lam, P.Y.S.; Clark, C.G.; Li, R.; Pinto, D.J.P.; Orwat, M.J.; Galemmo, R.A.; Fevig, J.M.; Teleha, C.A.; Alexander, R.S.; Smallwood, A.M.; Rossi, K.A.;

Wright, M.R.; Bai, S.A.; He, K.; Luettgen, J.M.; Wong, P.C.; Knabb, R.M.; Wexler, R.R. *J. Med. Chem.* **2003**, *46*, 4405.
(23) Qiao, J.X.; Lam, P.Y.S. *Synthesis* **2011**, *6*, 829.
(24) Qua, M.L.; Lam, P.Y.S.; Han, Q.; Pinto, D.J.P.; He, M.Y.; Li, R.; Ellis, C.D.; Clark, C.G.; Teleha, C.A.; Sun, J.H.; Alexander, R.S.; Bai, S.; Luettgen, J.M.; Knabb, R.M.; Wong, P.C.; Wexler, R.R. *J. Med. Chem.* **2005**, *48*, 1729.
(25) Wong, P.C.; Crain, E.J.; Watson, C.A.; Wexler, R.R.; Lam, P.Y.S.; Quan, M.L.; Knabb, R.M. *J. Thromb. Thrombolysis* **2007**, *24*, 43.
(26) Varnes, J.G.; Wacker, D.A.; Jacobson, I.C.; Quan, M.L.; Ellis, C.D.; Rossi, K.A.; He, M.Y.; Luettgen, J.M.; Knabb, R.M.; Bai, S.; He, K.; Lam, P.Y.S.; Wexler, R.R. *Bioorg. Med. Chem. Lett.* **2007**, *17*, 6481.
(27) Varnes, J.G.; Wacker, D.A.; Pinto, D.J.P.; Orwat, M.J.; Theroff, J.P.; Wells, B.; Galemmo, R.A.; Luettgen, J.M.; Knabb, R.M.; Bai, S.; He, K.; Lam, P.Y.S.; Wexler, R.R. *Bioorg. Med. Chem. Lett.* **2008**, *18*, 749.
(28) Fevig, J.M.; Cacciola, J.; Buriak, J.; Rossi, K.A.; Knabb, R.M.; Luettgen, J.M.; Wong, P.C.; Bai, S.A.; Wexler, R.R.; Lam, P.Y.S. *Bioorg. Med. Chem. Lett.* **2006**, *16*, 3755.
(29) Li, Y.L.; Fevig, J.M.; Cacciola, J.; Buriak, J.; Rossi, K.A.; Jona, J.; Knabb, R.M.; Luettgen, J.M.; Wong, P.C.; Bai, S.A.; Wexler, R.R.; Lam, P.Y.S. *Bioorg. Med. Chem. Lett.* **2006**, *16*, 5176.
(30) Pinto, D.J.P.; Orwat, M.J.; Quan, M.L.; Han, Q.; Galemmo, R.A.; Amparo, E.; Wells, B.; Ellis, C.D.; He, M.Y.; Alexander, R.S.; Rossi, K.A.; Smallwood, A.; Wong, P.C.; Luettgen, J.M.; Rendina, A.R.; Knabb, R.M.; Mersinger, L.; Kettner, C.; Bai, S.; He, K.; Wexler, R.R.; Lam, P.Y.S. *Bioorg. Med. Chem. Lett.* **2006**, *16*, 4141.
(31) Pinto, D.J.P.; Orwat, M.J.; Koch, S.; Rossi, K.A.; Alexander, R.S.; Smallwood, A.; Wong, P.C.; Rendina, A.R.; Luettgen, J.M.; Knabb, R.M.; He, K.; Xin, B.; Wexler, R.R.; Lam, P.Y.S. *J. Med. Chem.* **2007**, *50*, 5339.
(32) He, K.; Luettgen, J.M.; Zhang, D.; He, B.; Grace, J.E.; Xin, B.; Pinto, D.J.P.; Wong, P.C.; Knabb, R.M.; Lam, P.Y.S.; Wexler, R.R.; Grossman, S. *Eur. J. Drug Metab. Pharmacokinet.* **2011**, *36*, 129.
(33) Luettgen, J.M.; Knabb, R.M.; He, K.; Pinto, D.J.P.; Rendina, A.R. *J. Enzyme Inhib. Med. Chem.* **2011**, *26*, 514.
(34) Wong, P.C.; Pinto, D.J.P.; Zhang, D. *J. Thromb. Thrombolysis* **2011**, *31*, 478.
(35) ^{Pr}ELIQUIS (apixaban) Product Monograph [online]. http://www.pfizer.ca/en/our_products/products/monograph/313. Accessed Feb 6, 2012.
(36) Lassen, M.R.; Davidson, B.L.; Gallus, A.; Pineo, G.; Ansell, J.; Deitchman, D. *J. Thromb. Haemost.* **2007**, *5*, 2368.
(37) Lassen, M.R.; Raskob, G.E.; Gallus, A.; Pineo, G.; Chen, D.; Hornick, P.; ADVANCE-2 investigators, *Lancet* **2010**, *375*, 807.
(38) Lassen, M.R.; Gallus, A.; Raskob, G.E.; Pineo, G.; Chen, D.; Ramirez, L.M.; ADVANCE-3 Investigators, *N. Engl. J. Med.* **2010**, *363*, 2487.
(39) Lassen, M.R.; Raskob, G.E.; Gallus, A.; Pineo, G.; Chen, D.; Portman, R.J. *N. Engl. J. Med.* **2009**, *361*, 594.
(40) Buller, H.; Deitchman, D.; Prins, M.; Segers, A., Botticelli Investigators, Writing Committee, *J. Thromb. Haemost.* **2008**, *6*, 1313.
(41) Alexander, J.H.; Becker, R.C.; Bhatt, D.L.; Cools, F.; Crea, F.; Dellborg, M.; Fox, K.A.; Goodman, S.G.; Harrington, R.A.; Huber, K.; Husted, S.; Lewis, B.S.; Lopez-Sendon, J.; Mohan, P.; Montalescot, G.; Ruda, M.; Ruzyllo, W.; Verheugt, F.; Wallentin, L., APPRAISE Steering Committee and Investigators, *Circulation* **2009**, *119*, 2877.

(42) Alexander, J.H.; Lopes, R.D.; James, S.; Kilaru, R.; He, Y.; Mohan, P.; Bhatt, D.L.; Goodman, S.; Verheugt, F.W.; Flather, M.; Huber, K.; Liaw, D.; Husted, S.E.; Lopez-Sendon, J.; De Caterina, R.; Jansky, P.; Darius, H.; Vinereanu, D.; Cornel, J.H.; Cools, F.; Atar, D.; Leiva-Pons, J.L.; Keltai, M.; Ogawa, H.; Pais, P.; Parkhomenko, A.; Ruzyllo, W.; Diaz, R.; White, H.; Ruda, M.; Geraldes, M.; Lawrence, J.; Harrington, R.A.; Wallentin, L., APPRAISE-2 Investigators, *N. Engl. J. Med.* **2011**, *365*, 699.

(43) Connolly, S.J.; Eikelboom, J.; Joyner, C.; Diener, H.C.; Hart, R.; Golitsyn, S.; Flaker, G.; Avezum, A.; Hohnloser, S.H.; Diaz, R.; Talajic, M.; Zhu, J.; Pais, P.; Budaj, A.; Parkhomenko, A.; Jansky, P.; Commerford, P.; Tan, R.S.; Sim, K.H.; Lewis, B.S.; Van Mieghem, W.; Lip, G.Y.; Kim, J.H.; Lanas-Zanetti, F.; Gonzalez-Hermosillo, A.; Dans, A.L.; Munawar, M.; O'Donnell, M.; Lawrence, J.; Lewis, G.; Afzal, R.; Yusuf, S., AVERROES Steering Committee and Investigators, *N. Engl. J. Med.* **2011**, *364*, 806.

CHAPTER TEN

AMPK Activation in Health and Disease

Iyassu K. Sebhat, Robert W. Myers
Departments of Medicinal Chemistry and Diabetes and Endocrinology, Merck Research Laboratories, Rahway, New Jersey, USA

Contents

1. INTRODUCTION

The maintenance of cellular energy levels is a fundamental biological process. The primary mechanism for detecting and responding to changes in energy state is through the 5′ adenosine monophosphate-activated protein kinase (AMPK) pathway. The enzyme is activated in response to exercise and following treatment with metformin, the first-line therapy for type 2 diabetes mellitus (T2DM). As might be expected from its role, AMPK effects are pleiotropic, impacting multiple metabolic pathways in order to spare and/or generate 5′ adenosine triphosphate (ATP).

The goal of this review is to provide a broad overview of reported AMPK activators and to update the reader on recent advances in the elucidation of AMPK function. Given the breadth of cellular processes impacted by AMPK, the authors have chosen to focus on impacts to metabolic function and potential indications around the treatment of metabolic syndrome and T2DM.

Annual Reports in Medicinal Chemistry, Volume 47
ISSN 0065-7743
http://dx.doi.org/10.1016/B978-0-12-396492-2.00010-2

2. AMPK—ENZYME STRUCTURE AND FUNCTION

AMPK is a Ser/Thr protein kinase consisting of one α-catalytic (2 isoforms), one β-"scaffold" (2 isoforms), and one γ-regulatory (3 isoforms) subunit.[1] Excluding splice variants, 12 distinct AMPK complexes exist. The detailed molecular structure of AMPK has been the subject of considerable investigation. While the field awaits higher resolution structures, available crystallographic studies provide considerable detail on the overall structure and regulation of the complex.[2,3]

AMPK is a "stress" kinase that spares/replenishes depleted cellular energy reserves. Its activity is principally controlled by α-subunit Thr172 phosphorylation. Four Ser/Thr kinases that phosphorylate AMPK (AMPKK) have been identified in mammalian cells: liver kinase B1 (LKB1), calcium/calmodulin-dependent protein kinase kinase β (CAMKKβ), ataxia telangiectasia mutated, and transforming growth factor β-activated kinase 1.[4]

AMPK was so-named based on the ability of 5′ adenosine monophosphate (AMP) to stimulate its activity. Recent studies have established that AMPK is more broadly an "adenylate charge regulated" kinase.[2,3,5] AMP binding to the γ-subunit activates AMPK in three distinct ways: (1) increasing the accessibility of Thr172 for phosphorylation (generating pThr172); (2) inducing a conformation that decreases pThr172 dephosphorylation, and (3) allosterically activating phosphorylated AMPK (pAMPK) up to 2.5-fold. 5′ Adenosine diphosphate (ADP) binding has similar effects, but it cannot intrinsically activate pAMPK. By contrast, Mg^{2+}-ATP binding leads to inactivation by increasing the accessibility of pThr172 to protein phosphatases and by competing for AMP and ADP binding.

AMPK activation triggers phosphorylation of downstream targets, leading to activation or inhibition of target function. Functional consequences can either be immediate or delayed through modulation of transcription factors.[6] On balance, the overriding effect of AMPK action is to inhibit ATP-requiring processes and activate ATP-producing, catabolic processes.

3. MAJOR AMPK-MEDIATED EFFECTS ON LIPID AND CARBOHYDRATE METABOLISM

AMPK activation has profound effects on lipid metabolism.[1,7–9] The enzyme is responsible for phosphorylation and inhibition of both acetyl CoA carboxylase (ACC) and HMG-CoA reductase (the target of the statin class of hypercholesterolemia drugs).[10] ACC isoforms 1 and 2 are key enzymes regulating lipid metabolism. AMPK-mediated phosphorylation of ACC inhibits fatty acid synthesis and elongation and increases fat oxidation.[11] AMPK also phosphorylates several transcription factors, reducing lipogenic gene expression, for example, sterol regulatory element-binding protein-1c and -2 and carbohydrate responsive element-binding protein.[12,13] Additional data demonstrate that AMPK modulates the synthesis and mobilization of triglycerides (TGs) and increases fatty acid transport into cells.[1,9,14,15] In summary, the net effect of AMPK activation is to decrease lipid synthesis and storage and increase lipid utilization for energy production.

AMPK activation also has major effects on carbohydrate metabolism. Among the most important of these is the AMPK-induced increase in skeletal muscle glucose uptake, which is observed following its activation by exercise and 5-amino-1-β-D-ribofuranosyl-imidazole-4-carboxamide (AICAR).[16–18] AMPK activation increases glucose uptake in part by increasing the plasma membrane content of glucose transporter 4 (Glut4).[19] This is triggered by phosphorylation of key proteins including tre-2/USP6, BUB2, cdc16 domain family member 1 (TBC1D1), and Akt substrate of 160 kDa (AS160), leading to displacement of Glut4 from its intracellular docking sites.[20] Insulin-stimulated Glut4 translocation is a critical component of the mechanism by which insulin acts to reduce postprandial plasma hyperglycemia.[19] In this manner, AMPK activation can be viewed as an insulin mimic, which has important therapeutic consequences.[21–24] The resulting increased tissue glucose is either stored as glycogen or utilized for glycolysis; both of these processes are stimulated by AMPK activation.[25,26]

During fasting, increased hepatic glucose production plays a key role in maintaining whole-body glucose homeostasis. Early studies using the AMPK activators AICAR and metformin demonstrated inhibition of hepatic gluconeogenesis, but these effects are likely not solely AMPK mediated.[27] AICAR is a potent inhibitor of the essential gluconeogenic enzyme fructose-1,6-bisphosphatase (FBPase). Moreover, inhibition of gluconeogenesis by metformin is manifest in livers totally lacking AMPK.[28,29] However, hepatic overexpression of a constitutively active AMPK α2-subunit led to plasma glucose lowering in several mouse models and decreased key gluconeogenic gene expression.[30] Conversely, hepatic deletion of the AMPK α2-subunit results in fasting hyperglycemia.[31] AMPK activation represses gluconeogenic gene expression via phosphorylation and cytosolic sequestration of CREB-regulated transcriptional coactivator 2 (CRTC2).[32]

4. THERAPEUTIC POTENTIAL OF AMPK ACTIVATION

A compelling case can be made for the therapeutic potential of AMPK activation for a variety of diseases. AMPK-mediated alterations in lipid metabolism should lead to reductions in whole-body lipid content. Ectopic (nonadipose) lipid deposits, particularly those in liver, muscle, and beta cells, lead to increased inflammation, lipotoxicity, and insulin resistance and, as such, contribute significantly to metabolic diseases.[33–35] Numerous AMPK activators have been shown to improve hepatosteatosis, including metformin and AICAR.[8,29,36,37] AMPK activation also modulates cardiomyocyte, smooth muscle cell, and endothelial cell function. Combined lipid effects are anticipated to improve plasma dyslipidemia, pathological lipoprotein metabolism, and atherosclerosis, thus providing major cardiovascular benefits.[38–42]

AMPK activation modulates adipocyte metabolism.[9] The enzyme inhibits and activates adipogenesis in white and brown fat cells, respectively, and the mechanism is of interest for antiobesity therapy.[14,43,44] In fact, AMPK appears to be a key mediator of the positive metabolic benefits of calorie restriction, which includes increased lifespan.[45,46] Given the proinflammatory nature of white adipose tissue, AMPK activation has potential applications in numerous inflammatory diseases.[47–49]

A major therapeutic focus for the application of AMPK activators is insulin resistance and T2DM.[50] The approved antihyperglycemic agents metformin, rosiglitazone, and pioglitazone, and the insulin sensitizing hormone adiponectin, activate AMPK both *in vitro* and *in vivo*.[51] T2DM is characterized by excessive hepatic glucose production, extreme insulin resistance resulting in reduced peripheral glucose uptake and utilization, and insufficient insulin secretion to reverse the hyperglycemia. Redressing these imbalances are among the most prominent effects of exercise contributing to glucose homeostasis. Exercise activates AMPK, and conversely, chemical activation of AMPK may mimic many of the effects of exercise, which has profound benefits in T2DM patients.[18] In addition to the direct reduction in plasma glucose, AMPK activation should also improve insulin sensitivity over time by reducing ectopic fat depots (see above).

Table 10.1 Some additional therapeutic indications for AMPK activators

Indication	**References**
Hypertension	52,53
Heart failure and ischemic injury	38–42, 58–60
Kidney disease	61
Osteoporosis	62
Cancer	54

Finally, AMPK activation is of interest for a variety of other pathological conditions. One such area is the treatment of hypertension. AMPK phosphorylates and activates endothelial nitric oxide synthase and also decreases vascular smooth muscle cell contractility.[52,53] An additional area to highlight is the treatment of various cancers, which has become a major focus of AMPK research over the past several years.[54] This is based largely on the tumor suppressor function of LKB1 and the well-established modulation of mammalian target of rapamycin complex 1 (mTORC1) and the p53 tumor suppressor by AMPK.[12,55–57] Other indications are listed in Table 10.1.

Importantly, the number of adverse effects as a consequence of AMPK activation is anticipated to be relatively low. Data convincingly demonstrate that activation of AMPK in the brain, by either chemical or genetic means, increases food intake.[63,64] The hormones leptin and ghrelin control food intake at least in part by modulating hypothalamic AMPK activity. Other functions of AMPK in brain, including its impact on Alzheimer's disease, are the subject of current investigation.[65,66] Another potential adverse effect involves the heart. Mutations in the $\gamma 2$-subunit are found in familial Wolff–Parkinson–White syndrome patients (5′-AMP-activated protein kinase subunit $\gamma 2$ (PRKAG2) cardiomyopathy).[67,68] Although somewhat controversial, these mutations appear to result in (at least intrinsic) activation of AMPK, presumably by ablating the binding of inhibitory ATP. While it remains to be established which of the clinical manifestations of PRKAG2 cardiomyopathy have developmental origins, cardiac safety is likely to be a major consideration in the development of compounds that activate AMPK in the heart.

5. PHARMACOLOGICAL AMPK ACTIVATORS

The central role that AMPK plays in glucose and lipid metabolism has aroused interest in developing pharmacological activators. This interest has been fueled by the possibility (discussed above) that AMPK activation may mediate some of the beneficial effects of exercise and calorie restriction and that it may contribute to the efficacies of certain currently used antidiabetic agents.[69] Despite this growing attention and an increase in the number of reported AMPK activators, the chemical space remains relatively narrow.

The endogenous function of AMPK as a stress-activated kinase needs to be considered when assessing reports in the area. Compounds that exhibit cytotoxicity or interact with pathways associated with cellular stressors (e.g., hypoxia, oxidative stress, etc.) may activate the enzyme indirectly through elevation of cellular AMP or calcium levels. This complicates interpretation of the mechanism underlying the effects of a particular pharmacological agent.

This section will provide a brief overview of the more important and widely reported indirect AMPK activators and a closer focus on disclosed direct AMPK activators.

Since a number of different assays are used to assess AMPK activity, structures with data refer to the following: (a) activation of unspecified AMPK complex, (b) activation of recombinant $\alpha1\beta1\gamma1$ AMPK complex, (c) activation of recombinant $\alpha1\beta1\gamma1$ and $\alpha2\beta1\gamma1$ AMPK complex, and (d) inhibition of cellular fatty acid synthesis.

5.1. Indirect AMPK activators

Metformin (**1**) is the first line of therapy for T2DM in most countries and it has been shown to be an LKB1-dependent indirect activator of AMPK both *in vitro* and *in vivo*.[70,71] The compound inhibits complex I of the mitochondrial respiratory chain with subsequent reduction in cellular energy charge.[29,72]

The degree to which AMPK activation contributes to the hypoglycemic efficacy of metformin is unknown. Recent data provide strong evidence that at least some effects on glucose homeostasis are AMPK-independent.[28]

The thiazolidinediones (TZDs), exemplified by rosiglitazone (**2**) and pioglitazone (**3**), are purported to have dual mechanisms of action.[73] The compounds exert transcriptional effects following PPARγ activation which result in adipogenesis, adiponectin release and reduced hepatic gluconeogenesis.[51] In addition, *in vitro* studies and studies in mice deficient in adiponectin suggest that the compounds exert adiponectin-independent effects on skeletal muscle glucose uptake coincident with increases in AMPK and ACC phosphorylation. While the mechanism for AMPK activation has yet to be fully determined, similar to metformin, it may involve inhibition of respiratory complex I.[74]

There are numerous reports of compounds derived from natural products that activate AMPK. Of these, the alkaloid berberine (**4**) and the polyphenol resveratrol (**5**) have been most widely investigated. The mechanism of AMPK activation for both is understood to be indirect. Berberine has been shown to inhibit mitochondrial respiratory complex I in much the same fashion as metformin and the TZDs.[75] Resveratrol may reduce ATP synthesis via inhibition of F_1-ATPase.[76]

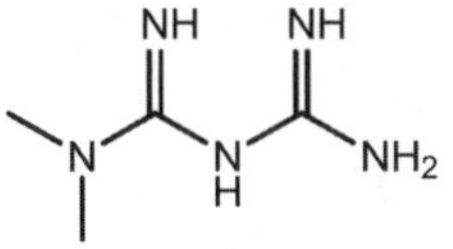

1

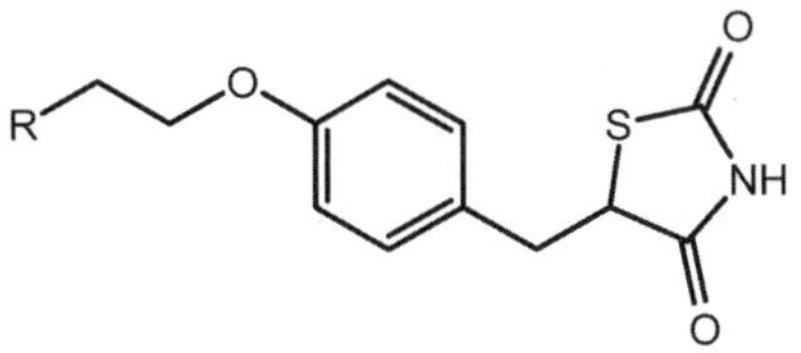

2: R =

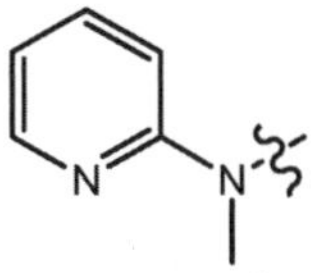

3: R =

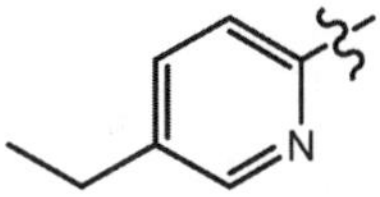

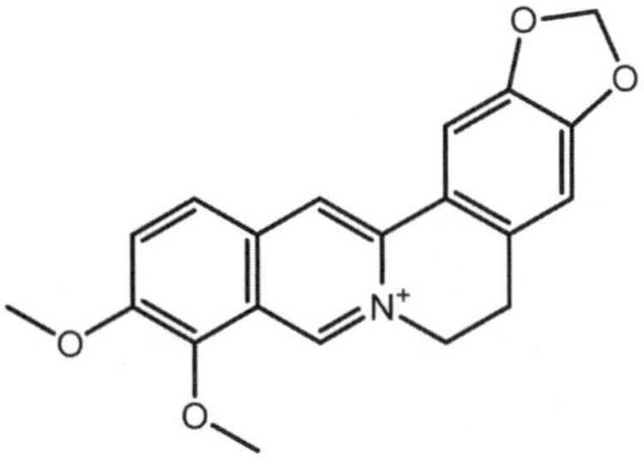

4

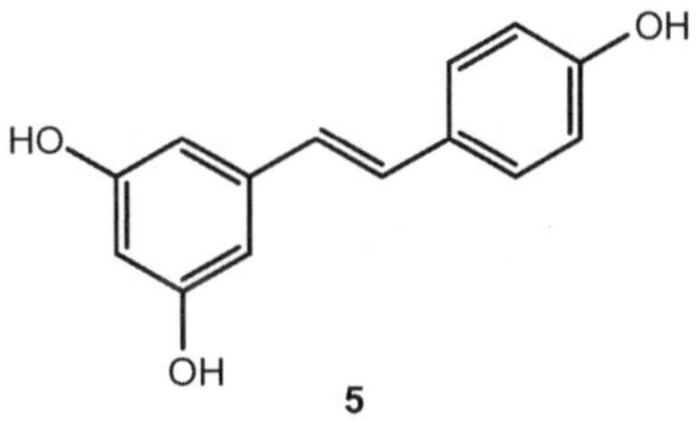

5

5.2. Direct AMPK activators

5.2.1 Synthetic nucleoside AMPK activators

Analogs of AMP have found wide use in the field as tools to investigate the effects of AMPK activation. AICAR (**6**) is a cell permeable nucleoside that undergoes intracellular phosphorylation to generate ZMP (**7**).[77] In various dysmetabolic rodent models, treatment with AICAR results in improvements in metabolic parameters.[78] Limitations of the compounds focus on their short half-lives and extensive off-target activities. In particular, ZMP inhibits FBPase and activates glycogen phosphorylase (GPPase)

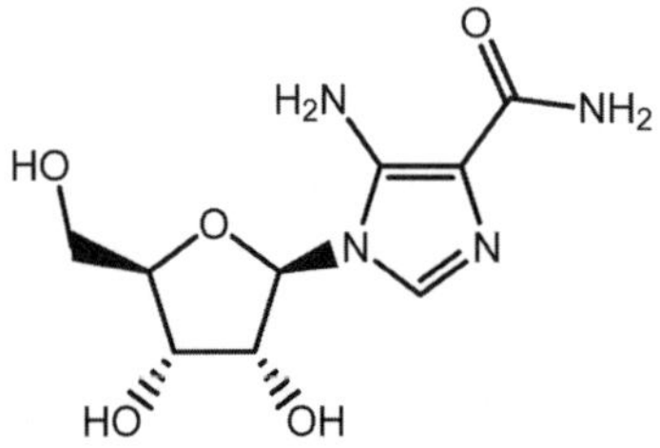

6: IC_{50} ~ 100 μM[d]

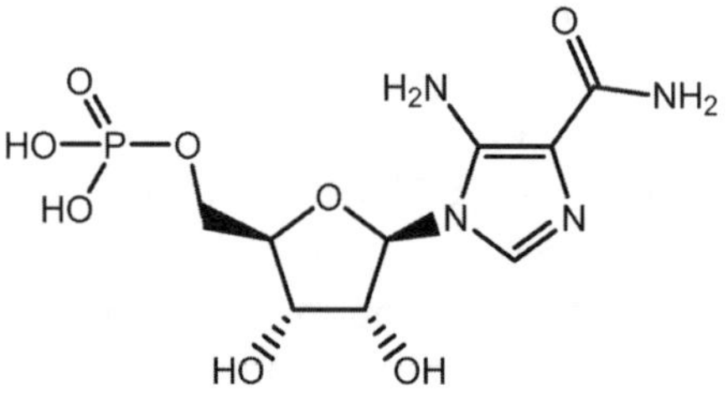

7: EC_{50} = 164 μM[a]

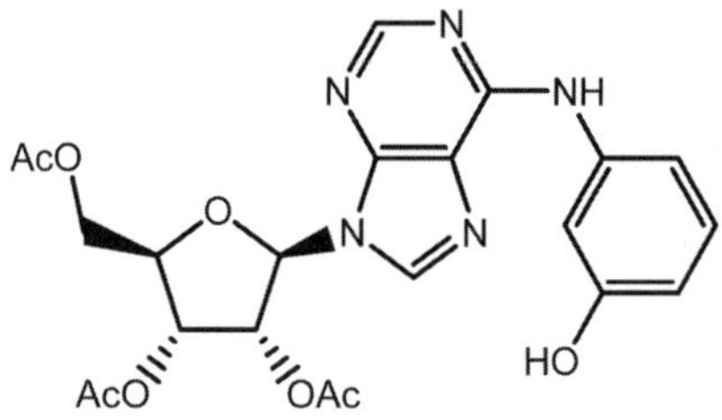

8

complicating interpretation of any AMPK-mediated hypoglycemic efficacy observed following AICAR administration. While poor physical properties and selectivity issues limit nucleosides as particularly promising leads for drug development, some efforts have been invested to identify new compounds. The most recently reported analog is WS010117 (**8**), which stimulates AMPK activity in HepG2 cells with consequent reductions in *de novo* lipogenesis and elevation of fatty acid oxidation. *In vivo*, the compound ($\leq$18 mg kg^{-1} QD PO, 2–8 weeks) reduced high-fat diet-induced plasma and liver lipid accumulation in hamsters.[79]

5.2.2 Nonnucleoside AMPK activators

Reports of small molecule direct AMPK activators that are not closely related to nucleosides comprise the largest area of growth in the past few years. Unfortunately, many of these reports emerge from patent applications where descriptions of data are limited. We have included generic structures and available data of disclosed series. A small number of compounds have been more widely published and we have included a broader discussion of their reported *in vitro* and *in vivo* properties.

Compounds containing a pyridone core form the largest structural class in the field. The first reports described the structure and activity of a number of AMPK activators bearing a thienopyridone core.[80,81] High-throughput screening of 700,000 compounds followed by a hit-to-lead effort identified A-769662 (**9**). Since its disclosure, the compound has become an important tool in efforts to further delineate the consequences of AMPK activation.

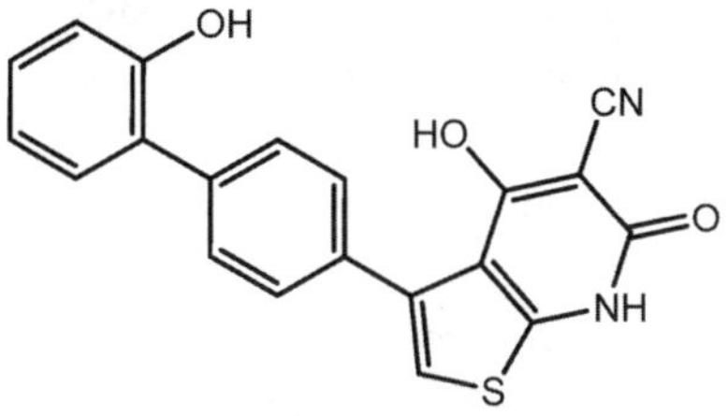

9: EC_{50} = 0.8 μM[a]

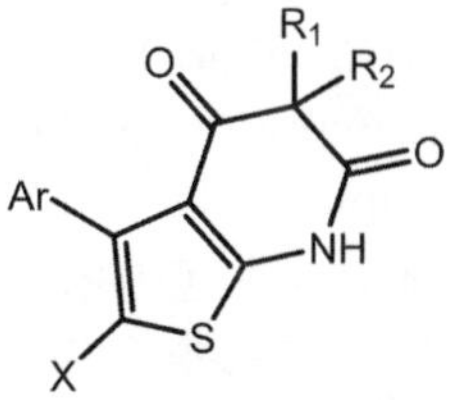

10: >75% activation vs. AMP @30 μM[a]
Ar=2-fluoro-4-methoxyphenyl;
X=Cl; R_1=Me; R_2=Ph

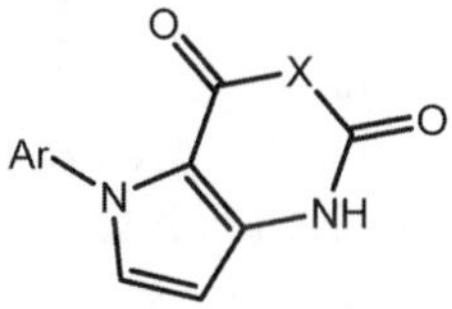

11: EC_{50} = 1.3 μM[c]
Ar=[1,1′-biphenyl]-2-ol;
X = NH

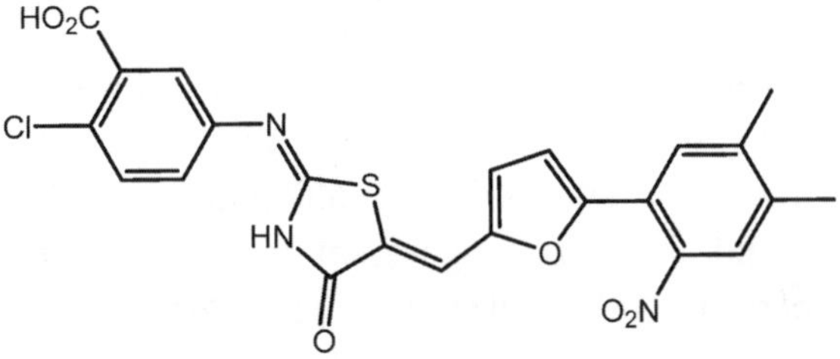

12: EC_{50} = 0.3 μM[b]

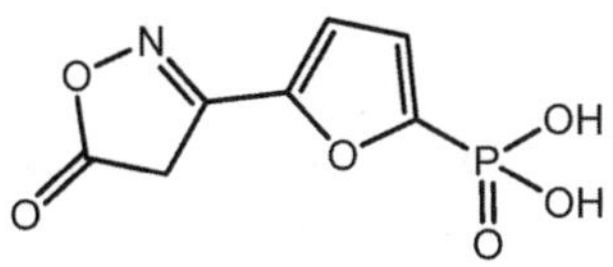

11: EC_{50} = 6.3 nM[a]

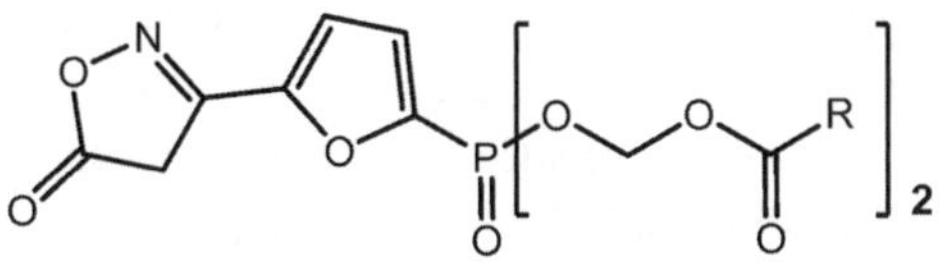

12: IC_{50} = 20 nM[d]
R=iso-propyl

A-769662 binds to a site on the enzyme distinct from the γ-subunit CBS domains and selectively activates β1-containing AMPK complexes (vs. β2-containing heterotrimers).[1] Similar to AMP, the compound is an allosteric activator of the enzyme (EC_{50} = 0.7–0.8 μM) and additionally protects pAMPK from dephosphorylation. While off-target activities have been documented for the compound,[82] A-769662 is considerably more selective than the nucleoside analogs with minimal effect on GPPase, FBPase, and a number of other receptor, ion channel, and kinase targets.

Incubation of rat hepatocytes with A-769662 dose-dependently increased ACC phosphorylation in an AMPKK (LKB1 or CaMKKβ) and AMPK-dependent manner with consequent inhibition of lipogenesis (IC_{50} = 3.2 μM) and without effecting intracellular adenine nucleotide levels. The compound also inhibits 3T3-L1 adipocyte differentiation.[43] At high concentrations, A-769662 induced mouse muscle glucose uptake in an AMPK-independent (P13-kinase-dependent) manner, highlighting an additional off-target activity and suggesting that activation by the compound is insufficient for AMPK-mediated glucose uptake.[83]

Due to poor oral pharmacokinetics (7% bioavailability in rats), *in vivo* study of A-769662 was conducted using IP dosing. Single acute doses (30 mg kg^{-1}) in rats were sufficient to induce a 33% reduction in hepatic malonyl CoA levels. While fatty acid oxidation was not measured directly, the compound did cause a reduction in respiratory exchange ratio. Acute effects on hepatic gluconeogenic and lipogenic gene expression were not evident, but reductions in PEPCK (37%), G6Pase (63%), and FAS (31%) mRNA expression were observed after 5 days of dosing.

Similar to effects seen with AICAR and metformin, A-769662 (30 mg kg^{-1} IP BID, 5 days) reduced fed glucose levels by 30–40% with concomitant reductions in TGs in plasma (63%) and liver (48%) in *ob/ob* mice. The compound also caused significant reductions in body weight (9% after 14 days of treatment). While the body weight effects may be mechanism based, they do introduce complications in understanding the mechanisms underlying glucose and TG reductions.

Finally, recent studies suggest that AMPK activation by pretreatment with A-769662 protects the heart from ischemia–reperfusion injury in mice[84] and reduces infarct size in isolated, perfused rat hearts.[85]

A number of additional pyridones have been disclosed in patent applications. This includes a series of substituted thienopyridones (e.g., **10**) and a novel series of pyrrolopyridones (e.g., **11**). While data are limited, the compounds appear to have similar *in vitro* potency in activating the $\alpha1\beta1\gamma1$ isoform of AMPK.[86]

Thiazolidinone-derived PT1 (**12**) was discovered in a screen of 3600 compounds.[87] The compound likely binds to and activates the catalytic α-subunit directly via suppression of autoinhibition ($EC_{50}=8\ \mu M$, ~eightfold activation above baseline). Activation of the intact $\alpha1\beta1\gamma1$ complex was also documented ($EC_{50}=0.3\ \mu M$, 1.5-fold maximal activation @ 5 μM). In cell cultures, PT1 caused a CaMKKβ-dependent increase in AMPK and ACC phosphorylation in L6 myotubes and HeLa cells without affecting the AMP:ATP ratio.

A screen of a ~1200-member library of AMP mimetics identified a hydroxy-isoxazole substituted furan phosphonic acid (**11**).[88] The compound is a full activator of human AMPK ($EC_{50}=6.3$ nM) and a partial activator of the rat enzyme ($EC_{50}=21$ nM, 51% of the maximal activation induced by AMP) with no activity against GPPase and FBPase and good selectivity over a panel of 64 other targets. The compound is charged at physiologic pH limiting permeability. When it was administered as an esterase-sensitive prodrug (e.g., **12**—itself unable to activate AMPK), the compound caused a significant reduction in *de novo* lipogenesis in plated rat hepatocytes ($IC_{50}=20$ nM) concomitant with increased ACC phosphorylation. A single acute dose of compound **12** (30 mg kg^{-1} IP) was sufficient to reduce hepatic *de novo* lipogenesis by 78% over the course of 1 h in mice. Much of the SAR described focuses on prodrug design and established that sterically unencumbered phosphonate esters and carbonates maintained similar potencies in *in vitro* assays suggesting similarly rapid conversion to active compound.

5.2.3 Other compounds

Some SAR is apparent in a series of benzimidazole compounds reported in five patent applications. Data suggest that replacement of a sulfur substituent in the 2-position of the benzimidazole with an oxygen significantly increases potency; with the latter (e.g., **13**) providing EC_{50}s in the low nanomolar range.[89]

The remaining structural series have been claimed in patent applications with more limited data. This includes a series of related carboxamide, sulfonamide, and amine compounds (e.g., **14**) with potencies ranging as low as $EC_{50}<0.1\ \mu M$,[90] a series of oxindoles (e.g., **15** $EC_{50}=0.66\ \mu M$),[91] and a series of tetrahydroquinolines (e.g., **16** $EC_{50}=1.24\ \mu M$).[92]

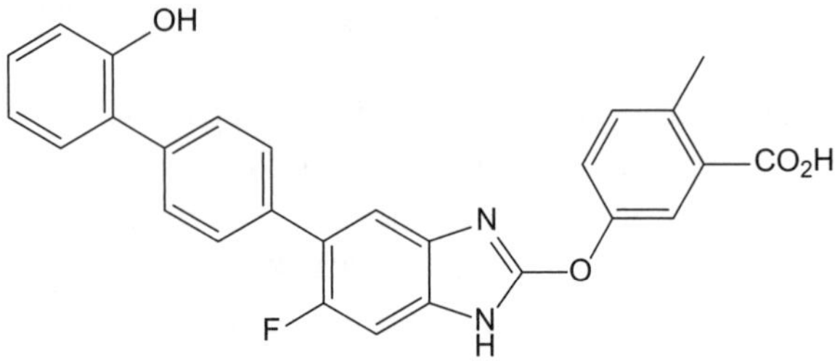

13: EC_{50} = 1 nM (Max Act = 187% vs. AMP)[a]

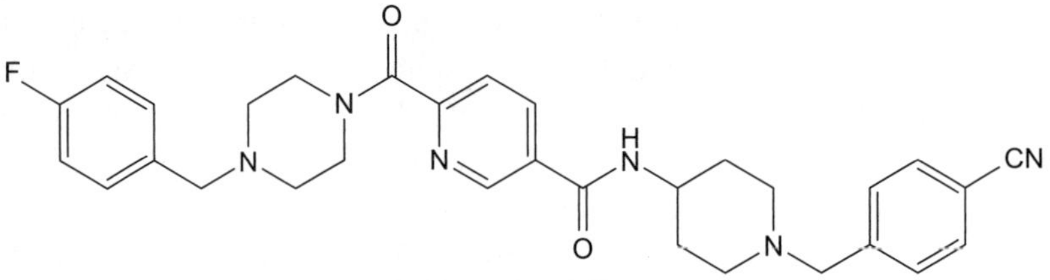

14: EC_{50} < 0.5 μM[a]

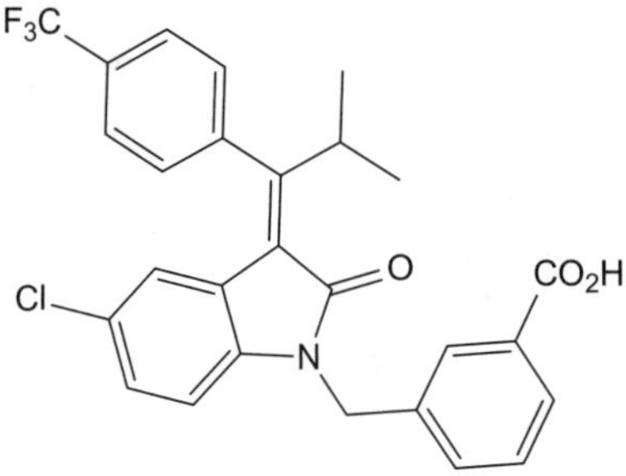

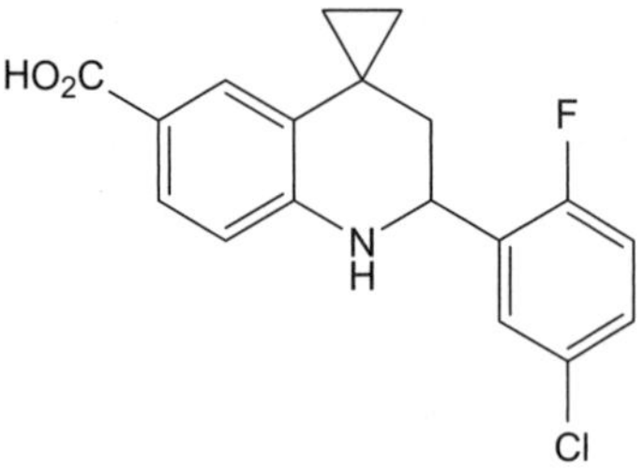

15: EC_{50} = 0.66 μM[c]

16: EC_{50} = 1.24 μM[c]

6. CONCLUSION

Five years ago, the development of chemotherapeutic AMPK activators would have been considered a remote possibility due to significant complexities emanating from the target's complex structure and pleiotropic function. This situation was compounded by the paucity of chemical matter capable of interacting with the enzyme. Increasing knowledge regarding the structure and mechanism of AMPK, coupled with the identification of new chemical matter in recent years, suggests that AMPK activation is indeed druggable. A significant remaining hurdle is to better understand the potential mechanism-based adverse effects of AMPK activation.

REFERENCES

(1) Steinberg, G.R.; Kemp, B.E. *Physiol. Rev.* **2009**, *89*, 1025.
(2) Scott, J.W.; Oakhill, J.S.; van Denderen, B.J. *Front. Biosci.* **2009**, *14*, 596.
(3) Carling, D.; Mayer, F.V.; Sanders, M.J.; Gamblin, S.J. *Nat. Chem. Biol.* **2011**, *7*, 512.
(4) Carling, D.; Sanders, M.J.; Woods, A. *Int. J. Obes.* **2008**, *32*, S55.
(5) Oakhill, J.S.; Scott, J.W.; Kemp, B.E. *Trends Endocrinol. Metab.* **2012**, *23*, 125.
(6) McGee, S.L.; Hargreaves, M. *Clin. Sci.* **2010**, *118*, 507.
(7) Thomson, D.M.; Winder, W.W. *Acta Physiol.* **2009**, *196*, 147.
(8) Viollet, B.; Guigas, B.; Leclerc, J.; Hébrard, S.; Lantier, L.; Mounier, R.; Andreelli, F.; Foretz, M. *Acta Physiol.* **2009**, *196*, 81.
(9) Daval, M.; Foufelle, F.; Ferré, P. *J. Physiol.* **2006**, *574*, 55–62.
(10) Carling, D.; Clarke, P.R.; Zammit, V.A.; Hardie, D.G. *Eur. J. Biochem.* **1989**, *186*, 129.
(11) Saggerson, D. *Annu. Rev. Nutr.* **2008**, *28*, 253.
(12) Mihaylova, M.M.; Shaw, R.J. *Nat. Cell Biol.* **2011**, *13*, 1016.
(13) Uyeda, K.; Repa, J.J. *Cell Metab.* **2006**, *4*, 107.
(14) Ahmadian, M.; Abbott, M.J.; Tang, T.; Hudak, C.S.; Kim, Y.; Bruss, M.; Hellerstein, M.K.; Lee, H.Y.; Samuel, V.T.; Shulman, G.I.; Wang, Y.; Duncan, R.E.; Kang, C.; Sul, H.S. *Cell Metab.* **2011**, *13*, 739.
(15) Nickerson, J.G.; Momken, I.; Benton, C.R.; Lally, J.; Holloway, G.P.; Han, X.X.; Glatz, J.F.; Chabowski, A.; Luiken, J.J.; Bonen, A. *Appl. Physiol. Nutr. Metab.* **2007**, *32*, 865.
(16) Hardie, D.G. *Proc. Nutr. Soc.* **2011**, *70*, 92.
(17) Jensen, T.E.; Wojtaszewski, J.F.; Richter, E.A. *Acta Physiol.* **2009**, *196*, 155.
(18) Richter, E.A.; Ruderman, N.B. *Biochem. J.* **2009**, *418*, 261.
(19) Leney, S.E.; Tavaré, J.M. *J. Endocrinol.* **2009**, *203*, 1.
(20) Chen, S.; Synowsky, S.; Tinti, M.; MacKintosh, C. *Trends Endocrinol. Metab.* **2011**, *22*, 429.
(21) Zhang, B.B.; Zhou, G.; Li, C. *Cell Metab.* **2009**, *9*, 407.
(22) Hegarty, B.D.; Turner, N.; Cooney, G.J.; Kraegen, E.W. *Acta Physiol.* **2009**, *196*, 129.
(23) Yamada, E.; Lee, T.W.; Pessin, J.E.; Bastie, C.C. *Future Med. Chem.* **2010**, *2*, 1785.
(24) Saha, A.K.; Xu, X.J.; Balon, T.W.; Brandon, A.; Kraegen, E.W.; Ruderman, N.B. *Cell Cycle* **2011**, *10*, 3447.
(25) Hunter, R.W.; Treebak, J.T.; Wojtaszewski, J.F.; Sakamoto, K. *Diabetes* **2011**, *60*, 766.
(26) Marsin, A.S.; Bouzin, C.; Bertrand, L.; Hue, L. *J. Biol. Chem.* **2002**, *277*, 30778.
(27) Viollet, B.; Athea, Y.; Mounier, R.; Guigas, B.; Zarrinpashneh, E.; Horman, S.; Lantier, L.; Hebrard, S.; Devin-Leclerc, J.; Beauloye, C.; Foretz, M.; Andreelli, F.; Ventura-Clapier, R.; Bertrand, L. *Front. Biosci.* **2009**, *14*, 19.
(28) Foretz, M.; Hébrard, S.; Leclerc, J.; Zarrinpashneh, E.; Soty, M.; Mithieux, G.; Sakamoto, K.; Andreelli, F.; Viollet, B. *J. Clin. Invest.* **2010**, *120*, 2355.
(29) Viollet, B.; Guigas, B.; Sanz Garcia, G.N.; Leclerc, J.; Foretz, M.; Andreelli, F. *Clin. Sci.* **2012**, *122*, 253.
(30) Foretz, M.; Ancellin, N.; Andreelli, F.; Saintillan, Y.; Grondin, P.; Kahn, A.; Thorens, B.; Vaulont, S.; Viollet, B. *Diabetes* **2005**, *54*, 1331.
(31) Andreelli, F.; Foretz, M.; Knauf, C.; Cani, P.D.; Perrin, C.; Iglesias, M.A.; Pillot, B.; Bado, A.; Tronche, F.; Mithieux, G.; Vaulont, S.; Burcelin, R.; Viollet, B. *Endocrinology* **2006**, *147*, 2432.
(32) Koo, S.H.; Flechner, L.; Qi, L.; Zhang, X.; Screaton, R.A.; Jeffries, S.; Hedrick, S.; Xu, W.; Boussouar, F.; Brindle, P.; Takemori, H.; Montminy, M. *Nature* **2005**, *437*, 1109.
(33) Musso, G.; Gambino, R.; Cassader, M. *Annu. Rev. Med.* **2010**, *61*, 375.
(34) Unger, R.H.; Clark, G.O.; Scherer, P.E.; Orci, L. *Biochim. Biophys. Acta* **2010**, *1801*, 209.

(35) Hill, M.J.; Metcalfe, D.; McTernan, P.G. *Clin. Sci.* **2009**, *116*, 113.
(36) Boyle, J.G.; Salt, I.P.; McKay, G.A. *Diabet. Med.* **2010**, *27*, 1097.
(37) Guigas, B.; Sakamoto, K.; Taleux, N.; Reyna, S.M.; Musi, N.; Viollet, B.; Hue, L. *IUBMB Life* **2009**, *61*, 18.
(38) Wong, A.K.; Howie, J.; Petrie, J.R.; Lang, C.C. *Clin. Sci.* **2009**, *116*, 607.
(39) Ewart, M.A.; Kennedy, S. *Pharmacol. Ther.* **2011**, *131*, 242.
(40) Zou, M.H.; Wu, Y. *Clin. Exp. Pharmacol. Physiol.* **2008**, *35*, 535.
(41) Kim, A.S.; Miller, E.J.; Young, L.H. *Acta Physiol.* **2009**, *196*, 37.
(42) Kim, M.; Tian, R. *J. Mol. Cell. Cardiol.* **2011**, *51*, 548.
(43) Zhou, Y.; Wang, D.; Zhu, Q.; Gao, X.; Yang, S.; Xu, A.; Wu, D. *Biol. Pharm. Bull.* **2009**, *32*, 993.
(44) Kola, B.; Grossman, A.B.; Korbonits, M. *Front. Horm. Res.* **2008**, *36*, 198.
(45) Cantó, C.; Auwerx, J. *Physiology* **2011**, *26*, 214.
(46) Mair, W.; Morantte, I.; Rodrigues, A.P.C.; Manning, G.; Montminy, M.; Shaw, R.J.; Dillin, A. *Nature* **2011**, *470*, 404.
(47) Salminen, A.; Hyttinen, J.M.; Kaarniranta, K. *J. Mol. Med.* **2011**, *89*, 667.
(48) Harford, K.A.; Reynolds, C.M.; McGillicuddy, F.C.; Roche, H.M. *Proc. Nutr. Soc.* **2011**, *70*, 408.
(49) Hardy, O.T.; Czech, M.P.; Corvera, S. *Curr. Opin. Endocrinol. Diabetes Obes.* **2012**, *19*, 81.
(50) Viollet, B.; Lantier, L.; Devin-Leclerc, J.; Hebrard, S.; Amouyal, C.; Mounier, R.; Foretz, M.; Andreelli, F. *Front. Biosci.* **2009**, *14*, 3380.
(51) Gu, W.; Li, Y. *BioDrugs* **2012**, *26*, 1.
(52) Fisslthaler, B.; Fleming, I. *Circ. Res.* **2009**, *105*, 114.
(53) Wang, S.; Liang, B.; Viollet, B.; Zou, M.H. *Hypertension* **2011**, *57*, 1010.
(54) Luo, Z.; Zang, M.; Guo, W. *Future Oncol.* **2010**, *6*, 457.
(55) Sebbagh, M.; Olschwang, S.; Santoni, M.J.; Borg, J.P. *Fam. Cancer* **2011**, *10*, 415.
(56) Inoki, K.; Kim, J.; Guan, K.L. *Annu. Rev. Pharmacol. Toxicol.* **2012**, *52*, 381.
(57) Alers, S.; Löffler, A.S.; Wesselborg, S.; Stork, B. *Mol. Cell. Biol.* **2012**, *32*, 2.
(58) Beauloye, C.; Bertrand, L.; Horman, S.; Hue, L. *Cardiovasc. Res.* **2011**, *90*, 224.
(59) Arad, M.; Seidman, C.E.; Seidman, J.G. *Circ. Res.* **2007**, *100*, 474.
(60) Dolinsky, V.W.; Dyck, J.R. *Am. J. Physiol. Heart Circ. Physiol.* **2006**, *291*, H2557.
(61) Hallows, K.R.; Mount, P.F.; Pastor-Soler, N.M.; Power, D.A. *Am. J. Physiol. Renal Physiol.* **2010**, *298*, F1067.
(62) Jeyabalan, J.; Shah, M.; Viollet, B.; Chenu, C. *J. Endocrinol.* **2012**, *212*, 277.
(63) Minokoshi, Y.; Shiuchi, T.; Lee, S.; Suzuki, A.; Okamoto, S. *Nutrition* **2008**, *24*, 786.
(64) Kola, B. *J. Neuroendocrinol.* **2008**, *20*, 942.
(65) Amato, S.; Man, H.Y. *Cell Cycle* **2011**, *10*, 3452.
(66) Salminen, A.; Kaarniranta, K.; Haapasalo, A.; Soininen, H.; Hiltunen, M. *J. Neurochem.* **2011**, *118*, 460.
(67) Luptak, I.; Shen, M.; He, H.; Hirshman, M.F.; Musi, N.; Goodyear, L.J.; Yan, J.; Wakimoto, H.; Morita, H.; Arad, M.; Seidman, C.E.; Seidman, J.G.; Ingwall, J.S.; Balschi, J.A.; Tian, R. *J. Clin. Invest.* **2007**, *117*, 1432.
(68) Moffat, C.; Ellen Harper, M. *IUBMB Life* **2010**, *62*, 739.
(69) Fryer, L.G.; Parbu-Patel, A.; Carling, D. *J. Biol. Chem.* **2002**, *277*, 25226.
(70) Musi, N.; Hirshman, M.F.; Nygren, J.; Svanfeldt, M.; Bavenholm, P.; Rooyackers, O.; Zhou, G.; Williamson, J.M.; Ljunqvist, O.; Efendic, S.; Moller, D.E.; Thorell, A.; Goodyear, L.J. *Diabetes* **2002**, *51*, 2074.
(71) Shaw, R.J.; Lamia, K.A.; Vasquez, D.; Koo, S.H.; Bardeesy, N.; Depinho, R.A.; Montminy, M.; Cantley, L.C. *Science* **2005**, *310*, 1642.
(72) El-Mir, M.Y.; Nogueira, V.; Fontaine, E.; Averet, N.; Rigoulet, M.; Leverve, X. *J. Biol. Chem.* **2000**, *275*, 223.

(73) Kubota, N.; Terauchi, Y.; Kubota, T.; Kumagai, H.; Itoh, S.; Satoh, H.; Yano, W.; Ogata, H.; Tokuyama, K.; Takamoto, I.; Mineyama, T.; Ishikawa, M.; Moroi, M.; Sugi, K.; Yamauchi, T.; Ueki, K.; Tobe, K.; Noda, T.; Nagai, R.; Kadowaki, T. *J. Biol. Chem.* **2006**, *281*, 8748.

(74) Phielix, E.; Szendroedi, J.; Roden, M. *Trends Pharmacol. Sci.* **2011**, *32*, 607.

(75) Turner, N.; Li, J.Y.; Gosby, A.; To, S.W.; Cheng, Z.; Miyoshi, H.; Taketo, M.M.; Cooney, G.J.; Kraegen, E.W.; James, D.E.; Hu, L.H.; Li, J.; Ye, J.M. *Diabetes* **2008**, *57*, 1414.

(76) Gledhill, J.R.; Montgomery, M.G.; Leslie, A.G.; Walker, J.E. *Proc. Natl. Acad. Sci. U.S. A.* **2007**, *104*, 13632.

(77) Corton, J.M.; Gillespie, J.G.; Hawley, S.A.; Hardie, D.G. *Eur. J. Biochem.* **1995**, *229*, 558.

(78) Pold, R.; Jensen, L.S.; Jessen, N.; Buhl, E.S.; Schmitz, O.; Flyvbjerg, A.; Fujii, N.; Goodyear, L.J.; Gotfredsen, C.F.; Brand, C.L.; Lund, S. *Diabetes* **2005**, *54*, 928.

(79) Guo, P.; Lian, Z.Q.; Sheng, L.H.; Wu, C.M.; Gao, J.; Li, J.; Wang, Y.; Guo, Y.S.; Zhu, H.B. *Life Sci.* **2012**, *90*, 1.

(80) Cool, B.; Zinker, B.; Chiou, W.; Kifle, L.; Cao, N.; Perham, M.; Dickinson, R.; Adler, A.; Gagne, G.; Iyengar, R.; Zhao, G.; Marsh, K.; Kym, P.; Jung, P.; Camp, H.S.; Frevert, E. *Cell Metab.* **2006**, *3*, 403.

(81) Zhao, G.; Iyengar, R.R.; Judd, A.S.; Cool, B.; Chiou, W.; Kifle, L.; Frevert, E.; Sham, H.; Kym, P.R. *Bioorg. Med. Chem. Lett.* **2007**, *17*, 3254.

(82) Moreno, D.; Knecht, E.; Viollet, B.; Sanz, P. *FEBS Lett.* **2008**, *582*, 2650.

(83) Treebak, J.T.; Birk, J.B.; Hansen, B.F.; Olsen, G.S.; Wojtaszewski, J.F. *Am. J. Physiol. Cell Physiol.* **2009**, *297*, C1041.

(84) Kim, A.S.; Miller, E.J.; Wright, T.M.; Li, J.; Qi, D.; Atsina, K.; Zaha, V.; Sakamoto, K.; Young, L.H. *J. Mol. Cell. Cardiol.* **2011**, *51*, 24.

(85) Paiva, M.A.; Rutter-Locher, Z.; Goncalves, L.M.; Providencia, L.A.; Davidson, S.M.; Yellon, D.M.; Mocanu, M.M. *Am. J. Physiol. Heart Circ. Physiol.* **2011**, *300*, H2123.

(86) Mirguet, O.; Bouillot, A.M.J. Patent Application WO 2011/138307-A1. **2011**.

(87) Pang, T.; Zhang, Z.S.; Gu, M.; Qiu, B.Y.; Yu, L.F.; Cao, P.R.; Shao, W.; Su, M.B.; Li, J.Y.; Nan, F.J.; Li, J. *J. Biol. Chem.* **2008**, *283*, 16051.

(88) Gomez-Galeno, J.E.; Dang, Q.; Nguyen, T.H.; Boyer, S.H.; Grote, M.P.; Sun, Z.; Chen, M.; Craigo, W.A.; van Poelje, P.D.; MacKenna, D.A.; Cable, E.E.; Rolzin, P.A.; Finn, P.D.; Chi, B.; Linemeyer, D.L.; Hecker, S.J.; Erion, M.D. *ACS Med. Chem. Lett.* **2010**, *1*, 478.

(89) Bookser, B.; Dang, Q.; Gibson, T.S.; Jiang, H.; Chung, D.M.; Bao, J.; Jiang, J.; Kassick, A.; Kekec, A.; Lan, P.; Lu, H.; Makara, G.M.; Romero, F.A.; Sebhat, I.; Wilson, D.; Wodka, D. Patent Application WO 2010/036613-A1. **2010**.

(90) Goff, D.; Payan, D.; Singh, R.; Shaw, S.; Carroll, D.; Hitoshi, Y. Patent Application WO 2012/016217-A1. **2012**.

(91) Chen, L.; Feng, L.; Huang, M.; Li, J.; Nan, F.; Pang, T.; Yu, L.; Zhang, M. Patent Application WO 2011/033099-A1. **2011**.

(92) Chen, L.; Feng, L.; He, Y.; Huang, M.; Liu, Y.; Yun, H.; Zhou, M. Patent Application WO 2012/001020-A1. **2011**.

CHAPTER ELEVEN

Type-2 Diabetes and Associated Comorbidities as an Inflammatory Syndrome

Juan C. Jaen, Jay P. Powers, Tim Sullivan
ChemoCentryx Inc., Mountain View, California, USA

Contents

Annual Reports in Medicinal Chemistry, Volume 47
ISSN 0065-7743
http://dx.doi.org/10.1016/B978-0-12-396492-2.00011-4

1. INTRODUCTION

Acute inflammation represents a protective response to injury or infection which normally resolves after the threat has been eliminated. Incomplete resolution or repeated attempts to neutralize nonexisting threats leads to chronic inflammation. Persistent low-level inflammation is involved in the etiology of many chronic human diseases, including type-2 diabetes (T2D), cardiovascular disease, and chronic kidney disease.[1] While certain cellular pathways are common to many of these, specific mechanisms are particularly important to each pathology. For this review, inflammation is defined broadly and includes the activity of immune cells and processes occurring in parenchymal cells.

T2D is characterized by pancreatic deficiency, insulin resistance, and hyperglycemia. Prominent among the consequences of chronic hyperglycemia are microvascular complications that include diabetic nephropathy, retinopathy, and neuropathy. This chapter reviews recent evidence on the clinical and preclinical effects of a wide range of anti-inflammatory compounds in these conditions. Due to space constraints, we will not review the inflammatory aspects of atherosclerosis (a macrovascular complication associated with diabetes), which are nevertheless equally important.[2]

While the role of inflammatory/immune pathways in the etiology of type-1 diabetes (T1D, autoimmune attack on pancreatic islets that produce insulin) is distinct from T2D, the microvascular complications resulting from uncontrolled hyperglycemia are similar across both diseases. Consequently, clinical studies in diabetic microvascular diseases often enroll both types of patients, and preclinical hyperglycemia-driven models of them are just as often induced by insulin resistance (T2D) as by pancreatic destruction (T1D).

2. TYPE-2 DIABETES

T2D is associated with multiple clinical markers of systemic inflammation (e.g., increased serum levels of acute-phase proteins and inflammatory cytokines).[3] Only a small number of the anti-inflammatory compounds described preclinically as potential antidiabetic agents have progressed into clinical trials. Due to the safety requirements for antidiabetic agents, non-specific immunosuppressive drugs have generally not been given much clinical consideration. However, although beyond the scope of this review, we note the various anti-IL-1β biological agents currently in late-stage T2D clinical trials.[3]

2.1. Salicylates

Salicylic acid (**1**) and its derivatives possess anti-inflammatory activity due to their inhibition of cyclooxygenases and of NF-κB activation. Following promising pilot studies with high-dose aspirin (**2**), salsalate (**3**), a prodrug of salicylic acid, produced a highly significant reduction in blood glucose, hemoglobin-A1c (−0.49% vs. placebo, $p=0.001$), and triglyceride levels when dosed to T2D patients for 14 weeks at doses up to 4 g/day.[4] A large Phase 3 trial (target > 560 patients) was initiated in 2008 and should complete soon.[5] While the mechanism of action for **3** is not entirely clear, the available clinical data are promising. Although it already has a long history of use in arthritic patients, its safety profile in T2D with respect to gastrointestinal bleeding and tinnitus will be critical to its therapeutic utility.

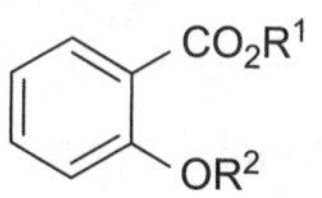

1: R^1 = H; R^2 = H
2: R^1 = H; R^2 = Ac
3: R^1 = (2-CO_2H)Ph; R^2 = H

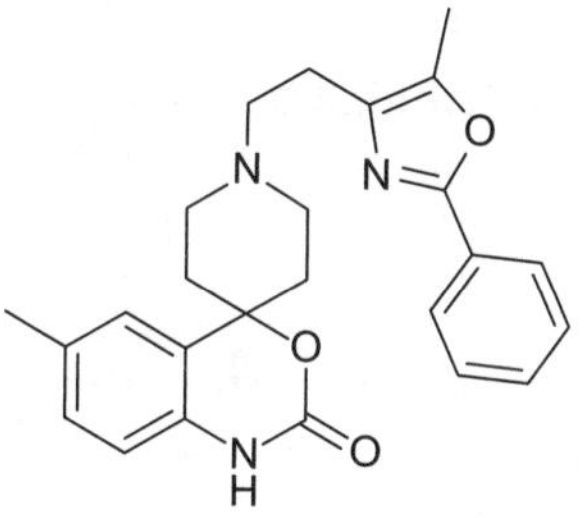

4

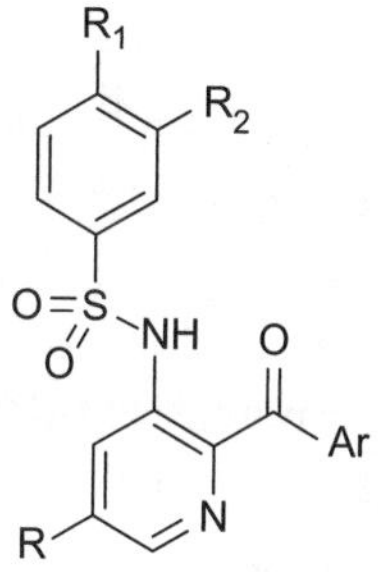

5

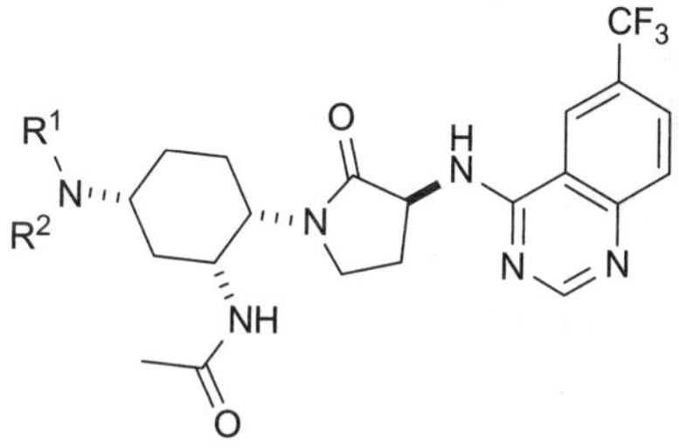

6a: R^1 = $C(CH_3)_3$, R^2 = H
6b: R^1 = $CH(CH_3)_2$, R^2 = CH_3

2.2. CCR2 antagonists

Much of the systemic inflammation in T2D appears to be driven by inflammation of adipose tissue (and perhaps also liver), a process recently linked to monocyte recruitment. Monocytes can differentiate into macrophages in these tissues, driving forward the vicious cycle of inflammation. C–C chemokine receptor 2 (CCR2) and its main ligand MCP-1 (monocyte chemoattractant protein 1) regulate inflammatory macrophage recruitment to adipose tissue and liver.[6,7] CCR2 antagonist RS504393 (**4**) inhibits MCP-1-induced chemotaxis of murine spleen cells with an IC_{50} of 0.8 μM. After 12 weeks of dosing in chow to *db/db* mice at 2 mg/kg/day, **4** ameliorated adipose tissue inflammation (MCP-1 expression and lipid hydroperoxide levels) and macrophage infiltration, concurrently with improved insulin sensitivity and hepatic steatosis.[8]

The orally active CCR2 antagonist CCX140-B potently ($IC_{50}=8$ nM) inhibits CCR2-mediated chemotaxis of human monocytes while lacking effects on any other chemokine receptor. CCX140-B improved hyperglycemia and insulin sensitivity in a diet-induced-obese mouse model, by a mechanism involving reduction in adipose macrophage content and improved insulin sensitivity.[9,10] CCX140-B (10 mg/day; 28 days) was well tolerated and reduced plasma glucose and hemoglobin-A1c levels (0.23% reduction from baseline; $p=0.045$ vs. placebo) in a 159-patient Phase 2 T2D trial.[11] Unlike other CCR2 antagonists described in the literature, CCX140-B did not alter circulating levels of monocytes or MCP-1, which would have been potentially detrimental to the desired therapeutic effect. CCX140-B was reported to belong to the class represented by formula **5**.[9,12] CCX140-B is undergoing further clinical study in DN (Section 3).

BMS-741672, an orally active CCR2 antagonist whose potency and structure have not been disclosed, but believed to be **6a** or **6b**,[13,14] was tested in T2D. The Phase 2 trial, completed in 2009, targeted 58 drug-naïve subjects, randomized to placebo or 50 mg BMS-741672 for 12 weeks. The primary end point was changes in hemoglobin-A1c from baseline.[15] Clinical results have not been disclosed, and there is no recent development activity for BMS-741672.

3. DIABETIC NEPHROPATHY

The first line of defense against microvascular diabetic complications is intensive control of hyperglycemia, dyslipidemia, and hypertension. However, this strategy is insufficient, as patients continue to develop these complications at a high rate.[16] Thus, therapies directed more proximally to the pathophysiology of these diabetic complications are needed. Pivotal diabetic nephropathy (DN) clinical trials generally aim to demonstrate clear benefits on number of renal failures (transplant or dialysis) or deaths, while Phase 2 studies often rely on clinical assessments such as estimated glomerular filtration rate (eGFR) and proteinuria, which generally correlate with the desired long-term outcomes. Proteinuria (increased urinary excretion of protein, mostly albumin) reflects glomerular injury and also contributes directly to tubular and interstitial damage by activation of renal tubular cells, production of proinflammatory chemokines and cytokines, and interstitial recruitment of leukocytes, mainly monocytes/macrophages.[17] Even though no animal model completely recapitulates all pathological features of human DN, certain aspects (particularly proteinuria, leukocyte infiltration, glomerular hypertrophy, renal hyperfiltration, and selected histopathological features) can be modeled preclinically.

3.1. AGE inhibitors

Advanced glycation end-products (AGEs) are formed by nonenzymatic reaction of amino groups in proteins with breakdown products of glucose, such as pyruvate. This process is exacerbated by high glucose levels and oxidative stress. AGEs are recognized by various receptors (known as RAGE), the physiological role of which is unknown. The AGE–RAGE axis stimulates cellular proinflammatory responses that are involved in chronic injury in all major diabetic complications.[18] These effects are mediated by activation of the transcription factor NF-κB. Inhibition of AGE formation or stimulation of AGE cross-link breakage provides renal protection in experimental models of DN.

Benfotiamine (**7**), a vitamin B1 analogue, was shown to block renal AGE accumulation and urinary excretion in a diabetic rat model in which **7** was administered orally at 70 mg/kg/day for 24 weeks.[19] Pyridoxamine (**8**) is a derivative of vitamin B6 that inhibits AGE formation and advanced lipoxidation end-products during lipid peroxidation reactions. Following encouraging preclinical and pilot clinical results, a recent trial, in which 317 proteinuric DN patients were randomized to placebo or pyridoxamine (150 or 300 mg twice daily) for 1 year, failed to demonstrate any benefit, either on eGFR or proteinuria.[20] LR-90 (**9**) was shown in experimental DN models (e.g., Zucker diabetic fatty rat) to decrease AGE accumulation in kidney glomeruli and nitrotyrosine deposition in the renal cortex. *In vitro*, LR-90 was capable of trapping reactive carbonyl compounds and was a potent metal chelator, which is thought to be important for inhibition of AGE formation. LR-90 also blocked the development of DN, based on its lowering of serum creatinine levels and proteinuria.[21] There are no clinical reports for **9**.

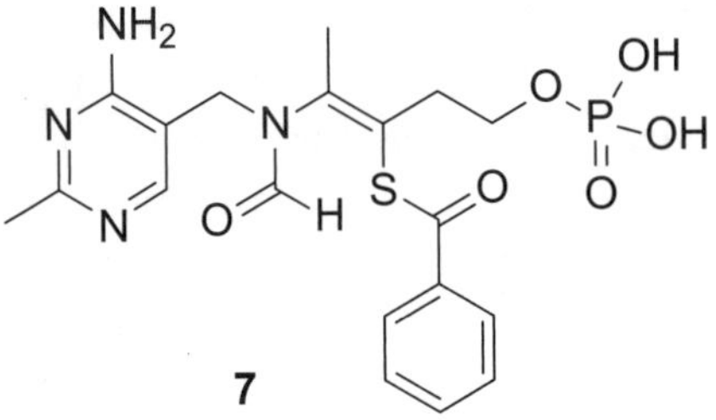

7

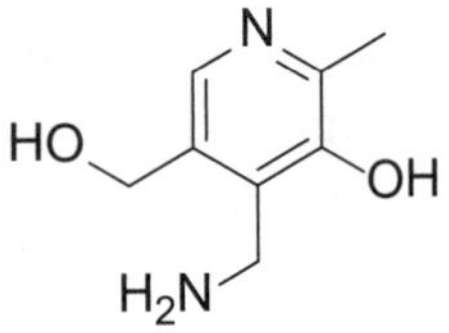

8

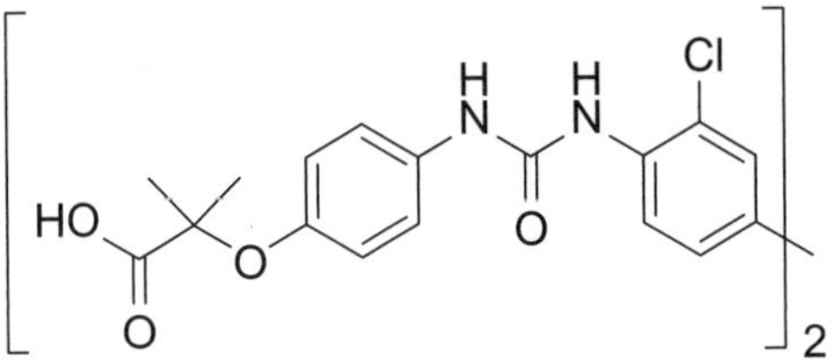

9

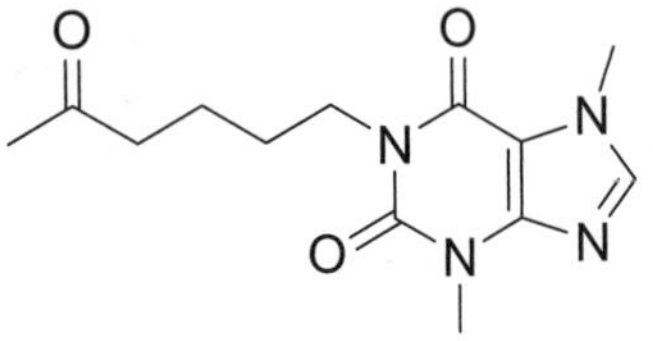

10

3.2. Inhibitors of TNF-α production

Pentoxifylline (**10**), a nonselective phosphodiesterase inhibitor, inhibits TNF-α synthesis and inflammation and is approved for the treatment of intermittent claudication in peripheral vascular disease. Over the years, pentoxifylline has been studied in a series of small clinical trials in DN, generally producing improvements in proteinuria and urinary markers of renal injury. For example, in one study involving 61 proteinuric T2D subjects, pentoxifylline (1.2 g/day) for 4 months reduced urinary albumin excretion from 900 to 791 mg/day ($p < 0.001$).[22,23] However, results have been less conclusive regarding eGFR, perhaps due to the small size and short duration of these clinical trials. Pentoxifylline continues to be studied clinically in DN in a randomized trial in which 169 DN subjects will be randomized to pentoxifylline (1200 mg/day) or placebo for 24 months. The primary outcome measure will be the difference in eGFR between the groups.[17]

3.3. Antifibrotic agents

Pirfenidone (**11**) is an orally active antifibrotic drug approved in Japan and Europe for idiopathic pulmonary fibrosis. While its molecular target is not known, pirfenidone decreases the production of the profibrotic cytokine TGF-β and scavenges radical oxygen species (ROS), among other activities, in cell culture and animal models of renal damage, such as the cyclosporine-induced chronic nephrotoxicity model. In a pilot Phase 2 trial involving 77 proteinuric DN subjects for 1 year, 1.2 g/day pirfenidone produced a small ($\Delta=5.5$ ml/min/1.73 m^2) but statistically significant improvement in eGFR relative to placebo; the higher dose of compound (2.4 g/day) did not reach significance due, in part, to the high drop-out rate. There was no improvement in proteinuria.[24] There are no reports of further development in DN.

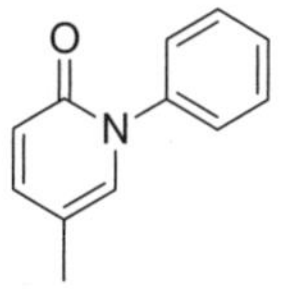

11

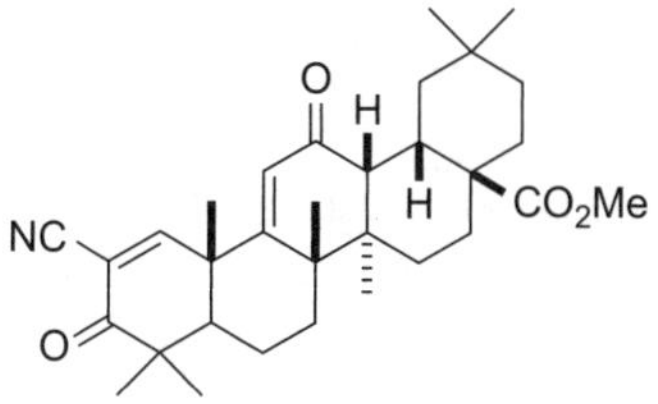

12

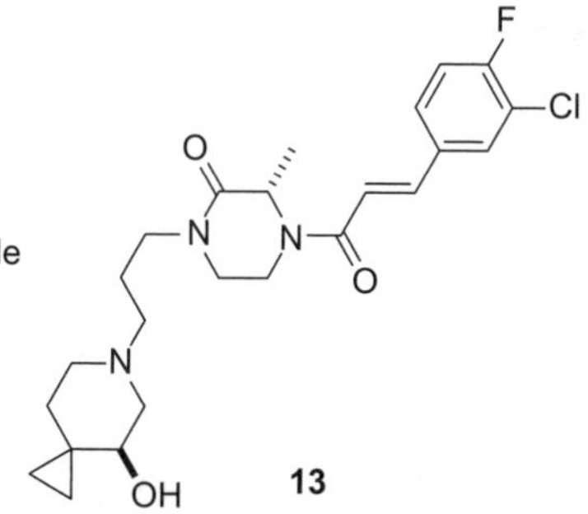

13

3.4. Antioxidant compounds

The transcription factor NF-E2-related factor 2 (Nrf2) regulates expression of over 300 genes involved in modulating oxidative stress and inflammation. Its activity is suppressed in DN animal models, and kidney tissue from Nrf2-knockout mice shows impaired antioxidant activity, increased oxidative damage, and multiple histopathological changes. Bardoxolone methyl (**12**) is an oral activator of the Nrf2 pathway by preventing Nrf2 interaction with the negative regulator KEAP1 (kelch-like ECH-associated protein 1). **12** inhibits NF-κB activation *in vitro* and increases production of antioxidant molecules.[25] In a Phase 2 DN trial, 227 subjects were randomized to placebo or various doses (25–150 mg) of **12** for 1 year. At the higher doses, eGFR increased rapidly by about 5–10 ml/min/1.73 m^2, an effect that was sustained throughout the study.[26] Since eGFR is calculated from serum creatinine levels, some controversy exists as to whether this calculation accurately reflects true GFR or whether changes to creatinine metabolism and/or blood pressure might account for the lowering of creatinine levels.[25] Using a new formulation of **12**, a Phase 3 trial (20 mg/day) is ongoing, targeting 1600 T2D subjects with significantly impaired renal function (eGFR: 15–30 ml/min/1.73 m^2). The primary outcome assessment will be time-to-composite end point of renal failure or cardiovascular death.[27] A number of other putative Nrf2 activators have been shown to possess renoprotective effects in experimental models.[25]

3.5. CCR2 antagonists

The CCR2/MCP-1 axis is critical for recruitment of monocytes from blood into diabetic kidneys, where they differentiate into macrophages. Administration of **4** to diabetic *db/db* mice decreased proteinuria and mesangial expansion, and suppressed profibrotic and proinflammatory cytokine synthesis.[8] In uninephrectomized *db/db* mice, which suffer a more pronounced loss of renal function, dosing of CCR2 antagonist RO5234444 (**13**) at 100 mg/kg in chow for 8 weeks resulted in improved glomerulosclerosis, albuminuria, and glomerular filtration rate.[28]

CCX140-B (**5**) is currently in two Phase 2 DN trials.[29,30] The larger trial targets 135 DN patients with mild–moderate proteinuria, reduction of which will be a key study outcome. Placebo or CCX140-B (5 or 10 mg/day) will be administered for 12 weeks. Given the high selectivity of CCX140-B for human versus murine CCR2, preclinical data were generated with CCX417 (structure undisclosed), a close analogue that blocks mouse CCR2 with an IC_{50} of 70 nM. Administration of CCX417 (30 mg/kg s.c.) to diabetic proteinuric *db/db* mice resulted in rapid (2 days) and sustained (2 weeks) reduction in proteinuria, as well as normalization of serum markers (creatinine, BUN) of renal function.[31]

3.6. PKC-β inhibitors

Hyperglycemia-induced activation of protein kinase C beta (PKC-β) and downstream signaling pathways has been implicated in the vascular injury underlying diabetic microvascular complications. PKC-β is upregulated and activated in diabetic kidney, a process leading to cell growth, fibrosis, and tissue injury. Although perhaps not entirely an anti-inflammatory mechanism, PKC-β inhibition deserves mention here because at least some of the diabetic stimuli (e.g., AGE) that activate PKC-β are inflammatory in nature and because ruboxistaurin (**14**) is the only compound that has received wide attention in all three diabetic complications reviewed here. Ruboxistaurin is an orally bioavailable PKC-β inhibitor shown to ameliorate DN in various animal models. For example, administration of ruboxistaurin (10 mg/kg/day) to diabetic *db/db* mice for 16 weeks significantly ameliorated their proteinuria by $>50\%$, reduced the extent of glomerular mesangial expansion, and prevented the enhanced expression of extracellular matrix proteins and TGF-β.[32] In a Phase 2 trial, 123 T2D patients with persistent albuminuria were randomized to ruboxistaurin (32 mg/day) or placebo for 1 year. Drug treatment resulted in a 24% reduction in albuminuria ($p = 0.02$) and a stabilization of eGFR.[33] Ruboxistaurin has been most extensively studied in DR (see Section 4).

3.7. Miscellaneous immunosuppressants

Strongly immunosuppressive drugs such as mycophenolate mofetil (**15**), a reversible inhibitor of IMPDH (inosine-5′-monophosphate dehydrogenase) that blocks lymphocyte proliferation, have demonstrated benefit on proteinuria, inflammatory cell infiltration, and fibrosis in experimental models of DN.[17,34] In spite of this, the clinical community has generally been reluctant to expose DN patients to strongly immunosuppressive agents, suggesting that clinical progress will be limited to agents that target specific pathways in the pathophysiology of DN.

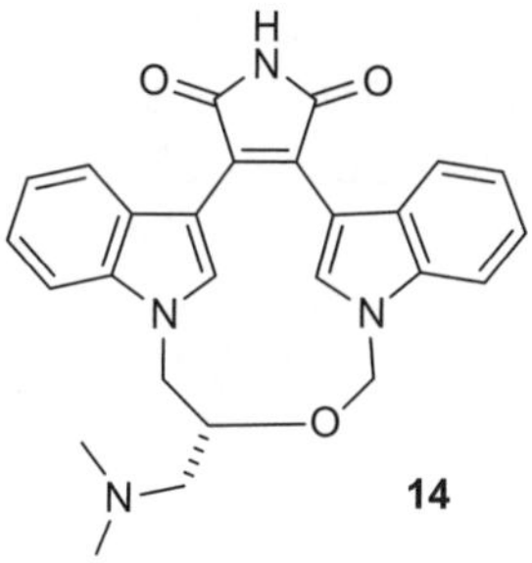

14

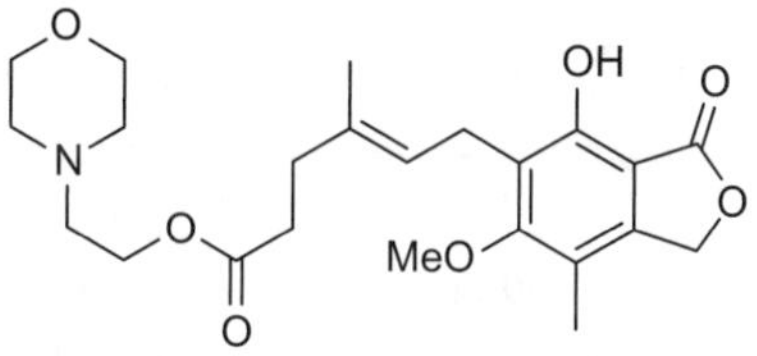

15

4. DIABETIC RETINOPATHY

Diabetic retinopathy (DR) affects ~75% of diabetics and is a leading cause of adult blindness. The earliest stages of DR involve altered retinal blood flow and vascular inflammation, characterized by the release of inflammatory cytokines and chemokines, oxidative stress, and leukocyte accumulation. An initial nonproliferative stage, characterized by retinal hemorrhage and macular edema (ME), may be followed by a final proliferative stage characterized by the ischemia-induced formation of new blood vessels. In ME, leakage of blood and fluids into the retinal *macula* (which provides fine resolution of images) reduces vision quality. New blood vessels formed during proliferative DR are fragile, bleed easily, and may undergo fibrosis/contraction, leading to retinal detachment and other complications resulting in vision loss. Laser photocoagulation of bleeding blood vessels, which also destroys neural tissues, remains the main standard of care.

DR displays signs of chronic subclinical inflammation. In fact, topical and intravitreal application of corticosteroids such as dexamethasone and triamcinolone, with their broad-spectrum anti-inflammatory effects, is a standard treatment option. The search for targeted anti-inflammatory treatments has gained attention in recent years, and several excellent reviews are available on this topic.[35–37]

4.1. Salicylates and NSAIDs

The benefit of aspirin on DR was first described 50 years ago.[38] Later prospective trials on its effects on the development of DN yielded contradictory results. However, these studies were conducted at doses deemed too low for anti-inflammatory effects.[35] Various salicylate-based drugs (salicylic acid, aspirin, sulfasalazine (**16**)), differing in their ability to inhibit prostaglandin synthesis by cyclooxygenases but sharing the ability to block activation of NF-κB, inhibited the development of retinal vascular pathology in diabetic rats and dogs.[39] In a small randomized trial, the NSAID sulindac (**17**) demonstrated a significant preventive effect on the development and progression of DR.[40] Various NSAIDs (nepafenac (**18**), bromfenac (**19**), diclofenac (**20**), and ketorolac (**21**)), either as topical or intravitreous formulations, are undergoing clinical evaluation in ME secondary to DR.[41–44] Ketorolac is also being studied in a Phase 2 trial in proliferative DR.[45]

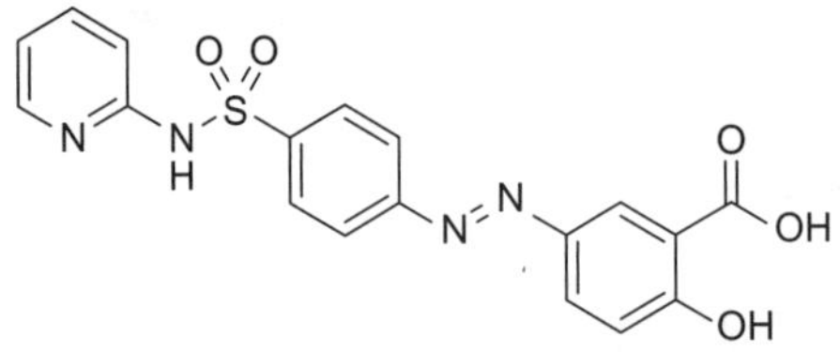

16

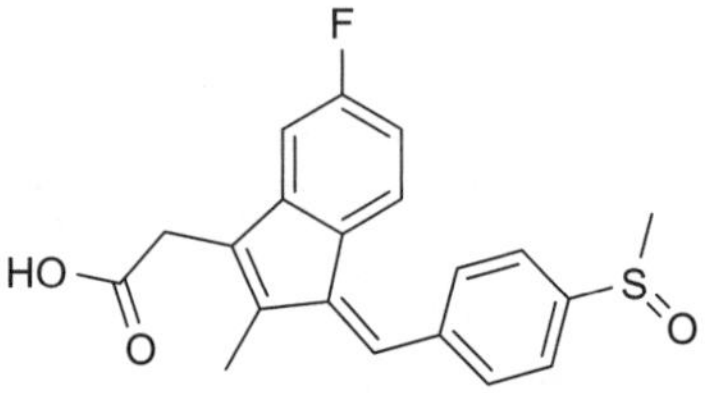

17

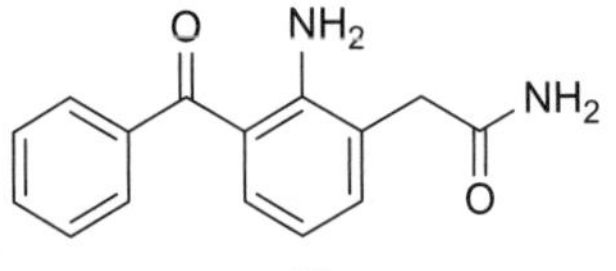

18

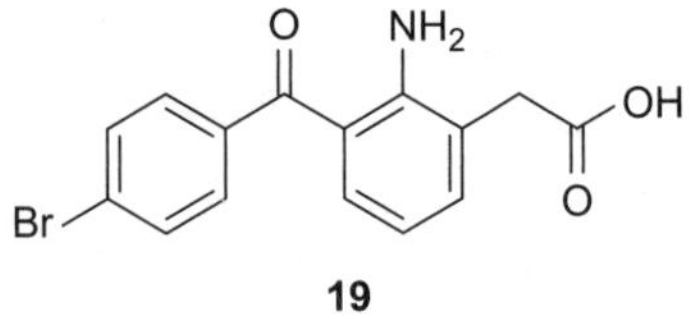

19

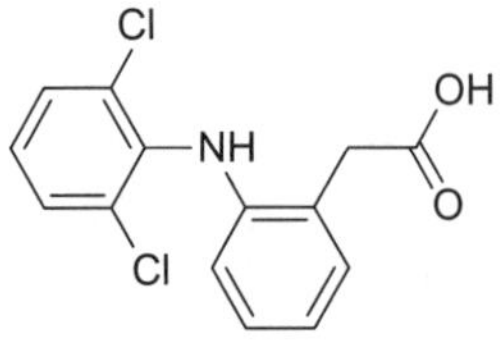

20

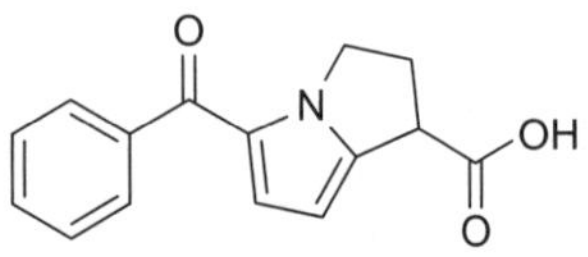

21

4.2. AGE inhibitors

AGE levels are greatly increased in diabetic vitreous and retinal vasculature. AGE receptors (RAGE) are expressed by numerous cells in the retina, including Müller (glia) cells, endothelial cells, and neurons. The AGE–RAGE system activates many downstream effects, including NF-κB activation, ROS formation, and leukocyte recruitment. The previously mentioned AGE inhibitor **9** administered in drinking water (50 mg/ml) to insulin-deficient diabetic rats for 32 weeks produced statistically significant benefits in the number of acellular capillaries and pericytes.[46]

4.3. Leukocyte adhesion inhibitors

Cell adhesion molecules are upregulated on endothelial cells in inflammatory settings and facilitate the adhesion and extravasation of circulating leukocytes. Leukocytes (neutrophils, monocytes/macrophages, T cells) accumulate in and around the abnormal vasculature in diabetic retina and choroid. These leukocytes disrupt vascular integrity and facilitate proliferative damage in DR. LFA-1 (lymphocyte function-associated antigen-1) is an integrin expressed on leukocytes which is important for leukocyte–endothelial cell interaction. SAR1118 (**22**) inhibits the binding of LFA-1 to ICAM-1 on Jurkat cells with an IC_{50} of 2 nM. When delivered via eye drops, **22** reduced leukostasis and retinal vascular leakage in a diabetic rat model.[47]

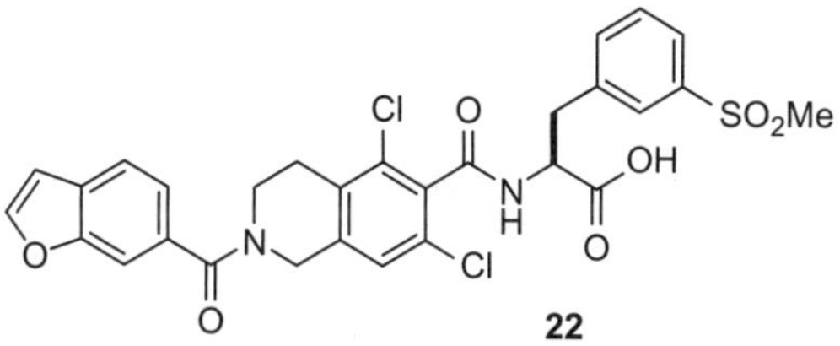

22

4.4. CCR2 antagonists

Certain common inflammatory pathways are involved in multiple vitreoretinal diseases, including DR, macular degeneration, and retinal detachment. For example, levels of IL-6, IL-8, and MCP-1 were highly correlated with each other in all vitreoretinal diseases evaluated.[48] MCP-1 levels in DR vitreous correlate with disease severity, which has been confirmed in a rat DR model.[49] It was recently shown that the CCR2/MCP-1 axis mediates recruitment of monocytes/macrophages and photoreceptor apoptosis in an experimental retinal detachment model.[50] INCB3344 (**23**) is a potent inhibitor of human and mouse MCP-1 binding (IC_{50}=7 and 10 nM for human and mouse, respectively) and is also a functional inhibitor of CCR2-mediated chemotaxis (human monocyte $IC_{50}=4$ nM). Intravitreal administration of **23** suppressed macrophage recruitment and choroidal neovascularization in a laser-induced mouse model of age-related macular vascularization.[51] To date, there have been no preclinical or clinical reports regarding evaluation of CCR2 antagonists in DR.

4.5. Miscellaneous anti-inflammatory agents

Among other anti-inflammatory drugs evaluated in preclinical DR models, minocycline (**24**), a tetracycline derivative with poorly defined anti-inflammatory effects (inhibition of microglia proliferation, activation, and apoptosis), was shown to inhibit neuronal cell death in the retinal ganglion cell layer of diabetic rats.[52] Long-term administration of minocycline also significantly inhibited the degeneration of retinal capillaries in diabetic mice.[53] An exploratory clinical trial is being conducted with **24** (100 mg b.i.d. for 24 months) in diabetic ME.[54]

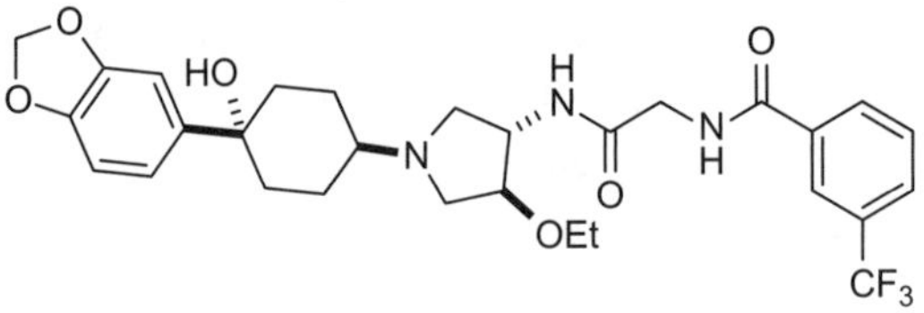

23

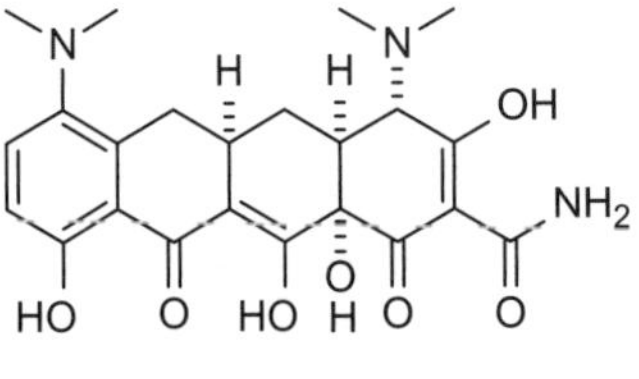

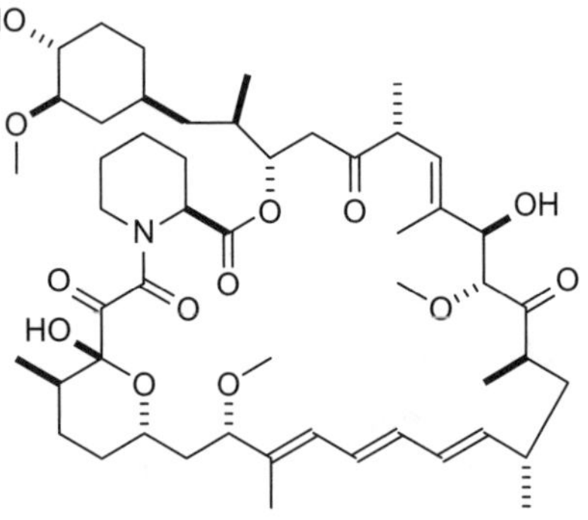

24 **25**

The immunosuppressive agent rapamycin (**25**), an inhibitor of mTOR (mammalian target of rapamycin), is currently in a placebo-controlled Phase 2 trial involving 120 subjects with diabetic ME secondary to DR.[55] The biological rationale behind this clinical trial has not been described in the literature; however, it appears to relate to the activation of PI3-kinase/AKT/mTOR (phosphatidylinositol 3-kinase/protein kinase B/mTOR) signaling pathways activated by growth factors involved in angiogenesis and vascular (endothelial) cell proliferation and survival.[56]

A number of agents that are effective in animal models of DR may, at least in part, do so by interfering with proinflammatory processes. For example, aldose reductase inhibition by sorbinil (**26**) blocked 55% and 71% of the genes involved in TGF-β signaling and oxidative stress, respectively, which were upregulated in an experimental rat DR model.[57]

4.6. PKC-β inhibitors

Ruboxistaurin showed promising efficacy in several large clinical trials in DR. In a 3-year study involving 685 patients with moderate-to-severe DR, **14** markedly reduced the risk of sustained vision loss compared with placebo.[58] An NDA was filed in 2006 and the drug received priority review by the FDA. Following a regulatory request for additional clinical data, the compound no longer appears to be under active clinical development.[59]

5. DIABETIC NEUROPATHY

Diabetic neuropathy (DNeu) represents a series of progressive axonopathies (nerve death that starts with axon degeneration) impacting ~50% of diabetics. The longest axons, those innervating the feet, are generally affected first. Typical symptoms include pain, numbness, and weakness. DNeu increases the risk of lower-limb ulcerations and amputations. In the absence of data on compounds aimed at prevention of nerve damage, this review covers anti-inflammatory approaches that ameliorate the neuropathic pain associated with such damage.

5.1. PKC-β inhibitors

Alterations in the microvasculature, similar to those noted in DN and DR, are also observed in DNeu.[60] However, unlike the efficacy documented in those other conditions, **14** only demonstrated efficacy on select end points (e.g., skin microvascular blood flow), while failing on others (e.g., sensory symptoms, neurological deficits, and nerve conduction studies) in a small 6-month Phase 2 DNeu trial.[61] Development of **14** in DNeu has been discontinued.[59]

5.2. Antioxidant agents

Neurons are highly susceptible to glucose-mediated injury. Excess glucose leads to intracellular ROS generation, lactate accumulation, NAD^+ depletion, and activation of back-up pathways such as the polyol and hexose pathways, both of which lead to increased oxidative burden and neuronal damage.[60] Despite this, very few antioxidant strategies have been carefully explored in DNeu, with the possible exception of α-lipoic acid (**27**), a standard of diabetic care in Germany. According to a meta-analysis comprising 1258 patients, infusions of **27** (600 mg/day) ameliorated neuropathic symptoms after 3 weeks. In a randomized trial, 460 DNeu patients were assigned to oral **27** (600 mg/day) or placebo. **27** was well tolerated and, after 4 years, a significant difference in neuropathic deficits was noted relative to placebo.[62]

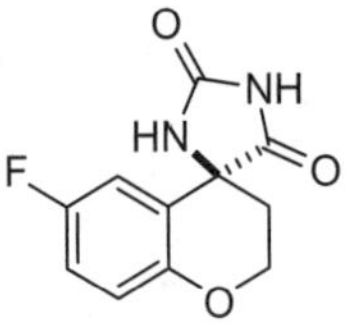

26

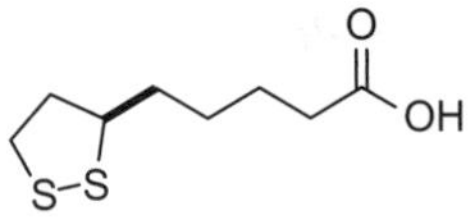

27

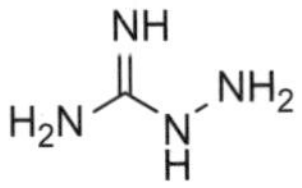

28

5.3. AGE inhibitors

In experimental diabetes models, RAGE expression is elevated in peripheral epidermal axons, sural axons, Schwann cells, and dorsal root ganglia (DRG) neurons, following the pattern of electrophysiological and structural abnormalities associated with neuropathy. The contribution of AGE to the development of DNeu has been studied preclinically using AGE inhibitors such as aminoguanidine (**28**) and benfotiamine and RAGE-knockout mice.[63]

5.4. CCR2 antagonists

CCR2 antagonists are worth noting because they may intervene at two distinct points in the pathology of DNeu, possibly slowing down peripheral nerve loss as well as ameliorating the resulting pain. Macrophages are recruited to areas of axon damage and DRG following damage to the afferent nerves. This process, which may accelerate nerve destruction and set in motion pain pathways, requires CCR2-mediated recruitment of circulating monocytes to the injury site.[64] CCR2 and its ligand MCP-1 are upregulated not only at the site of nerve injury but also distally in the DRG and spinal cord, where it regulates microglia–neuron communication.[65] AZ889 (undisclosed structure) is a potent antagonist of mouse CCR2, blocking mCCL2-induced calcium mobilization with an IC_{50} of 1.3 nM in HEK293 cells, and is also a potent antagonist of rat CCR5, blocking MIP-1α-induced calcium mobilization with an IC_{50} of 79 nM in CHO cells. Single intrathecal injections (1–30 nmol/rat) of AZ889 to rats who had undergone sciatic nerve ligation (a model of neuropathic pain) resulted in a significant increase in their threshold to mechanical and thermal pain, suggesting that at least some of the antinociceptive effects of CCR2 antagonists are centrally (spinal cord) mediated.[66] Treatment (0.05–1 mg/kg; 7 days) with the orally active CCR2/CCR5 antagonist peptide RAP-103 (undisclosed structure) blocked mechanical allodynia and development of thermal hyperalgesia after partial ligation of the sciatic nerve in rats. RAP-103 also reduced spinal microglial activation and monocyte infiltration. RAP-103 was reported to be a potent antagonist of both CCR2- (IC_{50}=4.2 pM) and CCR5- (IC_{50}=0.18 pM) mediated monocyte chemotaxis.[67]

BMS-741672 (**6**) was tested in a Phase 2 DNeu trial.[68] The study, completed in 2009, targeted 50 diabetics with painful sensory-motor neuropathy secondary to diabetes. Subjects received either placebo or 100 mg BMS-741672 daily for 3 weeks. Results of this trial have not been disclosed.

6. CONCLUSIONS

Multiple anti-inflammatory compounds, with various degrees of selectivity for a well-defined pathway, continue to be tested clinically in T2D and associated microvascular complications. At this time, DN appears to have the largest number of active programs moving toward large-scale clinical trials.

REFERENCES

(1) Manabe, I. *Circ. J.* **2011**, *75*, 2739.
(2) Libby, P.; Okamoto, Y.; Rocha, V.Z.; Folco, E. *Circ. J.* **2010**, *74*, 213.
(3) Donath, M.; Shoelson, S. *Nat. Rev. Immunol.* **2011**, *11*, 98.
(4) Goldfine, A.; Fonseca, V.; Jablonski, K.; Pyle, L.; Staten, M.; Shoelson, S. *Ann. Intern. Med.* **2010**, *152*, 346.
(5) http://clinicaltrials.gov/ct2/show/NCT00799643.
(6) Oh-da, Y.; Morinaga, H.; Talukdar, S.; Bae, E.; Olefsky, J. *Diabetes* **2012**, *61*, 346.
(7) Obstfeld, A.; Sugaru, E.; Thearle, M.; Francisco, A.; Gayet, C.; Ginsberg, H.; Ables, E.; Ferrante, A. *Diabetes* **2010**, *59*, 916.
(8) Kang, Y.; Lee, M.; Song, H.; Ko, G.; Kwon, O.; Lim, T.; Kim, S.; Han, S.; Han, K.; Lee, J.; Han, J.; Kim, H.; Cha, D. *Kidney Int.* **2010**, *78*, 883.
(9) Hanefeld, M.; Schell, E.; Gouni-Berthold, I.; Melichar, M.; Vesela, I.; Sullivan, T.; Miao, S.; Johnson, D.; Jaen, J.; Bekker, P.; Schall, T.J. *Diabetologia* **2011**, *54*(S1), S87.
(10) Jaen, J.; Sullivan, T.; Miao, Z.; Zhao, N.; Berahovich, R.; Ert, L.; Baumgart, T.; Krasinski, A.; Greenman, K.; Bekker, P.; Powers, J.; Schall, T. *Diabetologia* **2011**, *54* (S1), S365.
(11) Hanefeld, M.; Schell, E.; Gouni-Berthold, I.; Melichar, M.; Vesela, I.; Sullivan, T.; Miao, S.; Johnson, D.; Jaen, J.; Bekker, P.; Schall, T. *Diabetes* **2011**, *60*(S1), A85.

(12) Basak, A.; Charvat, T.; Chen, W.; Dairaghi, D.; Hansen, D.; Jin, J.; Krasinski, A.; Moore, J.; Pennell, A.; Punna, S.; Ungashe, S.; Wang, Q.; Wei, Z.; Wright, J.; Zeng, Y. Patent Application WO 008431-A2, **2008**.
(13) Carter, P.; Duncia, J.; Mudryk, B.; Randazzo, M.; Xiao, Z.; Yang, M.; Zhao, R. Patent Application WO 014360-A2, **2008**.
(14) Yang, M.; Cherney, R.; Eastgate, M.; Muslehiddinoglu, J.; Prasad, S.; Xiao, Z. Patent Application WO 014381-A2, **2008**.
(15) http://clinicaltrials.gov/ct2/show/NCT00699790.
(16) Gaede, P.; Vedel, P.; Larsen, N.; Jensen, G.V.; Parving, H.; Pedersen, O. *N. Engl. J. Med.* **2003**, *348*, 383.
(17) Navarro-Gonzalez, J.; Mora-Fernandez, C.; Muros-de-Fuente, M.; García-Pérez, J. *Nat. Rev. Immunol.* **2011**, 7, 327.
(18) Bierhaus, A.; Nawroth, P. *Diabetologia* **2009**, *52*, 2251.
(19) Karachalias, N.; Babaei-Jadidi, R.; Rabbani, N.; Thornalley, P. *Diabetologia* **2010**, *53*, 1506.
(20) Lewis, E.; Greene, T.; Spitalewiz, S.; Blumenthal, S.; Berl, T.; Hunsicker, L.; Pohl, M.; Rohde, R.; Raz, I.; Yerushalmy, Y.; Yagil, Y.; Herskovits, T.; Atkins, R.; Reutens, A.; Packham, D.; Lewis, J. *J. Am. Soc. Nephrol.* **2012**, *23*, 131.
(21) Figarola, J.; Loera, S.; Weng, Y.; Shanmugam, N.; Natarajan, R.; Rahbar, S. *Diabetologia* **2008**, *51*, 882.
(22) McCormick, B.; Sydor, A.; Akbari, A.; Fergusson, D.; Doucette, S.; Knoll, G. *Am. J. Kidney Dis.* **2008**, *52*, 454.
(23) Badri, S.; Dashti-Khavidaki, S.; Lessan-Pezeshki, M.; Abdollahi, M. *J. Pharm. Pharm. Sci.* **2011**, *14*, 128.
(24) Sharma, K.; Ix, J.; Mathew, A.; Cho, M.; Pflueger, A.; Dunn, S.; Francos, B.; Sharma, S.; Falkner, B.; McGowan, T.; Donohue, M.; Ramachandrarao, S.; Xu, R.; Fervenza, F.; Kopp, J. *J. Am. Soc. Nephrol.* **2011**, *22*, 1144.
(25) Thomas, M.; Cooper, M. *Nat. Rev. Nephrol.* **2011**, 7, 552.
(26) Pergola, P.; Raskin, P.; Toto, R.; Meyer, C.; Huff, J.; Grossman, E.; Krauth, M.; Ruiz, S.; Audhya, P.; Christ-Schmidt, H.; Wittes, J.; Warnock, D. *N. Engl. J. Med.* **2011**, *365*, 327.
(27) http://clinicaltrials.gov/ct2/show/NCT01351675.
(28) Sayyed, S.; Ryu, M.; Kulkarni, O.; Schmid, H.; Lichtnekert, J.; Gruner, S.; Green, L.; Mattei, P.; Hartmann, G.; Anders, H. *Kidney Int.* **2011**, *80*, 68.
(29) http://clinicaltrials.gov/ct2/show/NCT01440257.
(30) http://clinicaltrials.gov/ct2/show/NCT01447147.
(31) Sullivan, T.; Miao, Z.; Zhao, N.; Berahovich, R.; Powers, J.; Ertl, L.; Dang, T.; Zhao, R.; Liu, S.; Miao, S.; Seitz, L.; Bekker, P.; Schall, T.; Jaen, J. *J. Am. Soc. Nephrol.* **2011**, *22*(S1), 236A.
(32) Koya, D.; Haneda, M.; Nakagawa, H.; Isshiki, K.; Sato, H.; Maeda, S.; Sugimoto, T.; Yasuda, H.; Kashiwagi, A.; Ways, D.; King, G.; Kikkawa, R. *FASEB J.* **2000**, *14*, 439.
(33) Tuttle, K.; Bakris, G.; Toto, R.; McGill, J.; Hu, K.; Anderson, P. *Diabetes Care* **2005**, *28*, 2686.
(34) Rodríguez-Iturbe, B.; Quiroz, Y.; Shahkarami, A.; Li, Z.; Vaziri, N. *Kidney Int.* **2005**, *68*, 1041.
(35) Tang, J.; Kern, T. *Prog. Retin. Eye Res.* **2011**, *30*, 343.
(36) Zhang, W.; Liu, H.; Al-Shabrawey, M.; Caldwell, R.; Caldwell, R. *J. Cardiovasc. Dis. Res.* **2011**, *2*, 96.
(37) Noda, K.; Nakao, S.; Ishida, S.; Ishibashi, T. *J. Ophthalmol.* **2012**. http://dx.doi.org/10.1155/2012/279037 ePub Nov 2, 2011.
(38) Powell, E.; Field, R. *Lancet* **1964**, *2*, 17.
(39) Zheng, L.; Howell, S.; Hatala, D.; Huang, K.; Kern, T. *Diabetes* **2007**, *56*, 337.

(40) Hattori, Y.; Hashizume, K.; Nakajima, K.; Nishimura, Y.; Naka, M.; Miyanaga, K. *Curr. Med. Res. Opin.* **2007**, *23*, 1913.
(41) http://clinicaltrials.gov/ct2/show/NCT01331005.
(42) http://clinicaltrials.gov/ct2/show/NCT00758628.
(43) http://clinicaltrials.gov/ct2/show/NCT00619034.
(44) http://clinicaltrials.gov/ct2/show/NCT00900887.
(45) http://clinicaltrials.gov/ct2/show/NCT00907114.
(46) Bhatwadekar, A.; Glenn, J.; Figarola, J.; Scott, S.; Gardiner, T.; Rahbar, S.; Stitt, A. *Br. J. Ophthalmol.* **2008**, *92*, 545.
(47) Rao, V.; Prescott, E.; Shelke, N.; Trivedi, R.; Thomas, P.; Struble, C.; Gadek, T.; O'Neill, C.; Kompella, U. *Invest. Ophthalmol. Vis. Sci.* **2010**, *51*, 5198.
(48) Brucklacher, R.; Patel, K.; VanGuilder, H.; Bixler, G.; Barber, A.; Antonetti, D.; Lin, C.; LaNoue, K.; GarDNeur, T.; Bronson, S.; Freeman, W. *BMC Med. Genomics* **2008**, *1*, 26.
(49) Yoshimura, T.; Sonoda, K.; Sugahara, M.; Mochizuki, Y.; Enaida, H.; Oshima, Y.; Ueno, A.; Hata, Y.; Yoshida, H.; Ishibashi, T. *PLoS One* **2009**, *4*, e8158.
(50) Nakazawa, T.; Hisatomi, T.; Nakazawa, C.; Noda, K.; Maruyama, K.; She, H.; Matsubara, A.; Miyahara, S.; Nakao, S.; Yin, Y.; Benowitz, L.; Hafezi-Moghadam, A.; Miller, J. *Proc. Natl. Acad. Sci. U.S.A.* **2007**, *104*, 2425.
(51) Xie, P.; Kamei, M.; Suzuki, M.; Matsumura, N.; Nishida, K.; Sakimoto, S.; Sakaguchi, H.; Nishida, K. *PLoS One* **2011**, *6*, e28933.
(52) Krady, J.; Basu, A.; Allen, C.; Xu, Y.; LaNoue, K.; GarDNeur, T.; Levison, S. *Diabetes* **2005**, *54*, 1559.
(53) Vincent, J.; Mohr, S. *Diabetes* **2007**, *56*, 224.
(54) http://clinicaltrials.gov/ct2/show/NCT01120899.
(55) http://clinicaltrials.gov/ct2/show/NCT00656643.
(56) Jacot, J.; Sherris, D. *J. Ophthalmol.* **2011**, *31*, 95S. http://dx.doi.org/10.1155/2011/589813 Article ID 589813.
(57) Gerhardinger, C.; Dagher, Z.; Sebastiani, P.; Park, Y.; Lorenzi, M. *Diabetes* **2009**, *58*, 1659.
(58) *Drugs R&D* **2007**, *8*, 193.
(59) Danis, R.; Sheetz, M. *Expert Opin. Pharmacother.* **2009**, *10*, 2913.
(60) Vincent, A.; Callaghan, B.; Smith, A.; Feldman, E. *Nat. Rev. Neurol.* **2011**, *7*, 573.
(61) Casellini, C.; Barlow, P.; Rice, A.; Casey, M.; Simmons, K.; Pittenger, G.; Bastyr, E.; Wolka, A.; Vinik, A. *Diabetes Care* **2007**, *30*, 896.
(62) Ziegler, D. *Diabetes Care* **2009**, *32*(S2), S414.
(63) Cameron, N.; Gibson, T.; Nangle, M.; Cotter, M. *Ann. N.Y. Acad. Sci.* **2005**, *1043*, 784.
(64) Abbadie, C.; Lindia, J.; Cumiskey, A.; Peterson, L.; Mudgett, J.; Bayne, E.; DeMartino, J.; MacIntyre, D.; Forrest, M. *Proc. Natl. Acad. Sci. U.S.A.* **2003**, *100*, 7947.
(65) Abbadie, C. *Trends Immunol.* **2005**, *26*, 529.
(66) Serrano, A.; Paré, M.; McIntosh, F.; Elmes, S.; Martino, G.; Jomphe, C.; Lessard, E.; Lembo, P.; Vaillancourt, F.; Perkins, M.; Cao, C. *Mol. Pain* **2010**, *6*, 90.
(67) Padi, S.; Shi, X.; Zhao, Y.; Ruff, M.; Baichoo, N.; Pert, C.; Zhang, J. *Pain* **2012**, *153*, 95.
(68) http://clinicaltrials.gov/ct2/show/NCT00683423.

CHAPTER TWELVE

Beyond PPARs and Metformin: New Insulin Sensitizers for the Treatment of Type 2 Diabetes

Philip A. Carpino, David Hepworth
Department of Worldwide Medicinal Chemistry, Pfizer PharmaTherapeutics, Cambridge, Massachusetts, USA

Contents

1. INTRODUCTION

Type 2 diabetes (T2D) is a chronic disease of elevated plasma glucose levels resulting from a failure of the pancreas to produce sufficient insulin and/or peripheral tissues to respond to insulin (i.e., insulin resistance). Prevalence of T2D has rapidly increased over the past decade in both the United States and the rest of the world, paralleling the rise of obesity, a known risk factor. In 2010, approximately 26-million Americans were considered diabetic, an 8% increase from 2008.[1] Major complications of T2D include nephropathy, retinopathy, and neuropathy. T2D doubles the risk for cardiovascular disease and stroke. The economic costs associated with T2D were estimated as $174 billion in 2007, the last year for which figures are available.[2]

T2D is managed through a combination of lifestyle interventions and/or pharmacological therapy. Bariatric surgery has been shown to provide

Annual Reports in Medicinal Chemistry, Volume 47
ISSN 0065-7743
http://dx.doi.org/10.1016/B978-0-12-396492-2.00012-6

remission from T2D but carries short-term risks.[3] Goals for T2D treatment have generally focused on reduction in glycosylated hemoglobin (HbA1c), a biomarker for plasma glucose. Lifestyle interventions, especially body weight loss and increased physical activity, can significantly lower HbA1c levels but are difficult to maintain over an extended period. Pharmacological treatment also results in lower HbA1c levels, but the effectiveness of such treatment can decrease over time. The choice of drugs is governed by a series of diabetes care algorithms which have been periodically updated, most recently in 2009.[4]

The past decade has seen the emergence of several new classes of anti-T2D drugs such as the dipeptidyl peptidase IV (DPPIV) inhibitors and incretin mimetics. The only insulin sensitizers currently on the market were launched in the 1990s when metformin and the thiazolidinediones (TZDs) pioglitazone and rosiglitazone received approval from the U.S. Food and Drug Administration. The TZDs are peroxisome proliferator-activated receptor γ (PPARγ) agonists that, when used as monotherapy, lower HbA1c levels more effectively than DPPIV inhibitors or glucagon-like peptide-1 mimetics and achieve greater durability than metformin,[4,5] However, the side effect and safety profiles of PPARγ agonists as a class have come into question.[6] The marketing of rosiglitazone was severely restricted in 2011. The non-TZD aleglitazar, a balanced PPARα/γ agonist in Phase III trials for the treatment of T2D and dyslipidemias, is the only current example of a full PPARγ agonist, albeit a nonselective PPAR modulator, in clinical development,[6,7] It has been proposed that selective and/or partial PPARγ modulators (SPPARγMs) such as INT131 might exhibit reduced side effects, but their efficacy is unproven.[8]

The discovery of non-PPARγ insulin sensitizers has attracted considerable interest. New insulin sensitizers would fill a much needed void in treatment options for T2D patients who no longer respond to metformin or cannot use pioglitazone due to its contraindications.[9] Only a few small molecule insulin sensitizers are currently in clinical development including imeglimin (a metformin analog), MSDC-0602 (a TZD analog with no PPARγ transactivation activity), VVP808 (a repurposed drug with ophthalmic activity), and AC-201 (a new formulation of diacerin),[10–13] Many other insulin sensitizers have been reported to show robust efficacy in animal models, although it is not known whether this efficacy will translate into humans. Herein, we review progress toward the next generation of insulin sensitizers, with a focus on synthetic agents that have demonstrated body weight-independent insulin-sensitizing efficacy in animal models.

2. INSULIN SENSITIZERS

Insulin is a pleiotropic hormone involved in carbohydrate and lipid metabolism and cell proliferation. An insulin sensitizer improves insulin's ability to stimulate glucose utilization, leading to reduced plasma glucose.[14] At the level of tissues such as the liver, adipose tissue, or muscle, an insulin sensitizer may enhance the ability of insulin to decrease gluconeogenesis, inhibit lipolysis, or stimulate glucose uptake, respectively. The molecular mechanisms by which insulin sensitizers enhance the insulin response are varied and described in considerable detail elsewhere.[15]

2.1. Insulin sensitizers that act within the insulin signaling pathway

The insulin/insulin receptor (IR) signaling cascade leading to activation of the phosphoinositide 3-kinase (PI3K)/Akt pathway is complex, involving many phosphorylation/dephosphorylation events and requiring an array of adaptor proteins and secondary signaling molecules.[16] In T2D, the system becomes dysregulated, with insulin losing its ability to inhibit gluconeogenesis or stimulate glycogenesis. Few opportunities for pharmacological modulation within this system have been identified to safely restore normal insulin signaling. Two novel classes of insulin sensitizers, protein tyrosine phosphatase-1b (PTP1b) inhibitors and type-II SH2-domain-containing inositol 5-phosphatase (SHIP2) inhibitors, are discussed below. The recently disclosed insulin sensitizers **1–3** (Fig. 12.1) also act within the IR pathway but are not discussed further here.[17–19]

2.1.1 PTP1b inhibitors

PTB1b negatively regulates insulin signaling by catalyzing the dephosphorylation of both the IR and insulin-receptor substrate-1 (IRS1). Genetic polymorphism data associate PTP1b with protection from and/or development of insulin resistance and diabetes in humans.[20] Ablation of PTP1b in mice improves insulin sensitivity and prevents weight gain on a high fat diet.[21]

The identification of small molecule PTP1b inhibitors with appropriate pharmacokinetic properties and selectivity has proven extremely challenging.[22] Many competitive PTP1b inhibitors are phosphotyrosine mimetics that contain a carboxylic or phosphonic acid (i.e., to drive binding affinity to the high-affinity catalytic site) and a large lipophilic tail seeking to

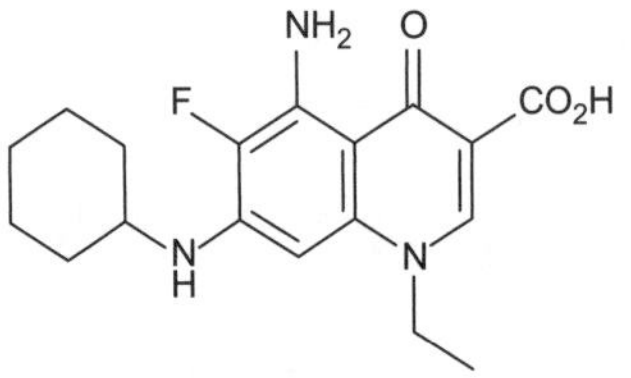

1 AS1842856

G6Pase mRNA IC50 = 0.13 μM (Fao cells)

PEPCK mRNA IC_{50} = 0.037 μM (Fao cells)

Forkhead transcription factor-01 (Fox01) modulators

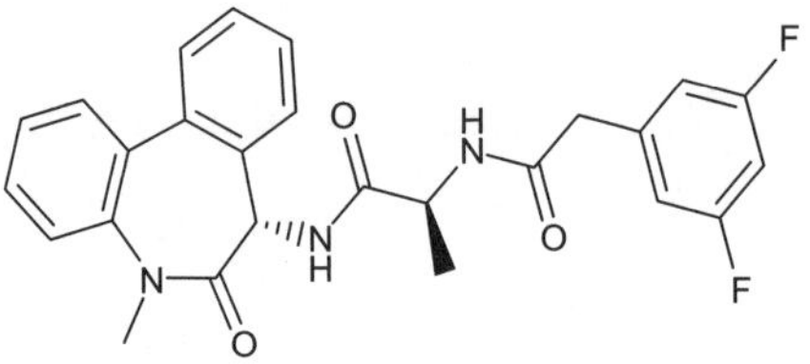

2

Notch/gamma-secretase inhibitors

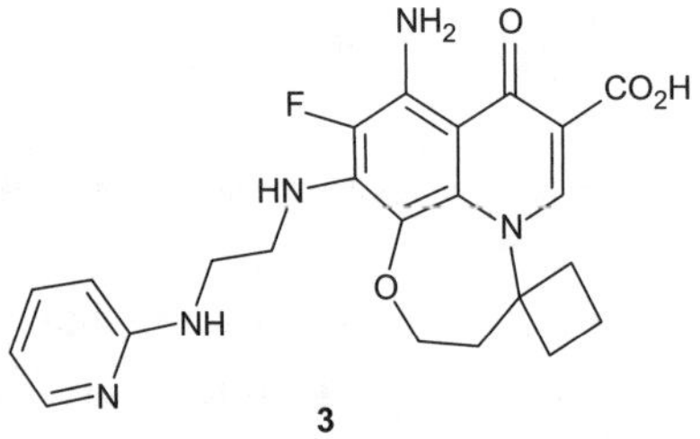

3

Cell free IC50 = 0.004 μM

Cell IC50 = 3 μM

Glycogen synthase kinase-3alpha/beta inhibitors

Figure 12.1 Structures of insulin sensitizers **1–3** that act within the insulin signaling pathway.

maximize the additional binding energy that can be obtained from the shallow binding pocket.[23] These PTP1b inhibitors generally exhibit low ligand efficiency, poor cell activity, and suboptimal pharmacokinetic properties. Application of acidic isosteres has led to some improvement in ADME properties, but not sufficient to produce clinical candidates against the target.[24] An allosteric binding site on the enzyme has been identified but not yet successfully exploited in the design of oral inhibitors.[25] The discovery of new PTP1b inhibitors remains an active area of research, especially in academia, but further breakthroughs will be required to deliver orally active agents.[26]

While PTP1b does not appear to be druggable from a small molecule perspective, PTP1b protein levels can be modulated using antisense oligonucleotides (ASOs) that bind mRNA and reduce protein transcription. A PTP1b ASO ISIS113715 was advanced into Phase II clinical trials and shown to reduce plasma glucose and LDL-cholesterol in diabetic patients without causing weight gain.[27] Development of ASO ISIS113715 has been reportedly suspended, replaced by a new ASO ISIS-PTP1B$_{RX}$ (structure unknown), currently in Phase I trials.[28]

2.1.2 SHIP2 inhibitors (SHIP2i)

In the insulin signaling pathway, phosphatidylinositol 3,4,5-triphosphate (PIP3), a product of PI3K, activates Akt and other downstream kinases to initiate a cascade that results in translocation of glucose transporter type 4 (GLUT4) to the plasma membrane and glucose uptake into tissues. SHIP2 acts as a negative regulator of this pathway by catalyzing dephosphorylation of PIP3, thereby blocking IR signaling.[29] SHIP2 knockout and transgenic mouse studies, tissue selective knockdowns, and human genetics studies linking polymorphisms in SHIP2 with susceptibility to T2D support a role for this target in enhancing insulin sensitivity.[29] A thiophene derivative AS 1949490 (**4**, $IC_{50}=0.6$ μM, Fig. 12.2), discovered by high-throughput screening (HTS), was shown to increase phosphorylation of Akt and glucose uptake in L6 myoblasts treated with insulin.[30] When dosed at 300 mg/kg bid to *db/db* mice, **4** reduced plasma glucose levels by 23%. Since **4** does not contain an acidic "warhead" as required by many phosphatase inhibitors, this class of compounds may exhibit improved pharmacokinetic properties compared to PTP1b inhibitors.

2.2. Insulin sensitizers that modulate lipid synthesis

Insulin resistance is often associated with accumulation of lipid derivatives in the liver and other tissues. Excess lipid deposition in the liver arises when hepatic uptake and synthesis of lipids exceeds hepatic clearance rates, via either oxidation or export as very low-density lipoprotein particles or as phospholipids into bile. Lipids that accrue in the liver come from dietary sources, elevated lipolytic activity in adipose tissue, or increased endogenous synthesis. It is not yet known whether fatty liver causes insulin resistance or insulin resistance causes the storage of excess fat in the liver. On the one hand, lipid intermediates are known to inhibit Akt signaling and activate serine kinases, with the latter able to phosphorylate IRS1, preventing its interaction with the IR.[31] On the other hand, the compensatory increases in plasma insulin arising from insulin resistance drive activation of the transcription factor sterol regulatory-element-binding protein-1c (SREBP-1c), which is a key regulator of FA synthesis.

Studies with transgenic animals and pharmacological tool compounds have shown that blocking the synthesis of saturated and (poly)unsaturated

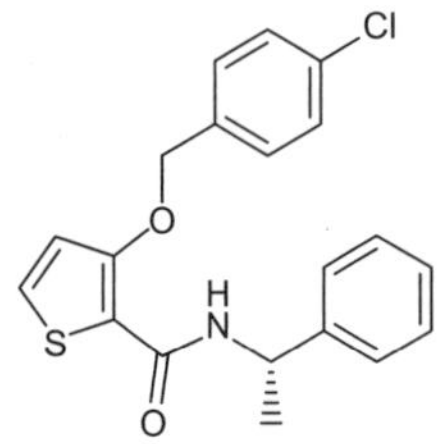

4 AS 1949490

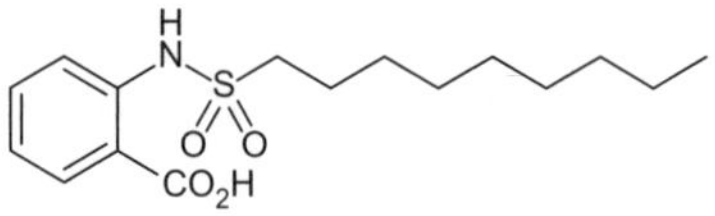

5 IC_{50} = 25 μM

Microsomal glycerol-3-phosphate acyltransferase inhibitor

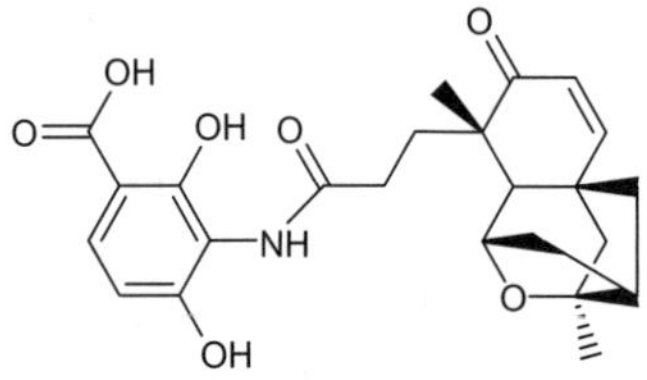

6 Fatty acid synthase IC_{50} = 0.1 μM

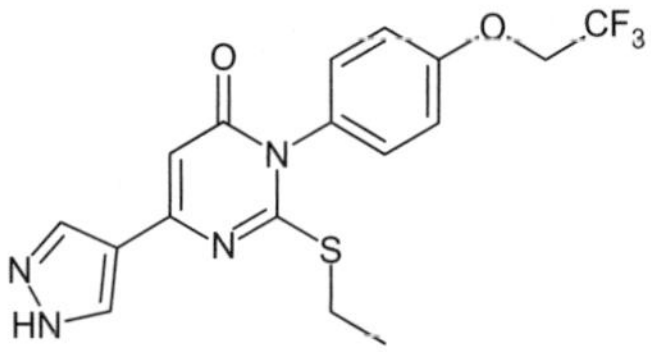

7 %inhibition @ 1 μM in HepG2 cells = 103%

Delta-5 desaturase inhibitor

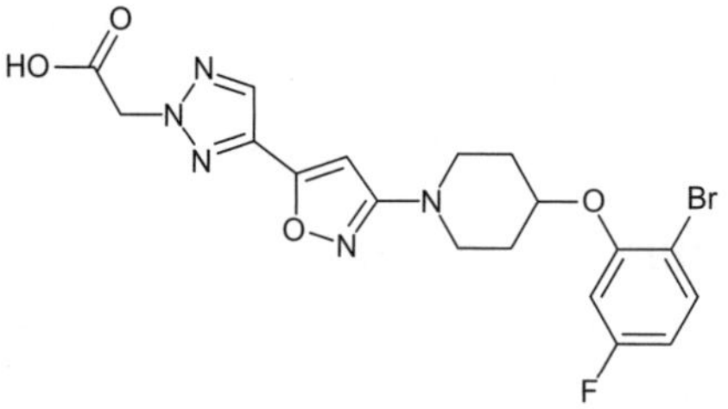

8 Stearoyl CoA desaturase IC_{50} = 0.003 μM

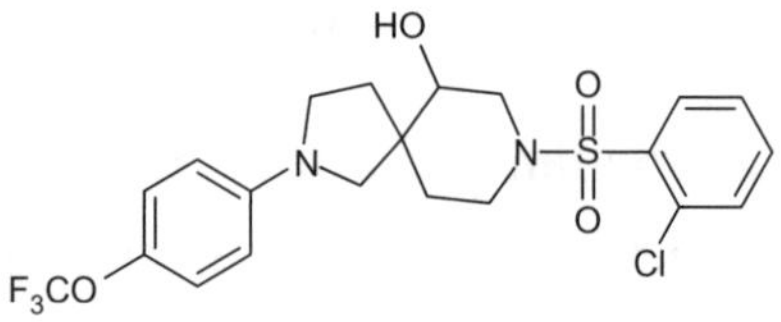

9 Hormone sensitive lipase IC_{50} = 0.01 μM

Figure 12.2 Structures of SHIP2i **4** and insulin sensitizers **5–9** that can potentially modulate lipid levels.

lipids in the liver (Sections 2.2.1 and 2.2.2), or changing the composition of lipid intermediates, especially lipokines, in the total lipid pool (Sections 2.2.3 and 2.2.4) can lead to significant improvements in insulin sensitivity.[15] Other recently disclosed insulin sensitizers **5–9** have also been shown to modulate lipid levels but are not discussed further in this review (Fig. 12.2).[32–36] Insulin sensitizers that activate AMP-activated kinase (AMPK) to decrease lipid synthesis are described in greater detail in this issue of *Annual Reports* while those that inhibit acetyl-CoA carboxylase (ACC), a direct target of AMPK activators, have been reviewed in a past issue.[37]

2.2.1 DGAT inhibitors

The final step in triacylglycerol (TAG) synthesis involves the acylation of diacylglycerols by acyl CoA:diacylglycerol acyltransferase (DGAT) enzymes. Two isoforms, DGAT1 and DGAT2, have been identified, with the former localized mainly in the intestines and thought to play a role in the reesterification of hydrolyzed TAGs, and the latter in liver and adipose tissue. *Dgat*1$^{-/-}$ knockout mice are resistant to diet-induced obesity and hepatic steatosis and exhibit improved insulin sensitivity.[38] ASO knockdown of DGAT2 in rodent models was shown to decrease hepatic lipids and improve insulin sensitivity, without appearing to cause any overt toxicity.[39] A DGAT2 ASO ISIS-$DGAT_{RX}$ is currently in preclinical development.[28]

DGAT1 has been the subject of extensive research and patent activity, reviewed in detail elsewhere.[40] Since 2010, a few novel DGAT1 chemotypes have been disclosed, such as **10** (IC_{50} = 0.089 μM; Fig. 12.3).[41] Two DGAT1 inhibitors, PF-04620110 and AZD7687, have advanced into clinical trials for the treatment of T2D and obesity, but no longer appear to be under development.[42] It remains to be determined whether other classes of DGAT1 inhibitors will demonstrate insulin-sensitizing activities in humans.[42] The only published example of a small molecule inhibitor of DGAT2 is niacin, a weak noncompetitive inhibitor in HepG2 cells (IC_{50} = 0.1 mM).[43]

2.2.2 Adiponectin receptor agonists

Adiponectin is a hormone secreted by adipocytes and believed to act in muscle and liver to improve insulin sensitivity by stimulating FA oxidation, inhibiting FA synthesis (via activation of AMPK), and decreasing gluconeogenesis.[44] Adiponectin signals, in part, through two 7-transmembrane (7TM) receptors, AdipoR1 and AdipoR2, which differ from classical 7TM G-protein-coupled receptors (GPCRs) in that the carboxyl terminal domain is extracellular.[44] A series of spirocyclic derivatives such as **11** (IC_{50} = 0.57 μM; Fig. 12.3) was disclosed that reportedly activate AdipoR1 and stimulate ACC phosphorylation in cells overexpressing AdipoR1.[45] Compounds illustrated by **12** (AMPK EC_{50}'s = 0.5–1 μM) are structurally related to a series of benzylpiperidine derivatives exemplified by **13** (AMPK EC_{50}'s < 1 μM) that were discovered from an adiponectin binding displacement assay.[46,47] A 10-mer peptide analog ADP 355 (H-*D*-Asn-Ile-Pro-Nva-Leu-Tyr-*D*-Ser-Phe-Ala-*D*-Ser-NH_2) derived from the loop-β-sheet

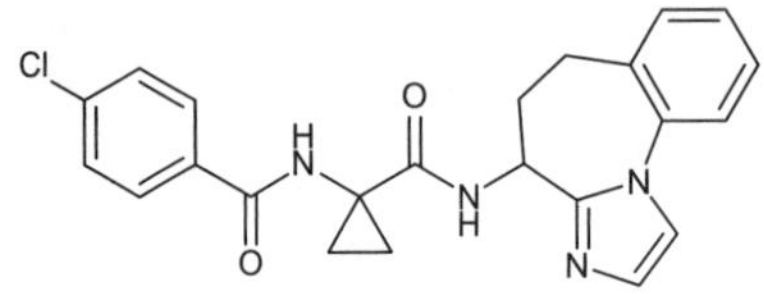

10

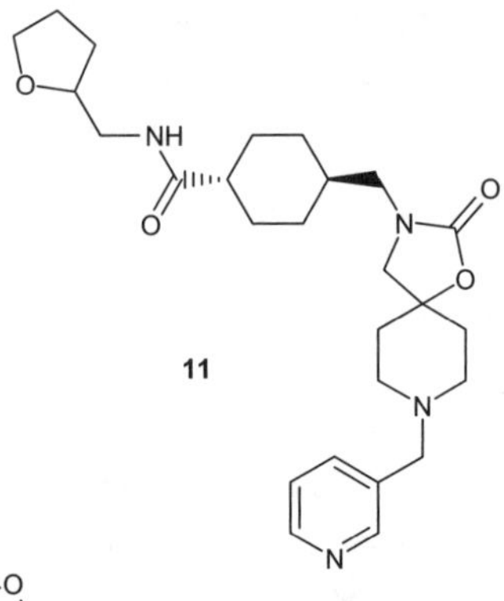

11

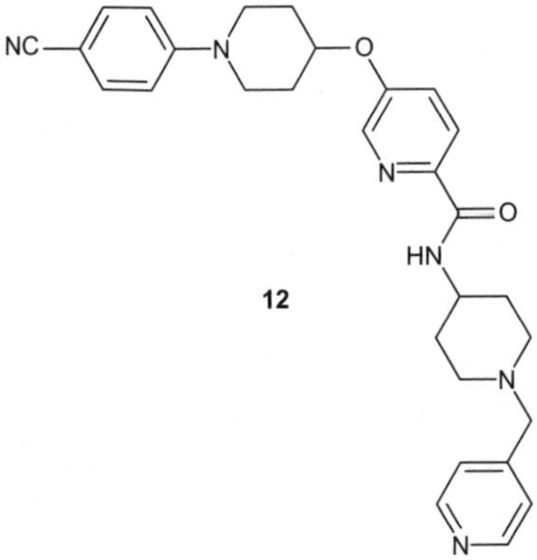

12

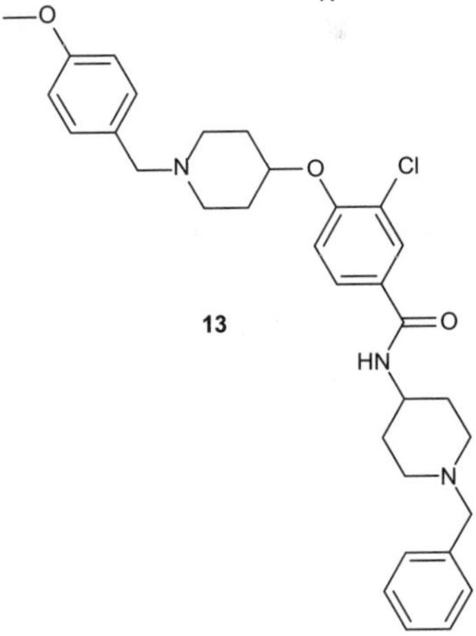

13

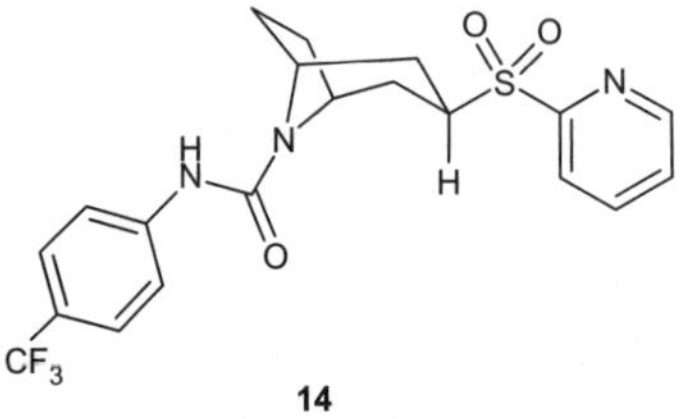

14

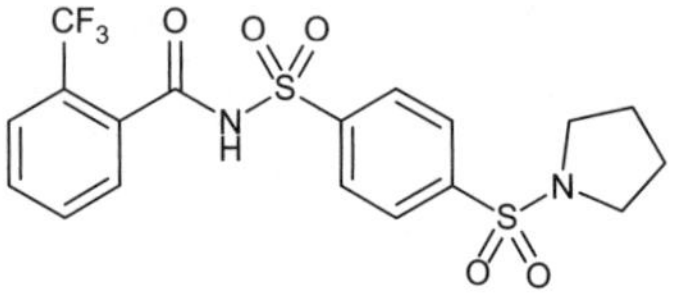

15

Figure 12.3 DGAT1 inhibitor **10**, adiponectin receptor mimetics **11–13**, and Elovl6 inhibitors **14–15**.

region of adiponectin has also been shown to activate adiponectin receptors, primarily AdipoR1, and stimulate downstream signaling, for example, AMPK.[48] The target specificity of ADP 355 was confirmed using siRNA technology. ADP 355 has been evaluated as an anticancer agent, but not as an insulin sensitizer.

2.2.3 Elovl6 inhibitors

Elongation of very long chain fatty acids protein 6 (Elovl6) catalyzes the formation of C14-, C16-, and C18- saturated and unsaturated FAs from C12-, C14-, and C16- fatty acids, respectively, using malonyl CoA as a two carbon donor. Recent evidence suggests that the FA palmitoleate may act as a lipokine to stimulate muscle insulin action.[49] Genetic knockdown of Elov6 does not protect mice on a high fat diet from obesity or severe hepatic steatosis but does prevent the development of insulin resistance. Pharmacological inhibition using the Elov6 inhibitor **14** ($IC_{50}=0.038$ μM; Fig. 12.3), however, did not recapitulate the insulin sensitivity phenotype observed in the knockout mice.[50] It was suggested that this may be due to selective inhibition of liver Elovl6 and a lack of inhibition in muscle and adipose tissue. No changes in the ratio of palmitoleate to palmitate were observed during the study. A novel series of Elov6 inhibitors exemplified by the acylsulfonamide **15** was recently disclosed (%inh @ 10 μM = 100%).[51]

2.2.4 GCS inhibitors (GCSi)

Excess production of sialic acid-containing glycosphingolipids such as ganglioside GM3 has been linked to insulin resistance through the formation of lipid rafts that disrupt the activity of the membrane-bound IR. Transgenic mice lacking GM3 due to the absence of GM3 synthase show improved insulin sensitivity.[52] Glucosylceramide synthase (GCS) is a glucosyltransferase that processes sphingolipid ceramides into glucosylceramides, intermediates in the production of GM3. GCSi Genz-123346 (**16**; $IC_{50}=0.016$ μM; Fig. 12.4) modulates GM3 levels and improves glycemic control and insulin sensitivity in animal models of type 2 diabetes.[53] Genz-112638 (**17**; $IC_{50}=0.009$ μM) is currently in Phase III clinical trials for type 1 Gaucher disease. A series of HTS-derived nonsphingosine inhibitors exemplified by **18** ($IC_{50}=0.016$ μM) has been recently described.[54] No GCSi's are yet believed to be under clinical evaluation for the treatment of T2D.

2.3. Insulin sensitizers that modulate inflammatory pathways

Proinflammatory cytokines secreted by adipocytes or resident macrophages can stimulate lipolysis (e.g., tumor necrosis factor α, TNF-α) or modulate the insulin signaling pathway, potentially by activating intracellular kinases such as c-Jun N-terminal kinase (JNK) and inhibitor of kappa B kinase (IKK) (e.g., TNF-α or interleukin 1, IL-1).[55,56] JNK can phosphorylate IRS1, preventing IRS1 binding to IR. JNK inhibitors such as **19** have been shown to improve insulin sensitivity (Fig. 12.4).[57] IKK activates

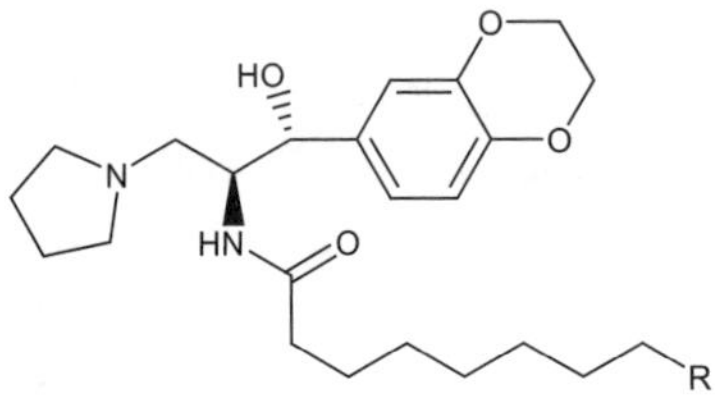

16 R = H; Genz-112638

17 R = Me; Genz-123346

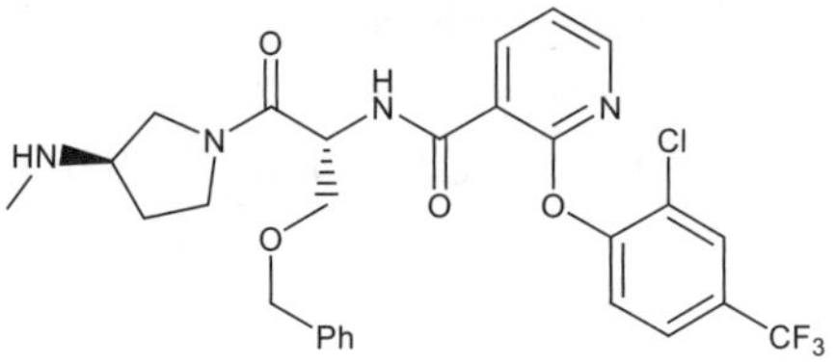

18

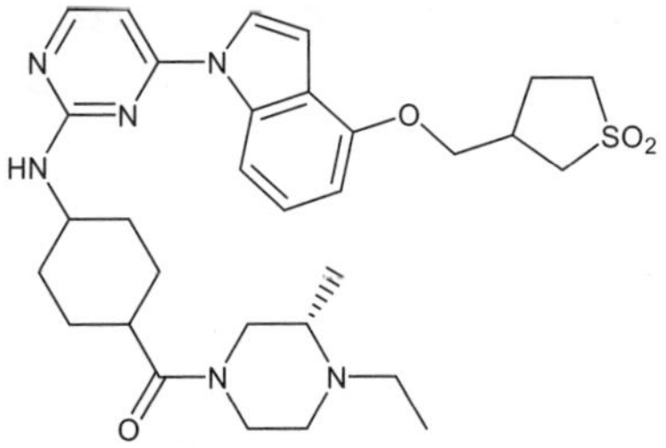

19 c-Jun N-terminal kinase IC_{50} = 0.019 μM

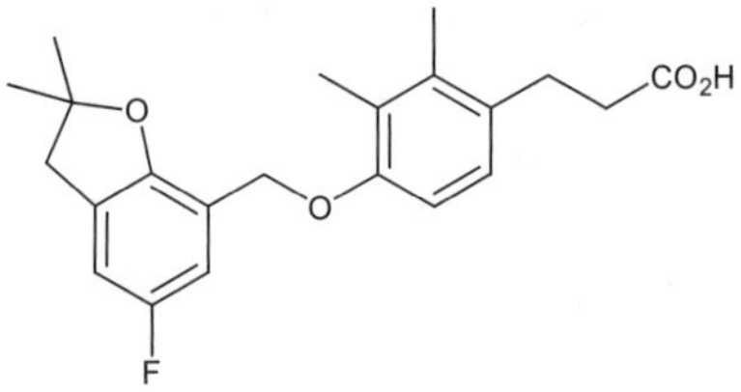

20

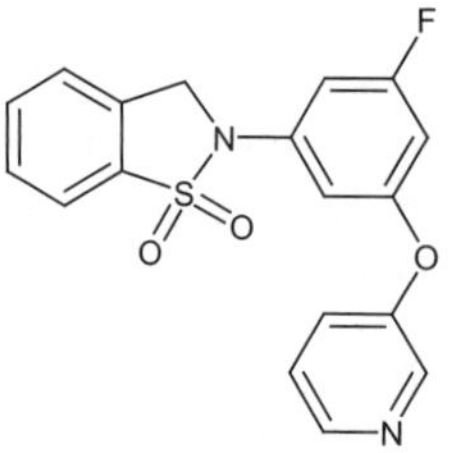

21

Figure 12.4 Structures of GCSi's **16–18**, JNK inhibitor **19**, and GPR120 agonists **20–21**.

translocation of the transcription factor nuclear factor-κB (NF-κB), which regulates the expression of many proinflammatory genes, including TNF-α and IL-1. Both the JNK and IKK/NF-κB pathways are upregulated in insulin-resistant states in animal models and humans,[58,59]

2.3.1 GPR120 agonists

GPR120 is a GPCR that responds to long-chain FAs. This receptor was recently shown to be expressed in macrophages, where it is coupled to the β-arrestin pathway.[60] FA activation leads to recruitment of β-arrestin-2, receptor internalization, and sequestration of proinflammatory proteins.[60] Omega-3 FAs such as docosahexaenoic acid are β-arrestin-biased GPR120 agonists (EC_{50}'s = 1–10 μM) that reduce inflammation in adipose tissue and improve insulin sensitivity in mice on a high fat diet.[60] Several GPR120 agonists, such as the carboxylic acid derivative **20** (EC_{50}'s = 1–10 μM; Fig. 12.4), have been disclosed.[61] These compounds appear to have been identified in assays measuring ligand-stimulated mobilization of intracellular calcium, another GPR120 signaling pathway thought to play a role in incretin secretion from the gastrointestinal tract. Compound **20** was shown to reduce glucose excursion in an IP glucose tolerance test (42% reduction @ 30 mg/kg po). A neutral series of GPR120 agonists, exemplified by **21** (EC_{50} = 0.18 μM), has also been identified. The activity of **21** for β-arrestin recruitment has not been reported.[62] No GPR120 agonists are reported to be clinical development.

2.4. Insulin sensitizers that act through other pathways

2.4.1 Ghrelin receptor antagonists

Ghrelin is an orexigenic peptide that contains a unique octanoyl group at Ser3. Its pharmacological activities are mediated via the ghrelin receptor (GHSR1a). Because of the role that ghrelin plays in stimulating food intake, GHSR1a antagonists were originally investigated as potential weight loss agents.[63] Recently, GHSR1a knockdown in mice and pharmacological inhibition using antagonists such as **22** (IC_{50} = 0.015 μM; Fig. 12.5) have been shown to enhance insulin sensitivity.[64,65] The ghrelin receptor is believed to possess high constitutive activity, suggesting that receptor modulation by an inverse agonist may lead to a differentiated profile compared to a neutral antagonist.[66] Neutral ghrelin antagonist **23** unexpectedly stimulated food intake in rodent models.[67] Recent disclosures include macrocyclic compounds such as **24** (IC_{50}'s = 0.001–0.01 μM),[68] and the inverse agonist **25**.[69] No ghrelin receptor antagonists are reported to be in clinical development.

2.4.2 CDK5 inhibitors

A novel mode of action was recently uncovered for PPARγ agonists such as rosiglitazone.[70] These agonists were shown to block cyclin-dependent kinase-5 (CDK5)-mediated phosphorylation of PPARγ, preventing PPARγ-mediated suppression of insulin responsive genes that regulate adipokines such as adiponectin. In human T2D patients treated with rosiglitazone, individual reductions in fasting plasma glucose and insulin

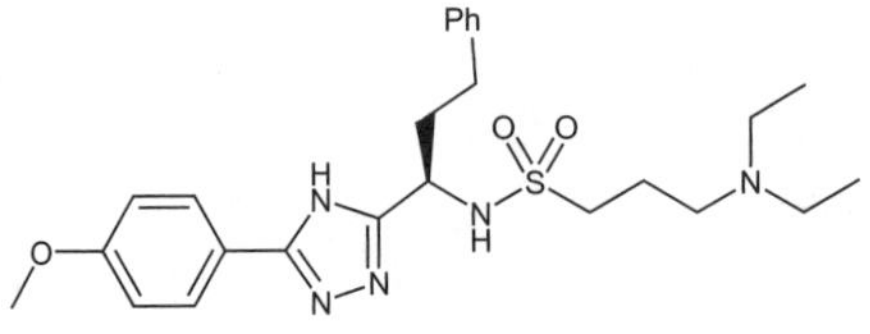

22

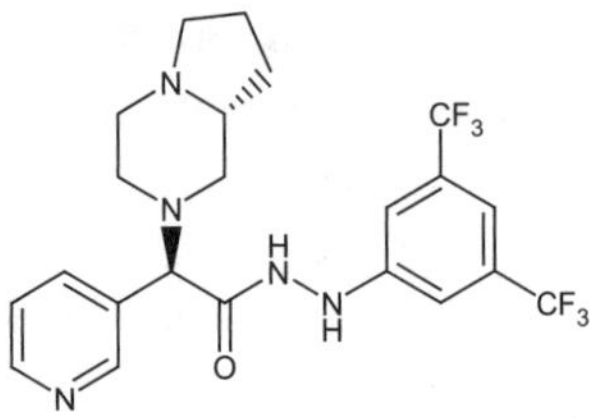

23

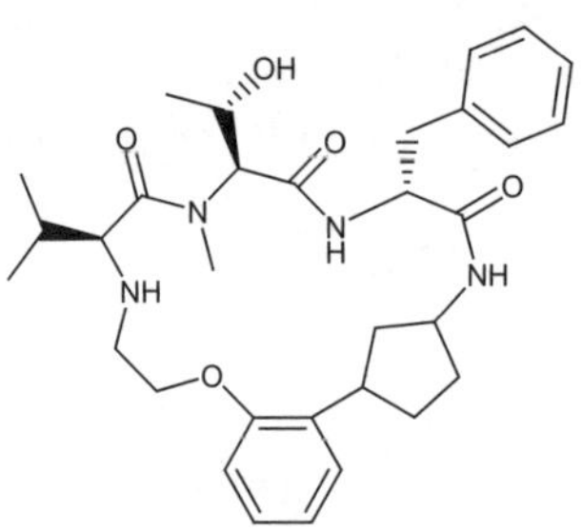

24

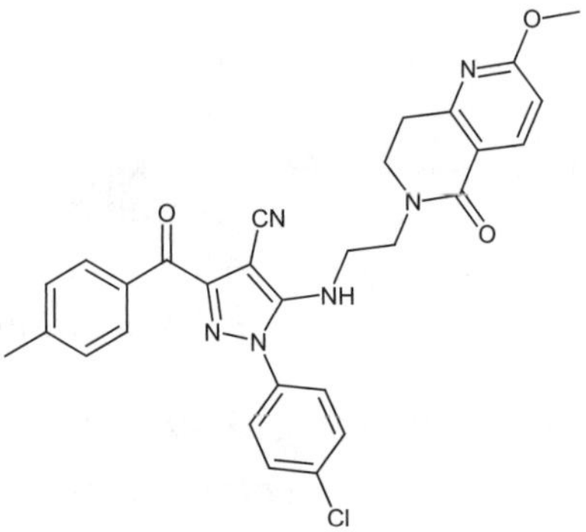

25

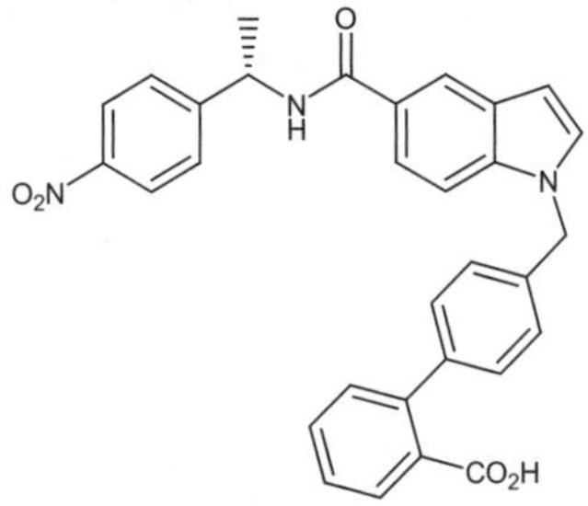

26, SR1664

Figure 12.5 Structures of ghrelin receptor antagonists **22–25** and compound **26** that blocks CDK5-mediated phosphorylation of PPARγ.

levels correlated with the degree of agonist-induced inhibition of PPARγ phosphorylation.[70] Novel inhibitors of CDK5-mediated phosphorylation of PPARγ such as SR1664 (**26**; Fig. 12.5) have been disclosed which are devoid of PPARγ transcriptional agonism.[71] Compound **26** demonstrated equivalent glucose-lowering activity to rosiglitazone in rodent models but exhibited an improved profile with respect to weight gain, edema, and bone formation. It remains to be determined whether the efficacy and safety profiles of agents such as **26** will translate to humans and differentiate from full PPARγ agonists and SPPARγMs.[7]

3. CONCLUSION

The rapid world-wide growth in the prevalence of T2D, combined with the limited durability of established antidiabetic medicines, suggests that a large medical need exists for new, safe, and effective treatments. While many gliptins and insulin secretagogues/incretin mimetics have been approved since 2000, and many others are in development, no new insulin sensitizers have reached the marketplace in recent years, attesting to the difficulty of discovering such drugs. The efficacy achieved with TZDs not only has set a high bar in terms of glucose lowering but also has illustrated the challenges and high demands of patients, prescribers, and payers on drug safety.

In the past decade, a greater understanding of the science underlying the molecular pathways that lead to insulin resistance has emerged. Accumulation of lipids in the liver and other tissues and the development of inflammation in adipose tissue and liver have been shown to play important roles in the development of T2D. While the number of druggable targets within the IR signaling cascade is limited, human genetic data and studies using transgenic rodent models have identified several promising targets involved in lipid metabolism and inflammatory pathways that could potentially be modulated to improve insulin sensitivity. Both small and large molecules with activities against some of these targets have shown promise in rodent models as insulin sensitizers. A formidable challenge in the coming years will be identifying targets for which rodent pharmacology, and equally importantly safety, will translate to robust and safe insulin sensitizing and glucose-lowering activities in humans to meet demands from the ever-growing population of T2D patients.

REFERENCES

(1) Centers for Disease Control Statistics 2011. Available at: http://www.cdc.gov/diabetes/pubs/references11.htm.

(2) Economic costs of diabetes in the US. In 2007. *Diabetes Care* **2008**, *31*, 596.

(3) Kohli, R.; Stefater, M.A.; Inge, T.H. *Rev. Endocr. Metab. Disord.* **2011**, *12*, 211.

(4) Nathan, D.M.; Buse, J.B.; Davidson, M.B.; Ferrannini, E.; Holman, R.R.; Sherwin, R.; Zinman, B. *Diabetes Care* **2009**, *32*, 193.

(5) Kahn, S.E.; Haffner, S.M.; Heise, M.A.; Herman, W.H.; Holman, R.R.; Jones, N.P.; Kravitz, B.G.; Lachin, J.M.; O'Neill, C.; Zinman, B.; Viberti, G. *N. Engl. J. Med.* **2006**, *355*, 2427.

(6) Tolman, K.G. *Expert Opin. Drug Saf.* **2011**, *10*, 419.
(7) Pirat, C.; Farce, A.; Lebegue, N.; Renault, N.; Furman, C.; Millet, R.; Yous, S.; Speca, S.; Berthelot, P.; Desreumaux, P.; Chavatte, P. *J. Med. Chem.* **2012**, *55*, 4027.
(8) Wang, C.; McCann, M.E.; Doebber, T.W.; Wu, M.; Chang, C.H.; McNamara, L.; McKeever, B.; Mosley, R.T.; Berger, J.P.; Meinke, P.T. *J. Med. Chem.* **2011**, *54*, 8541.
(9) Khanderia, U.; Pop-Busui, R.; Eagle, K.A. *Ann. Pharmacother.* **2008**, *42*, 1466.
(10) http://www.poxel.com/pipeline/imeglimin.
(11) http://www.msdrx.com/4_2.php.
(12) http://www.vervapharma.com/research-development/diabetes/.
(13) http://www.twipharma.com/ndd.htm.
(14) Cefalu, W.T. *Exp. Biol. Med.* **2001**, *226*, 13.
(15) Samuel, V.T.; Shulman, G.I. *Cell* **2012**, *148*, 852.
(16) Cheng, Z.; Tseng, Y.; White, M.F. *Trends Endocrinol. Metab.* **2010**, *21*, 589.
(17) Kim, J.J.; Li, P.; Huntley, J.; Chang, J.P.; Arden, K.C.; Olefsky, J.M. *Diabetes* **2009**, *58*, 1275.
(18) Pajvani, U.B.; Shawber, C.J.; Samuel, V.T.; Birkenfeld, A.L.; Shulman, G.I.; Kitajewski, J.; Accili, D. *Nat. Med.* **2011**, *17*, 961.
(19) Seto, S.; Yumoto, K.; Okada, K.; Asahina, Y.; Iwane, A.; Iwago, M.; Terasawa, R.; Shreder, K.R.; Murakami, K.; Kohno, Y. *Bioorg. Med. Chem.* **2012**, *20*, 1188.
(20) Bento, J.L.; Palmer, N.D.; Mychaleckyj, J.C.; Lange, L.A.; Langefeld, C.D.; Rich, S.S.; Freedman, B.I.; Bowden, D.W. *Diabetes* **2004**, *53*, 3007.
(21) Elchebly, M.; Payette, P.; Michaliszyn, E.; Cromlish, W.; Collins, S.; Loy, A.L.; Normandin, D.; Cheng, A.; Himms-Hagen, J.; Chan, C.-C.; Ramachandran, C.; Gresser, M.J.; Tremblay, M.L.; Kennedy, B.P. *Science* **1999**, *283*, 1544.
(22) Wan, Z.-K.; Follows, B.; Kirincich, S.; Wilson, D.; Binnun, E.; Xu, W.; Joseph-McCarthy, D.; Wu, J.; Smith, M.; Zhang, Y.-L.; Tam, M.; Erbe, D.; Tam, S.; Saiah, E.; Lee, J. *Bioorg. Med. Chem. Lett.* **2007**, *17*, 2913.
(23) Popov, D. *Biochem. Biophys. Res. Commun.* **2011**, *410*, 377.
(24) Fukuda, S.; Ohta, T.; Sakata, S.; Morinaga, H.; Ito, M.; Nakagawa, Y.; Tanaka, M.; Matsushita, M. *Diabetes Obes. Metab.* **2010**, *12*, 299.
(25) Wiesmann, C.; Barr, K.J.; Kung, J.; Zhu, J.; Erlanson, D.A.; Shen, W.; Fahr, B.J.; Zhong, M.; Taylor, L.; Randal, M.; McDowell, R.S.; Hansen, S.K. *Nat. Struct. Mol. Biol.* **2004**, *11*, 730.
(26) Patel, D.; Jain, M.; Shah, S.R.; Bahekar, R.; Jadav, P.; Joharapurkar, A.; Dhanesha, N.; Shaikh, M.; Sairam, KVVM; Kapadnis, P. *Bioorg. Med. Chem. Lett.* **2012**, *22*, 1111.
(27) Geary, R.S.; Bradley, J.D.; Watanabe, T.; Kwon, Y.; Wedel, M.; van Lier, J.J.; van Vliet, A.A. *Clin. Pharmacokinet.* **2006**, *45*, 789.
(28) http://www.isispharm.com/Pipeline/index.htm.
(29) Suwa, A.; Kurama, T.; Shimokawa, T. *Expert Opin. Ther. Targets* **2010**, *14*, 727.
(30) Suwa, A.; Yamamoto, T.; Sawada, A.; Minoura, K.; Hosogai, N.; Tahara, A.; Kurama, T.; Shimokawa, T.; Aramori, I. *Br. J. Pharmacol.* **2009**, *158*, 879.
(31) Morino, K.; Petersen, K.F.; Shulman, G.I. *Diabetes* **2006**, *55*, S9.
(32) Wydysh, E.A.; Medghalchi, S.M.; Vadlamudi, A.; Townsend, C.A. *J. Med. Chem.* **2009**, *52*, 3317.
(33) Wu, M.; Singh, S.B.; Wang, J.; Chung, C.C.; Salituro, G.; Karanam, B.V.; Lee, S.H.; Powles, M.; Ellsworth, K.P.; Lassman, M.E.; Miller, C.; Myers, R.W.; Tota, M.R.; Zhang, B.B.; Li, C. *Proc. Natl. Acad. Sci. U.S.A.* **2011**, *108*, 5378.
(34) Matsunaga, N.; Igawa, H.; Suzuki, H.; Okamoto, R.; Furukawa, H.; Murayama, K. *Patent Application WO 2012/011592*, 2012.
(35) Oballa, R.M.; Belair, L.; Black, W.C.; Bleasby, K.; Chan, C.C.; Desroches, C.; Du, X.; Gordon, R.; Guay, J.; Guiral, S.; Hafey, M.J.; Hamelin, E.; Huang, Z.; Kennedy, B.;

Lachance, N.; Landry, F.; Li, C.S.; Mancini, J.; Normandin, D.; Pocai, A.; Powell, D.A.; Ramtohul, Y.K.; Skorey, K.; Sorensen, D.; Sturkenboom, W.; Styhler, A.; Waddleton, D.M.; Wang, H.; Wong, S.; Xu, L.; Zhang, L. *J. Med. Chem.* **2011**, *54*, 5082.

(36) Ackermann, J.; Conte, A.; Hunziker, D.; Neidhart, W.; Nettekoven, M.; Schulz-Gasch, T.; Wertheimer, S. *Patent Application WO 2010/130665*, 2010.

(37) Bourbeau, M.P.; Allen, J.G.; Gu, W. *Annu. Rep. Med. Chem.* **2010**, *45*, 95.

(38) Chen, H.C.; Smith, S.J.; Ladha, Z.; Jensen, D.R.; Ferreira, L.D.; Pulawa, L.K.; McGuire, J.G.; Pitas, R.E.; Eckel, R.H.; Farese, R.V., Jr. *J. Clin. Invest.* **2002**, *109*, 1049.

(39) Choi, C.S.; Savage, D.B.; Kulkarni, A.; Yu, X.X.; Liu, Z.X.; Morino, K.; Kim, S.; Distefano, A.; Samuel, V.T.; Neschen, S.; Zhang, D.; Wang, A.; Zhang, X.M.; Kahn, M.; Cline, G.W.; Pandey, S.K.; Geisler, J.G.; Bhanot, S.; Monia, B.P.; Shulman, G.I. *J. Biol. Chem.* **2007**, *282*, 22678.

(40) Birch, A.M.; Buckett, L.K.; Turnbull, A.V. *Curr. Opin. Drug Discov. Devel.* **2010**, *13*, 489.

(41) Arnold, M.B.; Beauchamp, T.J.; Canada, E.J.; Hembre, E.J.; Lu, J.; Rizzo, J.R.; Schaus, J.M.; Shi, Q. Patent Application WO 2011/071840, 2011.

(42) Dow, R.L.; Li, J.-C.; Pence, M.P.; Gibbs, E.M.; LaPerle, J.L.; Litchfield, J.; Piotrowski, D.W.; Munchhof, M.J.; Manion, T.B.; Zavadoski, W.J.; Walker, G.S.; McPherson, R.K.; Tapley, S.; Sugarman, E.; Guzman-Perez, A.; DaSilva-Jardine, P. *ACS Med. Chem. Lett.* **2011**, *2*, 407.

(43) Ganji, S.H.; Tavintharan, S.; Zhu, D.; Xing, Y.; Kamanna, V.S.; Kashyap, M.L. *J. Lipid Res.* **2004**, *45*, 1835.

(44) Kadowaki, T.; Yamauchi, T.; Kubota, N.; Hara, K.; Ueki, K.; Tobe, K. *J. Clin. Invest.* **2006**, *116*, 1784.

(45) Nakano, S.; Takahashi, K.; Takada, J.; Iwamoto, T.; Nagae, K.; Maruyama, Y.; Shintani, Y.; Okada, T.; Ito, Y.; Kadowaki, T.; Yamauchi, T.; Iwabu, M.; Iwabu, M. Patent Application WO 2011/142359, 2011.

(46) Singh, R.; Yu, J.; Hong, H.; Thota, S.; Xu, X. Patent Application WO 2008/083124, 2008.

(47) Payan, D.; Hitoshi, Y.; Kinsella, T. Patent Application WO 2011/123681. **2011**.

(48) Otvos, L., Jr.; Haspinger, E.; La Russa, F.; Maspero, F.; Graziano, P.; Kovalszky, I.; Lovas, S.; Nama, K.; Hoffmann, R.; Knappe, D.; Cassone, M.; Wade, J.; Surmacz, E. *BMC Biotechnol.* **2011**, *11*, 90.

(49) Cao, H.; Gerhold, K.; Mayers, J.R.; Wiest, M.M.; Watkins, S.M.; Hotamisligil, G.S. *Cell* **2008**, *134*, 933.

(50) Shimamura, K.; Nagumo, A.; Miyamoto, Y.; Kitazawa, H.; Kanesaka, M.; Yoshimoto, R.; Aragane, K.; Morita, N.; Ohe, T.; Takahashi, T.; Nagase, T.; Sato, N.; Tokita, S. *Eur. J. Pharmacol.* **2010**, *630*, 34.

(51) Tawaraishi, T. Patent Application WO 2011/102514, 2011.

(52) Yamashita, T.; Hashiramoto, A.; Haluzik, M.; Mizukami, H.; Beck, S.; Norton, A.; Kono, M.; Tsuji, S.; Daniotti, J.L.; Werth, N.; Sandhoff, R.; Sandhoff, K.; Proia, R.L. *Proc. Natl. Acad. Sci. U.S.A.* **2003**, *100*, 3445.

(53) Zhao, H.; Przybylska, M.; Wu, I.H.; Zhang, J.; Siegel, C.; Komarnitsky, S.; Yew, N.S.; Cheng, S.H. *Diabetes* **2007**, *56*, 1210.

(54) Koltun, E.; Richards, S.; Chan, V.; Nachtigall, J.; Du, H.; Noson, K.; Galan, A.; Aay, N.; Hanel, A.; Harrison, A.; Zhang, J.; Won, K.-A.; Tam, D.; Qian, F.; Wang, T.; Finn, P.; Ogilvie, K.; Rosen, J.; Mohan, R.; Larson, C.; Lamb, P.; Nuss, J.; Kearney, P. *Bioorg. Med. Chem. Lett.* **2011**, *21*, 6773.

(55) Schenk, S.; Saberi, M.; Olefsky, J.M. *J. Clin. Invest.* **2008**, *118*, 2992.

(56) Shoelson, S.E.; Lee, J.; Goldfine, A.B. *J. Clin. Invest.* **2006**, *116*, 2308.

(57) Bogoyevitch, M.A. *Bioessays* **2006**, *28*, 923.
(58) Masharani, U.B.; Maddux, B.A.; Li, X.; Sakkas, G.K.; Mulligan, K.; Schambelan, M.; Goldfine, I.D.; Youngren, J.F. *PLoS One* **2011**, *6*, e19878.
(59) Shoelson, S.E.; Lee, J.; Yuan, M. *Int. J. Obes.* **2003**, *27*, S49.
(60) Oh, D.Y.; Talukdar, S.; Bae, E.J.; Imamura, T.; Morinaga, H.; Fan, W.Q.; Li, P.; Lu, W.J.; Watkins, S.M.; Olefsky, J.M. *Cell* **2010**, *142*, 687.
(61) Shi, D.F.; Song, J.; Ma, J.; Novack, A.; Pham, P.; Nashashibi, I.F.; Rabbat, C.J.; Chen, X. Patent Application US 2011/0313003, 2011.
(62) Arakawa, K.; Nishimura, T.; Sugimoto, Y.; Takahashi, H.; Shimamura, T. Patent Application WO 2010/104195, 2010.
(63) Seim, I.; El-Salhy, M.; Hausken, T.; Gundersen, D.; Chopin, L. *Curr. Pharm. Des.* **2012**, *18*, 768.
(64) Longo, K.A.; Govek, E.K.; Nolan, A.; McDonagh, T.; Charoenthongtrakul, S.; Giuliana, D.J.; Morgan, K.; Hixon, J.; Zhou, C.; Kelder, B.; Kopchick, J.J.; Saunders, J.O.; Navia, M.A.; Curtis, R.; DiStefano, P.S.; Geddes, B.J. *J. Pharmacol. Exp. Ther.* **2011**, *339*, 115.
(65) Longo, K.A.; Charoenthongtrakul, S.; Giuliana, D.J.; Govek, E.K.; McDonagh, T.; Qi, Y.; DiStefano, P.S.; Geddes, B.J. *Regul. Pept.* **2008**, *150*, 55.
(66) Els, S.; Beck-Sickinger, A.G.; Chollet, C. *Methods Enzymol.* **2010**, *485*, 103.
(67) Xin, Z.; Serby, M.D.; Zhao, H.; Kosogof, C.; Szczepankiewicz, B.G.; Liu, M.; Hutchins, C.W.; Sarris, K.A.; Hoff, E.D.; Falls, H.D.; Lin, C.W.; Ogiela, C.A.; Collins, C.A.; Brune, M.E.; Bush, E.N.; Droz, B.A.; Fey, T.A.; Knourek-Segel, V.E.; Shapiro, R.; Jacobson, P.B.; Beno, D.W.A.; Turner, T.M.; Sham, H.L.; Liu, G. *J. Med. Chem.* **2006**, *49*, 4459.
(68) Hoveyda, H.; Marsault, E.; Thomas, H.; Fraser, G.; Beaubien, S.; Mathieu, A.; Beignet, J.; Bonin, M.-A.; Phoenix, S.; Drutz, D.; Peterson, M.; Beauchemin, S.; Brassard, M.; Vezina, M. Patent Application WO 2011/053821, 2011.
(69) Ge, M.; Cline, E.; Yang, L.; Mills, S.G. Patent Application WO 2008/008286, 2008.
(70) Choi, J.H.; Banks, A.S.; Estall, J.L.; Kajimura, S.; Bostrom, P.; Laznik, D.; Ruas, J.L.; Chalmers, M.J.; Kamenecka, T.M.; Bluher, M.; Griffin, P.R.; Spiegelman, B.M. *Nature* **2010**, *466*, 451.
(71) Choi, J.H.; Banks, A.S.; Kamenecka, T.M.; Busby, S.A.; Chalmers, M.J.; Kumar, N.; Kuruvilla, D.S.; Shin, Y.; He, Y.; Bruning, J.B.; Marciano, D.P.; Cameron, M.D.; Laznik, D.; Jurczak, M.J.; Schurer, S.C.; Vidovic, D.; Shulman, G.I.; Spiegelman, B.M.; Griffin, P.R. *Nature* **2011**, *477*, 477.

PART 3

Inflammation Pulmonary GI

Editor: David S. Weinstein
Bristol-Myers Squibb R&D, Princeton, New Jersey

CHAPTER THIRTEEN

Recent Advances in the Discovery and Development of Sphingosine-1-Phosphate-1 Receptor Agonists

Alaric J. Dyckman
Bristol-Myers Squibb R&D, Princeton, New Jersey, USA

Contents

1. INTRODUCTION

Sphingosine-1-phosphate (S1P, **3**) is a zwitterionic lysophospholipid metabolite of sphingosine (Sph, **2**), which in turn is derived from enzymatic cleavage of ceramides (**1**). Enzymatic phosphorylation of Sph by two kinases (SphK1 and SphK2) leads to the production of S1P largely from erythrocytes.[1] Originally thought to operate solely as an intracellular signaling molecule, S1P was subsequently determined to be a high-affinity ligand for five members of the endothelial differentiation gene class of G-protein-coupled receptors subsequently renamed S1P1–5.[2] The interaction of S1P with the S1P receptors plays a fundamental physiological role in a large number of processes including proliferation, vascular development, and lymphocyte trafficking.[3] Tissue levels of S1P are maintained lower than circulating levels through the action of degrading enzymes (S1P-phosphatases and S1P-lyase). The resulting gradient of S1P is sensed by lymphocytes through interaction with cell surface S1P1 receptors to promote their migration out of the lymphatic system.[4]

Annual Reports in Medicinal Chemistry, Volume 47
ISSN 0065-7743
http://dx.doi.org/10.1016/B978-0-12-396492-2.00013-8

Research efforts around S1P receptors intensified after they were linked to the efficacy observed with the potent immunomodulatory agent fingolimod (**4**, FTY720). Administration of fingolimod was known to induce a significant, but reversible, reduction in circulating lymphocytes through a novel mechanism in which the affected cells were found to be trapped in the thymus and secondary lymphoid organs, thus preventing their access to sites of inflammation or tissue graft.[5] As such, fingolimod was found to be active in a wide array of animal models of inflammatory diseases and solid organ transplant.[6] The precise molecular mode of action of fingolimod was unknown until the discovery that it was stereospecifically monophosphorylated *in vivo* by SphK2, analogous to the conversion of Sph to S1P.[7,8] The resulting metabolite **5** (fingolimod-P) was found to act as a potent agonist of four of the five S1P receptors (S1P1, 3–5). Further genetic and pharmacological experiments linked the lymphocyte sequestration ability of fingolimod specifically to agonism of the S1P1 receptor by **5**. Agonism of S1P1 by **5** results in prolonged internalization and degradation of the receptor, in effect depleting it from the cell surface.[9,10] In this "functional antagonist" mode of operation, the affected lymphocytes can no longer sense the S1P gradient from tissue to blood, thereby losing their ability to exit into circulation. Consistent with this mechanistic hypothesis, recent disclosures have reported the ability of synthetic *antagonists* of S1P1 to also elicit lymphopenia.[11]

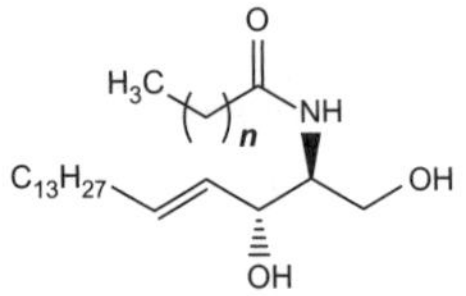

1 Ceramides

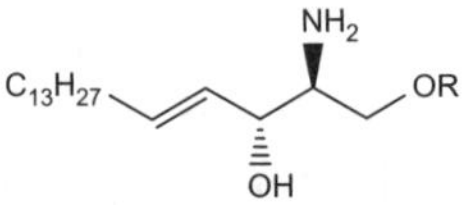

2 R = H **Sphingosine**
3 R = PO_3H_2 **Sphingosine-1-phosphate**

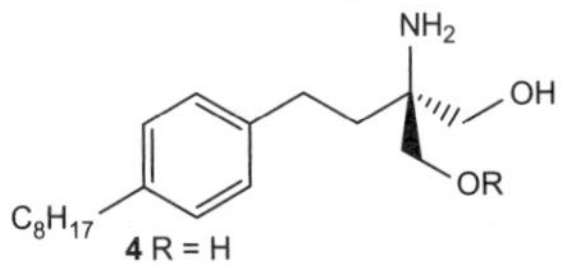

4 R = H
5 R = PO_3H_2

2. RECENT CLINICAL DEVELOPMENTS OF S1P1 AGONISTS

Fingolimod was initially advanced into the clinic for the prevention of graft rejection in renal transplant patients, and while mid-stage trials demonstrated proof of confidence, with results comparable to standard of care, the overall safety and efficacy profile that emerged from larger trials evaluating 2.5–5.0 mg QD fingolimod was not sufficient to continue development for this indication.[12–14] Concurrent trials initiated for the treatment of relapsing remitting multiple sclerosis (RRMS) provided more favorable results, with efficacy improved over a standard of care that was maintained even with substantially lower doses than those required in the transplant trials (0.5 mg QD).[15] In 2010, fingolimod (Gilenya™) was approved in the United States by the FDA as the first oral disease-modifying treatment for RRMS. Concerns noted with fingolimod include cardiovascular effects (e.g., heart rate reduction, blood pressure elevation), macular edema, teratogenic effects, decline in pulmonary function, elevation of liver enzymes, and a long half-life.[16] Follow-on efforts have been directed toward improving upon this profile primarily in two ways. First, identification of compounds with a reduced pharmacokinetic (PK) half-life would mitigate concerns over the slow recovery from lymphocyte suppression observed upon drug withdrawal. Second, by improving receptor selectivity (eliminating interaction with S1P3), it was hoped to avoid the cardiovascular effects which were clearly tied to agonism of S1P3 in rodents.[17] Recent information from clinical trials of selective S1P1 agonists, however, indicates that the cardiovascular liabilities are not eliminated with S1P3-sparing agonists.[18] Nevertheless, as the activity on S1P3 does not appear to contribute to efficacy, the emphasis on identifying agonists with improved selectivity for S1P1 remains a consistent theme in nearly all recent reports.

Additional S1P1 agonists that have entered clinical trials and for which a structure has been disclosed include two prodrug compounds KRP-203 (**6**) and CS-0777 (**8**).[19,20] Their corresponding phosphates demonstrated improved S1P1 versus S1P3 selectivity relative to fingolimod-P (P3/P1 = 10 for **5** vs. >300 for **9** and >1000 for **7**).[21,22] KRP-203 is under evaluation for the treatment of ulcerative colitis as well as subacute cutaneous lupus erythematosus.[23] CS-0777 was evaluated in a single ascending dose study at 0.1–2.5 mg, with dose-related lymphocyte reductions ranging from 7% to 85%.[24] CS-0777 was found to have a long human half-life of 171–211 h, similar to that of fingolimod. Repeat dosing of CS-0777 (0.1–0.6 mg QW or Q2W) in patients with MS revealed a dose-dependent reduction in absolute lymphocyte counts and reduction in heart rate.[25]

Diol **10** (ponesimod) is a direct-acting agonist of S1P1; unlike the previously described compounds, it does not require phosphorylation for its pharmacological activity. Ponesimod displays slightly improved receptor selectivity over fingolimod (P3/P1 = 18) and a shorter human half-life of 22–33 h.[26] In addition to completing a Phase 2b trial for RRMS, ponesimod is under evaluation in separate trials for psoriasis.[27] Additional direct-acting agonists include amino-cyclobutane carboxylic acid **11** (PF-991) and azetidine carboxylic acid **12**. Advanced into early clinical trials as a potential treatment for rheumatoid arthritis, selective direct-acting agonist **11** (P3/P1 > 3000) demonstrated 50% or 60% reduction of lymphocytes at 24 h from doses of 1 and 3 mg, respectively.[28] Compound **12** was administered to healthy subjects in a 1-month repeat-dose study across a dose range from 0.3 to 20 mg QD.[29] The mean change in absolute lymphocyte counts ranged from 42% to 85% with reductions in heart rate also being observed within this dose range. These data have been separately associated with BAF312 (structure not disclosed), a compound reported to demonstrate excellent receptor selectivity (P3/P1 > 3000) and a human half-life of approximately 30 h.[30] An initial trial in RRMS patients was recently completed for BAF312, and a separate ongoing trial is evaluating its efficacy in patients with the chronic inflammatory diseases, polymyositis or dermatomyositis.[31]

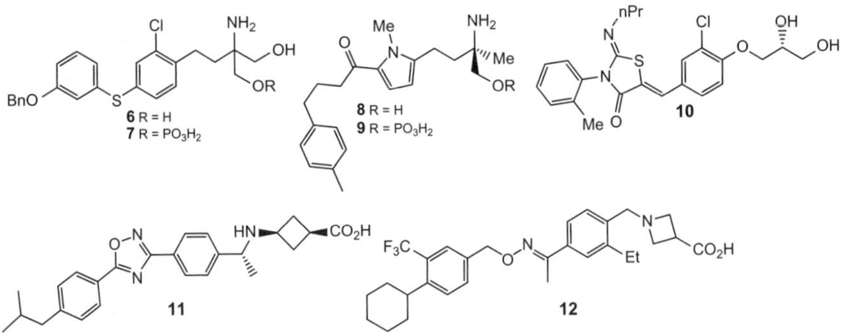

3. RECENT PRECLINICAL DEVELOPMENTS OF S1P1 AGONISTS

The medicinal chemistry efforts that were initiated shortly after the identification of S1P1 as the molecular target of fingolimod-P emphasized two main branches. In one, the focus turned to modifications of the amino diol prodrug compounds, optimizing for potency, selectivity, and *in vivo* properties, such as reduced half-life and increased formation of circulating phosphate metabolite. In the other area, direct acting agonists were explored, with early examples focusing on ionic phosphate mimetics. Early design efforts relied on homology models, refined by mutagenesis, which defined a lipophilic-binding pocket and pointed to key ion-pair interactions of the ligand amino with specific residues (including Arg120 and Glu121) within S1P1.[32] A recently disclosed cocrystal structure of an antagonist bound to S1P1 is a major advance for the field and will undoubtedly provide the basis for further structure-based design.[33] Stabilized phosphate surrogates such as phosphonates, thiophosphates, and carboxylic acids were designed to retain the putative interaction of the phosphate group with Arg120 of S1P1.[34] This was met with the greatest success in the case of the amino acid analogs, in particular, when azetidine-3-carboxylate was utilized in conjunction with a 3,5-diaryl-1,2,4-oxadiazole scaffold, first identified from high-throughput screening (HTS).[35] The resulting compounds (e.g., **16**) were potent agonists of S1P1, possessed good PK properties, and were able to induce lymphopenia in multiple species upon oral dosing.[36] Variations around this triaryl amino acid template have been explored by numerous research groups over the past few years, with modifications to the aryl core, the amino acid, and the lipophilic region of the molecules.[37,38] This is evidenced directly in **12**, and the influence is also apparent in the amino-cyclobutane of **11**.

In designing analogs of fingolimod, the only conserved element required for bioactivation has been the aminoethanol fragment. An amide was introduced as a rigidifying element in **13**, which was selective (phosphate: P3/P1 = 230) and active in eliciting lymphopenia in mice, but exhibited low conversion to the phosphate *in vivo*.[39] Elaboration of the amides to heterocycles was first realized in imidazole analogs such as **14** where ortho-CF3 substitution enhanced selectivity (phosphate: P3/P1 > 1800). Oral administration of **14** to mice led to maximal lymphopenia at doses as low as 1 mg/kg, with improved phosphorylation.[40,41] The compound had good PK in mouse ($t_{1/2}$ = 20 h) and was active in models of transplant and multiple sclerosis.[42] A similar heterocyclic constraint and ortho-substitution were retained in the structure of development candidate **15** (GSK1842799).[43]

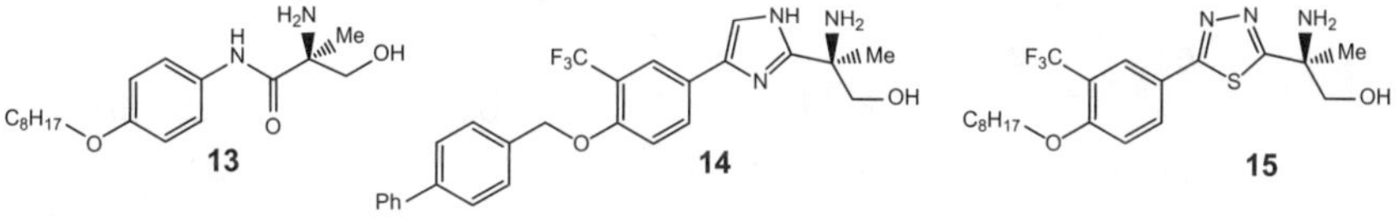

Among direct-acting agonists, the azetidine carboxylic acid originally described for **16** has been by far the most replicated surrogate for the amino phosphates. While in **17** an isoxazole was found to replace the phenyl ring of **16**, compound **18** utilizes a benzamido-thiadiazole to presumably occupy the same lipophilic pocket.[44,45] In **19**, the orientation of the azetidine and oxadiazole was changed from *para* to *meta*, yet good potency and selectivity were maintained (S1P1 pEC_{50} = 9.5; S1P3 pEC_{50} < 5.5).[46]

Alternatively, in CS-2100 (**20**), an ethyl-thiophene was found to be an effective phenyl replacement (S1P1 $EC_{50}=4.0$ nM; S1P3 $EC_{50}>20{,}000$ nM) leading to its selection as a clinical candidate.[47] However, development of CS-2100 was discontinued, reportedly due to concerns over the formation of 4-phenoxy benzoic acid via enterobacterial intestinal metabolism of the oxadiazole.[48] Efforts to identify metabolically stable replacements for the oxadiazole have been described.[49]

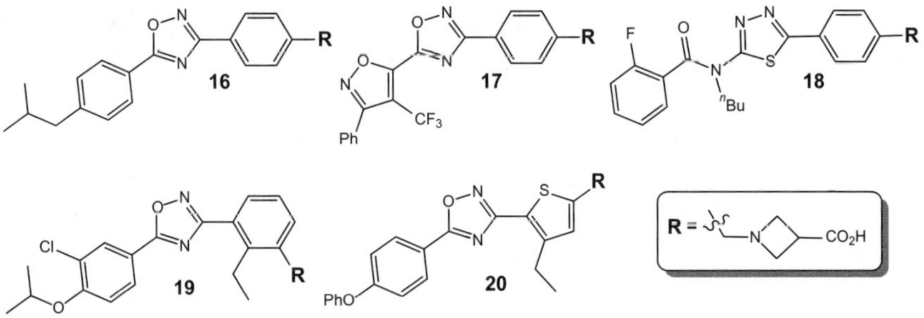

The azetidine carboxylate appears intact in several other noteworthy reports.[50–53] Toxicological observations of an undisclosed nature were noted after 1-month evaluation of benzofuran **21** as well as a close benzothiazole analog. Modifications aimed at increasing polarity were introduced to avoid off-target toxicity. Ultimately, an aza-benzothiazole core along with incorporation of a cyclopropyl constraint in AMG369 (**22**) led to an increase in potency of more than 10-fold (S1P1 $EC_{50}=2$ nM; P3/P1 = 444; rat lymphopenia $EC_{50}=1.6$ ng/mL). After demonstrating an adequate safety margin in 1-month toxicological evaluations, AMG369 was selected for further development.

A dihydronaphthalene core provided the basis for another series of azetidine carboxylate S1P1 agonists (**23–25**).[54–56] Compound **23** bearing a 4-phenylbutoxy side chain was potent (S1P1 $EC_{50}=2.9$ nM) and selective (S1P3 $EC_{50}>10{,}000$ nM) with a long half-life in rodent (rat $t_{1/2}=16.7$ h) resulting in a low efficacious dose that maintained lymphocyte suppression through 24 h (mouse $ED_{50}=1.9$ mg/kg). Truncated analog **24** provided a 20-fold improvement to *in vivo* potency (mouse 24 h lymphopenia $ED_{50}=0.095$ mg/kg) while maintaining good selectivity (P3/P1 = 15,000). Further truncation afforded benzylic ether **25**.

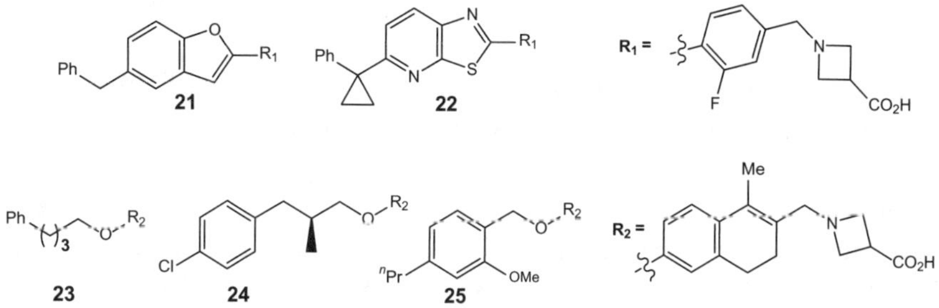

A large number of recent examples utilize indane or indole frameworks, the genesis of which traces back to early examples of S1P1 direct-acting agonists.[57] Indane **26** maintained lymphopenia in rat at a low oral dose (24 h $ED_{50}=0.25$ mg/kg).[58] Demonstrating good receptor selectivity (S1P1 $pEC_{50}>11$; S1P3 $pEC_{50}<5$), **27** was central nervous system (CNS) penetrant and efficacious in a rodent model of multiple sclerosis despite a relatively short half-life (mouse $t_{1/2}=3.8$ h).[59] Substituted tetrahydroisoquinolines (THIQs) have also provided effective S1P1 agonists including advanced compounds **28** and **29**.[60,61] Carboxylic acid-substituted THIQ **28** was able to induce maximal lymphopenia in rat through 12 h with doses as low as 0.1 mg/kg. Partial recovery from lymphopenia was evident by 24 h with doses up to 1.0 mg/kg, consistent with the compound's short half-life (rat $t_{1/2}=3.0$ h). Significant efficacy in the rat adjuvant arthritis model was demonstrated with a 3 mg/kg once-daily oral dose of **28**, and PK/PD modeling suggested a human dose of <10 mg would provide the desired target of 60% reduction in circulating lymphocytes.[62] As **28** was both zwitterionic and a substrate for P-glycoprotein, it was expected to have low CNS penetration, a factor believed to limit its utility in multiple sclerosis for which additional beneficial effects of S1P agonism in the CNS have been postulated.[63,64] Efforts to identify analogs of **28** with improved CNS penetration culminated in the discovery of **29**, which provided a steady-state brain-to-blood concentration ratio of 1.85:1. The diol motif of **29** is reminiscent of **10** and is likewise not phosphorylated *in vivo*. This potent and selective direct-acting agonist had an estimated *in vivo* EC_{50} of less than 0.1 nM representing an unexpectedly significant improvement over the *in vivo* potency of **28** ($EC_{50}=9$ nM), a difference that was attributed to a more favorable tissue distribution.

A series of disclosures illustrate the evolution of an indoline framework.[65–69] *N*-acylated indoline **30**, derived from lead optimization efforts around an indazole–oxadiazole template, was able to drive significant lymphopenia in mouse with an oral dose of 1.0 mg/kg, although the effects were not sustained through 24 h. Side chain migration led to C-linked carboxylate **31** (S1P1 $EC_{50}=8$ nM), which was then used in the design of constrained tricyclic indole analogs through annulation resulting in **32** or **33**. Although **33** was nearly fivefold more potent than **32** *in vitro* (mouse S1P1 $EC_{50}=0.52$ and 2.6 nM, respectively), their *in vivo* potencies were nearly identical (mouse lymphopenia $EC_{50}=750$–760 ng/mL).

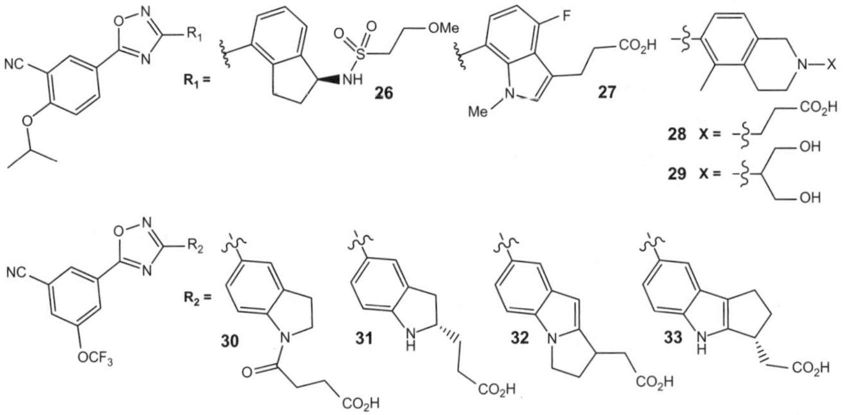

The number of examples appearing in recent literature that offer structural diversity significantly different than the initially described S1P1 agonists is quite limited. Pyrazole **34**, identified through HTS, is notable for its lack of polar head group, and it served as the basis for the proposal of a pharmacophore model for S1P1 agonism.[70,71] Also identified through HTS were coumarin-based agonists such as **35** (S1P1 $EC_{50}=3$ nM; S1P3 $EC_{50}=790$ nM).[72,73] The potency of the series was enhanced through modification of the carboxamide as in **36** (S1P1 $EC_{50}=0.3$ nM), however, the compound suffered from poor aqueous solubility and the formation of a long-lived *N*-acylated metabolite. Installation of gem-dimethyl (**37**) prevented the formation of the metabolite, while aqueous solubility was then improved through conversion of the C8 propyl group to ethoxy (**38**). Coumarin **38** demonstrated a long half-life in rat (34.8 h), was efficacious in a rat model of multiple sclerosis, and was selective across the S1P receptor family (S1P1 EC_{50} = 2 nM; S1P5 EC_{50} = 560 nM; S1P3,4 > 10,000 nM).

An additional series of structurally atypical agonists was generated from an HTS lead, optimization of which gave acylurea **39** as a monoselective agonist (S1P1 EC_{50} = 35 nM; > 100-fold vs. S1P2–5) with an extended pharmacodynamic effect in rat (78% lymphopenia at 24 h after 1 mg/kg oral dose) consistent with its long half-life (19 h).[74] The optimized trifluoromethyl-substituted biphenyl of **39** is reminiscent of the lipophilic groups in previously described S1P1 agonists, including **12** above, suggesting similar orientation of these compounds in the ligand-binding pocket of S1P1. Subsequent production of conformationally constrained analogs followed by a systematic N-scan SAR strategy and installation of a hydroxymethyl group afforded compound **40**.[75,76] In a rat lymphopenia assay, oral administration of **40** afforded 60% reduction at 24 h (corresponding to 84 ng/mL plasma concentration).

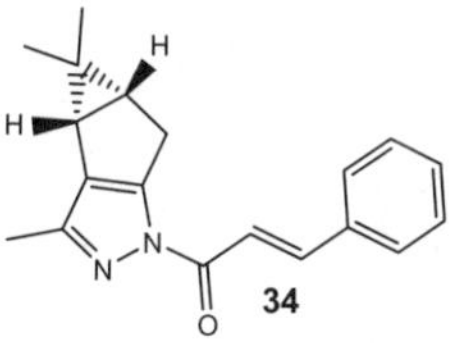
34

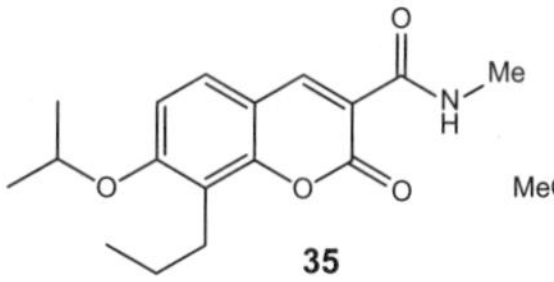
35

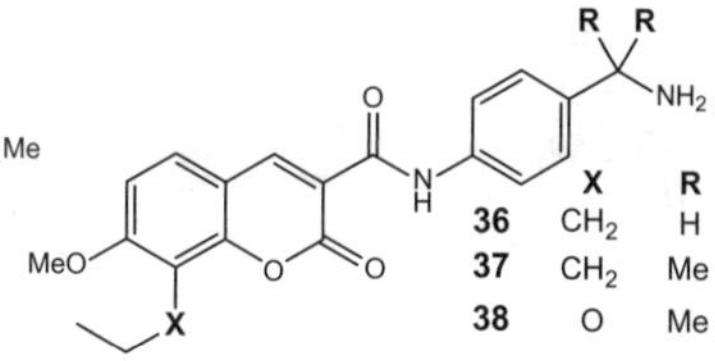

	X	R
36	CH_2	H
37	CH_2	Me
38	O	Me

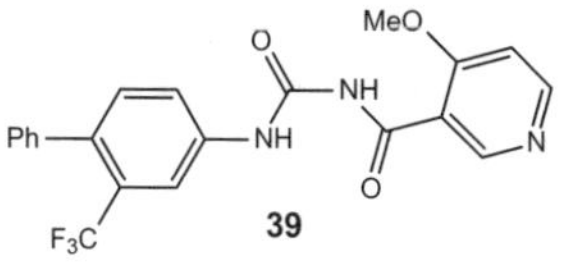
39

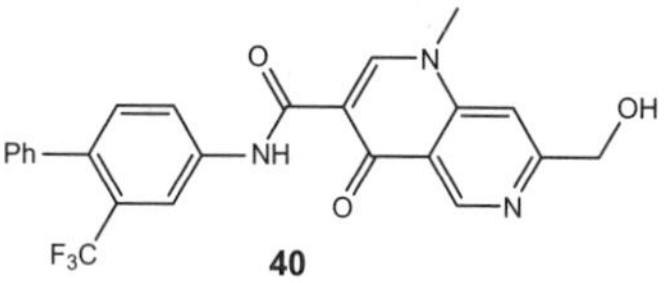
40

4. CONCLUSIONS

Following the identification of its molecular target, the impressive *in vivo* biological activity and continued clinical success of fingolimod have spurred tremendous and sustained efforts in the identification of novel agonists of S1P1 with improved profiles. The focus of optimization has been the identification of compounds with reduced half-life to permit a more rapid restoration of normal lymphocyte trafficking as well as improving upon the receptor selectivity profile of fingolimod-P. Although initial modifications retained the prodrug nature of fingolimod, direct-acting S1P1 agonists that do not require bioactivation were identified and have since come to form the majority of new disclosures. While this review focused only on the past few years of literature, it is clear that the foundational studies to define stable phosphate mimics of fingolimod such as **16** have had a lasting impact on the field, with elements of those early agonists being readily identifiable in the great majority of recently disclosed compounds described above. The continued evaluation of fingolimod in patients with RRMS will define the long-term significance of the identified liabilities. Multiple examples of both classes of S1P1 agonists (direct-acting and prodrug) have now entered human clinical trials, and it remains to be determined if they can provide a similar therapeutic benefit to fingolimod with an improved safety profile.

REFERENCES

(1) Pappu, R.; Schwab, S.R.; Cornelissen, I. *Science* **2007**, *316*, 295.
(2) Chun, J.; Hla, T.; Lynch, K.R.; Spiegel, S.; Moolenaar, W.H. *Pharmacol. Rev.* **2010**, *62*, 579.
(3) Olivera, A.; Rivera, J. *Adv. Exp. Med. Biol.* **2011**, *716*, 123.
(4) Matloubian, M.; Lo, C.G.; Cinamon, G.; Lesneski, M.J.; Xu, Y.; Brinkmann, V. *Nature* **2004**, *427*, 355.
(5) Chiba, K.; Yanagawa, Y.; Masubuchi, Y.; Kataoka, H.; Kawaguchi, T.; Ohtsuki, M. *J. Immunol.* **1998**, *160*, 5037.
(6) Dumont, F.J. *IDrugs* **2005**, *8*, 236.
(7) Mandala, S.; Hajdu, R.; Bergstrom, J.; Quackenbush, E.; Xie, J.; Milligan, J.; Thornton, R.; Shei, G.J.; Card, D.; Keohane, C.; Rosenbach, M.; Hale, J.; Lynch, C.L.; Rupprecht, K.; Parsons, W.; Rosen, H. *Science* **2002**, *296*, 346.
(8) Brinkmann, V.; Davis, M.D.; Heise, C.E.; Albert, R.; Cottens, S.; Hof, R.; Bruns, C.; Prieschl, E.; Baumruker, T.; Hiestand, P.; Foster, C.A.; Zollinger, M.; Lynch, K.R. *J. Biol. Chem.* **2002**, *277*, 21453.
(9) Billich, A.; Bornancin, F.; Devay, P.; Mechtcheriakova, D.; Urtz, N.; Baumruker, T. *J. Biol. Chem.* **2003**, *278*, 47408.
(10) Paugh, S.W.; Payne, S.G.; Barbour, S.E.; Milstien, S.; Spiegel, S. *FEBS Lett.* **2003**, *554*, 189.
(11) Fujii, Y.; Hirayama, T.; Ohtake, H.; Ono, N.; Inoue, T.; Sakurai, T.; Takayama, T.; Matsumoto, K.; Tsukahara, N.; Hidano, S. *J. Immunol.* **2012**, *188*, 206.
(12) Mulgaonkar, S.; Tedesco, H.; Oppenheimer, F.; Walker, R.; Kunzendorf, U.; Russ, G.; Knoflach, A.; Patel, Y.; Ferguson, R. *Am. J. Transplant.* **1848**, *2006*, 6.
(13) Tedesco-Silva, H.; Pescovitz, M.D.; Cibrik, D.; Rees, M.A.; Mulgaonkar, S.; Kahan, B.D.; Gugliuzza, K.K.; Rajagopalan, P.R.; de M. Esmeraldo, R.; Lord, H.; Salvadori, M.; Slade, J.M. *Transplantation* **2006**, *82*, 1689.
(14) Salvadori, M.; Budde, K.; Charpentier, B.; Klempnauer, J.; Nashan, B.; Pallardo, L.M.; Eris, J.; Schena, F.P.; Eisenberger, U.; Rostaing, L.; Hmissi, A.; Aradhye, S. *Am. J. Transplant.* **2006**, *6*, 2912.
(15) Warnke, C.; Stüve, O.; Hartung, H.-P.; Fogdell-Hahn, A.; Kieseier, B.C. *Neuropsychiatr. Dis. Treat.* **2011**, *7*, 519.
(16) Pelletier, D.; Hafler, D.A. *N. Engl. J. Med.* **2012**, *366*, 339.
(17) Hale, J.J.; Doherty, G.; Toth, L.; Mills, S.G.; Hajdu, R.; Keohane, C.A.; Rosenbach, M.; Milligan, J.; Shei, G.-J.; Chrebet, G.; Bergstrom, J.; Card, D.; Forrest, M.; Sun, S.-Y.; West, S.; Xie, H.; Nomura, N.; Rosen, H.; Mandala, S. *Bioorg. Med. Chem. Lett.* **2004**, *14*, 3501.

(18) Gergely, P.; Wallstroüm, E.; Nuesslein-Hildesheim, B.; Bruns, C.; Zécri, F.; Cooke, N.; Traebert, M.; Tuntland, T.; Rosenberg, M.; Saltzman, M. *Mult. Scler.* **2009**, *15*, S125.
(19) Shimizu, H.; Takahashi, M.; Kaneko, T.; Murakami, T.; Hakamata, Y.; Kudou, S.; Kishi, T.; Fukuchi, K.; Iwanami, S.; Kuriyama, K.; Yasue, T.; Enosawa, S.; Matsumoto, K.; Takeyoshi, I.; Morishita, Y.; Kobayashi, E. *Circulation* **2005**, *111*, 222.
(20) Nishi, T.; Miyazaki, S.; Takemoto, T.; Suzuki, K.; Iio, Y.; Nakajima, K.; Ohnuki, T.; Kawase, Y.; Nara, F.; Inaba, S.; Izumi, T.; Yuita, H.; Oshima, K.; Doi, H.; Inoue, R.; Tomisato, W.; Kagari, T.; Shimozato, T. *ACS Med. Chem. Lett.* **2011**, *2*, 368.
(21) Song, J.; Matsuda, C.; Kai, Y.; Nishida, T.; Nakajima, K.; Mizushima, T.; Kinoshita, M.; Yasue, T.; Sawa, Y.; Ito, T. *J. Pharmacol. Exp. Ther.* **2007**, *324*, 276.
(22) Zahir, H.; Moberly, J. B.; Ford, D.; Truitt, K. E.; Bar-Or, A.; Vollmer, T. L. AAN Annual Meeting, **2010**; P04.196.
(23) http://www.clinicaltrials.gov/ct2/results?term=KRP203.
(24) Moberly, J.B.; Rohatagi, S.; Zahir, H.; Hsu, C.; Noveck, R.J.; Kenneth, E. *J. Clin. Pharmacol.* **2012**, *52*, 996.
(25) Moberly, J.B.; Ford, D.M.; Zahir, H.; Chen, S.; Mochizuki, T.; Truitt, K.E.; Vollmer, T.L. *J. Neuroimmunol.* **2012**, *246*, 100.
(26) Brossard, P.; Hofmann, S.; Cavallaro, M.; Halabi, A.; Dingemanse, J. *Clin. Pharmacol. Ther.* **2009**, *85*, S63 PII-87.
(27) http://www.clinicaltrials.gov/ct2/show/NCT00852670?term=ACT-128800&rank=3.
(28) Walker, J. K.; Huff, R. M.; Cornicelli, J.; Mickelson, J. W.; Brown, M. F.; Wendling, J. M.; Funckes-Shippy, C. L.; Hiebsch, R. R.; Radi, Z.; Lawson, J. P.; Nagiec, M. M.; Strohbach, J. W.; Beebe, J. S.; Bradley, J.; Blinn, J. R. 239th ACS National Meeting, San Francisco, CA, March 21-25, **2010**; Abstracts of Papers, MEDI-39.
(29) Legangneux, E. US20110039818, **2011**.
(30) Nuesslein-Hildesheim, B.; Gergely, P.; Wallström, E.; Luttringer, O.; Groenewegen, A.; Howard, L.; Pan, S.; Gray, N.; Chen, Y. A.; Bruns, C.; Zécri, F. Abstr ECTRIMS, **2009**; P438.
(31) http://www.clinicaltrials.gov/ct2/results?term=baf312.
(32) Parrill, A.L.; Wang, D.; Bautista, D.L.; Van Brocklyn, J.R.; Lorincz, Z.; Fischer, D.J.; Baker, D.L.; Liliom, K.; Spiegeli, S.; Tigyi, G. *J. Biol. Chem.* **2000**, *275*, 39379.
(33) Hanson, M.A.; Roth, C.B.; Jo, E.; Griffith, M.T.; Scott, F.L.; Reinhart, G.; Desale, H.; Clemons, B.; Cahalan, S.M.; Schuerer, S.C.; Sanna, M.G.; Han, G.W.; Kuhn, P.; Rosen, H.; Stevens, R.C. *Science* **2012**, *335*, 851.
(34) Fujiwara, Y.; Osborne, D.A.; Walker, M.D.; Wang, D.; Bautista, D.A.; Liliom, K.; Van Brocklyn, J.R.; Parrill, A.L.; Tigyi, G. *J. Biol. Chem.* **2007**, *282*, 2374.
(35) Hale, J.J.; Lynch, C.L.; Neway, W.; Mills, S.G.; Hajdu, R.; Keohane, C.A.; Rosenbach, M.J.; Milligan, J.A.; Shei, G.-J.; Parent, S.A.; Chrebet, G.; Bergstrom, J.; Card, D.; Ferrer, M.; Hodder, P.; Strulovici, B.; Rosen, H.; Mandala, S. *J. Med. Chem.* **2004**, *47*, 6662.
(36) Li, Z.; Chen, W.; Hale, J.J.; Lynch, C.L.; Mills, S.G.; Hajdu, R.; Keohane, C.A.; Rosenbach, M.J.; Milligan, J.A.; Shei, G.-J.; Chrebet, G.; Parent, S.A.; Bergstrom, J.; Card, D.; Forrest, M.; Quackenbush, E.J.; Wickham, L.A.; Vargas, H.; Evans, R.M.; Rosen, H.; Mandala, S. *J. Med. Chem.* **2005**, *48*, 6169.
(37) Bolli, M.H.; Abele, S.; Binkert, C.; Bravo, R.; Buchmann, S.; Bur, D.; Gatfield, J.; Hess, P.; Kohl, C.; Mangold, C.; Mathys, B.; Menyhart, K.; Muller, C.; Nayler, O.; Scherz, M.; Schmidt, G.; Sippel, V.; Steiner, B.; Strasser, D.; Treiber, A.; Weller, T. *J. Med. Chem.* **2010**, *53*, 4198.
(38) Schmidt, G.; Reber, S.; Bolli, M.H.; Abele, S. *Org. Process Res. Dev.* **2012**, *16*, 595–604.
(39) Evindar, G.; Bernier, S.G.; Kavarana, M.J.; Doyle, E.; Lorusso, J.; Kelley, M.S.; Halley, K.; Hutchings, A.; Wright, A.D.; Saha, A.K.; Hannig, G.; Morgan, B.A.; Westlin, W.F. *Bioorg. Med. Chem. Lett.* **2009**, *19*, 369.

(40) Evindar, G.; Satz, A.L.; Bernier, S.G.; Kavarana, M.J.; Doyle, E.; Lorusso, J.; Taghizadeh, N.; Halley, K.; Hutchings, A.; Kelley, M.S.; Wright, A.D.; Saha, A.K.; Hannig, G.; Morgan, B.A.; Westlin, W.F. *Bioorg. Med. Chem. Lett.* **2009**, *19*, 2315.
(41) Evindar, G.; Bernier, S.G.; Doyle, E.; Kavarana, M.J.; Satz, A.L.; Lorusso, J.; Blanchette, H.S.; Saha, A.K.; Hannig, G.; Morgan, B.A.; Westlin, W.F. *Bioorg. Med. Chem. Lett.* **2010**, *20*, 2520.
(42) Evindar, G.; Bernier, S. G.; Satz, A. L.; Kavarana, M. J.; Doyle, E.; Lorusso, J.; Taghizadeh, N.; Blanchette, H. S.; Halley, K.; Hutchings, A. 238th ACS National Meeting, Washington, DC, August 16–20, **2009**; Abstracts of Papers, MEDI-002.
(43) Anson, M.S.; Graham, J.P.; Roberts, A.J. *Org. Process Res. Dev.* **2011**, *15*, 649.
(44) Watterson, S. H.; Dyckman, A. J.; Pitts, W. J.; Spergel, S. PCT International Application WO201008558, **2010**.
(45) Aguilar, N.; Mir, M.; Grima, P. M.; Lopez, M.; Godessart, N.; Tarrason, G.; Domenech, T.; Vilella, D.; Armengol, C.; Cordoba, M.; Sabate, M.; Casals, D.; Dominguez, M. 240th ACS National Meeting, Boston, MA, August 22–26, **2010**; Abstracts of Papers, MEDI-164.
(46) Deng, G.; Lin, X.; Ren, F.; Zhao, B. PCT International Application WO2011134280, **2011**.
(47) Nakamura, T.; Asano, M.; Sekiguchi, Y.; Mizuno, Y.; Tamaki, K.; Kimura, T.; Nara, F.; Kawase, Y.; Shimozato, T.; Doi, H. *Bioorg. Med. Chem. Lett.* **2012**, *22*, 1788.
(48) Nakamura, T.; Asano, M.; Sekiguchi, Y.; Mizuno, Y.; Tamaki, K.; Nara, F.; Kawase, Y.; Yabe, Y.; Nakai, D.; Kamiyama, E.; Urasaki-Kaneno, Y.; Shimozato, T.; Doi-Komuro, H.; Kagari, T.; Tomisato, W.; Inoue, R.; Nagasaki, M.; Yuita, H.; Oguchi-Oshima, K.; Kaneko, R.; Nishi, T. *Eur. J. Med. Chem.* **2012**, *51*, 92.
(49) Asano, M.; Nakamura, T.; Sekiguchi, Y.; Mizuno, Y.; Yamaguchi, T.; Tamaki, K.; Shimozato, T.; Doi-Komuro, H.; Kagari, T.; Tomisato, W.; Inoue, R.; Yuita, H.; Oguchi-Oshima, K.; Kaneko, R.; Nara, F.; Kawase, Y.; Masubuchi, N.; Nakayama, S.; Koga, T.; Namba, E.; Nasu, H.; Nishi, T. *Bioorg. Med. Chem. Lett.* **2012**, *22*, 3083.
(50) Saha, A.K.; Yu, X.; Lin, J.; Lobera, M.; Sharadendu, A.; Chereku, S.; Schutz, N.; Segal, D.; Marantz, Y.; McCauley, D.; Middleton, S.; Siu, J.; Burli, R.W.; Buys, J.; Horner, M.; Salyers, K.; Schrag, M.; Vargas, H.M.; Xu, Y.; McElvain, M.; Xu, H. *ACS Med. Chem. Lett.* **2011**, *2*, 97.
(51) Cee, V. J. 241st ACS National Meeting, Anaheim, CA, March 27–31, **2011**; Abstracts of Papers, MEDI-25.
(52) Lanman, B.A.; Cee, V.J.; Cheruku, S.R.; Frohn, M.; Golden, J.; Lin, J.; Lobera, M.; Marantz, Y.; Muller, K.M.; Neira, S.C. *ACS Med. Chem. Lett.* **2011**, *2*, 102.
(53) Cee, V.J.; Frohn, M.; Lanman, B.A.; Golden, J.; Muller, K.; Neira, S.; Pickrell, A.; Arnett, H.; Buys, J.; Gore, A. *ACS Med. Chem. Lett.* **2011**, *2*, 107.
(54) Kurata, H.; Kusumi, K.; Otsuki, K.; Suzuki, R.; Kurono, M.; Takada, Y.; Shioya, H.; Komiya, T.; Mizuno, H.; Ono, T. *Bioorg. Med. Chem. Lett.* **2011**, *21*, 3885.
(55) Kurata, H.; Kusumi, K.; Otsuki, K.; Suzuki, R.; Kurono, M.; Tokuda, N.; Takada, Y.; Shioya, H.; Mizuno, H.; Komiya, T. *Bioorg. Med. Chem. Lett.* **2012**, *22*, 144.
(56) Habashita, H.; Kurata, H.; Nakade, S. PCT International Application WO2006064757, **2006**.
(57) Colandrea, V. J.; Doherty, G. A.; Hale, J. J.; Huo, P.; Legiec, I. E.; Toth, L.; Vachal, P.; Yan, L. PCT International Application WO2005058848, **2005**.
(58) Martinborough, E.; Boehm, M. F.; Yeager, A. R.; Tamiya, J.; Huang, L.; Brahmachary, E.; Moorjani, M.; Timony, G. A.; Brooks, J. L.; Peach, R. PCT International Application WO2011060392, **2011**.

(59) Meng, Q.; Zhao, B.; Xu, Q.; Xu, X.; Deng, G.; Li, C.; Luan, L.; Ren, F.; Wang, H.; Xu, H. *Bioorg. Med. Chem. Lett.* **2012**, *22*, 2794.
(60) Demont, E.H.; Andrews, B.I.; Bit, R.A.; Campbell, C.A.; Cooke, J.W.B.; Deeks, N.; Desai, S.; Dowell, S.J.; Gaskin, P.; Gray, J.R.J. *ACS Med. Chem. Lett.* **2011**, *2*, 444.
(61) Demont, E.H.; Arpino, S.; Bit, R.A.; Campbell, C.A.; Deeks, N.; Desai, S.; Dowell, S.J.; Gaskin, P.; Gray, J.R.J.; Harrison, L.A. *J. Med. Chem.* **2011**, *54*, 6724.
(62) Taylor, S.; Gray, J.R.J.; Willis, R.; Deeks, N.; Haynes, A.; Campbell, C.; Gaskin, P.; Leavens, K.; Demont, E.; Dowell, S.; Cryan, J.; Morse, M.; Patel, A.; Garden, H.; Witherington, J. *Xenobiotica* **2012**, *16*, 671.
(63) Brinkmann, V. *Br. J. Pharmacol.* **2009**, *158*, 1173.
(64) Noguchi, K.; Chun, J. *Crit. Rev. Biochem. Mol. Biol.* **2011**, *46*, 2.
(65) Buzard, D.J.; Han, S.; Thoresen, L.; Moody, J.; Lopez, L.; Kawasaki, A.; Schrader, T.; Sage, C.; Gao, Y.; Edwards, J. *Bioorg. Med. Chem. Lett.* **2011**, *21*, 6013.
(66) Kawasaki, A.; Thoresen, L.; Buzard, D. J.; Moody, J.; Lopez, L.; Ullman, B.; Lehmann, J.; Zhu, X.; Edwards, J.; Barden, J. 241st ACS National Meeting, Anaheim, CA, March 27–31, **2011**; Abstracts of Papers, MEDI-254.
(67) Han, S.; Thoresen, L.; Moody, J.; Buzard, D. J.; Lopez, L.; Sage, C.; Edwards, J.; Barden, J.; Kawasaki, A.; Ullman, B. 241st ACS National Meeting, Anaheim, CA, March 27–31, **2011**; Abstracts of Papers, MEDI-98.
(68) Jones, R. M.; Buzard, D. J.; Han, S.; Kim, S. H.; Lehmann, J.; Ullman, B.; Moody, J. V.; Zhu, X.; Stirn, S. PCT International Application WO2010011316, **2010**.
(69) Montalban, A. G.; Buzard, D. J.; Demattei, J. A.; Gharbaoui, T.; Johannsen, S. R.; Krishnan, A. M.; Kuhlman, Y. M.; Ma, Y. -A.; Martinelli, M. J.; Sato, S. M. PCT International Application WO2011094008, **2011**.
(70) Baenteli, R.; Cooke, N. G.; Zecri, F.; Alexander, B. PCT International Application WO2008092930, **2008**.
(71) Zecri, F.J.; Albert, R.; Landrum, G.; Hinterding, K.; Cooke, N.G.; Guerini, D.; Streiff, M.; Bruns, C.; Nuesslein-Hildesheim, B. *Bioorg. Med. Chem. Lett.* **2010**, *20*, 35.
(72) Zecri, F. J.; Albert, R.; Baenteli, R.; Landrum, G.; Cooke, N. G.; Beerli, C.; Bruns, C.; Guerini, D.; Streiff, M.; Nusslein-Hildesheim, B. 239th ACS National Meeting, San Francisco, CA, March 21–25, **2010**; Abstracts of Papers, MEDI-31.
(73) Baenteli, R.; Cooke, N. G.; Weiler, S.; Zecri, F. PCT International Application WO2007115820, **2007**.
(74) Pennington, L.D.; Sham, K.K.C.; Pickrell, A.J.; Harrington, P.E.; Frohn, M.J.; Lanman, B.A.; Reed, A.B.; Croghan, M.D.; Lee, M.R.; Xu, H.; Zhang, X.; McElvain, M.; Xu, Y.; Zhang, X.; Fiorino, M.; Horner, M.; Morrison, H.G.; Arnett, H.A.; Fotsch, C.; Wong, M; Cee, V.J. *ACS Med. Chem. Lett.* **2011**, *2*, 752.
(75) Pennington, L.D.; Croghan, M.D.; Sham, K.K.C.; Pickrell, A.J.; Harrington, P.E.; Frohn, M.J.; Lanman, B.A.; Reed, A.B.; Lee, M.R.; Xu, H.; McElvain, M.; Xu, Y.; Zhang, X.; Fiorino, M.; Horner, M.; Morrison, H.G.; Arnett, H.A.; Fotsch, C.; Tasker, A.S.; Wong, M.; Cee, V.J. *Bioorg. Med. Chem. Lett.* **2012**, *22*, 527.
(76) Harrington, P.E.; Croghan, M.D.; Fotsch, C.; Frohn, M.; Lanman, B.A.; Pennington, L.D.; Pickrell, A.J.; Reed, A.B.; Sham, K.K.C.; Tasker, A.S.; Arnett, H.A.; Fiorino, M.; Lee, M.R.; McElvain, M.; Morrison, H.G.; Xu, H.; Xu, Y.; Zhang, X.; Wong, M.; Cee, V.J. *ACS Med. Chem. Lett.* **2012**, *3*, 74.

CHAPTER FOURTEEN

Bifunctional Compounds for the Treatment of COPD

Gary Phillips* and Michael Salmon†
*Medicinal chemistry, Gilead Sciences, Inc., Seattle, Washington, USA
†Respiratory and Immunology Franchise, Merck Research Laboratories, Boston, Massachusetts, USA

Contents

1. INTRODUCTION

1.1. COPD pathophysiology and current therapies

Chronic obstructive pulmonary disease (COPD) is a chronic lung disease considered to be of high unmet medical need. The disease is characterized by airflow obstruction which is only partially reversible and is often progressive in nature.[1] The pulmonary symptoms are caused by pathological changes in the lungs particularly associated with small airways disease (obstructive bronchiolitis) and emphysema (alveolar tissue destruction). In addition to a compromised lung function, the disease may also be associated with chronic cough and sputum production and systemic components such as cachexia and depression. COPD is often associated with a number of comorbidities, most notably cardiovascular disease. Long-term cigarette smoking is the most common risk factor for COPD, but exposure of the lungs to other environmental noxious particles or gases has also been

Annual Reports in Medicinal Chemistry, Volume 47
ISSN 0065-7743
http://dx.doi.org/10.1016/B978-0-12-396492-2.00014-X

implicated. Such exposures are believed to contribute to a chronic inflammatory process that underlies the disease progression in predisposed individuals. Chronic inflammatory processes lead to structural remodeling including narrowing of the small airways and parenchymal destruction and loss of lung elastic recoil which causes the loss of lung function.[1] COPD is of increasing concern as a major public health problem with increasing rates of morbidity and mortality. It is projected to be the fourth major burden of disease and the third leading cause of death by 2030.[2,3]

Inhaled bronchodilators are central to symptomatic relief in COPD and two major classes exist: β_2-adrenoceptor agonists (beta agonists) and anticholinergics (Fig. 14.1). Beta agonists act on β_2 receptors, which are seven-transmembrane domain-spanning G-protein-coupled receptors (GPCRs) situated on the

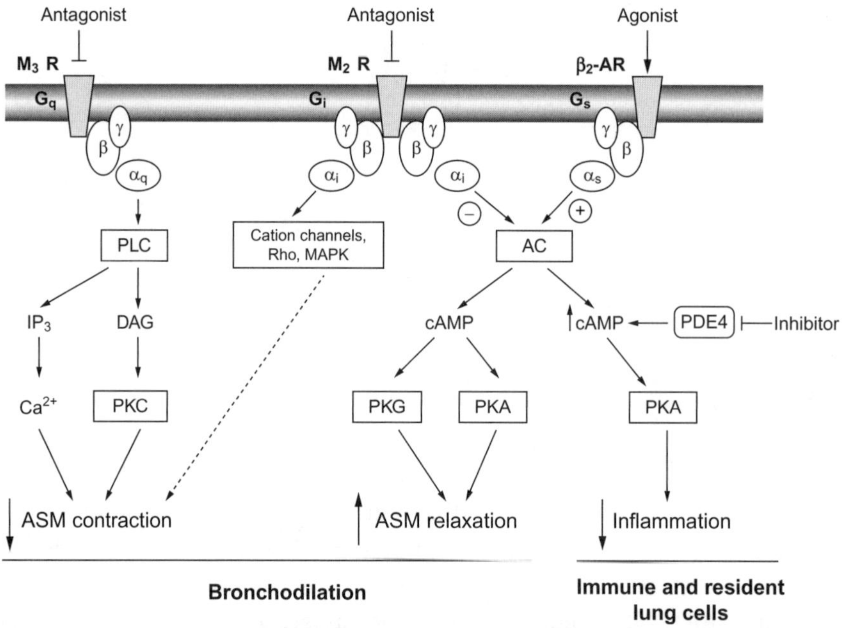

Figure 14.1 Interaction of muscarinic M_2 and M_3 receptor- and β_2-adrenoceptor-mediated intracellular signaling pathways and their modulation by muscarinic antagonists, β_2 agonists, and PDE4 inhibitors to elicit bronchodilation and anti-inflammatory activity in the lungs. (Adapted with kind permission from Springer Science+Business Media from Ref. 4.) Abbreviations: M_2, muscarinic 2 receptor; M_3, muscarinic 3 receptor; β_2, β_2 adrenoceptor; G_q, G_i, G_s: G proteins; α: alpha, β: beta, γ: gamma, G-protein subunits; PLC, phospholipase C; IP3, inositol 1,4,5-triphosphate; DAG, sn-1,2-diacylglycerol; Ca^{2+}, calcium ions; PKC, protein kinase C; AC, adenylyl cyclase; cAMP, cyclic adenosine 3′,5′ monophosphate; PKG, protein kinase G; PKA, protein kinase A; ASM, airway smooth muscle; PDE4, phosphodiesterase-4; −, negative regulator of cAMP; +, positive regulator of cAMP; Rho, rho-associated kinases; MAPK, mitogen-activated protein kinases.

smooth muscle cells in the airways. There are three subtypes of β adrenoceptors (β_1–β_3) which mediate the actions of adrenaline and noradrenaline. β_1 Receptors are present in the heart, and agonist binding can elicit both increases in heart rate and force of contraction. Functional selectivity of agonists for β_2 over β_1 receptors is therefore desirable as it could increase the therapeutic index for β_2-mediated bronchodilation over unwanted β_1-mediated increases in heart rate. β_2 Receptors are coupled to G_s proteins which activate adenylyl cyclase leading to formation of cyclic AMP. This elevation of intracellular cyclic AMP leads to relaxation of the smooth muscle and bronchodilation.[5] Muscarinic receptors are seven-transmembrane-spanning GPCRs which mediate acetylcholine (ACh) signaling from the cell surface. There are a total of five muscarinic receptor subtypes (M_1–M_5) of which M_1, M_2, and M_3 are present in the lungs. M_1 and M_3 receptors are coupled to G_q proteins and utilize calcium as a second messenger through the action of phospholipase C and inositol triphosphate, whereas M_2 receptors are G_i linked which decreases cellular cyclic AMP levels and inhibits voltage-gated calcium channels.[6] Muscarinic receptor antagonists act to block ACh signaling, thereby inhibiting the airway smooth muscle contraction which leads to bronchoconstriction. Muscarinic M_1 receptors are located on parasympathetic nerve ganglia and are responsible for facilitation of nerve transmission. M_2 receptors are located on postganglionic parasympathetic nerves and are the predominant receptor subtype on airway smooth muscle. The function of M_2 receptors is autoinhibitory which serves to maintain a tight regulation of ACh release. Muscarinic M_3 receptors are also located on airway smooth muscle cells and mediate airway smooth muscle contraction. Blockade of both M_1 and M_3 muscarinic receptor subtypes inhibits cholinergic-mediated bronchoconstriction. Presently, long-acting β_2 agonists (LABA) and long-acting muscarinic antagonists (LAMA) are used as the standard of care for symptomatic control in COPD.[7]

Recently, roflumilast, a novel oral anti-inflammatory agent that inhibits phosphodiesterase 4 (PDE4) enzyme activity, was approved for treatment of moderate and severe COPD associated with chronic bronchitis in patients at risk of exacerbations.[8] While the clinical effect of roflumilast and related PDE4 inhibitors are as anti-inflammatory agents rather than as bronchodilators, PDE4 is present in airway smooth muscle cells and is responsible for the hydrolysis of cyclic AMP, which is important in bronchodilation. This compound has been shown to be effective in improving lung function as an add-on to the LABA, salmeterol and the LAMA, tiotropium bromide.[9] To date, to the authors' knowledge, no inhaled PDE4 inhibitors have progressed to registration trials in pulmonary disease. The pursuit of PDE4 subtype selective inhibitors has also been an area of high pharmaceutical industry interest as a way to achieve clinical efficacy and improve the systemic adverse effects associated with the compound drug class.[10] There is also emerging preclinical evidence that dual PDE3/PDE4 inhibitors can combine bronchodilatory effects via direct PDE3 inhibition and act as an anti-inflammatory, which could be an alternative novel therapeutic approach.[11]

Inhaled glucocorticoids have a broad range of anti-inflammatory actions through binding and activating cytosolic glucocorticoid receptor (GR)-α. Upon glucocorticoid binding with GR-α, the complex formed translocates to the cellular nucleus where it can bind with a glucocorticoid response element in the promoter region of target genes to modulate gene expression. Glucocorticoids can mediate both gene repression (transrepression) and gene induction (transactivation) to exert their anti-inflammatory effects.[12] The level of efficacy achieved by inhaled glucocorticoids in COPD remains a controversial topic but they do improve symptoms, lung function, and reduce exacerbations in more severe patients.[1] The combination of an inhaled β_2 agonist and glucocorticoid therapies has been shown to be superior to individual treatments alone with respect to lung function and health status as well as reducing disease-related exacerbations in patients with moderate to severe disease.[13]

1.2. Bifunctional compounds: Concepts and advantages

Combination products consisting of two medications and two modes of action have been highly successful. Seretide®/Advair®, a combination of the LABA, salmeterol, and the inhaled corticosteroid, fluticasone propionate, used for both COPD and asthma, was the prescription drug with the third highest sales in 2010.[14] Coadministration of two mechanistically distinct chemical entities is one approach for combination drug therapy, but it is also possible to use a "bifunctional compound," also referred as a "dual selective pharmacology molecule," which is a single chemical entity with two distinct pharmacophores covalently bonded. While the design seems conceptually simple, connection points and linkers must be chosen carefully, as the choices can significantly impact the activity of either pharmacophore. The compounds, being the combination of two pharmacophores, tend to have high molecular weights, but the net molecular size may offer an advantage as an inhaled therapeutic, because it can result in greater lung retention and lower oral bioavailability, thereby reducing systemic exposure and related downstream toxicology issues.[15] Due to the size of the molecules and the inherent flexibility of many of the linkers, it may also be a challenge to obtain crystalline compounds for development. However, a single chemical entity with two activities would offer multiple advantages, including matched pharmacokinetics, simplified formulation, and simplified clinical development.[16] Significantly, there is potential for a combination product consisting of a novel bifunctional compound and a separate additional medication in a single device to deliver three mechanisms. The alternative development and engineering of an inhaler capable of delivering these separate drugs, with unique disease-related mechanisms, is yet to be realized.

Described herein are novel bifunctional molecules designed through linking of pharmacophores of approved therapeutic targets: muscarinic antagonists, β agonists, and PDE4 inhibitors. Additional combination examples, such as neutrophil elastase inhibitor/muscarinic antagonist[17] or epithelial sodium channel inhibitor/β agonist,[18] have also been disclosed.

2. BIFUNCTIONAL STRATEGIES AND COMPOUNDS

2.1. Muscarinic receptor antagonist–β_2 agonist

Clinical trials have demonstrated that the combination of an individually dosed muscarinic antagonist along with a β agonist provides greater bronchodilation than either component alone.[19] The two mechanisms are complementary, with the β_2 agonist increasing cyclic AMP, leading to smooth muscle relaxation, while the muscarinic antagonist blocks ACh-mediated bronchoconstriction. Furthermore, the addition of a β_2 agonist actually decreases the amount of ACh released, thereby amplifying the effect induced by the muscarinic antagonist. Combivent®, an inhaled combination product used for the relief of symptoms, contains the short-acting β agonist, albuterol, and the short-acting muscarinic antagonist, ipratropium. Combining long-acting agents with each individual mechanism is recommended therapy for more severe COPD,[1] yet no single commercial inhalation device is currently marketed which delivers both therapeutics, although clinical trials are underway.[20]

Bifunctional compounds comprised of pharmacophores representing each class of bronchodilators have been reviewed.[20] One such compound (GSK961081, TD5959, structure not disclosed) has been reported to be in clinical studies both as a stand-alone drug and in combination with fluticasone propionate, an inhaled corticosteroid.[21] The structure of GSK961081 has been speculated to be **1**, as developable salts of this compound have been reported.[22–25] Compound **1** connects the common quinolinone head group of indacaterol (**2**) with the biphenyl of the muscarinic receptor antagonist, YM-46303 (**3**).[26] Once-daily administration of GSK961081 has shown equivalent changes in FEV_1 to once-daily tiotropium (LAMA) plus twice-daily salmeterol (LABA) after 14 days in COPD patients.[16]

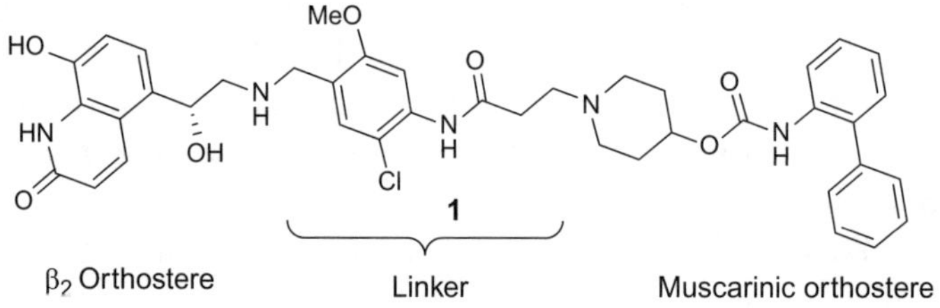

1

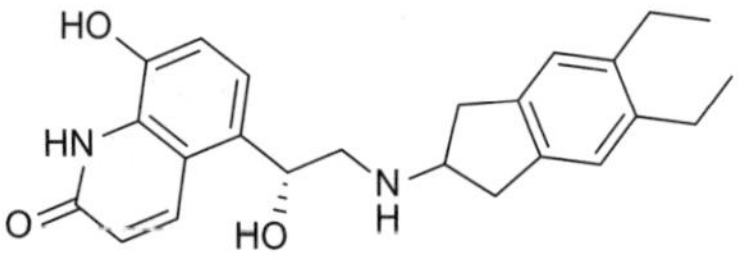

2

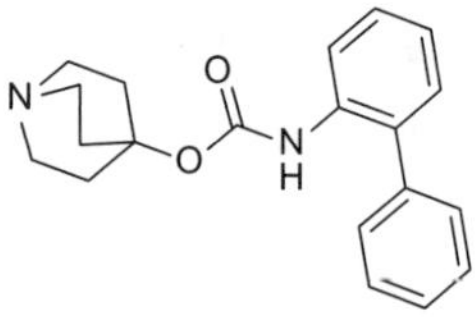

3

A detailed characterization of a related compound with the same key pharmacophores, THRX-198321 (**4**) has been reported.[22,27–29] Compound **4** is reported to have a K_i for the β_2 receptor of 3.5 nM and for the M_3 receptor of 0.01 nM with greater than eightfold selectivity for β_2 over β_1.[22,30] A systematic study of chain length resulted in the identification of the nine-carbon spacer between the two pharmacophores as being optimal for highest activity on both receptors. Representative β_2-active moieties were also compared, with the quinolinone moiety providing the most potent β_2-agonist activity and concomitant potentiation of M_3 antagonist activity. Increased binding affinity at each receptor was reported for **4** over the respective fragments, which was proposed to be due to simultaneous binding of the alternate pharmacophore to an allosteric binding site in each protein.[27]

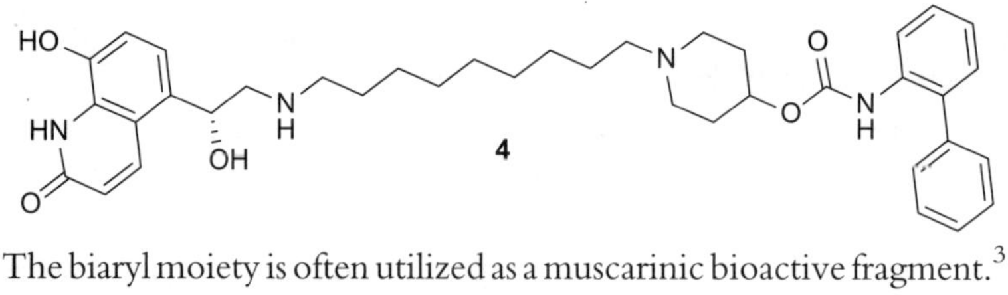

4

The biaryl moiety is often utilized as a muscarinic bioactive fragment.[31–37] Compound **5** is described to have an EC_{50} as a β_2 agonist of 0.13 nM and an IC_{50} on the M_3 receptor of 0.72 nM.[31] A similar series with the same pharmacophores separated by a diamide linker has also been disclosed.[32]

A similar compound, **6** is reported to have a K_i of less than 30 nM against the β_2 receptor with greater than 10-fold β_2/β_1 selectivity and a K_i of less than 10 nM against the M_3.[33] A related patent application discloses various alternatives to the distal aryl group of the biphenyl, including thiophene, thiazole, and pyridine, with the thiophene **7** having activity against both receptors less than 10 nM (K_i) and greater than fivefold selectivity for β_2 over β_1.[34] Another application describes an acyclic linker replacement for the piperidine, with the activity of **8** being less than 300 nM against M_3 and β_2 receptors.[35] Compounds similar to **1** with a substituted phenyl in the linker and a formoterol-like β_2-agonist head group as exemplified by **9**, exhibit activity against both receptors less than 10 nM (K_i) and greater than 100-fold selectivity for β_2 over β_3 and β_1 subtypes.[36,37]

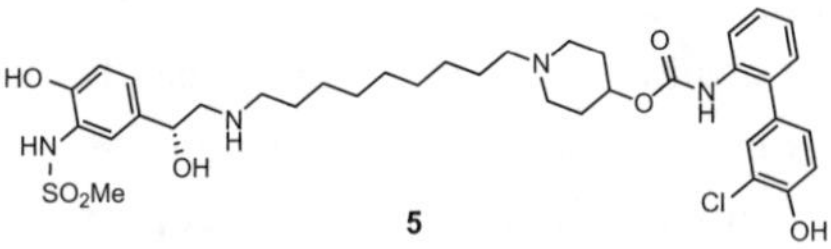

5

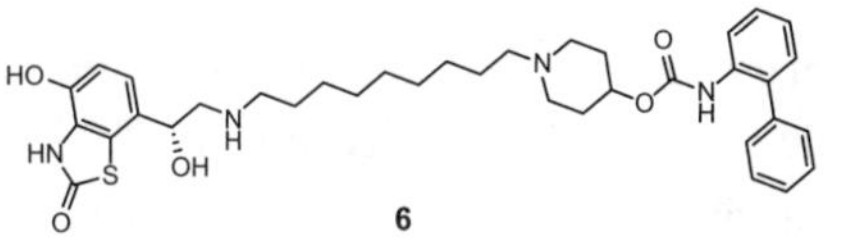

6

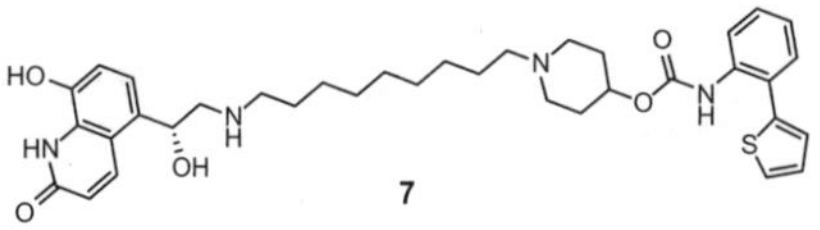

7

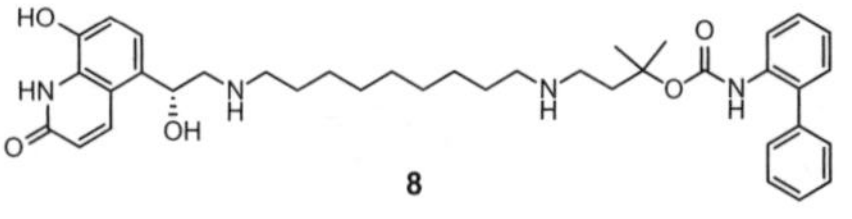

8

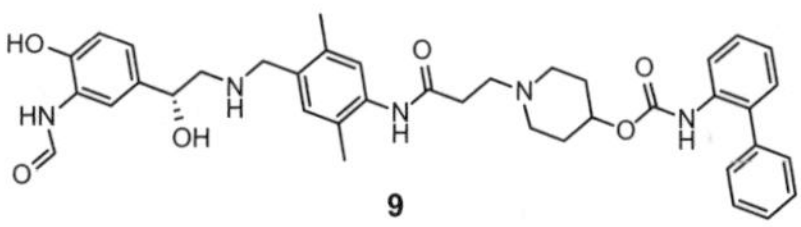

9

Conjugation of the muscarinic antagonist tolterodine (**10**) to β_2 agonists have resulted in the discovery of new muscarinic receptor antagonist–β_2 agonists (MABAs) such as **11** (M_3 K_i 0.3 nM, β_2 EC_{50} 2.4 nM).[38] Various examples with common β_2 moieties and tethers were compared, with most compounds showing good potency against both targets, with the quinolinone-containing fragment being the most potent. The compounds, being optimized for inhaled delivery, were designed to have poor oral availability and low metabolic stability to minimize systemic exposure. A patent application covering MABAs with the headpiece containing **10** has appeared.[39]

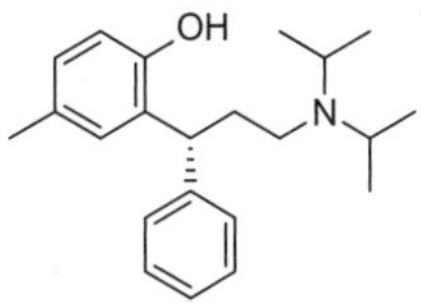

10

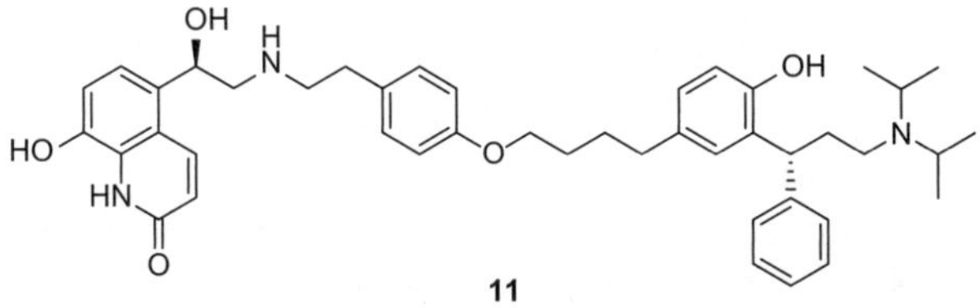

11

Another often appearing muscarinic receptor antagonist moiety utilized in bifunctional compound design is based on the biarylmethylene of tiotropium bromide (**12**). Linking to a quinolinone provided **13** (K_i under 50 nM against β_2 and M_3 receptors).[40] Related compounds have been described with the tertiary amine (**14**) reported to be less than 100 nM against the β_2 receptor and less than 1 nM on the M_3 receptor.[41] The naphthalene-1,5-disulfonate salt of **14** has also been described as well as additional related compounds.[42–45] The bicyclic moieties have been replaced with an oxazole to give **15** with a K_i less than 100 nM against the β_2 receptor and less than 5 μM against the M_3 receptor.[46] Additional five-membered ring heterocycles in the linker have also been exemplified with **16** having an IC_{50} of 4 nM on M_3 binding and an EC_{50} of 1.2 nM[47] and **17** having an IC_{50} of 0.1 nM on M_3 binding and an EC_{50} of 2 nM.[48] Related triazoles have also been reported.[49] Further work around the biarylmethylenes, including quaternary amines in the linker, has been described.[50] The alcohol has been replaced with a carboxamide as in darifenacin, **18**,[51] to give **19** with a K_i less than 50 nM against both receptors.[52]

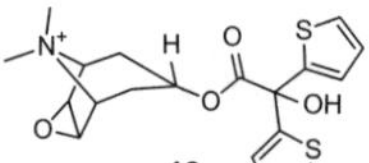

12

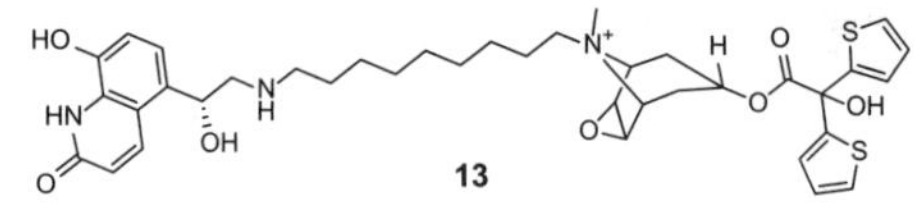

13

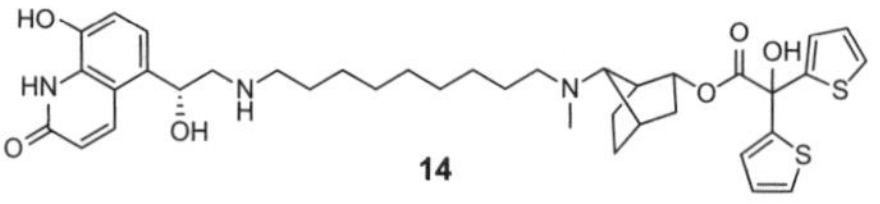

14

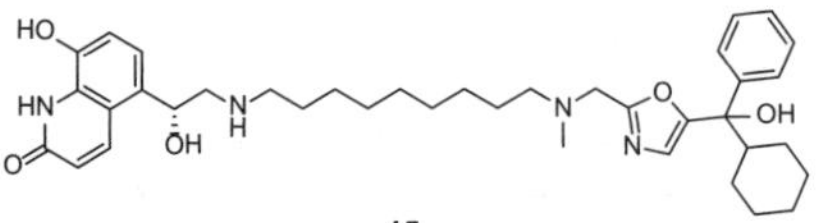

15

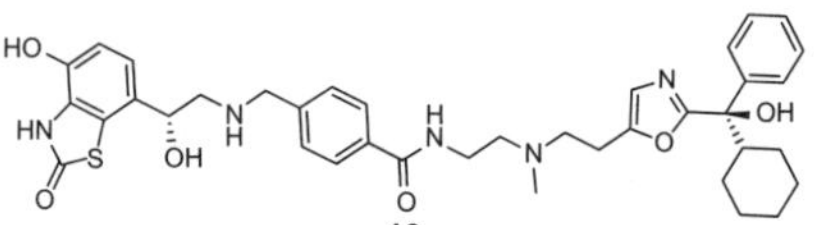

16

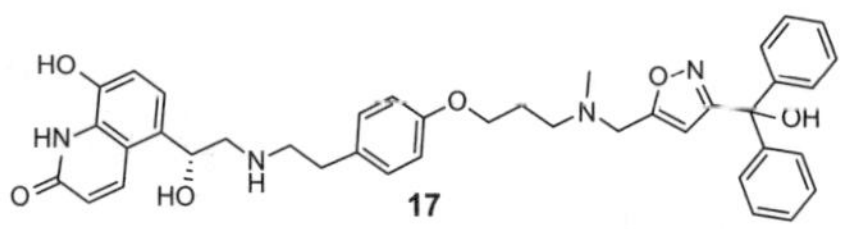

17

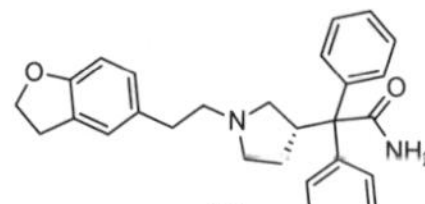

18

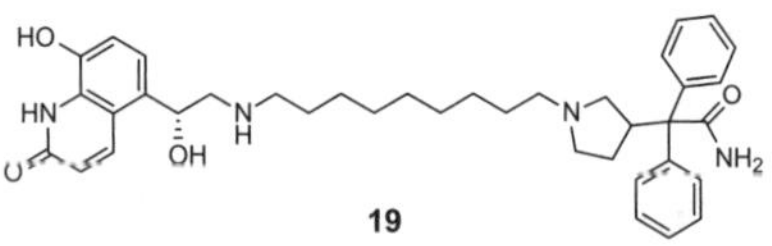

19

Additional examples with unique muscarinic pharmacophores have also been reported.[53–55] One such example (**20**) began with a novel, selective M_3 antagonist (**21**) which evolved from a nonselective norepinephrine reuptake inhibitor.[56] Compound **20** has high clearance in microsomes and poor membrane permeability, a favorable profile for an inhaled therapeutic.

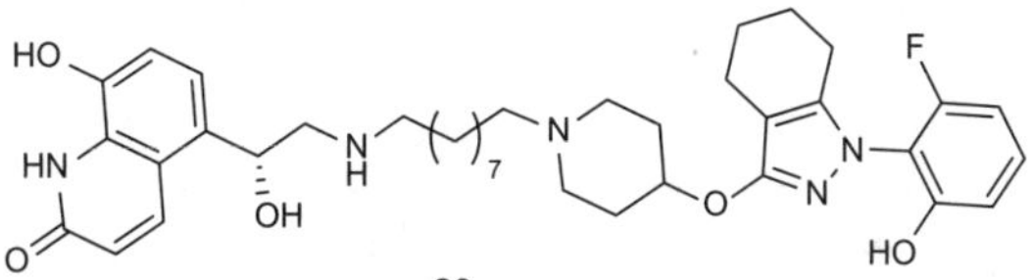

20

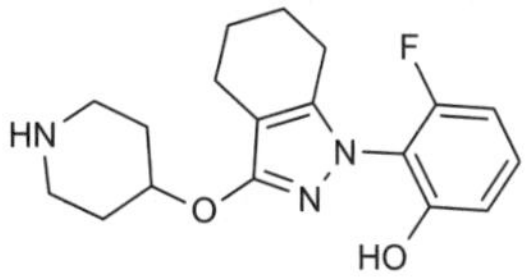

21

2.2. Muscarinic receptor antagonist–PDE4 inhibitor

Addition of an oral PDE4 inhibitor with an inhaled muscarinic receptor antagonist has clinical benefit in improving lung function in COPD beyond the effect of muscarinic antagonist treatment alone.[57] An inhaled bifunctional molecule against both these targets, therefore, has the potential for improved efficacy as the systemic side effects, which are dose limiting to oral PDE4 inhibitors, are less likely to be of concern. Additionally, a PDE4 inhibitor and antimuscarinic bifunctional molecule could be codosed with either a β_2 agonist or glucocorticoid to target two bronchodilator or anti-inflammatory mechanisms. More conventional small molecules with such dual activity have been identified.[58,59]

A series of four related applications have appeared that describe bifunctional compounds in which a PDE4 inhibitor is connected to a muscarinic receptor antagonist.[60–63] All utilize a pyrazolopyridine (**22**)[64] as the PDE4 inhibitor and a biaryl-containing muscarinic antagonist, but differ in the linker. A muscarinic receptor antagonist such as **23**[65] used in **24**, with only a urea separating the pharmacophores, is reported to be active against both targets, although the specific activities are not reported.[63] Bifunctional compounds of this class tend to have shorter linkers, suggesting that the activity of each pharmacophore is less sensitive to the presence of the other.

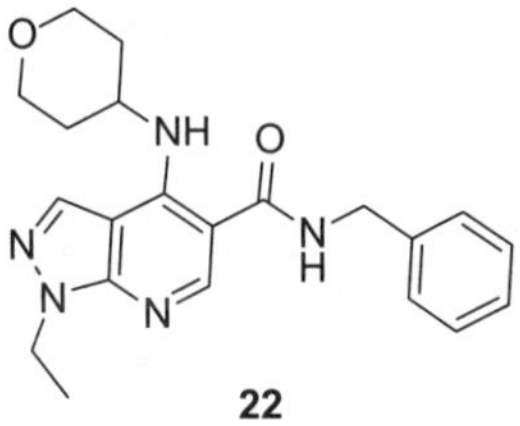

22

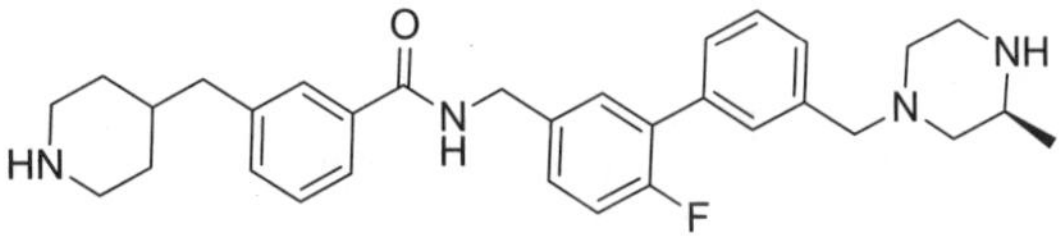

23

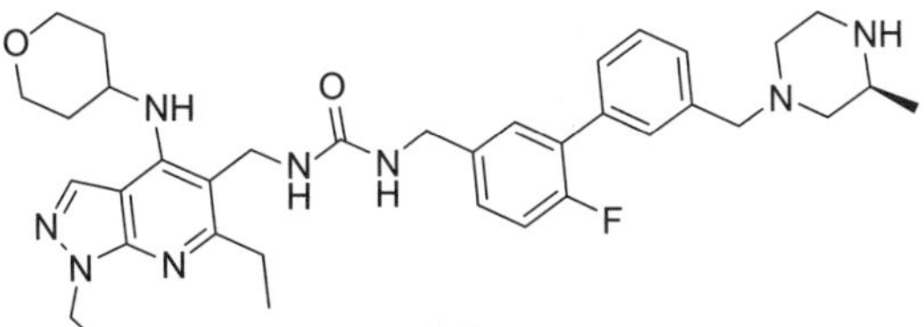

24

2.3. PDE4 inhibitor–β_2 agonist

Conceptually, the conjugation of a β_2 agonist to a PDE4 inhibitor could be an alternative way for developing a single molecule with both bronchodilator and anti-inflammatory properties. In support of this approach, the addition of the oral PDE4 inhibitor roflumilast to an inhaled β_2 agonist has demonstrated clinical efficacy for improved lung function over β_2-agonist treatment alone.[57] A potential advantage of this combination is that both β_2 agonists and PDE4 inhibitors rely on modulation of the secondary messenger cyclic AMP to elicit their effects, and it is possible that the combination could provide additive or synergistic pulmonary anti-inflammatory activity. Studies reporting positive anti-inflammatory interactions through modulation of cyclic AMP using a combination of β_2 agonist and PDE4 inhibitor have been reported in human immune cells and lung fibroblasts.[66,67]

Three applications have appeared which describe the combination of a PDE4 inhibitor with a β_2 agonist.[37,68,69] The applications differentiate from one another by the PDE4 inhibitor pharmacophore that is utilized. In one, the standard set of β_2-agonist head groups is connected with a pyrazolopyridine PDE4 inhibitor (**22**) to give compounds exemplified by **25** (IC_{50}s against both targets are less than 100 nM).[68] Another example connects GSK256066 (**26**)[70] with a variety of similar linkers and a standard set of β_2-agonist head groups of which **27** is an example (IC_{50} against both targets less than 100 nM).[69] Another study describes the connection shown in **28**, which has an EC_{50} of 0.1 nM for relaxation of guinea pig tracheal smooth muscle in an *ex vivo* assay and an IC_{50} of 260 nM as a PDE4 inhibitor.[71] The authors describe the β_2-agonist activity to be sensitive to chain length, with little effect on PDE4 activity.

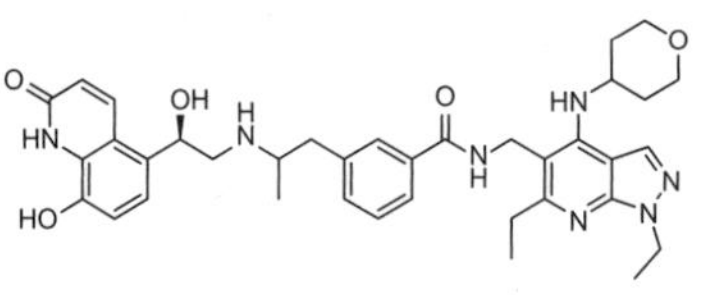

25

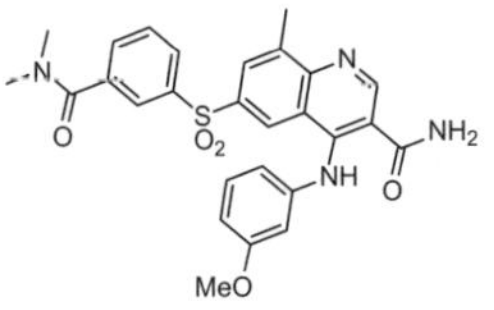

26

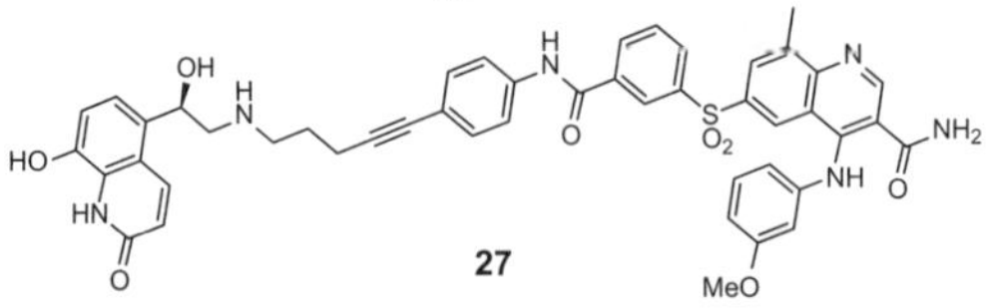

27

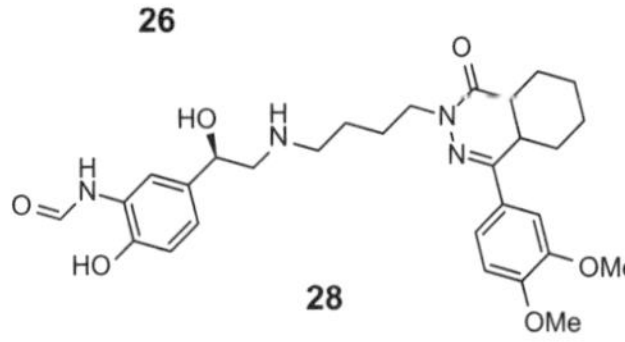

28

3. CONCLUSIONS

There remains great need for the discovery and development of more efficacious and novel treatments for COPD. The development of bifunctional molecules may be ideally suited toward inhaled delivery, given their physical chemical properties, which should favor lung retention while minimizing systemic adverse effects and toxicity. In the absence of more sophisticated delivery devices that could deliver multiple combinations, the utilization of bifunctional compounds has been predicted to be a potential future basis of a therapeutic regimen that affects three targets and which would consist of the combination of a bifunctional compound with another agent.[16] It is also tempting to speculate that in the future, a quadruple target treatment may also be possible based on a combination of two bifunctional molecules delivered in currently available oral inhalation devices.

REFERENCES

(1) Global Initiative for Chronic Obstructive Lung Disease. Global strategy for the diagnosis, management and prevention of COPD. Revised 2011. Available from http//www.goldcopd.org/uploads/users/files/GOLD_Report_2011Feb21.pdf.
(2) Mathers, C.C.; Loncar, D. *PLoS Med.* **2006**, *3*, e442.
(3) Chapman, K.R.; Mannino, D.M.; Soriano, J.B.; Vermeire, P.A.; Buist1, A.S.; Thune, M.J.; Connelle, C.; Jemale, A.; Lee, T.A.; Miravitlles, M.; Aldington, S.; Beasley, R. *Eur. Respir. J.* **2006**, *27*, 188.
(4) Meurs, H.; Roffel, A.F.; Elzinga, C.R.; Zaagsma, J. In *Muscarinic Receptors in Airways Disease*; Zaagsma, J., Meurs, H., Roffel, A.F., Eds.; Birkhauser-Verlag: Basel, 2001; pp 121–157.
(5) Johnson, M. *J. Allergy Clin. Immunol.* **2006**, *117*, 18.
(6) Gosens, R.; Zaagsma, J.; Meurs, H.; Halayko, A.J. *Respir. Res.* **2006**, *7*, 73.
(7) Lainé, D.I. *Expert Rev. Clin. Pharmacol.* **2010**, *3*, 43.
(8) http://www.fda.gov/downloads/AdvisoryCommittees/CommitteesMeetingMaterials/Drugs/Pulmonary-AllergyDrugsAdvisoryCommittee/UCM207377.pdf.
(9) Gross, N.J.; Giembycz, M.A.; Rennard, S.I. *COPD* **2010**, *7*, 141.
(10) Page, C.P.; Spina, D. *Curr. Opin. Pharmacol.* **2012**, *12*, 1.
(11) Boswell-Smith, V.; Cazzola, M.; Page, C.P. *J. Allergy Clin. Immunol.* **2006**, *117*, 1237.
(12) Newton, R.; Holden, N. *Mol. Pharmacol.* **2007**, *72*, 799.
(13) Calverley, P.M.; Anderson, J.A.; Celli, B.; Ferguson, G.T.; Jenkins, C.; Jones, P.W.; Yates, J.C.; Vestbo, J. *N. Engl. J. Med.* **2007**, *356*, 775.
(14) Arrowsmith, J. *Nat. Rev. Drug Dis.* **2012**, *11*, 17.
(15) Robinson, C.; Zhang, J.; Garrod, D.R.; Newton, G.K.; Jenkins, K.; Perrior, T.R. *Future Med. Chem.* **2011**, *3*, 1567.
(16) Matera, M.G.; Page, C.P.; Cazzola, M. *Trends Pharm. Sci.* **2011**, *32*, 495.
(17) Finch, H.; Ray, N. C.; Edwards, C. Patent Application WO 060203, **2009**.
(18) Johnson, M. R.; Hirsch, A. J.; Boucher, R. C.; Zhang, J. Patent Application WO 146869, **2007**.

(19) Cazzola, M.; Molimard, M. *Pulm. Pharmacol. Ther.* **2010**, *23*, 257.
(20) Hughes, A.D.; Jones, L.H. *Future Med. Chem.* **2011**, *3*, 1585.
(21) Theravance. Press Release October 27, **2011**. www.theravance.com.
(22) Mammen, M.; Dunham, S.; Hughes, A.; Lee, T. W.; Husfeld, C.; Stangeland, E. Patent Application WO 074246, **2004**.
(23) Ray, N.C.; Alcaraz, L. *Expert Opin.* **2009**, *19*, 1.
(24) Chudasama, R.; Kennedy, A.; Kindon, L. J.; Mallet, F. P. Patent Application WO 090859, **2007**.
(25) Chao, R. S.; Rapta, M.; Colson, P. Patent Application WO 023454, **2006**.
(26) Naito, R.; Takeuchi, M.; Morihira, K.; Hayakawa, M.; Ikeda, K.; Shibanuma, T.; Isomura, Y. *Chem. Pharm. Bull.* **1998**, *46*, 1286.
(27) Steinfeld, T.; Hughes, A.D.; Klein, U.; Smith, J.A.M.; Mammen, M. *Mol. Pharmacol.* **2011**, *79*, 389.
(28) Mammen, M.; Dunham, S.; Hughes, A. Patent Application WO 074812, **2004**.
(29) Mammen, M.; Dunham, S.; Hughes, A.; Lee, T. W. Patent Application WO 074276, **2004**.
(30) Hughes, A.D.; Chin, K.H.; Dunham, S.L.; Jasper, J.R.; King, K.E.; Lee, T.W.; Mammen, M.; Martin, J.; Steinfeld, T. *Bioorg. Med. Chem. Lett.* **2011**, *21*, 1354.
(31) Jones, L. H.; Lunn, G.; Price, D. A. Patent Application WO 041095, **2008**.
(32) Hughes, A.; Byun, D.; Chen, Y.; Fleury, M.; Jacobsen, J. R.; Strangeland, E. R.; Wilson, R. D.; Yen, R. Patent Application WO 123766, **2010**.
(33) Mammen, M.; Dunham, S. Patent Application WO 051946, **2005**.
(34) Mammen, M.; Mischki, T.; Hughes, A.; Ji, Y. Patent Application WO 023460, **2006**.
(35) Mammen, M.; Mischki, T. Patent Application WO 023457, **2006**.
(36) Colson, P.; Hughes, A.; Husfeld, C.; Mammen, M.; Rapta, M. Patent Application WO 127196, **2007**.
(37) Collingwood, S. P.; Fairhurst, R. A.; Press, N. J. Patent Application WO 000483, **2008**.
(38) Jones, L.H.; Baldock, H.; Bunnage, M.E.; Burrows, J.; Clarke, N.; Coghlan, M.; Entwistle, D.; Fairman, D.; Feeder, N.; Fulton, C.; Hilton, L.; James, K.; Jones, R.M.; Kenyon, A.S.; Marshal, S.; Newman, S.D.; Osborne, R.A.; Patel, S.; Selby, M.D.; Stuart, E.F.; Trevethick, M.A.; Wright, K.N.; Price, D.A. *Bioorg. Med. Chem. Lett.* **2011**, *21*, 2759.
(39) James, K.; Jones, L. H.; Price, D. A. Patent Application WO 107828, **2007**.
(40) Mammen, M.; Hughes, A. Patent Application WO 106333, **2004**.
(41) Finch, H.; Ray, N. C.; Bull, R. J.; Neil, M. B.; Jennings, A. S. R. Patent Application WO 017670, **2007**.
(42) Finch, H.; Bull, R. J. Patent Application WO 017824, **2008**.
(43) Alcaraz, L.; Kindon, N.; Sutton, J. Patent Application WO 096127, **2008**.
(44) Alcaraz, L.; Kindon, N.; Sutton, J. Patent Application WO 149110, **2008**.
(45) Bourssou, T.; Schnapp, A.; Lustenberger, P.; Rudolf, K.; Pieper, M. P.; Grauert, M.; Breitfelder, S.; Speck, G. Patent Application WO 111004, **2005**.
(46) Ray, N. C.; Bull, R. J.; Finch, H.; Van Den Heuvel, M.; Bravo, J. A. Patent Application WO 017669, **2008**.
(47) Edwards, C.; Kindon, N.; Ray, N. C.; Sutton, J. M.; Alcaraz, L. Patent Application WO 006129, **2009**.
(48) Sutton, J. M.; Roussel, F.; Van Den Heuvel, M.; Ray, N. C.; Alcaraz, L. Patent Application WO 015792, **2010**.
(49) Jones, L. H.; Roberts, D. F.; Strang, R. S. Patent Application WO 004517, **2010**.
(50) Finch, H.; Bull, R. J.; Sutton, J. M. Patent Application WO 017827, **2008**.
(51) Haab, F. *Drugs Today* **2005**, *41*, 441.
(52) Mammen, M.; Hughes, A. Patent Application WO 089892, **2004**.

(53) Jennings, A. S. R.; Ray, N. C.; Roussel, F.; Sutton, J. M. Patent Application WO 154678, **2011**.
(54) Alcaraz, L.; Bailey, A.; Bull, R. J.; Johnson, T.; Kindon, N. D.; Lister, A. S.; Robbins, A. J.; Stocks, M. J.; Tobald, B. Patent Application WO 008448, **2009**.
(55) Alcaraz, L.; Bailey, A.; Kindon, N. Patent Application WO 012896, **2011**.
(56) Osborne, R.; Clark, N.; Glossop, P.; Kenyon, A.; Liu, H.; Patel, S.; Summerhill, S.; Jones, L.H. *J. Med. Chem.* **2011**, *54*, 6998.
(57) Fabbri, L.M.; Calverley, P.M.; Izquierdo-Alonso, J.L.; Bundschuh, D.S.; Brose, M.; Martinez, F.J.; Rabe, K.F. *Lancet* **2009**, *374*, 695.
(58) Provins, L.; Christophe, B.; Danhaive, P.; Dulieu, J.; Durieu, V.; Gillard, M.; Lebon, F.; Lengele, S.; Que´re, L.; van Keulen, B. *Bioorg. Med. Chem. Lett.* **2006**, *16*, 1834.
(59) Provins, L.; Christophe, B.; Danhaive, P.; Dulieu, J.; Gillard, M.; Que´re´, L.; Stebbins, K. *Bioorg. Med. Chem. Lett.* **2007**, *17*, 3077.
(60) Callahan, J. F.; Lin, G.; Wan, Z.; Yan, H. Patent Application WO 100166, **2009**.
(61) Callahan, J. F.; Lin, G.; Wan, Z.; Yan, H. Patent Application WO 100167, **2009**.
(62) Callahan, J. F.; Li, T.; Wan, Z.; Yan, H. Patent Application WO 100169, **2009**.
(63) Callahan, J. F.; Wan, Z.; Yan, H. Patent Application WO 100170, **2009**.
(64) Hamblin, J.N.; Angell, T.D.R.; Ballantine, S.P.; Cook, C.M.; Cooper, A.W.J.; Dawson, J.; Delves, C.J.; Jones, P.S.; Lindvall, M.; Lucas, F.S.; Mitchell, C.J.; Neu, M.Y.; Ranshaw, L.E.; Solanke, Y.E.; Somers, D.O.; Wiseman, J.O. *Bioorg. Med. Chem. Lett.* **2008**, *18*, 4237.
(65) Jin, J.; Budzik, B.; Wang, Y.; Shi, D.; Wang, F.; Xie, H.; Wan, Z.; Zhu, C.; Foley, J.J.; Webb, E.F.; Berlanga, M.; Burman, M.; Sarau, H.M.; Morrow, D.M.; Moore, M.L.; Rivero, R.A.; Palovich, M.; Salmon, M.; Belmonte, K.E.; Laine, D.I. *J. Med. Chem.* **2008**, *51*, 5915.
(66) Tannheimer, S.L.; Sorensen, E.A.; Haran, A.C.; Mansfield, C.N.; Wright, C.D.; Salmon, M. *Pulm. Pharmacol. Ther.* **2012**, *25*, 178.
(67) Tannheimer, S.L.; Wright, C.D.; Salmon, M. *Respir. Res.* **2012**, *13*, 28.
(68) Baker, W. R.; Cai, S.; Kaplan, J. A.; Kim, M.; Phillips, G.; Purvis, L.; Stasiak, M.; Stevens, K. L.; Van Veldhuizen, J. Patent Application WO 143106, **2011**.
(69) Baker, W. R.; Cai, S.; Kaplan, J. A.; Kim, M.; Loyer-Drew, J. A.; Perrault, S.; Phillips, G.; Purvis, L.; Stasiak, M.; Stevens, K. L.; Van Veldhuizen, J. Patent Application WO 143105, **2011**.
(70) Tralau-Stewart, C.J.; Williamson, R.A.; Nials, A.T.; Gascoigne, M.; Dawson, J.; Hart, G.J.; Angell, A.D.R.; Solanke, Y.E.; Lucas, F.S.; Wiseman, J.; Ward, P.; Ranshaw, L.E.; Knowles, R.G. *J. Pharmacol. Exp. Ther.* **2011**, *337*, 145.
(71) Shan, W.; Huang, L.; Zhou, Q.; Jiang, H.; Luo, Z.; Lai, K.; Li, X. *Bioorg. Med. Chem. Lett.* **2012**, *22*, 1523.

CHAPTER FIFTEEN

Inflammatory Targets for the Treatment of Atherosclerosis

Robert O. Hughes, Alessandra Bartolozzi, Hidenori Takahashi
Boehringer Ingelheim, Ridgefield Connecticut, USA

Contents

1. INTRODUCTION

Atherosclerosis is a complex disease of the arterial wall and is the primary cause of coronary artery disease, peripheral vascular disease, and stroke. The prevalence of these conditions is a leading cause of death worldwide.[1] Atherosclerosis is characterized by the formation of lipid-rich arterial plaques that may eventually rupture causing major adverse cardiovascular events such as myocardial infarct and stroke. Against a backdrop of genetic risk factors,[2] formation of these plaques and the pathogenesis of atherosclerosis derive from unbalanced lipid metabolism and an inappropriate response from the immune system.[3–5] Treatments targeting the modulation of key

Annual Reports in Medicinal Chemistry, Volume 47
ISSN 0065-7743
http://dx.doi.org/10.1016/B978-0-12-396492-2.00015-1

lipids (aimed at increasing or decreasing levels) have been successful at improving outcomes in at risk patients, although significant residual risk remains.[6] For example, treatment aimed at improving the distribution of lipid fractions in patients diagnosed with acute coronary syndrome has had a significant clinical benefit. Yet, the residual risk for major adverse cardiovascular events in this same patient population has been estimated at 70–80%. There is clearly an opportunity for improved treatment paradigms. Recently, an appreciation of the role of inflammation in the pathogenesis of atherosclerosis has led to the identification of various targets suitable for small molecule intervention.[3–5] The development of agents targeted against inflammatory targets in conjunction with lipid modulating therapies should present new treatment modalities for improving outcomes in patients. In this review, clinical and preclinical evidence for anti-inflammatory targets in cardiovascular diseases amenable to small molecule modulation will be presented. Specific examples of compounds either in development or that have helped validate these targets will be discussed. Agents for which the primary mode of action appears to be modulation of lipid metabolism, but which may also impart an anti-inflammatory effect, will not be covered; these agents have been recently reviewed.[5]

2. PHOSPHOLIPASE A_2

Phospholipase A_2 (PLA_2) is a superfamily of lipolytic enzymes comprised of 15 groups and multiple subgroups.[7,8] There are four main types of PLA_2s: the secreted PLA_2s ($sPLA_2$s), the cytosolic PLA_2s ($cPLA_2$s), the calcium-independent PLA_2s ($iPLA_2$s), and the lipoprotein-associated PLA_2s (Lp-PLA_2s, also known as platelet activating factor acetylhydrolase, PAF-AH). PLA_2s catalyze the hydrolysis of membrane phospholipids at the *sn*-2 position to generate free fatty acids and lysophospholipids that are subsequently converted to prostaglandins and leukotrienes (LTs). All three end products have been recognized for their inflammatory roles; therefore, PLA_2s have been targeted for the treatment of multiple disease indications.[9–12] Among the PLA_2s, $sPLA_2$s and Lp-PLA_2 have been specifically targeted for the treatment of cardiovascular diseases. The most recent advances achieved for the inhibition of $sPLA_2$s and Lp-PLA_2s and the relevance of these two targets to the treatment of cardiovascular diseases are described below.

2.1. Lp-PLA$_2$ inhibitors

Lp-PLA$_2$s are calcium-independent enzymes produced by inflammatory cells (macrophages and monocytes). The role played by Lp-PLA$_2$s in atherosclerosis has been controversial, with initial evidence suggesting an atheroprotective effect due to hydrolysis of the proinflammatory PAF.[13] Subsequent investigations demonstrated that Lp-PLA$_2$s do not play a predominant role in the hydrolysis of PAF. Additionally, it has been demonstrated that Lp-PLA$_2$s accumulate in human atherosclerotic lesions with evidence suggesting a proatherogenic role in the vessel wall.[14] Elevated levels of circulating Lp-PLA$_2$s are associated with plaque vulnerability and endothelial dysfunction, and it has been determined that patients with acute coronary syndromes have higher levels of Lp-PLA$_2$s than normal.[15]

Lp-PLA$_2$s are delivered to lesion-prone segments of the arterial wall by binding to apolipoprotein B (ApoB) on low-density lipoprotein (LDL). Oxidation of LDL leads to the formation of truncated phospholipids in the *sn*-2 position that are subsequently subjected to hydrolysis by Lp-PLA$_2$s. Lysophosphatidylcholine (lysoPC) and oxidized nonesterified fatty acids (oxNEFA) are the products of this hydrolysis and are believed to play a prominent role in the local increase of inflammatory mediators in lesion-prone areas. It has been demonstrated that expression of Lp-PLA$_2$s is highly upregulated in macrophages undergoing apoptosis within the necrotic core and fibrous cap of vulnerable and ruptured plaque.[16] Based on this evidence, Lp-PLA$_2$ inhibitors have been targeted to decrease plaque instability as a treatment for atherosclerosis. Multiple pyrimidone- and pyridinone-based inhibitors (e.g., **1**, **2**, **3**, and **4**) of Lp-PLA$_2$ have been reported.[17–19] SB-568859 (structure not reported) and **3** reached phase I and phase II clinical trials for atherosclerosis, respectively, while **4** is currently under evaluation in a phase III trial.[20] Three ongoing large outcome trials are currently evaluating **4** for primary and secondary prevention.[2,21,22]

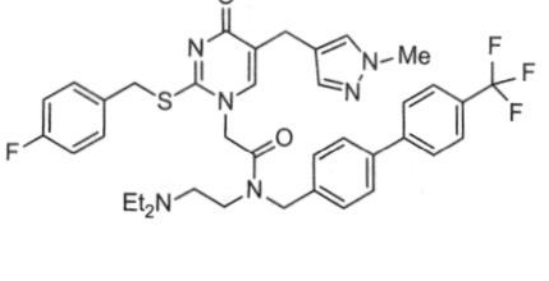

1

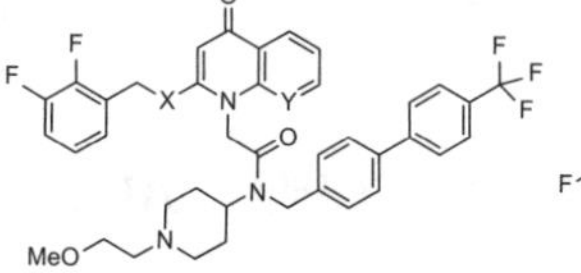

2 X = CH$_2$, Y = N
3 X = S, Y = CH

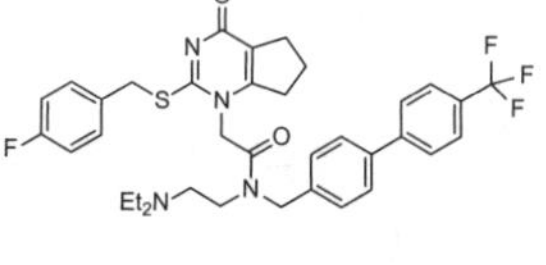

4

Additional small molecule Lp-PLA$_2$ inhibitors have been reported. Examples include oximes **5**,[23] amides of xanthurenic acid **6**,[24] and tricycle compounds **7**.[25]

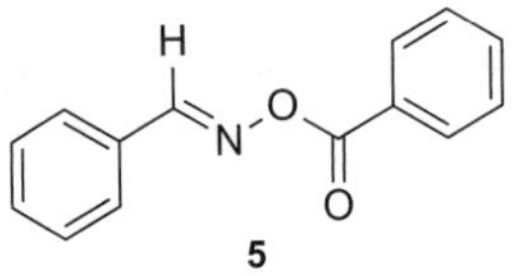
5

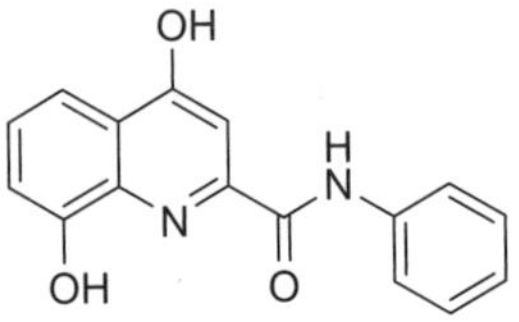
6

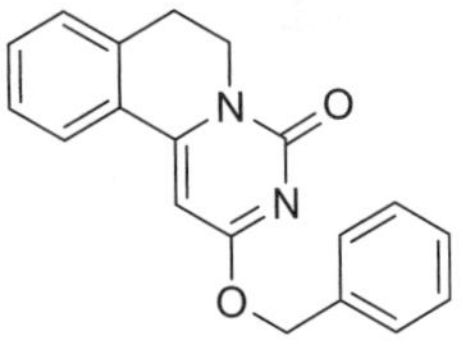
7

2.2. S-PLA$_2$ inhibitors

S-PLA$_2$s are calcium-dependent enzymes that catalyze the hydrolysis of phospholipids at the *sn*-2 position to release arachidonic acid (AA). Coronary atherosclerotic lesions of patients with acute myocardial infarction contain S-PLA$_2$s which are detected within intimal macrophages, vascular smooth muscle cells, and in extracellular deposits. Levels and activity of S-PLA$_2$s in acute coronary syndrome patients were found to correlate well with disease severity.[26] Multiple sPLA$_2$ inhibitors have been reported and have been reviewed.[11,12] **8** and its methyl ester prodrug, **9**, are sPLA$_2$ inhibitors that have advanced the furthest into development.[27–32] **8** has been evaluated in two phase II clinical trials.[28–30] **9** was studied in phase II[31] and phase III clinical trials for acute chest syndrome.[32] Recent publications have highlighted related analogs including **10**.[33] Additionally, molecular docking studies on 2-oxoamides, **11** and peptide inhibitors have been reported.[34,35]

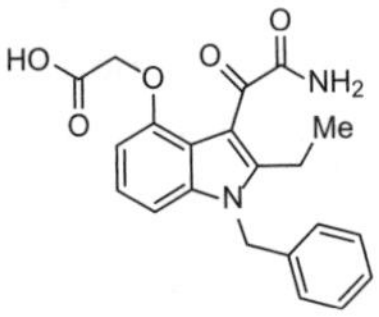
8

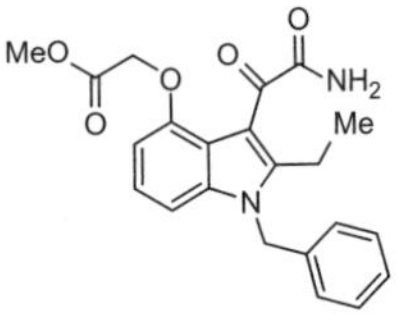
9

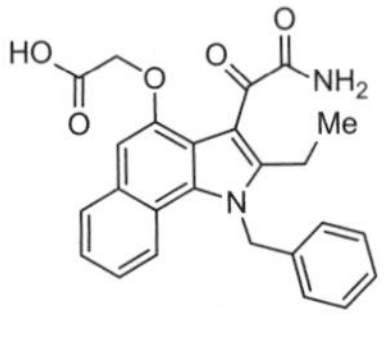
10

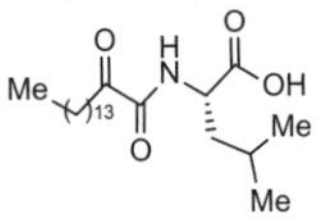
11

3. LEUKOTRIENE PATHWAY AND ATHEROSCLEROSIS

The LT pathway constitutes a fundamental series of events underlying the inflammatory component of atherosclerosis.[36–38] The pathway is initiated after release of AA from cell membranes by cPLA$_2$ (Fig. 15.1). Oxidation of AA catalyzed by 5-lipoxygenase (5-LO) generates LTs, lipid mediators that trigger proinflammatory signaling through activation of the

G protein-coupled transmembrane receptors (GPCRs) BLT (for LTB_4) and CysLT (for LTC_4, D_4, and E_4). LT signaling is involved in several processes associated with atherosclerosis, including lipid retention, accumulation of foam cells, and development of intimal plaque. The LT pathway is also believed to contribute to the rupture of plaque, resulting in ischemic injuries, such as myocardial infarct and stroke.

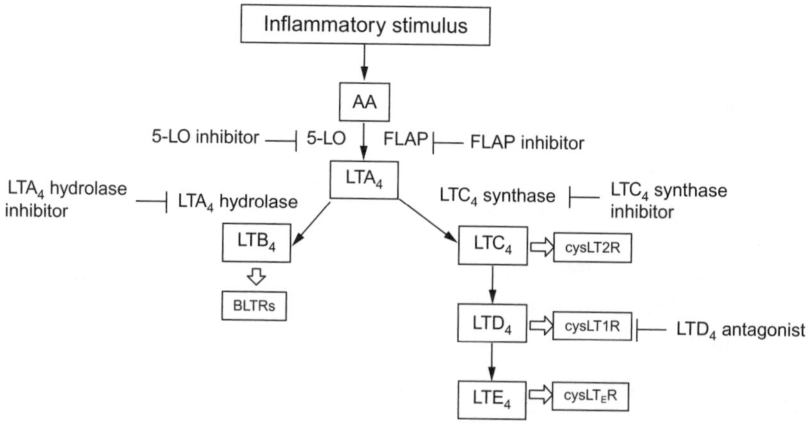

Figure 15.1 Drug targets of the leukotriene (LT) pathway involved in atherogenesis.

3.1. 5-LO inhibitors

5-LO catalyzes the oxidative metabolism of AA aided by 5-lipoxygenase-activating protein (FLAP) and is a rate-limiting enzyme in LT biosynthesis. Zileuton, **12**, is a marketed 5-LO inhibitor indicated for the treatment of asthma.[39] Although a number of 5-LO inhibitors have been targeted for inflammatory indications such as asthma, allergic rhinitis, rheumatoid arthritis, and inflammatory bowel disease, only atreleuton, **13**, has been developed to treat atherosclerosis.[40] A phase II clinical trial of **13** in patients who had experienced recent acute coronary syndrome has completed, with results showing a reduction in LT production, and imaging data suggestive of a positive effect on atherosclerosis.[41,42] A non-*N*-hydroxyurea containing, selective nonredox 5-LO inhibitor, **14**, showing an IC_{50} of 130 nM in a human whole blood assay has been reported.[43]

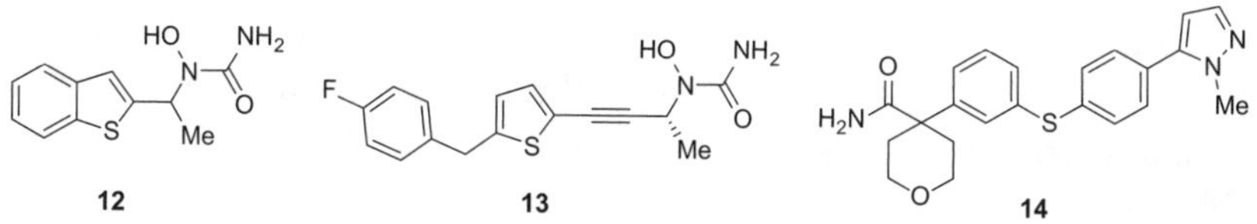

12 **13** **14**

3.2. 5-Lipoxygenase-activating protein inhibitors

AA released from phospholipids is transferred to 5-LO via FLAP for oxidative conversion to LTA_4. Four FLAP inhibitors representing three structural classes have advanced into clinical trials: indoles (**15** and **16**), indoloquinolines (**17**), and quinolines (**18**). Phase I and II clinical trials of compounds **15**, **17**, and **18** for the treatment of asthma have completed, with a reduction of urinary LTE_4 excretion being reported for all three compounds.[44–46] Detailed SAR around indoles related to **16** has been reported.[47] **18** advanced to phase III clinical trials for the prevention of myocardial infarction.[48] However, development of this compound has been discontinued due to a problem with its formulation.[49] Patent applications claiming FLAP inhibitors of different scaffold classes, such as benzimidazoles (e.g., **19**) or oxadiazoles (e.g., **20**), have also published.[50–52]

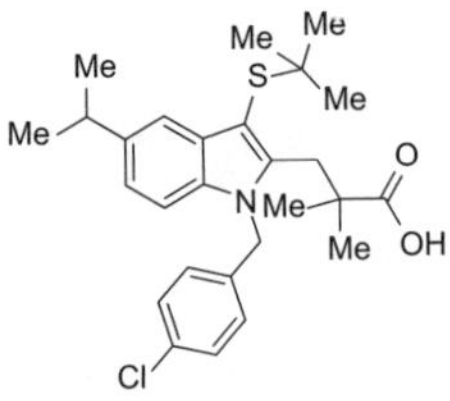

15

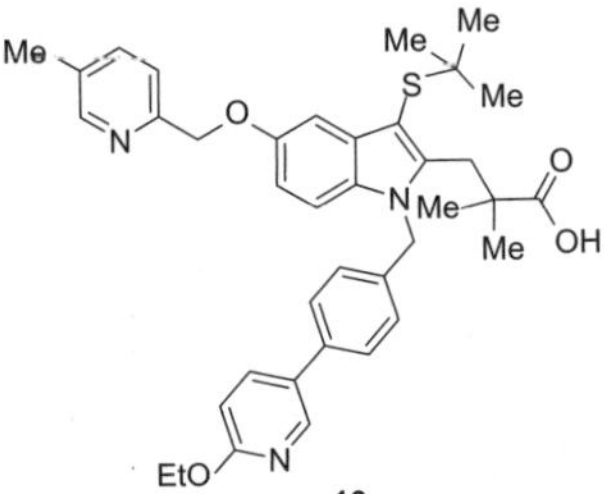

16

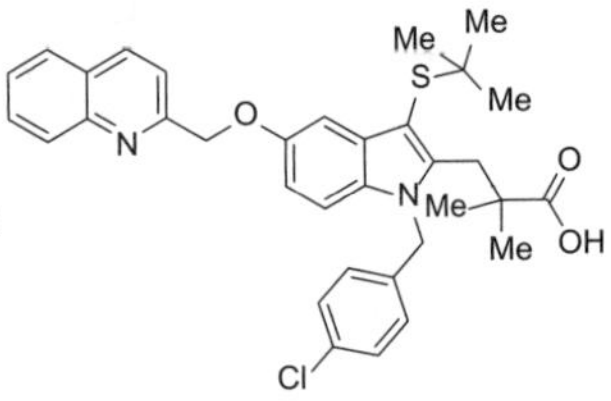

17

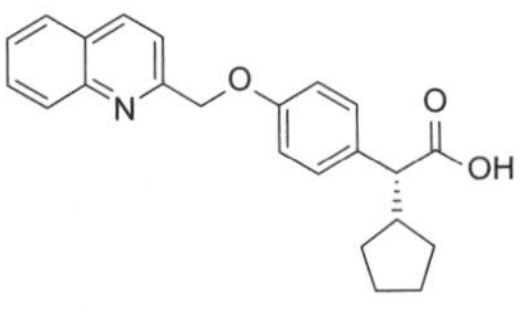

18

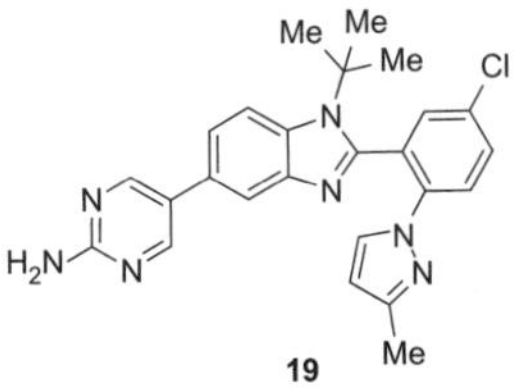

19

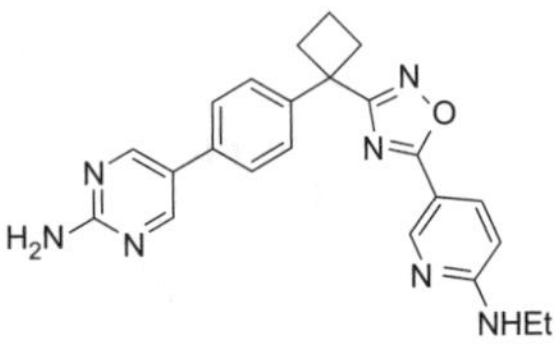

20

3.3. LTA_4 hydrolase inhibitors and leukotriene C_4 inhibitors

An unstable epoxide, LTA_4, is further transformed to either LTB_4 by LTA_4 hydrolase or leukotriene C_4 (LTC_4) by LTC_4 synthase. Two LTA_4 hydrolase inhibitors, JNJ-40929837 and CTX-4430, have reportedly been in development for the treatment of COPD and/or MS, but structures of these compounds have not been disclosed.[53,54] Only **21** advanced to phase II clinical trials (for treatment of myocardial infarction), but its development has been discontinued.[55] A patent application disclosing benzoic acid-containing potent LTC_4 synthase inhibitors (e.g., **22**, $IC_{50} = 18$ nM) has published.[56]

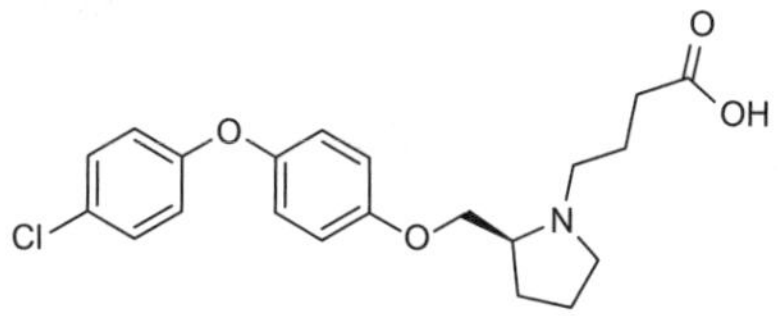

21

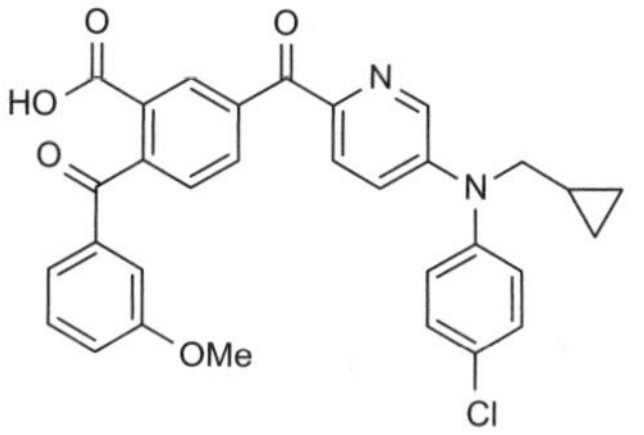

22

3.4. LTD_4 antagonists

Three cysteinyl leukotriene (CysLT) receptor blockers (montelukast, **23**, zafirlukast, **24**, and pranlukast **25**) have reached market for the treatment of respiratory disorders including allergy, allergic rhinitis, or asthma. Although no CysLT receptor antagonists have been reported to be in development for cardiovascular disease, a retrospective analysis of Swedish asthmatic patients treated with montelukast indicated its potential utility for secondary prevention.[57] These data could support CysLT receptor as a potential target for cardiovascular diseases.

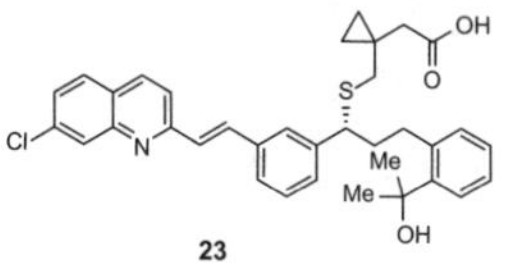

23

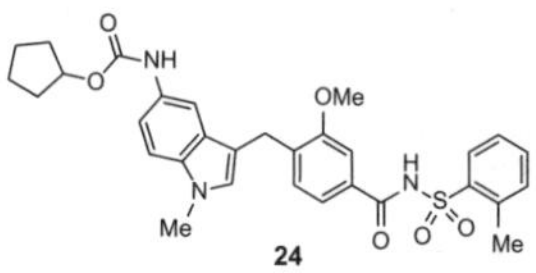

24

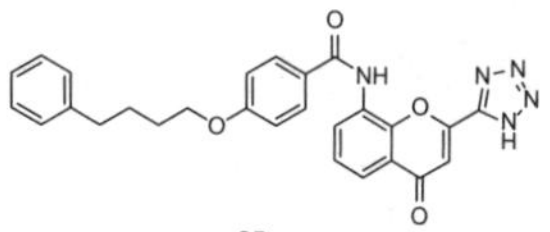

25

4. CHEMOKINES AND ATHEROSCLEROSIS

The chemokines are secreted proteins that interact with cognate GPCRs to regulate the trafficking and activation of leukocytes. They are named according to the location and number of conserved cysteine residues at the N-terminus (CC, CXC and CX3C). There are more than 60 known chemokines and chemokine receptors (CCRs). Both chemokines and their receptors have been targeted for pharmacologic intervention, Chemokine receptors are more amenable to small molecule intervention than the chemokines themselves and are the focus of this section. Recent studies have shown the role of chemokine signaling in the recruitment of inflammatory cells, including monocytes and T-cells, to developing plaques. CCR2 and CCR5 in particular have been implicated in atherogenesis.[58,59] Antagonists of both receptors are described below. Chemokine receptor antagonists often contain basic amines which pose challenges to improving oral bioavailability, reducing Pgp-mediated efflux, and achieving selectivity over hERG and other ion channels.

4.1. CCR2 antagonists

There has been significant effort directed toward the optimization of CCR2 antagonists, with several compounds entering clinical trials for indications such as rheumatoid arthritis, multiple sclerosis, neuropathic pain, diabetes mellitus, and allergic rhinitis. The field has recently been reviewed.[60] Several structural classes of CCR2 antagonists have yielded clinical candidates and these are summarized below. The amino cyclopentane carboxamide scaffold has yielded five clinical candidates: **24**, **25**,[61–63] **26**, **27**, and **28**.[64–66] Both **25** and **28** are reported to behave as insurmountable antagonists, displaying long, dissociative half-lives from CCR2.[63,66] A series of cyclohexanols, **29** and **30**, were recently disclosed as clinical candidates.[67,68] The quaternary salt, **31**, administered as a nasal spray, completed a phase II clinical trial.[69,70] BMS-741672 is reported to be in clinical trials.[71] The structure of BMS-741672 has not been disclosed, but may be that of lactam **32**, a compound for which significant characterization was disclosed in a patent application.[72] Clinical development of CCX140 for Type 2 diabetes has been reported.[73] A recent disclosure characterizing the pharmacology of CCX140 and a patent referenced therein suggest the compound is related to sulfonamides such as **33**.[74,75]

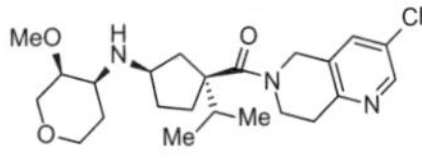

24

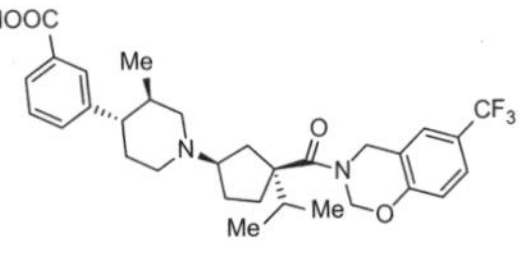

25

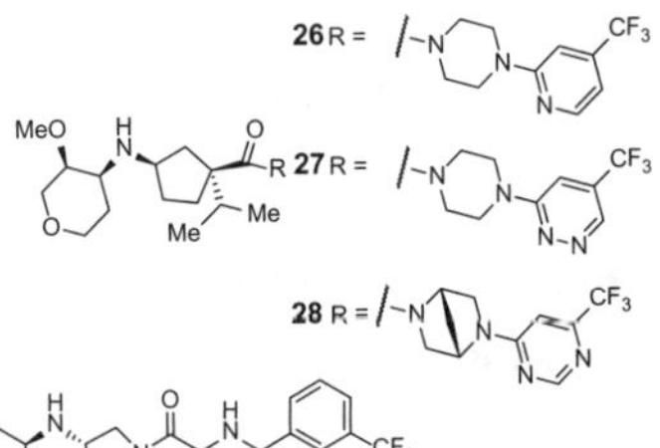

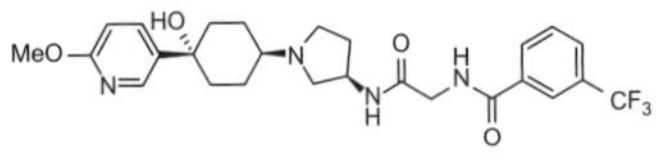

29

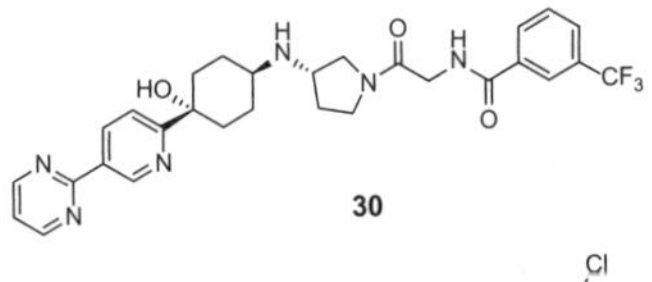

30

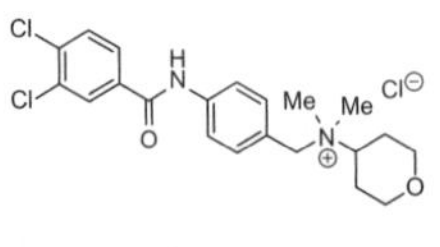

31

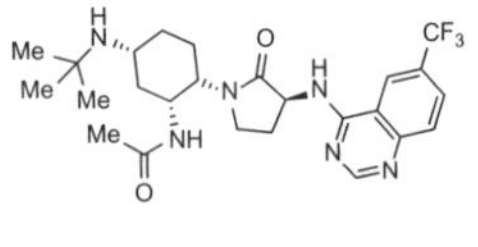

32

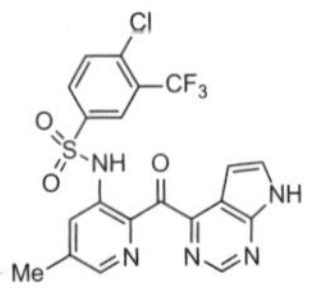

33

4.2. CCR5 antagonists

While emerging evidence points to the role of CCR5 in atherosclerosis,[76] initial interest in the target stemmed from its role as a coreceptor for HIV-1 transmission. Accordingly, there has been significant effort directed toward the optimization of CCR5 antagonists. These efforts have recently been reviewed.[77] Several compounds, which are highlighted below, have entered clinical trials and maraviroc, **34**, has been approved for clinical use. The lead optimization of **34** has been described.[78] A related compound, **35**, has advanced to phase II clinical trials.[79] Amino cyclopentane **36** is under development as a topical agent.[80] A pyrimidinyl amide, vicriviroc, **37**, which has entered late-stage clinical trials, has been described.[81,82] A conformationally restricted analog, **38**, has demonstrated an improved pharmacokinetic profile in phase I.[83] Clinical trials of the spirocyclic carboxylic acid, aplaviroc **39**, were halted due to hepatotoxicity.[84] Biaryl **40** and piperidine **41** have progressed into clinical trials as has the sulfonamide **42**.[85–87]

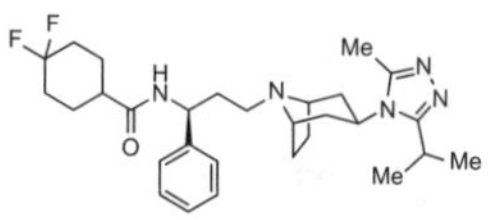
34

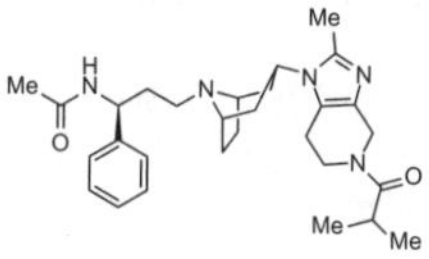
35

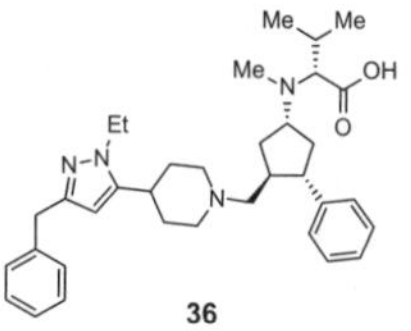
36

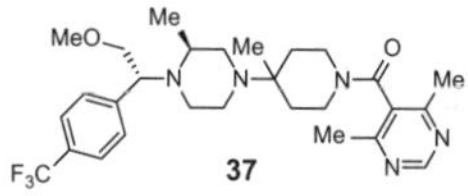
37

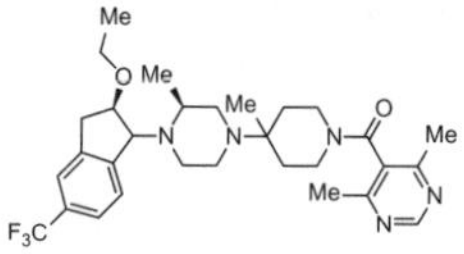
38

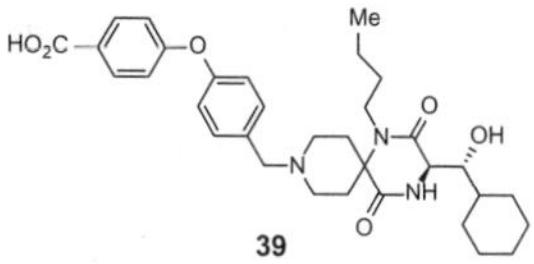
39

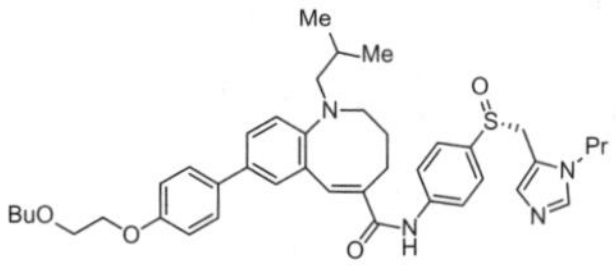
40

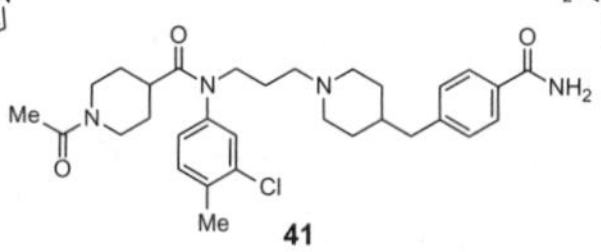
41

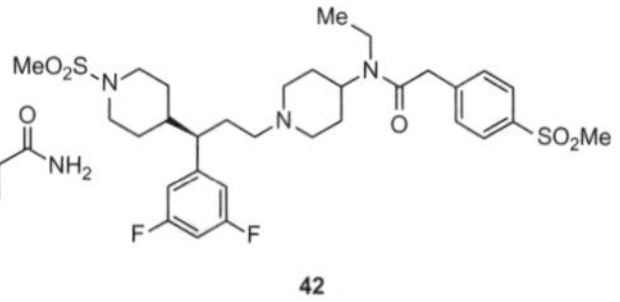
42

5. CONCLUSIONS

The key role played by inflammatory mediators in the pathogenesis of atherosclerosis has presented a wealth of targets for small molecule intervention. While lead optimization and clinical programs against inflammatory targets may not have been initially directed toward the potential treatment of atherosclerosis, the availability of high-quality chemical matter should allow for clinical proof-of-concept studies to be conducted and hypotheses tested.

REFERENCES

(1) Weber, C.; Noels, H. *Nat. Med.* **2011**, *17*, 1410.
(2) White, H.; Held, C.; Stewart, R.; Watson, D.; Harrington, R.; Budaj, A.; Steg, P.G.; Cannon, C.P.; Krug-Gourley, S.; Wittes, J.; Trivedi, T.; Tarka, E.; Wallentin, L. *Am. Heart J.* **2010**, *160*, 655.
(3) Tousoulis, D.; Kampoli, A.-M.; Papageorgiou, N.; Androulakis, E.; Antoniades, C.; Toutouzas, K.; Stefanadis, C. *Curr. Pharm. Design* **2011**, *17*, 4089.
(4) Libby, P.; Okamoto, Y. *Circ. J.* **2010**, *74*, 213.
(5) Charo, I.F.; Taub, R. *Nat. Rev.* **2011**, *10*, 365.
(6) Cannon, C.P.; Braunwald, E.; McCabe, C.H. *N. Engl. J. Med.* **2004**, *350*, 1495.
(7) Burke, J.E.; Dennis, E.A. *Cardiovasc. Drugs Therapy* **2009**, *23*, 49.
(8) Schaloske, R.H.; Dennis, E.A. *Biochim. Biophys. Acta* **2006**, *1761*, 1246.
(9) Lehr, M. *Anti-Inflammatory & Anti-Allergy Agents in Medicinal Chemistry* **2006**, *5*, 149.
(10) Magrioti, V.; Kokotos, G. *Anti-Inflammatory & Anti-Allergy Agents in Medicinal Chemistry* **2006**, *5*, 189.
(11) Mouchlis, V.D.; Barbayianni, E.; Mavromoustakos, T.M.; Kokotos, G. *Curr. Med. Chem.* **2011**, *18*, 2566.
(12) Reid, R.C. *Curr. Med. Chem.* **2005**, *12*, 3011.
(13) Chauffe, R.J.; Wilensky, R.L.; Mohler, E.R., III *Curr. Atheroscler. Rep.* **2010**, *12*, 43.
(14) Zalewski, A.; Macphee, C. *Arterioscler. Thromb. Vasc. Biol.* **2005**, *25*, 923.
(15) Anderson, J.L. *Am. J. Cardiol.* **2008**, *101*, 23F.
(16) Kolodgie, F.D.; Burke, A.P.; Skorija, K.S.; Ladich, E.; Kutys, R.; Makuria, A.T.; Virmani, R. *Arterioscler. Thromb. Vasc. Biol.* **2006**, *26*, 2523.
(17) Blackie, J.A.; Bloomer, J.C.; Brown, M.J.B.; Cheng, H.Y.; Elliott, R.L.; Hammond, B.; Hickey, D.M.B.; Ife, R.J.; Leach, C.A.; Lewis, V.A.; Macphee, C.H.; Milliner, K.J.; Moores, K.E.; Pinto, I.L.; Smith, S.A.; Stansfield, I.G.; Stanway, S.J.; Taylor, M.A.; Theobald, C.J.; Whittaker, C.M. *Bioorg. Med. Chem. Lett.* **2002**, *12*, 2603.
(18) Hickey, D.M.B.; Ife, R.J.; Leach, C.A.; Liddle, J.; Pinto, I.L.; Smith, S.A.; Stanway, S.J. Patent Application WO 2002/030904. **2002**.
(19) Blackie, J.A.; Bloomer, J.C.; Brown, M.J.B.; Cheng, H.Y.; Hammond, B.; Hickey, D.M.B.; Ife, R.J.; Leach, C.A.; Lewis, V.A.; Macphee, C.H.; Milliner, K.J.; Moores, K.E.; Pinto, I.L.; Smith, S.A.; Stansfield, I.G.; Stanway, S.J.; Taylor, M.A.; Theobald, C.J. *Bioorg. Med. Chem. Lett.* **2003**, *13*, 1067.
(20) ClinicalTrials.gov NCT: 00470145, NCT: 00695305.
(21) Clinical Trials.gov NCT: 01067339.
(22) O'Donoghue, M.L.; Braunwald, E.; White, H.D.; Serruys, P.; Steg, P.G.; Hochman, J.; Maggioni, A.P.; Bode, C.; Weaver, D.; Johnson, J.L.; Cicconetti, G.; Lukas, M.A.; Tarka, E.; Cannon, C.P. *Am. Heart J.* **2011**, *162*, 613.
(23) Jeong, T.S.; Kim, M.J.; Yu, H.; Kim, K.S.; Choi, J.K.; Kim, S.S.; Lee, W.S. *Bioorg. Med. Chem. Lett.* **2005**, *15*, 1525.
(24) Lin, E.C.K.; Hu, Y.; Amantea, C.M.; Pham, L.M.; Cajica, J.; Okerberg, E.; Brown, H.E.; Fraser, A.; Du, L.; Kohno, Y.; Ishiyama, J.; Kozarich, J.W.; Shreder, K.R. *Bioorg. Med. Chem. Lett.* **2012**, *22*, 868.
(25) Fukumoto, S.; Sasaki, M.; Kanra, T.; Hasui, T.; Fujimoto, J.; Kato, T. Patent Application JP 2011/088847. **2011**.

(26) Hartford, M.; Wiklund, O.; Hulten, L.M.; Perers, E.; Person, A.; Herlitz, J.; Hurt-Camejo, E.; Karlsson, T.; Caidahl, K. *Int. J. Card.* **2006**, *108*, 55.
(27) Draheim, S.E.; Bach, N.J.; Dillard, R.D.; Berry, D.R.; Carlson, D.G.; Chirgadze, N.Y.; Clawson, D.K.; Hartley, L.W.; Johnson, L.M.; Jones, N.D.; McKinney, E.R.; Mihelich, E.D.; Olkowski, J.L.; Schevitz, R.W.; Smith, A.C.; Snyder, D.W.; Sommers, C.D.; Wery, J.P. *J. Med. Chem.* **1996**, *39*, 5159.
(28) Bradley, J.D.; Dmitrienko, A.A.; Kivitz, A.J.; Gluck, O.S.; Weaver, A.L.; Wiesenhutter, C.; Myers, S.L.; Sides, G.D. *J. Rheumatol.* **2005**, *32*, 417.
(29) Abraham, E.; Naum, C.; Bandi, V.; Gervich, D.; Lowry, S.F.; Wunderink, R.; Schein, R.M.; Macias, W.; Skerjanec, S.; Dmitrienko, A.; Farid, N.; Forgue, S.T.; Jiang, F. *Crit. Care Med.* **2003**, *31*, 718.
(30) Rosenson, R.S.; Fraser, H.; Goulder, M.A.; Hislop, C. *Cardiovasc. Drugs Therapy* **2011**, *25*, 539.
(31) Rosenson, R.S.; Hislop, C.; Elliott, M.; Stasiv, Y.; Goulder, M.; Waters, D. *J. Am. Coll. Cardiol.* **2010**, *56*, 1079.
(32) Dzavik, V.; Lavi, S.; Thorpe, K.; Yip, P.M.; Plante, S.; Ing, D.; Overgaard, C.B.; Osten, M.D.; Lan, J.; Robbins, K.; Miner, S.E.; Horlick, E.M.; Cantor, W.J. *Circulation* **2010**, *122*, 2411.
(33) Oslund, R.C.; Cermak, N.; Gelb, M.H. *J. Med. Chem.* **2008**, *51*, 4708.
(34) Mouchlis, V.D.; Magrioti, V.; Barbayianni, E.; Cermak, N.; Oslund, R.C.; Mavromoustakos, T.M.; Gelb, M.H.; Kokotos, G. *Bioorg. Med. Chem. Lett.* **2011**, *19*, 735.
(35) Thwin, M.M.; Satyanarayanajois, S.D.; Nagarajarao, L.M.; Sato, K.; Arjunan, P.; Ramapatna, S.L.; Kumar, P.V.; Gopalakrishnakone, P. *J. Med. Chem.* **2007**, *50*, 5938.
(36) Back, M. *Cardiovasc. Drugs Ther.* **2009**, *23*, 41.
(37) Radmark, O. *Arterioscler. Thromb. Vasc. Biol.* **2003**, *23*, 1140.
(38) Riccioni, G.; Zanasi, A.; Vitulano, N.; Mancini, B.; D'Orazio, N. *Mediators Inflamm.* **2009**, *17*, 1.
(39) McGill, K.A.; Busse, W.W. *Lancet* **1996**, *348*, 519.
(40) Brooks, C.D.W.; Stewart, A.O.; Basha, A.; Bhatia, P.; Ratajczyk, J.D.; Martin, J.G.; Craig, R.A.; Kolasa, T.; Bouska, J.B.; Lanni, C.; Harris, R.R.; Malo, P.E.; Carter, G.W.; Bell, R.L. *J. Med. Chem.* **1995**, *38*, 4768.
(41) Back, M. *Curr. Pharm. Des.* **2009**, *15*, 3116.
(42) Tardif, J.-C.; L'Allier, P.L.; Ibrahim, R.; Gregoire, J.C.; Nozza, A.; Cossette, M.; Kouz, S.; Lavoie, M.-A.; Paquin, J.; Brotz, T.M.; Taub, R.; Presacco, J. *Circ. Cardiovasc. Imag.* **2010**, *3*, 298.
(43) Masferrer, J.L.; Zweifel, B.S.; Hardy, M.; Anderson, G.D.; Dufield, D.; Cortes-Burgos, L.; Pufahl, R.A.; Graneto, M. *J. Pharmacol. Exp. Ther.* **2010**, *334*, 294.
(44) Friedman, B.S.; Bel, E.H.; Tanaka, W.; Han, Y.H.; Shingo, S.; Spector, R.; Sterk, P. *Am. Rev. Respir. Dis.* **1993**, *147*, 839.
(45) Diamant, Z.; Timmers, M.C.; van der Veen, H.; Friedman, B.S.; De Smet, M.; Depre, M.; Hilliard, D.; Bel, E.H.; Sterk, P.J. *J. Allergy Clin. Immunol.* **1995**, *95*, 42.
(46) Dahlen, B.; Kumlin, M.; Ihre, E.; Zetterstrom, O.; Dahlen, S.E. *Thorax* **1997**, *52*, 342.
(47) Stock, N.S.; Bain, G.; Zunic, J.; Li, Y.; Ziff, J.; Roppe, J.; Santini, A.; Darlington, J.; Prodanovich, P.; King, C.D.; Baccei, C.; Lee, C.; Rong, H.; Chapman, C.; Broadhead, A.; Lorrain, D.; Correa, L.; Hutchinson, J.H.; Evans, J.F.; Prasit, P. *J. Med. Chem.* **2011**, *54*, 8013.
(48) Hakonarson, H.; Thorvaldsson, S.; Helgadottir, A.; Gudbjartsson, D.; Zink, F.; Andersdottir, M.; Manolescu, A.; Arnar, D.O.; Andersen, K.; Sigurdsson, A.; Thorgeirsson, G.; Jonsson, A.; Agnarsson, U.; Bjornsdottir, H.; Gottskalksson, G.; Einarsson, A.; Gudmundsdottir, H.; Adalsteinsdottir, A.E.; Gudmundsson, K.; Kristjansson, K.; Hardarson, T.; Kristinsson, A.; Topol, E.J.; Gulcher, J.; Kong, A.; Gurney, M.; Thorgeirsson, G.; Stefansson, K. *JAMA* **2005**, *293*, 2245.

(49) deCODE genetics, Inc., Press Release 2006: October 5.
(50) Chen, Z.; Hao, M.H.; Liu, W.; Lo, H.Y.; Loke, P.L.; Man, C.C.; Morwick, T.M.; Nemoto, P.A.; Takahashi, H.; Tye, H.; Wu, L. Patent Application, WO 2011/068821. **2011**.
(51) Bartolozzi, A.; Bosanac, T.; Chen, Z.; De Lombaert, S.; Huber, J.; Lo, H.Y.; Loke, P.L.; Liu, W.; Morwick, T.M.; Olague, A.; Riether, D.; Tye, H.; Wu, L.; Zindell, R. Patent Application, WO 2012/024150. **2012**.
(52) Bartolozzi, A.; Bosanac, T.; Chen, Z.; De Lombaert, S.; Dines, J.A.; Huber, J.D.; Liu, W.; Lo, H.Y.; Loke, P.L.; Morwick, T.M.; Nemoto, P.A.; Olague, A.; Riether, D.; Tye, H.; Wu, L.; Zindell, R.M. Patent Application, WO 2012/027322. **2012**.
(53) Rao, N.L.; Dunford, P.J.; Xue, X.; Jiang, X.; Lundeen, K.A.; Coles, F.; Riley, J.P.; Williams, K.N.; Grice, C.A.; Edwards, J.P.; Karlsson, L.; Fourie, A.M. *J. Pharmacol. Exp. Ther.* **2007**, *321*, 1154.
(54) Celtaxsys Inc., Press Release, 2012: January 05.
(55) deCODE genetics, Inc., Press Release, 2010: January 21.
(56) Nilsson, P.; Pelcman, B.; Roenn, M.; Krog-Jensen, C. Patent Application, WO 2011/110824. **2011**.
(57) Ingelsson, E.; Yin, L.; Baeck, M. *J. Allergy Clin. Immunol.*, **2012**, *129,* 702.
(58) Barlic, J.; Murphy, P.M. *J. Leukoc. Biol.* **2007**, *27*, 226.
(59) Zernecke, A.; Shagdarsuren, E.; Weber, C. *Arterioscler. Thromb. Vasc. Biol.* **2008**, *28*, 1897.
(60) Struthers, M.; Pasternak, A. *Curr. Top. Med. Chem.* **2010**, *10*, 1278.
(61) Yang, L.; Jiao, R.X.; Moyes, C.; Morriello, G.; Butora, G.; Shankaran, K.; Pasternak, A.; Goble, S.; Zhou, C.; MacCoss, M.; Cumiskey, A.-M.; Peterson, L.; Forrest, M.; Ayala, J.M.; Jin, H.; DeMartion, J.; Mills, S.G. 233rd ACS national meeting, Chicago, IL, March, **2007**.
(62) Pasternak, A.; Goble, S.D.; Struthers, M.; Vicario, P.P.; Ayala, J.M.; Di Salvo, J.; Kilburn, R.; Wisniewski, T.; De Martino, J.A.; Mills, S.G.; Yang, L. *ACS Med. Chem. Lett.* **2010**, *1*, 14.
(63) Yang, L.; Zhou, C.; Guo, L.; Morriello, G.; Butora, G.; Pasternak, A.; Parsons, W.H.; Mills, S.G.; MacCoss, M.; Vicario, P.P.; Zweerink, H.; Ayala, J.M.; Goyal, S.; Hanlon, W.A.; Cascierind, M.A.; Springer, M.S. *Bioorg. Med. Chem. Lett.* **2006**, *16*, 3735.
(64) Zheng, C.; Cao, G.; Xia, M.; Feng, H.; Glenn, J.; Anand, R.; Zhang, K.; Huang, T.; Wang, A.; Kong, L.; Li, M.; Galya, L.; Hughes, R.O.; Devraj, R.; Morton, P.A.; Rogier, J.D.; Covington, M.; Baribaud, F.; Shin, N.; Scherle, P.; Diamond, S.; Yeleswaram, S.; Vaddi, K.; Newton, R.; Hollis, G.; Friedman, S.; Metcalf, B.; Xue, C.-B. *Bioorg. Med. Chem. Lett.* **2011**, *21*, 1442.
(65) Hughes, R.O.; Rogier, D.J.; Devraj, R.; Zheng, C.; Cao, G.; Feng, H.; Xia, M.; Anand, R.; Xing, L.; Glenn, J.; Zhang, K.; Convington, M.; Morton, P.A.; Hutzler, M.J.; Davis, J.W., II; Scherle, P.; Baribaud, F.; Bahinski, A.; Mo, Z.-L.; Newton, R.; Metcalf, B.; Xue, C.-B. *Bioorg. Med. Chem. Lett.* **2011**, *21*, 2626.
(66) Hughes, R.O.; Rogier, D.J.; Devraj, R.; Xue, C.-B.; Cao, G.; Turner, S.R.; Morton, P.A.; Keys, K.; Covington, M.; Bond, B.R.; Yu, Y.; Meade, H.; Hood, W.F.; Roeberds, S.; Newton, R.; Metcalf, B. 242nd ACS National Meeting & Exposition, Denver, CO, **2011**.
(67) Xue, C.-B.; Feng, H.; Cao, G.; Huang, T.; Glenn, J.; Anand, R.; Meloni, D.; Zhang, K.; Kong, L.; Wang, A.; Zhang, Y.; Zheng, C.; Xia, M.; Chen, L.; Tanaka, H.; Han, Q.; Robinson, D.J.; Modi, D.; Storace, L.; Shao, L.; Sharief, V.; Li, M.; Galya, L.G.; Covington, M.; Scherle, P.; Diamond, S.; Emm, T.; Yeleswaram, S.; Contel, N.; Caddi, K.; Newton, R.; Hollis, G.; Friedman, S.; Metcalf, B. *ACS Med. Chem. Lett.* **2011**, *2*, 450.

(68) Xue, C.-B.; Wang, A.; Han, Q.; Zhang, Y.; Cao, G.; Feng, H.; Huang, T.; Zheng, C.; Xia, M.; Zheng, K.; Kong, L.; Glenn, J.; Anand, R.; Meloni, D.; Robinson, D.J.; Shao, L.; Storace, L.; Li, M.; Hughes, R.O.; Devraj, R.; Morton, P.A.; Rogier, J.D.; Covington, M.; Scherle, P.; Diamond, S.; Emm, T.; Yeleswaram, S.; Contel, N.; Vaddi, K.; Newton, R.; Hollis, G.; Metcalf, B. *ACS Med. Chem. Lett.* **2011**, *2*, 913.
(69) Clinical Trials.gov NCT:00604123.
(70) Lagu, B.; Gerchak, C.; Pan, M.; Hou, C.F.; Singer, M.; Malaviya, R.; Matheis, M.; Olini, G.; Cavender, D.; Wachter, M. *Bioorg. Med. Chem. Lett.* **2007**, *17*, 4382.
(71) Clinical Trials.gov NCT:00683423, NCT:00699790.
(72) Carter, P.H.; Dunica, J.V.; Mudryk, B.M.; Randazzo, M.E.; Xiao, Z.; Yang, M.G.; Zang, R. Patent Application WO 2008/014360, **2008**.
(73) Clinical Trials.gov NCT:01440257.
(74) Sullivan, T.J.; Daiaghi, D.J.; Krasinki, A.; Miao, Z.; Wang, Y.; Zhao, B.N.; Baumgart, T.; Berahovich, R.; Ertl, L.S.; Pennell, A.; Seitz, L.; Miao, S.; Ungashe, S.; Wei, Z.; Johnson, D.; Boring, L.; Tsou, C.-L.; Charo, I.F.; Bekker, P.; Schall, T.J.; Jaen, J.C. *J. Pharm. Exp. Ther. Fast Forward* **2012**. http://jpet.aspetjournals.org/content/early/2012/06/04/jpet.111.190918 (accessed Mar 1, 2012).
(75) Dairaghi, D.; Charvat, T.T.; Moore, J.; Hansen, D.; Chen, W.; Jin, J.; Krasinski, A.; Basak, A; Wei, Z.; Wang, Q.; Ungashe, S.; Wright, J.J.; Punna, S.; Pennel, S.; Zeng, Y. Patent Application WO2008/008431, **2008**.
(76) Jones, K.L.; Maguire, J.J.; Davenport, A.P. *Br. J. Pharmacol.* **2011**, *162*, 1453.
(77) Pyrde, D.C.; Barber, C.G. *Chemokine Recept. Drug Targets* **2011**, 209.
(78) Barber, C.; Pyrde, D. *RSC Drug Disc. Series* **2011**, *4*, 183.
(79) Stupple, P.A.; Batcelor, D.V.; Corless, M.; Dorr, P.K.; Fenwick, D.R.; Galan, S.R.G.; Jones, R.M.; Mason, H.J.; Middleton, D.S.; Perros, M.; Perruccio, F.; Platts, M.Y.; Pryde, D.C.; Rodrigues, D.; Smith, N.N.; Stephenson, P.T.; Webster, R.; Westby, M.; Wood, A. *J. Med. Chem.* **2011**, *54*, 67.
(80) Veazey, R.S.; Klasse, P.J.; Ketas, T.J.; Reeves, J.D.; Patrick, M., Jr.; Kunstman, K.; Kuhmann, S.E.; Marx, P.A.; Lifson, J.D.; Dufour, J.; Mefford, M.; Pandrea, I.; Wolinsky, S.M.; Doms, R.W.; DeMartino, J.A.; Siciliano, S.J.; Lyons, K.; Springer, M.S.; Moore, J.P. *J. Exp. Med.* **2003**, *198*, 1551.
(81) Clinical Trials.gov NCT: 00686829.
(82) Tagat, J.R.; McCombie, S.W.; Nazareno, D.; Labroli, M.A.; Xiao, Y.; Steensma, R.W.; Strizki, J.M.; Baroudy, B.M.; Cox, K.; Lachowicz, J.; Varty, H.; Watkins, R. *J. Med. Chem.* **2004**, *47*, 2405.
(83) Xue, C.-B.; Chen, L.; Cau, G.; Zheng, K.; Wang, A.; Meloni, D.; Glenn, J.; Anand, R.; Xia, M.; Kong, L.; Huang, T.; Feng, H.; Zheng, C.; Li, M.; Galya, L.; Zhou, J.; Shin, N.; Baribaud, F.; Solomon, K.; Scherle, P.; Zhao, B.; Diamond, S.; Emm, T.; Keller, D.; Contel, N.; Yeleswaram, S.; Vaddi, K.; Hollis, G.; Newton, R.; Friedman, S.; Metcalf, B. *ACS Med. Chem. Lett.* **2010**, *1*, 483.
(84) Nichols, W.G.; Steel, H.M.; Bonny, T.; Adkison, K.; Curtis, L.; Millard, J.; Kabeya, K.; Clumeck, N. *Antimicrob. Agents Chemother.* **2008**, *52*, 858.
(85) Baba, M.; Takashima, K.; Miyake, H.; Kanzaki, N.; Teshima, K.; Wang, X.; Shiraishi, M.; Iizawa, Y. *Antimicrob. Agents Chemother.* **2005**, *49*, 4584.
(86) Tremblay, C.L.; Giguel, F.; Gaun, Y.; Chou, T.-C.; Takashima, K.; Hirsch, M.S. *Antimicrob. Agents Chemother.* **2005**, *49*, 3483.
(87) Cumming, J.C.; Tucker, H.; Oldfield, J.; Fielding, C.; Highton, A.; Faull, A.; Wild, M.; Brown, D.; Wells, S.; Shaw, J. *Bioorg. Med. Chem. Lett.* **2012**, *22*, 1655.

PART 4

Oncology

Editor: Shelli R. McAlpine
School of Chemistry, University of New South Wales,
Sydney, Australia

CHAPTER SIXTEEN

Nanotechnology Therapeutics in Oncology—Recent Developments and Future Outlook

Paul F. Richardson
Pfizer Inc., La Jolla, California, USA

Contents

1. INTRODUCTION

There has been intense interest in nanotechnology for the treatment of cancer, and numerous recent reviews have been published covering work in the field.[1,2] Understanding the underlying causes of cancer has lead to a new generation of chemotherapeutics.[3] However, these new "molecularly targeted" therapeutics suffer the drawback that only a very small fraction

Annual Reports in Medicinal Chemistry, Volume 47
ISSN 0065-7743
http://dx.doi.org/10.1016/B978-0-12-396492-2.00016-3

of the compound reaches the tumor site. Often following administration, the compounds are rapidly cleared and in a number of cases, even if the drug reaches the tumor, its residence time at the site of action is not sufficient to exercise a strong therapeutic effect. This drug delivery issue brings us to the key driver for using nanoparticulate formulations: improving small molecule drug levels at the tumor site (site-specific delivery), while keeping the drug away from healthy tissue (site-avoidance delivery).[4] Furthermore, the development of this technology offers the opportunity to reevaluate and expand the clinical use of classical chemotherapeutic agents. These potentially very useful cytotoxic drugs act by targeting fast growing cells, though differentiation between diseased and healthy cells is a critical issue, which contributes to the severe side effects observed, and leads to a narrow therapeutic index.

Nanoparticles for medical applications are typically defined as microscopic particles with a size between 1 and 1000 nm. They can be formulated from numerous materials and are optimally formulated to carry a variety of substances in a controlled and targeted manner. Nanoparticle distribution within the body is based on various parameters such as size, and surface coating, which can, in turn, be carefully controlled. In addition to this, nanoparticles offer considerable scope for engineering, thus enabling optimization of physicochemical characteristics.

2. PASSIVE TARGETING—THE EPR EFFECT

The most general method to target tumors relies on the enhanced permeability and retention (EPR) effect discovered by Maeda.[5] Most solid tumors have architecturally defective blood vessels and produce vascular permeability factors.[6] Therefore, they exhibit enhanced vascular permeability, which will ensure a sufficient supply of nutrients and oxygen to tumor tissues for rapid growth. In tumors, this defective architecture allows macromolecules (>40 kDa) to selectively leak out from tumor vessels and accumulate in the tumor tissue. In contrast, this does not occur in normal tissues. As such, this unique phenomenon is considered one of the cornerstones in tumor-targeting chemotherapy. All the currently approved nanomedicines, and the majority in clinical trials, rely on this passive accumulation within tumors.[7] Compared with conventional anticancer drugs (MW *ca.* 500), these macromolecular drugs have far superior pharmacokinetics (prolonged plasma half-life) and greater tumor selectivity. However, despite a focus on utilizing the EPR effect as a targeting mechanism, there are drawbacks with this approach.[8] In particular, the random nature of the EPR effect makes it difficult to control the process and can induce multiple-drug resistance leading to the drug being expelled from the tumor tissue. Furthermore, many large tumors display a high degree of pathophysiological heterogeneity, which limits the penetration of the nanoparticulate formulation, and finally although often the formulation is effectively delivered to the tumor, internalization of the drug substance into the cell is an issue.

3. IDEAL CHARACTERISTICS OF NANOPARTICLES

Size and surface characteristics have long been recognized as key characteristics in the design and development of nanoparticles.[9] Effective drug delivery to tumor tissue requires that nanoparticles remain in the bloodstream for a considerable time. General trends on spherical nanoparticles have emerged. Particles less than 5 nm are rapidly cleared from the circulation through extravasation or renal clearance. The behavior of nanoparticles between 10 nm and 15 μm varies widely in terms of both biodistribution and cell uptake with the RES (reticuloendothelial system) being the primary mechanism for removal from the circulation. Particles of ~15 μm accumulate in the liver, spleen, and bone marrow. The ideal size of nanoparticles was found to be between 70 and 200 nm to reach tumor tissues through the EPR effect while avoiding first pass elimination by cells of the RES.[10]

However, size alone is not sufficient to ensure long circulation times in the bloodstream. The major issue observed is uptake by macrophages in the RES. The chief strategy to avoiding macrophage capture and increasing circulation time is incorporating a hydrophilic surface on the particle. This can be achieved either by coating the surface with a hydrophilic polymer (PEG) in order to protect it from opsonization or by formulating nanoparticles from block polymers with both hydrophilic and hydrophobic domains.

Designing nanoparticles that have either slightly negative or positive surface charges is also advantageous.[11] Charges serve to minimize self–self and self–non-self interactions. However, as the surface charge becomes larger, macrophage scavenging is increased and can lead to greater clearance by the RES. Thus, minimizing nonspecific interactions *via* steric stabilization and control of surface charge helps to prevent nanoparticle loss.

One additional characteristic that impacts nanoparticle delivery is the particle's shape.[12] The effect of shape and geometry has been investigated using both spherical and nonspherical polystyrene microparticles during phagocytosis.[13] With elliptical disk-shaped microparticles, if the macrophage first contacted particles along the major axis, the particles were rapidly internalized (<6 min). However, when first contact was along the minor axis, even after 12 h, the particles were not internalized. Spherical particles were rapidly and uniformly internalized because of their symmetry.[14]

4. LOADING OF NANOPARTICLES

Most oncology nanoparticles are highly engineered drug delivery devices. The formulation is utilized to improve the biodistribution and pharmacokinetic profile of the chemotherapeutic agent.[15] Given this, strategies can be divided into two broad categories. The first involves physically entrapping or absorbing the active substance onto the nanoparticle through noncovalent interactions. The second method involves directly attaching the drug substance to the nanoparticulate matrix by degradable or nondegradable covalent bonds. For both methods, it is key that the chemotherapeutic agent remains retained with the matrix until it reaches its ultimate site of action *in vivo*. Current research focuses on elucidating methods for stimulating release of the active drug at the tumor site. The bulk composition of the engineered nanoparticle must be judiciously chosen to take into account biocompatibility, immunotoxicity, and ability to solubilize or sequester effectively the drug substance.[16]

5. TYPES OF NANOPARTICLES

The most common types of nanocarriers are shown in Figure 16.1.

5.1. Liposomes

Liposomes are the most studied colloidal particles applied in antitumor therapy,[17] and their potential as drug carriers was rapidly recognized for anticancer treatment.[18] Two major issues emerged preventing the development of liposomes: (1) problems with robust loading and (2) preventing rapid clearance.[19] However, the development of technology to efficiently encapsulate ionizable products within liposomes (conventional liposomes) enabled the clinical application of liposomal doxorubicin (i.e., Myocet).[20] Myocet was approved in Europe and Canada in 2000 and is currently undergoing a phase 3 global trial. Myocet provides a limited degree of prolonged circulation compared to free drug (2–3 vs. 0.2 h), but this innovative encapsulation technology did not afford long-term stability in the bloodstream.

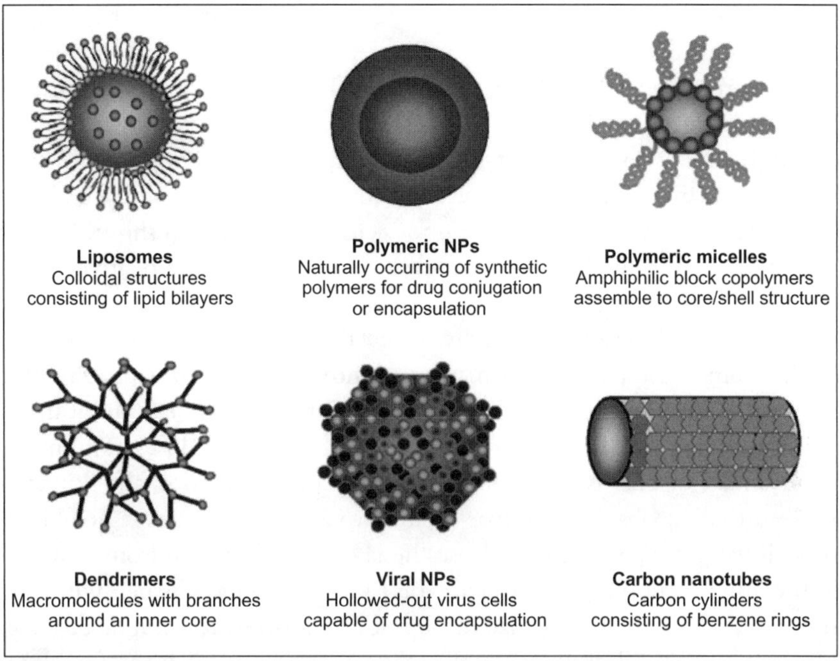

Figure 16.1 Common nanocarriers for drug delivery (reprinted by permission from the American Association for Cancer Research: Cho, K.; Wang, X.; Nie, S.; Chen, Z.; Shin, D.M. Therapeutic Nanoparticles for Drug Delivery in Cancer, *Clin. Cancer Res.* **2008**, *14* (5), 1310–1316. http://dx.doi.org/10.1158/1078-0432.CCR-07-1441).

Second-generation liposomes are referred to as "long circulating liposomes," and achieve this property through two key modifications to the liposome structure. First, their size is tightly controlled (<100 nm), and second, saturated lipids are more stable than unsaturated lipids in plasma. DaunoXome (a liposomal formulation of daunorubicin) was approved in 1996 for the treatment of Kaposi's sarcoma.[21] DaunoXome provides extended circulation (half-life of 5 h against 45 min for the free drug) due to its small size (~ 45 nm) and is efficacious against numerous tumors while limiting cardiotoxicity observed with free drug.

Another branch of liposome research emerged during this time which investigated the effects of coating liposome surfaces with inert materials. A major advance in the surface-modified liposomes was the development of the coated liposomes using polymers like polyethylene glycol (PEG) which has many attractive properties. Pegylated-liposomes circulated for remarkably long times after intravenous administration and evaded interception by the immune system, thus terming them "stealth" liposomes.[22] Currently, the only approved stealth liposome is pegylated liposomal doxorubicin (PLD, Doxil), which treats Kaposi's sarcoma, ovarian, and breast cancer.[23] The PEG surface coating substantially extends the half-life of doxorubicin *in vivo* with a half-life of approximately 55 h in humans versus 0.2 h for the free drug. Intravenous injection of doxorubicin allows less than 1% of the free drug to reach the tumor cells, whereas Doxil shows 10-fold higher levels in cells. Total plasma levels of doxorubicin are relatively high for several days after PLD administration; however, the majority of the dose is not available to healthy tissues thus protecting these from any toxic effects. Presumably, this is because the majority of doxorubicin in plasma (95–99%) remains encapsulated within the liposomes. Patients on PLD exhibit minimal cardiotoxicity even after receiving high cumulative doses. The major disadvantage of PLD is that acute infusion reaction, mucositis, and skin toxicities are often observed as dose-related toxicities.[24] The most notable of these is palmar-plantar erythrodyesesthesia ("hand-foot" syndrome), which is the usual dose-limiting toxicity of PLD, and causes issues with any substantial dose escalation that the liposomal formulation might provide. However, PLD still displays a higher therapeutic index than the free drug. Interestingly, these skin toxicities are not observed with Myocet, and as such are associated with the PEG surface coating.

5.2. Polymeric nanoparticles (polymer–drug conjugates)

The development of polymeric nanoparticles from concept to a commercial product for oncology has mirrored that of liposomes in that, despite initial rapid progress, the first protein-based nanoparticle did not receive commercial approval until 2005.[25] Numerous naturally occurring and synthetic polymers have been evaluated as materials for the delivery of drugs. Poly-L-glutamic acid, which was the first biodegradable polymer to be used for conjugate synthesis, is commonly used in drug delivery because the free γ-carboxylic acid enhances water solubility and can be utilized as a chemical handle for derivatization.[26] To be successfully employed in controlled drug delivery formulations, a material must be (1) chemically inert, (2) nontoxic, (3) free of leachable impurities, (4) biodegradable, and (5) degraded into molecules that can be metabolized and eliminated *via* normal metabolic pathways. During drug delivery, the hydrophobic core of the polymeric nanoparticle encapsulates a wide range of drug molecules with high loading efficiency, while the hydrophilic shell provides steric protection. Numerous techniques are available to formulate polymeric nanoparticles, with PRINT (particle replication in nonwetting templates) being of interest as it controls the precise size, composition, and shape of particles.[27]

In 2005, the FDA approved Abraxane (ABI-007) for the treatment of metastatic breast cancer.[28] Abraxane is a cremophor-free protein-stabilized nanoparticle formulation of paclitaxel. Paclitaxel is traditionally formulated with the micelle-forming vehicle cremophor EL (polyoxyethylated castor oil) and ethanol. The nanoparticle albumin-bound platform (nab) uses albumin as a therapeutic carrier for the delivery of hydrophobic chemotherapeutics.[29] Abraxane is a colloidal suspension of albumin-bound paclitaxel in which the particles are approximately 130 nm. Direct comparison of efficacy demonstrated that Abraxane had significantly higher response rates compared with conventional paclitaxel (33% vs. 19%), and longer time to disease progression (23 vs. 16.9 weeks). Overall, side effects were lower and no premedications were required using Abraxane. This is the first demonstration of a higher therapeutic efficacy for a nanoparticle therapeutic over the normal formulation.[30]

5.3. Polymeric micelles

The core/shell segments of polymeric micelles (PM) control the functionalities of the nanoparticle structure.[31] The hydrophilic outer shell region stabilizes the hydrophobic core region and controls the *in vivo* pharmacokinetic behavior by rendering the polymers water soluble to enable facile intravenous administration. The hydrophobic inner core is responsible for drug loading capacity, stability, and drug release behavior. Size control of the PM is simple through control of the block copolymer chain lengths. PM can also alter the drug internalization route through interaction and binding of the block copolymer to the cell membrane to facilitate endocytosis. The polymers also have been shown to avoid P-glycoprotein (P-gp) expressed in membranes of multidrug resistant cells, thus increasing drug absorption. In addition, they possess a higher drug loading capacity as well as improved stability compared to other nanoparticle systems.[32] Due to these and other favorable characteristics, interest in PM has been rapidly increasing and several formulations are currently undergoing clinical evaluation (NK105, a PM formulation of Paclitaxel is currently in phase III). Previously, NK105 has demonstrated complete tumor effacement after a single dose in HT-29 colon cancer cells in female nude mice.[33]

5.4. Dendrimers

Dendrimers are macromolecular compounds that comprise branches around an inner core. Their key advantage is that they can be tailored for specific applications and are ideally suited as drug delivery systems as it is easy to tailor their topology, functionality, and size.[34] The terminal groups can be utilized for bioconjugation, as targeting moieties, for signaling purposes, or simply to modify the dendrimer surface. Through judicious selection of the surface functional groups, biodistribution, permeability, and other pharmacokinetic properties can be modulated. No dendrimer-based products are currently in clinical development for oncology, but several promising studies have been carried out both *in vitro* and *in vivo*.

5.5. Viral nanoparticles

Viral nanoparticles are hollowed out virus cells, which can be utilized to target drugs directly to cancer cells.[35] Numerous viruses have been developed for nanotechnology applications, and in general, plant viruses are utilized as they are the easiest to produce in large quantities. These virus cells readily self-assemble into a nanoparticle with approximately a 10-nm core capable of encapsulating a high load of drug substance.[36] Researchers have demonstrated *in vivo* the enhanced recognition of prostate tumors by a PEG-coated cowpea mosaic virus particle modified with bombesin.[37] The major challenges for the development of viral nanoparticle-based systems are enhancement of physicochemical properties to prevent rapid clearance and limiting toxicity.

5.6. Carbon nanotubes

Carbon nanotubes are tubular materials consisting of benzene rings with nanometer-sized diameters, and axial symmetries, which leads to unique properties in the treatment of cancer. The high surface area of carbon nanotubes allows multiple functionalization sites to attach therapeutic agents and targeting ligands, or to aid solubility.[38] However, toxicity problems related to chronic insolubility have inhibited development, though functionalization can enhance solubility and alleviate these effects.[39] Interest in this drug delivery modality is focused on their ability to cross biological barriers.[40] However, despite the fact that modified carbon nanotubes have been evaluated with numerous therapeutic agents, the issue of biodegradability remains unresolved.

6. THE NEXT GENERATION OF NANOMEDICINES FOR ONCOLOGY

The major downside to current nanotechnology delivery systems is that the efficacy of cancer treatment is reduced because nanoparticles release their payload *via* a slow and passive mechanism. Although passive targeting has succeeded in reducing off-target toxicity effects (and as a result the formulations often show increased therapeutic index), polymer–drug carriers have shown limited increases in therapeutic efficacy. Large tumors are not homogeneous and nanoparticles cannot disperse uniformly through the tumor. In addition, the negative pressure gradient within the tumor limits the nanoparticle's effective range and can force them out of the tumor. Finally, although the EPR effect localizes the nanoparticles to the tumor, release of the drug substance and its subsequent internalization into the cells is still an issue. Thus, the next generation of nanoparticles has focused on incorporation of "smart" technologies that are responsive to environmental stimuli. Two major areas have emerged. The first involves active-targeting, whereby particles search for and subsequently attach themselves to diseased cells.[41] The second involves active drug payload that is triggered to release at the desired site of action owing to chemical or physical changes in the environment.[42,43] The ultimate goal in nanooncology would involve combining site-targeting and site-specific release into the same nanoparticle.[44]

6.1. Active targeting of nanoparticles

The most common approach to achieve this goal is to attach targeting ligands to the surface of the particle, taking advantage of receptor overexpression on specific tumor cell surfaces compared to healthy cells. Coupled with the EPR effect, targeted particles should increase the interaction time between the particles and the tumor and increase the likelihood of the particles being taken up by the tumor cells *via* endocytosis. Numerous ligands have been utilized to take advantage of targeting mechanisms,[45] and a tabulated summary of the more common approaches has been compiled.[46]

The folate receptor, which is overexpressed in many tumors, has been widely exploited for this purpose.[47] Attaching folic acid and related derivatives to the particle surface creates a targeted drug delivery vehicle.[48] Folic acid-functionalized dendrimers conjugated with methotrexate show a 10-fold increased efficacy and significantly lower toxicity when compared to the free drug at an equivalent cumulative dose in a mouse model of human epithelial cancer. This approach shows great promise as the folic acid particles can be produced at low cost.

Monoclonal antibodies (mAbs) have also been aggressively pursued as targeting moieties owing to their exquisite ability to target specific disease processes.[49] One key problem with mAbs is that they are targeted by the immune system. In addition, mAbs are large, complex molecules, expensive to manufacture, and require significant engineering at the molecular level to be effective.[50] One potential solution involves using antibody fragments as targeting molecules. Antibody fragments retain high affinity for the target antigens but have less immunogenicity (due to the lack of an Fc region, and relatively short circulation times); being smaller in size, they are better suited for molecular targeting.[51]

Aptamers are small nucleic acid ligands that bind to targets with both high sensitivity and specificity. One major advantage of aptamers is that they are identified using an *in vitro* evolutionary process called systemic evolution of ligands by exponential enrichment (SELEX).[52] This process uses a library of 10^{15} random oligonucleotides, and enrichment identifies aptamers that bind with high affinity and specificity to the target.[53] This approach allows scale-up without batch to batch variation. Further advantages are their small size ($\sim$15 kDa) and low immunogenicity which leads to controlled biodistribution. Several aptamer-targeted nanoparticles are in preclinical development. Unfortunately, aptamers are unstable in serum and have relatively high production costs.

Peptides represent an extremely attractive targeting moiety because of their small size, low immunogenicity, high stability, and relative ease of manufacture. Development of peptide phage libraries ($\sim 10^{11}$ different peptide sequences) and efficient screening technologies has made selection of potential targeting ligands extremely facile.[54] They also have the ability to bind to their targets with high affinity and specificity.[55]

6.2. Target-controlled release

Efficient release of drugs from nanoparticles is critical to bioavailability. Controlling the timing and degree of release are the two major challenges. First, stability is key to avoiding discharge of the cargo into healthy tissues. Second, specificity and efficiency of the trigger are essential so that nanocarrier instability and release of cargo are done in the desired tumor environment. Triggers can be divided into two categories based on where the stimuli originates. Internal triggers rely on exploiting subtle differences within the tumor environment such as pH or specific enzymes, while external triggers release the drug cargo prompted by stimuli from outside the body such as temperature, ultrasound, or radiation.

Of the potential triggers, pH is the most common.[56] Mildly acidic conditions exist in tumor and inflammatory tissues due to hypoxia and cell death, where the pH in tumor tissues is ~6.5 versus physiological pH of 7.4. A more pronounced pH change is encountered if the nanoparticles enter into cancer cells where the pH is ~5. Development of acid-labile linkers and acid-sensitive nanocarrier shells releases their cargo on a drop in pH. Numerous variations on these themes have also been investigated. Several *in vivo* studies have demonstrated the viability of this approach, but these have yet to be translated into the clinic.

Enzyme-triggered release uses localization of enzymes in specific areas. The most exploited enzymes are matrix metalloproteinases (MMPs) which are often overexpressed in tumor sites. Researchers have incorporated a MMP cleavage site within both liposomes and dendrimers to trigger release of the encapsulated cargo.[57] The main disadvantage of enzyme-triggered release is that no enzyme is solely expressed in the target region; as such nonspecific release occurs throughout the body. Work utilizing dendrimers bearing cell-penetrating peptides that were exposed upon cleavage to MMP indicated 4- to 15-fold higher uptake than the corresponding dendrimers without the peptides.

Two other approaches are being investigated utilizing heat[58] or ultrasound waves to trigger release. The heat-activated liposome formulation, Thermodox, is currently undergoing clinical evaluation. Use of ultrasound as a triggering mechanism is still at an early stage of development, though proof of concept has been achieved.[59]

6.3. In development

Numerous nanoparticle formulations are currently in development. A selection of these with comments regarding the key differentiating features is presented in Table 16.1[60,61]

7. CONCLUSIONS

There continues to be explosive growth in oncology research focused on utilizing nanoparticles as drug delivery vehicles. Although the development of both a commercial liposomal and a polymeric product within the field took place 30 years ago, there are many areas that still need

Table 16.1 Selection of nanoparticle formulations currently in clinical development

Name	Company	Indication	Status	Comments	Improvement
Genexol-PM	Samyang	Pancreatic cancer/metastatic breast cancer/NSCLC	Phase 2/3 (US)	Approved in Korea for treatment of metastatic breast cancer. Polymeric micelle cremophor-free formulation of Paclitaxel	38% partial response versus 24% with Paclitaxel. Formulation enables higher MTD
Thermodox	Celsion	Hepatocellular carcinoma	Phase 3	Encapsulated doxorubicin in a heat-activated liposome. Activated by radio frequency ablation (RFA)	Significant tumor size decrease. 26.1% with RFA as opposed to 12.1% without
CDP 791	UCB Pharma	NSCLC	Phase 2	A PEGylated, humanized di-Fab fragment specifically inhibiting VEGFR-2 activation and blocks all signaling through this receptor	Six ascending doses once every 3 weeks. Well tolerated and met safety endpoint. No dose limiting toxicity or immunogenicity
CALAA-01[62]	Calando Pharm.	Solid tumors	Phase 1	Polymer formulation of siRNA using transferrin as a targeting ligand for binding to transferrin receptors	Active safety study trial
Lipoplatin	Regulon	NSCLC and pancreatic cancer	Phase 2/3	Liposomal formulation of cisplatin with significantly reduced toxicity	Half-life observed to be 60–117 h as opposed to 6 h for cisplatin. 11.1% partial response in patients, who had failed all previous chemotherapy
BIND-014[63]	BIND bioscience	Prostate cancer/solid tumors	Phase 1	Prostate-specific membrane antigen-targeted docetaxel-encapsulated polymeric formulation. First targeted nanoparticle to enter clinical trials. Also displays controlled release of drug	Evidence for antitumor activity in cancers for which conventional docetaxel is ineffective. Well tolerated

improvement. The lessons learned from the first generation of products, investment in the field, and our increased understanding should lead to rapid progress.

REFERENCES

(1) Wang, A.Z.; Langer, R.; Farokhzad, O.C. *Annu. Rev. Med.* **2012**, *63*, 185.
(2) Alexis, F.; Pridgen, E.M.; Langer, R.; Farokhzad, O.C. *Handb. Exp. Pharmacol.* **2010**, *197*, 55.
(3) Hanahan, D.; Weinberg, R.A. *Cell* **2011**, *144*, 646.
(4) Farokhzad, O.C.; Langer, R. *ACS Nano* **2009**, *3*, 16.
(5) Ang, J.; Nakamura, H.; Maeda, H. *Adv. Drug Deliv. Rev.* **2011**, *63*, 136.
(6) Ruoslahti, E.; Bhatia, S.N.; Sailor, M.J. *J. Cell Biol.* **2010**, *188*, 759.
(7) Torchillin, V.P. *Handb. Exp. Pharmacol.* **2010**, *197*, 3.
(8) Jain, R.K. *Sci. Am.* **1994**, *271*, 58.
(9) Petros, R.A.; DeSimone, J.M. *Nat. Rev. Drug Discov.* **2010**, *9*, 615.
(10) Storm, G.; Belliot, S.O.; Daemen, T.; Lasic, D.D. *Adv. Drug Deliv. Rev.* **1995**, *17*, 31.
(11) Arvizot, R.R.; Miranda, O.R.; Thompson, M.A.; Pabelick, C.M.; Bhattacharya, R.; Robertson, J.D.; Rotello, V.M.; Prakash, Y.S.; Mukherjee, P. *Nano Lett.* **2010**, *10*, 2543.
(12) Champion, J.A.; Katare, Y.K.; Mitragotri, S. *J. Control. Release* **2007**, *121*, 3.
(13) Champion, J.A.; Mitragotri, S. *Proc. Natl. Acad. Sci. U.S.A.* **2006**, *103*, 4930.
(14) Champion, J.A.; Mitragotri, S. *Pharm. Res.* **2009**, *26*, 244.
(15) Alexis, F.; Pridgen, E.; Molnar, L.K.; Farokhzad, O.C. *Mol. Pharm.* **2008**, *5*, 505.
(16) Dobrovolskaia, M.A.; McNeil, S.E. *Nat. Nanotechnol.* **2007**, *2*, 469.
(17) Maurer, N.; Fenske, D.B.; Cullis, P.R. *Expert Opin. Biol. Ther.* **2001**, *1*, 923.
(18) Gregoriadis, G.; Wills, E.J.; Swain, C.P.; Tavill, A.S. *Lancet* **1974**, *303*, 1313.
(19) Gabizon, A.A.; Shmeeda, H.; Zalipsky, S. *J. Liposome Res.* **2006**, *16*, 175.
(20) Batist, G.; Barton, J.; Chaikin, P.; Swenson, C.; Welles, L. *Expert Opin. Pharmacother.* **2002**, *3*, 1739.
(21) Petre, C.E.; Dittmer, D.P. *Int. J. Nanomedicine* **2007**, *2*, 277.
(22) Immordino, M.L.; Dosio, F.; Cattel, L. *Int. J. Nanomedicine* **2006**, *1*, 297.
(23) Gabizon, A.A. *Cancer Invest.* **2001**, *19*, 424.
(24) Gordon, K.B.; Tajuddin, A.; Guitart, J.; Kuzel, T.M.; Eramo, L.R.; Vonroenn, J. *Cancer* **1995**, *75*, 2169.
(25) Kreuter, J. *Int. J. Pharm.* **2007**, *331*, 1.
(26) Sinha, R.; Kim, G.J.; Nie, S.; Shin, D.M. *Mol. Cancer Ther.* **1909**, *2006*, 5.
(27) Gratton, S.E.; Napier, M.E.; Ropp, P.A.; Tian, S.M.; DeSimone, J.M. *Pharm. Res.* **2008**, *25*, 2845.
(28) Gradishar, W.J. *Expert Opin. Pharmacother.* **2006**, 7, 1041.
(29) Elsedek, B.; Kratz, F. *J. Control. Release* **2012**, *157*, 4.
(30) Henderson, I.C.; Bhatia, V. *Expert Rev. Anticancer Ther.* **2007**, 7, 919.
(31) Gong, J.; Chen, M.; Zheng, Y.; Wang, S.; Wang, Y. http://dx.doi.org/10.1016/j.jconrel.2011.12.012.
(32) Mikhal, A.S.; Allen, C. *J. Control. Release* **2009**, *138*, 214.
(33) Hamaguchi, T.; Matsumura, Y.; Suzuki, M.; Shimizu, K.; Goda, R.; Nakamura, I.; Nakatomi, I.; Yokoyama, M.; Kataoka, K.; Kakizoe, T. *Br. J. Cancer* **2005**, *92*, 1240.
(34) Gajbhiye, V.; Kumar, P.V.; Tekade, R.K.; Jain, N.K. *Curr. Pharm. Des.* **2007**, *13*, 415.
(35) Franzen, S.; Lommel, S.A. *Nanomedicine* **2009**, *4*, 575.
(36) Ren, Y.; Wong, S.M.; Lim, L.Y. *Bioconjug. Chem.* **2007**, *18*, 836.

(37) Steinmetz, N.F.; Ablack, A.L.; Hickey, J.L.; Ablack, J.; Manocha, B.; Mymryk, J.S.; Luyt, L.G.; Lewis, J.D. *Small* **2011**, 7, 1664.
(38) Fabbro, C.; Ali-Boucetta, H.; Da Ros, T.; Kostarelos, K.; Bianco, A.; Prato, M. *Chem. Commun.* **2012**, *48*, 3911.
(39) Pastrorin, G. *Pharm. Res.* **2009**, *26*, 746.
(40) Wu, P.; Chen, X.; Hu, N.; Tan, U.C.; Blixt, O.; Zettl, A.; Bertozzi, C.R. *Angew. Chem. Int. Ed Engl.* **2008**, *47*, 5022.
(41) Byrne, J.D.; Betancourt, T.; Brannon-Pappas, L. *Adv. Drug Deliv. Rev.* **2008**, *60*, 1615.
(42) Ganta, S.; Devalpally, H.; Shahiwala, A.; Amiji, M. *J. Control. Release* **2008**, *126*, 187.
(43) Oh, K.T.; Yin, H.O.; Lee, E.S.; Bae, Y.H. *J. Mater. Chem.* **2007**, *17*, 3987.
(44) Rahman, M.; Ahmad, M.Z.; Kazmi, I.; Akhter, S.; Afzal, M.; Gupta, G.; Jalees Ahmed, F.; Anwar, F. *Expert Opin. Drug Deliv.* **2012**, *9*, 367.
(45) Wang, A.Z.; Gu, F.; Zhang, L.; Chan, J.M.; Radovic-Moreno, A.; Shaikh, M.R.; Farokhzad, O.C. *Expert Opin. Biol. Ther.* **2008**, *8*, 1063.
(46) Loomis, K.; McNeely, K.; Bellamkonda, R.V. *Soft Matter* **2011**, 7, 839.
(47) Gabizon, A.; Shmeeda, H.; Baabur-Cohen, H.; Saatchi-Fainaro, R. In *Targeted Drug Strategies for Cancer and Inflammation*; Jackman, A.L., Leamon, C.P., Eds.; Springer: New York, 2011; p 217.
(48) Kularatne, S.A.; Low, P.S. *Methods Mol. Biol.* **2010**, *624*, 249.
(49) McCarron, P.A.; Marouf, W.A.; Quinn, D.J.; Fay, F.; Burden, R.E.; Olwill, S.A.; Scott, C.J. *Bioconjug. Chem.* **2008**, *19*, 1561.
(50) Pasquetto, M.V.; Vecchia, L.; Covini, D.; Digilio, R.; Scotti, C. *J. Immunother.* **2011**, *34*, 611.
(51) Lu, R.-M.; Chang, Y.-M.; Chen, M.-S.; Wu, H.-C. *Biomaterials* **2011**, *32*, 3265.
(52) Fang, X.; Tan, W. *Acc. Chem. Res.* **2010**, *43*, 48.
(53) He, X.; Hai, L.; Su, J.; Wang, K.; Wu, X. *Nanoscale* **2011**, *3*, 2936.
(54) Aina, O.H.; Liu, R.; Sutcliffe, J.L.; Marik, J.; Pan, C.X.; Lam, K.S. *Mol. Pharm.* **2007**, *4*, 631.
(55) Raha, S.; Paunesku, T.; Woloschak, G. *WIREs Nanomed. Nanobiotechnol.* **2011**, *3*, 269.
(56) Gao, W.; Chen, J.M.; Farokhzad, O.C. *Mol. Pharm.* **2010**, 7, 1913.
(57) Olson, E.S.; Jiang, T.; Aguilera, T.A.; Nguyen, Q.T.; Ellies, L.G.; Scadeng, M.; Tsien, R.Y. *Proc. Natl. Acad. Sci. U.S.A.* **2010**, *107*, 4311.
(58) Li, G.Y.; Guo, L.; Ma, S.M. *J. Appl. Polym. Sci.* **2009**, *113*, 1364.
(59) Liu, Y.; Miyoshi, H.; Nakamura, M. *J. Control. Release* **2006**, *114*, 89.
(60) Duncan, R.; Gaspar, R. *Mol. Pharm.* **2011**, *8*, 2101.
(61) Kamaly, N.; Xiao, Z.; Valencia, P.M.; Radovic-Moreno, A.F.; Farokhzad, O.C. *Chem. Soc. Rev.* **2012**, *41*, 2971.
(62) Davis, M.E. *Mol. Pharm.* **2009**, *6*, 659.
(63) Shi, J.; Xiao, Z.; Kamaly, N.; Farokhzad, O.C. *Acc. Chem. Res.* **2011**, *44*, 1123.

CHAPTER SEVENTEEN

Small-Molecule Antagonists of Bcl-2 Family Proteins

Sean P. Brown, Joshua P. Taygerly
Amgen Inc., South San Francisco, California, USA

Contents

1. INTRODUCTION

The Bcl-2 family of proteins is central to regulating apoptosis, or programmed cell death. Alterations to the native expression levels of Bcl-2 family proteins have been shown to correlate with cancer progression, chemotherapy and radiotherapy resistance, and overall poor clinical outcomes. Apoptosis may be initiated by two different pathways. The extrinsic pathway is initiated through activation of death receptors on the cell surface, while the intrinsic pathway is initiated by increases in mitochondrial outer membrane permeability (MOMP). Mitochondrial membrane integrity is regulated through a balance of pro- and antiapoptotic Bcl-2 family proteins. The multidomain proapoptotic proteins Bax, Bak, and Bok share sequence homology in three α-helical Bcl-2 homology domains (BH1–BH3), while the antiapoptotic proteins Bcl-2, Bcl-x_L, Mcl-1, Bcl-w, Bcl-B, and A1 (Bfl-1) share sequence homology in four domains (BH1–BH4) and similar tertiary structure. In a healthy cell, antiapoptotic proteins deactivate the

Annual Reports in Medicinal Chemistry, Volume 47
ISSN 0065-7743
http://dx.doi.org/10.1016/B978-0-12-396492-2.00017-5

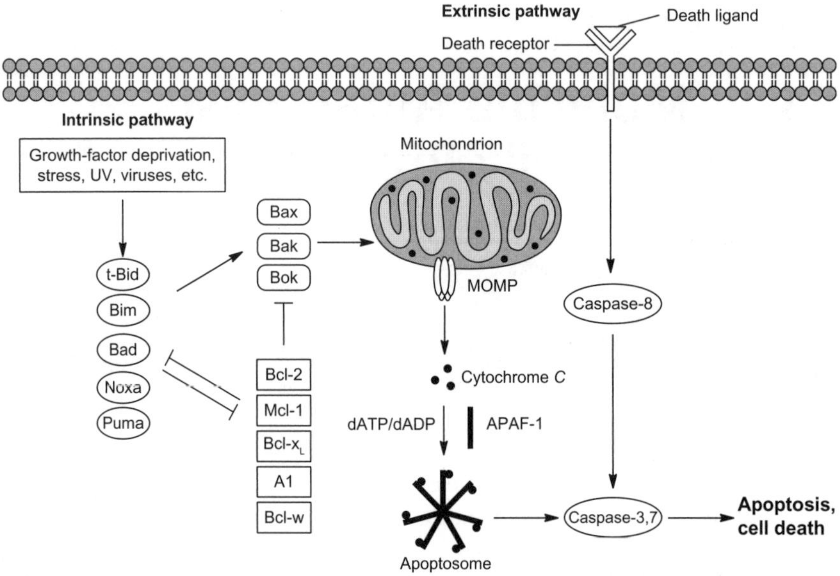

Figure 17.1 Scheme depicting intrinsic and extrinsic pathways of apoptosis.

multidomain proapoptotic proteins on the mitochondrion outer membrane (Fig. 17.1). Damaged cells receive death signals from the activated proapoptotic BH3-only proteins Bad, Bim, Puma, Bid, Bik, Noxa, Hrk, and Bmf. This initiates a cascade in which the proapoptotic proteins Bax, Bak, and Bok are activated and then homooligomerize on the outer mitochondrial membrane. This oligomerization forms pores in the mitochondrion that cause MOMP, the key step in commitment to apoptosis. Permeabilization of the mitochondrion releases cytochrome *c* into the cytosol where it complexes with APAF-1, caspase-9, and dATP/dADP to form the apoptosome that ultimately leads to caspase activation and cell death.

Although the precise mechanism of interaction between the Bcl-2 proteins is still under debate, it is widely agreed upon that the antiapoptotic proteins bind and neutralize the proapoptotic members through protein–protein interactions.[1] A hydrophobic groove displayed by the antiapoptotic Bcl-2 proteins accommodates the α-helical BH3 domain of both the proapoptotic BH3-only proteins and the multidomain proteins, creating a high-affinity interface that can potentially be disrupted with small molecules. Antagonizing the antiapoptotic Bcl-2 family proteins should result in the release of BH3-only proteins and thus induce apoptosis in the cell. Additionally, it has been shown that tumors expressing high levels of Bcl-2, Bcl-x_L, and Mcl-1 are resistant to radiation and chemotherapy, suggesting that antagonizing the antiapoptotic

Bcl-2 proteins is a feasible and promising strategy for the treatment of cancer.[2,3] In this chapter, we present the different strategies applied to small-molecule lead generation and the progress toward the discovery of antiapoptotic Bcl-2 protein antagonists for the treatment of cancer.

2. SMALL-MOLECULE INHIBITORS OF Bcl-2 Family Proteins

2.1. Fragment-based approaches

To date, the most selective and potent inhibitors of Bcl-2 family proteins have been discovered using fragment screening followed by synthetic optimization. This flexible technique enables the discovery of leads for various targets with vastly different binding pockets. It also has advantages over other lead discovery techniques for finding inhibitors of protein–protein interactions, since these are challenging targets that do not fit the classical definition of "druggable."[4] Multiple leads and several clinical candidates have been discovered using this technique.

In 2006, a fragment screening effort was described that led to the discovery of potent Bcl-x_L inhibitors using ^{15}N HSQC NMR.[5] Initial screens identified low-affinity binders **1** ($K_d \sim 300$ μM) and **2** ($K_d \sim 6000$ μM) that were chemically linked to generate ligand **3**, a modest inhibitor of Bcl-x_L ($K_i = 1.4$ μM). Further synthetic efforts identified compound **4**, a potent inhibitor ($K_i = 0.036$ μM) of Bcl-x_L that binds to the BH3-domain. A liability of ligand **4** was low affinity in the presence of human serum (HS) ($K_i = 2.50$ μM with 1%HS). Further modifications resulted in A-385358 (**5**), an inhibitor with increased affinity in serum ($K_i = 0.36$ μM with 10%HS) as well as affinity for Bcl-2 ($K_i = 0.067$ μM). As a single agent, A-385358 showed little activity on cancer cell viability. However, it potentiated the activity of paclitaxel in A549 NSCLC cells (up to 25-fold compared to monotherapy) and significantly enhanced A549 tumor shrinkage when codosed with paclitaxel in a mouse xenograft model.[6,7]

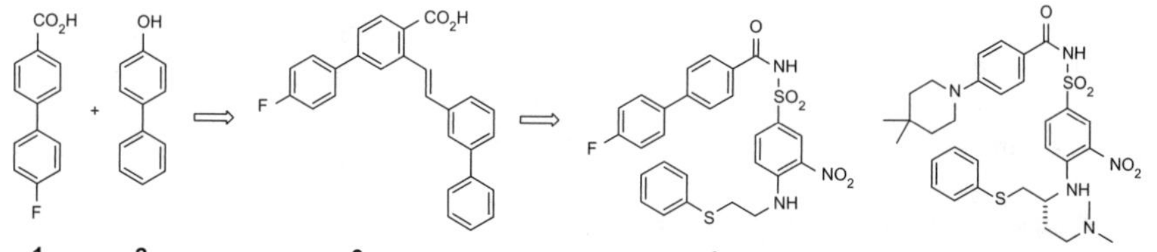

1 **2** **3** **4** A-385358 (**5**)

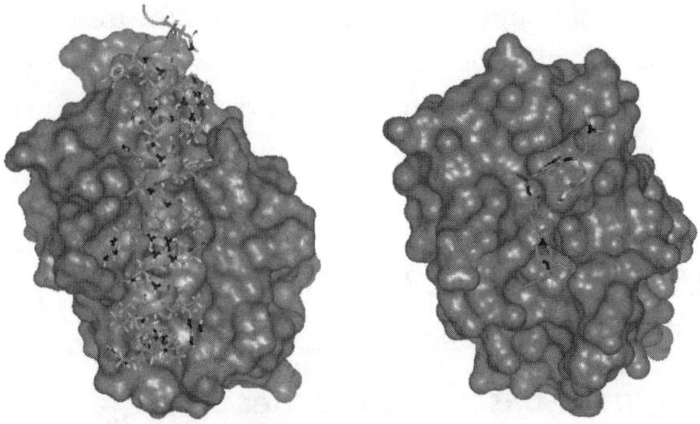

Figure 17.2 X-ray crystal structure of antiapoptotic protein Bcl-x_L bound to the Bim BH3-domain peptide (left), and the small-molecule BH3 mimetic ABT-737 (right). (See Color Plate 17.2 in Color Plate Section.)

Further optimization led to the discovery of ABT-737 (**6**), an inhibitor with high affinity for Bcl-x_L, Bcl-2, and Bcl-w ($K_i < 0.001$ μM) but not Mcl-1, Bcl-B, or A1.[8] The increased affinities are rationalized by the chlorobiphenyl substituent occupying hydrophobic regions common to the three proteins.[9] ABT-737 was cocrystallized with Bcl-x_L, demonstrating that it mimics the BH3 helix and binds to the hydrophobic BH3-binding domain (Fig. 17.2). ABT-737 demonstrates efficacy as a single agent in various cancer cells lines, such as NCI-H889 SCLC cells ($IC_{50} = 20$ nmol/L). Cancer cell lines sensitive to ABT-737 generally show high expression levels of Bcl-2 and Bcl-x_L, while high levels of Mcl-1 correlate with resistance.[10] ABT-737 also potentiates the proapoptotic effects of various chemotherapeutic agents such as paclitaxel (fourfold over monotherapy in A549 cells). Additionally, in H146 SCLC tumor xenograft models, ABT-737 caused complete regression, and tumors did not regrow for 58 days after therapy cessation.[8] However, the poor oral bioavailability and solubility profile of ABT-737 hampered the progression of this molecule through clinical studies.[11]

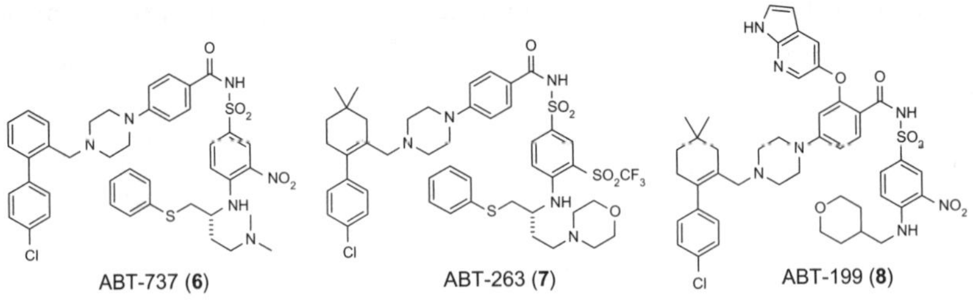

ABT-737 (**6**) ABT-263 (**7**) ABT-199 (**8**)

Optimization of ABT-737 led to ABT-263 (navitoclax, **7**), an inhibitor with improved solubility and pharmacokinetic profiles.[12] ABT-263 displays high affinity for Bcl-x_L (K_i=0.0042 μM) and Bcl-2 (K_i=0.0056 μM) and shows oral bioavailability in various animal species. It selectively kills SCLC H146 cells as a single agent (EC_{50}=87 nM with 10% HS) and has shown efficacy in other lymphoma and chronic lymphocyte leukemia lines. ABT-263 potentiates the activity of a broad range of chemotherapeutic agents over a wide range of cell types. For example, it synergizes with erlotinib in NSCLC NCI-H1650 cells with a combination index value of 0.44.[13] ABT-263 induces complete tumor regression when dosed at 100 mg/kg/day for 21 days in a SCLC H1963 xenograft model.[14] Tumor sensitivity positively correlates with Bcl-2 and Bcl-x_L expression and negatively correlates with Mcl-1 expression.[10,15] ABT-263 is currently in phase I/II clinical trials for the treatment of various solid tumors, hematologic malignancies, chronic lymphocytic leukemia, and lymphoid malignancies.[16] However, phase II results in patients with relapsed SCLC demonstrated limited efficacy (partial response in 2.6% of patients and stable disease in 23%), while dosing has been limited by Bcl-x_L-mediated thrombocytopenia.[17]

To overcome the Bcl-x_L-related platelet toxicity, ABT-199 (**8**) was designed as a selective Bcl-2 inhibitor. ABT-199 is a potent Bcl-2 inhibitor (K_i<0.001 μM) but has significantly less affinity for Bcl-x_L (K_i>1.0 μM). ABT-199 has a similar sensitivity profile to ABT-263 across various cells, but exhibits an approximate 10-fold increased potency in all sensitive cell lines. Importantly, ABT-199 has little effect on platelets *in vitro* compared to ABT-263, which kills platelets with an EC_{50} of 80 nM.[18] ABT-199 is currently in phase I clinical trials.[19]

Other inhibitors against the various Bcl-2 family members include quinazoline sulfonamide **9**.[20] The quinazoline sulfonamide core is a suitable isostere for the phenyl acylsulfonamide in ABT-263, and X-ray crystallographic data show that both inhibitors bind to Bcl-x_L with similar binding modes. Unlike ABT-263, quinazoline sulfonamide displays high affinity for Bcl-x_L (IC_{50}=0.007 μM) and Bcl-2 (IC_{50}=0.0082 μM) while displaying significantly lower affinity for Bcl-w (IC_{50}=0.44 μM). These analogs currently remain in preclinical development. Selective inhibitors of Mcl-1, exemplified by indole **10**, have been disclosed although biological data have yet to be reported.[21,22]

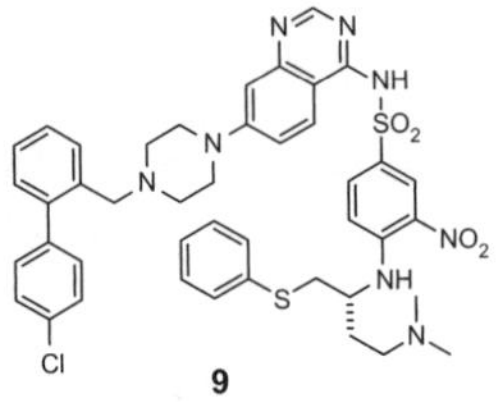

9

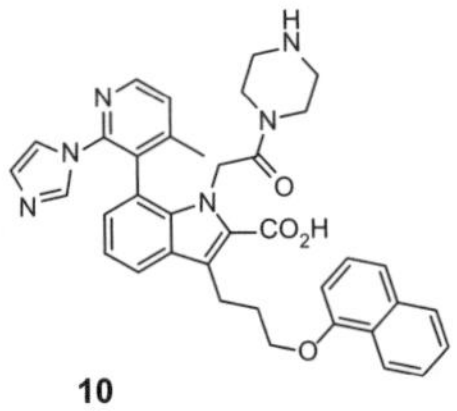

10

Finally, a fragment-based lead discovery approach reported a dual Bcl-x_L and Mcl-1 inhibitor.[23] NMR screening revealed weak-binding fragments **11** (K_i = 690 μM) and **12** (K_i = 380 μM). Fragment linking and optimization gave acylsulfonamide **13**, with affinity for both Bcl-x_L (IC_{50} = 0.086 μM) and Mcl-1 (IC_{50} = 0.14 μM).

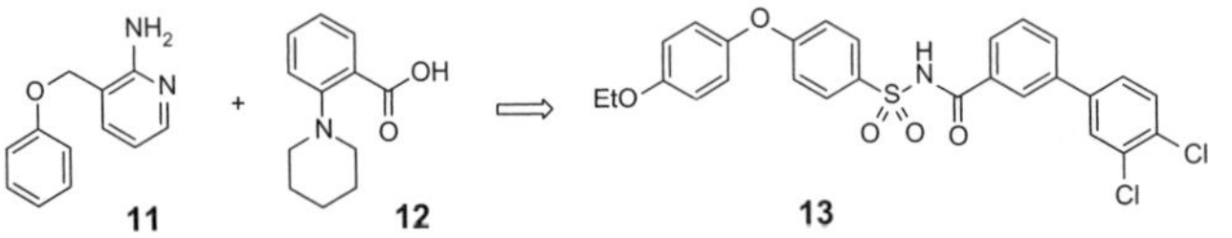

11 12 13

2.2. Natural product-based approaches

The discovery of several classes of cytotoxic natural products that may function by antagonizing the antiapoptotic Bcl-2 proteins has provided several small-molecule leads and multiple clinical candidates. In contrast to the potent inhibitors discovered by fragment screening, however, most of the natural product inhibitors display only modest affinity for Bcl-2 family proteins, and in several cases, additional biological activity has been reported.

Obatoclax (GX15-070, **14**), an analog of cycloprodigiosin, inhibits all six antiapoptotic Bcl-2 proteins with IC_{50} values ranging from 1 to 7 μM. Obatoclax displays 87% growth inhibition of C33A tumors cells compared to vehicle in a mouse xenograft model and is currently in phase I/II clinical trials for solid and hematological malignancies.[24] Although obatoclax displays clinical efficacy, reducing circulating lymphocytes in 18 of 26 patients with a median reduction of 24%,[25] this efficacy may not be driven solely by antagonism of Bcl-2 proteins. In fact, obatoclax induces apoptosis in the absence of Bax and Bak and also induces an S-G2 cell cycle block, suggesting that it acts through multiple targets.[26]

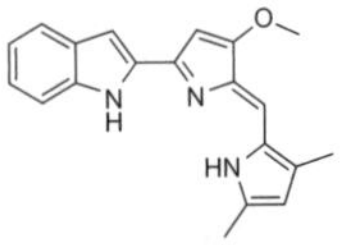

Obatoclax (GX15-070, **14**)

Gossypol is a natural polyphenol isolated from cotton seeds and roots. (±)-Gossypol (BL-193) is a mixture of atropisomers where (−)-gossypol (AT101, **15**) displays greater potency of the two isoforms. AT101 exhibits affinity for Bcl-2, Bcl-x_L, and Mcl-1 ($IC_{50}=0.32$, 0.48, and 0.18 μM, respectively), although evidence suggests that the cytotoxicity results from additional mechanisms, such as DNA cleavage or the generation of reactive oxygen species that promote apoptosis.[27] AT101 demonstrates activity in a WSU-DLCL$_2$ mouse xenograft model, providing 51% tumor growth inhibition compared to vehicle.[28] This single-agent efficacy has prompted the therapeutic potential of AT101 to be tested in phase I/II clinical trials, but dosing has been limited by gastrointestinal toxicity.[29]

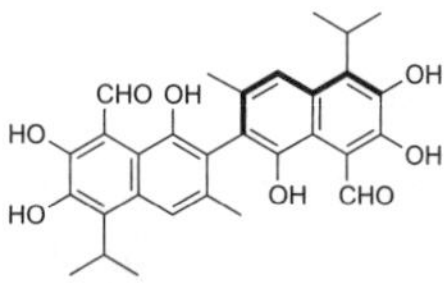

(−)-Gossypol (AT101, **15**)

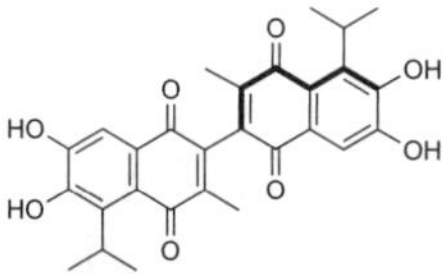

Apogossypolone (ApoG2, **16**)

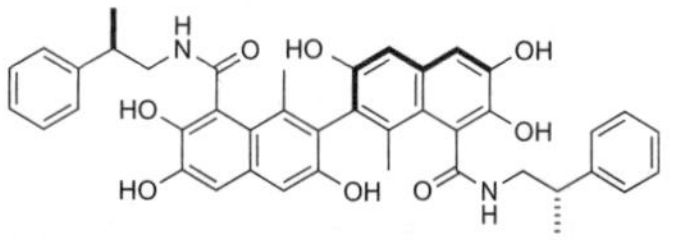

Sabutoclax (BI-97C1, **17**)

In an effort to reduce the toxicity observed with AT101 treatment, apogossypolone (ApoG2, **16**) was synthesized within a series of gossypol derivatives designed without the electrophilic aldehyde groups. ApoG2 displays increased binding affinity toward Bcl-2 ($IC_{50}=0.040$ μM), Bcl-x_L ($IC_{50}=0.088$ μM), Mcl-1 ($IC_{50}=0.056$ μM), and A1 ($IC_{50}=0.211$ μM). Further optimization culminated in the identification of BI-97C1 (sabutoclax, **17**). A pan-Bcl-2 family inhibitor, sabutoclax exhibits submicromolar affinity for Bcl-2 ($IC_{50}=0.32$ μM), Bcl-x_L ($IC_{50}=0.31$ μM), Mcl-1 ($IC_{50}=0.20$ μM), and A1 ($IC_{50}=0.62$ μM) while displaying *in vitro* cytotoxicity toward prostate cancer (PC3 cells; $EC_{50}=0.13$ μM), lung cancer (H460 cells; $EC_{50}=0.42$ μM), and lymphoma (BP3 cells; $EC_{50}=0.049$ μM) cell lines. Sabutoclax has minimal effect on $Bax^{-/-}Bak^{-/-}$ cells providing evidence that its cytotoxicity is mitochondrion mediated.[30]

Efforts to design (−)-gossypol mimetics resulted in the identification of TW-37 (**18**), a pan-Bcl-2 family inhibitor.[31] TW-37 exhibits affinity for Bcl-2 ($IC_{50}=0.29$ μM), Mcl-1 ($IC_{50}=0.26$ μM), and Bcl-x_L ($IC_{50}=1.11$ μM) and inhibits growth of PC3 cells with an IC_{50} of 200 nM, although its cytotoxic effects have also been attributed to other mechanisms.[32] A structure-based design effort utilizing the gossypol/Bcl-x_L structure revealed BI-33 (**19**), a modest inhibitor of Bcl-x_L ($K_i=1.2$ μM) but a potent inhibitor of both Bcl-2 ($K_i=0.017$ μM) and Mcl-1 ($K_i=0.017$ μM). A related effort revealed dihydroisoquinoline TM-1206 (**20**), an inhibitor of Bcl-x_L ($K_i=0.64$ μM), Bcl-2 ($K_i=0.11$ μM), and

Mcl-1 ($K_i = 0.15$ μM). Both BI-33 and TM-1206 induce death in MDA-MB-231 cells with IC_{50} values of 0.11 and 0.10 μM, respectively.[33,34]

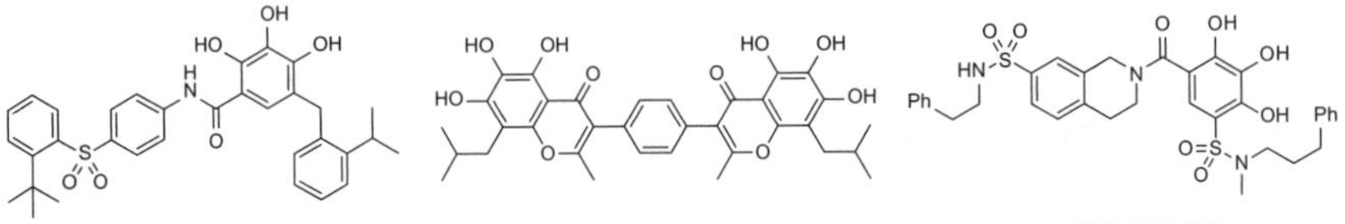

TW-37 (**18**) BI-33 (**19**) TM-1206 (**20**)

Several classes of natural polyphenols have been identified as Bcl-x_L inhibitors in screening campaigns.[35] Purpurogallin (**21**), isolated from edible oils, is a modest inhibitor of Bcl-x_L ($K_i = 2.2$ μM). Black tea component theaflavanin (**22**) has submicromolar affinities for Bcl-x_L ($K_i = 0.25$ μM) and Bcl-2 ($K_i = 0.28$ μM), and green tea component (−)-catechin-3 gallate (**23**) has similar Bcl-x_L ($K_i = 0.12$ μM) and Bcl-2 ($K_i = 0.40$ μM) affinities.[36] Both these compounds induce apoptosis in various cancer cells, but the exact mechanisms are a matter of debate.[37]

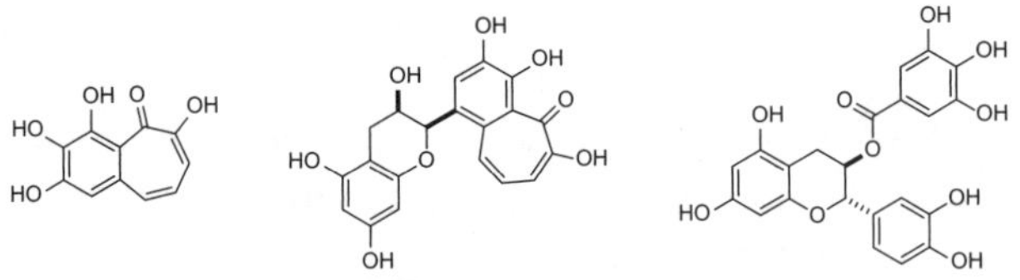

Purpurogallin (**21**) Theaflavanin (**22**) (−)-Catechin-3 gallate (**23**)

Other structurally diverse natural products with moderate affinity for Bcl-2 family proteins have been reported. Antimycin A (**24**), isolated from *Streptomyces*, was shown to selectively induce apoptosis in cells with high Bcl-x_L expression levels, although superoxide release from the mitochondria was also observed, suggesting that multiple apoptotic mechanisms may be occurring.[38] Chelerythrine (**25**), isolated from *Bocconia vulcanica*, demonstrates modest Bcl-x_L affinity ($IC_{50} = 1.5$ μM) and was shown to disrupt Bak/Bcl-x_L interactions in an immunoprecipitation assay. However, activity against Bak/Bax-deficient cell lines suggests that additional biological mechanisms may contribute to this cytotoxicity.[39] Cryptosphaerolide (**26**), isolated from *Cryptosphaeria*, is a modest binder of Mcl-1 ($IC_{50} = 11.4$ μM) and kills HCT-116 cells ($EC_{50} = 4.5$ μM) *in vitro*.[40] Marinopyrrole A (**27**), isolated from marine *Streptomycetes*, exhibits modest affinity for Mcl-1 ($IC_{50} = 10$ μM) but not for Bcl-x_L. Marinopyrrole A enhances the cytotoxicity of ABT-737 by 60-fold compared to monotherapy in K562 cells.[41] Finally, tetrocacin A (**28**), isolated from *Actinomycete*, sensitizes cells overexpressing Bcl-2 and Bcl-x_L to radiation, but again evidence suggests that this activity may result from other pathways.[42]

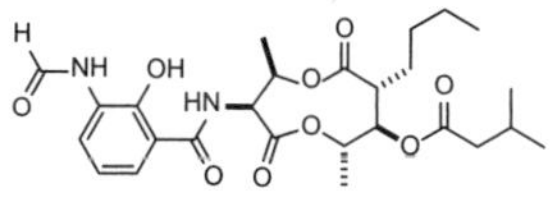

Antimycin A (**24**)

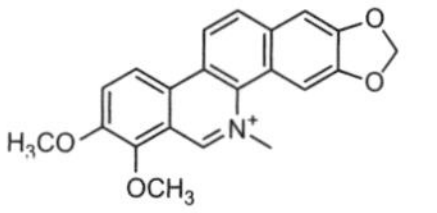

Chelerythrine (**25**)

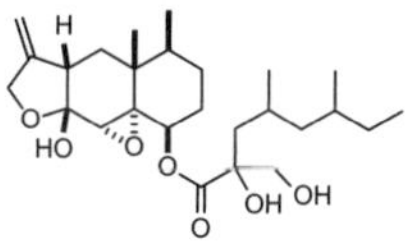

Cryptosphaerolide (**26**)

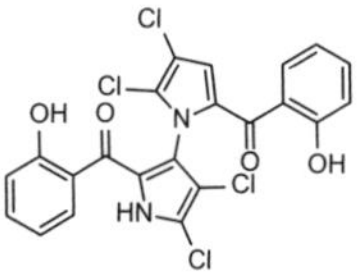

Marinopyrrole A (**27**)

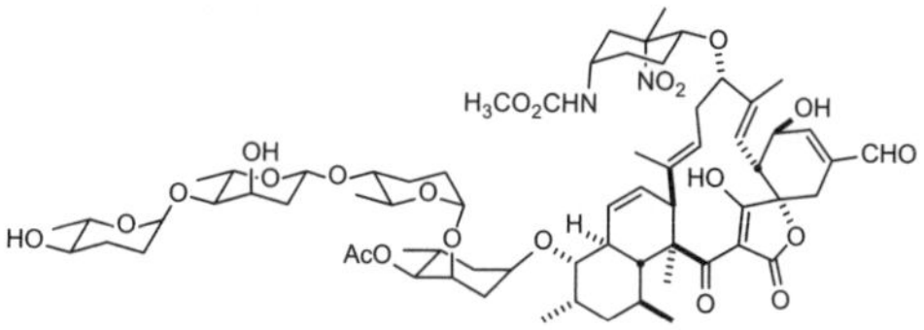

Tetrocarcin A (**28**)

2.3. *In silico* screening approaches

Several inhibitors with modest affinity for Bcl-2 family proteins have been discovered utilizing *in silico* screens of commercial 3D libraries. The first reported Bcl-2 inhibitor, dihydrochroman HA14-1 (**29**), was discovered through a virtual screen of the MDL/ACD-3D database and displays weak affinity for Bcl-2 (IC_{50} = 9.0 μM) and efficacy against HL-60 cells (EC_{90} = 50 μM). However, HA14-1 was shown to decompose to generate reactive oxygen species that may induce apoptosis.[43] A virtual screen of the Maybridge chemical library revealed BI-21C5 (**30**, IC_{50} = 5.1 μM) and BI-21C6 (**31**, IC_{50} = 0.5 μM) with modest affinity for Bcl-x_L.[44] BI-21C5 kills ZR-75-1 cells (EC_{50} = 11.7 μM), while BI-21C6 is inactive. A virtual screen of the NCI-3D database led to BL-11 (**32**), with modest Bcl-x_L (IC_{50} = 9.0 μM) and Bcl-2 (IC_{50} = 10.4 μM) affinities.[45] However, BL-11 demonstrated similar activity in Bak/Bax-deficient cells, suggesting that activity may be off mechanism.[46] A virtual screen of the NCI-3D database revealed MNB (**33**), a Bcl-2 inhibitor (IC_{50} = 0.7 μM) with cytotoxic activity (EC_{65} = 5 μM) against HL-60 cells.[47] Finally, a virtual screen of the SPECS database revealed tetracycle **34**, an inhibitor with moderate Bcl-x_L (IC_{50} = 3.4 μM), Bcl-2 (IC_{50} = 3.1 μM), and Mcl-1 (IC_{50} = 6.4 μM) affinities.[46] Tetracycle **34** showed dose-dependent cell killing, but reactive oxygen species were observed that may be responsible for initiating apoptosis. Compounds **29–34** have not been evaluated with *in vivo* tumor models.

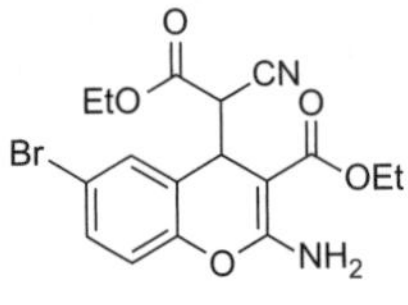

HA-14-1 (**29**)

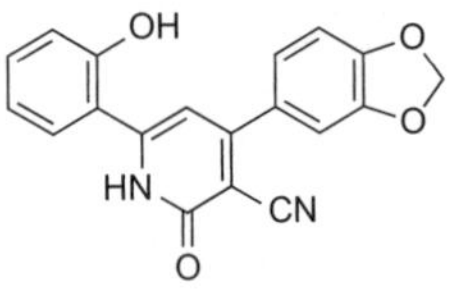

BI-21C5 (**30**)

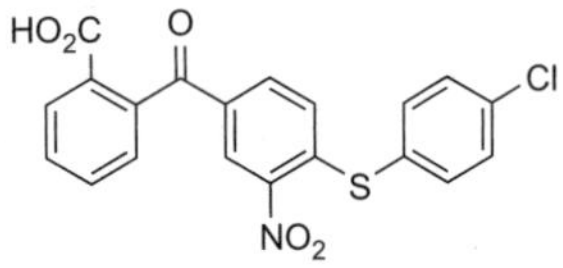

BI-21C6 (**31**)

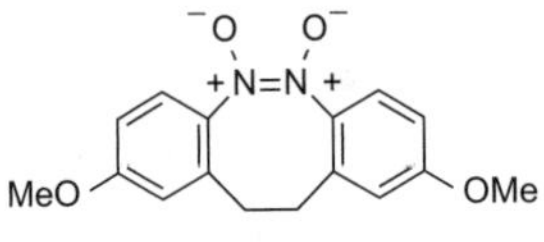

BL-11 (**32**)

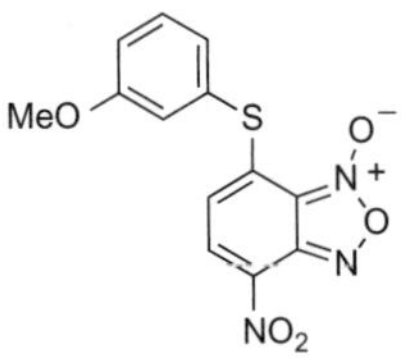

MNB (**33**)

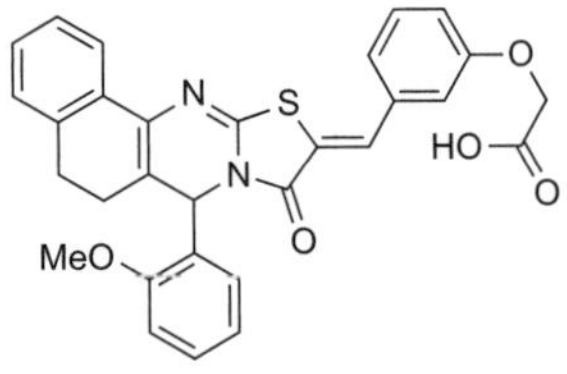

34

2.4. Library screening approaches

Several reports of Bcl-2 family inhibitors discovered from library high-throughput screens (HTSs) have also been published, but thus far, this approach has only led to the discovery of low-affinity inhibitors. One report describes an HTS campaign that resulted in phenylpyrazole **35**, a dual Bcl-2 ($IC_{50}=0.16$ μM) and Bcl-x_L ($IC_{50}=0.25$ μM) inhibitor.[48] Modest Bcl-x_L inhibitors rhodanine **36** ($K_i=2.4$ μM) and diarylsulfone **37** ($K_i=4.1$ μM) were discovered using an HTS of commercial compounds. However, cytotoxicity in Bak/Bax-deficient cell lines suggests that activity may be independent from Bcl-x_L binding.[49] An HTS produced compound **38** as the only hit in the screen to induce >50% inhibition of Bcl-x_L.[50] Inhibitor **38** disrupted the Bax/Bcl-x_L interaction in an immunoprecipitation assay and induced apoptosis at high concentrations ($EC_{50}\sim0.5$ mM) in MCF7 cells.

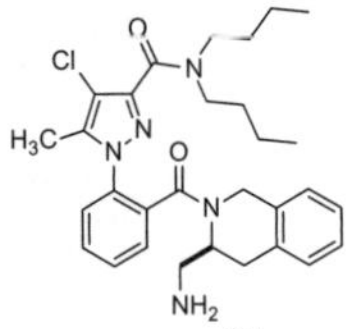

35

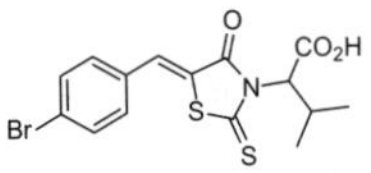

BH3I-1 (**36**)

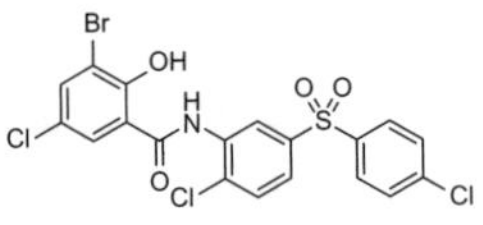

BH3I-2 (**37**)

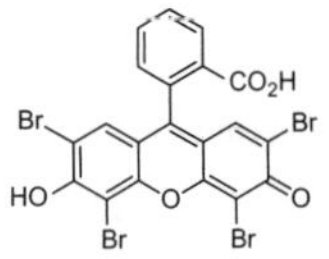

38

2.5. Other approaches

Efforts to mimic the α-helical BH3 domain of BH3-only proteins with a small molecule led to the discovery of terphenyl **39**, a weak inhibitor ($K_i=114$ μM) of Bcl-x_L.[51] All attempts to improve the poor physical properties of this compound proved unsuccessful. A series of isooxazolidines represented by structure **40** were discovered using a diversity-oriented synthesis approach.[52] These isooxazolidines are dual Bcl-2 ($K_i < 0.001$ μM) and Bcl-x_L ($K_i < 1.0$ μM) inhibitors and demonstrate dose-dependent killing of various cell lines. An effort to design DNA intercalating agents revealed rigid chromophore S1 (**41**), a compound that reportedly lacks DNA intercalation ability but induces apoptosis in a variety of Bcl-2-sensitive cell lines. S1 was claimed to be a dual inhibitor of Bcl-2 ($IC_{50} = 0.285$ μM) and Mcl-1 ($IC_{50} = 0.035$ μM).[53] Mipralden (**42**) is a weak dual inhibitor of Bcl-2 ($K_d = 70$ μM) and Mcl-1 ($K_d = 25$ μM) that was discovered through a *de novo* structure-based approach.[54] Compounds **39–42** are currently in the preclinical evaluation stage.

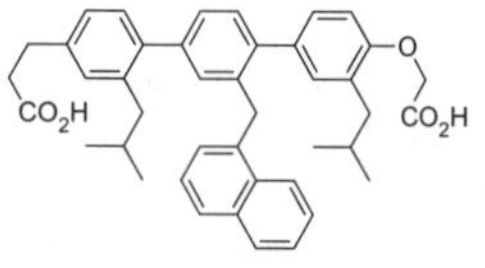

39

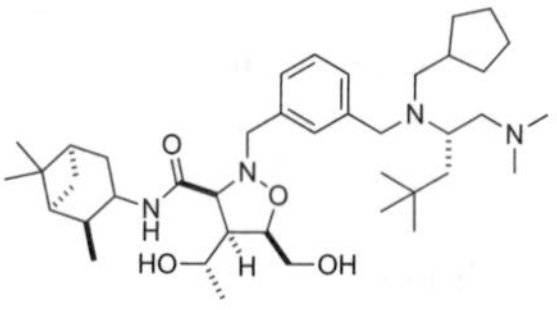

40

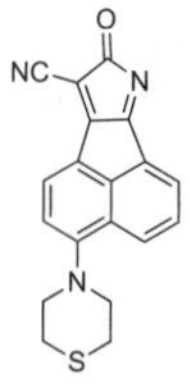

S1 (**41**)

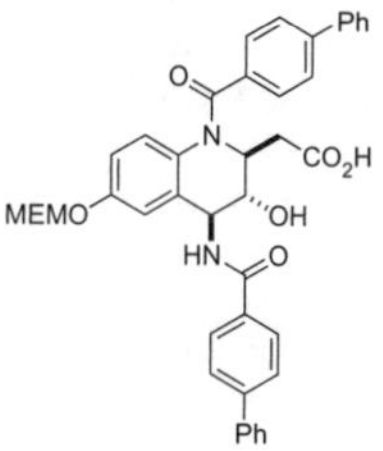

Mipralden (**42**)

3. CONCLUSIONS

The work described in this review demonstrates that inhibition of Bcl-2 family proteins with high-affinity small molecules is possible and that Bcl-2 proteins are valid targets for clinical oncology indications. The structural diversity of chemical inhibitors suggests that a wide range of chemotypes are useful, although the physiochemical properties of these molecules may fall outside the range traditionally considered "drug-like." Development of clinical candidates with a range of selectivity profiles for the Bcl-2 family proteins will provide clarity on their role in cancer and should ultimately lead to new oncology therapies.

REFERENCES

(1) Leber, B.; Lin, J.; Andrews, D. *Oncogene* **2010**, *29*, 5221–5230.
(2) Bajwa, N.; Liao, C.; Nikolovska-Coleska, Z. *Expert Opin. Ther. Pat.* **2012**, *22*, 37–55.
(3) Lessene, G.; Czabotar, P.; Colman, P. *Nat. Rev. Drug Discov.* **2008**, *7*, 989–1000.
(4) Keller, T.; Pichota, A.; Yin, Z. *Curr. Opin. Chem. Biol.* **2006**, *4*, 357–361.
(5) Petros, A.; Dinges, J.; Augeri, D.; Baumeister, S.; Betebenner, D.; Bures, M.; Elmore, S.; Hajduk, P.; Joseph, M.; Landis, S.; Nettesheim, D.; Rosenberg, S.;

Shen, W.; Thomas, S.; Wang, X.; Zanze, I.; Zhang, H.; Fesik, S. *J. Med. Chem.* **2006**, *49*, 656–663.
(6) Wendt, M.; Shen, W.; Kunzer, A.; McClellan, W.; Bruncko, M.; Oost, T.; Ding, H.; Joseph, M.; Zhang, H.; Nimmer, P.; Ng, S.; Shoemaker, A.; Petros, A.; Oleksijew, A.; Marsh, K.; Bauch, J.; Oltersdorf, T.; Belli, B.; Martineau, D.; Fesik, S.; Rosenberg, S.; Elmore, S. *J. Med. Chem.* **2006**, *49*, 1165–1181.
(7) Shoemaker, A.; Oleksijew, A.; Bauch, J.; Belli, B.; Borre, T.; Bruncko, M.; Deckwirth, T.; Frost, D.; Jarvis, K.; Joseph, M.; Marsh, K.; McClellan, W.; Nellans, H.; Ng, S.; Nimmer, P.; O'Connor, J.; Oltersdorf, T.; Qing, W.; Shen, W.; Stavropoulos, J.; Tahir, S.; Wang, B.; Warner, R.; Zhang, H.; Fesik, S.; Rosenberg, S.; Elmore, S. *Cancer Res.* **2006**, *66*, 8731–8739.
(8) Oltersdorf, T.; Elmore, S.; Shoemaker, A.; Armstrong, R.; Augeri, D.; Belli, B.; Bruncko, M.; Deckwerth, T.; Dinges, J.; Hajduk, P.; Joseph, M.; Kitada, S.; Korsmeyer, S.; Kunzer, A.; Letai, A.; Li, C.; Mitten, M.; Nettesheim, D.; Ng, S.; Nimmer, P.; O'Connor, J.; Oleksijew, A.; Petros, A.; Reed, J.; Shen, W.; Tahir, S.; Thompson, C.; Tomaselli, K.; Wang, B.; Wendt, M.; Zhang, H.; Fesik, S.; Rosenberg, S. *Nature* **2005**, *435*, 677–681.
(9) Bruncko, M.; Oost, T.; Belli, B.; Ding, H.; Joseph, M.; Kunzer, A.; Martineau, D.; McClellan, W.; Mitten, M.; Ng, S.; Nimmer, P.; Oltersdorf, T.; Park, C.; Petros, A.; Shoemaker, A.; Song, X.; Wang, X.; Wendt, M.; Zhang, H.; Fesik, S.; Rosenberg, S.; Elmore, S. *J. Med. Chem.* **2007**, *50*, 641–662.
(10) Tahir, S.; Yang, X.; Anderson, M.; Morgan-Lappe, S.; Sarthy, A.; Chen, J.; Warner, R.; Ng, S.; Fesik, S.; Elmore, S.; Rosenberg, S.; Tse, C. *Cancer Res.* **2007**, *67*, 1176–1183.
(11) Tse, C.; Shoemaker, A.; Adickes, J.; Anderson, M.; Chen, J.; Jin, S.; Johnson, E.; Marsh, K.; Mitten, M.; Nimmer, P.; Roberts, L.; Tahir, S.; Xiao, Y.; Yang, X.; Zhang, H.; Fesik, S.; Rosenberg, S.; Elmore, S. *Cancer Res.* **2008**, *68*, 3421–3428.
(12) Park, C.; Bruncko, M.; Adickes, J.; Bauch, J.; Ding, H.; Kunzer, A.; Marsh, K.; Nimmer, P.; Shoemaker, A.; Song, X.; Tahir, S.; Tse, C.; Wang, X.; Wendt, M.; Yang, X.; Zhang, H.; Fesik, S.; Rosenberg, S.; Elmore, S. *J. Med. Chem.* **2008**, *51*, 6902–6915.
(13) Chen, J.; Jin, S.; Abraham, V.; Huang, X.; Liu, B.; Mitten, M.; Nimmer, P.; Lin, X.; Smith, M.; Shen, Y.; Shoemaker, A.; Tahir, S.; Zhang, H.; Ackler, S.; Rosenberg, S.; Maecker, H.; Sampath, D.; Leverson, J.; Tse, C.; Elmore, S. *Mol. Cancer Ther.* **2011**, *12*, 2340–2349.
(14) Shoemaker, A.; Mitten, M.; Adickes, J.; Ackler, S.; Refici, M.; Ferguson, D.; Oleksijew, A.; O'Connor, J.; Wang, B.; Frost, D.; Bauch, J.; Marsh, K.; Tahir, S.; Yang, X.; Tse, C.; Fesik, S.; Rosenberg, S.; Elmore, S. *Clin. Cancer Res.* **2008**, *14*, 3268–3277.
(15) Lam, L.; Lu, X.; Zhang, H.; Lesniewski, R.; Rosenberg, S.; Semizarov, D. *Mol. Cancer Ther.* **2010**, *9*, 2943–2950.
(16) http://clinicaltrials.gov/ct2/results?term=abt-263.
(17) Rudin, C.; Hann, C.; Garon, E.; de Oliveira, M.; Bonomi, P.; Camidge, D.; Chu, Q.; Giaccone, G.; Khaira, D.; Ramalingam, S.; Ranson, M.; Dive, C.; McKeegan, E.; Chyla, B.; Dowell, B.; Chakravartty, A.; Nolan, C.; Rudersdorf, N.; Busman, T.; Mabry, M.; Krivoshik, A.; Humerickhouse, R.; Shapiro, G.; Gandhi, L. *Clin. Cancer Res.* **2012**, *18*, 3163–3169.
(18) Elmore, S. ABT-199: A potent and selective inhibitor of Bcl-2. Oral Presentation at the AACR Annual Meeting, Apr **2012**, Chicago, IL.
(19) http://clinicaltrials.gov/ct2/results?term=abt-199.
(20) Sleebs, B.; Czabotar, P.; Fairbrother, W.; Fairlie, W.; Flygare, J.; Huang, D.; Kersten, W.; Koehler, M.; Lessene, G.; Lowes, K.; Parisot, J.P.; Smith, B.; Smith, M.; Souers, A.; Street, I.; Yang, H.; Baell, J. *J. Med. Chem.* **2011**, *54*, 1914–1926.

(21) Song, X.; Ding, H.; Elmore, S.; Bruncko, M.; Madar, D.; Souers, A.; Park, C.; Tao, Z.; Wang, X.; Kunzer, A. U.S. Patent Application Publication 7981888, **2011**.
(22) Elmore, S.; Souers, A.; Bruncko, M.; Song, X.; Wang, X.; Hasvold, L.; Wang, L.; Kunzer, A.; Park, C.; Wendt, M.; Tao, Z.; Madar, D. PCT Publication WO 131000, **2008**.
(23) Rega, M.; Wu, B.; Wei, J.; Zhang, Z.; Cellitti, J.; Pellecchia, M. *J. Med. Chem.* **2011**, *54*, 6000–6013.
(24) http://clinicaltrials.gov/ct2/results?term=obatoclax.
(25) Attardo, G.; Lavallee, J. -F.; Rioux, E.; Doyle, T. U.S. Patent Application Publication 7425553, **2008**.
(26) Konopleva, M.; Watt, J.; Contractor, R.; Tsao, T.; Harris, D.; Estrov, Z.; Bornmann, W.; Kantarjian, H.; Viallet, J.; Samudio, I.; Andreeff, M. *Cancer Res.* **2008**, *68*, 3413–3420.
(27) Balakrishnan, K.; Wierda, W.; Keating, M.; Gandi, V. *Blood* **2008**, *112*, 1971–1980.
(28) Mohammed, R.; Wang, S.; Aboukameel, A.; Chen, B.; Wu, X.; Chen, J.; Al-Katib, A. *Mol. Cancer Ther.* **2005**, *4*, 13–21.
(29) Liu, G.; Kelly, W.; Wilding, G.; Leopold, L.; Brill, K.; Somer, B. *Clin. Cancer Res.* **2009**, *15*, 3172–3176.
(30) Dash, R.; Azab, B.; Quinn, B.; Shen, X.; Wang, X.; Das, S.; Rahmani, M.; Wei, J.; Hedvat, M.; Dent, P.; Dmitriev, I.; Curiel, D.; Grant, S.; Wu, B.; Stebbins, J.; Pellecchia, M.; Reed, J.; Sarkar, D.; Fisher, P. *Proc. Natl. Acad. Sci. U.S.A.* **2011**, *108*, 8785–8790.
(31) Wang, G.; Nikolovska-Coleska, Z.; Yang, C.; Wang, R.; Tang, G.; Guo, J.; Shangary, S.; Qiu, S.; Gao, W.; Yang, D.; Meagher, J.; Stuckey, J.; Krajewski, K.; Jiang, S.; Roller, P.; Abaan, H.; Tomita, Y.; Wang, S. *J. Med. Chem.* **2006**, *49*, 6139–6142.
(32) Zeitlin, B.; Joo, E.; Dong, Z.; Warner, K.; Wang, G.; Nikolovska-Coleska, Z.; Wang, S.; Nör, J. *Cancer Res.* **2006**, *66*, 8698–8706.
(33) Tang, G.; Ding, K.; Nikolovska-Coleska, Z.; Yang, C.; Qiu, S.; Shangary, S.; Wang, R.; Guo, J.; Gao, W.; Meagher, J.; Stuckey, J.; Krajewski, K.; Jiang, S.; Roller, P.; Wang, S. *J. Med. Chem.* **2007**, *50*, 3163–3166.
(34) Tang, G.; Yang, C.; Nikolovska-Coleska, Z.; Guo, J.; Qiu, S.; Wang, R.; Gao, W.; Wang, G.; Stuckey, J.; Krajewski, K.; Jiang, S.; Roller, P.; Wang, S. *J. Med. Chem.* **2007**, *50*, 1723–1726.
(35) Pellecchia, M.; Reed, J. *Curr. Pharm. Des.* **2004**, *10*, 1387–1398.
(36) Leone, M.; Zhai, D.; Sareth, S.; Kitada, S.; Reed, J.; Pellecchia, M. *Cancer Res.* **2003**, *63*, 8118–8121.
(37) Lahiry, L.; Saha, B.; Chakraborty, J.; Bhattacharyya, S.; Chattopadhyay, S.; Banerjee, S.; Choudhuri, T.; Mandal, D.; Bhattacharyya, A.; Sa, G.; Das, T. *Apoptosis* **2008**, *11*, 771–781.
(38) Tzung, S.; Kim, K.; Basanez, G.; Giedt, C.; Simon, J.; Zimmerberg, J.; Zhang, K.; Hockenbery, D. *Nat. Cell Biol.* **2001**, *3*, 183–191.
(39) Wan, K.; Chan, S.; Sukumaran, S.; Lee, M.; Yu, V. *J. Biol. Chem.* **2008**, *283*, 8423–8433 and references therein.
(40) Oh, H.; Jensen, P.; Murphy, B.; Fiorilla, C.; Sullivan, J.; Ramsey, R.; Fenical, W. *J. Nat. Prod.* **2010**, *73*, 998–1001.
(41) Doi, K.; Li, R.; Sung, S.; Wu, H.; Liu, Y.; Manieri, W.; Krishnegowda, G.; Awwad, A.; Dewey, A.; Liu, X.; Amin, S.; Cheng, C.; Qin, Y.; Schonbrunn, E.; Daughdrill, G.; Loughran, T., Jr.; Sebti, S.; Wang, H. *J. Biol. Chem.* **2012**, *287*, 10224–10235.
(42) van Delft, M.; Wei, A.; Mason, K.; Vandenberg, C.; Chen, L.; Czabatar, P.; Willis, S.; Scott, C.; Day, C.; Cory, S.; Adams, J.; Roberts, A.; Huang, D. *Cancer Cell* **2006**, *10*, 389–399 and references therein.

(43) Wang, J.; Liu, D.; Zhang, Z.; Shan, S.; Han, X.; Srinivasula, S.; Croce, C.; Alnemri, E.; Huang, Z. *Proc. Natl. Acad. Sci. U.S.A.* **2000**, *97*, 7124–7129.
(44) Raga, M.; Leone, M.; Jung, D.; Cotton, N.J.; Stebbins, J.; Pellecchia, M. *Bioorg. Chem.* **2007**, *35*, 344–353.
(45) Enyedy, I.; Ling, Y.; Nacro, K.; Tomita, Y.; Wu, X.; Cao, Y.; Guo, R.; Li, B.; Zhy, X.; Huang, Y.; Lont, Y.; Roller, P.; Yang, D.; Wang, S. *J. Med. Chem.* **2001**, *44*, 4313–4324.
(46) Feng, Y.; Ding, X.; Chen, T.; Chen, L.; Liu, F.; Jia, X.; Luo, X.; Shen, X.; Chen, K.; Jiang, H.; Wang, H.; Liu, H.; Liu, D. *J. Med. Chem.* **2010**, *53*, 3465–3479.
(47) Zhang, M.; Ling, Y.; Yang, C.; Liu, H.; Wang, R.; Wu, X.; Ding, K.; Zhu, F.; Griffith, B.; Mohammad, R.; Wang, S.; Yang, D. *Ann. Hematol.* **2007**, *86*, 471–481.
(48) Porter, J.; Payne, A.; de Candole, B.; Ford, D.; Hutchinson, B.; Trevitt, G.; Turner, J.; Edwards, C.; Watkins, C.; Whitcombe, I.; Davis, J.; Stufferfield, C. *Bioorg. Med. Chem. Lett.* **2009**, *19*, 230–233.
(49) Degterev, A.; Lugovskoy, A.; Cardone, M.; Mulley, B.; Wagner, G.; Mitchison, T.; Yuan, J. *Nat. Cell Biol.* **2001**, *3*, 173–182.
(50) Tan, Y.; Teng, E.; Ting, A. *J. Cancer Res. Clin. Oncol.* **2003**, *129*, 437–448.
(51) Kutzki, O.; Park, H.; Ernst, J.; Orner, B.; Yin, H.; Hamilton, A. *J. Am. Chem. Soc.* **2002**, *124*, 11838–11839.
(52) Castro, A.; Holson, E.; Hopkins, B.; Koney, N.; Snyder, D.; Tibbitts, T. U.S. Patent Application Publication 7842815, **2010**.
(53) Zhang, Z.; Wu, G.; Xie, F.; Song, T.; Chang, X. *J. Med. Chem.* **2011**, *54*, 1101–1105.
(54) Prakesch, M.; Denisov, A.; Naim, M.; Gehring, K.; Arya, P. *Bioorg. Med. Chem.* **2008**, *16*, 7443–7449.

CHAPTER EIGHTEEN

Notch Pathway Modulators as Anticancer Chemotherapeutics

Vibhavari Sail, M. Kyle Hadden
Department of Pharmaceutical Sciences, University of Connecticut School of Pharmacy, Storrs, Connecticut, USA

Contents

1. INTRODUCTION

Targeting developmental signaling pathways (Hedgehog, Wnt, and Notch) that control cellular growth and tissue differentiation has emerged as a viable strategy for the identification and development of new anticancer chemotherapeutics.[1] These pathways serve numerous functions during embryogenesis and in self-renewing adult tissues; including, stem cell maintenance, cellular fate, proliferation, and apoptosis. In general, pathway dysregulation, through a variety of different mechanisms, leads to constitutive activation and tumor formation. However, the Notch pathway has

Annual Reports in Medicinal Chemistry, Volume 47
ISSN 0065-7743
http://dx.doi.org/10.1016/B978-0-12-396492-2.00018-7

demonstrated both oncogenic and tumor suppressor activity, depending on the cell type, suggesting that both agonists and antagonists of Notch signaling hold potential for future development as anticancer drugs.[2,3]

2. MECHANISM OF NOTCH SIGNAL TRANSDUCTION

In multicellular organisms, Notch signaling is an evolutionarily conserved pathway that has a fundamental role in regulating stem cell maintenance and cell fate decisions during development. The Notch gene was first described by Thomas H. Morgan in 1917. In the fruit fly *Drosophila melanogaster*, the partial loss of function of this gene caused a notched wing phenotype.[4] The mammalian Notch system consists of four receptors (Notch1–4) and five ligands (delta-like 1, delta-like 3, delta-like 4, jagged1, and jagged2).[2] Notch receptors are single-pass transmembrane proteins consisting of an extracellular domain, transmembrane domain, and an intracellular domain (Fig. 18.1). Upon binding to one of the ligands on an adjacent cell, the cytoplasmic domain of the Notch receptor undergoes a two-step metalloprotease- and γ-secretase-mediated cleavage. The cleaved domain, termed Notch intracellular domain (NICD), recruits various coactivator proteins (CoA, MAML1) to the nucleus and converts the CSL (CBF/Su (H)/Lag-1) repressor complex to a transcriptional activator. The ultimate result of canonical Notch signaling is the transcription of pathway target genes, including the Hairy Enhancer of Split (HES), Hairy-related (HRT: also known as HEY) families, Notch receptors and ligands, cyclin D, MYC, and Bcl-2.[2,3] These proteins serve a number of cellular functions, and dysregulation of Notch signaling can drive tumorigenesis through several mechanisms (described below).

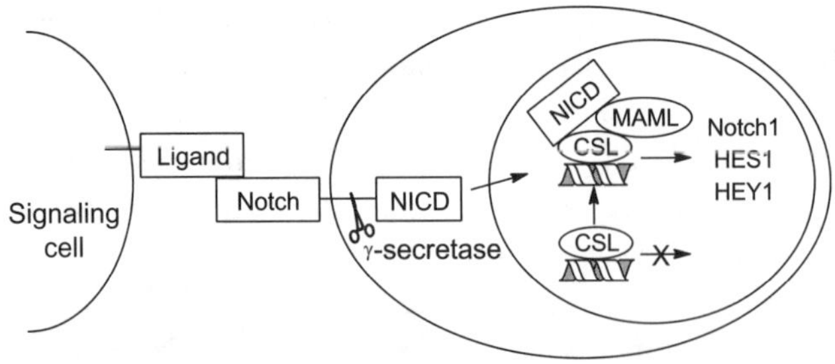

Figure 18.1 Primary components of the Notch signaling pathway.

3. ROLE IN TUMORIGENESIS

3.1. Notch as an oncogene

The oncogenic nature of aberrant Notch signaling was originally identified in human T-cell acute lymphoblastic leukemia (T-ALL).[5] T-ALL accounts for approximately 15% of all diagnosed ALL patients and was initially linked to dysregulated Notch signaling through the identification of a specific *NOTCH1* mutation that resulted in overexpression of NICD. A subsequent study demonstrated that 56% of T-ALL patients contained activating *NOTCH1* mutations, establishing a clear oncogenic role for Notch signaling in T-ALL development.[6] Of note, recent evidence suggests that the activating mutations present in T-ALL may not be sufficient to completely induce the disease and that cellular effects downstream of *NOTCH1* mutations are imperative for T-ALL development.[7] Since the initial discovery of its involvement in T-ALL, oncogenic Notch signaling has been implicated in a variety of solid tumors, including, breast, colorectal, pancreatic, and lung cancer.[2] Ligand overexpression, loss of function of negative pathway regulators, and activating mutations have all been linked to increased Notch activity in solid tumors. The ultimate result of this dysregulation and the main mechanism through which Notch signaling appears to drive tumorigenesis is promoting cell cycle progression and inhibiting apoptosis through target gene overexpression.[2,3]

3.2. Notch as a tumor suppressor

Recently, growing evidence supports a tumor suppressor role for Notch signaling in a variety of human tissues, including, skin, liver, and neuroendocrine tumors (NETs).[8] Although the mechanisms through which Notch exerts its tumor suppressor activity are not completely understood, two potential roles have been identified: (1) modulation of cell growth through expression of transcriptional repressor target genes (HES1 and HEY1) and (2) cross talk with other signaling pathways.

The best example of Notch acting as a tumor suppressor through transcriptional repression can be found in NETs (also termed carcinoid tumors).[9] NETs are characterized by the secretion of excessive levels of various bioactive hormones that can result in a variety of debilitating effects (asthma, cardiac disease, and dehydration) depending upon the tumor origin. Notch signaling is inactive in several specific types of NETs, including gastrointestinal (GI) carcinoids and small cell lung cancer (SCLC).[9–11] These types of NETs are also characterized by increased levels of achaete-scute complex-like 1 (ASCL-1), a transcription factor that governs progenitor cell differentiation and is controlled by Notch/HES1 signaling.[10,11] The mechanisms through which ASCL-1 mediates NET onset and progression are poorly understood; however, upregulation of HES1 through activation of Notch signaling in NCI-H209 SCLC cells results in growth inhibition, cell cycle arrest, and apoptosis.[11] In addition, a gene expression profile analysis of patients with hepatocellular carcinoma demonstrated that increased *NOTCH1* and *HES1* expression correlated with a significantly better survival rate.[12]

The second mechanism through which Notch exerts its tumor suppressor activity is through cross talk with other oncogenic signaling pathways (Hh, Wnt, PI3K/AKT, and NF-κB; Fig. 18.2). Loss of function of *NOTCH1* resulted in spontaneous basal cell carcinoma (BCC) formation.[13] As Hh and Wnt are known mediators of BCC, these results suggest Notch suppression of Hh and/or Wnt signaling in skin as a mechanism of regulating cell growth. The Notch pathway has also been implicated as a regulator of PI3K/Akt signaling network through its control of a transcriptional network that regulates PTEN expression and PI3K/Akt signaling activity in normal thymocytes and leukemic T cells.[14] Finally, vascular endothelial growth factor (VEGF) controls angiogenesis by increasing delta-like 4 levels to drive vessel formation. Ligand activation of Notch receptors on adjacent cells downregulates VEGF2, modulating vessel growth through a negative feedback loop.[15] Disruption of this cross talk with delta-like 4-specific antibodies resulted in the development of vascular tumors, suggesting a tumor suppressor role for Notch in this context.[16]

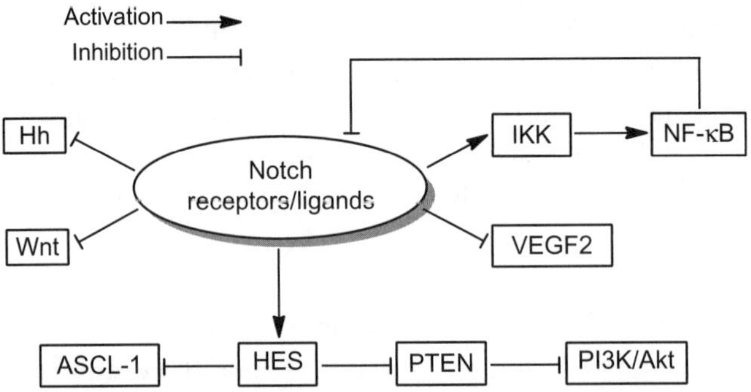

Figure 18.2 Examples of cross talk between Notch and oncogenic signaling pathways.

4. INHIBITORS OF NOTCH SIGNALING

4.1. Gamma secretase inhibitors

Gamma secretase inhibitors (GSIs) have long been investigated for their potential as Alzheimer's disease (AD) treatments due to their ability to block secretion of amyloid β-peptide, the precursor of amyloid plaques associated with AD. The integral role that γ-secretase plays in regulating NICD levels has attracted increasing interest for GSIs as anticancer therapeutics that function through Notch pathway inhibition. This section focuses on the early analysis of GSIs as Notch pathway inhibitors and the most recent preclinical evaluation of compounds that have entered or are approaching clinical trials.

Initial studies on the ability of these compounds to inhibit Notch signaling and exhibit anticancer effects focused on the early di- and tripeptide GSIs. The tripeptide GSI-1, z-Leu-Leu-Nle-CHO, and the peptidomimetic LY-411,575 (**1**) blocked Notch activation and induced apoptosis in *in vitro* and *in vivo* models of Kaposi's sarcoma.[17] Treatment of murine T-ALL cells with the dipeptide DAPT (**2**, 1 μM) resulted in Notch inhibition and induced G_0—G_1 cell cycle arrest and apoptosis.[18] Compound **2** also showed a dose-dependent downregulation of HES1, inhibited cell growth and induced G1 cell cycle arrest in A2780 ovarian cancer cells.[19] These early reports led to more detailed preclinical studies aimed at evaluating the effects of several peptidomimetic and small molecule GSIs as anticancer chemotherapeutics.

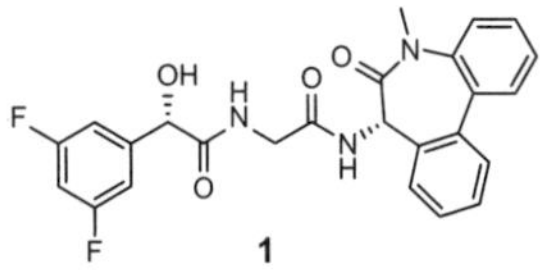
1

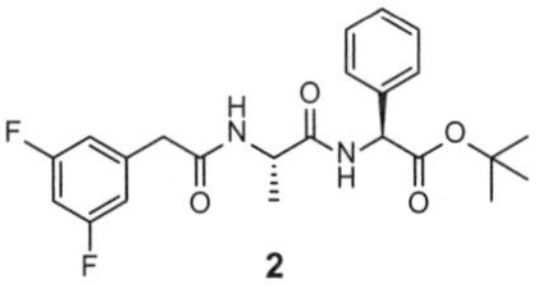
2

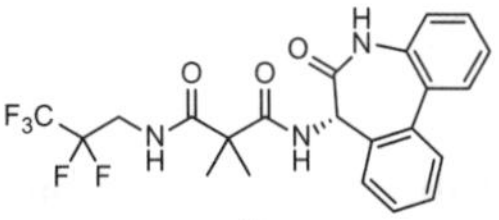
3

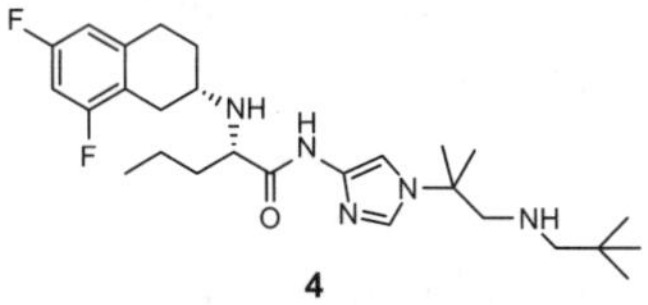
4

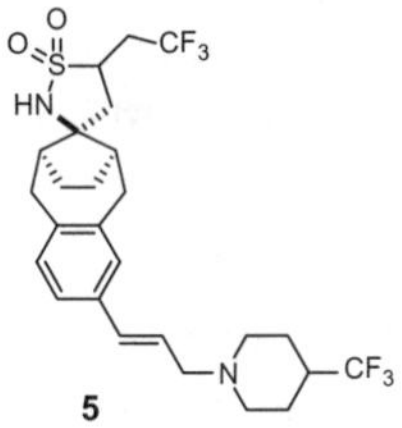
5

The peptidomimetic RO4929097 (**3**) was derived from **1** as a GSI with improved potency, selectivity, and stability.[20] Compound **3** inhibited Notch processing in a dose-dependent fashion ($IC_{50}=5$ nM) in a cell-based reporter gene assay and reduced NICD formation with a concomitant reduction in HES1 mRNA in A549 non-small cell lung carcinoma cells at concentrations as low as 100 nM. Interestingly, *in vitro* studies did not demonstrate inhibition of cellular growth or induction of apoptosis; rather, treatment with the compound produced a less transformed, slower growing phenotype.[21] In an A549 xenograft model, once-daily dosing (3–60 mg/kg) for 7, 14, or 21 days resulted in significant tumor growth inhibition that correlated well with HES1 downregulation. Similar effects were seen in human primary melanoma cells and in a melanoma xenograft model,[22] and **3** has entered multiple Phase I and II trials as a single agent or combination therapy for the treatment of human cancers.[23] A Phase I trial in patients with advanced solid tumors demonstrated encouraging anticancer activity in patients with melanoma, sarcoma, and ovarian cancers with only mild side effects reported.

In a recent Phase II study in patients with refractory metastatic colorectal cancer, **3** showed no clinical benefit at the 20-mg oral dose.[24] Further published reports of the clinical activity of **3** have not been disclosed. Phase II clinical trials for **3** as monotherapy are currently being conducted in advanced and metastatic stages in human cancers such as glioblastoma, breast cancer, nonsmall cell lung cancer, and stage IV melanoma. In addition, numerous Phase II combination trials are underway including, combination therapy (FOLFOX regimen) and Bevacizumab in metastatic colorectal cancer; cisplatin, vinblastine, and temozolomide in metastatic melanoma; and with the Hh pathway inhibitor GDC-0449 in metastatic sarcoma.

In HPB-ALL cancer cells at concentrations as low as 10 nM, PF-03084014 (**4**) lowers NICD and HES1 levels and induces apoptosis in a dose-dependent fashion.[25] Treatment with **4** (75–150 mg/kg) in multiple Notch-dependent xenograft models demonstrated robust NICD reduction, tumor growth inhibition, and apoptosis induction. In addition, a 7 days on/7 days off dosing scheduled reduced GI toxicity typically associated with Notch inhibition and reduced tumor growth at levels comparable to twice-daily dosing for 14 days.[25] This reduction in toxicity is notable because GSI treatment is known to cause significant accumulation of goblet cells in the small intestine leading to GI distress and patient withdrawal from the trial.[26,27] The establishment of a dosing regimen that could overcome this side effect is a key to the advancement of GSIs in the clinic.

MRK-003 (**5**) inhibits cellular proliferation in multiple Notch-dependent T-ALL cell lines (GI_{50} values, 5–320 nM), resulting in G_0/G_1 cell cycle arrest and apoptosis.[28] Treatment of HPB-ALL cells with **5** resulted in a dose-dependent reduction of NICD ($IC_{50} \sim 50$ nM). In an *in vivo* mouse model of breast cancer, **5** specifically targeted cancer stem cells and completely abrogated tumor growth in all animals at 5 μM.[29] Tumors did not recur in any treated mice up to 1 year after administration, suggesting γ-secretase/Notch inhibition as a potential strategy for treating breast cancer. Similar effects were seen in *in vitro* models of multiple myeloma and non-Hodgkin's lymphoma[30] and combination therapy with gemcitabine prolonged survival in a murine model of pancreatic cancer.[31] Although no clinical trials with compound **5** are underway in the United States, the promising results demonstrated for the combination therapy on pancreatic cancer have prompted a trial in the United Kingdom to assess the combination of **5** and gemcitabine in humans.

4.2. Other small molecule inhibitors

4.2.1 Genistein

Genistein (**6**) is a natural isoflavone primarily found in soybeans and soybean-enriched products. Many studies have reported a wide range of biological effects such as antioxidant, antiangiogenic, anthelmintic, and anticancer activity.[32] Based on the interrelations between genistein/NF-κB and NF-κB/Notch, research studies were undertaken to determine whether the effects of genistein on NF-κB signaling were mediated by the Notch pathway in BxPC-3 pancreatic cancer cells. Genistein treatment of BxPC-3 cells inhibited cell growth in a dose-dependent fashion, induced apoptosis, and downregulated Notch1 protein expression.[33] Genistein-induced downregulation of Notch1 protein correlated directly with the downregulation of Notch target genes HES1, Bcl-X_L, and cyclin D1.[33] Finally, genistein inhibited NF-κB DNA-binding activity, providing further evidence that NF-κB is a downstream signaling regulator of the Notch pathway in pancreatic cancer and suggesting the potential of genistein in pancreatic cancer therapy.

More recently, Notch1 downregulation by genistein has been suggested as an antitumor and antimetastatic approach for the treatment of prostate cancer.[34] Developmental pathways such as Notch and FoxM1 are overexpressed in prostate cancer cell lines, and based on its ability to regulate Notch signaling in pancreatic cells, the anticancer effects of genistein on prostate cancer cells were explored.[34] Genistein downregulated Notch1, p-Akt, and FoxM1 (mRNA and protein) in a dose-dependent manner, leading to cell growth inhibition and apoptosis in three distinct prostate cancer cell lines PC-3, LNCaP, and C4-2B. The *in vitro* combination treatment of genistein (30 μM) and taxotere (docetaxel, 1 nM) showed a synergistic effect in inhibiting PC-3 cell growth and inducing apoptotic cell death. Finally, genistein treatment in two *in vivo* mouse models of prostate cancer (PC-3 and C4-2B xenograft) demonstrated a significant reduction in Notch1-mediated signaling and a concomitant decrease in tumor volume.[34,35] Although continued studies to deconvolute the multiple biological activities attributed to genistein are needed, the *in vitro* and *in vivo* activities of genistein suggest that its inhibition of Notch/Akt/FoxM1 signaling pathways holds promise as a novel therapeutic approach in the treatment of both prostate and pancreatic cancer.

4.2.2 Withaferin A

Withaferin A (**7**) is a natural product isolated from the medicinal plant *Withania somnifera*, an important ingredient in Indian Ayurvedic medicine, and has been widely studied for its anti-inflammatory properties.[36] Recently, studies characterizing the biological activities of withaferin A have expanded to include the determination of its anticancer properties. Withaferin A was initially reported to inhibit cell growth and induce apoptosis in *in vitro* and *in vivo* models of breast cancer.[37] Subsequent studies of the anticancer activity of withaferin A in colon cancer cells demonstrated its ability to modulate Notch1.[38] Withaferin A treatment significantly inhibited Notch1 signaling as measured by downregulation of pathway target genes HES1 and HEY1. In addition, it downregulated numerous prosurvival pathways such as the Akt/NF-κB/Bcl-2 signaling network and the mTOR pathway components pS6K and p4E-BP1. Finally, withaferin A activated c-Jun N-terminal kinase-mediated apoptosis in colon cancer cells.[38]

4.2.3 Curcumin and related analogues

The anticancer activity of curcumin (**8**) has been extensively studied and has been attributed to the regulation of multiple signaling pathways, including the NF-κB network.[39] The known cross talk between Notch and NF-κB led researchers to explore the ability of curcumin to regulate Notch signaling in pancreatic cancer cells.[40] Curcumin treatment of BxPC-3 and PANC-1 pancreatic cancer cells resulted in a dose-dependent reduction in cell growth and induction of apoptosis ($IC_{50}s \sim 10\ \mu M$).[40] Moreover, these effects directly correlated with the downregulation of target proteins Notch1, HES1, and cyclin D1. Curcumin treatment resulted in similar effects in esophageal cancer cells.[41] Of note, these studies suggest that curcumin inhibits Notch signaling by inhibiting the γ-secretase complex and preventing NICD formation.[41] Finally, the curcumin analogue DiFiD (**9**) inhibited Notch signaling and cell growth in pancreatic cancer cell lines, and these results were also attributed to disruption of the γ-secretase complex.[42]

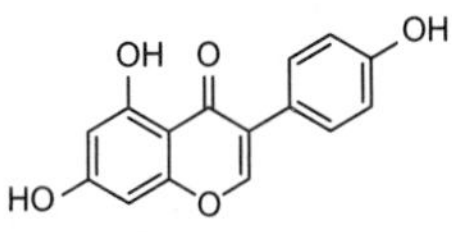
6

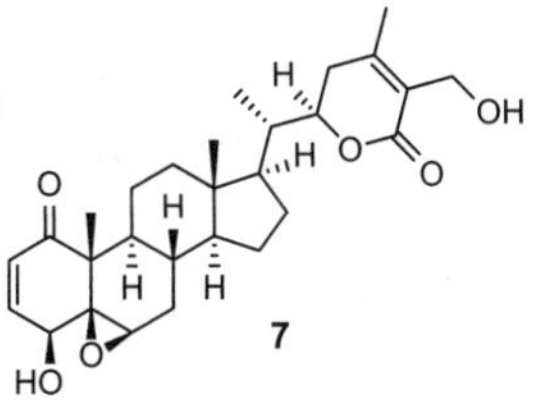
7

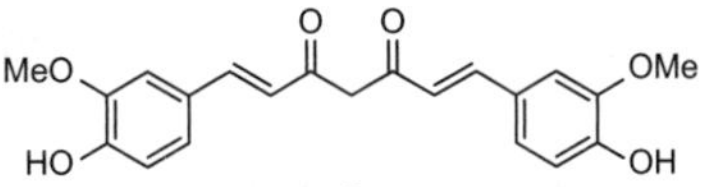
8

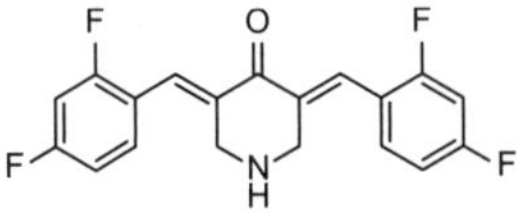
9

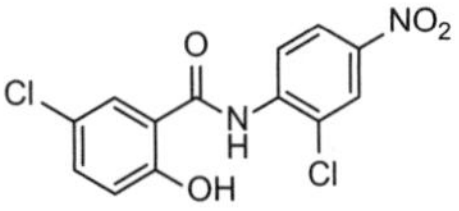
10

4.2.4 Niclosamide

Niclosamide (**10**) has been FDA approved as an anthelmintic for 50 years; however; the past 5 years have seen increasing interest in its use as an anticancer agent. A recent screening effort in K562 human erythroleukemia cells identified niclosamide as a potent inhibitor of Notch signaling (IC_{50} ~ 0.75 μM). Niclosamide treatment (1.25 μM) of K562 cells resulted in the downregulation of functional HES1, NICD, and cyclin D1.[43] Of note, niclosamide also induced HEY2 expression in K562 cells, providing further evidence of differential Notch-mediated effects, depending on the cell type.

5. ACTIVATORS OF NOTCH SIGNALING

5.1. Resveratrol

Resveratrol (**11**), a polyphenol found in the skins of grapes, berries, and peanuts, has been reported to have anti-inflammatory, antioxidant, and anticancer properties.[44] Currently, resveratrol is being evaluated in multiple Phase I and II clinical trials for the treatment of a variety of human diseases. A recent high-throughput screening effort in BON cells, a pancreatic carcinoid cell line, identified resveratrol as an activator of Notch signaling.[45] This activity was further validated by analysis of resveratrol-treated BON and NCI H727 pulmonary carcinoid tumor cells. Resveratrol exhibited antiproliferative effects in both cell lines (IC_{50} values ~50 μM), which correlated well to a dose-dependent suppression of the neuroendocrine markers ASCL1,

CgA, and serotonin. Resveratrol treatment also inhibited growth of carcinoid tumor cells *in vivo*. Intraperitoneal injection of resveratrol (5 mg/kg, 15 days) reduced tumor volume by 65% compared to the control in a NCI H727 xenograft.[45] In addition, resveratrol has been reported as a potential drug candidate in the treatment of medullary thyroid cancer (MTC). Resveratrol treatment in human MTC cells has effects similar to that observed in carcinoid cells characterized by growth inhibition and dose-dependent reduction of neuroendocrine markers.[46] Based on these results, a Phase I clinical trial is currently recruiting patients to study the effects of resveratrol on Notch signaling and NET growth in a human population.

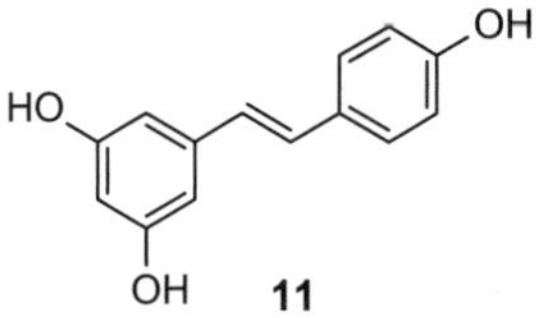
11

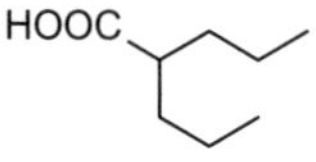
12

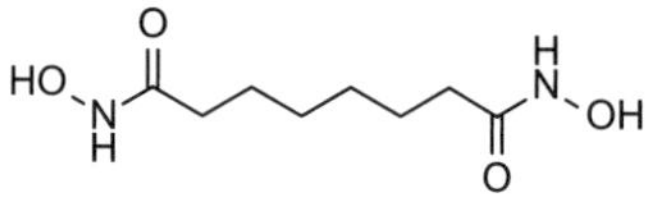
13

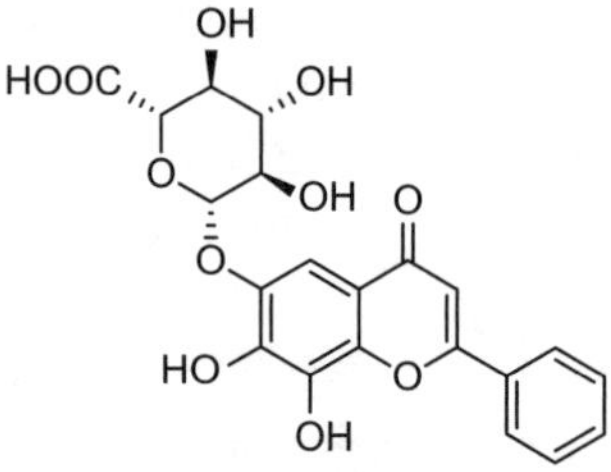
14

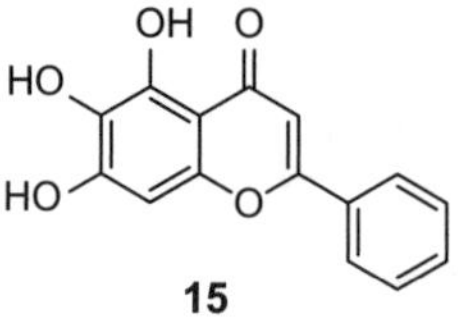
15

5.2. Histone deacetylase inhibitors

Histone deacetylase (HDAC) inhibitors are a class of small molecules that have long been pursued as anticancer agents for their ability to regulate gene expression and cellular growth through histone modification. More recently, two HDAC inhibitors, valproic acid (VPA, **12**) and suberoyl bishydroxamic acid (SBHA, **13**), have been reported to activate Notch1 signaling in NETs such as carcinoid tumor and MTC cells. VPA is a well-established HDAC inhibitor and has been studied in numerous Phase I/II clinical trials for a variety of hematologic and solid tumor malignancies. To date, these clinical studies have primarily focused on dual therapy with the DNA methyltransferase inhibitor 5-azacytidine (myelodysplastic syndrome) or as an addition to standard therapy in myeloma, melanoma, and solid tumor patients.[47] Reports from these trials suggest that the addition of VPA to standard regimens holds potential; however, more in-depth studies are warranted to determine its efficacy in relation to the FDA-approved HDAC inhibitor vorinostat. With respect to its induction of Notch signaling *in vitro*, VPA treatment in NETs such as carcinoid tumors[48] and SCLC[49] induces full-length Notch1 and NICD expression, resulting in suppression of NET markers (ASCL1and CgA) and inhibition of NET cell growth.

By contrast, SBHA is a relatively new HDAC inhibitor. Similar to VPA, SBHA inhibits cell growth and apoptosis in carcinoid tumor and MTC cells.[50] In addition, SBHA treatment (200 mg/kg, i.p.) for 12 days in an MTC tumor xenograft model resulted in 55% inhibition of tumor growth and upregulation of NICD with a concomitant decrease in ASCL-1 and CgA expression.[51] Combination therapy consisting of an HDAC inhibitor and lithium chloride, a known pharmacologic inhibitor of glycogen synthase kinase-3β (GSK-3β), has been suggested as a novel therapeutic approach for the treatment of carcinoid tumors. Combination treatment of either HDAC inhibitor (VPA, 3 mM or SBHA, 20 μM) and lithium chloride (20 mM) demonstrated significant Notch1 activation and suppressed the expression of NET markers and GSK-3β thus targeting two signaling pathways.[52]

5.3. Baicalin and baicalein

Baicalin (**14**) and its aglycone baicalein (**15**) are flavones that exhibit a variety of biological effects, including anti-inflammatory and anticancer properties. Both natural products inhibit cancer cell growth and induce apoptosis in a variety of human cancer cell lines. The same screening effort in K562 cells that identified niclosamide as a pathway inhibitor demonstrated the ability of both flavones to induce Notch pathway activity (baicalin, $EC_{50} \sim 20$ μM and baicalein, $EC_{50} \sim 2.5$ μM).[43] Both compounds upregulated Notch1, HES1, and HEY1 mRNA and significantly reduced the number and size of K562 colonies *in vitro*.

6. CONCLUSIONS

The dual role that Notch signaling plays in cellular growth provides evidence that pathway agonists and antagonists hold potential as future chemotherapeutics. The most advanced class of pathway modulators, the GSIs, has seen several compounds advance into clinical trials for the treatment of a variety of Notch-dependent cancers. Their potential clinical use could be limited by several factors, including extreme GI distress and the nonselective nature of GSIs given that the γ-secretase complex processes multiple substrates throughout the human body. To date, other natural products and small molecules identified as Notch pathway modulators (agonists and antagonists) are nonselective compounds that exert a myriad of physiological activities, potentially complicating their future clinical development and highlighting the necessity of identifying and characterizing specific small molecule modulators of Notch signaling.

REFERENCES

(1) Muller, J.-M.; Chevrier, L.; Cochaud, S.; Meunier, A.-C.; Chadeneau, C. *Drug Discov. Today Dis. Mech.* **2007**, *4*, 285.

(2) Ranganathan, P.; Weaver, K.L.; Capobianco, A.J. *Nat. Rev. Cancer* **2011**, *11*, 338.

(3) Radtke, F.; Raj, K. *Nat. Rev. Cancer* **2003**, *3*, 756.

(4) Morgan, T.H. *Am. Nat.* **1917**, *51*, 513.

(5) Ellisen, L.W.; Bird, J.; West, D.C.; Soreng, A.L.; Reynolds, T.C.; Smith, S.D.; Sklar, J. *Cell* **1991**, *66*, 649.

(6) Weng, A.P.; Ferrnado, A.A.; Lee, W.; Morris, J.P.; Silverman, L.B.; Sanchez-Irizarry, C.; Blacklow, S.C.; Look, A.T.; Aster, J.C. *Science* **2004**, *306*, 269.

(7) Chiang, M.Y.; Xu, L.; Shestova, O.; Histen, G.; L'heurexu, S.; Romany, C.; Childs, M.E.; Gimotty, P.A.; Aster, J.C.; Pear, W.S. *J. Clin. Invest.* **2008**, *118*, 3181.

(8) South, A.P.; Cho, R.J.; Aster, J.C. *Semin. Cell Dev. Biol.* **2012**, *23*, 458–464.

(9) Kunnimalaiyaan, M.; Chen, H. *Oncologist* **2007**, *12*, 535.

(10) Nakakura, E.K.; Sriuranpong, V.R.; Kunnimalaiyaan, M.; Hsiao, E.C.; Schuebel, K.E.; Borges, M.W.; Jin, N.; Collins, B.J.; Nelkin, B.D.; Chen, H.; Ball, D.W. *J. Clin. Endocrinol. Metab.* **2005**, *90*, 4350.

(11) Sriuranpong, V.; Borges, M.W.; Ravi, R.K.; Arnold, D.R.; Nelkin, B.D.; Baylin, S.B.; Ball, D.W. *Cancer Res.* **2001**, *61*, 3200.

(12) Viatour, P.; Ehmer, U.; Saddic, L.A.; Dorrell, C.; Andersen, J.B.; Lin, C.; Zmoos, A.-F.; Mazur, P.K.; Schaffer, B.E.; Ostermeier, A.; Vogel, H.; Sylvester, K.G.; Thorgeirsson, S.S.; Grompe, M.; Sage, J.J. *J. Exp. Med.* **2011**, *10*, 1963.

(13) Nicholas, M.; Wolfer, A.; Raj, K.; Kummer, J.A.; Mill, P.; van Noort, M.; Hui, C.C.; Clevers, H.; Dotto, G.P.; Radtke, F. *Nat. Genet.* **2003**, *33*, 416.

(14) Palomero, T.; Sulis, M.L.; Cortina, M.; Real, P.J.; Barnes, K.; Ciofani, M.; Caparros, E.; Buteau, J.; Brown, K.; Perkins, S.L.; Bhagat, G.; Agarwal, A.M.; Basso, G.; Castillo, M.; Nagase, S.; Cordon-Cardo, C.; Parsons, R.; Zuniga-Pflucker, J.; Dominguez, M.; Ferrando, A.A. *Nat. Med.* **2007**, *13*, 1203.

(15) Suchting, S.; Frietas, C.; le Noble, F.; Benedito, R.; Breant, C.; Duarte, A.; Eichmann, A. *Proc. Natl. Acad. Sci. U.S.A.* **2007**, *104*, 3225.

(16) Yan, M.; Callahan, C.A.; Beyer, J.C.; Allamneni, K.P.; Zhang, G.; Ridgeway, J.B.; Niessen, K.; Plowman, G.D. *Nature* **2010**, *463*, E6.

(17) Curry, C.L.; Reed, L.L.; Golde, T.E.; Miele, L.; Nickoloff, B.J.; Foreman, K.E. *Oncogene* **2005**, *24*, 6333.

(18) O'Neill, J.; Calvo, J.; McKenna, K.; Krishnamoorthy, V.; Aster, J.C.; Bassing, C.H.; Alt, F.W.; Kelliher, M.; Look, A.T. *Blood* **2006**, *107*, 781.

(19) Wang, M.; Wu, L.; Wang, L.; Xin, X. *Biochem. Biophys. Res. Commun.* **2010**, *393*, 144.

(20) Peters, J.-U.; Galley, G.; Jacobsen, H.; Czech, C.; David-Pierson, P.; Kitas, E.A.; Ozmen, L. *Bioorg. Med. Chem. Lett.* **2007**, *17*, 5918.

(21) Luistro, L.; He, W.; Smith, M.; Packman, K.; Vilenchik, M.; Carvajal, D.; Roberts, J.; Cai, J.; Berkofsky-Fessler, W.; Hilton, H.; Linn, M.; Flohr, A.; Jakob-Rotne, R.; Jacobsen, H.; Glenn, K.; Heimbrook, D.; Boylan, J.F. *Cancer Res.* **2009**, *69*, 672.

(22) Huynh, C.; Poliseno, L.; Segura, M.F.; Medicherla, R.; Haimovic, A.; Menendez, S.; Shang, S.; Pavlick, A.; Shao, Y.; Darvishian, F.; Boylan, J.F.; Osman, I.; Hernando, E. *PLoS One* **2011**, *6*, e25264.
(23) A complete list of current clinical trials including RO4929097 can be found at http://clinicaltrials.gov/ct2/results?term=RO4929097.
(24) Strosberg, J.R.; Yeatman, T.; Weber, J.; Coppola, D.; Schell, M.J.; Han, G.; Almhanna, K.; Kim, R.; Valone, T.; Jump, H.; Sullivan, D. *Eur. J. Cancer* **2012**, *48*, 997.
(25) Wei, P.; Walls, M.; Qiu, M.; Ding, R.; Denlinger, R.H.; Wong, A.; Tsaparikos, K.; Jani, J.P.; Hosea, N.; Sands, M.; Randolph, S.; Smeal, T. *Mol. Cancer Ther.* **2010**, *9*, 1618.
(26) Milano, J.; McKay, J.; Dagenais, C.; Foster-Brown, L.; Pognan, F.; Gadient, R.; Jacobs, R.T.; Zacco, A.; Greenberg, B.; Ciaccio, P.J. *Toxicol. Sci.* **2004**, *82*, 341.
(27) Deangelo, D.J.; Stone, R.M.; Silverman, L.B.; Stock, W.; Arttar, E.C.; Fearen, I.; Dallob, A.; Matthews, C.; Stone, J.; Freedman, S.J.; Aster, J. *J. Clin. Oncol.* **2006**, *24*, 6585 (abstract).
(28) Lewis, H.D.; Leveridge, M.; Strack, P.R.; Haldon, C.D.; O'Neill, J.; Kim, H.; Madi, A.; Hannam, J.C.; Look, A.T.; Kohl, N.; Draetta, G.; Harrison, T.; Kerby, J.A.; Shearman, M.S.; Beher, D. *Chem. Biol.* **2007**, *14*, 209.
(29) Kondratyev, M.; Kreso, A.; Hallett, R.M.; Girgis-Gabardo, A.; Barcelon, M.E.; Ilieva, D.; Ware, C.; Majumder, P.K.; Hassell, J.A. *Oncogene* **2012**, *31*, 93.
(30) Ramakrishnan, V.; Ansell, S.; Haug, J.; Grote, D.; Kimlinger, T.; Stenson, M.; Timm, M.; Wellik, L.; Halling, T.; Rajkumar, S.V.; Kumar, S. *Leukemia* **2012**, *26*, 340.
(31) Cook, N.; Frese, K.K.; Bapiro, T.E.; Jacobetz, M.A.; Gopinathan, A.; Miller, J.L.; Rao, S.S.; Demuth, T.; Howat, W.J.; Jodrell, D.I.; Tuveson, D.A. *J. Exp. Med.* **2012**, *209*, 437–444.
(32) Polkowski, K.; Mazurek, A.P. *Acta Pol. Pharm.* **2000**, *57*, 135.
(33) Wang, Z.; Zhang, Y.; Li, Y.; Banerjee, S.; Liao, J.; Sarkar, F.H. *Mol. Cancer Ther.* **2006**, *5*, 483.
(34) Wang, Z.; Li, Y.; Ahmad, A.; Banerjee, S.; Azmi, A.S.; Kong, D.; Wojewoda, C.; Miele, L.; Sarkar, F.H. *J. Cell. Biochem.* **2011**, *112*, 78.
(35) Wang, Z.; Li, Y.; Banerjee, S.; Kong, D.; Ahmad, A.; Nogueira, V.; Hay, N.; Sarkar, F.H. *J. Cell. Biochem.* **2010**, *109*, 726.
(36) Misra, L.; Mishra, P.; Pandey, A.; Sangwan, R.S.; Sangwan, N.S.; Tuli, R. *Phytochemistry* **2008**, *69*, 1000.
(37) Stan, S.D.; Hahm, E.; Warin, R.; Singh, S.V. *Cancer Res.* **2008**, *68*, 7661.
(38) Koduru, S.; Kumar, R.; Srinivasan, S.; Evers, M.B.; Damodaran, C. *Mol. Cancer Ther.* **2010**, *9*, 202.
(39) Ravindran, J.; Prasad, S.; Aggarwal, B.B. *AAPS J.* **2009**, *11*, 495.
(40) Wang, Z.; Zhang, Y.; Banerjee, S.; Li, Y.; Sarkar, F.H. *Cancer* **2006**, *106*, 2503.
(41) Subramaniam, D.; Ponnurangam, S.; Ramamoorthy, P.; Standing, D.; Battafarano, R.J.; Anant, S.; Sharma, P. *PLoS One* **2012**, 7, e30590.
(42) Subramaniam, D.; Nicholes, N.D.; Dhar, A.; Umar, S.; Awasthi, V.; Welch, D.R.; Jensen, R.A.; Anant, S. *Mol. Cancer Ther.* **2011**, *10*, 2146.
(43) Wang, A.-M.; Ku, H.-H.; Liang, Y.-C.; Chen, Y.-C.; Hwu, Y.-M.; Yeh, T.-S. *J. Cell. Biochem.* **2009**, *106*, 682.
(44) Fremont, L. *Life Sci.* **2000**, *66*, 663.
(45) Pinchot, S.N.; Jaskula-Sztul, R.; Ning, L.; Peters, N.R.; Cook, M.R.; Kunnimalaiyaan, M.; Chen, H. *Cancer* **2011**, *117*, 1386.
(46) Truong, M.; Cook, M.; Pinchot, S.; Kunnimalaiyaan, M.; Chen, H. *Ann. Surg. Oncol.* **2011**, *18*, 1506.

(47) Wagner, J.M.; Hackanson, B.; Lubbert, M.; Jung, M. *Clin. Epigenetics* **2010**, *1*, 117.
(48) Greenblatt, D.Y.; Vaccaro, A.M.; Jaskula-Sztul, R.; Ning, L.; Haymart, M.; Kunnimalaiyaan, M.; Chen, H. *Oncologist* **2007**, *12*, 942.
(49) Platta, C.S.; Greenblatt, D.Y.; Kunnimalaiyaan, M.; Chen, H. *J. Surg. Res.* **2008**, *148*, 31.
(50) Ning, L.; Greenblatt, D.Y.; Kunnimalaiyaan, M.; Chen, H. *Oncologist* **2008**, *13*, 98.
(51) Ning, L.; Jaskula-Sztul, R.; Kunnimalaiyaan, M.; Chen, H. *Ann. Surg. Oncol.* **2008**, *15*, 2600.
(52) Adler, J.; Hottinger, D.; Kunnimalaiyaan, M.; Chen, H. *Ann. Surg. Oncol.* **2009**, *16*, 481.

CHAPTER NINETEEN

Anaplastic Lymphoma Kinase Inhibitors for the Treatment of ALK-Positive Cancers

Kazutomo Kinoshita, Nobuhiro Oikawa, Takuo Tsukuda
Research Division, Chugai Pharmaceutical Co., Ltd., Kamakura, Japan

Contents

1. INTRODUCTION

Anaplastic lymphoma kinase (ALK) is a receptor tyrosine kinase belonging to the insulin receptor superfamily. *ALK* was originally identified as a part of the fusion oncogene nucleophosmin *(NPM)-ALK* resulting from a *t*(2;5) chromosomal translocation in anaplastic large cell lymphomas (ALCL).[1] The fusion gene *NPM-ALK* is detected in approximately 75% of all *ALK*-positive ALCL and is implicated in the pathogenesis of ALCL.[2] The function of ALK in normal human tissues is unclear but it appears to play a role in physiological development and function of the nervous system.[3] Importantly, *ALK* knockout mice indicated no overt developmental, anatomical, or locomotor deficiencies; therefore, inhibition of ALK does not lead to serious disorders.

The discovery of echinoderm microtubule-associated protein-like 4 *(EML4)-ALK* fusion gene in nonsmall cell lung cancer (NSCLC) led to ALK emerging as a novel drug target for cancer therapy.[4] This was the first

Annual Reports in Medicinal Chemistry, Volume 47
ISSN 0065-7743
http://dx.doi.org/10.1016/B978-0-12-396492-2.00019-9

report of an oncogenic fusion gene in a solid tumor. Thus, just as breakpoint cluster region-Abelson *(BCR-ABL)* is a key oncogenic factor in chronic myeloid leukemias (CML),[5] *EML4-ALK* is the key oncogenic factor in *EML4-ALK*-positive NSCLC.[6] *EML4-ALK* has been detected in approximately 5% of NSCLC patients and is mutually exclusive for known oncogenic mutated endothelial growth factor receptor *(EGFR)* and v-Ki-ras2 Kirsten rat sarcoma viral oncogene homolog *(K-RAS)*, which are identified as oncogenic drivers in NSCLC.[7] One study revealed that *EML4-ALK*-positive patients had a lower response rate to platinum-based chemotherapy than patients who have an EGFR mutation.[8] Further, 2,4-pyrimidinediamine ALK inhibitor **1** inhibits the growth of EML4-ALK tumors in transgenic mice.

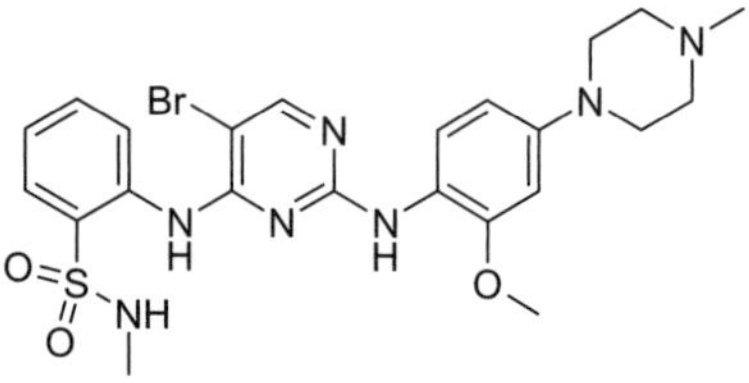

1

Other fusion genes of ALK have been identified not only in ALCL and NSCLC but also in inflammatory myofibroblastic tumors[9] and diffuse large B-cell lymphomas.[10] Furthermore, in childhood neuroblastoma, it was found that genetic mutations R1275Q and F1174L were activated, and the ALK gene was amplified.[11]

Thus, ALK is an attractive drug target for the treatment of various *ALK*-positive cancers in both blood and solid tumors, and a number of ALK inhibitors have been developed. Carrying out structural studies facilitated rational drug design efforts, and consequently seven crystal structures of human ALK in complex with small molecule inhibitors have been published thus far. Five inhibitors, namely crizotinib **2** (PDB ID 2XP2),[12] CH5424802 **3** (PDB ID 3AOX),[13] NVP-TAE684 **4** (PDB ID 2XB7),[14] PHA-E429 **5** (PDB ID 2XBA), and piperidine carboxamide **6** (PDB ID 4DCE),[15] have been designed to achieve target selectivity (Fig. 19.1). They are all ATP-competitive inhibitors and contain hinge-binding moieties that compete with the adenine base of ATP in order to interact with the ATP-binding site, commonly referred to as the hinge region. These inhibitors form one to three hydrogen bonds with the hinge region (marked by dotted lines), where the number of hydrogen bonds presumably affects their kinase selectivity.[16]

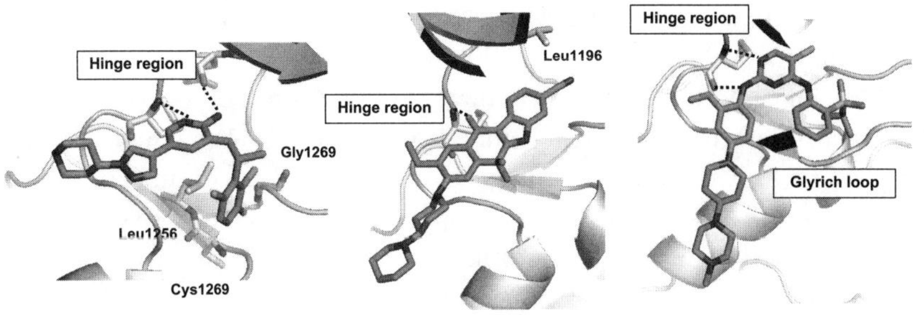

Crizotinib (2XP2) CH5424802 (3AOX) NVP-TAE684 (2XB7)

Figure 19.1 Binding modes of representative ALK inhibitors. (See Color Plate 19.1 in Color Plate Section.)

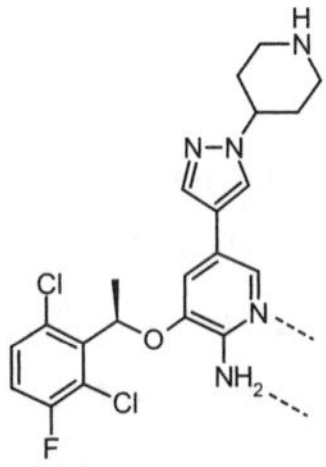

2

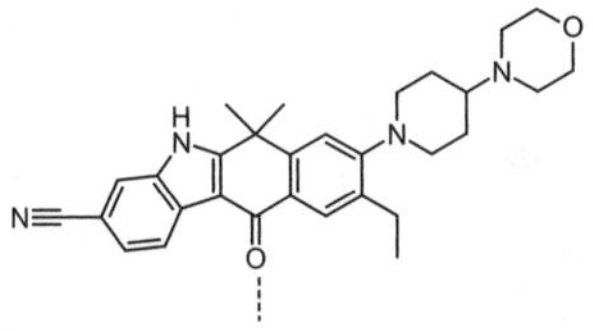

3

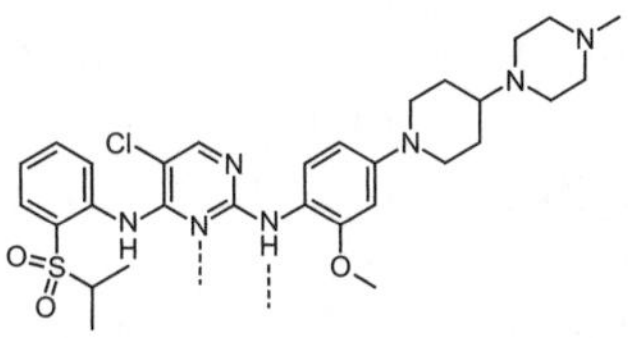

4

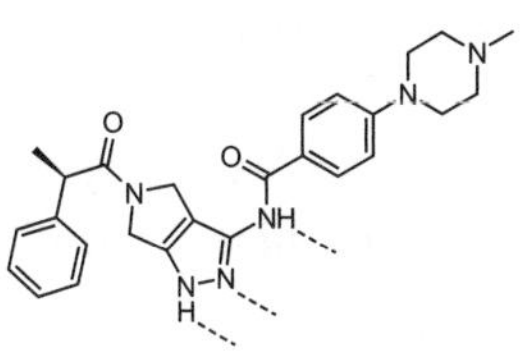

5

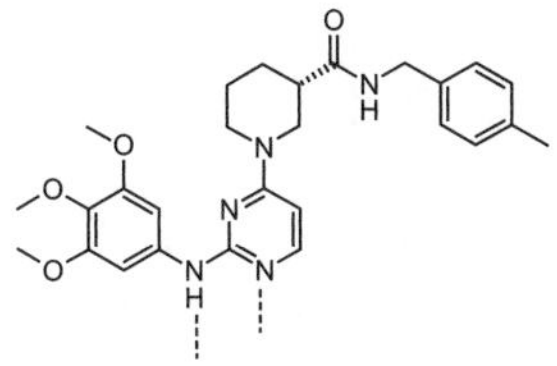

6

2. CRIZOTINIB (XALKORI®)

Crizotinib (PF-02341066) by Pfizer is the first marketed ALK inhibitor, which was originally designed in a c-Met inhibitor program. Pfizer's researchers screened kinase inhibitors, 3-substituted indoline-2-ones, to identify a potent and selective c-Met inhibitor PHA-665752 (**7**). In order to improve physicochemical properties of **7**, they re-engineered the core scaffold by structure-based drug design (SBDD) and produced the novel 5-aryl-3-benzyloxy-2-aminopyridine core. Further, optimization of the substituents led to crizotinib as a potent c-Met inhibitor, which also demonstrated potent inhibitory activity against the ALK enzyme and an *NPM-ALK*-positive cell line, with IC_{50} values of < 1.0 and 20 nM, respectively.

One of the remarkable structural characteristics of crizotinib is an L-shaped conformation in the binding state (Fig. 19.1), which is driven by a methyl group attached to the chiral benzylic position. The hinge-binding motif of crizotinib, 2-aminopyridine, is located above Leu1256 and forms two hydrogen bonds with the hinge region, with the attached 2,6-dichloro-3-fluorophenyl moiety bending down into a hydrophobic binding pocket formed by amino acid residues including Cys1255, Leu1256, and Gly1269. The hydrophobic binding pocket of ALK and c-Met (PDB ID 2WGJ) are formed by a unique activation loop, which seems to contribute to the selectivity observed with ALK inhibitors.

When the *EML4-ALK* fusion oncogene was reported, Pfizer was conducting a Phase I study of crizotinib for c-Met-targeted cancer therapy. In a new clinical study, they immediately tested the same inhibitor for the treatment of *EML4-ALK*-positive NSCLC. The Phase I clinical trial of crizotinib for *EML4-ALK*-positive NSCLC patients demonstrated an overall response rate of 57%, and a 72% probability of progression-free survival (PFS) for 6 months.[17] The U.S. Food and Drug Administration approved crizotinib for the treatment of patients with locally advanced or metastatic *ALK*-positive NSCLC in August 2011.[18] Thus, time between discovery of the novel drug target and approval was only 4 years, which is surprisingly short.

It was also noteworthy that crizotinib was concurrently approved with a fluorescence *in situ* hybridization (FISH) diagnostic test, the Vysis® ALK Break-Apart FISH Probe Kit (Abbott Molecular, Inc.) to help identify the minor population of patients with the ALK fusion gene (ca. 5%) from the total NSCLC patients. Other diagnostic modalities such as immunohistochemistry and reverse transcriptase-polymerase chain reaction are also being explored in order to establish a reliable and cost-effective *ALK*-positive NSCLC screening.[19]

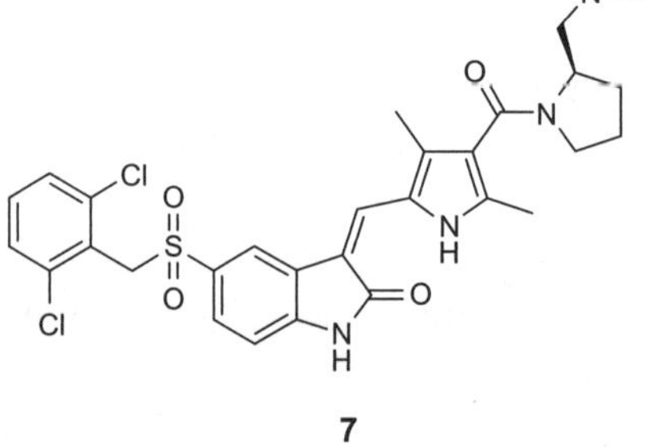

7

3. ACQUIRED CRIZOTINIB RESISTANCE

One of the main issues of tyrosine kinase inhibitors is acquired drug resistance. This means that although many patients with *ALK*-positive NSCLC obtain clinical benefit from crizotinib treatment, the effect is relatively short-lasting and most patients treated with crizotinib develop acquired resistance eventually. In a Phase I clinical trial of crizotinib, the median duration of response is 48 weeks and PFS is 10 months, although median overall survival has not been reached as of the data cut off.[20]

A secondary point mutation is a well-known mechanism of acquired drug resistance against kinase inhibitors. A so-called gatekeeper of kinases, the amino acid residue adjacent to the hinge region, and point mutations of the gatekeeper often diminish the inhibitory activity of kinase inhibitors. T315I mutation in BCR-ABL[21] and T790M in EGFR[22] are typical examples of gatekeeper mutations. The gatekeeper of ALK is Leu1196, and L1196M mutation not only hampers crizotinib from interacting with the kinase domain but also enhances catalytic efficiency.[23]

In addition to the mutation L1196M, various types of secondary mutations within the kinase domain from *ALK*-positive NSCLC tumors have been identified from patients who were treated with crizotinib. Cys1156 and Lys1152 are located far from the ATP-contact surface, and these residues are unlikely to interact with crizotinib directly. Therefore, C1156Y and L1152R mutations appear to induce resistance by modifying the kinetics of ALK's conformation change.[24] The presence of a secondary mutation, G1269A, reduces the binding affinity of crizotinib by sterically hindering the binding event.[25] In addition, G1202R, S1206Y, and 1151Tins mutations were recently identified in crizotinib-resistant patients.[26] Gly1202 and Ser1206 residues are located at the solvent front of the ATP-binding pocket, and the mutation of these amino acids may also lead to steric hindrance of the crizotinib binding event. The T1151 insertion located far from the crizotinib contact surface is speculated to change the affinity of ALK for ATP. Amplification of the ALK fusion gene, aberrant activation of other oncogenic drivers by K-RAS mutation, amplification of KIT, and increased autophosphorylation of EGFR are other mechanisms of crizotinib resistance. In a subset of patients, multiple resistance mechanisms develop simultaneously, creating a formidable problem.

4. CLINICAL CANDIDATES

Four ALK inhibitors, namely LDK378 (Phase I),[27] AP26113 (Phase I/II),[28] ASP3026 (Phase I, **8**)[29], and CH5424802 (Phase I/II, **3**), are in clinical trials for the treatment of tumors. Only the structures of CH5424802 and ASP3026 have been disclosed.

LDK378, created by Novartis, is a selective ALK inhibitor with weak inhibitory activity against c-Met. In the *EML4-ALK*-positive cell line, NCI-H2228 mouse xenograft model, oral administration of LDK378 resulted in tumor regression, and after 14 days of treatment with LDK378, tumor regrowth did not occur for the monitored period of 4 months. LDK378 also demonstrated *in vivo* efficacy against the crizotinib-resistant mouse model (developed by treating the NCI-H2228 xenograft model with crizotinib).

AP26113, created by Ariad, is approximately fivefold more potent *in vitro* than crizotinib: IC_{50} values of AP26113 and crizotinib are 0.62 and 3.6 nM, respectively. AP26113 demonstrated *in vivo* antitumor efficacy in the crizotinib-resistant EML4-ALK mutant mouse xenograft models.[30]

ASP3026 (**8**), created by Astellas, is an analog of NVP-TAE684 (**4**), which was generated by Novartis.[31] The only structural difference between these compounds is a hinge-binding motif (1,3,5-triazine vs. 5-chloropyrimidine, respectively). In the NCI-H2228 mouse xenograft model, oral administration of ASP3026 resulted in tumor regression.

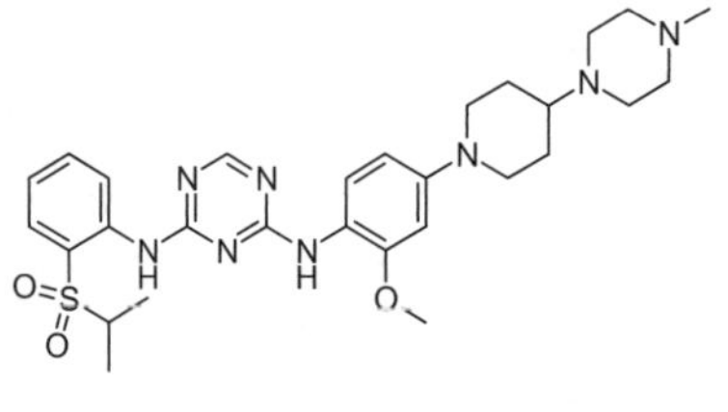

8

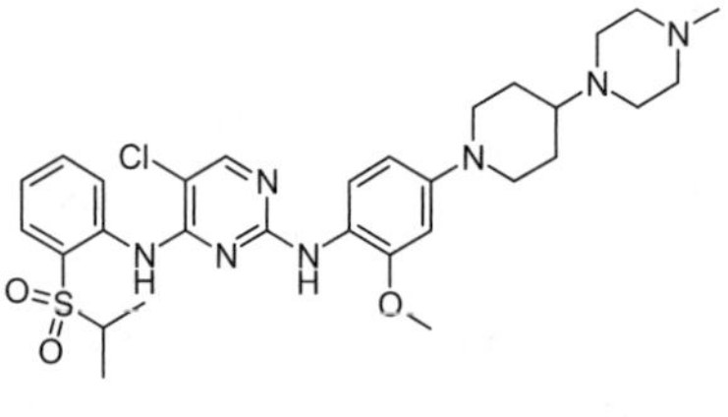

4

A highly selective ALK inhibitor created by Chugai, CH5424802 (**3**), is designed by SBDD from a high-throughput screening hit compound.[32] CH5424802 showed potent inhibitory activity against ALK (IC_{50} of 1.9 nM) and the *NPM-ALK*-positive cell line, KARPAS-299 (IC_{50} of 3.0 nM), but demonstrated weak or no inhibition against 24 other protein kinases. In an NCI-H2228 mouse xenograft model, once-daily oral administration of CH5424802 for 11 days resulted in dose-dependent tumor growth inhibition and tumor regression (168% tumor growth inhibition at 20 mg/kg/day). Further, tumor regrowth did not occur throughout the following 4-week drug-free period. In a Ba/F3 cell growth assay, CH5424802 demonstrated cell growth inhibition against L1196M and C1156Y, which are both reported crizotinib mutants. In addition, in a mouse model bearing Ba/F3 EML4-ALK harboring L1196M, oral administration of CH5424802 showed tumor growth inhibition at 60 mg/kg once daily for 8 days. CH5424802's high ALK selectivity is a result of steric repulsion between the 9-ethyl group of CH5424802 and non-ALK kinases, as ALK has a wide-open surface at the entrance of its ATP-binding pocket.[33]

5. PRECLINICAL CANDIDATES

NVP-TAE684 (**4**) was identified before *EML4-ALK* was discovered. The compound was developed to treat ALCL by targeting NPM-ALK and identified in a cellular assay that screened for compounds with selective cytotoxicity against *NPM-ALK*-transformed Ba/F3 cells. The compound showed good antiproliferative activity against Ba/F3 NPM-ALK cells (IC_{50} of 3 nM) and very weak activity against Ba/F3 cells that had constitutively activated tyrosine kinases (IC_{50} values of >500 nM).

Interestingly, comparison of NVP-TAE684 to crizotinib in the two *EML4-ALK*-positive NSCLC cell lines, NCI-H2228 and NCI-H3122, revealed that NVP-TAE684 was a more potent inhibitor than crizotinib in the NSCLC models.[34] In the NCI-H2228 xenograft model, NVP-TAE684 showed complete tumor regression within a week (at 10 mg/kg), whereas crizotinib had no effect on the tumor growth at the same dose. In the NCI-H3122 xenograft mouse model, treatment with NVP-TAE684 at 50 mg/kg resulted in tumor regression, whereas treatment with crizotinib at the same dose demonstrated marginal tumor growth inhibition.

Xcovery reported the discovery of two ALK inhibitors bearing the aminopyridazine scaffold, X-376 (**9**) and X-396 (structure not disclosed).[35] X-376 has the same hydrophobic moiety as that of crizotinib (2,6-dichloro-3-fluoro-phenylethoxy group) but a unique hinge-binding motif (6-aminopyridazine-3-carboxamide). The significant difference between crizotinib and X-376 is in the hydrophilic side chains (circled). The binding model of X-376 with ALK suggests that the compound forms two more hydrogen bonds with ALK than crizotinib, using the pyridazine nitrogen and amide nitrogen (marked by dotted lines), which led a significant increase in affinity of X-376 over crizotinib for ALK. In addition, X-376 and X-396 demonstrated more potent antiproliferative activity against the *EML4-ALK*-positive cell line, H3122, than crizotinib (IC_{50}=77, 15, and 180 nM, respectively). In the H3122 xenograft mouse model, twice-daily oral administration of either X-376 (50 mg/kg) or X-396 (25 mg/kg) resulted in tumor growth inhibition without body weight loss. The efficacy is comparable to that of crizotinib at 50 mg/kg, bid.

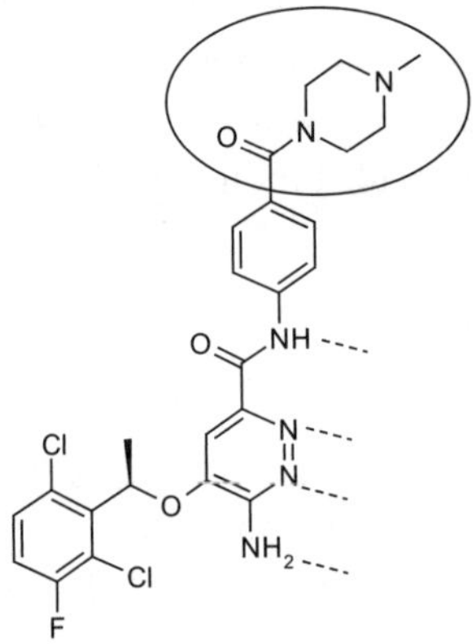

9

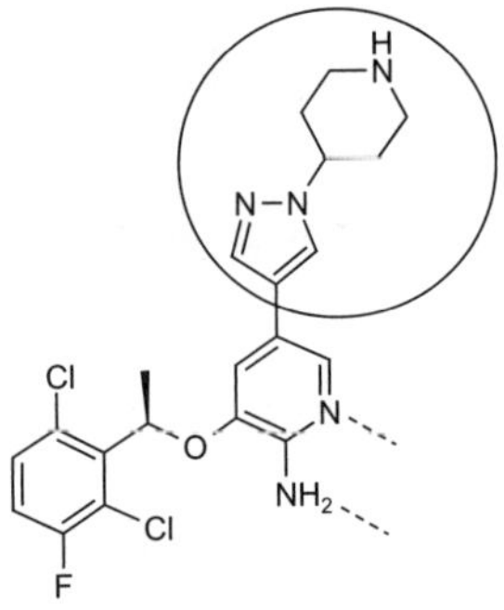

2

Cephalon recently disclosed three unique inhibitors derived from NVP-TAE684 (**4**). One of the inhibitors (**10**) is a compound bearing a bicyclo [2.2.1]hept-5-ene ring system instead of an aryl moiety.[36] The compound showed potent inhibitory activity against ALK (IC_{50}=14 nM) and antiproliferative activity against the *NPM-ALK*-positive cell line, KARPAS-299 (IC_{50}=45 nM). On the other hand, compound **10** had very weak inhibitory activity against the structurally related insulin receptor kinase (InsR) (IC_{50}=597 nM), achieving >40-fold selectivity for ALK. Furthermore, in *ALK*-positive ALCL xenografts (SUP-M2) in SCID mice, oral administration of compound **10** resulted in dose-dependent antitumor efficacy (10–55 mg/kg, bid) without overt toxicity. In this study, complete/near-complete tumor regressions were observed at 55 mg/kg, bid.

The second unique inhibitor, **11**, has a 2,7-disubstituted-pyrrolo[2,1-f][1,2,4]triazine scaffold that mimics the bioactive conformation of a diaminopyrimidine inhibitor such as NVP-TAE684 (**4**).[37] Compound **11** also had potent inhibitory activity against ALK (IC_{50} of 10 nM) and KARPAS-299 (IC_{50} of 60 nM). On the other hand, **11** showed more than 100-fold weaker inhibitor activity against InsR than that of ALK. In addition, in the same xenograft models as those used to test compound **10**, twice-daily oral administration of **11** resulted in 35%, 81%, and 98% tumor growth inhibition at 10, 30, and 55 mg/kg, respectively.

The third compound (**12**) is a macrocyclic ALK inhibitor designed to lock the diaminopyrimidine scaffold into its bioactive conformation.[38] Compound **12** demonstrated strong inhibitory activity against ALK (IC_{50} of 0.51 nM) and cellular activity with inhibition of NPM-ALK phosphorylation (IC_{50} = 10 nM). Currently, no *in vivo* data of **12** have been reported.

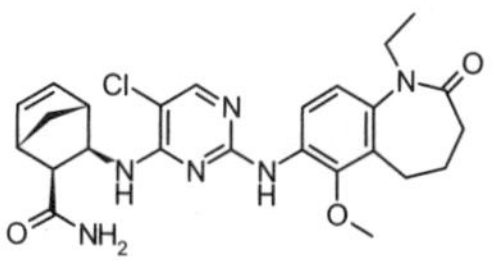

10

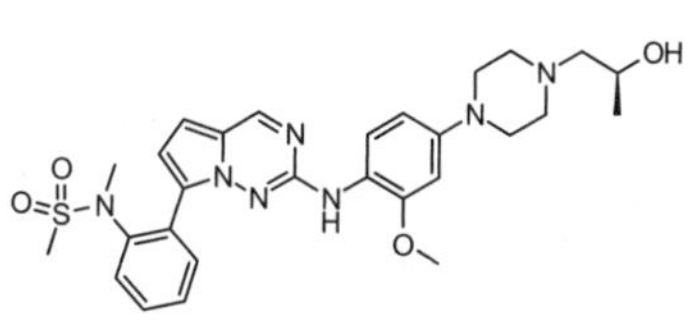

11

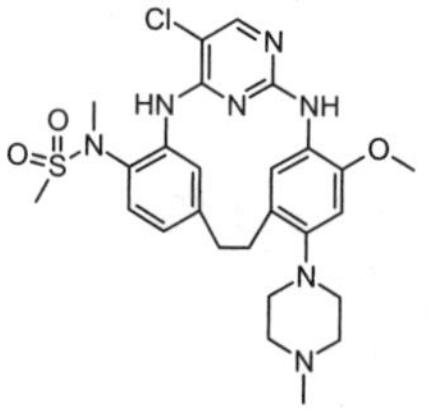

12

Recently, Cephalon disclosed the structure of CEP-28122 (**13**), which is an analog of compound **10**, but with a difference in the solvent accessible moiety (morpholino group).[39] The *in vitro* profile of CEP-28122 was improved over that for compound **10**,[40] where CEP-28122 showed potent inhibitory activity against ALK (IC_{50} = 1.9 nM) and very weak inhibition toward InsR (IC_{50} = 1257 nM). In ALK-positive ALCL xenografts (SUP-M2) in SCID mice, CEP-28122 demonstrated dose-dependent antitumor activity (3–30 mg/kg, po, bid) without overt toxicity. Furthermore, in a 4-week antitumor study using the same xenograft model, treatment with CEP-28122 (55 and 100 mg/kg, po, bid) gave complete tumor regression and no tumor regrowth for 60 days after the cessation of treatment. In other *ALK*-positive xenograft models (NCI-H2228, NCI-H3122, and NB-1), CEP-28122 also demonstrated dose-dependent antitumor efficacy (30 and 55 mg/kg, po, bid in each study).

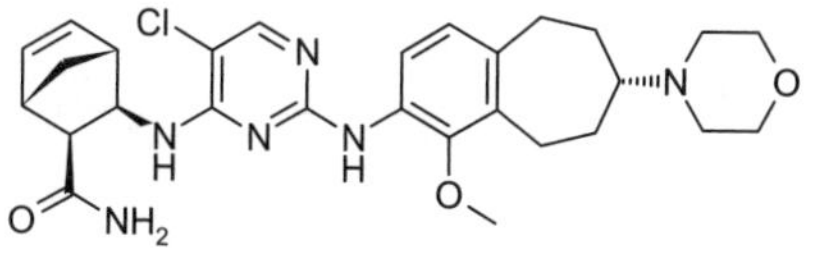

13

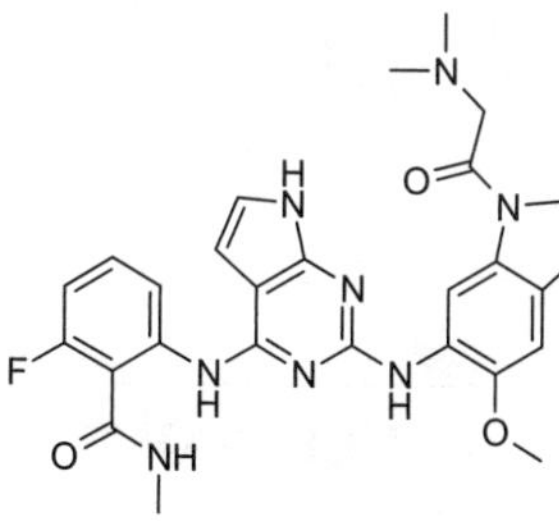

14

GSK1838705A (**14**) created by GSK has a similar scaffold to NVP-TAE684. The compound has a potent inhibitory activity against ALK, insulin-like growth factor-1 receptor (IGF-1R), and InsR (IC_{50} of 0.5, 2.0, and 1.6 nM, respectively).[41] Despite strong inhibitory activity against IGF-1R and InsR, **14** showed a transient and modest effect on blood glucose levels. Once-daily oral administration of **14** in *ALK*-positive ALCL xenograft mouse models (KARPAS-299 and SR-786) resulted in dose-dependent antitumor effects. Tumor growth inhibition against established KARPAS-299 xenograft was 22% and 93% at 10 and 30 mg/kg/day, respectively. Tumor growth inhibition against established SR-786 xenograft was 63% and 93% at 30 and 60 mg/kg/day, respectively. In addition, excellent oral bioavailability (98%) was observed in a rat pharmacokinetic study.[42]

Amgen reported piperidine carboxamides as ALK inhibitors. High-throughput screening showed that compound **6** had moderate inhibitory activity against ALK (IC_{50}=0.174 μM) with selectivity over IGF-1R (IC_{50}=4.61 μM).[43] X-ray crystallographic analysis of the co-crystal of **6** with ALK (PDB ID 4DCE) revealed that compound **6** has a unique binding interaction resulting from the DFG sequence shift, which is observed in type I 1/2 inhibitors.[44] Type I 1/2 inhibitors recognize DFG-in conformation of a target kinase in the same manner as type I inhibitors and additionally access to the extended hydrophobic pocket created by conformation shift of DFG motif. Utilizing a parallel synthesis approach, rapid SAR development designed to improve ALK selectivity over other kinases resulted in potent ALK inhibitor **15** (ALK, IGF-1R, and KARPAS-299 with IC_{50} values=31, 7390, and 28 nM, respectively).

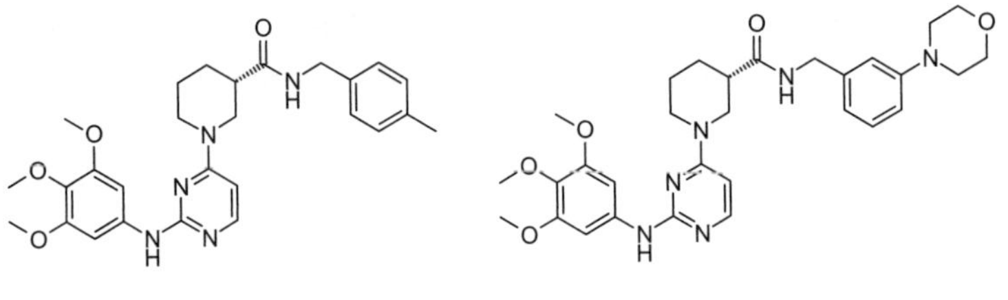

6 **15**

6. CONCLUSIONS

ALK is one of the most highly competitive drug targets in oncology research. The development of ALK inhibitors was accelerated by the discovery of an oncogenic driver *EML4-ALK*, reminiscent of *BCR-ABL* in CML.

A number of research groups have identified promising drugs, where Pfizer succeeded in launching a potent ALK inhibitor, crizotinib, for treatment of *EML4-ALK*-positive NSCLC. However, acquired drug resistance caused by mutations of ALK has been identified in patients who were treated with crizotinib; therefore, many clinical and preclinical second-generation ALK inhibitors are now required to have efficacy against ALK mutants. The patient population of ALK-related cancers is not large; however, an ALK inhibitor is expected to be critically effective for particular patients who harbor a genetic ALK mutation. Therefore, genetic diagnostics play an important role in ALK-related cancer therapy and the combination of the diagnostic test and crizotinib is one of the most successful milestones in personalized therapy.

REFERENCES

(1) Morris, S.W.; Kirstein, M.N.; Valentine, M.B.; Dittmer, K.G.; Shapiro, D.N.; Saltman, D.L.; Look, A.T. *Science* **1994**, *263*, 1281.

(2) Webb, T.R.; Slavish, J.; George, R.E.; Look, A.T.; Xue, L.; Jiang, Q.; Cui, X.; Rentrop, W.B.; Morris, S.W. *Expert Rev. Anticancer Ther.* **2009**, *9*, 331.

(3) Pulford, K.; Lamant, L.; Morris, S.W.; Butler, L.H.; Wood, K.M.; Stroud, D.; Delsol, G.; Mason, D.Y. *Blood* **1997**, *89*, 1394.

(4) Soda, M.; Choi, Y.L.; Enomoto, M.; Takada, S.; Yamashita, Y.; Ishikawa, S.; Fujiwara, S.; Watanabe, H.; Kurashina, K.; Hatanaka, H.; Bando, M.; Ohno, S.; Ishikawa, Y.; Aburatani, H.; Niki, T.; Sohara, Y.; Sugiyama, Y.; Mano, H. *Nature* **2007**, *448*, 561.

(5) Capdeville, R.; Buchdunger, E.; Zimmermann, J.; Matter, A. *Nat. Rev. Drug Discovery* **2002**, *1*, 493.

(6) Soda, M.; Takada, S.; Takeuchi, K.; Choi, Y.L.; Enomoto, M.; Ueno, T.; Haruta, H.; Hamada, T.; Yamashita, Y.; Ishikawa, Y.; Sugiyama, Y.; Mano, H. *Proc. Natl. Acad. Sci. U.S.A.* **2008**, *105*, 19893.

(7) Takahashi, T.; Sonobe, M.; Kobayashi, M.; Yoshizawa, A.; Menju, T.; Nakayama, E.; Mino, N.; Iwakiri, S.; Sato, K.; Miyahara, R.; Okubo, K.; Manabe, T.; Date, H. *Ann. Surg. Oncol.* **2010**, *17*, 889.

(8) Shaw, A.T.; Yeap, B.Y.; Mino-Kenudson, M.; Digumarthy, S.R.; Costa, D.B.; Heist, R.S.; Solomon, B.; Stubbs, H.; Admane, S.; McDermott, U.; Settleman, J.; Kobayashi, S.; Mark, E.J.; Rodig, S.J.; Chirieac, L.R.; Kwak, E.L.; Lynch, T.J.; Iafrate, A.J. *J. Clin. Oncol.* **2009**, *27*, 4247.

(9) Griffin, C.A.; Hawkins, A.L.; Dvorak, C.; Henkle, C.; Ellingham, T.; Perlman, E.J. *Cancer Res.* **1999**, *59*, 2776.

(10) Gascoyne, R.D.; Lamant, L.; Martin-Subero, J.I.; Lestou, V.S.; Harris, N.L.; Muller-Hermelink, H.K.; Seymour, J.F.; Campbell, L.J.; Horsman, D.E.; Auvigne, I.; Espinos, E.; Siebert, R.; Delsol, G. *Blood* **2003**, *102*, 2568.

(11) Mossé, Y.P.; Laudenslager, M.; Longo, L.; Cole, K.A.; Wood, A.; Attiyeh, E.F.; Laquaglia, M.J.; Sennett, R.; Lynch, J.E.; Perri, P.; Laureys, G.; Speleman, F.; Kim, C.; Hou, C.; Hakonarson, H.; Torkamani, A.; Schork, N.J.; Brodeur, G.M.; Tonini, G.P.; Rappaport, E.; Devoto, M.; Maris, J.M. *Nature* **2008**, *455*, 930.

(12) Cui, J.J.; Tran-Dube, M.; Shen, H.; Nambu, M.; Kung, P.P.; Pairish, M.; Jia, L.; Meng, J.; Funk, L.; Botrous, I.; McTigue, M.; Grodsky, N.; Ryan, K.; Padrique, E.;

Alton, G.; Timofeevski, S.; Yamazaki, S.; Li, Q.; Zou, H.; Christensen, J.; Mroczkowski, B.; Bender, S.; Kania, R.S.; Edwards, M.P. *J. Med. Chem.* **2011**, *54*, 6342.
(13) Sakamoto, H.; Tsukaguchi, T.; Hiroshima, S.; Kodama, T.; Kobayashi, T.; Fukami, T.A.; Oikawa, N.; Tsukuda, T.; Ishii, N.; Aoki, Y. *Cancer Cell* **2011**, *19*, 679.
(14) Bossi, R.T.; Saccardo, M.B.; Ardini, E.; Menichincheri, M.; Rusconi, L.; Magnaghi, P.; Orsini, P.; Avanzi, N.; Borgia, A.L.; Nesi, M.; Bandiera, T.; Fogliatto, G.; Bertrand, J.A. *Biochemistry* **2010**, *49*, 6813.
(15) Bryan, M.C.; Whittington, D.A.; Doherty, E.M.; Falsey, J.R.; Cheng, A.C.; Emkey, R.; Brake, R.L.; Lewis, R.T. *J. Med. Chem.* **2012**, *55*, 1698.
(16) Morphy, R. *J. Med. Chem.* **2010**, *53*, 1413.
(17) Kwak, E.L.; Bang, Y.J.; Camidge, D.R.; Shaw, A.T.; Solomon, B.; Maki, R.G.; Ou, S.H.; Dezube, B.J.; Jänne, P.A.; Costa, D.B.; Varella-Garcia, M.; Kim, W.H.; Lynch, T.J.; Fidias, P.; Stubbs, H.; Engelman, J.A.; Sequist, L.V.; Tan, W.; Gandhi, L.; Mino-Kenudson, M.; Wei, G.C.; Shreeve, S.M.; Ratain, M.J.; Settleman, J.; Christensen, J.G.; Haber, D.A.; Wilner, K.; Salgia, R.; Shapiro, G.I.; Clark, J.W.; Iafrate, A.J. *N. Engl. J. Med.* **2010**, *363*, 1693.
(18) Shaw, A.T.; Yasothan, U.; Kirkpatrick, P. *Nat. Rev. Drug Discovery* **2011**, *10*, 897.
(19) Shaw, A.T.; Solomon, B.; Kenudson, M.M. *J. Natl. Compr. Canc. Netw.* **2011**, *9*, 1335.
(20) Camidge, D.R.; Bang, Y.; Kwak, E.L.; Shaw, A.T.; Iafrate, A.J.; Maki, R.G.; Solomon, B.J.; Ou, S.I.; Salgia, R.; Wilner, K.D.; Costa, D.B.; Shapiro, G.; LoRusso, P.; Stephenson, P.; Tang, Y.; Ruffner, K.; Clark, J.W. *J. Clin. Oncol.* **2011**, *29*(Suppl.) Abstract 2501.
(21) Yamamoto, M.; Kurosu, T.; Kakihana, K.; Mizuchi, D.; Miura, O. *Biochem. Biophys. Res. Commun.* **2004**, *319*, 1272.
(22) Bell, D.W.; Gore, I.; Okimoto, R.A.; Godin-Heymann, N.; Sordella, R.; Mulloy, R.; Sharma, S.V.; Brannigan, B.W.; Mohapatra, G.; Settleman, J.; Haber, D.A. *Nat. Genet.* **2005**, *37*, 1315.
(23) Choi, Y.L.; Soda, M.; Yamashita, Y.; Ueno, T.; Takashima, J.; Nakajima, T.; Yatabe, Y.; Takeuchi, K.; Hamada, T.; Haruta, H.; Ishikawa, Y.; Kimura, H.; Mitsudomi, T.; Tanio, Y.; Mano, H. *N. Engl. J. Med.* **2010**, *363*, 1734.
(24) Sasaki, T.; Okuda, K.; Zheng, W.; Butrynski, J.; Capelletti, M.; Wang, L.; Gray, N.S.; Wilner, K.; Christensen, J.G.; Demetri, G.; Shapiro, G.I.; Rodig, S.J.; Eck, M.J.; Jänne, P.A. *Cancer Res.* **2011**, *71*, 6051.
(25) Doebele, R.C.; Pilling, A.B.; Aisner, D.L.; Kutateladze, T.G.; Le, A.T.; Weickhardt, A.J.; Kondo, K.L.; Linderman, D.J.; Heasley, L.E.; Franklin, W.A.; Varella-Garcia, M.; Camidge, D.R. *Clin. Cancer Res.* **2012**, *18*, 1472.
(26) Katayama, R.; Shaw, A.T.; Khan, T.M.; Mino-Kenudson, M.; Solomon, B.J.; Halmos, B.; Jessop, N.A.; Wain, J.C.; Yeo, A.T.; Benes, C.; Drew, L.; Saeh, J.C.; Crosby, K.; Sequist, L.V.; Iafrate, A.J.; Engelman, J.A. *Sci. Transl. Med.* **2012**, *4*, 120ra17.
(27) Li, N.; Michellys, P.-Y.; Kim, S.; Pferdekamper, A.C.; Li, J.; Kasibhatla, S.; Tompkins, C.S.; Steffy, A.; Li, A.; Sun, F.; Sun, X.; Hua, S.; Tiedt, R.; Sarkisova, Y.; Marsilje, T.H.; McNamara, P.; Harris, J. In B232, 23rd AACR-NCI-EORTC International Conference on Molecular Targets and Cancer Therapeutics, San Francisco, CA, Nov 2011, 2011.
(28) Katayama, R.; Khan, T.M.; Benes, C.; Lifshits, E.; Ebi, H.; Rivera, V.M.; Shakespeare, W.C.; Iafrate, A.J.; Engelman, J.A.; Shaw, A.T. *Proc. Natl. Acad. Sci. U.S.A.* **2011**, *108*, 7535.
(29) Kuromitsu, S.; Mori, M.; Shimada, I.; Kondoh, Y.; Shindoh, N.; Soga, T.; Furutani, T.; Konagai, S.; Sakagami, H.; Nakata, M.; Ueno, Y.; Saito, R.; Sasamata, M.; Mano, H.; Kudou, M. In #2821, 102nd American Association for Cancer Research Annual Meeting, Orlando, FL, Apr 2011, **2011**.

(30) Zhang, S.; Wang, F.; Keats, J.; Ning, Y.; Wardwell, S.D.; Moran, L.; Mohemmad, Q.K.; Ye, E.; Anjum, R.; Wang, Y.; Zhu, X.; Miret, J.J.; Dalgarno, D.; Narasimhan, N.I.; Clackson, T.; Shakespeare, W.C.; Rivera, V.M. In LB-298, 101st American Association for Cancer Research Annual Meeting, Washington, DC, Apr 2010, **2010**.
(31) Galkin, A.V.; Melnick, J.S.; Kim, S.; Hood, T.L.; Li, N.; Li, L.; Xia, G.; Steensma, R.; Chopiuk, G.; Jiang, J.; Wan, Y.; Ding, P.; Liu, Y.; Sun, F.; Schultz, P.G.; Gray, N.S.; Warmuth, M. *Proc. Natl. Acad. Sci. U.S.A.* **2007**, *104*, 270.
(32) Kinoshita, K.; Asoh, K.; Furuichi, N.; Ito, T.; Kawada, H.; Hara, S.; Ohwada, J.; Miyagi, T.; Kobayashi, T.; Takanashi, K.; Tsukaguchi, T.; Sakamoto, H.; Tsukuda, T.; Oikawa, N. *Bioorg. Med. Chem.* **2012**, *20*, 1271.
(33) Kinoshita, K.; Kobayashi, T.; Asoh, K.; Furuichi, N.; Ito, T.; Kawada, H.; Hara, S.; Ohwada, J.; Hattori, K.; Miyagi, T.; Hong, W.S.; Park, M.J.; Takanashi, K.; Tsukaguchi, T.; Sakamoto, H.; Tsukuda, T.; Oikawa, N. *J. Med. Chem.* **2011**, *54*, 6286.
(34) Li, Y.; Ye, X.; Liu, J.; Zha, J.; Pei, L. *Neoplasia* **2011**, *13*, 1.
(35) Lovly, C.M.; Heuckmann, J.M.; de Stanchina, E.; Chen, H.; Thomas, R.K.; Liang, C.; Pao, W. *Cancer Res.* **2011**, *71*, 4920.
(36) Ott, G.R.; Tripathy, R.; Cheng, M.; McHugh, R.; Anzalone, A.V.; Underiner, T.L.; Curry, M.A.; Quail, M.R.; Lu, L.; Wan, W.; Angeles, T.S.; Albom, M.S.; Aimone, L.D.; Ator, M.A.; Ruggeri, B.A.; Dorsey, B.D. *ACS Med. Chem. Lett.* **2010**, *1*, 493.
(37) Ott, G.R.; Wells, G.J.; Thieu, T.-V.; Quail, M.R.; Lisko, J.G.; Mesaros, E.F.; Gingrich, D.E.; Ghose, A.K.; Wan, W.-H.; Lu, L.-H.; Cheng, M.-G.; Albom, M.S.; Angeles, T.S.; Huang, Z.-Q.; Aimone, L.D.; Ator, M.A.; Ruggeri, B.A.; Dorsey, B.D. *J. Med. Chem.* **2011**, *54*, 6328.
(38) Breslin, H.J.; Lane, B.M.; Ott, G.R.; Ghose, A.K.; Angeles, T.S.; Albom, M.S.; Cheng, M.; Wan, W.; Haltiwanger, R.C.; Wells-Knecht, K.J.; Dorsey, B.D. *J. Med. Chem.* **2012**, *55*, 449.
(39) Allwein, S.P.; Roemmele, R.C.; Haley, J.J.; Mowrey, D.R.; Petrillo, D.E.; Reif, J.J.; Gingrich, D.E.; Bakale, R.P. *Org. Process Res. Dev.* **2012**, *16*, 148.
(40) Cheng, M.; Quail, M.R.; Gingrich, D.E.; Ott, G.R.; Lu, L.; Wan, W.; Albom, M.S.; Angeles, T.S.; Aimone, L.D.; Cristofani, F.; Machiorlatti, R.; Abele, C.; Ator, M.A.; Dorsey, B.D.; Inghirami, G.; Ruggeri, B.A. *Mol. Cancer Ther.* **2012**, *11*, 670.
(41) Sabbatini, P.; Korenchuk, S.; Rowand, J.L.; Groy, A.; Liu, Q.; Leperi, D.; Atkins, C.; Dumble, M.; Yang, J.; Anderson, K.; Kruger, R.G.; Gontarek, R.R.; Maksimchuk, K.R.; Suravajjala, S.; Lapierre, R.R.; Shotwell, J.B.; Wilson, J.W.; Chamberlain, S.D.; Rabindran, S.K.; Kumar, R. *Mol. Cancer Ther.* **2009**, *8*, 2811.
(42) Chamberlain, S.D.; Redman, A.M.; Wilson, J.W.; Deanda, F.; Shotwell, J.B.; Gerding, R.; Lei, H.; Yang, B.; Stevens, K.L.; Hassell, A.M.; Shewchuk, L.M.; Leesnitzer, M.A.; Smith, J.L.; Sabbatini, P.; Atkins, C.; Groy, A.; Rowand, J.L.; Kumar, R.; Mook, R.A., Jr.; Moorthy, G.; Patnaik, S. *Bioorg. Med. Chem. Lett.* **2009**, *19*, 360.
(43) Deng, X.; Wang, J.; Zhang, J.; Sim, T.; Kim, N.D.; Sasaki, T.; Luther, W., II; George, R.E.; Jänne, P.A.; Gray, N.S. *ACS Med. Chem. Lett.* **2011**, *2*, 379.
(44) Zuccotto, F.; Ardini, E.; Casale, E.; Angiolini, M. *J. Med. Chem.* **2010**, *53*, 2681.

PART 5

Infectious Diseases

Editor: John Primeau
Westford, Massachusetts

CHAPTER TWENTY

Recent Advances in the Discovery of Dengue Virus Inhibitors

Jeremy Green, Upul Bandarage, Kate Luisi, Rene Rijnbrand
Vertex Pharmaceuticals, Cambridge, Massachusetts, USA

Contents

ABBREVIATIONS

ADE antibody-dependent enhancement
BHK21 baby hamster kidney-21 cells
BVDV bovine viral diarrhea virus
DENV dengue virus
DHF dengue hemorrhagic fever
DSS dengue shock syndrome
ER endoplasmic reticulum
EC_{50} half maximal effective concentration
HCV hepatitis C virus
IC_{50} half maximal inhibitory concentration
JEV Japanese encephalitis virus

Annual Reports in Medicinal Chemistry, Volume 47
ISSN 0065-7743
http://dx.doi.org/10.1016/B978-0-12-396492-2.00020-5

K_i inhibitory dissociation constant
NOAEL no observable adverse event level
NS2B/NS3pro dengue NS3 serine protease with NS2B cofactor
RdRp RNA-dependent RNA polymerase
Vero African green monkey kidney epithelial cells
VSV vesicular stomatitis virus
WEEV Western equine encephalitis virus
WNV West Nile virus
YFV yellow fever virus

1. INTRODUCTION

Dengue virus (DENV) is a mosquito-borne infection that causes significant morbidity and mortality throughout the tropical and subtropical areas of the world.[1] Severe dengue disease (dengue hemorrhagic fever (DHF) or dengue shock syndrome (DSS)) was first recognized in the 1950s following epidemics in the Philippines and Thailand.[1] Prior to 1970, only nine countries had experienced severe dengue outbreaks, but in recent decades, the incidence of dengue has dramatically increased, particularly in urban and suburban areas. Today, the disease is endemic in over 100 countries, with severe dengue disease affecting Southeast Asia, Africa, the Caribbean, the South Pacific, and Central and South America.[2] Currently, over 2.5 billion people worldwide are thought to be at risk of contracting this disease and it is estimated that between 50 and 100 million dengue infections occur every year. According to the World Health Organization (WHO), dengue is now a leading cause of hospitalization and death among children in endemic regions, such as Asia and South America, and is now considered a major international health concern.[3]

After malaria, dengue is the most significant mosquito-borne human pathogen and is transmitted by the *Aedes* family of mosquito. The main vector is the urban-adapted *Aedes aegypti* mosquito, but *Aedes albopictus* and *Aedes polynesiensis* have also been implicated in some outbreaks. The global resurgence of dengue has been attributed to an increase in the *A. aegypti* population and its adaptation to urban areas.[1] Mosquito population control has been employed as a mechanism to prevent dengue outbreaks. These efforts have included the eradication of habitat,[4] the use of bacterial pathogens,[5] and the release of sterilized mosquitoes into the natural population.[6] Although these efforts have been shown to reduce the mosquito population, they have not permanently eradicated the mosquitoes.[1,4] In addition, the termination of mosquito control efforts leads to a rebounding of the mosquito population along with the recurrence of dengue disease.[7] This was observed in Central and South America where the reemergence of dengue during the 1970s was ascribed to the cessation of mosquito control measures.[7]

There are four similar, yet antigenically distinct, serotypes of DENV: DENV1, DENV2, DENV3, and DENV4. Initial infection with any of the serotypes is typically self-limiting with relatively mild symptoms. Resolution of infection occurs within 4–7 days and is associated with a robust innate and adaptive immune response. Recovery provides lifelong immunity against that particular serotype, but cross-immunity with other serotypes is only partial and temporary. A subsequent infection with a different serotype is more likely to result in severe disease,[1] a result of a phenomenon known as antibody-dependent enhancement (ADE). This occurs when cross-reactive non-neutralizing antibodies generated during the primary infection recognize a heterologous DENV during a second infection resulting in an increase in virus uptake into Fc-receptor-bearing cells. This infection of T-cells increases viral replication and triggers the release of proinflammatory cytokines that are thought to be involved in the development of plasma leakage and DHF/DSS.

DENV is a member of the *Flaviviridae* family of viruses, which also includes West Nile virus (WNV), yellow fever virus (YFV), and Japanese encephalitis virus (JEV). Flaviviruses are small-enveloped viruses that contain a single molecule of positive-strand RNA. The RNA genome is approximately 11 kb in length and encodes for three structural proteins (capsid [C], premembrane [PrM], and envelope [E] proteins) and seven nonstructural proteins (NS1, NS2A, NS2B, NS3, NS4A, NS4B, and NS5) (Fig. 20.1).

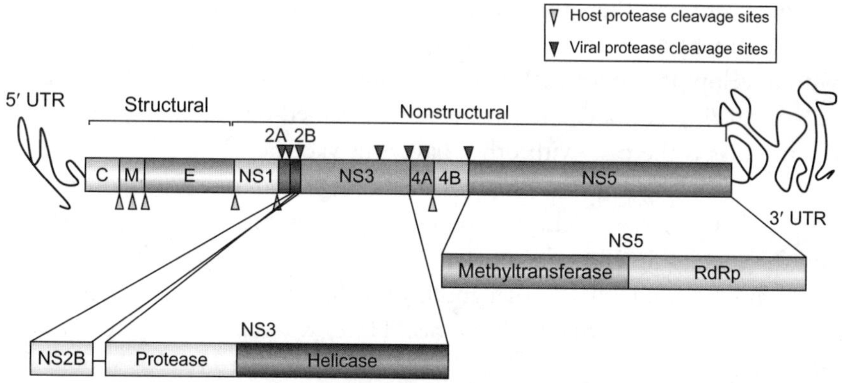

Figure 20.1 Schematic of the DENV genome. (See Color Plate 20.1 in Color Plate Section.)

The structural proteins are responsible for the formation and maturation of new viral particles. The nonstructural proteins are required for the replication of the genome, virion assembly[8], and evasion of the innate immune system.[9] Of these proteins, there are only two that are known to have enzymatic activities, NS3 and NS5, making these proteins obvious targets for the development of DENV targeting antivirals. The N-terminal domain of NS3 carries the catalytic domain of the viral serine protease, which, together with the cofactor NS2B, cleaves the viral polyprotein.[10] The C-terminal domain of NS3 contains the helicase,[11] nucleotide triphosphatase (NTPase)[12], and RNA triphosphatase (RTPase)[13] activity of the virus. NS5 contains methyltransferase activity[14] in the N-terminal domain and the RNA-dependent RNA polymerase (RdRp)[15] function in the C-terminal domain. Although these are the only two viral proteins with enzymatic activity, the other nonstructural proteins are also required for proper formation of the replication complex and are also potential antiviral targets.[16–18]

There are currently no therapies or vaccines available against DENV. Multiple efforts are in progress to develop a dengue vaccine, but these vaccines face significant challenges due to targeting multiple serotypes and the cost of producing the vaccine.[19] Currently, the most advanced efforts to develop a dengue vaccine consist of either the combination of live-attenuated strains of each of the four serotypes or the use of chimeric viruses (based on existing licensed vaccines) designed to generate an immune response to either the prM or E protein. Studies have shown that these approaches can generate neutralizing antibody titers to all four serotypes and that the vaccine is relatively safe with low levels of viremia. Interference of the four virus components of the vaccines does occur and it typically requires three vaccinations over a period of several months to achieve neutralizing antibody titers to each of the four serotypes. This raises concerns over whether these vaccines may prove too costly for use in endemic regions with developing economies. It also remains to be seen if the observed levels of neutralizing antibody titers can be maintained or whether boosters are required, as is the case with other flavivirus vaccines.[20]

Drug discovery efforts directed toward DENV have focused on both viral and host targets. The increasing understanding of flavivirus biology and the increasing clinical successes in HCV drug discovery have inspired research efforts applying virtual and high-throughput screening approaches to DENV drug discovery programs. Here we report the most recent advances in anti-DENV medicinal chemistry. Interested readers should also refer to other recent reviews in this rapidly evolving area.[21,22]

2. VIRAL STRUCTURAL PROTEIN TARGETS

2.1. E protein

The DENV glycoprotein E forms the outer shell of the flavivirus particle and is responsible for virus–host receptor interactions. Following attachment, the particles are internalized through endocytosis. In the low pH environment of the endosome, the E protein undergoes structural rearrangements that drive the fusion of viral and endosomal membranes prior to release of the viral genome into the cytoplasm.[23,24] Preventing attachment to the host receptor or proper deployment and activation of the E protein fusion domain will prevent infection of permissive cells.

The DENV E protein consists of three structural domains (I, II, and III). The *n*-octyl-β-D-glucoside (β-OG) pocket is a channel buried at the hinge between domains I and II that undergoes a structural change during the fusion activation process. It is highly conserved among flaviviruses and has received attention from several groups.[25]

Virtual screening of 135,000 commercial molecules directed toward the β-OG pocket led to the selection of five molecules for evaluation in Vero cells infected with DENV2.[26] Compound **1** inhibited DENV ($IC_{50}=1.2$ μM), as well as WNV ($IC_{50}=3.8$ μM), and YFV ($IC_{50}=1.6$ μM). Activity in an E protein and pH-specific c6/36 insect cell fusion assay provides evidence that the mechanism of **1** is through inhibition of E protein-mediated cell fusion.

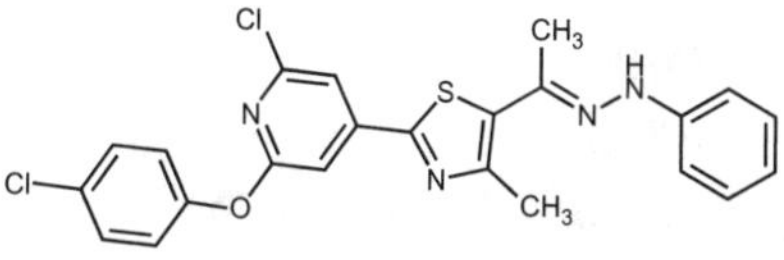

1

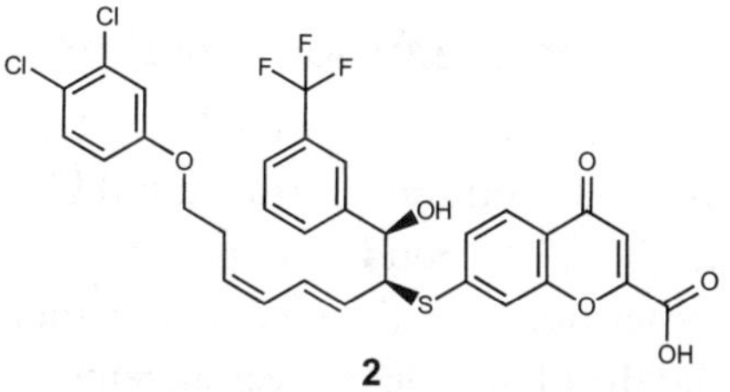

2

A similar virtual screening approach based on a corporate collection, including natural products, led to the identification of **2**. Compound **2** demonstrated activity in a quantitative DENV2 fusion assay ($IC_{50}=6.8$ μM) and the antiviral activity was confirmed in DENV2-infected BHK21 fibroblasts ($IC_{50}=9.8$ μM).[27]

Using a docking study of the β-OG pocket with a 586,829 compound library, two molecules were identified that inhibited DENV2 infection of BHK21 cells.[28] Further SAR development resulted in **3** ($EC_{50}=0.07$ μM), which is active against all four serotypes of DENV

($EC_{50}s = 0.068–0.496\ \mu M$), as well as YFV, JEV, and WNV ($EC_{50}s = 0.47–1.42\ \mu M$). Biological assays confirmed that **3** bound to viral particles and acted on an early step in the viral life cycle. Additionally, the virus appeared arrested in endosomes, where the E protein normally undergoes a pH-induced conformational change that drives the membrane fusion process.

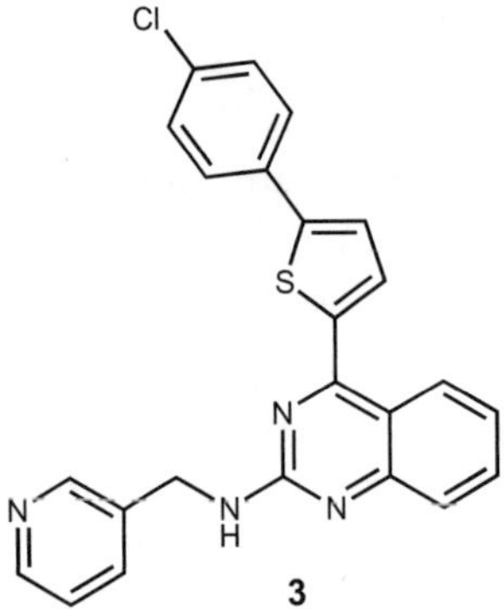

3

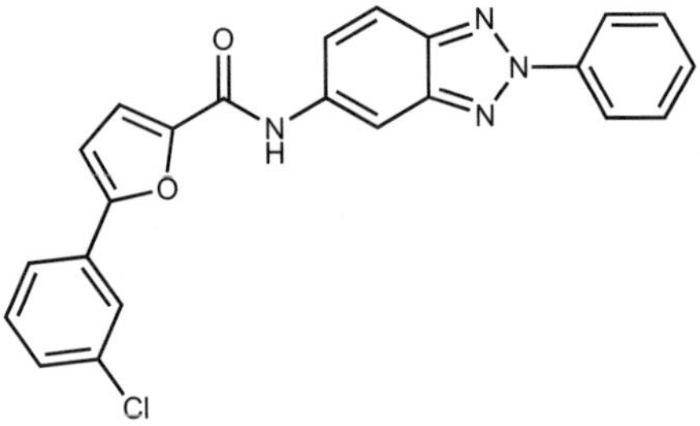

4

An alternate approach, comparing the pre- and postfusion forms of the E protein, identified two sites suitable for small-molecule inhibitor binding.[29] An *in silico* screen was performed using several publicly and commercially available chemical databases. Seven molecules from this screen were tested in DENV2-infected Vero cells and **4** was identified ($IC_{50} = 4\ \mu M$).

2.2. Peptide fusion inhibitors

A 29-amino acid peptide derived from the DENV2 stem (aa 419–447) targets the postfusion E trimer and blocks pore formation.[30] Surprisingly, the $DV2^{419–447}$ peptide seems to be introduced into the endosome by initially associating with the viral membranes and only associates with the stem during the pH-mediated rearrangement in the endosome. This is driven in part by a relatively hydrophobic C-terminal domain.

Using structural information, the E domain II hinge region was targeted with energy-minimized and structurally stabilized peptides.[31] The two most active peptides identified in this screen contained 20 and 28 amino acids and inhibited DENV2 with $IC_{50}s$ of 7 and 8 μM, respectively. The interaction of the peptides with virions led to surface changes and loss of icosahedral symmetry prior to internalization of the particles. It should be noted, however, that these changes can occur after virus attachment to the cells.

The ability of pre-existing antibodies to enhance DENV uptake into cells was used to develop an assay to test peptide inhibitors for their ability to prevent ADE in K562 cells.[32] Two previously characterized peptides of 33 amino acids[33] and 20 amino acids[31] in length were found to prevent ADE with IC_{50}s of 3 and 6 μM, respectively.

A particular obstacle for the development of DENV entry inhibitors is the possibility that the virus might have multiple paths to enter cells.[24,34,35] This is supported by the observation that carbohydrate-binding agents were able to inhibit infection of some cell types but not others.[34,36] A further complication is that these molecules will have no beneficial effect on already infected cells.

3. VIRAL NONSTRUCTURAL PROTEIN TARGETS

3.1. NS2B/NS3 protease

The full-length DENV NS3 protein contains both protease and helicase domains. The N-terminal 180-amino acid residue domain encodes for a serine protease (NS3pro), which requires the NS2B cofactor to be stably folded and to exhibit proteolytic activity.[37] The NS2B/NS3pro complex mediates cleavage of the viral polyprotein (see Fig. 20.1) and possesses a classic serine protease catalytic triad of His51, Asp75, and Ser135. Since NS2B/NS3pro plays a critical role in viral polyprotein processing and replication, it is a promising target for therapeutic intervention for DENV infection.[38]

3.1.1 Competitive inhibitors

By analogy to the successful approaches to inhibit the HCV NS3 protease,[39,40] binding of inhibitors to the nonprime subsites of DENV protease provides an opportunity to develop effective small-molecule inhibitors.[21] However, it should be noted that the substrate specificity of NS2B/NS3pro is markedly different to that of HCV NS3/4A protease, while maintaining the drug development challenge of a shallow, solvent-exposed active-site surface. DENV NS2B/NS3pro possesses trypsin-like substrate preference for basic residues (Lys, Arg) at P1, as well as at P2.[41,42]

A series of substrate-based, tetrapeptide inhibitors containing various functional groups that can covalently bond with the catalytic Ser has been synthesized and tested against DENV2 NS2B/NS3pro.[43–45] Tetrapeptide boronic acid **5** and trifluoromethylketone **6** have shown good inhibitory potencies toward NS2B/NS3pro with K_is = 0.043 and 0.85 μM, respectively.[43] The substrate-based tetrapeptide aldehyde **7** has also been reported to have inhibitory potency for NS2B/NS3pro with $K_i = 5.8\ \mu M$,[43] while a truncated but more basic aldehyde **8** was slightly more potent ($K_i = 1.5\ \mu M$).[44]

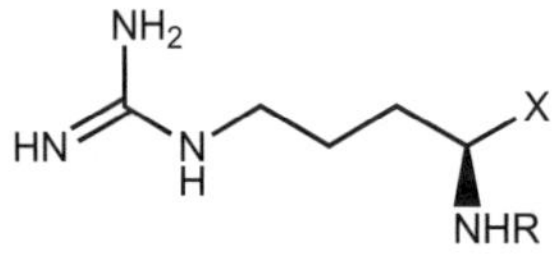

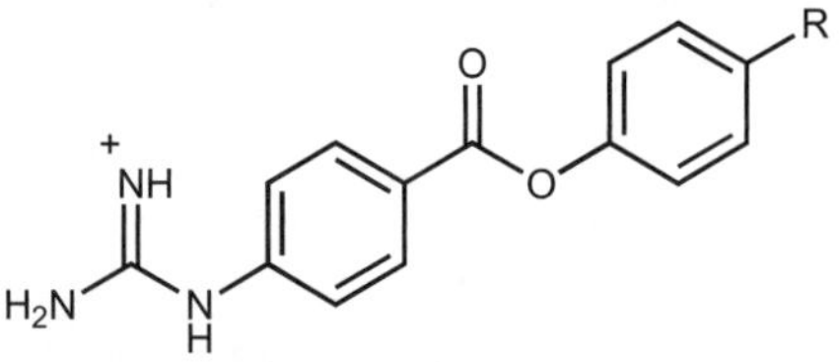

R = Bz-Nle-Lys-Arg   X = $B(OH)_2$ **5**
R = Bz-Nle-Lys-Arg   X = $COCF_3$ **6**
R = Bz-Nle-Arg-Arg   X = CHO   **7**
R = Bz-Lys-Arg   X = CHO   **8**

R = 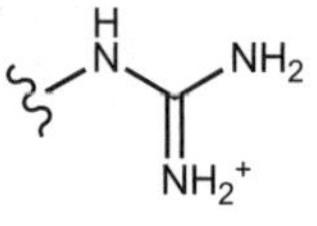

9

R = CN **10**

An *in silico*, structure-guided, fragment-based design approach to identify DENV protease inhibitors from commercially available molecules led to the identification of compounds **9** and **10**, which inhibited DENV2 NS2B/NS3pro with $IC_{50}s = 7.7$ and $37.9\ \mu M$, respectively.[46]

Arylcyanoacrylamides, better known as early receptor tyrosine kinase inhibitors,[47] have been recently reported as DENV and WNV NS2B/NS3pro inhibitors.[48] The electron density of the aryl group and the central double bond are crucial for the activity while substitution of the amide group did not improve potency. Compound **11** was found to be the most potent inhibitor in this series with $K_i = 35.7\ \mu M$. Possessing low molecular weight and high ligand efficacy,[49] compound **11** was identified as a candidate for further SAR exploration.[48] Combining the arylcyanoacrylamides with retropeptide-based inhibitors has led to **12**, but structural information on binding mode is lacking.[50]

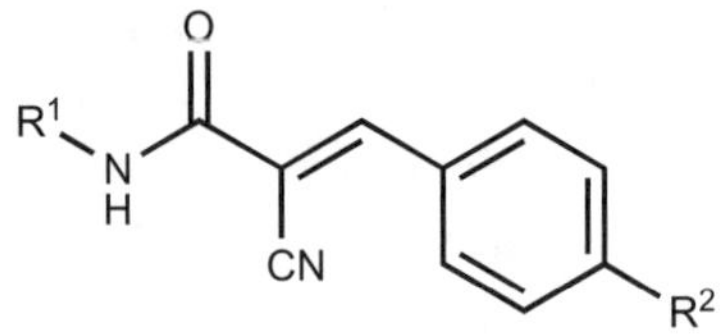

11 $R^1 = H$; $R^2 = OH$
12 $R^1 = cPr$; $R^2 = COArg\text{-}Lys\text{-}Nle\text{-}NH_2$

A series of triazole-containing benz[d]isothiazol-3(2H)-one derivatives has been identified as inhibitors of DENV2 NS2B/NS3pro.[51] Several compounds in this series displayed noteworthy inhibitory activities including compound **13** ($K_i = 4.77$ μM). Molecular docking suggests that **13** spans the substrate-binding subsites of the enzyme, yet is not in direct contact with the catalytic triad.

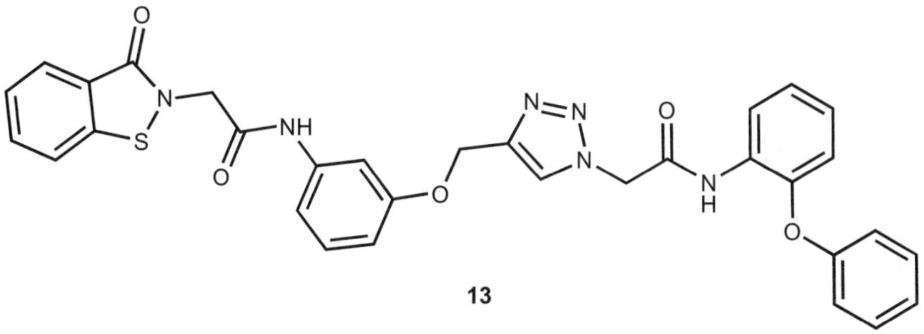

13

3.1.2 Noncompetitive inhibitors

To identify allosteric inhibitors of the protease activity of the related flavivirus WNV that bind the NS2B/NS3pro cofactor-binding site, a virtual screen of the NCI database of 275,000 compounds was performed. Two hits, **14** and **15**, inhibited DENV2 NS2B/NS3pro ($IC_{50} = 2.75$ and 2.04 μM, respectively).[52]

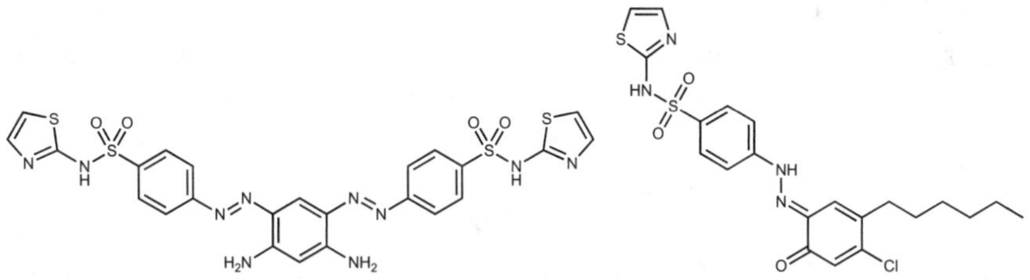

14 **15**

A high-throughput screen of DENV2 NS2b/NS3pro identified the dihydroxyfluoren-9-one derivative, **16** ($IC_{50} = 15.4$ μM), which showed comparable activity against all four DENV serotypes in infected BHK21 cells. Resistance mutations identified in the NS2B region conferred a 10-fold reduction in compound potency in enzyme inhibition and a 74-fold potency reduction in a replicon assay, suggestive of compound **16** binding to NS2B and interfering with the NS2B–NS3 interactions.[53]

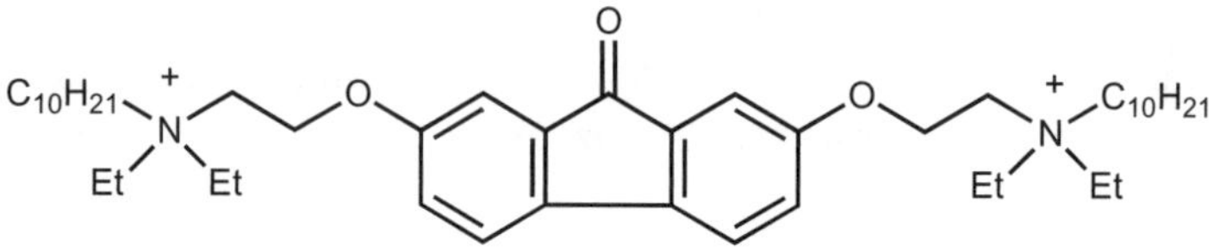

16

3.2. NS3 helicase

The C-terminal 440 residues of NS3 contains the RNA helicase, NTPase, and RTPase activities.[13] Few reports of inhibition of the C-terminal domain of NS3 (NTPase/helicase) of DENV have appeared, though available structural information[10,54–56] suggests the presence of a pocket for possible drug interaction.[56] Synthetic nucleoside derivatives tested against DENV NTPase/helicase showed only limited activity.[57]

3.3. NS4B

Though poorly understood, the NS4B protein is a small, hydrophobic, transmembrane protein that serves to anchor the viral replication machinery to the endoplasmic reticulum. Though lacking in intrinsic enzymatic activity, NS4B may represent a potential therapeutic target in the manner that the nonenzymic HCV NS5A protein has been exploited for therapeutic intervention.[58,59] Compound **17** was identified from a high-throughput screen of 1.8 million compounds against DENV2 replicon in A549 cells ($EC_{50}=1.0\ \mu M$).[60] Compound **17** was shown to have antiviral activity against all four DENV serotypes ($EC_{50}=1.5$, 1.6, and 4.1 μM for DENV1, 3, and 4, respectively), but was not active against WNV or YFV. Mechanism of action studies indicated that **17** acts by suppression of viral RNA synthesis, while resistance selection studies identified mutations in the NS4B protein, notably NS4B P104L and A119T. These residues are conserved among all four DENV serotypes, but not in other flaviviruses. Furthermore, P104 and A119 are believed to be in a region of NS4B that locates in the ER membrane.[18]

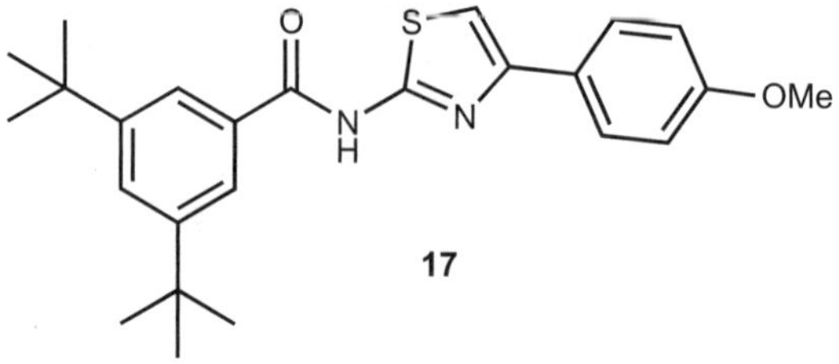

17

3.4. NS5 polymerase

The C-terminus of the NS5 protein contains the RdRp (NS5pol), which, together with both viral and host proteins, synthesizes both positive- and negative-strand RNA. Again, available structural information and clinical success in targeting the HCV NS5 RdRp provide impetus for the discovery of both active-site and allosteric inhibitors.[61–64]

3.4.1 Nucleoside inhibitors of NS5pol

Inhibition of viral polymerases by nucleosides is an established, successful approach to treatment of viral infection in a variety of diseases, with nucleosides representing the largest single class of antiviral drugs. Nucleoside drugs exert their therapeutic efficacy following processing to the corresponding nucleotide triphosphate which then competes with endogenous nucleotides for incorporation into viral DNA or RNA.

2′-Methyl-7-deazaadenosine (MK0608) (**18**), a known inhibitor of the HCV NS5B polymerase, has also been shown to inhibit DENV in cell culture and shows efficacy in animal models of infection.[65] This result led to an expanded evaluation of 2′-modified nucleosides for anti-DENV activity.[66,67] Studies on 2′-modified adenosines demonstrated that only methyl (**19**) and ethynyl (**22**) substituents showed activity in DENV2-infected A549 cells (IC_{50}s = 1.1 and 1.4 μM, respectively). Ethyl (**20**), ethenyl (**21**), and propargyl (**23**) substituents were all effectively inactive (>50 μM), as were 2′-ethynyl cytidine and guanines. To preclude deamination of 2′-C-substituted adenines, leading to inactive inosine derivatives, the C-7 purine nitrogen was replaced with carbon (compounds **24–27**). Compound **24** has emerged as a promising nucleoside lead, with submicromolar activity in DENV2-infected A549 cells (EC_{50} = 0.7 μM). Mechanism of action studies indicate that **24** inhibits DENV by blocking RNA synthesis. Further, both the 5′-*O*-mono- and triphosphates exhibit strong DENV inhibition in both cell culture and enzyme inhibition assays,[67] indicating **24** acts through blocking RNA chain elongation. The triphosphate of **24** demonstrates an NS5 polymerase K_i = 0.060 μM.[68] In addition, although unmodified **24** is observed in rat plasma, the mono-, di-, and triphosphate metabolites have been identified in blood cells. Pharmacokinetic studies indicated that **24** is orally bioavailable and has parameters consistent with twice-daily dosing.

In DENV2-infected AG129 mice, a single dose of **24** was administered (25, 75, 150, and 300 mg/kg, p.o.) 12 h postinfection with DENV2.[67] At day 3, peak viremia was reduced in all dose groups, though the highest doses were not significantly superior to 75 mg/kg. In addition, a single 75-mg/kg dose of **24** provided complete protection to mice infected with the lethal DENV2 D2S10 strain.[67] However, following 14-day toxicology studies in rats and dogs, a NOAEL could not be established for **24** in either species. Even at 10 mg/kg/day in rats and 1 mg/kg/day in dogs, notable toxicities were observed, including irreversible corneal damage.[66]

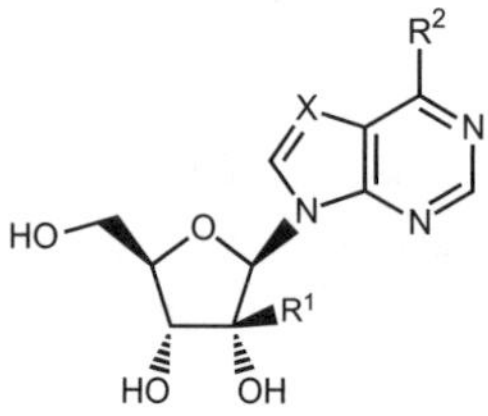

Compound	X	R^1	R^2
18	C-H	CH_3	NH_2
19	N	CH_3	NH_2
20	N	CH_2CH_3	NH_2
21	N	$CH{=}CH_2$	NH_2
22	N	$C{\equiv}CH$	NH_2
23	N	$C{\equiv}CCH_3$	NH_2
24	C-H	$C{\equiv}CH$	NH_2
25	C-F	$C{\equiv}CH$	NH_2
26	C-CN	$C{\equiv}CH$	NH_2
27	$C\text{-}CONH_2$	$C{\equiv}CH$	NH_2
28	N	H	CH_3
29	C-H	H	CH_3

Compound **27**, though slightly less potent in DENV2-infected A549 cells (EC_{50}: 2.6 μM), was also studied *in vivo*.[69] However, pharmacokinetic studies indicated very low oral bioavailability (*F* 1%) in mice and rats (*F* 2%). By analogy with the HCV polymerase inhibitor **30** (RG7128),[70] the di-isobutyryl derivative of **27** was evaluated.[69] Compound **31** exhibits enhanced cellular efficacy ($EC_{50}=0.69$ μM), as well as improved oral bioavailability of **27** (*F* 32% and 10–13% in mice and rats, respectively). The prodrug **31** was not detected in plasma. Dosing 25 mg/kg of **31** in the mouse viremia model resulted in a 30-fold reduction in peak viremia; however, toxicological studies demonstrated weight loss and death in Wistar rats at doses of 30 and 75 mg/kg/day.[69]

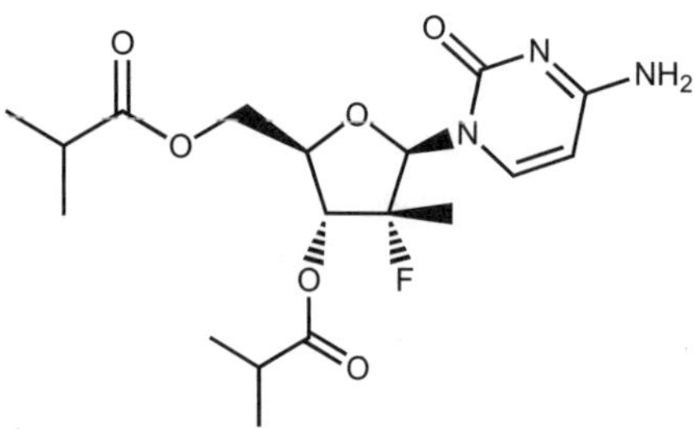

30

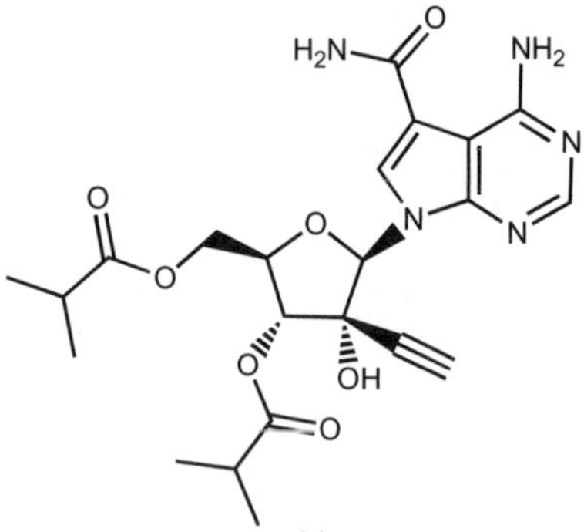

31

The 6-methyl adenosine analogs, compounds **28** and **29**, have been reported as inhibitors of the DENV2 replicon system in BHK21 cells ($EC_{50}=5.5$ and 0.9 μM, respectively) and **29** showed even greater potency in DENV2-infected Vero cells ($EC_{50}=0.039$–0.062 μM), though cytotoxicity was noted in HeLa cells.[71] Thus, while anti-DENV efficacy and promising pharmacokinetics have been attained, further investigation of this class of antiviral is warranted in order to identify a viable drug candidate.

3.4.2 Non-nucleoside inhibitors of NS5pol

Non-nucleoside, allosteric inhibitors of HCV NS5B polymerase have garnered significant attention, with numerous molecules in clinical evaluation[63,64] and, by analogy, provide inspiration for analogous approaches to inhibit DENV NS5pol.[21]

A high-throughput screen of more than 1 million compounds against the full-length NS5 protein reportedly resulted in a hit rate of 0.7%, from which compound **32** (IC_{50} = 7.2 μM) was identified as the basis for further optimization.[72,73] This optimization effort led to **33** (IC_{50} = 0.7 μM), which showed no activity toward human DNA polymerases, nor HCV NS5B, or WNV NS5 polymerases. Neither compound showed significant activity in infected cells. Further SAR around this series suggested that the carboxylate at the 2-position is essential for activity and that minor improvements might be attained by inclusion of an electron-withdrawing group at the aryl 5-position. A slight improvement in potency was obtained by replacing the *N*-benzyl group with *N*-2-naphthylmethyl, compound 34 (IC_{50} = 0.26 μM).[72] To better understand the binding location of this series, a photoaffinity experiment was performed with compound **35** (IC_{50} = 1.5 μM), which irreversibly inhibited NS5pol following UV irradiation. On the basis of the labeled residue (methionine 320/343) and structural analysis of the available X-ray structure and docking experiments, a binding mode was proposed at a site between the "finger" and "thumb" regions of the polymerase,[62] which would cause the occlusion of the RNA template tunnel. Compound **35** was notably more potent when added to the enzyme prior to the RNA template and weaker when added after. Cellular efficacy of these compounds was lacking.[73]

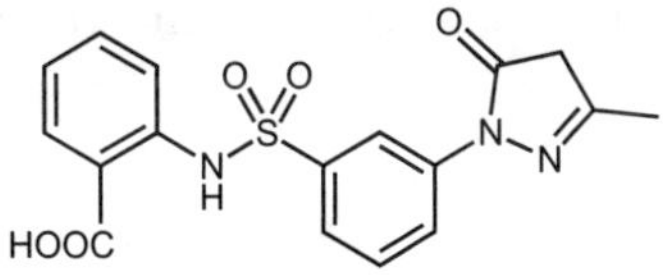

32

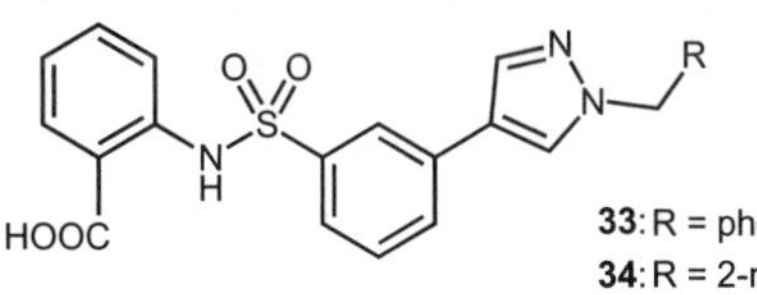

33: R = phenyl
34: R = 2-naphthyl

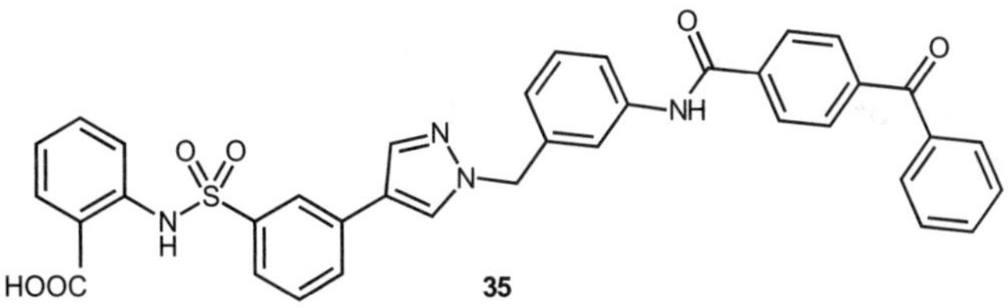

35

3.5. Methyltransferase inhibitors

The N-terminus of the NS5 protein contains the methyl transferase domain (NS5mt) which serves to sequentially methylate both guanine N7 and adenosine ribose 2′-OH of the viral RNA cap. This is accomplished by methyl transfer from *S*-adenosyl methionine (SAM, **40**), affording *S*-adenosyl homocysteine (SAH, **41**) as the by-product.[74] X-ray crystal structures of DENV NS5mt have been reported[14,75–79] and have served as templates for virtual screening approaches to the discovery of new inhibitors.[80] Docking experiments at both the SAM and RNA sites led to the identification of confirmed inhibitors, two acting at the SAM site (**36** and **37** $IC_{50}s = 9.5$ and 4.4 μM, respectively) and two acting at the RNA site (**38** and **39** $IC_{50}s = 7.1$ and 4.9 μM, respectively).

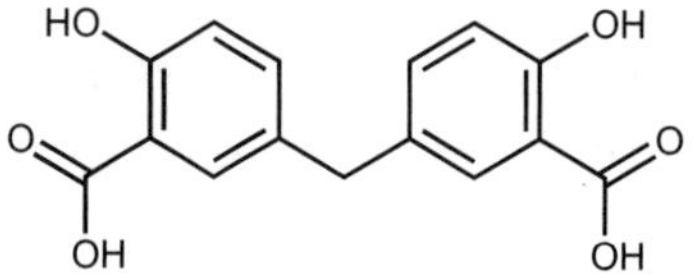

36

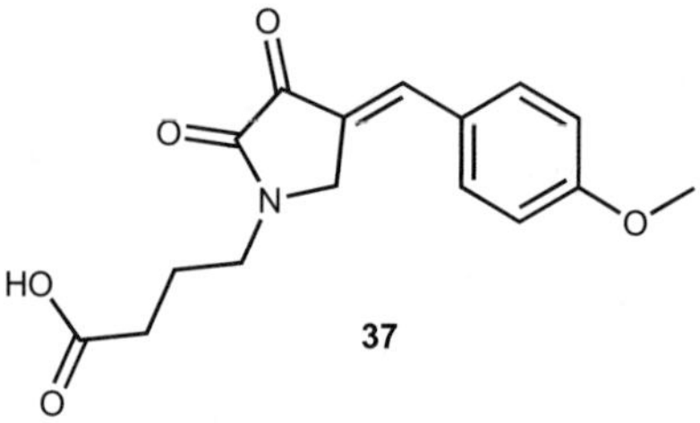

37

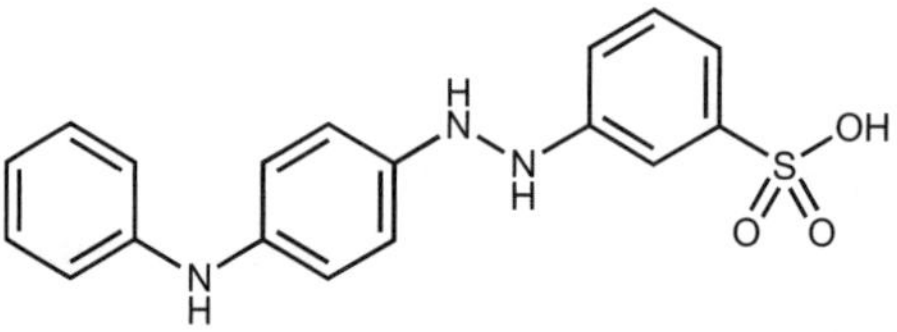

38

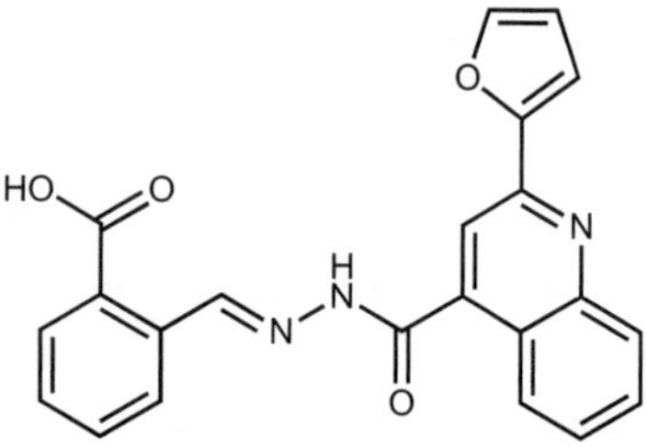

39

The RNA site has been observed to be shallow and solvent-exposed and may not represent a suitable site for drug interaction. In contrast, the SAM site is well defined, though the substrate is ubiquitous in methyltransferases and presents a potential selectivity challenge. Recently, a detailed analysis of flavivirus NS5 structures identified a conserved cavity adjacent to the adenine-binding pocket.[81] Synthesis of a series of SAH analogs identified a number of inhibitors of DENV3 NS5mt-mediated N7 and 2′-O-methylation, which also had activity against WNV NS5mt. Furthermore, N6-benzyl derivatives of **41** (compounds **42–44**) showed no inhibition toward human RNA guanine-7-methyltransferase nor human DNA methyltransferase 1A, suggesting opportunities for selective inhibitor design.

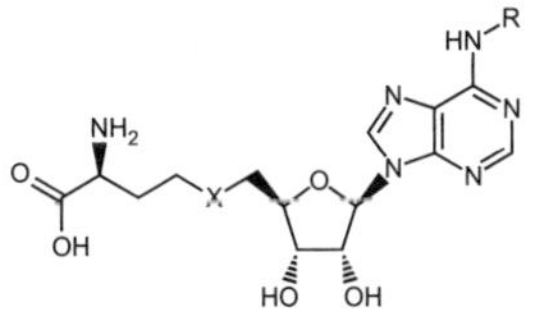

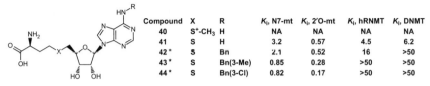

Compound	X	R	K_i, N7-mt	K_i, 2′O-mt	K_i, hRNMT	K_i, DNMT
40	S^+-CH_3	H	NA	NA	NA	NA
41	S	H	3.2	0.57	4.5	6.2
42 *	S	Bn	2.1	0.52	16	>50
43 *	S	Bn(3-Me)	0.85	0.28	>50	>50
44 *	S	Bn(3-Cl)	0.82	0.17	>50	>50

* epimeric at amino acid

4. HOST TARGETS

4.1. Maturation inhibitors

The viral E protein is a glycoprotein that is modified by the cellular sugar modifying machinery in the ER and Golgi complex. The sugar groups attached to the protein play a critical role during several steps in the viral life cycle.[82,83] Inhibition of the cellular enzymes responsible for the modifications on the E glycoprotein might be expected to reduce the production of new infectious virions and has a broad spectrum potential. Iminosugars have been used as competitive inhibitors for the ER α-glucosidases I and II which trim N-linked glycans on glycoproteins.[84] Until recently, the potential of iminosugars was limited by their low level of potency in combination with relatively high toxicity. However, deoxynojirimycin derivative **45** has improved activity ($IC_{50}=2$ μM) and reduced cytotoxicity.[85] Derivatives **46** and **47** inhibited DENV2 with IC_{50}s of 0.075–0.1 μM. These molecules are also active against WNV and BVDV, albeit with lower potency.[86]

The iminosugar derivative **48** (DENV2 $IC_{50}=1.1$ μM) was tested in the AG129 mouse DENV2 infection model.[87] When dosed p.o. at 100 mg/kg, the plasma concentration of **48**, 11 h postdosing, was approximately 5 μM ($5\times IC_{50}$). Under these conditions, viral loads dropped approximately two-fold. **48** was synergistic with ribavirin; however, α-glucosidase I activity was not detected suggesting that the activity might be mostly due to α-glucosidase II activity.[87] **48** also showed a survival benefit in a lethal DENV mouse challenge model.[88]

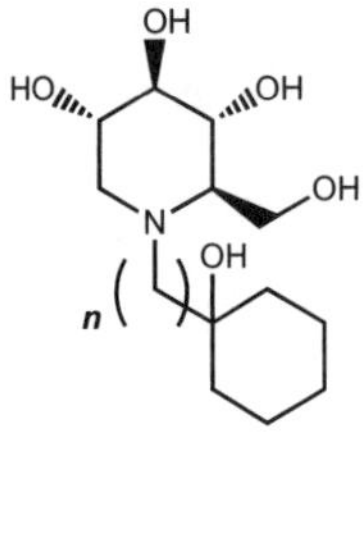

45 $n = 5$

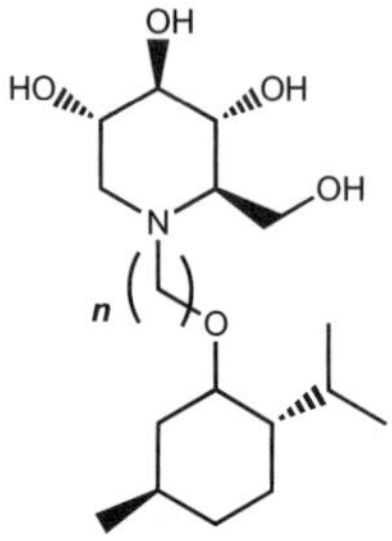

46 $n = 6$

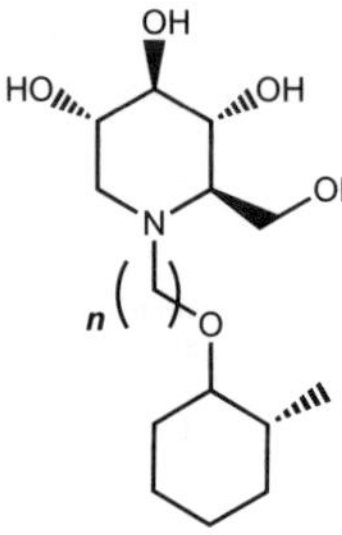

47 $n = 6$

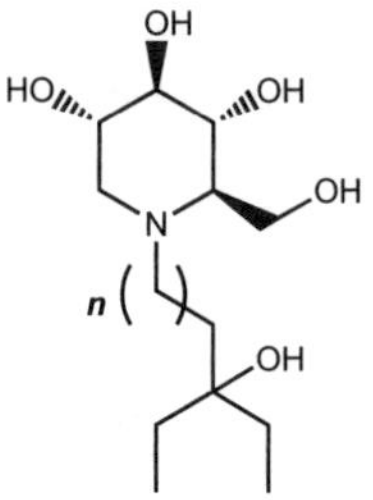

48 $n = 5$

Celgosivir (6-*O*-butanoylcastanospermine, **49**), an inhibitor of α-glucosidase I and II, is known to affect folding of N-glycosylated proteins. DENV proteins prM, E, and NS1 are all believed to be N-glycosylated and **49** showed strong inhibition of DENV2-infected BHK21 fibroblasts (IC_{50} = 0.20 μM), with similar antiviral activity across all serotypes (IC_{50} = 0.65, 0.68, and 0.31 μM for DENV1, 3, and 4, respectively).[89] Further studies demonstrated both an effect on NS1 post-translational modification as well as effective inhibition of DENV2 replication in a subgenomic replicon system (EC_{50} = 2.2 μM). Compound **49** was evaluated in both a primary model of infection and in the lethal ADE model, in each case dosed at 50 mg/kg bid. In the latter model, mice are administered a DENV E protein cross-reactive antibody and challenged after 24 h with virus, resulting in 100% mortality after 5 days in the absence of treatment. Compound **49** proved effective in both models, providing complete protection when dosed from the time of infection. Furthermore, 50% survival in the lethal ADE model was obtained even when dosing of **49** was delayed 2 days postinfection.

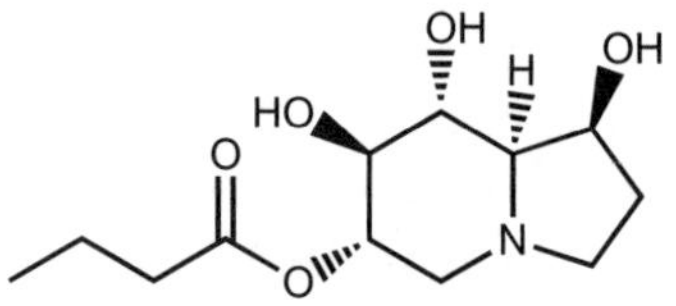

49

4.2. Translation inhibitors

Compound **50** is a molecule that was identified following a phenotypic screen with a 1.8 million compound library on DENV2-infected Huh-7 cells (EC_{50} = 0.55 μM).[90] The molecule inhibited the related YFV and WNV with EC_{90}s of 4.9 and 4.5 μM, respectively, while the EC_{90}s against the nonrelated WEEV and VSV were ~20 and >20 μM, respectively.

SAR efforts to improve the chemical characteristics of **50** resulted in **51**. Compound **51**, however, was found to nonselectively inhibit both viral and cellular protein translation. In short-term tissue culture studies, toxicity was limited; however, mice dosed at 75 mg/kg had severe side effects.

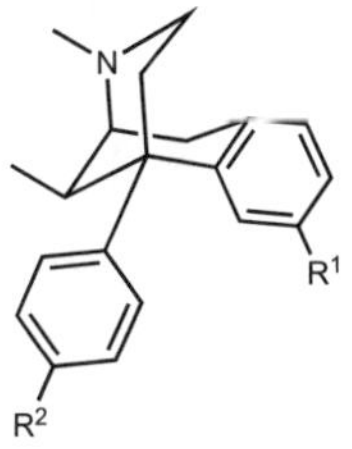

Compound	R^1	R^2
50*	OH	F
51	CN	H

* racemic

4.3. Other mechanisms

Other reportedly effective approaches to blocking DENV replication in cells have been demonstrated with the dihydroorotate dehydrogenase inhibitors brequinar, **52** ($EC_{50}=0.078\ \mu M$),[91,92] and compound **53**,[93] the cholesterol uptake inhibitor **54** ($EC_{50}=6.2\ \mu M$),[94] the deubiquinating enzyme inhibitor **55** ($EC_{50}=40\ \mu M$),[95] the NTRK1/MAPKAPK5 kinase inhibitor **56** ($EC_{50}=0.4\ \mu M$),[96] and compound **57** ($EC_{50}=0.9\ \mu M$), with an unknown mechanism of action.[97]

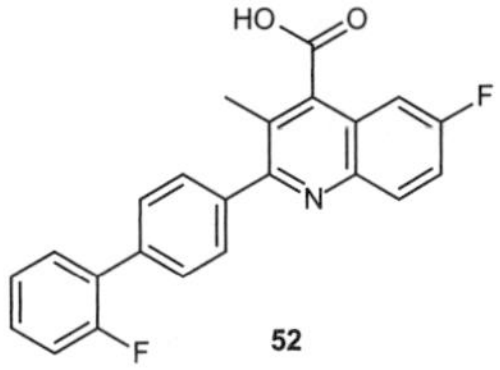

52

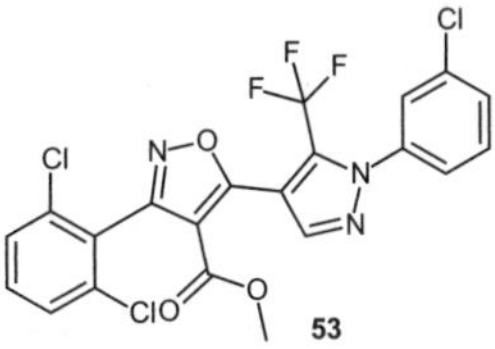

53

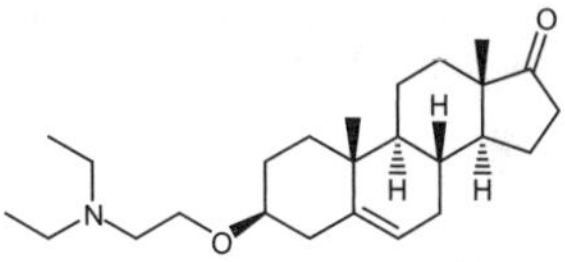

54

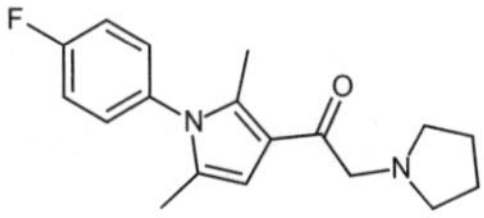

55

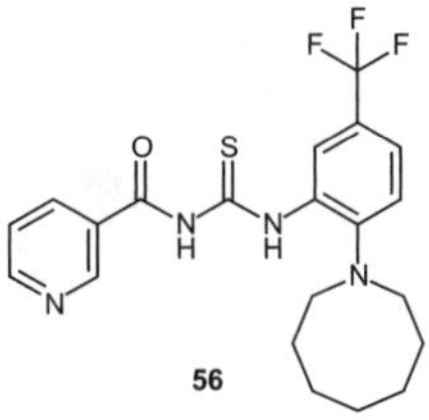

56

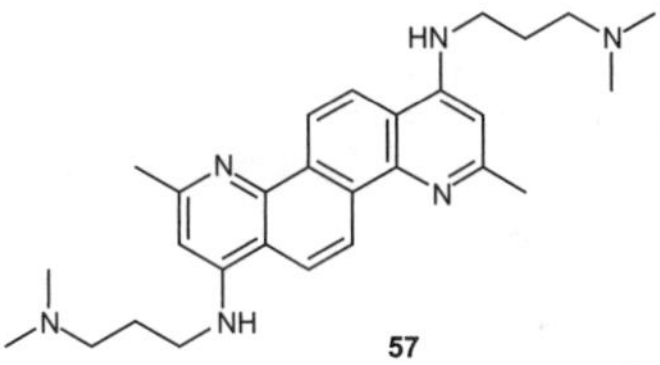

57

5. CONCLUSIONS

The rapid geographic spread of DENV over the past half century represents an emergence of an alarming global health threat. Recent clinical successes in the treatment of the related hepatitis C virus suggest that similar approaches to the treatment of DENV infection may represent fruitful opportunities for drug discovery exploration. The well-established understanding of virus biology, coupled with the accessibility of viral protein target structures, has led to a significant application of virtual as well as high-throughput screens to identify chemical matter. Successes in animal models of infection also give cause for optimism, yet ideal candidate molecules for clinical development have yet to be reported.

REFERENCES

(1) Gubler, D.J. *Clin. Microbiol. Rev.* **1998**, *11*, 480.
(2) Guzman, A.; Istúriz, R.E. *Int. J. Antimicrob. Agents* **2010**, *36S*, S40.
(3) WHO. Dengue and severe dengue. Fact sheet number 117, January 2012. http://www.who.int/mediacentre/factsheets/fs117/en/ (accessed May 4, 2012).
(4) Gubler, D.J. *Am. J. Trop. Med. Hyg.* **1989**, *40*, 571.
(5) Iturbe-Ormaetxe, I.; Walker, T.; O'Neill, S.L. *EMBO Rep.* **2011**, *12*, 508.
(6) Harris, A.F.; Nimmo, D.; McKerney, A.R.; Kelly, N.; Scaife, S.; Donnelly, C.A.; Beech, C.; Petrie, W.D.; Alphey, L. *Nat. Biotechnol.* **2011**, *29*, 1034.
(7) Reiter, P.; Gubler, D.J. In *Dengue and Dengue Hemorrhagic Fever*; Gubler, D.J., Kuno, G., Eds.; CAB International: New York, 1997; p 425.
(8) Kummerer, B.M.; Rice, C.M. *J. Virol.* **2002**, *76*, 4773.
(9) Guo, J.T.; Hayashi, J.; Seeger, C. *J. Virol.* **2005**, *79*, 1343.
(10) Luo, D.; Xu, T.; Hunke, C.; Gruber, G.; Vasudevan, S.G.; Lescar, J. *J. Virol.* **2008**, *82*, 173.
(11) Wang, C.C.; Huang, Z.S.; Chiang, P.L.; Chen, C.T.; Wu, H.N. *FEBS Lett.* **2009**, *583*, 691.
(12) Li, H.; Clum, S.; You, S.; Ebner, K.E.; Padmanabhan, R. *J. Virol.* **1999**, *73*, 3108.
(13) Bartelma, G.; Padmanabhan, R. *Virology* **2002**, *299*, 122.
(14) Egloff, M.P.; Benarroch, D.; Selisko, B.; Romette, J.L.; Canard, B. *EMBO J.* **2002**, *21*, 2757.
(15) Ackermann, M.; Padmanabhan, R. *J. Biol. Chem.* **2001**, *276*, 39926.
(16) Lindenbach, B.D.; Rice, C.M. *J. Virol.* **1997**, *71*, 9608.
(17) Miller, S.; Kastner, S.; Krijnse-Locker, J.; Bühler, S.; Bartenschlager, R. *J. Biol. Chem.* **2007**, *282*, 8873.
(18) Miller, S.; Sparacio, S.; Bartenschlager, R. *J. Biol. Chem.* **2006**, *281*, 8854.
(19) Murphy, B.R.; Whitehead, S.S. *Annu. Rev. Immunol.* **2011**, *29*, 587.
(20) Durbin, A.P.; Whitehead, S.S. *Viruses* **1800**, *2011*, 3.
(21) Parkinson, T.; Pryde, D.C. *Future Med. Chem.* **2010**, *2*, 1181.
(22) Stevens, A.J.; Gahan, M.E.; Mahalingam, S.; Keller, P.A. *J. Med. Chem.* **2009**, *52*, 7911.
(23) Kaufmann, B.; Rossmann, M.G. *Microbes Infect.* **2011**, *13*, 1.
(24) Smit, J.M.; Moesker, B.; Rodenhuis-Zybert, I.; Wilschut, J. *Viruses* **2011**, *3*, 160.
(25) Modis, Y.; Ogata, S.; Clements, D.; Harrison, S.C. *Proc. Natl. Acad. Sci. U.S.A.* **2003**, *100*, 6986.
(26) Kampmann, T.; Yennamalli, R.; Campbell, P.; Stoermer, M.J.; Fairlie, D.P.; Kobe, B.; Young, P.R. *Antiviral Res.* **2009**, *84*, 234.
(27) Poh, M.K.; Shui, G.; Xie, X.; Shi, P.Y.; Wenk, M.R.; Gu, F. *Antiviral Res.* **2011**, *93*, 191.
(28) Wang, Q.-Y.; Patel, S.J.; Vangrevelinghe, E.; Xu, H.Y.; Rao, R.; Jaber, D.; Schul, W.; Gu, F.; Heudi, O.; Ma, N.L.; Poh, M.K.; Phong, W.Y.; Keller, T.H.; Jacoby, E.; Vasudevan, S.G. *Antimicrob. Agents Chemother.* **1823**, *2009*, 53.

(29) Yennamalli, R.; Subbarao, N.; Kampmann, T.; McGeary, R.P.; Young, P.R.; Kobe, B. *J. Comput. Aided Mol. Des.* **2009**, *23*, 333.
(30) Schmidt, A.G.; Yang, P.L.; Harrison, S.C. *J. Virol.* **2010**, *84*, 12549.
(31) Costin, J.M.; Jenwitheesuk, E.; Lok, S.-M.; Hunsperger, E.; Conrads, K.A.; Fontaine, K.A.; Rees, C.R.; Rossmann, M.G.; Isern, S.; Samudrala, R.; Michael, S.F. *PLoS Negl. Trop. Dis.* **2010**, *4*, e721.
(32) Nicholson, C.O.; Costin, J.M.; Rowe, D.K.; Lin, L.; Jenwitheesuk, E.; Samudrala, R.; Isern, S.; Michael, S.F. *Antiviral Res.* **2011**, *89*, 71.
(33) Hrobowski, Y.M.; Garry, R.F.; Michael, S.F. *Virol. J.* **2005**, *2*, 49.
(34) Alen, M.M.F.; Kaptein, S.J.F.; De Burghgraeve, T.; Balzarini, J.; Neyts, J.; Schols, D. *Virology* **2009**, *387*, 67.
(35) Acosta, E.G.; Castilla, V.; Damonte, E.B. *Cell. Microbiol.* **2009**, *11*, 1533.
(36) Kato, D.; Era, S.; Watanabe, I.; Arihara, M.; Sugiura, N.; Kimata, K.; Suzuki, Y.; Morita, K.; Hidari, K.I.P.J.; Suzuki, T. *Antiviral Res.* **2010**, *88*, 236.
(37) Erbel, P.; Schiering, N.; D'Arcy, A.; Renatus, M.; Kroemer, M.; Lim, S.P.; Yin, Z.; Keller, T.H.; Vasudevan, S.G.; Hommel, U. *Nat. Struct. Mol. Biol.* **2006**, *13*, 372.
(38) Lescar, J.; Luo, D.; Xu, T.; Sampath, A.; Lim, S.P.; Canard, B.; Vasudevan, S.G. *Antiviral Res.* **2008**, *80*, 94.
(39) Chen, K.X.; Njoroge, F.G. *Curr. Opin. Investig. Drugs* **2009**, *10*, 821.
(40) Mani, N.; Rao, B.G.; Kieffer, T.L.; Kwong, A.D. In *Antiviral Drug Strategies*; de Clercq, E., Ed.; Wiley-VCH Verlag, Weinheim 2011; p 307.
(41) Li, J.; Lim, S.P.; Beer, D.; Patel, V.; Wen, D.; Tumanut, C.; Tully, D.C.; Williams, J.A.; Jiricek, J.; Priestle, J.P.; Harris, J.L.; Vasudevan, S.G. *J. Biol. Chem.* **2005**, *280*, 28766.
(42) Schechter, I.; Berger, A. *Biochem. Biophys. Res. Commun.* **1967**, *27*, 157.
(43) Yin, Z.; Patel, S.J.; Wang, W.-L.; Wang, G.; Chan, W.-L.; Rao, K.R.R.; Alam, J.; Jeyaraj, D.A.; Ngew, X.; Patel, V.; Beer, D.; Lim, S.P.; Vasudevan, S.G.; Keller, T.K. *Bioorg. Med. Chem. Lett.* **2006**, *16*, 36.
(44) Yin, Z.; Patel, S.J.; Wang, W.-L.; Wang, G.; Chan, W.-L.; Rao, K.R.R.; Alam, J.; Jeyaraj, D.A.; Ngew, X.; Patel, V.; Beer, D.; Lim, S.P.; Vasudevan, S.G.; Keller, T.K. *Bioorg. Med. Chem. Lett.* **2006**, *16*, 40.
(45) Schüller, A.; Yin, Z.; Brian Chia, C.S.; Doan, D.N.P.; Kim, H.-K.; Shang, L.; Loh, T.P.; Hill, J.; Vasudevan, S.G. *Antiviral Res.* **2011**, *92*, 96–101.
(46) Knehans, T.; Schüller, A.; Doan, D.N.; Nacro, K.; Hill, J.; Güntert, P.; Madhusudhan, M.S.; Weil, T.; Vasudevan, S.G. *J. Comput. Aided Mol. Des.* **2011**, *25*, 263.
(47) Levitzki, A.; Mishani, E. *Annu. Rev. Biochem.* **2006**, *75*, 93.
(48) Nitsche, C.; Steuer, C.; Klein, C.D. *Bioorg. Med. Chem.* **2011**, *19*, 7318.
(49) Perola, E. *J. Med. Chem.* **2010**, *53*, 2986.
(50) Nitsche, C.; Behnam, M. A. M.; Steuer, C.; Klein, C. D. *Antiviral Res.* **2012**, *94*, 72.
(51) Tiew, K.-C.; Dou, D.; Teramoto, T.; Lai, H.; Alliston, K.R.; Lushington, G.H.; Padmanabhan, R.; Groutas, W.C. *Bioorg. Med. Chem.* **2012**, *20*, 1213.
(52) Shiryaev, S.A.; Cheltsov, A.V.; Gawlik, K.; Ratnikov, B.I.; Strongin, A.Y. *Assay Drug Dev. Technol.* **2011**, *9*, 69.
(53) Yang, C.-C.; Hsieh, Y.-C.; Lee, S.-J.-H.; Wu, S.-H.; Liao, C.-L.; Tsao, C.-L.; Chao, Y.-S.; Chern, J.-H.; Wu, C.-P.; Yueh, A. *Antimicrob. Agents Chemother.* **2011**, *55*, 229.
(54) Luo, D.; Wei, N.; Doan, D.N.; Paradkar, P.N.; Chong, Y.; Davidson, A.D.; Kotaka, M.; Lescar, J.; Vasudevan, S.G. *J. Biol. Chem.* **2010**, *285*, 18817.
(55) Xu, T.; Sampath, A.; Chao, A.; Wen, D.; Nanao, M.; Chene, P.; Vasudevan, S.G.; Lescar, J. *J. Virol.* **2005**, *79*, 10278.
(56) Sampath, A.; Xu, T.; Chao, A.; Luo, D.; Lescar, J.; Vasudevan, S.G. *J. Virol.* **2006**, *80*, 6686.

(57) Bretner, M.; Schalinski, S.; Hang, A.; Lang, M.; Schmitz, H.; Baier, A.; Behrens, S.-E.; Kulikowski, T.; Borowski, P. *Antiviral Chem. Chemother.* **2004**, *15*, 35.
(58) Gao, M.; Nettles, R.E.; Belema, M.; Snyder, L.B.; Nguyen, V.N.; Fridell, R.A.; Serrano-Wu, M.H.; Langley, D.R.; Sun, J.-H.; O'Boyle, D.R., II; Lemm, J.A.; Wang, C.; Knipe, J.O.; Chien, C.; Colonno, R.J.; Grasela, D.M.; Meanwell, N.A.; Hamann, L.G. *Nature* **2010**, *465*, 96.
(59) Lemm, J.A.; Leet, J.E.; O'Boyle, D.R.; Romine, J.L.; Huang, X.S.; Schroeder, D.R.; Alberts, J.; Cantone, J.L.; Sun, J.-H.; Nower, P.T.; Martin, S.W.; Serrano-Wu, M.H.; Meanwell, N.A.; Snyder, L.B.; Gao, M. *Antimicrob. Agents Chemother.* **2011**, *55*, 3795.
(60) Xie, X.; Wang, Q.-Y.; Xu, H.Y.; Qing, M.; Kramer, L.; Yuan, Z.; Shi, P.-Y. *J. Virol.* **2011**, *85*, 11183.
(61) Rawlinson, S.M.; Pryor, M.J.; Wright, P.J.; Jans, D.A. *Curr. Drug Targets* **2006**, 7, 1623.
(62) Choi, K.H.; Rossmann, M.G. *Curr. Opin. Struct. Biol.* **2009**, *19*, 746.
(63) Beaulieu, P.L. *Curr. Opin. Investig. Drugs* **2007**, *8*, 614.
(64) Deore, R.R.; Chern, J.W. *Curr. Med. Chem.* **2010**, *17*, 3806.
(65) Schul, W.; Liu, W.; Xu, H.-Y.; Flamand, M.; Vasudevan, S.G. *J. Infect. Dis.* **2007**, *195*, 665.
(66) Yin, Z.; Chen, Y.L.; Schul, W.; Wang, Q.Y.; Gu, F.; Duraiswamy, J.; Kondreddi, R.R.; Niyomrattanakit, P.; Lakshminarayana, S.B.; Goh, A.; Xu, H.Y.; Liu, W.; Liu, B.; Lim, J.Y.H.; Ng, C.Y.; Qing, M.; Lim, C.C.; Yip, A.; Wang, G.; Chan, W.L.; Tan, H.P.; Lin, K.; Zhang, B.; Zou, G.; Bernard, K.A.; Garrett, C.; Beltz, K.; Dong, M.; Weaver, M.; He, H.; Pichota, A.; Dartois, V.; Keller, T.H.; Shi, P.-Y. *Proc. Natl. Acad. Sci. U.S.A.* **2009**, *106*, 20435.
(67) Chen, Y.-L.; Yin, Z.; Duraiswamy, J.; Schul, W.; Lim, C.C.; Liu, B.; Xu, H.Y.; Qing, M.; Yip, A.; Wang, G.; Chan, W.L.; Tan, H.P.; Lo, M.; Liung, S.; Kondreddi, R.R.; Rao, R.; Gu, H.; He, H.; Keller, T.H.; Shi, P.-Y. *Antimicrob. Agents Chemother.* **2010**, *54*, 2932.
(68) Latour, D.R.; Jekle, A.; Javanbakht, H.; Henningsen, R.; Gee, P.; Lee, I.; Tran, P.; Ren, S.; Kutach, A.K.; Harris, S.F.; Wang, S.M.; Lok, S.J.; Shaw, D.; Li, J.; Heilek, G.; Klumpp, K.; Swinney, D.C.; Deval, J. *Antiviral Res.* **2010**, *87*, 213.
(69) Chen, Y.-L.; Yin, Z.; Lakshminarayana, S.B.; Qing, M.; Schul, W.; Duraiswamy, J.; Kondreddi, R.R.; Goh, A.; Xu, H.Y.; Yip, A.; Liu, B.; Weaver, M.; Dartois, V.; Keller, T.H.; Shi, P.-Y. *Antimicrob. Agents Chemother.* **2010**, *54*, 3255.
(70) Cole, P.; Castaner, R.; Bolos, J. *Drugs Future* **2009**, *34*, 282.
(71) Wu, R.; Smidansky, E.D.; Oh, H.S.; Takhampunya, R.; Padmanabhan, R.; Cameron, C.E.; Peterson, B.R. *J. Med. Chem.* **2010**, *53*, 7958.
(72) Yin, Z.; Chen, Y.-L.; Kondreddi, R.R.; Chan, W.L.; Wang, G.; Ng, R.H.; Lim, J.Y.H.; Lee, W.Y.; Jeyaraj, D.A.; Niyomrattanakit, P.; Wen, D.; Chao, A.; Glickman, J.F.; Voshol, H.; Mueller, D.; Spanka, C.; Dressler, S.; Nilar, S.; Vasudevan, S.G.; Shi, P.-Y.; Keller, T.H. *J. Med. Chem.* **2009**, *52*, 7934.
(73) Niyomrattanakit, P.; Chen, Y.-L.; Dong, H.; Yin, Z.; Qing, M.; Glickman, J.F.; Lin, K.; Mueller, D.; Voshol, H.; Lim, J.Y.H.; Nilar, S.; Keller, T.H.; Shi, P.-Y. *J. Virol.* **2010**, *84*, 5678.
(74) Liu, L.; Dong, H.; Chen, H.; Zhang, J.; Ling, H.; Li, Z.; Shi, P.Y.; Li, H. *Front. Biol.* **2010**, *5*, 286.
(75) Benarroch, D.; Egloff, M.-P.; Mulard, L.; Guerreiro, C.; Romette, J.-L.; Canard, B. *J. Biol. Chem.* **2004**, *279*, 35638.
(76) Zhou, Y.; Ray, D.; Zhao, Y.; Dong, H.; Ren, S.; Li, Z.; Guo, Y.; Bernard, K.A.; Shi, P.-Y.; Li, H. *J. Virol.* **2007**, *81*, 3891.
(77) Egloff, M.-P.; Decroly, E.; Malet, H.; Selisko, B.; Benarroch, D.; Ferron, F.; Canard, B. *J. Mol. Biol.* **2007**, *372*, 723.
(78) Geiss, B.J.; Thompson, A.A.; Andrews, A.J.; Sons, R.L.; Gari, H.H.; Keenan, S.M.; Peersen, O.B. *J. Mol. Biol.* **2009**, *385*, 1643.

(79) Yap, L.J.; Luo, D.; Chung, K.Y.; Lim, S.P.; Bodenreider, C.; Noble, C.; Shi, P.-Y.; Lescar, J. *PLoS One* **2010**, *5*, e12836.
(80) Podvinec, M.; Lim, S.P.; Schmidt, T.; Scarsi, M.; Wen, D.; Sonntag, L.-S.; Sanschagrin, P.; Shenkin, P.S.; Schwede, T. *J. Med. Chem.* **2010**, *53*, 1483.
(81) Lim, S.P.; Sonntag, L.S.; Noble, C.; Nilar, S.H.; Ng, R.H.; Zou, G.; Monaghan, P.; Chung, K.Y.; Dong, H.; Liu, B.; Bodenreider, C.; Lee, G.; Ding, M.; Chan, W.L.; Wang, G.; Jian, Y.L.; Chao, A.T.; Lescar, J.; Yin, Z.; Vedananda, T.R.; Keller, T.H.; Shi, P.-Y. *J. Biol. Chem.* **2011**, *286*, 6233–6240.
(82) Mondotte, J.A.; Lozach, P.-Y.; Amara, A.; Gamarnik, A.V. *J. Virol.* **2007**, *81*, 7136.
(83) Bryant, J.E.; Calvert, A.E.; Mesesan, K.; Crabtree, M.B.; Volpe, K.E.; Silengo, S.; Kinney, R.M.; Huang, C.Y.-H.; Miller, B.R.; Roehrig, J.T. *Virology* **2007**, *366*, 415.
(84) Courageot, M.-P.; Frenkiel, M.-P.; Duarte Dos Santos, C.; Deubel, V.; Desprès, P. *J. Virol.* **2000**, *74*, 564.
(85) Gu, B.; Mason, P.; Wang, L.; Norton, P.; Bourne, N.; Moriarty, R.; Mehta, A.; Despande, M.; Shah, R.; Block, T. *Antivir. Chem. Chemother.* **2007**, *18*, 49.
(86) Chang, J.; Wang, L.; Ma, D.; Qu, X.; Guo, H.; Xu, X.; Mason, P.M.; Bourne, N.; Moriarty, R.; Gu, B.; Guo, J.-T.; Block, T.M. *Antimicrob. Agents Chemother.* **2009**, *53*, 1501.
(87) Chang, J.; Schul, W.; Butters, T.D.; Yip, A.; Liu, B.; Goh, A.; Lakshminarayana, S.B.; Alonzi, D.; Reinkensmeier, G.; Pan, X.; Qu, X.; Weidner, J.M.; Wang, L.; Yu, W.; Bourne, N.; Kinch, M.A.; Rayahin, J.E.; Moriarty, R.; Xu, X.; Shi, P.-Y.; Guo, J.-T.; Block, T.M. *Antiviral Res.* **2011**, *89*, 26.
(88) Chang, J.; Schul, W.; Yip, A.; Xu, X.; Guo, J.-T.; Block, T.M. *Antiviral Res.* **2011**, *92*, 369–371.
(89) Rathore, A.P.S.; Paradkar, P.N.; Watanabe, S.; Tan, K.H.; Sung, C.; Connolly, J.E.; Low, J.; Ooi, E.E.; Vasudevan, S.G. *Antiviral Res.* **2011**, *92*, 453.
(90) Wang, Q.-Y.; Kondreddi, R.R.; Xie, X.; Rao, R.; Nilar, S.; Xu, H.Y.; Qing, M.; Chang, D.; Dong, H.; Yokokawa, F.; Lakshminarayana, S.B.; Goh, A.; Schul, W.; Kramer, L.; Keller, T.H.; Shi, P.-Y. *Antimicrob. Agents Chemother.* **2011**, *55*, 4072.
(91) Tan, Y. H.; Driscoll, J. S.; Mui Mui, S. Patent Application WO 2001/024785, **2001**.
(92) Qing, M.; Zou, G.; Wang, Q.-Y.; Xu, H.Y.; Dong, H.; Yuan, Z.; Shi, P.-Y. *Antimicrob. Agents Chemother.* **2010**, *54*, 3686.
(93) Wang, Q.-Y.; Bushell, S.; Qing, M.; Xu, H.Y.; Bonavia, A.; Nunes, S.; Zhou, J.; Poh, M.K.; Florez de Sessions, P.; Niyomrattanakit, P.; Dong, H.; Hoffmaster, K.; Goh, A.; Nilar, S.; Schul, W.; Jones, S.; Kramer, L.; Compton, T.; Shi, P.-Y. *J. Virol.* **2011**, *85*, 6548.
(94) Poh, M.K.; Shui, G.; Xie, X.; Shi, P.Y.; Wenk, M.R.; Gu, F. *Antiviral Res.* **2012**, *93*, 191.
(95) Nag, D.K.; Finley, D. *Virus Res.* **2012**, *165*, 103.
(96) Anwar, A.; Hosoya, T.; Leong, K.M.; Onogi, H.; Okuno, Y.; Hiramatsu, T.; Koyama, H.; Suzuki, M.; Hagiwara, M.; Garcia-Blanco, M.A. *PLoS One* **2011**, *6*, e23246.
(97) Aman, M.J.; Kinch, M.S.; Warfield, K.; Warren, T.; Yunus, A.; Enterlein, S.; Stavale, E.; Wang, P.; Chang, S.; Tang, Q.; Porter, K.; Goldblatt, M.; Bavari, S. *Antiviral Res.* **2009**, *83*, 245.

CHAPTER TWENTY-ONE

Nonfluoroquinolone-Based Inhibitors of Mycobacterial Type II Topoisomerase as Potential Therapeutic Agents for TB

Pravin S. Shirude, Shahul Hameed
Department of Medicinal Chemistry, AstraZeneca India Pvt. Ltd, Avishkar, Bangalore, India

Contents

1. INTRODUCTION

Tuberculosis (TB) is more prevalent in the world today than at any other time in human history.[1] *Mycobacterium tuberculosis* (Mtb), the pathogen responsible for TB, uses diverse strategies to survive in a variety of host lesions and to evade immune surveillance.[1] Few novel drug targets for Mtb have been identified, in spite of the genomic sequence having been known for more than a decade and the proliferation of new genetic technologies.[2] Extensive research has elucidated the Mtb *genes* that are required for growth and pathogenicity, but very few new drug targets have been validated using small-molecule inhibitors.[2] The development of novel treatments for TB is further complicated by the requirement that a new drug

Annual Reports in Medicinal Chemistry, Volume 47
ISSN 0065-7743
http://dx.doi.org/10.1016/B978-0-12-396492-2.00021-7

must provide an improved standard of care over the current multidrug, directly observed treatment short course regimen, which has proven to be quite effective.[3] Thus, it is imperative to validate new targets with potent small-molecule inhibitors and demonstrate their potential for therapeutic value in the context of new TB treatment regimens.[3]

One promising target for antimycobacterials is DNA gyrase, which belongs to a class of enzymes known as topoisomerases which are involved in the vital processes of DNA replication, transcription, translation, and recombination in prokaryotic and eukaryotic cells.[4,5] Two types of the topoisomerases are known: type I topoisomerases change the degree of supercoiling of DNA by causing single-strand breaks and religation and type II topoisomerases (including bacterial DNA gyrase) cause double-strand breaks.[4,5] DNA gyrase, unique to prokaryotes, binds DNA as a tetramer in which two A and two B subunits (GyrA and GyrB, respectively) combine with an appropriately displayed DNA leading strand that becomes cleaved. Subsequent passage of a lagging DNA strand through the interior of the enzyme complex and through the DNA cleavage site is followed by religation with hydrolysis of ATP to drive the catalytic cycles and produce a negative supercoil in the DNA.[6–8]

DNA gyrase has many of the ideal attributes required for an attractive antibiotic target. It is an essential gene for bacterial viability, it is present in a single copy, and its inhibition results in significant cell death because there are no viable alternative mechanisms for performing this function.[2] Gyrase genes from among a variety of Mtb isolates that have been sequenced are nearly 99.9% homologous, further signifying its broad applicability as a drug target.[2] The B subunit of DNA gyrase (GyrB) contains the ATP-binding pocket. Tetrameric topoisomerase IV (topoIV) is closely related to DNA gyrase and consists of two subunits of ParC and two of ParE, with ParE containing the ATP-binding pocket (Fig. 21.1) analogous to GyrB. The A subunit of DNA gyrase (GyrA) is responsible for DNA cleavage where covalent bonds to each of the cleaved DNA strands are made into active-site tyrosine residues. Analogously, the ParC subunits of topoIV execute DNA strand cleavage and religation.

To date, no advanced inhibitors of bacterial type I topoisomerases have advanced into the clinic.[9] In contrast, clinically valuable inhibitors of type II topoisomerases are abundant, from the well-established fluoroquinolone class to a variety of emerging classes in various stages of clinical or preclinical evaluation. Fluoroquinolones principally bind GyrA near the intersection of the GyrB subunits and the associated DNA strand of the functional DNA

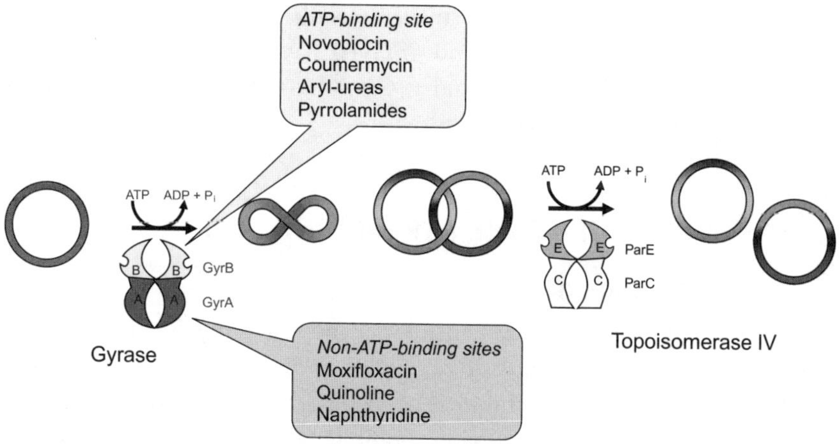

Figure 21.1 Schematic diagram of bacterial type II topoisomerases. (See Color Plate 21.1 in Color Plate Section.)

gyrase heterotetramer (A2B2) complex. They analogously bind to the ParC units of topoIV, at the interface with ParE and DNA.[5] Resistance to fluoroquinolones is typically (though not exclusively) generated by point mutations in the *gyrA* gene, which gives rise to resistance against the entire class of fluoroquinolones, thus embodying class resistance.[5,6,10] As fluoroquinolones [e.g., in a recent Phase II clinical trial, moxifloxacin (**1**) was used in place of ethambutol (in combination with isoniazid, rifampicin, and pyrazinamide) and demonstrated improved culture conversion in the initial phase of TB treatment[11]] become incorporated into clinical TB drug treatment regimens, alternative drug combinations will be required to combat the emergence of fluoroquinolone-resistant Mtb strains. However, if GyrB is targeted, it still exerts the same phenotypic effects on bacterial viability as do the fluoroquinolones. Compounds such as the coumarin antibiotics novobiocin (**2**), chlorobiocin, and coumermycin that target GyrB and ParE by binding to the ATP sites also kill bacteria, thus establishing both the GyrA and GyrB subunits of DNA gyrase complex as worthwhile objectives for inhibitor and drug development against TB.

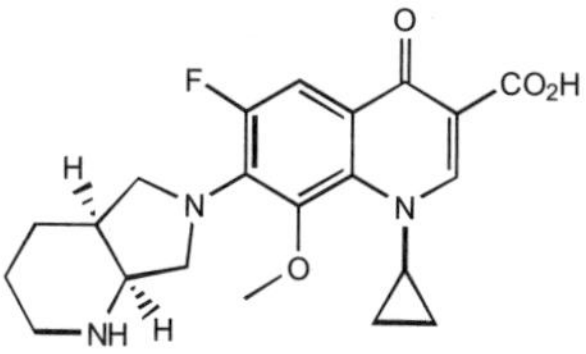

1

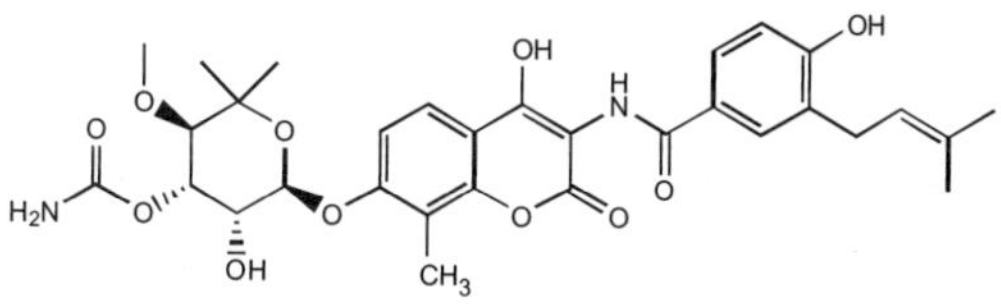

2

This review largely focuses on small-molecule inhibitors of mycobacterial type II topoisomerase that bind at GyrB having ATP sites and non-fluoroquinolone inhibitors that bind at GyrA having other than ATP sites within the enzyme tetramers.

2. INHIBITION AT THE ATP-BINDING SITE

2.1. Novobiocin

As mentioned, the natural antibiotic novobiocin is member of aminocoumarin class which was described in 1955,[1] and its mode of action was demonstrated to be the inhibition of bacterial topoisomerase, specifically DNA gyrase, in 1970.[4,5] Novobiocin inhibits DNA gyrase and topoIV by binding to the ATP pocket of GyrB and ParE, respectively.[2,12–19] The *Streptomyces* strain that produces this and related antibiotics, and its biosynthetic pathways have been elucidated and characterized.[20] GyrB has been genetically demonstrated to be essential for Mtb viability,[2] but there have not been any effective drug developed against this target for TB. Novobiocin has been shown to be a potent inhibitor of GyrB, with enzyme inhibition (K_i) and binding (K_d) constants in the low nanomolar range (7–15 nM) as well as the ability to inhibit DNA supercoiling *in vitro*.[2] It was originally approved for the treatment of methicillin-resistant *Staphylococcus aureus* (MRSA) infections, but has since been withdrawn from the market due to poor pharmacological properties and safety concerns.[2,12–19]

2.2. Coumermycin analogs

The antibiotic coumermycin Al (**3**) was first isolated from the fermentation broths of *Streptomyces rishiriensis* by Kawaguchi and co-workers.[21] It is an acidic antibiotic with *in vitro* activity against Gram-positive as well as some Gram-negative bacteria.[22,23] The spectrum of microorganisms inhibited by this compound includes *streptococci*, *pneumococci*, *bacillus*, and *mycobacteria* species, as well as a number of *enterobacteriaceae* strains. Kawaguchi *et al.* demonstrated *in vitro* activity of coumermycin A1 against three

Mycobacterium species; however, in mice experimentally infected with the $H_{37}Rv$ strain of Mtb, *in vivo* activity was not observed.[24] Although the antibiotic's subcutaneous LD_{60} of 250–380 mg/kg and oral LD_{50} of >2000 mg/kg[24,25] indicate only a moderate level of toxicity, the compound does have several undesirable characteristics. Among these are low oral bioavailability and an irritating effect on tissues at the site of parenteral administration. Coumermycin has several structural moieties in common with novobiocin[26,27] (**1**). Structural modification of the latter antibiotic by Walton *et al.*[28] failed to increase antibacterial potency or spectrum.

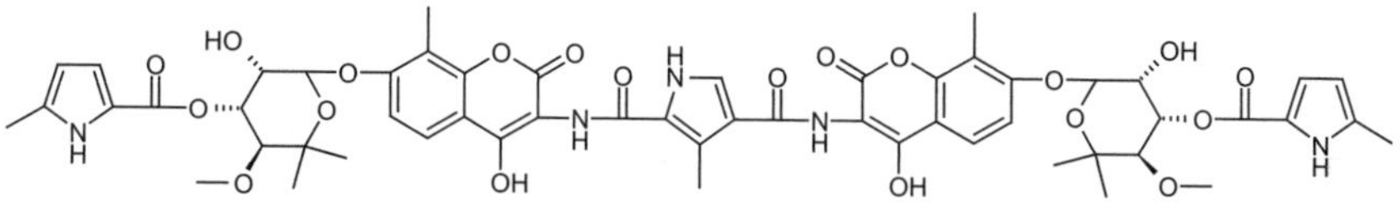

3

2.3. Aryl-ureas

As reported in the literature, aryl-ureas containing an ethyl-urea pharmacophore represent a novel class of GyrB inhibitors. This class was first identified in a high-throughput assay targeting the ATPase activity of the *S. aureus* GyrB.[29] The potent activity of this class has generated considerable attention by a variety of pharmaceutical companies through a series of scaffold-hopping approaches. Herein we describe some of the derivatives of aryl-urea scaffold that have demonstrated Mtb activity.

2.3.1 Aminobenzimidazole derivatives

The aminobenzimidazole (e.g., **4**) was developed as a GyrB inhibitors for the treatment of MRSA.[29,30] This and related compounds are much more potent and less toxic as compared to novobiocin.[2] Recently, a representative from this class was compared with novobiocin to validate the GyrB target in Mtb[2] as an opportunity for a first-line drug therapy. The enzyme potency [inhibition (K_i) and binding (K_d) constants] of aminobenzimidazole to GyrB is in the low nanomolar range, making them more potent than novobiocin. The aminobenzimidazoles also inhibit DNA supercoiling activity *in vitro*[29,31] and demonstrate cidality with an excellent activity against drug-resistant Mtb strains, including fluoroquinolone-resistant strains. These GyrB inhibitors do not exhibit antagonism against rifampicin and isoniazid[2,29,30] supporting their use in anti-TB multidrug cocktails. With GyrB as the target, aminobenzimidazole exhibits potent activity against nonreplicating Mtb expanding the attractiveness of the compounds for the treatment of TB.[32,33]

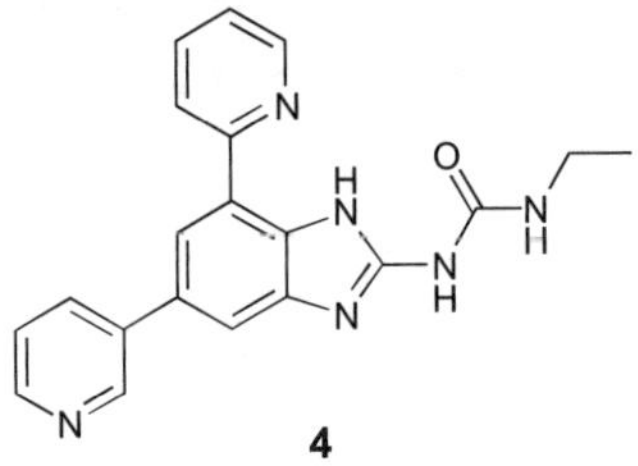

4

Based on the structural work with the aminobenzimidazoles, the urea portion of the molecule makes a critical hydrogen bonding interactions with an aspartic acid residue and thus the urea is essential for activity.[29,30] Therefore, any changes to the residues surrounding the urea region of the GyrB-binding pocket would result in a significant loss of activity. Because of mutations to these residues, not all mycobacteria are equally susceptible to these aminobenzimidazoles.[2] Another notable hydrogen bonding interaction is seen from the structural work between the pyridine ring of aminobenzimidazoles and an arginine residue similar to that seen for coumarin hydroxyl of aminocoumarins.[29] Mutation of the arginine residue, which is located on the edge of the ATP-binding pocket, resulted in resistance to novobiocin. However, aminobenzimidazoles do not significantly lose their potency against Mtb strains containing this novobiocin-resistant mutant, perhaps due to dual targeting of the topoIV DNA gyrase by aminobenzimidazoles.[2]

2.3.2 Other aryl ureas

The thiazolopyridine urea series was discovered at AstraZeneca through a scaffold-hopping approach combining benzimidazole and benzothiazole urea cores reported in the literature.[29] Compound (**5**) from this series displayed excellent biochemical potency by inhibiting *Mycobacterium smegmatis* (Msm) GyrB isozymes at 2.5 nM and good antimycobacterial activity (Mtu MIC 0.06 μg/ml).[34]

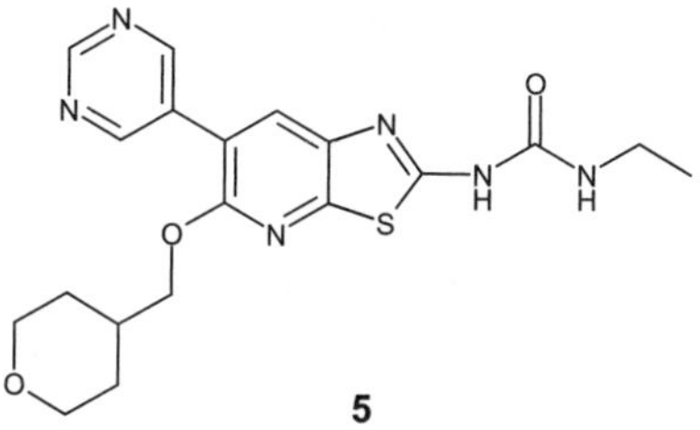

5

2.4. Miscellaneous compounds or inhibitors

2.4.1 Pyrrolamides

A novel class of bacterial DNA GyrB inhibitors, the pyrrolamides were discovered using fragment-based screening reported earlier.[35] Our group at AstraZeneca has profiled this novel class of inhibitors for their antimycobacterial properties with an objective of developing an orally active anti-TB agent. Initial screening and subsequent lead optimization lead to pyrrolamide (**6**), which has excellent enzyme and cellular activity against the Msm GyrB enzyme and Mtb $H_{37}Rv$ respectively.[36] Compound (**6**) also shows >70% oral bioavailability and efficacy of 1 log kill in an acute mouse model.

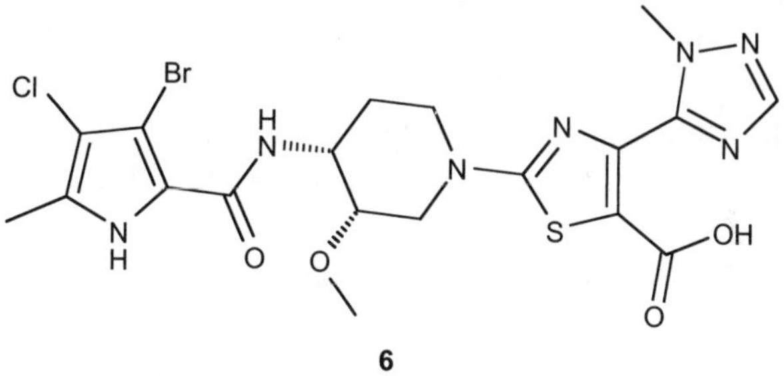

6

3. INHIBITION AT THE NON-ATP-BINDING SITE

3.1. Piperidinyl quinoline and naphthyridines

Without any indication of mode of action, a series of piperidinyl alkyl quinoline derivatives as antibacterials was first reported by GSK in 1999.[37] Compound **7** from this initial disclosure was reported to have an MIC of 4 μg/ml against *E. coli*. Subsequently, there has been a large body of work reported around this scaffold in an effort to develop novel antibacterial agents and establish new IP space. Novexel reported the target information for a clinical candidate **8** (NXL101), as inhibition of type II topoisomerases (both topoIV and gyrase) in *E. coli* and *S. pneumoniae*. NXL101 also showed good activity against fluoroquinolone-resistant strains of *S. aureus* with known mutations in the quinolone-resistance determining region (QRDR). This indicates that mechanism of inhibition of topoisomerases by NXL-101 is very different as that of fluoroquinolones mechanism.[38] Compound **8** was advanced to Phase I clinical studies, but was discontinued due to QT prolongation signals in the healthy volunteers.[39] Despite numerous patents covering quinoline- and naphthyridine-based gyrase inhibitors as broad spectrum antibacterial agents, only GSK has reported antimycobacterial activity.[40–44] Analogs from quinolone (**9,10**), pyridopyrazinone (**11**), and pyridopyrazinedione (**12**) series with bicyclic right-hand side (RHS) fragments showed Mtb MICs ranging from 0.3 to 2 μg/ml, though there is no data related to antimycobacterial mode of action reported for this set of compounds. It is likely that the Mtb activity is due to gyrase inhibition with the structural resemblance to **8** and the reported inhibition of type II topoisomerases in *E. coli* and *S. pneumonia*.[38] Further medicinal chemistry optimization for Mtb activity led to additional series with monocyclic RHS fragments as seen in compounds **13–16**, reported to demonstrate MICs of 0.01–0.3 μg/ml against Mtb.[45–47] Compound **14** and its closely related analogs maintain wild-type Mtb MICs and improve MICs versus fluoroquinolone-resistant strains with well-characterized mutations in the QRDR region (S91P, A90V, and D94G).[48,49] These compounds also retained Mtb MICs against extensively drug-resistant clinical isolates, which are completely resistant to first-line and second-line TB agents including ciprofloxacin and moxifloxacin. The above study reinforces that these antimycobacterial mechanisms of gyrase inhibition are distinctly different from the fluoroquinolone mode of inhibition.

Preliminary pharmacokinetic profiling of compound **13** in mouse showed 28% bioavailability and also demonstrated excellent efficacy with clear dose response (limit of quantification at 75 mg/kg BID, s.c.) in an acute mouse model with BID dosing through a subcutaneous route of administration. However, the series was reported to inhibit the hERG cardiac ion channel (hERG), and optimization of the series toward reducing the hERG liability is underway.[48,49]

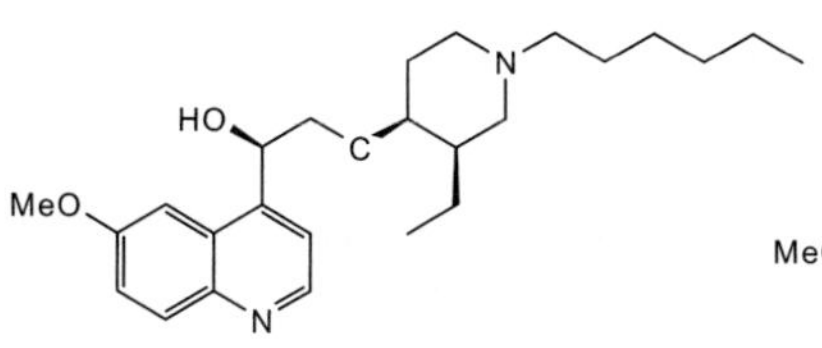

7

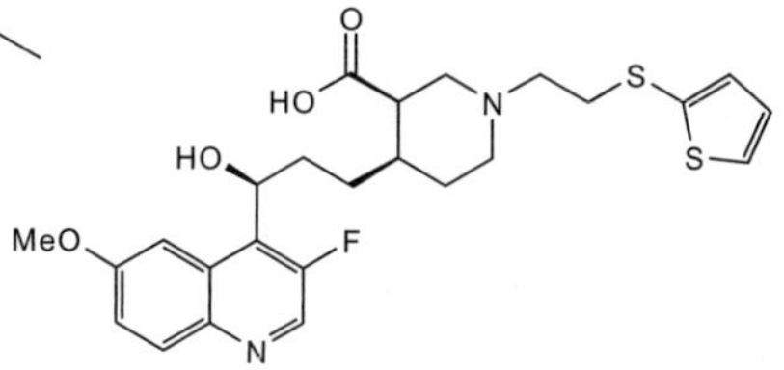

8

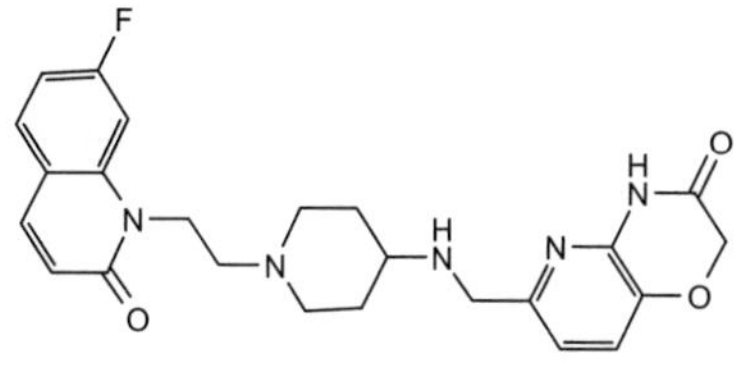

9

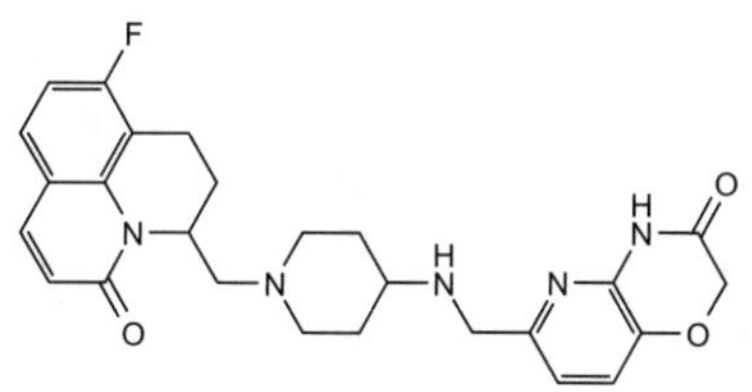

10

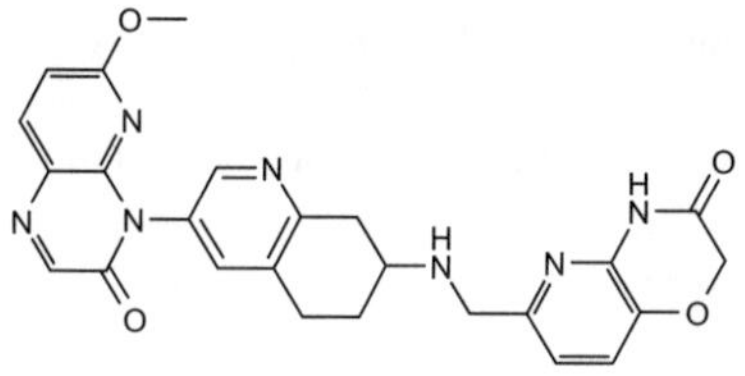

11

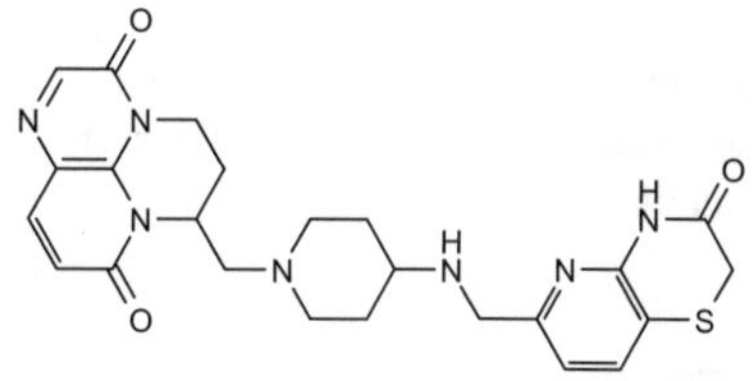

12

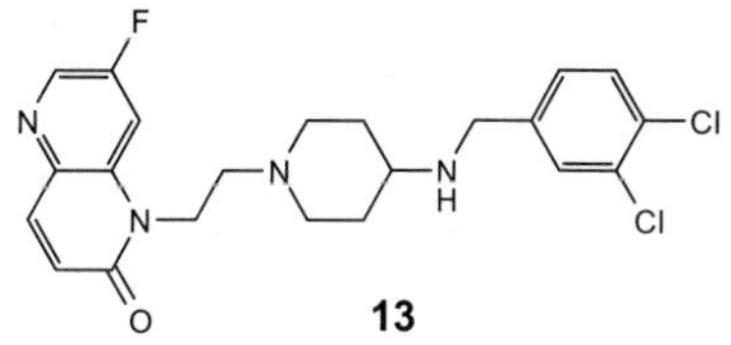

13

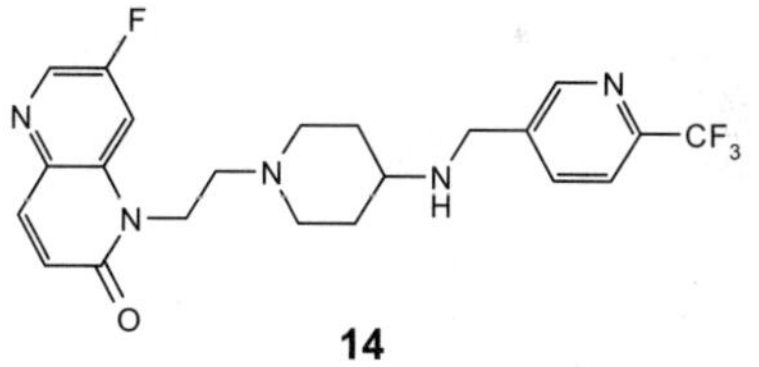

14

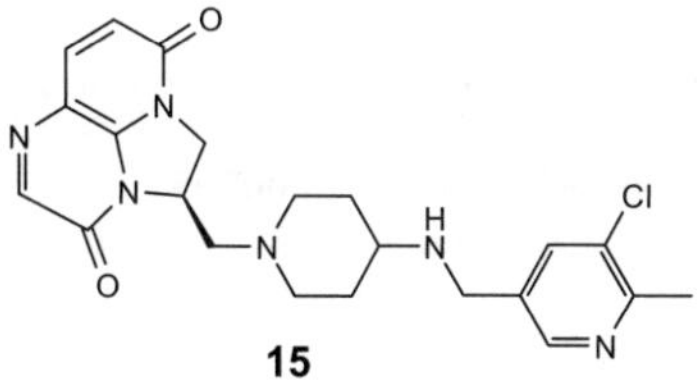

15

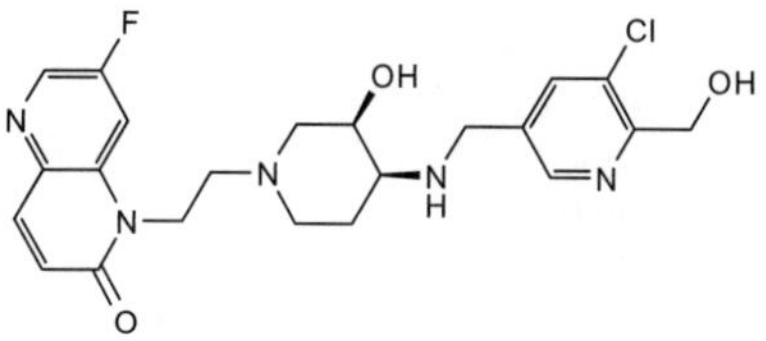

16

4. CONCLUSIONS

In today's advanced multidisciplinary drug discovery, TB still remains a challenging endeavor at every level. DNA gyrase remains sole target for quinolones and continues to be pharmaceutically effective target for drug discovery against Mtb. The ATPase activity of bacterial DNA gyrase that resides in the B subunit is emerging as a novel target and lot of efforts are being put by various groups with encouraging results. Based on this progress, if we can successfully leverage the opportunities in this target, there is hope that we will be able to raise novel gyrase inhibitor in earnest in the long struggle against TB.

ACKNOWLEDGMENTS

We deeply acknowledge Dr. Gregory Bisacchi for his guidance. We express thanks to our AZ Boston colleagues for proofreading of this manuscript. We also thank Dr. Bheemarao Ugarkar and Dr. Tanjore Balganesh for their constant inspiration, encouragement, and support.

REFERENCES

(1) Koul, A.; Arnoult, E.; Lounis, N.; Guillemont, J.; Andries, K. *Nature* **2011**, *469*, 483.
(2) Chopra, S.; Matsuyama, K.; Tran, T.; Malerich, J.P.; Wan, B.; Franzblau, S.G.; Lun, S.; Guo, H.; Maiga, M.C.; Bishai, W.R.; Madrid, P.B. *J. Antimicrob. Chemother.* **2012**, *67*, 415.
(3) Ginsberg, A.M.; Spigelman, M. *Nat. Med.* **2007**, *13*, 290.
(4) Wang, J.C. *Annu. Rev. Biochem.* **1996**, *65*, 635.
(5) Pommier, Y.; Pourquier, P.; Fan, Y.; Strumberg, D. *Biochim. Biophys. Acta* **1998**, *1400*, 83.
(6) Gellert, M.; O'Dea, M.H.; Itoh, T.; Tomizawa, J.-I. *Proc. Natl. Acad. Sci. U.S.A.* **1976**, *73*, 4474.
(7) Sugino, A.; Peebles, C.L.; Kreuzer, K.N.; Cozzarelli, N.R. *Proc. Natl. Acad. Sci. U.S.A.* **1977**, *74*, 4767.
(8) Watanabe, J.; Nakada, N.; Sawairi, S.; Shimada, H.; Oshima, S.; Kamiyama, T.; Arisawa, M. *J. Antibiot.* **1994**, *47*, 32.
(9) Cheng, B.; Liu, I.-F.; Tse-Dinh, Y.-C. *J. Antimicrob. Chemother.* **2007**, *59*, 640.
(10) Fernandes, P.B.; Menzel, R.; Hardy, D.J.; Tse-Ding, Y.-C.; Warren, A.; Elsemore, D.A. *Med. Res. Rev.* **1999**, *19*, 559.
(11) Conde, M.B.; Efron, A.; Loredo, C. *Lancet* **2009**, *373*, 1183.
(12) Maxwell, A. *Mol. Microbiol.* **1993**, *9*, 681.
(13) Ueda, Y.; Chuang, J.M.; Crast, L.B., Jr.; Partyka, R.A. *Antibiotics* **1989**, *42*, 1379.
(14) Ueda, Y.; Chuang, J.M.; Fung-Tomc, J.; Partyka, R.A. *Bioorg. Med. Chem. Lett.* **1994**, *4*, 1623.
(15) Bell, W.; Block, M.H.; Cook, C.; Grant, A.; Timms, D. *J. Chem. Soc. Perkin Trans.* **1997**, *1*, 2789.
(16) Klich, M.; Laurin, P.; Musicki, B.; Schio, L. Patent Application WO 9747634, 1998.
(17) Chartreaux, F.; Klich, M.; Schio, L. Patent Application EP 894805, 1999.

(18) Laurin, P.; Ferroud, D.; Klich, M.; Dupuis-Haelin, C.; Mauvais, P.; Lassaigne, P.; Bonnefoy, A.; Musicki, B. *Bioorg. Med. Chem. Lett.* **1999**, *9*, 2079.
(19) Laurin, P.; Ferroud, D.; Schio, L.; Klich, M.; Dupuis-Haelin, C.; Mauvais, P.; Lassaigne, P.; Bonnefoy, A.; Musicki, B. *Bioorg. Med. Chem. Lett.* **1999**, *9*, 2875.
(20) Heide, L.; Gust, B.; Anderle, C.; Li, S.-M. *Curr. Top. Med. Chem.* **2008**, *8*, 667.
(21) Kawaguchi, H.; Tsukiura, H.; Okanishi, M.; Miyaki, T.; Ohmori, T.; Fujisawa, K.; Koshiyama, H. *J. Antibiot. Ser. A* **1965**, *18*, 10.
(22) Price, K.E.; Chisholm, D.R.; Godfrey, J.C.; Misiek, M.; Gourevitch, A. *Appl. Microbiol.* **1970**, *19*, 14.
(23) Duma, R.J.; Warner, J.F. *Appl. Microbiol.* **1969**, *18*, 404.
(24) Grunberg, E.; Bennett, M. *Antimicrob. Agents Chemother.* **1966**, 786.
(25) Grunberg, E.; Cleeland, R.; Titsworth, E. *Antimicrob. Agents Chemother.* **1967**, 397.
(26) Berger, J.; Schocher, A.J.; Batcho, A.D.; Pecherer, B.; Keller, O.; Maricq, J.; Karr, A.E.; Vaterlaus, B.P.; Furlenmeier, A.; Spiegelberg, H. *Antimicrob. Agents Chemother.* **1965**, *5*, 778.
(27) Kawaguchi, H.; Takayuki, N.; Tsukiura, H. *J. Antibiot. Ser. A* **1965**, *18*, 11.
(28) Walton, R.B.; McDaniel, L.E.; Woodruff, H.B. *Dev. Ind. Microbiol.* **1962**, *3*, 370.
(29) Charifson, P.S.; Grillot, A.-L.; Grossman, T.H.; Parsons, J.D.; Badia, M.; Bellon, S.; Deininger, D.D.; Drumm, J.E.; Gross, C.H.; LeTiran, A.; Liao, Y.; Mani, N.; Nicolau, D.P.; Perola, E.; Ronkin, S.; Shannon, D.; Swenson, L.L.; Tang, Q.; Tessier, P.R.; Tian, S.-K.; Trudeau, M.; Wang, T.; Wei, Y.; Zhang, H.; Stamos, D. *J. Med. Chem.* **2008**, *51*, 5243.
(30) Grossman, T.H.; Bartels, D.J.; Mullin, S. *Antimicrob. Agents Chemother.* **2007**, *51*, 657.
(31) Glaser, B.T.; Malerich, J.P.; Duellman, S.J. *J. Biomol. Screen.* **2011**, *16*, 230.
(32) Piton, J.; Petrella, S.; Delarue, M. *PLoS One* **2010**, *5*, 12245.
(33) Sassetti, C.M.; Boyd, D.H.; Rubin, E.J. *Proc. Natl. Acad. Sci. U.S.A.* **2001**, *98*, 12712.
(34) Ghorpade, S. R.; Kale, M. G.; McKinney, D. C.; Peer Mohamed, S. H.; Raichurkar, A. K. Patent Application WO 2009/147431 A1, **2009**.
(35) Eakin, A.E.; Green, O.; Hales, N.; Walkup, G.K.; Bist, S.; Singh, A.; Mullen, G.; Bryant, J.; Embrey, K.; Gao, N.; Breeze, A.; Timms, D.; Andrews, B.; Uria-Nickelsen, M.; Demeritt, J.; Loch, J.T.; Hull, K.; Blodgett, A.; Illingworth, R.N.; Prince, B.; Boriack-Sjodin, P.A.; Hauck, S.; MacPherson, L.J.; Ni, H.; Sherer, B. *Antimicrob. Agents Chemother.* **2012**, *56*, 1240.
(36) Peer Mohamed, S. H.; Waterson, D. Patent Application WO 2010/067125, **2010**.
(37) Coates, W.J.; Gwynn, M. N.; Hatton, I. K.; Masters, P.J.; Pearson, N. D.; Rahman, S. S.; Slocombe, B.; Warrack, J. D. Patent Application WO 99/37635 A1, **1999**.
(38) Black, M.T.; Stachyra, T.; Platel, D.; Girard, A.M.; Claudon, M.; Bruneau, J.M.; Miossec, C. *Antimicrob. Agents Chemother.* **2008**, *52*, 3339.
(39) Press release, June 30, 2008. http://www.novexel.com/.
(40) Ballell, L.; Barros, D.; Brooks, G.; Castro Pichel, J.; Dabbs, S.; Daines, R. A.; Davies, D. T.; Fiandor Roman, J. M.; Giordano, I.; Hennessy, A. J.; Hoffman, J. B.; Jones, G. E.; Miles, T. J.; Pearson, N. D.; Pendrak, I.; Remuinan Blanco, M. J.; Rossi, J. A.; Zhang, L. Patent Application WO 2008/009700 A1, **2008**.
(41) Brown, P.; Dabbs, S.; Davies, D. T.; Pearson, N. D. Patent Application WO 2008/116815 A1, **2008**.
(42) Barfoot, C.; Davies, D. T.; Miles, T.; Pearson, N. D. Patent Application WO 2008/152603, **2008**.
(43) Brown, P.; Dabbs, S.; Hennessy, A. J. Patent Application WO 2009/087153 A1, **2009**.
(44) Giordina, I.; Hennessy, A. J. Patent Application WO 20010/043714, A1, **2010**.
(45) Alemparte-Gallardo, C.; Ballell-Pages, L.; Barros-Aguirre, D.; Cacho-Izquierdo, M.; Castro-Pichel, J.; Fiandor Roman, J. M.; Hennessy, A. J.; Pearson, N. D.; Remuinan-Blanco, J. M. Patent Application WO 2009/090222 A1, **2009**.

(46) Alemparte-Gallardo, C.; Barfoot, C.; Barros-Aguirre, D.; Cacho-Izquierdo, M.; Fiandor Roman, J. M.; Hennessy, A. J.; Pearson, N. D.; Remuinan-Blanco, M. J. Patent Application WO 2009/141398 A1, **2009**.
(47) Alemparte-Gallardo, C.; Barros-Aguirre, D.; Cacho-Izquierdo, M.; Fiandor-Roman, J. M.; Lavandera Diaz, J. L.; Remuinan-Blanco, M. J. Patent Application WO 2010/081874 A1, **2010**.
(48) Barrows, D. Recent advances in TB drug development, 40th IUATLD Meeting, Cancun, Mexico, Dec **2009**.
(49) Presentations, Dec 2009. http://www.newtbdrugs.org/eventfiles/p4/Novel%20Mtb%20DNA%20Gyrase%20inhibitors_The%20Union_40th.pdf.

CHAPTER TWENTY-TWO

HCV Inhibition Mediated Through the Nonstructural Protein 5A (NS5A) Replication Complex

Robert Hamatake, Andrew Maynard, Wieslaw M. Kazmierski
GlaxoSmithKline, Research Triangle Park, Durham, North Carolina, USA

Contents

1. INTRODUCTION

Hepatitis C virus (HCV) affects more than 170 million individuals worldwide. Despite increased monitoring of blood supply and prevention efforts, up to 4.7 million new infections take place each year.[1] While about 20% of all HCV infections clear spontaneously with time, the remaining ones become chronic[2] and potentially lead to steatosis, cirrhosis, and hepatocellular carcinoma.[3]

A combination of an injectable peginterferon and oral ribavirin has been the standard of care for the past decade. This 48-week, often grueling treatment, can be associated with very unpleasant side effects such as flu-like symptoms and fatigue, which, in turn, often leads to treatment discontinuation. The treatment options for HCV have expanded with the recent approval of HCV protease inhibitors (PI), boceprevir and telaprevir. New PI-inclusive treatment regimens have resulted in increased rates of sustained viral response (SVR) for both treatment naïve and experienced patients[4] and

Annual Reports in Medicinal Chemistry, Volume 47
ISSN 0065-7743
http://dx.doi.org/10.1016/B978-0-12-396492-2.00022-9

brought an all-oral HCV therapy closer to realization. To this end, other HCV targets[5] as well as small molecule inhibitors targeting host factors utilized by the virus for replication[6] have been pursued. Particularly noteworthy is the recently discovered class of NS5A inhibitors, due to their novel mechanism of action, high potency, and the promising clinical progression of several compounds.[7,8] Small-molecule NS5A drug discovery has been reviewed.[9,10] This chapter provides a brief historical perspective and a further update of recent activities in the field.

2. FIRST-GENERATION NS5A INHIBITORS

Initial inhibitors that targeted NS5A were peptides derived from an amphiphilic α-helical sequence in NS5A-termed amphiphilic α-helix (AH), thought to be responsible for NS5A binding to the membrane.[11] First-generation small molecule NS5A inhibitors included 4-aminoquinazolines **1**[12,13] from Arrow Therapeutics. Arrow and Astra-Zeneca subsequently advanced AZD-2836 and AZD-7295 to the clinic. Other early NS5A molecules included acetylene derivatives **2** from Presidio and XTL biopharmaceuticals[14] and piperazine-based inhibitors from Merck[15] (Fig. 22.1). Compound **3** in Merck series had replicon $EC_{50} = 160$ nM and caused mutations in NS5A, implicating it as a potential macromolecular target. Early compounds were generally not reported to be potent in the gt1a replicon assay.

3. CURRENT-GENERATION NS5A INHIBITORS

BMS scientists disclosed iminothiazolidinone HCV inhibitors,[16] which were susceptible to NS5A mutations. Optimization of the screening hit *BMS-858* led to *BMS-824* which showed a ~100-fold potency increase. *BMS-824* was found to undergo a spontaneous dimerization to potent compound **4**.[17] Subsequent structural explorations designed to define the minimum pharmacophore in **4** resulted in a picomolar inhibitor

1 2 3

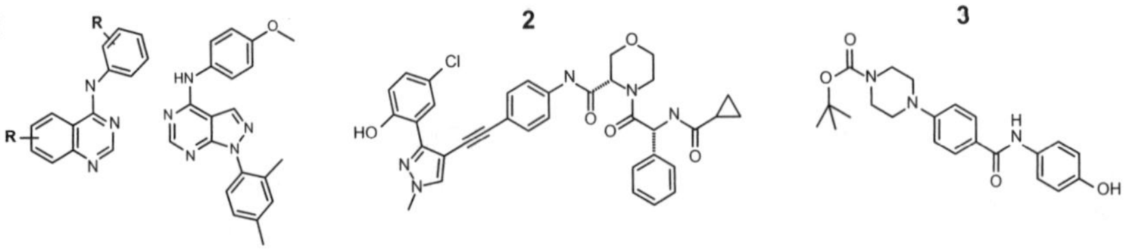

Figure 22.1 Structures of early NS5A small molecule inhibitors.

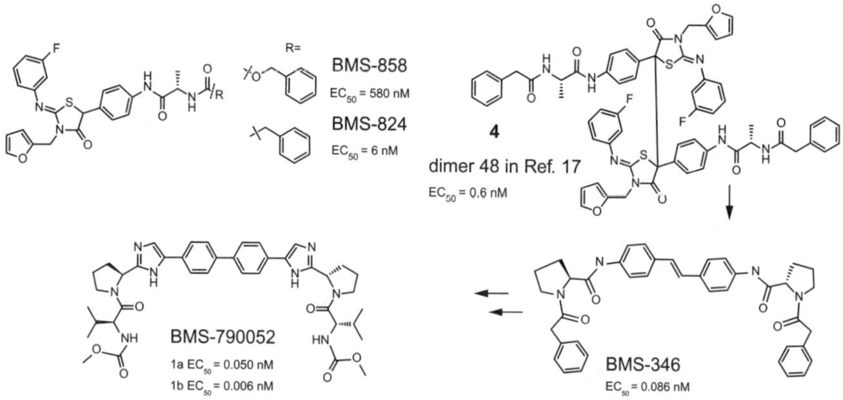

Figure 22.2 BMS progression from screening hit BMS-858 to clinical compound BMS-790052.

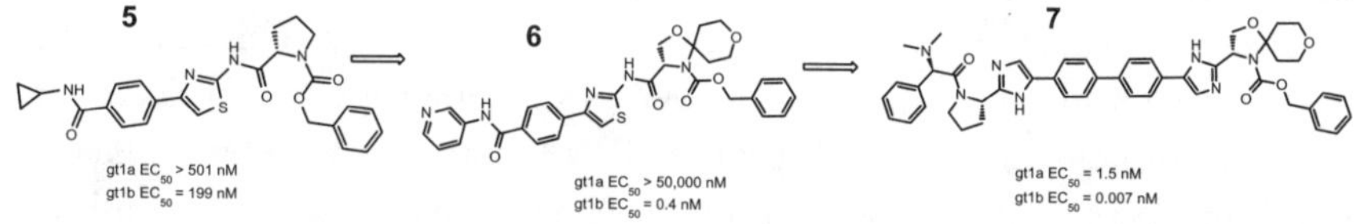

Figure 22.3 Genelabs and GSK discovery of spirocyclic NS5A inhibitor series.

BMS-346,[17,18] leading to BMS-790052. BMS-790052 was potent in both gt1b and gt1a replicon assays and was thus advanced to the clinic.[19] The structure of BMS-790052 (Fig. 22.2), anti-HCV properties, and single-dose monotherapy data in HCV-infected patients were described in 2010.[7]

Genelabs and GSK focused on optimizing novel thiazoles **5**.[20] Replacement of proline by the spiro tetrahydropyran-oxazolidine motif in **6** resulted in potency improvement in gt1b replicon.[21] Further extension of this motif to a pseudosymmetric biphenyl series (exemplified by compound **7**) improved both gt1b and, in particular, gt1a potencies[22–25] (Fig. 22.3).

4. NS5A STRUCTURAL BIOLOGY AND CURRENT INHIBITOR DESIGN

HCV NS5A is one of the six nonstructural viral proteins that coordinate intracellular viral replication. NS5A is very unique, having no known host or viral homolog, other than the NS5A homolog of the closely related GB virus B. It is reported to have multiple functions in the HCV life cycle and interact with a variety of viral and host proteins.[10] Recent studies indicate that NS5A recruits lipid kinase PI4Ka to support the integrity of the membranous replication complex through increased local concentration of PI(4)P.[26,27] Though NS5A is required for viral RNA replication and is essential to virion production, much remains unknown about how it functions mechanistically at the molecular level.

NS5A contains three distinct domains, separated by relatively disordered segments. The C-terminal domain III regulates viral assembly and fitness[28] through a conserved cluster of Ser/Thr residues via phosphorylation,[29] while domain II contains elements that antagonize innate immune defenses[30] and is a site of cyclophilin inhibitors.[31] The N-terminal domain I contains two features: a conserved structural Zn-finger (ZnF) motif that is essential for viral RNA replication[32] and the AH terminus that is implicated in membrane association.

Two X-ray crystallographic structures of the ZnF unit (residues 33–245) have been reported.[33,34] While both structures suggest NS5A dimerization, the relative orientation of the ZnF units differs significantly. This discrepancy has been interpreted as multiple "active" NS5A orientations associated with different protein–protein and protein–RNA interaction roles.[34] Recent trends in NS5A inhibitor design utilize a dimeric pharmacophore, consistent with the premise that a dimeric form of NS5A is inhibited. NS5A inhibitors feature a linearly conjugated bis-aryl imidazole core, symmetrically terminated by peptidic caps, as found in BMS-790052. A variety of spiro and fused-ring variants of proline are also exemplified, as well as alcohol and ether derivatives of valine.

NS5A inhibitor-induced mutations arise predominantly at residue positions 28, 30, 31, and 93 in domain I, color-coded blue in Fig. 22.4. In HCV1b, Y93 is located at edge of the ZnF dimer interface, flanked by residues L28, R30, and L31, located in a flexible loop that connects the AH and ZnF units of domain 1. Modeling[10] suggests that this locus of residues functions as a "hinge" that regulates the orientation and large-amplitude motion of the AH relative to the ZnF unit. The O—O distance between Y93 residues in the monomer units is 19 Å, which serves as an estimate of the width of the dimer interface. This distance is roughly the length of dimeric NS5A inhibitor cores, defined by the intramolecular distance between the imidazole–proline junctions. Assuming resistant mutations are due to direct ligand contacts, symmetry implies dimeric inhibitors interact with NS5A by straddling the ZnF dimer interface, positioning the peptidic caps in proximity to either NS5A loop where mutational resistance is prevalent, modeled in Fig. 22.4. As independent fragments, the caps

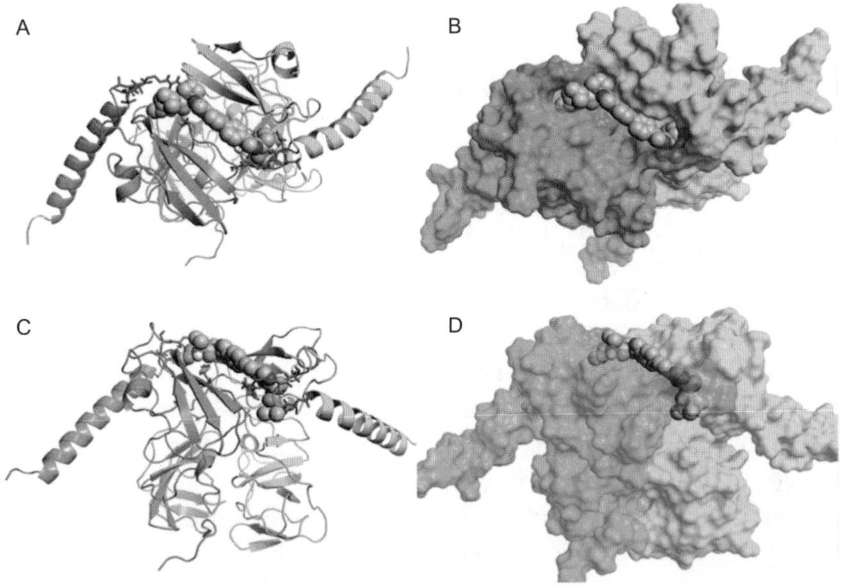

Figure 22.4 Putative binding mode of BMS-790052 with NS5A gt1b domain I, based on the crystal structure of the ZnF domain[34] and homology modeling of the AH and ZnF-AH loop units (residues 1–36).[10] Monomer units are color-coded red and cyan. BMS-790052-resistant mutant sites (gt1a: Met28, Gln30, Leu31, Tyr93) are color-coded blue. (A) Dimeric interaction mode of BMS-790052 (CPK spheres) looking down the C2-symmetry axis of the NS5A homodimer. The bis-phenyl imidazole core of BMS-790052 spans the NS5A dimer interface, positioning the Pro-Val-carbamate caps at the hinge loops, between the ZnF and AH units. (B) Transparent surface rendering of (A). (C) View of the dimer interface, perpendicular to the C2 axis. (D) Transparent surface rendering of (C). (See Color Plate 22.4 in Color Plate Section.)

are assumed to interact only weakly with the loop region ($K_{d,mono} \sim 1\ \mu M$), but by linking the caps to bind cooperatively, inhibition grows exponentially ($K_{d,dimer} \sim K_{d,mono} \cdot K_{d,mono} \sim 1\ pM = \exp[-\Delta G_{dimer}/RT] \sim \exp[-2\Delta G_{mono}/RT]$). This may explain the picomolar potency of dimeric NS5A inhibitors and the abrupt loss in potency that can be observed as dimeric NS5A inhibitors are truncated into monomeric analogues.

Notably, neither NS5A inhibitor cocrystal structures nor labeling studies have been published to date. Thus, the direct binding modes of NS5A inhibitors, dimeric or monomeric, remain elusive. Conceivably, the absence of the connecting loop between helix and ZnF domains, where mutations arise, could preclude the formation of a cocrystal complex with truncated protein constructs used thus far in structural biology. Given the variety of NS5A functions reported in the literature, multiple inhibitory mechanisms of action could also complicate the interpretation of SAR. Future experimental structure-based studies, as well as NS5A functional studies, will hopefully provide clearer insight into the SAR and functions of NS5A.

Since recent NS5A inhibitors are dimeric, their physical properties can exceed optimal lead-like metrics for small molecules by roughly a factor of two. The median MW, clog*P*, and clog*D* are approximately 787, 4.9, and 4.6, respectively, which can present a challenge for achieving *in vivo* oral bioavailability. Nonetheless, encouraging clinical results for BMS-790052 have demonstrated the viability of this dimeric NS5A inhibitor for treating HCV, as well as the caveats of strictly imposing widely accepted, but crude, cutoffs that can subvert discovery and development of promising therapeutics.

Figure 22.5 captures the distribution of patents exemplifying dimeric NS5A inhibitors. Following BMS-790052, a number of companies are in the clinic with reported NS5A inhibitors. While the chemical structures of most NS5A clinical candidates have not been disclosed, Fig. 22.6 illustrates the diversity of several inhibitors reported in the NS5A patent space

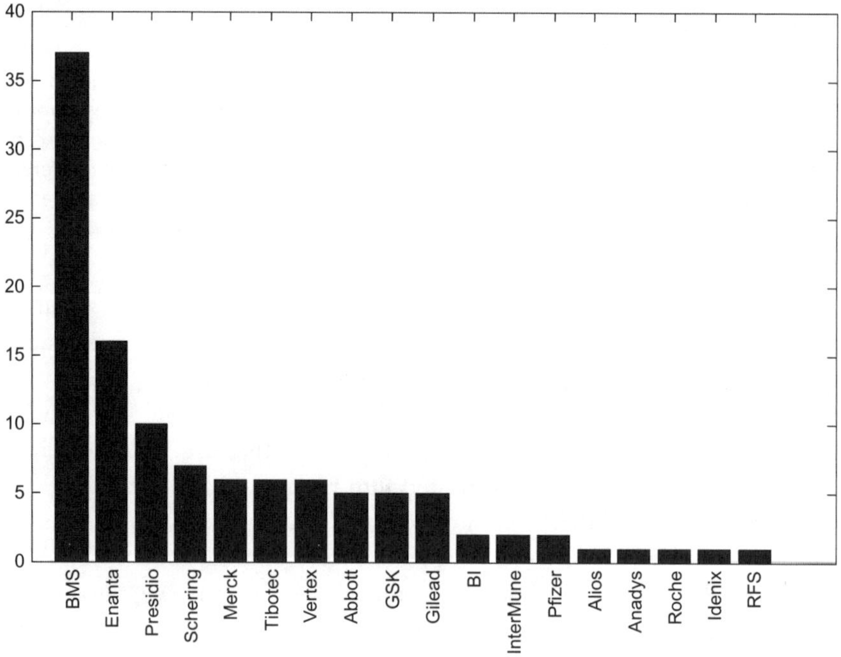

Figure 22.5 The distribution of patent applications associated with dimeric HCV NS5A inhibitors. (For color version of this figure, the reader is referred to the online version of this chapter.)

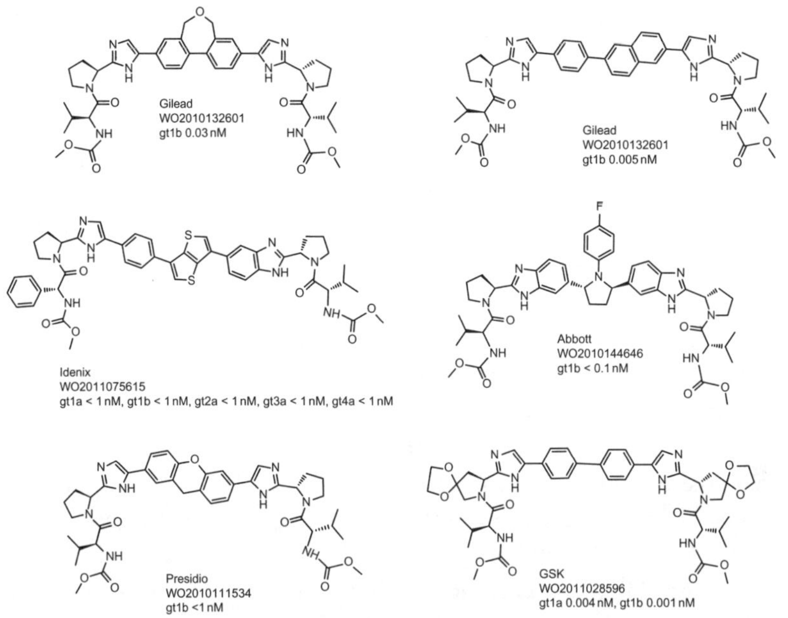

Figure 22.6 Selected diversity of NS5A inhibitors from the recent patent literature.

from companies currently pursing HCV clinical trials. The Abbott core motif is notable in that it suggests an orthogonal dimension for scaffold modification, along the C_2 axis of the core.

5. CLINICAL PROGRESS OF NS5A INHIBITORS

NS5A inhibitors have been evaluated in single and multiple dose studies in healthy volunteers to assess pharmacokinetics and safety. Evaluation in HCV-infected subjects has been done in monotherapy studies using single dose or multiple daily doses or in combination with other regimens in longer-term studies. Endpoints that are commonly used in HCV clinical trials are rapid virological response (RVR, undetectable HCV RNA after 4 weeks), extended RVR (eRVR, undetectable HCV RNA at weeks 4 and 12), and SVR (undetectable HCV RNA 12 weeks after treatment cessation, SVR12, or 24 weeks after treatment cessation, SVR24). SVR24 is considered a virological cure for HCV although regulatory agencies are now accepting SVR12 as an endpoint.

Bristol–Myers Squibb's Daclatasvir (BMS-790052) is the most advanced NS5A inhibitor in the clinic. It was first disclosed at the 59th annual meeting of the American Association for the Study of Liver Diseases and generated substantial interest in this class of inhibitors due to the very rapid and profound viral load reductions resulting from administration of single dose to HCV-infected patients.[7] A 14-day monotherapy trial has confirmed the potency of daclatasvir with a maximum mean log viral load reduction of 2.8 logs achieved with a 1-mg dose administered once a day.[35] However, a low genetic barrier to resistance was displayed by daclatasvir as viral breakthroughs occurred by day 7 at doses as high as 100 mg QD necessitating its combination with an inhibitor with a different mechanism of action. BMS has initiated multiple clinical trials with daclatasvir in combination with other regimens, and the early results are encouraging although these studies are small and have not reached their final endpoints.

In a Phase 2a study,[36] 48 treatment naïve genotype 1 HCV-infected patients were randomized to receive placebo, 3, 10, or 60 mg daclatasvir for 48 weeks in combination with pegylated interferon α2a and ribavirin (PR). In the 3-mg group, 5/12 patients had an RVR and SVR12. In the 10-mg group, 11/12 patients had an RVR and SVR12. In the 60-mg group, 10/12 patients had an RVR and SVR12. These responses were much better than the PR plus placebo group who had RVR in 1/12 patients and SVR12 in 3/12 patients. There were 7, 3, 2, and 9 virologic failures in the 3, 10, 60, and placebo groups, respectively, which highlights the ability of robust combination regimens to reduce virologic failures.

In a study in treatment naïve genotype 1-infected Japanese patients,[37] daclatasvir at 10 or 60 mg QD was given in combination with pegylated interferon α2a and ribavirin for 12 weeks. Patients who had a protocol-defined response (PDR) of HCV RNA $<$ LOQ at week 4 and undetectable at week 12 had an additional 12 weeks of triple therapy, while those who did not achieve PDR had an additional 36 weeks of triple therapy. 7/9 patients in the 10-mg group achieved PDR as did 8/8 patients in the 60-mg group. All patients who achieved PDR and who received 24 weeks of total treatment had an SVR24. The combination of daclatasvir and PR was also tested in patients who had previously undergone treatment with PR but were partial or null responders. 8/9 patients in the 10-mg group achieved PDR and four of these achieved SVR24. In the 60-mg group, 7/9 patients achieved PDR and 6 of these had SVR24. SVR24 rates on the patients who did not achieve PDR and went on to a longer course of treatment are not yet available.

In a similar study using pegylated interferon α2b and ribavirin in combination with 10 or 60 mg daclatasvir,[38] 7/9 treatment naïve patients in the 10-mg group and 10/10 in the 60-mg group achieved PDR and received the shorter treatment course. Of these, six patients in the 10-mg group and nine in the 60-mg group had an SVR24. This study also looked at prior PR nonresponders. The efficacy of daclatasvir in combination with pegylated interferon α2b and ribavirin was not as substantial in this difficult to treat group. Only 5/9 patients in the 10-mg group achieved PDR and only 2 of these had SVR24. In the 60-mg group, 3/9 patients achieved PDR and 2 of these had SVR24. For the overall nonresponder population, SVR24 was achieved in 22% of the 10-mg group and 33% of the 60-mg group.[39]

The SVR24 rates for the treatment naïve patients who had PDR for these two studies (30/32 patients) were similar to the SVR24 rate of 89% for treatment naïve patients who received telaprevir in combination with PR for 12 weeks and, using response guided therapy (undetectable HCV RNA at weeks 4 and 12), an additional 12 weeks PR.[40]

Daclatasvir is currently in a phase 2b study in combination with pegylated interferon α2a and ribavirin in prior PR nonresponders. Interim week 12 data was presented at the 2012 International Liver Congress.[41] In the 20-mg daclatasvir group, 18% of null responders and 26% of partial responders met the primary endpoint of eRVR. In the 60-mg daclatasvir group, 20% of null responders and 36% of partial responders had eRVR. Although the SVR rates are not yet available, these eRVR rates are not encouraging for the use of NS5A inhibitors as an add-on to retreatment with PR of prior nonresponders.

Daclatasvir has also been investigated in combination with other direct acting antivirals (DAAs) in interferon α free regimens. Daclatasvir has been combined with GS-7977, a nucleotide NS5B inhibitor, in treatment naïve patients infected with HCV genotype 1, 2, or 3. In genotype 1a/1b patients, 100% of patients receiving 60 mg daclatasivr and 400 mg GS-7977 for 24 weeks had SVR4. An SVR4 rate of 100% was also achieved in genotype 2 or 3 infected patients. When ribavirin was added on to the combination of daclatasvir and GS-7977, an SVR4 of 100% was obtained in the genotype 1/1b patients and an SVR4 of 86% in genotype 2/3 patients.[42]

An exploratory study in genotype 1 HCV-infected patients who were prior PR nonresponders dosed 11 patients with 60 mg daclatasvir once daily and 600 mg NS3 PI asunaprevir twice daily for 24 weeks. Another group of 10 patients received daclatasvir and asunaprevir in combination with PR for 24 weeks.[43] Four of 11 patients in the dual therapy group achieved the primary endpoint of SVR12, while 10/10 patients in the quadruple therapy group had SVR12. This was the first indication that an interferon free, all DAA regimen can effect a cure. The primary reason for treatment failure in the all DAA regimen was viral breakthrough which occurred in six patients who were infected with genotype 1a HCV. The two genotype 1b-infected patients in this group both had SVR12 consistent with the higher genetic barrier to resistance of NS5A inhibitors against genotype 1b HCV. A recently published study has confirmed the impressive efficacy that can be achieved against genotype 1b.[44] Ten genotype 1b HCV-infected patients who were prior PR nonresponders were treated with 60 mg daclatasvir once daily and initially 600 mg asunaprevir twice daily for 24 weeks. The dose of asunaprevir was reduced to 200 mg twice daily during the study after hepatic enzyme elevations occurred in an asunaprevir plus PR clinical trial. 9/10 patients completed 24 weeks of treatment and all 9 had SVR12 and SVR24. Additional null responder patients were added to this study resulting in an SVR24 for 19 of 21 patients.[45] This is an unprecedented response in the difficult to treat nonresponder population which has a 41% SVR rate when treated with the approved PI telaprevir in combination with PR.[46]

Gilead Sciences' GS-5885 is an NS5A inhibitor that showed clinical proof of concept in a 3-day monotherapy trial in genotype 1 patients.[47] HCV-infected subjects dosed with 1, 3, 10, 30, or 90 mg had median maximal reductions >3 logs in dose groups ≥ 3 mg. The dosing period was too short to detect viral breakthroughs during treatment, but genotypic and phenotypic analysis detected the presence by population sequencing of variants resistant to GS-5885 in all patients dosed ≥ 3 mg at day 4 or 14. Interestingly, three patients whose maximal viral load reductions were <1.6 logs had viral variants at baseline with resistance to NS5A inhibitors (Q30E and L31M and Y93C in genotype 1a). GS-5885 was well tolerated in this study and has progressed to Phase 2 studies. GS-5885 has also been investigated in an interferon free regimen in treatment naïve genotype 1-infected patients. Patients in Arm 1 received 30 mg QD GS-5885, 200 mg QD GS-9451, an NS3 PI, 30 mg BID GS-9190, an NS5B nonnucleoside inhibitor, and ribavirin for 24 weeks. In Arm 2, the dose of GS-5885 was increased to 90 mg QD and patients were randomized after 12 weeks to stop treatment or continue treatment through week 24. Not all patients have completed the study, but interim results presented at the 2012 International Liver Congress are promising.[48] In Arm 1, 80% of the patients who completed the study had an SVR12. In Arm 2, 81% of the patients who received 12 weeks of treatment had SVR12 and 100% of the patients who received 24 weeks of treatment had SVR12. These numbers could change as there are still a number of patients in follow-up, but the prospects for replacing interferon with an all-oral regimen are encouraging based on these early results.

GlaxoSmithKline is developing an NS5A inhibitor with a preclinical profile that is more potent on genotype 1b Y93N and Y93H mutants than the BMS compound daclatasvir.[49] In a Phase 1 study, GSK2336805 showed good PK in single and multiple ascending dose studies and was well tolerated up to 60 mg as a single dose and 75 mg for 14 days in multiple doses. Genotype 1 HCV-infected patients were dosed with 1, 10, 30, 60, and 120 mg of GSK2336805 and HCV viral load was monitored. 2.0–3.9 log reductions were achieved after single dose of $\geq$10 mg of GSK2336805.[50] Genotypic analysis of posttreatment samples detected the emergence of variants associated with resistance to GSK2336805 in 12 of 14 subjects dosed with $\geq$10 mg GSK2336805.

ABT-267 is Abbott Laboratories' NS5A inhibitor in clinical development for HCV. Single and multiple ascending dose studies in healthy volunteers have shown good PK and tolerability.[51] In a 3-day monotherapy study conducted in HCV genotype 1-infected patients, doses of 5, 50, and 200 mg QD of ABT-267 resulted in mean maximal log reductions in viral loads of 2.89, 2.77, and 3.1, respectively.[52] These same doses were given for 12 weeks in combination with PR followed by PR alone for an additional 36 weeks. One of the primary endpoints was RVR which was achieved in 33% of the 5-mg group, 56% of the 50-mg group, and 70% of the 200-mg group. SVR24 is not yet available from this ongoing study.[53]

Presidio Pharmaceuticals, Inc. is developing two NS5A inhibitors. PPI-461 is their lead compound which has shown proof of concept in the clinic, and PPI-668 is their back-up compound currently in a Phase 1b study. PPI-461, like many of the NS5A inhibitors in development, is an extremely potent inhibitor of genotype 1b HCV with slightly lower activity against genotype 1a. PPI-668 is more active against replicons containing the NS5A gene from genotypes 3a and 6a than PPI-461. PPI-461 was dosed at 50, 100, and 200 mg daily for 3 days and achieved mean log viral load reductions of 2.65, 3.65, and 3.62 logs, respectively.[54] One patient in the 50-mg cohort only had a 0.4-log reduction, and genotypic analysis showed the presence at baseline of HCV with four linked NS5A-resistance mutations. Resistance substitutions in NS5A were detected in 17/18 subjects after 3 days of monotherapy.[55]

Achillion Pharmaceuticals, Inc. has ACH-2928 in clinical development. ACH-2928 was dosed at 10 and 60 mg QD for 3 days in a monotherapy study in HCV genotype 1-infected patients. The 10-mg group had a mean maximal log viral load reduction of 2.79 logs, and the 60-mg group had a mean maximal log viral load reduction of 3.68 logs.[56] Achillion also has another NS5A inhibitor in development. ACH-3102 has improved potency against HCV replicons containing mutations associated with resistance to other NS5A inhibitors. Unlike ACH-2928 and daclatasvir, ACH-3102 also retains potency against chimeric replicons containing NS5A from genotype 2a and 2b patient isolates.[57]

6. FUTURE PROSPECTS

The chemical exploration of small molecules targeting NS5A has increased in the past 2–3 years. The clinical progression of NS5A inhibitors has also increased in terms of the number of candidates in the clinic and in the number of novel regimens that are being investigated with them. The addition of NS5A inhibitors to the clinician's armamentarium will improve SVR rates in difficult to treat populations such as interferon nonresponders. More importantly, the success seen in combinations of these very potent inhibitors with other DAAs shows the potential to completely replace interferon containing therapies with all oral drugs. The current NS5A inhibitors in clinical development are very potent inhibitors of genotype 1 HCV but are less potent against some of the other genotypes. Designing NS5A inhibitors with potent pan-genotype activity will allow all oral regimens to be extended to all HCV-infected patients regardless of genotype.

REFERENCES

(1) Alter, M.J. *World J. Gastroenterol.* **2007**, *13*, 2436.

(2) Kamal, S.M.; Fouly, A.E.; Kamel, R.R.; Hockenjos, B.; Al Tawil, A.; Khalifa, K.E.; He, Q.; Koziel, M.J.; El Naggar, K.M.; Rasenack, J.; Afdhal, N.H. *Gastroenterology* **2006**, *130*, 632.

(3) Tong, M.J.; el-Farra, N.S.; Reikes, A.R.; Co, R.L. *N. Engl. J. Med.* **1995**, *332*, 1463.

(4) Adeel, A.B.; Kanwal, F. *Clin. Infect. Dis.* **2012**, *54*, 96.

(5) Lemon, S.M.; McKeating, J.A.; Pietschmann, T.; Frick, D.N.; Glenn, J.S.; Tellinghuisen, T.L.; Symons, J.; Furman, P.A. *Antiviral Res.* **2010**, *86*, 79.

(6) Poenisch, M.; Bartenschlager, R. *Semin. Liver Dis.* **2010**, *30*, 333.

(7) Gao, M.; Nettles, R.E.; Belema, M.; Snyder, L.B.; Nguyen, V.N.; Fridell, R.A.; Serrano-Wu, M.H.; Langley, D.R.; Sun, J.-H.; O'Boyle, D.R., II; Lemm, J.A.; Wang, C.; Knipe, J.O.; Chien, C.; Colonno, R.J.; Grasela, D.M.; Meanwell, N.; Hamann, L.G. *Nature* **2010**, *465*, 96.

(8) Fridell, R.A.; Qiu, D.; Wang, C.; Valera, L.; Gao, M. *Antimicrob. Agents Chemother.* **2010**, *54*, 3641.

(9) Schmitz, U.; Tan, S.-L. *Recent Pat. Antiinfect. Drug Discov.* **2008**, *3*, 77.

(10) Cordek, D.G.; Bechtel, J.T.; Maynard, A.T.; Kazmierski, W.M.; Cameron, C.E. *Drug Future* **2011**, *36*, 691.

(11) Glenn, J.S.; Myers, T.M.; Glass, J.I. WO Patent Application 2002089731, **2002**.

(12) Tiberghien, N.; Lumley, J.; Reynolds, K.; Angell, R.M.; Matthews, N.; Cockerill, G.S.; Barnes, M.C. Patent Application WO2005047288, **2005**.

(13) Dennison, H.; Mathews, N.; Barnes, M.; Chana, S. Patent Application WO2005105761, **2005**.
(14) Li, G.; Fathi, R.; Yang, Z.; Liao, Y.; Zhu, Q.; Lam, A.; Sandrasagra, A.; Nawoschik, K.; Cho, H.-J.; Cao, J.; Ruoqiu, W.; Wobbe, C.R. Patent Application WO2008048589, **2008**.
(15) Conte, I.; Giuliano, C.; Ercolani, C.; Narjes, F.; Koch, U.; Rowley, M.; Altamura, S.; De Francesco, R.; Neddermann, P.; Migliaccio, G.; Stansfield, I. *Bioorg. Med. Chem. Lett.* **2009**, *19*, 1779.
(16) Colonno, R.; Lemm, J.; O'Boyle, D.; Gao, M.; Romine, J.L.; Martin, S.; Snyder, L.B.; Serrano-Wu, M.; Deshpande, M.; Whitehouse, D. Patent Application WO2004014313, **2004**.
(17) Romine, J.L.; St. Laurent, D.R.; Leet, J.E.; Martin, S.W.; Serrano-Wu, M.H.; Yang, F.; Gao, M.; O'Boyle, D.R., II; Lemm, J.A.; Sun, J.-H.; Nower, P.T.; Huang, X.; Deshpande, M.S.; Meanwell, N.A.; Snyder, L.B. *ACS Med. Chem. Lett.* **2011**, *2*, 224.
(18) Lemm, J.A.; Leet, J.E.; O'Boyle, D.R., II; Romine, J.L.; Huang, X.S.; Schroeder, D.R.; Alberts, J.; Cantone, J.L.; Sun, J.-H.; Nower, P.T.; Martin, S.W.; Serrano-Wu, M.H.; Meanwell, N.A.; Snyder, L.B.; Gao, M. *Antimicrob. Agents Chemother.* **2011**, *55*, 3795.
(19) Gao, M.; Fridell, R.; O'Boyle II, D.; Qiu, D.; Sun, J.H.; Lemm, J.; Nower, P.; Valera, L.; Voss, S.; Liu, M.; Belema, M.; Nguyen, V.; Romine, J.L.; Martin, S.W.; Serrano-Wu, M.; St. Laurent, D.; Snyder, L.B.; Colonno, R.C.; Hamann, L.G.; Meanwell, N.A. 15th international symposium on hepatitis C virus & related viruses, San Antonio, Texas, Oct 5–9, **2008**
(20) Schmitz, F.U.; Roberts, C.D.; Abadi, A.D.M.; Griffith, R.C.; Leivers, M.R.; Slobodov, I.; Rai, R. Patent Application WO2007070600, **2007**.
(21) Leivers, M.R.; Schmitz, F.U.; Griffith, R.C.; Roberts, C.D.; Dehghani Mohammad Abadi, A.; Chan, S.A.; Rai, R.; Slobodov, I.; Ton, T.L.; Patent Application WO2008064218, **2008**.
(22) Schmitz, F.U.; Rai, R.; Roberts, C.D.; Kazmierski, W.; Grimes, R. Patent Application WO2010062821, **2010**.
(23) Baskaran, S.; Botyanszki, J.; Cooper, J.P.; Duan, M.; Kazmierski, W.M.; McFadyen, R.B.; Redman, A. Patent Application WO2011050146A1, **2011**.
(24) Baskaran, S.; Dickerson, S.H.; Duan, M.; Kazmierski, W.M.; McFadyen, R.B. Patent Application WO2011091446 A1, **2010**.
(25) Cooper, J.P.; Duan, M.; Grimes, R.M.; Kazmierski, W.M.; Tallant, M.D. Patent Application WO2010088394A1, **2010**.
(26) Berger, K.L.; Cooper, J.D.; Heaton, N.; Yoon, R.; Oakland, T.; Jordan, T.; Mateu, G.; Grakoui, A.; Randall, G. *Proc. Natl. Acad. Sci. U.S.A.* **2009**, *106*, 7577.
(27) Reiss, S.; Rebhan, I.; Backes, P.; Romero-Brey, I.; Erfle, H.; Matula, P.; Kaderali, L.; Poenisch, M.; Blankenburg, H.; Hiet, M.S.; Longerich, T.; Diehl, S.; Ramirez, F.; Balla, T.; Rohr, K.; Kaul, A.; Bühler, S.; Pepperkok, R.; Lengauer, T.; Albrecht, M.; Eils, R.; Schirmacher, P.; Lohmann, V.; Bartenschlager, R. *Cell Host Microbe* **2011**, *9*, 32.
(28) Appel, N.; Zayas, M.; Miller, S.; Krijnse-Locker, J.; Schaller, T.; Friebe, P.; Kallis, S.; Engel, U.; Bartenschlager, R. *PLoS Pathog.* **2008**, *4*, 1.
(29) Kim, S.; Welsch, C.; Yi, M.; Lemon, S.M. *J. Virol.* **2011**, *85*, 6645.
(30) Toroney, R.; Nallagatla, S.; Boyer, J.; Cameron, C.; Bevilacqua, P. *J. Mol. Biol.* **2010**, *400*, 393.
(31) Yang, F.; Robotham, J.; Grise, H.; Frausto, S.; Madan, V.; Zayas, M.; Bartenschlager, R.; Robinson, M.; Greenstein, A.; Nag, A.; Logan, T.; Bienkiewicz, E.; Tang, H. *PLoS Pathog.* **2010**, *6*, 1.

(32) Tellinghuisen, T.L.; Marcotrigiano, J.; Gorbalenya, A.F.; Rice, C.M. *J. Biol. Chem.* **2004**, *279*, 48576.
(33) Tellinghuisen, T.L.; Marcotrigiano, J.; Rice, C.M. *Nature* **2005**, *435*, 374.
(34) Love, R.A.; Brodsky, O.; Hickey, M.J.; Wells, P.A.; Cronin, C.N. *J. Virol.* **2009**, *83*, 4395.
(35) Nettles, R.E.; Gao, M.; Bifano, M. *Hepatology* **2011**, *54*, 1956.
(36) Pol, S.; Ghalib, R.H.; Rustgi, V.K.; Martorell, C.; Everson, G.T.; Tatum, H.A.; Hezode, C.; Lim, J.K.; Bronowicki, J.-P.; Abrams, G.A.; Brau, N.; Morris, D.W.; Thuluvath, P.; Reindollar, R.; Yin, P.D.; Diva, U.; Hindes, R.; McPhee, F.; Gao, M.; Thiry, A.; Schnittman, S.; Hughes, E.A. EASL **2011**, Poster 1373.
(37) Izumi, N.; Asahina, Y.; Yokosuka, O.; Imazeki, F.; Kawada, N.; Tamori, A.; Osaki, Y.; Kimura, T.; Yamamoto, K.; Takaki, A.; Sata, M.; Ide, T.; Ishikawa, H.; Ueki, T.; Yang, R.; McPhee, F.; Hughes, E.A. *Hepatology* **2011**, *54*, 1439A.
(38) Suzuki, F.; Chayama, K.; Kawakami, Y.; Toyota, J.; Karino, Y.; Mochida, S.; Nagashi, S.; Hayashi, N.; Takehara, T.; Ishikawa, H.; Miyagoshi, H.; Yang, R.; McPhee, F.; Hughes, E.A.; Kumada, H. *Hepatology* **2011**, *54*, 1441A.
(39) Suzuki, F.; Chayama, K.; Kawakami, Y.; Toyota, J.; Karino, Y.; Mochida, S.; Nagoshi, S.; Hayashi, N.; Takehara, T.; Kumada, H. APASL **2012**.
(40) Jacobson, I.M.; McHutchison, J.G.; Dusheiko, G.; Di Bisceglie, A.M.; Reddy, K.R.; Bzowej, N.H.; Marcellin, P.; Muir, A.J.; Ferenci, P.; Flisiak, R.; George, J.; Rizzetto, M.; Shouval, D.; Sola, R.; Terg, R.A.; Yoshida, E.M.; Adda, N.; Bengtsson, L.; Sankoh, A.J.; Kieffer, T.L.; George, S.; Kauffman, R.S.; Zeuzem, S.; ADVANCE Study Team, *N. Engl. J. Med.* **2011**, *364*, 2405.
(41) Ratzui, V.; Gadano, A.; Pol, S.; Hezode, C.; Ramji, A.; Cheng, W.; Sulkowski, M.; Everson, G.; Diva, U.; McPhee, F.; Wind-Rotolo, M.; Hughes, E.A.; Yin, P.D.; Schnittman, S. *J. Hepatol.* **2012**, *56*, S478.
(42) Sulkowski, M.; Gardiner, D.; Lawitz, E.; Hinestrosa, F.; Nelson, D.; Thuluvath, P.; Rodriguez-Torres, M.; Lok, A.; Schwartz, H.; Reddy, K.R.; Eley, T.; Wind-Rotolo, M.; Huang, S.-P.; Gao, M.; McPhee, F.; Hindes, R.; Symonds, W.; Pasquinelli, C.; Grasela, D. EASL **2012**, poster 1422.
(43) Lok, A.; Gardiner, D.F.; Lawitz, E.; Gardiner, D.F.; Lawitz, E.; Martorell, C.; Everson, G.T.; Ghalib, R.; Reindollar, R.; Rustgi, V.; McPhee, F.; Wind-Rotolo, M.; Persson, A.; Zhu, K.; Dimitrova, D.I.; Eley, T.; Guo, T.; Grasela, D.M.; Pasquinelli, C. *New Engl. J. Med.* **2012**, *366*, 216.
(44) Chayama, K.; Takahashi, S.; Toyota, J.; Karino, Y.; Ikeda, K.; Ishikawa, H.; Watanabe, H.; McPhee, F.; Hughes, E.; Kumada, H. *Hepatology* **2012**, *55*, 742.
(45) Suzuki, F.; Ikeda, K.; Toyota, J.; Karino, Y.; Ohmura, T.; Chayama, K.; Takahashi, S.; Kawakami, Y.; Ishikawa, H.; Watanabe, H.; Hu, W.; McPhee, F.; Hughes, E.; Kumada, H. *J. Hepatol.* **2012**, *56*, S7.
(46) Zeuzem, S.; Andreone, P.; Pol, S.; Lawitz, E.; Diago, M.; Roberts, S.; Focaccia, R.; Younossi, Z.; Foster, G.R.; Horban, A.; Ferenci, P.; Nevens, F.; Müllhaupt, B.; Pockros, P.; Terg, R.; Shouval, D.; van Hoek, B.; Weiland, O.; Van Heeswijk, R.; De Meyer, S.; Luo, D.; Boogaerts, G.; Polo, R.; Picchio, G.; Beumont, M.; REALIZE Study Team, *New Engl. J. Med.* **2011**, *364*, 2417.
(47) Lawitz, E.J.; Gruener, D.; Hill, J.M.; Marbury, T.; Moorehead, L.; Mathias, A.; Cheng, G.; Link, J.O.; Wong, K.A.; Mo, H.; McHutchison, J.G.; Brainard, D.M. *J. Hepatol.* **2012**, *54*, S481. http://dx.doi.org/10.1016/j.jhep.2011.12.029.
(48) Sulkowski, M.; Rodriguez-Torres, M.; Lawitz, E.; Shiffman, M.; Pol, S.; Herring, R.; McHutchison, J.; Pang, P.; Brainard, D.; Wyles, D.; Habersetzer, F. *J. Hepatol.* **2012**, *56*, S560.
(49) Bechtel, J.; Crosby, R.; Wang, A.; Woldu, E.; Van Horn, S.; Horton, J.; Creech, K.; Caballo, L.H.; Voitenleitner, C.; Vamathevan, J.; Duan, M.; Spaltenstein, A.; Kazmierski, W.; Roberts, C.; Hamatake, R. *J. Hepatol.* **2011**, *54*, S307.

(50) Spreen, W.; Wilfret, D.; Bechtel, J.; Adkison, K.K.; Lou, Y.; Jones, L.; Willsie, S.K.; Glass, S.J.; Roberts, C.D. *Hepatology* **2011**, *54*, 400A.
(51) Dumas, E.O.; Lawal, A.; Menon, R.M.; Podsadecki, T.; Awni, W.; Dutta, S.; Williams, L. *J. Hepatol.* **2011**, *54*, S475.
(52) Lawitz, E.; Marbury, T.; Campbell, A.; Dumas, E.; Kapoor, M.; Pilot-Matias, T.; Krishnan, P.; Setze, C.; Xie, W.; Podsadecki, T.; Bernstein, B.; Williams, L. *J. Hepatol.* **2012**, *56*, S469.
(53) Sullivan, G.J.; Rodrigues-Torres, M.; Lawitz, E.; Poordad, F.; Kapoor, M.; Campbell, A.; Setze, C.; Xie, W.; Khatri, A.; Dumas, E.; Krishnan, P.; Pilot-Matias, T.; Williams, L.; Bernstein, B. *J. Hepatol.* **2012**, *56*, S480.
(54) Lalezari, J.; Agarwal, K.; Dusheiko, G.M.; Brown, A.S.; Weis, N.; Christensen, P.B.; Laursen, A.L.; Asmuth, D.M.; Vig, P.; Ruby, E.; Huang, N.; Huang, Q.; Colonno, R.; Harding, G.D.; Brown, N.A. *Hepatology* **2011**, *54*, 400A.
(55) Huang, N.; Huang, Q.; Lau, M.; Bencsik, M.; Huq, A.; Peng, E.; Agarwal, K.; Lalezari, J.; Vig, P.; Brown, N.A.; Colonno, R. *Hepatology* **2011**, *54*, 536A.
(56) Vince, B.; Lawitz, E.; Searle, S.; Marbury, T.; Robison, H.; Robarge, L.; Olek, E. *J. Hepatol.* **2012**, *56*, S480.
(57) Yang, G.; Wiles, J.; Patel, D.; Zhao, Y.; Fabrycki, J.; Weinheimer, S.; Marlor, C.; Rivera, J.; Wang, Q.; Gadhachanda, V.; Hashimoto, A.; Chen, D.; Pais, G.; Wang, X.; Deshpande, M.; Stauber, K.; Huang, M.; Phadke, A. *J. Hepatol.* **2012**, *56*, S330.

PART 6

Topics in Biology

Editor: John Lowe
JL3Pharma LLC, Stonington, Connecticut

CHAPTER TWENTY-THREE

Antibody–Drug Conjugates for Targeted Cancer Therapy

Victor S. Goldmacher, Thomas Chittenden, Ravi V.J. Chari, Yelena V. Kovtun, John M. Lambert
ImmunoGen, Inc., Waltham, Massachusetts, USA

Contents

ABBREVIATIONS

ADC antibody–drug conjugate
DM1 and DM4 thiol-containing maytansinoids
FDA U.S. Food and Drug Administration
MMAE monomethyl auristatin E
MMAF monomethyl auristatin F
PSMA prostate-specific membrane antigen
vc valine-citrulline dipeptide

Annual Reports in Medicinal Chemistry, Volume 47
ISSN 0065-7743
http://dx.doi.org/10.1016/B978-0-12-396492-2.00023-0

1. INTRODUCTION

An antibody–drug conjugate (ADC) for cancer therapy consists of a tumor-targeting antibody chemically attached (through a "linker") to a cytotoxic effector molecule (also called the "payload" or "drug").

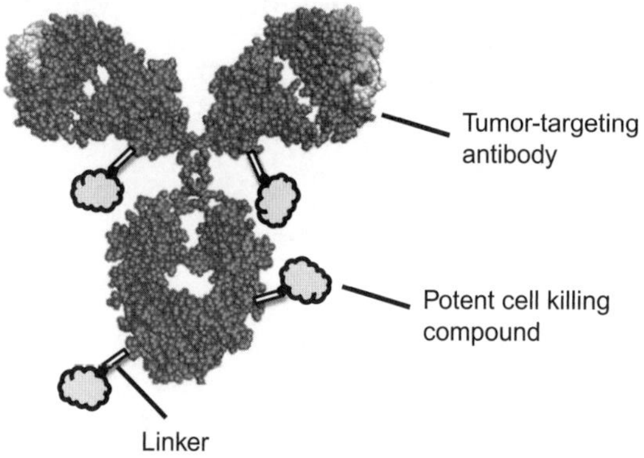

Mechanistically, an ADC acts by binding to the target antigen on the cell surface, followed by its internalization *via* antigen-mediated endocytosis, trafficking into the lysosome, and the release of the payload through the proteolytic degradation of the antibody moiety and/or cleavage of the linker.[1] The rationale for developing an ADC is that linking a cytotoxic agent to a tumor-targeting antibody will enable its selective targeting to cancer cells, leading to their eradication while sparing cells in normal tissues. Compelling clinical results with ADCs in both hematological malignancies and solid tumors has prompted renewed interest in the field and expanded efforts to exploit ADCs as a validated platform for developing new targeted anticancer compounds. ADCs represent a novel class of targeted therapy in oncology, with a distinct mode of action. Most targeted cancer therapies utilize small molecules or naked antibodies that inhibit key signal transduction molecules/pathways required for tumor cell growth or survival. Successful development of such agents requires intricate understanding of tumor cell biology, oncogenic driver mutations, and signaling pathways, which can vary widely even within a given tumor type. In the ADC approach, a potent cytotoxic agent is delivered selectively to cancer cells, exerting an antitumor mechanism of action that is not inherently dependent on particular oncogenic signaling pathways. ADC specificity is thus provided by the targeting antibody rather than by blocking key signal transduction nodes uniquely required for tumor growth or survival.

2. TARGET SELECTION

Selective expression of the target antigen on tumors relative to normal tissues provides the basis for ADC targeting to tumors and minimizes the potential for targeted toxicity in normal tissues. The abundance of the target antigen on the cell surface and its distribution in the tumor (homogeneous vs. heterogeneous expression) can be important determinants of ADC efficacy.[2,3] Attachment to secreted or shed antigen can alter the pharmacokinetics of an ADC and potentially limit effective targeting of tumors.[4] ADC effectiveness ultimately reflects a combination of target antigen properties including antigen density, internalization and intracellular processing, and sensitivity of the target cell to the cytotoxic payload. Most tumor-associated antigens are also expressed to some extent on normal tissues. Such antigens can be considered if expression is restricted to tissues that do not present a toxicity concern. For example, ADCs for prostate cancer have targeted PSMA or Prostate Stem Cell antigen (PSMA),[5,6] which are expressed on normal prostate tissue; gemtuzumab ozogamicin, **1**, targets CD33, which is expressed on both malignant and normal cells of myeloid lineage. Certain markers specific for B-cell malignancies targeted by ADCs, CD22, CD19, CD20, CD79b, and CD37,[5,7–9] are also expressed on normal B-cells, but their temporary depletion can be tolerated.

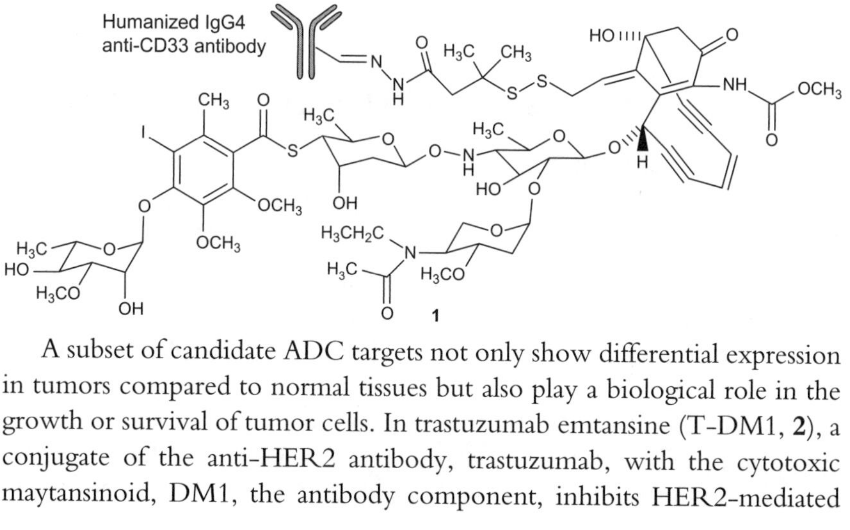

1

A subset of candidate ADC targets not only show differential expression in tumors compared to normal tissues but also play a biological role in the growth or survival of tumor cells. In trastuzumab emtansine (T-DM1, **2**), a conjugate of the anti-HER2 antibody, trastuzumab, with the cytotoxic maytansinoid, DM1, the antibody component, inhibits HER2-mediated signal transduction and is used as naked antibody therapy for metastatic breast cancer.

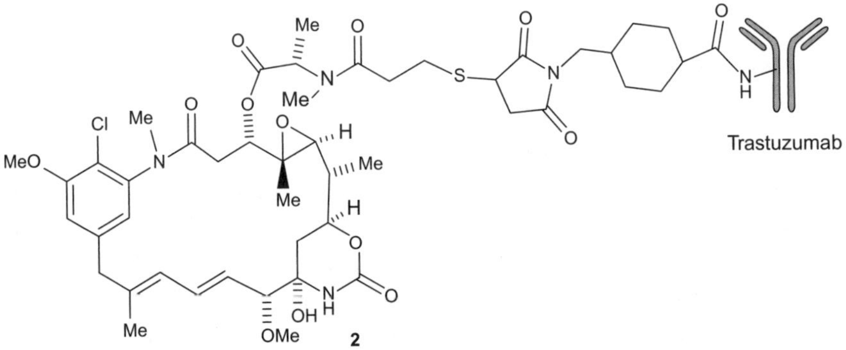

2

All of the functional properties of trastuzumab are retained in **2**, including HER2 binding affinity, inhibition of HER2 signaling, and antibody-dependent cellular cytotoxicity.[10] Moreover, based on clinical experience, **2** can be dosed in patients at levels that approach those used for trastuzumab and other naked "functional" antibodies. This raises the attractive possibility of developing ADCs that combine intrinsic antitumor activity(s) from the antibody component with the potent cytotoxic activity provided by the payload agent. IMGN529 is an example of an ADC designed with this rationale, comprising an anti-CD37 antibody with direct antitumor activities (proapoptotic, complement-dependent cytotoxicity, and antibody-dependent cellular cytotoxicity), conjugated to DM1 using the same linker and conjugate design as in **2**.[8] IMGN529 exhibits potent activity against B-cell tumors in preclinical studies.[8]

3. CYTOTOXIC AGENTS AND LINKERS USED IN ADCs

3.1. Considerations

Favorable parameters for the cytotoxic effectors include potency,[11] selectivity of impact on proliferating cells, low-molecular weight and the related lack of immunogenicity, reproducible and cost-effective production, attachability, and sufficient solubility and stability in water to facilitate preparation and storage of the ADC.[12] The cytotoxic molecules used in the ADCs currently in clinical testing are either microtubule-impacting agents or DNA-interacting agents. In addition, there are several ADCs with other cytotoxic agents in preclinical evaluation.[11,13,14] Delivery of the payload by an antibody requires that they are stably linked. However, once the antibody has bound to the antigen on the surface of cancer cells and the antibody–antigen complex is internalized into the cell, the cytotoxic agent needs to be released to enable it to efficiently arrive at the target and to inactivate it.

3.2. Maytansinoids

Maytansine **3** and its derivatives, maytansinoids, interfere with microtubule dynamics.[15]

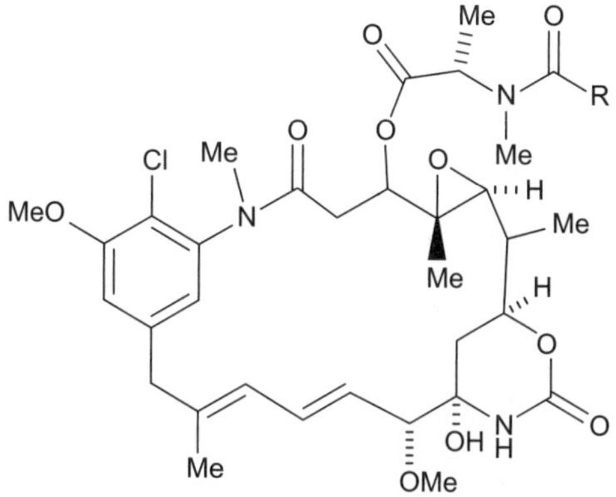

3: R = CH_3
4: R = CH_2CH_2SH
5: R = $CH_2CH_2C(CH_3)_2SH$

Proliferating cells are more sensitive to maytansine than are non-proliferating cells.[16] Appropriately designed maytansinoid agents can have potent cytotoxic activity, and, at the same time, good aqueous stability and reasonable aqueous solubility.[17] Structure–activity relationship studies of maytansinoids showed that the C3 ester side chain is required for biological activity but that the structure of the side chain can be modified without a significant loss of activity. Incorporation of methyldisulfide substituents into the C3 ester side chain has resulted in proprietary maytansinoid derivatives that retained or exceeded the *in vitro* potency of the parent compound. Reduction of the methyldisulfide led to the formation of a thiol group that enables linkage to antibodies. Two of these thiol-containing maytansinoids, $N^{2'}$-deacetyl-$N^{2'}$-(3-mercapto-1-oxopropyl)-maytansine (DM1, **4**) and $N^{2'}$-deacetyl-$N^{2'}$-(4-mercapto-4-methyl-1-oxopentyl)-maytansine (DM4, **5**), were linked to antibodies *via* cleavable disulfide bonds to provide antibody-DM1 (**6**) and antibody-DM4, SAR3419, (**7**).[1,18] DM1 was also linked to antibodies *via* a noncleavable thioether linker as in **2**.[19] Newer thiol-containing maytansinoids and linkers have been synthesized with one or more methyl groups (substituting hydrogen) on the carbon atom geminal to the sulfhydryl group, allowing for the generation of conjugates with varying degrees of steric hindrance around the disulfide bond, allowing a fine control of the rate of bond cleavage.[1,13]

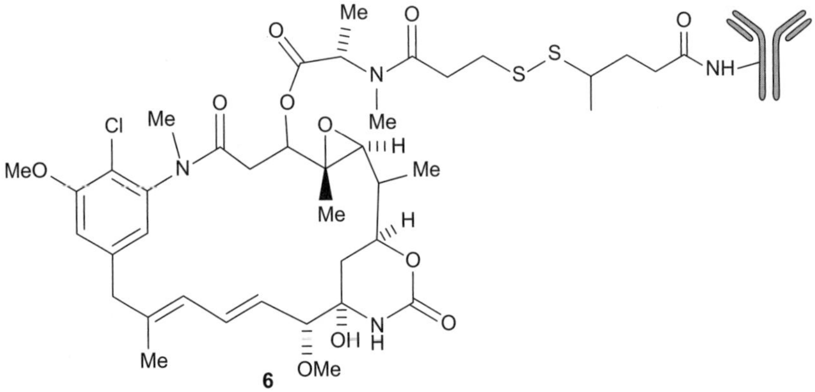

6

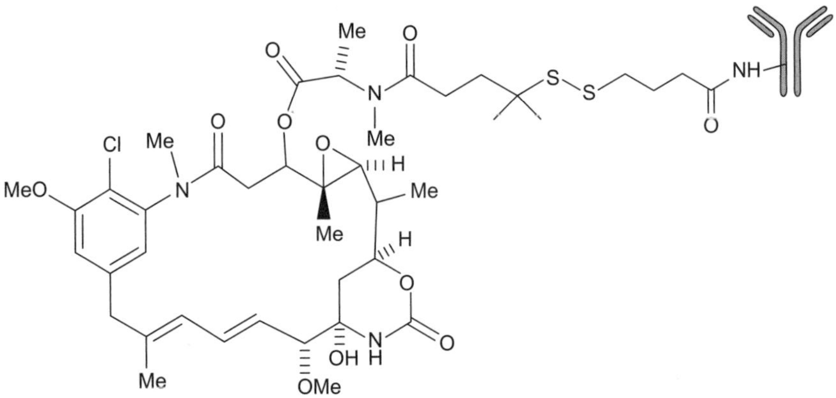

7

3.3. Auristatins

The auristatins, such as monomethyl auristatin E (MMAF) **8a** and monomethyl auristatin F (MMAF) **8b**, are analogs of dolastatin 10, a microtubule-impacting agent.[20] To prepare auristatin conjugates (MMAE conjugate **8c** shown), native disulfide bonds within the antibody are reduced to generate cysteine residues that are subsequently linked to a maleimido derivative of MMAE or MMAF, *via* a dipeptide linker.[20–22] New auristatin ADC linkers have been generated by replacing the maleimide with a halo-acetamide and have shown improved *in vivo* stability in preclinical studies.[23] Novel dipeptide linkers and new auristatins linked through the C-terminus, and some of these new conjugates showed improved therapeutic windows over the original valine-citrulline (vc)-para-aminobenzyl carbamate–MMAF conjugate.[24]

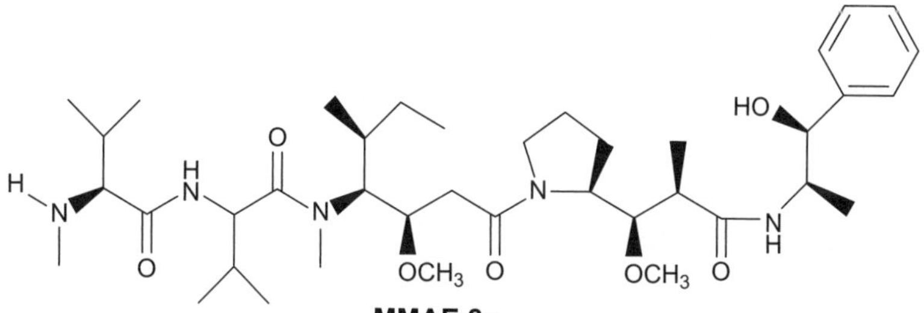

MMAE 8a

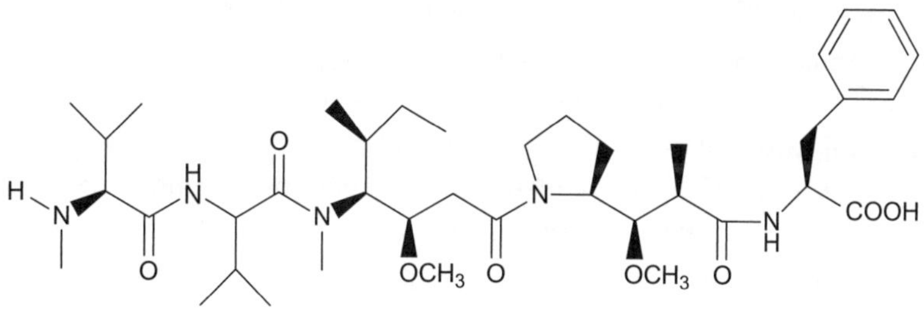

MMAF 8b

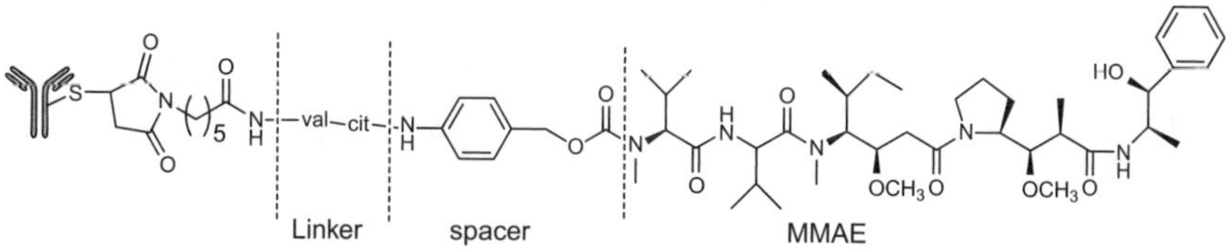

8c

3.4. Calicheamicins

Calicheamicins[25] induce DNA double-strand breaks. *N*-acetyl γ−calicheamicin, **9** was linked to lysine residues of antibodies through an acid-labile hydrazone bond or a noncleavable linker.[26] In gemtuzumab ozogamicin, *N*-acetyl-γ-calicheamicin dimethyl hydrazide is conjugated using a 4-(4′-acetylphenoxy) butanoic acid linker, which is stable at physiologic near-neutral pH, but hydrolyses at the acidic pH (~4) of lysosomes.[27]

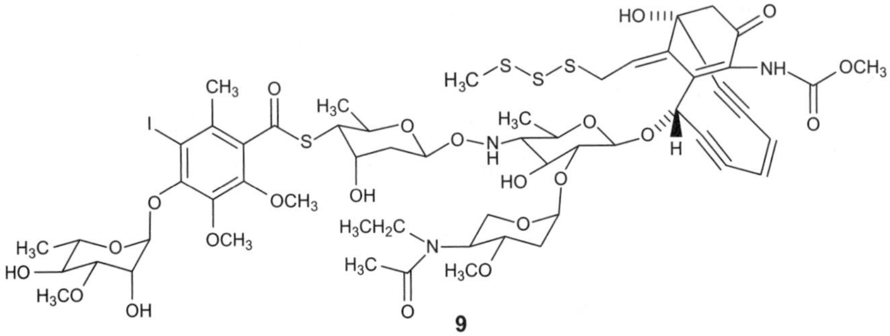

9

3.5. CC-1065 analogues

The parent compound CC-1065, **10**, bearing a cyclopropapyrroloindole pharmacophore alkylates *N*-3 guanine residues of DNA. Analogues containing another alkylating subunit, cyclopropabenzindole of CC-1065, proved chemically more stable, biologically more potent, and synthetically accessible.[28] A cyclopropabenzindole-based CC-1065 analogue, DC1, **11a** has been conjugated to antibodies *via* disulfide bonds. The resulting ADCs were highly cytotoxic in an antigen-specific manner but were not developed because of their instability in water and poor solubility. These problems were overcome by conversion of DC1 into the phenolic phosphate prodrug, DC4, **11b**.[29]

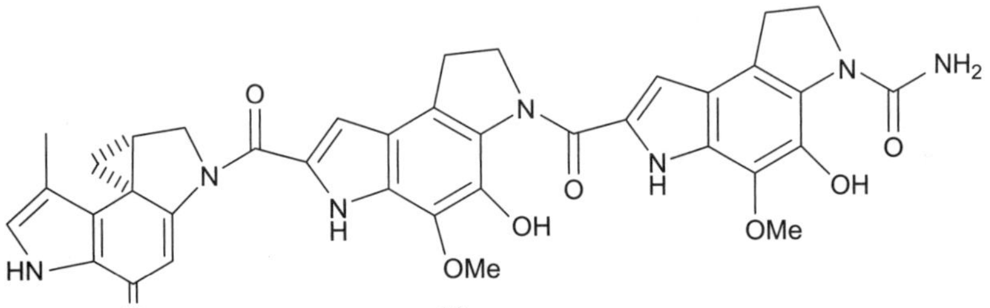

10

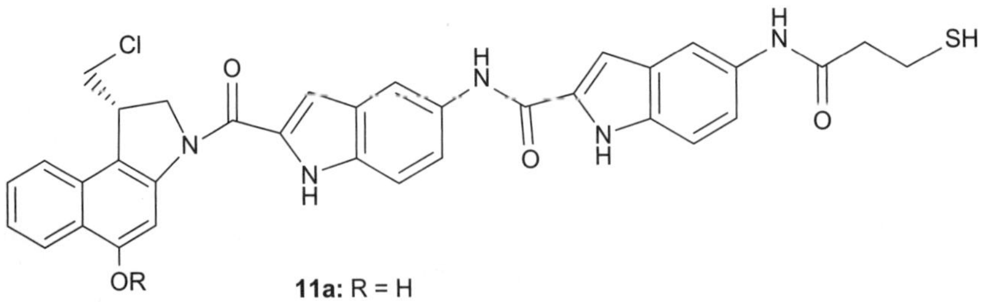

11a: R = H
11b: R = PO_3H

Duocarmycin is also structurally similar to CC-1065, except that it contains only one DNA-binding indolyl unit. In the MDX-1203 ADC **12**, a duocarmycin analogue was linked to an antibody *via* a dipeptide linker.[30]

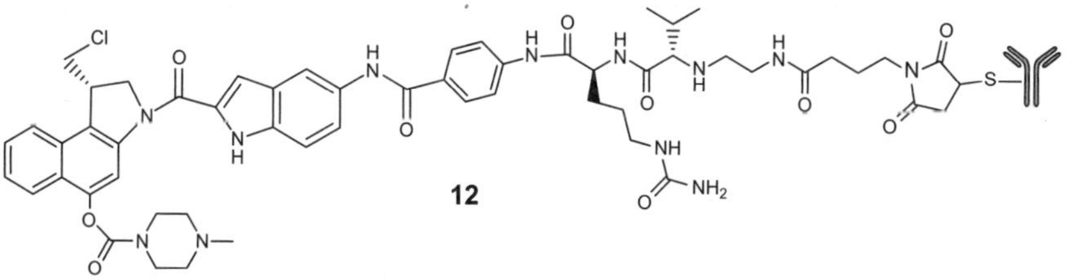

12

Conjugates currently in clinical testing (regardless of which cytotoxic compound, antibody, or linker was used) are mixtures of molecules with various drug-per-antibody ratio; most have, on average, 3–4 cytotoxic molecules per antibody molecule, randomly linked to lysine or cysteine residues on the antibody. Methods for site-specific conjugation are also being explored.[31]

4. INTRACELLULAR CATABOLISM OF ADCs

Regardless of whether the linkage is cleavable or noncleavable, initial degradation of auristatin- and maytansinoid-based ADCs apparently takes place in lysosomes. Calicheamicin-based hydrazone-linked ADCs are possibly hydrolyzed in the low-pH endosomal compartment. MMAE-based ADC, such as **8c** releases **8a** in cells *via* protease-(probably, cathepsin)-mediated cleavage of the amide bond between the peptide and the aromatic amine, followed by elimination of the para-amino benzyl moiety and carbon dioxide.[20,23] An auristatin derivative with a negatively charged C-terminal phenylalanine residue, **8b**, is much less cytotoxic as an unconjugated compound than **8a** likely due its reduced plasma membrane permeability.[21] Nonetheless, antibody–MMAF conjugates, conjugated to the antibody through a noncleavable linker, are comparable in their potency with antibody–MMAE conjugates.[21] Maytansinoid conjugates are degraded in lysosomes yielding metabolites consisting of a lysine adduct of the maytansinoid and linker, and the efficacy of this process is important for ADC activity.[1] Metabolites derived from a disulfide-linked maytansinoid conjugate are processed further *via* reduction of the disulfide bond and subsequent *S*-methylation, to produce lipophilic *S*-methyl-maytansinoid metabolites, which have been found to be highly cytotoxic when added exogenously. These observations may help explain both the phenomenon of target cell-activated killing of bystander cells and the superior efficacy of disulfide-linked conjugates over thioether-linked conjugates seen in some xenograft models.[32] The cytotoxic agents that are currently used in ADCs are substrates for at least one of the three multidrug-resistance transporters Multidrug Resistance Protein 1 (MDR1), Multidrug Resistance-associated Protein 1 (MRP1), or Breast Cancer Resistance Protein (BCRP).[18] Upon intracellular processing, conjugates are processed into cytotoxic metabolites,[1,21] which can be susceptible to the transporter-mediated efflux.[18] Hydrophilic linkers **13** and **14** that enable antibody–maytansinoid conjugates to overcome this MDR1-mediated resistance have been recently developed.[18,33]

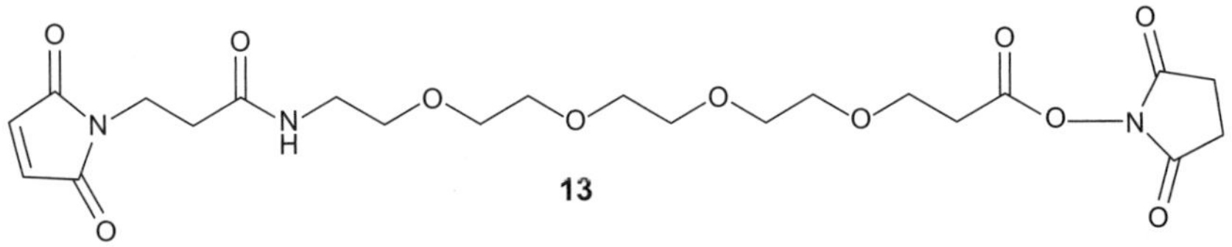

13

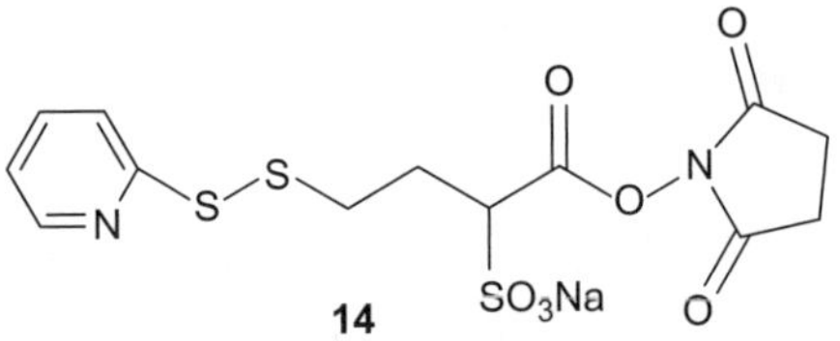

14

5. ADCs IN CLINICAL DEVELOPMENT

The ADCs that are in clinical development at the time of preparation of this review are listed in Table 23.1. One of those listed is gemtuzumab ozogamicin (Mylotarg), an anti-CD33-calicheamicin conjugate that was approved under an accelerated-approval process by U.S. Food and Drug Administration (FDA) in 2000 for treatment of acute myeloid leukemia, but withdrawn from the market in 2011 after an intended confirmatory trial showed no improvement in clinical benefit and an unfavorable toxicity. Brentuximab vedotin, **8c**, comprising an anti-CD30 antibody linked to MMAE, was granted conditional approval by FDA in 2011 for treating Hodgkin's lymphoma and ATCL. Several clinical trials are underway or in development to further evaluate the activity and safety of this ADC. Trastuzumab emtansine, described earlier in this review, exhibited robust antitumor activity and excellent tolerability in Phase I and Phase II trials,[34,35] suggesting that this agent may change the paradigm for the treatment of

Table 23.1 ADCs in clinical development

ADC	Target antigen	Linker-cytotoxic compound, class	Antibody	Tumor type(s)	Developer	Status
Gemtuzumab ozogamicin (Mylotarg®)	CD33 (Siglec-3)	Hydrazone, AcBu *N*-acetyl-γ calicheamicin Calicheamicin	hP67.6 Humanized IgG4	Acute myeloid leukemia	Pfizer	US FDA conditional approval 5/2000. Withdrawn 8/2011
Brentuximab vedotin (Adcetris®, SGN35)	CD30	Dipeptide, vc-MMAE Auristatin	Brentuximab Chimeric IgG1	Relapsed/refractory Hodgkin's lymphoma and systemic anaplastic large cell lymphoma	Seattle Genetics Millenium-Takeda	US FDA conditional approval 8/2011. Phase III and PhI/II combinations
Trastuzumab emtansine (T-DM1)	HER2 (ErbB2)	Thioether SMCC-DM1 Maytansinoid	Trastuzumab Humanized IgG1	HER2-positive breast cancer	Genentech-Roche	Ph II and Ph III and Ph I/II combinations
Inotuzumab ozogamicin (CMC-544)	CD22 (Siglec-2)	Hydrazone, AcBu *N*-acetyl-γ calicheamicin Calicheamicin	G5/44 Humanized IgG4	B-cell lymphomas	Pfizer	Ph I and Ph I/II Ph III combination

Continued

Table 23.1 ADCs in clinical development—cont'd

ADC	Target antigen	Linker-cytotoxic compound, class	Antibody	Tumor type(s)	Developer	Status
Glembatumumab vedotin (CR011-vc-MMAE, CDX-011)	Glycoprotein NMB (osteoactivin)	Dipeptide, vc-MMAE Auristatin	Glembatumumab Fully human IgG1	Metastatic breast cancer and melanoma	Celldex Therapeutics	Ph II
Lorvotuzumab mertansine (IMGN901, huN901-DM1, BB10901)	CD56 (NCAM)	Hindered disulfide SPP-DM1 Maytansinoid	Lorvotuzumab Humanized IgG1	SCLC and other CD56-positive solid tumors, multiple myeloma	ImmunoGen	Ph I Ph I/II combinations
SAR3419 (huB4-DM4)	CD19	Highly hindered disulfide SPDB-DM4 Maytansinoid	huB4 Humanized IgG1	B-cell malignancies	Sanofi	Ph II
BT-062	CD138 Syndecan-1	Highly hindered disulfide SPDB-DM4 Maytansinoid	Anti-CD138 chimeric IgG4	Multiple myeloma	Biotest	Ph I/II
SAR566658	CA6	Highly hindered disulfide SPDB-DM4 Maytansinoid	DS6 Humanized IgG1	CA6-positive solid tumors	Sanofi	Ph I

BAY 94-9343	Mesothelin	Highly hindered disulfide SPDB-DM4 Maytansinoid	Antimesothelin Fully human IgG1	Mesothelin-positive solid tumors	Bayer Healthcare	Ph I
SGN-75 (h1F6-mcMMAF)	CD70	Non-cleavable mcMMAF Auristatin	SGN-70 Humanized IgG1	Non-Hodgkin's lymphoma and renal cell carcinoma	Seattle Genetics	Ph I
PSMA ADC	Prostate-specific membrane antigen	Dipeptide, vc-MMAE Auristatin	Anti-PSMA Fully human IgG1	Metastatic, hormone-refractory prostate cancer	Progenics Pharmaceuticals	Ph I
ASG-5ME	SLC44A4 (AGS-5)	Dipeptide, vc-MMAE Auristatin	Anti-ASG-5 Fully human IgG2	Pancreatic cancer and prostate cancer	Astellas (Agensys) Seattle Genetics	Ph I
ASG-22ME	AGS-22 Nectin-4	Dipeptide, vc-MMAE Auristatin	Anti-Nectin Fully human IgG	Solid tumors	Astellas (Agensys) Seattle Genetics	Ph I
Anti-AGS-16 ADC (AGS-16M8F)	AGS-16 (ENPP3)	Auristatin	Anti-AGS-16 Fully human IgG2	Renal cell carcinoma and liver cancer	Astellas (Agensys)	Ph I
RG7593	CD22	Auristatin	Anti-CD22 human IgG	B-cell lymphoma	Genentech-Roche	Ph I

Continued

Table 23.1 ADCs in clinical development—cont'd

ADC	Target antigen	Linker-cytotoxic compound, class	Antibody	Tumor type(s)	Developer	Status
MDX-1203	CD70	Dipeptide (vc), prodrug of duocarmycin analog (DNA minor-groove binder and alkylator)	Anti-CD70 Fully human IgG	Non-Hodgkin's lymphoma and renal cell carcinoma	Bristol-Myers Squibb (Medarex)	Ph I
Milatuzumab-Doxorubicin hLL1-Dox	CD74	Thioether (SMCC) Doxorubicin	IMMU-110 Humanized IgG1	Multiple myeloma	Immunomedics	Ph I
IMGN529	CD37	Thioether SMCC-DM1 Maytansinoid	K7153A Humanized IgG1	B-cell malignancies	ImmunoGen	IND active
5 ADCs	Not identified	Auristatin	Not identified	Various solid and liquid cancers	Genentech-Roche	Ph I
2 ADCs	Not identified	Maytansinoid	Not identified	Undisclosed cancers	Amgen	IND

HER2-positive breast cancer, with promise for improved outcomes for patients. It is now being evaluated for treatment of HER2+ metastatic breast cancer in a number of Phase III trials. Besides **8c** and **2**, several other ADCs made with potent microtubule-impacting cytotoxic agents have entered phase II studies, or are in combination studies. Glembatumumab vedotin (see structure **8c** above, wherein the antibody is anti-NMB) combines an antiglycoprotein NMB fully human antibody with vc-MMAE.[36] The target, also known as osteoactivin, is highly expressed in melanoma[36] and breast cancer.[37] This ADC was active in preclinical xenograft models of melanoma.[36] Two Phase I/II trials were conducted in patients with advanced metastatic cancers: one in patients with melanoma, where antitumor activity was reported,[38] and the second in patients with metastatic breast cancer.[39] A Phase II trial in metastatic breast cancer is ongoing.

Lorvotuzumab mertansine, **6**, comprises a humanized version of the N901 antibody that targets CD56/NCAM, conjugated to DM1 *via* a hindered disulfide linker.[40] CD56 is expressed on a variety of cancers of hematopoietic and neuroendocrine origin, including multiple myeloma[40] and small-cell lung cancer.[41] Phase I trials showed encouraging evidence of antitumor activity coupled with an acceptable tolerability profile, in particular, a lack of clinically meaningful myelosuppression, both in CD56-positive solid tumors[42] and in multiple myeloma.[43] Clinical studies of **6** in combination with lenalidomide and dexamethasone in multiple myeloma, and in combination with carboplatin and etoposide in SCLC, have been initiated.[44]

SAR3419 comprises a humanized anti-CD19 antibody attached to DM4, linked *via* a highly hindered disulfide linker.[7,45] Phase I studies demonstrated good tolerability and promising activity in a variety of lymphoma subtypes, especially notable considering the heavy pretreatment of these patients and the mixed histology of those enrolled.[7,46,47] Three Phase II trials evaluating this ADC are underway in diffuse large B-cell lymphoma and acute lymphoblastic leukemia.

Inotuzumab ozogamicin, a conjugate of an anti-CD22 antibody with calicheamicin (see structure **1**, wherein the antibody is anti-CD22 instead of anti-CD33),[25,48] is being evaluated in a Phase III study in non-Hodgkin's lymphoma. The only other ADC with a DNA-acting payload is the duocarmycin conjugate **12**, wherein the antibody is anti-CD70; it is in a Phase I trial in CD70-positive renal cell cancer and non-Hodgkin's lymphoma.

There are several more ADCs in early clinical development (Table 23.1). Building upon the excitement around **8c** and **2**, nearly all of them utilize one of the two classes of potent tubulin-acting agents as payload.

6. CONCLUSION

The clinical development of ADCs, especially those employing potent microtubule-impacting agents, has changed the outlook for ADC technology. ADCs hold the promise of having an important role in cancer treatment, providing active therapeutics that reduces the severe toxicities associated with nontargeted cytotoxic chemotherapy.

REFERENCES

(1) Erickson, H.K.; Park, P.U.; Widdison, W.C.; Kovtun, Y.V.; Garrett, L.M.; Hoffman, K.; Lutz, R.J.; Goldmacher, V.S.; Blattler, W.A. *Cancer Res.* **2006**, *66*, 4426.
(2) Carter, P.; Smith, L.; Ryan, M. *Endocr. Relat. Cancer. Cancer* **2004**, *11*, 659.
(3) Xie, H.; Blattler, W.A. *Expert Opin. Biol. Ther.* **2006**, *6*, 281.
(4) Qin, A.; Mastico, R.A.; Lutz, R.J.; O'Keeffe, J.; Zildjian, S.; Mita, A.C.; Phan, A.T.; Tolcher, A.W. American Society of clinical oncology annual meeting proceedings, 2008, 3066.
(5) Vater, A.V.; Goldmacher, V.S. In *Macromolecular Anticancer Therapeutics*; Reddy, L.H., Couvreur, P., Eds.; Springer: New York, NY, 2009; p 331.
(6) Ross, S.; Spencer, S.D.; Holcomb, I.; Tan, C.; Hongo, J.; Devaux, B.; Rangell, L.; Keller, G.A.; Schow, P.; Steeves, R.M.; Lutz, R.J.; Frantz, G.; Hillan, K.; Peale, F.; Tobin, P.; Eberhard, D.; Rubin, M.A.; Lasky, L.A.; Koeppen, H. *Cancer Res.* **2002**, *62*, 2546.
(7) Blanc, V.; Bousseau, A.; Caron, A.; Carrez, C.; Lutz, R.J.; Lambert, J.M. *Clin. Cancer Res.* **2011**, *17*, 6448.
(8) Deckert, J.; Chicklas, S.; Yi, Y.; Li, M.; Pinkas, J.; Chittenden, T.; Lutz, R.J.; Park, P.U. *ASH Annu. Meeting Abstr.* **2011**, *118*, 3726.
(9) Dornan, D.; Bennett, F.; Chen, Y.; Dennis, M.; Eaton, D.; Elkins, K.; French, D.; Go, M.A.; Jack, A.; Junutula, J.R.; Koeppen, H.; Lau, J.; McBride, J.; Rawstron, A.; Shi, X.; Yu, N.; Yu, S.F.; Yue, P.; Zheng, B.; Ebens, A.; Polson, A.G. *Blood* **2009**, *114*, 2721.
(10) Junttila, T.T.; Li, G.; Parsons, K.; Lewis Phillips, G.; Sliwkowski, M.X. *Breast Cancer Res. Treat.* **2011**, *128*, 347.
(11) Goldmacher, V.S.; Blattler, W.A.; Lambert, J.M.; Chari, R.V.J. In *Biomedical Aspects of Drug Targeting*; Muzykantov, V., Torchilin, V., Eds.; Kluwer Academic Publishers: Boston/Dordrecht/London, 2002; p 291.
(12) Singh, R.; Erickson, H.K. Dimitrov, A.S., Ed.; Humana Press: Totowa, NJ, 2009; Vol. 525, p 445.
(13) Chari, R.V. *Acc. Chem. Res.* **2008**, *41*, 98.
(14) Burke, P.J.; Senter, P.D.; Meyer, D.W.; Miyamoto, J.B.; Anderson, M.; Toki, B.E.; Manikumar, G.; Wani, M.C.; Kroll, D.J.; Jeffrey, S.C. *Bioconjugate Chem.* **2009**, *20*, 1242.
(15) Oroudjev, E.; Lopus, M.; Wilson, L.; Audette, C.; Provenzano, C.; Erickson, H.; Kovtun, Y.; Chari, R.; Jordan, M.A. *Mol. Cancer Ther.* **2010**, *9*, 2700.
(16) Drewinko, B.; Patchen, M.; Yang, L.Y.; Barlogie, B. *Cancer Res.* **1981**, *41*, 2328.

(17) Widdison, W.C.; Wilhelm, S.D.; Cavanagh, E.E.; Whiteman, K.R.; Leece, B.A.; Kovtun, Y.; Goldmacher, V.S.; Xie, H.; Steeves, R.M.; Lutz, R.J.; Zhao, R.; Wang, L.; Blattler, W.A.; Chari, R.V. *J. Med. Chem.* **2006**, *49*, 4392.

(18) Kovtun, Y.V.; Audette, C.A.; Mayo, M.F.; Jones, G.E.; Doherty, H.; Maloney, E.K.; Erickson, H.K.; Sun, X.; Wilhelm, S.; Ab, O.; Lai, K.C.; Widdison, W.C.; Kellogg, B.; Johnson, H.; Pinkas, J.; Lutz, R.J.; Singh, R.; Goldmacher, V.S.; Chari, R.V. *Cancer Res.* **2010**, *70*, 2528.

(19) Lewis Phillips, G.D.; Li, G.; Dugger, D.L.; Crocker, L.M.; Parsons, K.L.; Mai, E.; Blattler, W.A.; Lambert, J.M.; Chari, R.V.; Lutz, R.J.; Wong, W.L.; Jacobson, F.S.; Koeppen, H.; Schwall, R.H.; Kenkare-Mitra, S.R.; Spencer, S.D.; Sliwkowski, M.X. *Cancer Res.* **2008**, *68*, 9280.

(20) Doronina, S.O.; Toki, B.E.; Torgov, M.Y.; Mendelsohn, B.A.; Cerveny, C.G.; Chace, D.F.; DeBlanc, R.L.; Gearing, R.P.; Bovee, T.D.; Siegall, C.B.; Francisco, J.A.; Wahl, A.F.; Meyer, D.L.; Senter, P.D. *Nat. Biotechnol.* **2003**, *21*, 778.

(21) Doronina, S.O.; Mendelsohn, B.A.; Bovee, T.D.; Cerveny, C.G.; Alley, S.C.; Meyer, D.L.; Oflazoglu, E.; Toki, B.E.; Sanderson, R.J.; Zabinski, R.F.; Wahl, A.F.; Senter, P.D. *Bioconjugate Chem.* **2006**, *17*, 114.

(22) Hamblett, K.J.; Senter, P.D.; Chace, D.F.; Sun, M.M.; Lenox, J.; Cerveny, C.G.; Kissler, K.M.; Bernhardt, S.X.; Kopcha, A.K.; Zabinski, R.F.; Meyer, D.L.; Francisco, J.A. *Clin. Cancer Res.* **2004**, *10*, 7063.

(23) Alley, S.C.; Benjamin, D.R.; Jeffrey, S.C.; Okeley, N.M.; Meyer, D.L.; Sanderson, R.J.; Senter, P.D. *Bioconjugate Chem.* **2008**, *19*, 759.

(24) Doronina, S.O.; Bovee, T.D.; Meyer, D.W.; Miyamoto, J.B.; Anderson, M.E.; Morris-Tilden, C.A.; Senter, P.D. *Bioconjugate Chem.* **2008**, *19*, 1960.

(25) DiJoseph, J.F.; Armellino, D.C.; Boghaert, E.R.; Khandke, K.; Dougher, M.M.; Sridharan, L.; Kunz, A.; Hamann, P.R.; Gorovits, B.; Udata, C.; Moran, J.K.; Popplewell, A.G.; Stephens, S.; Frost, P.; Damle, N.K. *Blood* **2004**, *103*, 1807.

(26) Hamann, P.R.; Hinman, L.M.; Beyer, C.F.; Greenberger, L.M.; Lin, C.; Lindh, D.; Menendez, A.T.; Wallace, R.; Durr, F.E.; Upeslacis, J. *Bioconjugate Chem.* **2005**, *16*, 346.

(27) Hamann, P.R.; Hinman, L.M.; Beyer, C.F.; Lindh, D.; Upeslacis, J.; Flowers, D.A.; Bernstein, I. *Bioconjugate Chem.* **2002**, *13*, 40.

(28) Boger, D.L.; Yun, W.; Han, N.; Johnson, D.S. *Bioorg. Med. Chem.* **1995**, *3*, 611.

(29) Zhao, R.Y.; Erickson, H.K.; Leece, B.A.; Reid, E.E.; Goldmacher, V.S.; Lambert, J.M.; Chari, R.V. *J. Med. Chem.* **2012**, *55*, 766.

(30) King, D.; Terrett, J.; Cardarelli, P.; Pan, C.; Rao, C.; Gangwar, S.; Deshpande, S.; Vangipuram, R.; Passmore, D.; Mirjolet, J.F.; Bichat, F. AACR meeting abstracts, Abstract 4057, San Diego, CA, April, **2008**.

(31) Junutula, J.R.; Raab, H.; Clark, S.; Bhakta, S.; Leipold, D.D.; Weir, S.; Chen, Y.; Simpson, M.; Tsai, S.P.; Dennis, M.S.; Lu, Y.; Meng, Y.G.; Ng, C.; Yang, J.; Lee, C.C.; Duenas, E.; Gorrell, J.; Katta, V.; Kim, A.; McDorman, K.; Flagella, K.; Venook, R.; Ross, S.; Spencer, S.D.; Lee Wong, W.; Lowman, H.B.; Vandlen, R.; Sliwkowski, M.X.; Scheller, R.H.; Polakis, P.; Mallet, W. *Nat. Biotechnol.* **2008**, *26*, 925.

(32) Kovtun, Y.V.; Goldmacher, V.S. *Cancer Lett.* **2007**, *255*, 232.

(33) Zhao, R.Y.; Wilhelm, S.D.; Audette, C.; Jones, G.; Leece, B.A.; Lazar, A.C.; Goldmacher, V.S.; Singh, R.; Kovtun, Y.; Widdison, W.C.; Lambert, J.M.; Chari, R.V.J. *J. Med. Chem.* **2011**, *54*, 3606.

(34) Hurvitz, S.; Dirix, L.; Kocsis, J.; Gianni, L.; Lu, J.; Vinholes, J.; Song, C.; Tong, B.; Chu, Y.W.; Perez, E.A. The 2011 European multidisciplinary cancer congress abstracts, Abstract 5001, 2011.

(35) Burris, H.A.; Rugo, H.S.; Vukelja, S.J.; Vogel, C.L.; Borson, R.A.; Limentani, S.; Tan-Chiu, E.; Krop, I.E.; Michaelson, R.A.; Girish, S.; Amler, L.; Zheng, M.; Chu, Y.W.; Klencke, B.; O'Shaughnessy, J.A. *J. Clin. Oncol.* **2011**, *29*, 398.
(36) Tse, K.F.; Jeffers, M.; Pollack, V.A.; McCabe, D.A.; Shadish, M.L.; Khramtsov, N.V.; Hackett, C.S.; Shenoy, S.G.; Kuang, B.; Boldog, F.L.; MacDougall, J.R.; Rastelli, L.; Herrmann, J.; Gallo, M.; Gazit-Bornstein, G.; Senter, P.D.; Meyer, D.L.; Lichenstein, H.S.; LaRochelle, W.J. *Clin. Cancer Res.* **2006**, *12*, 1373.
(37) Rose, A.A.; Grosset, A.A.; Dong, Z.; Russo, C.; Macdonald, P.A.; Bertos, N.R.; St-Pierre, Y.; Simantov, R.; Hallett, M.; Park, M.; Gaboury, L.; Siegel, P.M. *Clin. Cancer Res.* **2010**, *16*, 2147.
(38) Hamid, O.; Sznol, M.; Pavlick, A.C.; Kluger, H.M.; Kim, K.B.; Boasberg, P.D.; Simantov, R.; Davis, T.A.; Crowley, E.; Hwu, P. *ASCO Meeting Abstr.* **2010**, *28*, 8525.
(39) Saleh, M.N.; Bendell, J.C.; Rose, A.; Siegel, P.; Hart, L.L.; Sirpal, S.; Jones, S.F.; Crowley, E.; Simantov, R.; Vahdat, L.T. *ASCO Meeting Abstr.* **2010**, *28*, 1095.
(40) Tassone, P.; Gozzini, A.; Goldmacher, V.; Shammas, M.A.; Whiteman, K.R.; Carrasco, D.R.; Li, C.; Allam, C.K.; Venuta, S.; Anderson, K.C.; Munshi, N.C. *Cancer Res.* **2004**, *64*, 4629.
(41) Roy, D.C.; Ouellet, S.; Le Houillier, C.; Ariniello, P.D.; Perreault, C.; Lambert, J.M. *J. Natl. Cancer Inst.* **1996**, *88*, 1136.
(42) Wall, P.J.; O'Brien, M.; Fossella, F.; Shah, M.H.; Clinch, Y.; O'Keeffe, J.; Qin, A.; O'Leary, J.; Lorigan, P. *Ann. Oncol.* **2010**, *21*, 536P.
(43) Chanan-Khan, A.; Wolf, J.L.; Garcia, J.; Gharibo, M.; Jagannath, S.; Manfredi, D.; Sher, T.; Martin, C.; Zildjian, S.H.; O'Leary, J.; Vescio, R. *ASH Annu. Meeting Abstr.* **2010**, *116*, 1962.
(44) Berdeja, J.G.; Ailawadhi, S.; Weitman, S.D.; Zildjian, S.; O'Leary, J.J.; O'Keeffe, J.; Guild, R.; Whiteman, K.; Chanan-Khan, A.A.A. *ASCO Meeting Abstr.* **2011**, *29*, 8013.
(45) Al-Katib, A.M.; Aboukameel, A.; Mohammad, R.; Bissery, M.C.; Zuany-Amorim, C. *Clin. Cancer Res.* **2009**, *15*, 4038.
(46) Coiffier, B.; Ribrag, V.; Dupuis, J.; Tilly, H.; Haioun, C.; Morschhauser, F.; Lamy, T.; Copie-Bergman, C.; Brehar, O.; Houot, R.; Lambert, J.M.; Morarui-Zamfir, R. *ASCO Meeting Abstr.* **2011**, *29*, 8017.
(47) Younes, A.; Gordon, L.; Kim, S.; Romaguera, J.; Copeland, A.R.; de Castro Farial, S.; Kwak, L.; Fayad, L.; Hagemeister, F.; Fanale, M.; Lambert, J.; Bagulho, T.; Morariu-Zamfir, R. 51st ASH annual meeting, New Orleans, LA, December, **2009**.
(48) DiJoseph, J.F.; Dougher, M.M.; Kalyandrug, L.B.; Armellino, D.C.; Boghaert, E.R.; Hamann, P.R.; Moran, J.K.; Damle, N.K. *Clin. Cancer Res.* **2006**, *12*, 242.

CHAPTER TWENTY-FOUR

3D Cell Cultures: Mimicking *In Vivo* Tissues for Improved Predictability in Drug Discovery

Indira Padmalayam, Mark J. Suto
Southern Research Institute, Birmingham, Alabama, USA

Contents

ABBREVIATIONS

2D two dimension
3D three dimension
ECM extracellular matrix
ZO zona occludens
MTS multicellular spheroids
HTS high-throughput screening
CSC cancer stem cell

Annual Reports in Medicinal Chemistry, Volume 47
ISSN 0065-7743
http://dx.doi.org/10.1016/B978-0-12-396492-2.00024-2

1. INTRODUCTION

The cells and tissues in our body exist in a highly complex three-dimensional (3D) milieu. However, we have traditionally used a two-dimensional approach to study the behavior of these cells. For more than 100 years, mammalian cells have been grown as single layers on flat, impermeable dishes. While this approach has been quite useful and yielded a wealth of information across many areas of cell biology and research in general, it is now widely recognized that this so-called flat biology represents a "minimalistic" approach that is highly limited in the extent to which it models cells and tissues *in vivo*. There are numerous reports in all fields of biology showing that cells grown as monolayers or in two dimensions (2D) fail to capture key *in vivo* characteristics of cells. This recognition has led to an interest in finding methods to grow cells in a manner that recapitulates their natural *in vivo* environment. We describe herein the advantages of 3D culture techniques and their application to drug discovery.

2. 3D CELL CULTURE: A PHYSIOLOGICALLY RELEVANT BIOLOGICAL TOOL TO INVESTIGATE CELLULAR BEHAVIOR

2.1. What is 3D cell culture?

The use of 3D cell culture refers to the process of growing cells in their native context or *in vivo* environment. This is done by the addition of a third dimension, which can mimic the complex interactions between cells as well as cells and the extracellular matrix (ECM). In contrast to monolayers or 2D cultures where cells are mostly exposed to the media and plastic in the cell culture dish, in 3D cell cultures, all cells in the culture are exposed either to other cells or to the ECM (Fig. 24.1). Cells cultured in simplistic monolayers lose their original phenotypic and functional characteristics, often becoming dedifferentiated and very distinct from the tissues that they were derived from. Therefore, the biological responses of these cells may not be physiologically relevant, which can be a major factor hampering their predictability, and thus utility in drug discovery. Due to their architectural similarity to *in vivo* tissues, 3D cell cultures retain the phenotypic and functional characteristics of their *in vivo* counterparts, thereby providing a more physiologically relevant *in vitro* system to evaluate biological responses. There is now an increasing interest in 3D cell

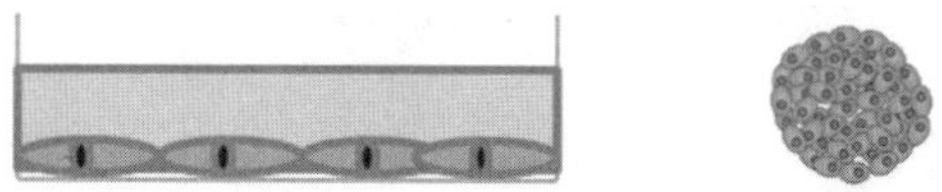

Figure 24.1 Culturing cells in 2D monolayers versus 3D. In 2D monolayer cultures, cells are grown on a flat, impermeable surface such as the plastic surface of a culture dish. These cells flatten out on the plastic, with exposure to plastic and medium and very little cell–cell contact. In 3D cell cultures, cells are completely exposed to other cells, or matrix. (See Color Plate 24.1 in Color Plate Section.)

cultures as an attractive alternative to the traditional 2D cell culture approach which is reflected by the exponential increase in the number of publications related to 3D cell cultures (www.3DCellCulture.com). In the world of drug discovery and development, 3D cell culture systems are fast replacing 2D cell cultures due to their reliability as an *in vitro* approach to predict *in vivo* responses.

2.2. History of evolution for 3D cell culture

The idea of 3D cell culture was originally conceived as early as 1907, when Ross Harrison from Johns Hopkins University conducted the first successful tissue culture experiment. He explanted neural tube fragments from frog embryos into fresh frog lymph on a cover slip, and once the lymph clotted, he inverted the cover slip onto a slide that had a depression. This was earliest demonstration of "hanging drop culture" which, interestingly, is one of the more popular 3D cell culture techniques today. Harrison observed the growth of neurite fibers using this simple culture system, thus establishing that animal cells could be grown outside the body.[1] This first "neurite outgrowth" experiment using the hanging drop method marked the advent of the mammalian cell culture technique. After further breakthroughs, animal cell culture became one of the most important and widely used tools in the field of biology. Ironically, the development of animal cell culture techniques led to oversimplification of the system and the generation of cells that were significantly different from the original tissues.

2.3. Biological differences between 2D and 3D cell cultures

Cellular context is a critical factor which governs tissue specificity, function, and homeostasis.[2] When cells are grown as monolayers on plastic, cellular context is compromised and minimal cell–cell and cell–ECM interactions occur. By contrast, cells grown in 3D reestablish cell–cell and cell–ECM interactions, maintaining cellular context and thus creating a physiological, "*in vivo*-like" environment. Differences in cellular context manifest as vast differences in the biology of cells grown in 2D versus 3D with respect to the following characteristics:

1. *Gene expression*: Striking differences in the expression of genes that play key roles in intercellular and cell–ECM interactions have been observed. For example, proteins associated with tight junctions, such as occludin, zona occludens (ZO)-1, E-cadherin, β-catenin, and cingulin, are abundantly expressed in cells cultured in 3D (3D cells), whereas they are either absent or minimally expressed when the same cells are cultured in 2D (2D cells).[3–6] Since tight junction proteins are critical for cell–cell communication, it is conceivable that 2D cells are inherently deficient in their ability to communicate with each other. Additionally, cell polarity, which is critical for the function of cell types such as endothelial and epithelial cells, is compromised in 2D cells due to a deficiency in the expression of proteins such as laminin and collagen IV, which are crucial for maintenance of cell polarity.[3–6] The expression levels of these proteins in 3D cells are higher, which preserves cellular polarity, and therefore their function. Conversely, 2D cultures of fibroblastic cells induce artificial polarity in these normally nonpolar cells, while 3D matrix-based cultures preserved their nonpolar characteristics.[7] Another feature which contributes to the nonphysiological nature of 2D cells is the incorrect localization of tight junction and polarity markers. In contrast, in 3D cells, these proteins are localized to the functionally relevant cellular compartments. For example, staining for ZO-1, E-cadherin, and β-catenin indicated that they were localized at cell–cell interfaces in A549 cells grown as 3D spheroids, while monolayers displayed a diffuse pattern of staining for these markers.[4,5] Similarly, staining for a polarity maker, collagen IV revealed that this protein was appropriately localized to the extracellular region in 3D cells, whereas it was retained in the perinuclear space in 2D cells.[4,5] Staining of human urothelial cells for cytokeratin 20 and uroplakin Ia revealed appropriate apical localization of these proteins in 3D organoids, with a more diffuse cytoplasmic staining in monolayers.[7] In addition, the expression and localization patterns seen in 3D cultures are consistent with the patterns observed in human tissues, suggesting that these cultures are more physiologically relevant.[3–6] Markers for differentiation of cells such as mucins and brush border proteins are also minimally expressed in 2D cells but abundantly expressed in

3D-cells.[4,5] This suggests that unlike 2D cells, 3D cells maintain their differentiated state in culture. Striking differences in gene expression are also observed between 2D and 3D cultures of proliferating cells such as tumor cell lines.[8–10] For example, genome-wide messenger RNA expression studies revealed that 3400 genes are differentially expressed in prostate cancer cell lines grown in 3D compared to 2D.[8]

2. *Cellular morphology*: Several studies have confirmed that the cellular morphology and architecture of cells in 3D more accurately represent *in vivo* tissues compared to those cultured in 2D. For example, intestinal epithelial cells, when cultured in 3D, arrange themselves in villus-like folds that closely resemble intestinal microvilli.[3] Ovarian and endometrial cancer cell lines when cultured in 3D display cytological hallmarks of primary tumor cells such as variation in cell size, nuclear polymorphism, and prominent nucleoli.[10] Nonmalignant breast epithelial cells grown in 3D form organized growth-arrested acinar structures that resemble normal human breast tissue.[11] In contrast, cells derived from breast carcinomas and grown in 3D behave like cancer cells *in vivo*, by forming disorganized structures that do not respond to growth inhibitory cues from the ECM.[11] These architectural differences are poorly captured in 2D cultures. Normal breast epithelial cells cease to be "normal" when grown in monolayers, because they are highly plastic and express the above-mentioned tumor-like characteristics, providing an example of the "nonphysiological" nature of 2D cell culture.[12]
3. *Drug sensitivities*: Striking differences in drug sensitivities have been observed between 3D and 2D cells. For example, whereas a panel of Food and Drug Administration-approved anticancer compounds (cisplatin, 5-azacytidine, emodin, mitoxantrone, and 5-fluorouracil) induced strong apoptosis of cells grown in 2D, they were ineffective when tested on 3D spheroids. Conversely, tetrandrine and miconazole induced apoptosis in 3D cultures but were less effective in 2D cultures.[13] Cell–ECM interactions and 3D morphology of breast cancer cells have a significant impact on the sensitivities to drugs targeting the HER2 pathway such as Trastuzumab, Pertuzumab, and Lapatinib.[14] Tung *et al.* made an interesting observation using the antiproliferative drug 5-fluoruracil (5-FU), and a hypoxia-activated cytotoxin, tirapazamine (TPZ), in 2D versus 3D spheroid cultures.[15] It was shown that 2D cultures were more sensitive than 3D spheroids to 5-FU, while 3D cultures were more sensitive to TPZ. These different responses are attributed to the multilayer structure of the spheroids, which accurately

recapitulates the tumor microenvironment. 5-FU can only kill the proliferating cells in the outer layers of the spheroids, but not the quiescent cells in the interior. In contrast, TPZ, which is activated by hypoxia, can kill the cells in the hypoxic interior of the spheroids.[15] These examples illustrate that 3D and 2D cell cultures are distinct in their responses to certain drugs, and since 3D cell cultures have been recognized to represent a more physiologically relevant system, it is conceivable that they would elicit more appropriate "*in vivo*-like" responses to drug treatment.

2.4. Methods of 3D cell culture

Typically, 3D cell culture methods are broadly classified into two approaches: (1) the scaffold-based approach where cells are cultured on natural or artificial matrices or scaffolds. These scaffolds provide a framework for the cells to adhere, grow, and establish cell–cell and cell–matrix interactions and adopt "*in vivo*-like" architecture. (2) The scaffold/matrix-free approach, which relies on the ability of cells to self-assemble, grow, and secrete ECM proteins. This approach perhaps more closely mimics the process of cell growth and tissue morphogenesis *in vivo*.

Scaffold/matrix-based methods: Various types of scaffolds ranging in complexity from simple agar or collagen matrices to sophisticated nanofibers have been used to provide cells with an environment that promotes intercellular communication.[16–18] These 3D matrix scaffolds are made from synthetic or natural materials and come in various forms such as membranes, sponges, gels, and microcarrier beads.[19] Matrigel™ is a mouse-tumor-derived basement membrane (BM) matrix which is frequently used to culture cancer cells in 3D. It contains all of the major BM constituents such as laminin, collagen IV, and heparin sulfate proteoglycans[17] and is therefore suitable for investigating the process of differentiation, migration, and invasion of tumor cells in response to signals from ECM components.

Scaffold/matrix-free methods: The most popular scaffold-free approach is the multicellular spheroid (MTS)-based 3D cell culture. MTS are naturally formed 3D aggregates of cells. Even though this approach is based on simple spheroidal geometry, it accurately models the complex 3D *in vivo* environment of normal tissue and is therefore physiologically relevant.[20–23] It also provides an approach to model dynamic processes such as growth and invasiveness by pathogenic agents. MTS from a wide range of cell types can be generated by the hanging drop technique[22] wherein cells aggregate within a droplet of medium in a static environment. Alternatively, MTS can be formed in rotary bioreactors such as the NASA-designed rotary wall vessel[4,5,22] or the biolevitator.[19] These bioreactors provide a simulated microgravity and low shear stress environment that is conducive to tissue-like self-aggregation and assembly of cells to form MTS structures.[24] Such an environment directly affects gene expression and promotes differentiation of cells and the formation of cell–cell contacts to facilitate cell signaling via key adhesion molecules.[24] Thus, the MTS approach provides an excellent *in vitro* system to study intercellular communication and its effects on cellular differentiation and function.

3. USE OF 3D CELL CULTURES IN DRUG DISCOVERY

3.1. Application of 3D cell cultures in drug screening: Improving predictability of *in vivo* drug efficacy

In the process of drug discovery, the use of 3D cell cultures, by virtue of their "*in vivo*-like" characteristics, can help bridge the gap between oversimplified, nonphysiological 2D systems and more complex and expensive *in vivo* systems. In addition, it should result in a reduction in the number of *in vivo* animal studies required. The need for a predictable *in vitro* system is particularly critical in the field of oncology where the current attrition rate of late-stage anticancer drugs is high (70% in Phase II and 59% in Phase III).[25] The failure of drugs at these stages is, in large part, attributed to a lack of efficacy and safety. Incorporating 3D cell culture-based *in vitro* screens in the early stages of drug discovery could potentially reduce efficacy-related issues that contribute to the proverbial "Valley of Death."[25] This recognition has fueled a recent surge of interest in testing the adaptability and feasibility of various types of 3D cell cultures in the early screening process. Two studies using matrix-based and matrix-free approaches to 3D cell cultures have compared the responses of anticancer drugs in 3D systems to the traditional 2D-based screens.[13,26] Both studies identified substantial differences in the potency and efficacy of the drugs between the 3D and 2D cell cultures (discussed in an earlier section), underscoring the dependence of compound activity on biological context. Since 3D cell cultures have now earned the recognition of being more physiologically relevant based upon the biological similarities with *in vivo* tissues, it would be fair to assume that drug responses in 3D systems are more predictive of *in vivo* responses. A 3D spheroid-based drug screen demonstrated the potential for adaptability to high-throughput screening (HTS),[13] confirming the feasibility of incorporating 3D systems into mainstream drug screening applications.

3.2. Application of 3D cell cultures in toxicology and drug metabolism: Early clues for drug safety

In vitro evaluation of drug metabolism and hepatoxicity is an essential stage in the drug discovery process. In 2D culture, hepatocytes rapidly dedifferentiate and lose their cuboidal morphology as they flatten and spread out on the plastic surface of the cell culture dish.[27] This is accompanied by reduced viability and striking changes in gene expression and metabolic activity. In particular, it has been reported that primary hepatocytes cultured in 2D have reduced activity of the genes involved in xenobiotic metabolism such as the cytochrome P450 enzymes as well as a loss of response to well-known inducers of these enzymes.[27–29] In contrast, culturing primary hepatocytes in 3D using scaffolds or scaffold-free, spheroid-based approaches improves their viability, retains their cuboidal morphology, and more closely resembles liver tissue *in vivo*.[27,28] Both of these 3D cell culture approaches maintain basal activity and inducibility of the xenobiotic enzymes associated with phase I–III drug metabolism. For example, in contrast to a lack of a response seen in 2D cell cultures, hepatocytes cultured in 3D and treated with 3-methylcholanthrene or a cocktail of rifampicin, dexamethosone, and β-naphtaflavone exhibited a dramatic increase in the activity of CYP-1A2, CYP-2B1, and CYP-3A2.[27] Similarly, 3D (scaffold-based) cultures, but not 2D cultures, displayed an increased sensitivity to acetaminophen, which was attributed to increased activity of phase I and phase II metabolic enzymes in the 3D system.[28] In other studies, hepatocyte spheroid cultures (hepatospheres) treated with drugs such as diclofenac, paracetamol, and galactosamine displayed dramatic changes in anchorage dependence, cellular morphology, energy metabolism, and biotransformation, which were consistent with hepatotoxicity.[29] Taken together, these findings suggest that hepatocytes cultured in 3D could serve as a predictable *in vitro* screening tool for hepatotoxicity of drug candidates.

4. APPLICATION OF 3D CELL CULTURES IN VARIOUS THERAPEUTIC AREAS

4.1. Oncology

In cancer drug discovery, there is currently a shift from antimitotic drugs, which kill proliferating tumor cells, to drugs that target persistent, non-proliferating cells such as cancer stem cells (CSCs). These CSCs promote cancer cell survival and tumor regrowth. While 2D cell cultures fail to provide a screening environment of sufficient complexity to identify compounds critical for CSC survival,[8] 3D cell cultures preserve the characteristics of *in vivo* tumors and are a more reliable and physiologically relevant *in vitro* model for cancer drug screening. Furthermore, 3D cocultures incorporating tumor-specific cells (epithelial cells, fibroblasts, endothelial cells) and matrix components model the structural architecture and establish interactions and biochemical networks[23,30,31] conducive to eliciting *in vivo*-like drug responses. The following examples suggest that 3D cell cultures have the characteristics of *in vivo* tumors that are relevant to cancer drug discovery:

Cellular interactions and architecture: Tumor epithelial cells grown as 3D cocultures with fibroblasts more accurately model the growth, invasion, and metastatic potential of breast tumors.[30,31] In another example, 3D cocultures of breast tumor cells with osteoblasts provided an *in vitro* model for bone metastasis.[32,33]

Tumor hypoxia and heterogeneity: In vivo, rapid proliferation of tumor cells forces blood vessels apart, creating heterogeneous populations of nutrient- and oxygen-deprived cells (depending upon their distance from blood vessels).[34] This heterogeneity is accurately reproduced by 3D tumor spheroids. Cells in the periphery of spheroids represent actively growing tumor cells adjacent to capillaries *in vivo*, and cells in the interior of the spheroids become quiescent and eventually die, resembling the hypoxic, necrotic regions of *in vivo* tumors.[34,35]

Tumor acidification: The acidification in the hypoxic interior of tumors due to a build-up of glycolytic products such as lactic acid is mimicked by 3D spheroids.[36] Since acidification is a factor that affects uptake of weakly basic anticancer drugs,[37] the preservation of this characteristic is critical for an anticancer drug screen.

Drug sensitivity: Good correlations have been observed between drug resistance *in vivo* and drug penetration in spheroid-based 3D cell cultures.[38,39] This suggests that the spheroid-based cultures accurately reproduce the complexity of the tumor microenvironment with respect to drug penetration characteristics.

4.2. Infectious diseases

The initial interaction between a pathogen and host is the defining event for most infections; hence the choice of an appropriate *in vitro* cell culture model is critical to derive appropriate *in vivo*-like responses. Due to their tissue-like characteristics, 3D cell cultures allow for more physiologically relevant host responses to infection and the pathways involved in the infectious process. Additionally, the use of 3D cell culture techniques allows for the study of pathogens that are not currently supported by 2D cell culture or animal models.[4] The following characteristics of 3D cultured cells enable them to more accurately model host–pathogen interactions, making them suitable for infectious disease drug discovery.

Architectural and functional: It was shown that 3D spheroid-based cultures display physiologically relevant expression and localization of proteins critical for pathogen invasion (receptors, adhesion, tight junction, differentiation, and polarity markers) in 3D cultures of A549 cells for *Pseudomonas aeruginosa* infection.[4] Hepatocytes cultured as spheroid aggregates express high levels of liver-specific markers and have increased expression and organization of hepatitis C virus (HCV) receptors. These features make 3D culture more permissive to HCV infection than 2D cultures.[40]

Infectivity: The characteristics of bacterial and viral infections in host cells cultured in 3D are more similar to *in vivo* infections. For example, during *P. aeruginosa* infection, 3D cultures of lung epithelial cells (A549) exhibit damage on the cell surface, but not the exaggerated excessive rounding and detachment response observed in 2D cell cultures.[4] Human bladder cells that are cultivated as organoids and infected with uropathogenic *Escherichia coli* display exfoliation, which is a natural host defense mechanism associated with clearing bacteria from the bladder.[6]

Inflammatory response: Studies have shown that 3D cell cultures exhibit *in vivo*-like production of inflammatory markers, which are more representative of the host response to infection, unlike the "exaggerated" inflammatory responses seen in 2D cell cultures.[4]

4.3. Neurodegenerative diseases

Central nervous system disorders such as Parkinson's disease represent a challenging area for the development of effective therapeutics, in part due to the lack of valid and reliable preclinical models.[41] The use of 3D cell cultures opens up an opportunity to establish long-term cultures of neuronal cells, which can be used for drug testing. For example, neuronal cells exhibit physiologically relevant growth in 3D collagen matrices compared to nonpermeable glass surfaces.[42] In another study, culturing midbrain-derived neural progenitor cells in 3D facilitated their differentiation into neurospheres, which were enriched in dopaminergic neurons.[41] Furthermore, the neurospheres were amenable to stable expression of a transgene for an extended period and thus provided a promising system for evaluating the efficacy of gene-based therapies.[41] A 3D coculture system of neuronal cells and astrocytes yielded a more clinically relevant model to study glial–neuronal interactions and the importance of astrocyte alignment as a possible future therapeutic intervention for spinal cord repair following injury.[43] It is conceivable that the 3D neurite outgrowth system that Ross Harrison used in the early 1900s will find its place in modern day neurodegenerative drug discovery programs.

5. CONCLUSION

In summary, there is sufficient evidence to suggest that 3D cell culture techniques will have a tremendous impact on drug discovery at all stages, including HTS, prioritizing compounds for further development and *in vivo* testing. In addition, 3D cultures can also be used to look at other parameters such as toxicity, cell penetration, and drug–drug interactions. The use of 3D cell cultures could also facilitate the use of routine drug sensitivity testing. For example, patient biopsy samples cultured in 3D would be tested to identify the drug sensitivity of the patient's own cancer cells. In the infectious disease arena, 3D cocultures of *Mycobacterium tuberculosis* with both alveolar epithelial cells and macrophages could provide an antimicrobial screen to identify compounds that block uptake of *M. tuberculosis* and prevent ensuing pathogenic events.

REFERENCES

(1) Harrison, R. *Anat. Rec.* **1907**, *1*, 116.
(2) Pampaloni, F.; Reynaud, E.G.; Stelze, E.H. *Nat. Rev. Mol. Cell Biol.* **2007**, *8*, 839.
(3) Carvalho, H.M.; Teel, L.D.; Goping, G.; O'Brien, A.D. *Cell. Microbiol.* **2005**, 7, 1771.
(4) Barrila, J.; Radke, A.L.; Crabbe, C.; Sarker, S.F.; Herbst-Kralovetz, M.M.; Ott, C.M.; Nickerson, C.A. *Nat. Rev. Microbiol.* **2010**, *8*, 791.
(5) Nickerson, C.A.; Richter, E.G.; Ott, C.M. *J. Neuroimmune Pharmacol.* **2007**, *2*, 26.
(6) Smith, Y.C.; Grande, K.K.; Rasmussen, S.B.; O'Brien, A.D. *Infect. Immun.* **2006**, *74*, 750.
(7) Cukierman, E.; Pankov, R.; Stevens, D.R.; Yamada, K.M. *Science* **2001**, *294*, 1708.
(8) Härmä, V.; Virtanen, J.; Mäkelä, R.; Happonen, A.; Mpindi, J.P.; Knuuttila, M.; Kohonen, P.; Lötjönen, J.; Kallioniemi, O.; Nees, M. *PLoS One* **2010**, *5*, e10431.
(9) Martin, K.J.; Patrick, D.R.; Bissell, M.J.; Fournier, M.V. *PLoS One* **2008**, *3*, e2994.
(10) Grun, B.; Benjamin, E.; Sinclair, J.; Timms, J.F.; Jacobs, I.J.; Gayther, S.A.; Dafou, D. *Cell Prolif.* **2009**, *42*, 219.
(11) Gupta, P.B.; Kuperwasser, C. *Drug Discov. Today* **2004**, *1*, 9.
(12) Kim, J.B.; Stein, R.; O'Hare, M.J. *Breast Cancer Res. Treat.* **2004**, *85*, 281.
(13) Herrmann, R.; Fayad, W.; Schwarz, S.; Berndtsson, M.; Linder, S. *J. Biomol. Screen.* **2008**, *13*, 1.

(14) Weigelt, B.; Lo, A.T.; Park, C.C.; Gray, J.W.; Bissell, M.J. *Breast Cancer Res. Treat.* **2010**, *122*, 35.
(15) Tung, Y.C.; Hsiao, A.Y.; Allen, S.G.; Torisawa, Y.S.; Ho, M.; Takayama, S. *Analyst* **2011**, *136*, 473.
(16) Tibbitt, M.W.; Anseth, K.S. *Biotechnol. Bioeng.* **2009**, *103*, 655.
(17) Gurski, L.A.; Jha, A.K.; Zhang, C.; Jia, X.; Farach-Carson, M.C. *Biomaterials* **2009**, *30*, 6076.
(18) Horning, J.L.; Sahoo, S.K.; Vijayaraghavalu, S.; Dimitrijevic, S.; Vasir, J.K.; Jain, T.K.; Panda, A.K.; Labhasetwar, V. *Mol. Pharm.* **2008**, *5*, 849.
(19) Justice, B.A.; Badr, N.A.; Felder, R.A. *Drug Discov. Today* **2009**, *14*, 102.
(20) Friedrich, J.; Seidel, C.; Ebner, R.; Kunz-Schughart, L.A. *Nat. Protoc.* **2009**, *4*, 309.
(21) Hirschhaeuser, F.; Menne, H.; Dittfeld, C.; West, J.; Mueller-Klieser, W.; Kunz-Schughart, L.A. *J. Biotechnol.* **2010**, *148*, 3.
(22) Kunz-Schughart, L.A.; Freyer, J.P.; Hofstaedter, F.; Ebner, R. *J. Biomol. Screen.* **2004**, *9*, 273.
(23) Friedrich, J.; Ebner, R.; Kunz-Schughart, L.A. *Int. J. Radiat. Biol.* **2007**, *83*, 849.
(24) Unsworth, B.R.; Lelkes, P.I. *Nat. Med.* **1998**, *8*, 901.
(25) Adams, D.J. *Trends Pharmacol. Sci.* **2012**, *33*, 173.
(26) Nirmalanandhan, V.S.; Duren, A.; Hendricks, P.; Vielhauer, G.; Sittampalam, G.S. *Assay Drug Dev. Technol.* **2010**, *8*, 581.
(27) Schutte, M.; Fox, B.; Baradez, M.O.; Devonshire, A.; Minguez, J.; Bokhari, M.; Przyborski, S.; Marshall, D. *Assay Drug Dev. Technol.* **2011**, *9*, 475.
(28) van Zijl, F.; Mikulits, W. *World J. Hepatol.* **2010**, *2*, 1.
(29) Xu, J.; Purcell, W.M. *Toxicol. Appl. Pharmacol.* **2006**, *216*, 293.
(30) Kalluri, R.; Zeisberg, M. *Nat. Rev. Cancer* **2006**, *6*, 392.
(31) Olsen, C.J.; Moreira, J.; Lukanidin, E.M.; Ambartsumian, N.S. *BMC Cancer* **2010**, *10*, 444.
(32) Dolznig, H.; Rupp, C.; Puri, C.; Haslinger, C.; Schweifer, N.; Wieser, E.; Kerjaschki, D.; Pilar, G. *Am. J. Pathol.* **2011**, *179*, 487.
(33) Krishnan, V.; Shuman, L.A.; Sosnoski, D.M.; Dhurjati, R.; Vogler, E.A.; Mastro, A.M. *J. Cell. Physiol.* **2011**, *226*, 2150.
(34) Sutherland, R.M. *Science* **1988**, *240*, 177.
(35) Minchinton, A.I.; Tannock, I.F. *Nat. Rev. Cancer* **2006**, *6*, 583.
(36) Khaitan, D.; Chandna, S.; Arya, M.B.; Dwarakanath, B.S. *J. Transl. Med.* **2006**, *2*, 4.
(37) Raghunand, N.; Gillies, R.J. *Drug Resist. Update* **2000**, *3*, 39.
(38) Ong, S.M.; Zhao, Z.; Arooz, T.; Zhao, D.; Zhang, S.; Du, T.; Wasser, M.; van Noort, D.; Yu, H. *Biomaterials* **2010**, *31*, 1180.
(39) Desoize, B.; Jardillier, J. *Crit. Rev. Oncol. Hematol.* **2000**, *36*, 193.
(40) Sainz, B., Jr.; TenCate, V.; Uprichard, S.L. *Virol. J.* **2009**, *6*, 103.
(41) Brito, C.; Simao, D.; Costa, I.; Malpique, R.; Pereira, C.I.; Fernandez, P.; Serra, M.; Schwarz, S.C.; Schwarz, J.; Kremer, E.J.; Alves, P.M. *Methods* **2012**, *56*, 452–460.
(42) Kofron, C.M.; Fong, V.J.; Hoffman-Kim, D. *J. Neural Eng.* **2009**, *6*, 016002.
(43) East, E.; Blum de Oliveira, D.; Golding, J.; Phillips, J. *Glia* **2009**, *57*, S159.

CHAPTER TWENTY-FIVE

Virally Encoded G Protein-Coupled Receptors: Overlooked Therapeutic Opportunities?

Nuska Tschammer
Department of Chemistry and Pharmacy, Emil Fischer Center, Friedrich Alexander University, Erlangen, Germany

Contents

ABBREVIATIONS

AIDS acquired immune deficiency syndrome
CCL17/TARC chemokine (C–C motif) ligand 17/thymus- and activation-regulated chemokine
CCL19/ELC chemokine (C–C motif) ligand 19/EBI 1 ligand chemokine
CCL2/MCP-1 chemokine (C–C motif) ligand 2/monocyte chemotactic protein 1
CCL21/SLC chemokine (C–C motif) ligand 21/secondary lymphoid tissue chemokine

Annual Reports in Medicinal Chemistry, Volume 47
ISSN 0065-7743
http://dx.doi.org/10.1016/B978-0-12-396492-2.00025-4

CCL22/MDC chemokine (C–C motif) ligand 22/macrophage-derived chemokine
CCL3/MIP-1α chemokine (C–C motif) ligand 3/macrophage inflammatory protein-1α
CCL4/MIP-1β chemokine (C–C motif) ligand 4/macrophage inflammatory protein-1β
CCL5/RANTES chemokine (C–C motif) ligand 5/regulated upon activation, normal T-cell expressed, and secreted
CCR1 chemokine (C–C motif) receptor 1
CD4 cluster of differentiation 4
CREB cAMP response element binding
CX_3CL1/Fractalkine chemokine (C–X_3–C motif) ligand 1/Fractalkine
CXCL1/GROα chemokine (C–X–C motif) ligand 1/growth-regulated alpha protein
CXCL10 chemokine (C–X–C motif) ligand 10
CXCL12 chemokine (C–X–C motif) ligand 12
CXCL3 chemokine (C–X–C motif) ligand 3
CXCL4 chemokine (C–X–C motif) ligand 4
CXCL5 chemokine (C–X–C motif) ligand 5
CXCL7 chemokine (C–X–C motif) ligand 7
CXCL8/IL-8 chemokine (C–X–C motif) ligand 8/interleukin-8
CXCR2 chemokine (C–X–C motif) receptor 2
CXCR4 chemokine (C–X–C motif) receptor 4
EBI2 Epstein–Barr virus-induced receptor 2
EBV Epstein–Barr virus
GPCR G protein-coupled receptor
GPR17 human G protein-coupled receptor 17
GPR39 human G protein-coupled receptor 39
HCMV human cytomegalovirus
HHV-7 human herpesvirus 7
HIV-1 human immunodeficiency virus-1
KSHV Kaposi's sarcoma herpesvirus
MC1R melanocortin 1 receptor
NFAT nuclear factor of activated T cells
ORF74 open reading frame 74
PLC phospholipase C
R33 rat cytomegalovirus G protein-coupled receptor 33
U12 G protein-coupled receptor encoded by HHV-7 *U12* gene
U51 G protein-coupled receptor encoded by HHV-7 *U51* gene
UL33 G protein-coupled receptor homolog UL33
UL78 G protein-coupled receptor homolog UL78
US27 G protein-coupled receptor homolog US27
US28 G protein-coupled receptor homolog US28
vGPCR virally encoded GPCR

1. INTRODUCTION

Viruses use a great variety of tools to invade a host organism. Large DNA viruses, such as Herpesviruses, encode proteins that mimic the G protein-coupled receptors (GPCRs) of the host. These virally encoded GPCRs (vGPCRs) are commonly homologues of mammalian chemokine receptors.[1–5] The vGPCRs' efficient interaction with the host's signaling machinery provides advantages to the virus. The expression of chemokine-like vGPCRs by viruses illustrates the adaptation of viruses to a complex system of cellular intercommunication in their host and even includes functions beyond immune evasion.[3]

The modulation of cellular functions by vGPCRs has been extensively reviewed.[1–5] This review focuses on (1) the main characteristics of cellular reprogramming caused by vGPCRs, (2) the current development of synthetic ligands that target vGPCRs, and (3) a discussion of the importance of these ligands as chemical probes used to dissect the signaling properties of vGPCRs and their involvement in oncogene transformation or development of atherosclerosis. GPCRs are the target of more than one-third of all drugs on the market,[6] and therefore, an understanding of the precise mechanism of signal transduction and negative regulation of vGPCRs by synthetic small weight inverse agonists should provide a strategy to block the actions of these vGPCRs and to develop practical therapies for diseases caused by Herpesviruses.

2. STRUCTURE, FUNCTION, AND PHYSIOLOGICAL CONSEQUENCES OF vGPCRs

vGPCRs are seven-transmembrane receptors with often striking sequence homology to host chemokine receptors, particularly in the helical regions. Host chemokine receptors are responsible for coordinating the immune system surveillance and the response to infection and inflammation.[7] Chemokine receptors bind chemokines (chemoattractant cytokines) that regulate the trafficking and effector functions of leukocytes. Each immune cell type carries a specific expression pattern of chemokine receptors. The induction of expression of particular chemokines determines which immune cells will migrate during infection and inflammation. vGPCRs may represent immune-evasion strategies to inactivate inflammatory cytokines, to redirect the immune response, and to improve the survival and spreading of the virus.[3,4]

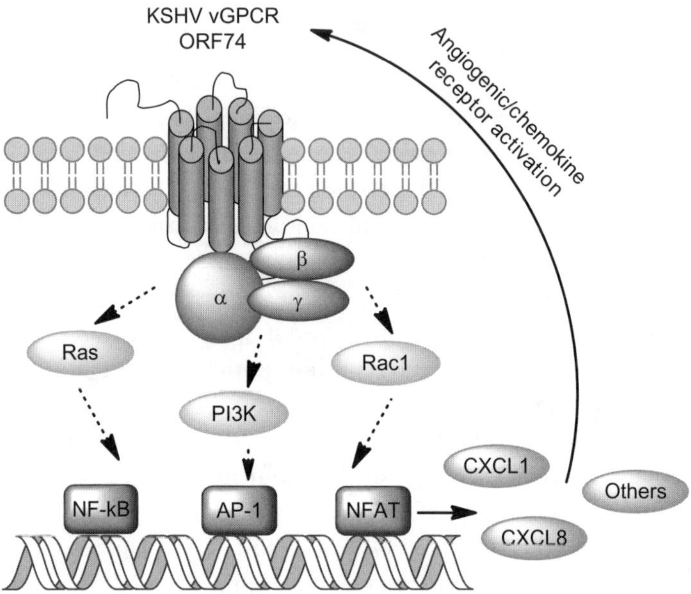

Figure 25.1 Interaction of vGPCRs with the host's signaling machinery, leading to reprogramming of cell signaling, activation of several transcription factors, and subsequent production of growth factors and chemokines. (See Color Plate 25.1 in Color Plate Section.)

vGPCRs interact efficiently with the signaling machinery of the host cell, which consequently often leads to pronounced biological response in the absence of a bound ligand (constitutive activity) (Fig. 25.1). The chemokine ligands of several vGPCRs are known.[1,3] The reprogramming of cell signaling by vGPCRs often results in prosurvival and angiogenic effects, inflammation, transformation, proliferation, and increased viral replication.[1,3,4,8]

2.1. Kaposi's sarcoma-associated herpesvirus

A γ-herpesvirus, Kaposi's sarcoma-associated herpesvirus (KSHV) is one of seven currently known human oncoviruses. It was first identified in AIDS patients suffering with the otherwise rare Kaposi's sarcoma.[9] KHSV is found in nearly 100% of tumors isolated from Kaposi's sarcoma patients. Moreover, it is associated with primary effusion lymphoma and multicentric Castleman's disease.[10,11] KSHV encodes one vGPCR, referred to as ORF74 (open reading frame 74).[9] ORF74, a homologue of the human chemokine CXC receptor 2, which binds a broad range of human chemokines from the CXC and CC family, is required for increased viral replication in response to chemokines and efficient reactivation from latency.[12,13] The sole expression of ORF74 is sufficient to induce the development of Kaposi's sarcoma-like lesions in transgenic mice.[14] Selective elimination of vGPCR-expressing cells in established allografts in nude mice resulted in tumor regression.[15]

Work from several laboratories has demonstrated that ORF74 promotes cell proliferation, enhances cell survival, modulates cell migration, stimulates angiogenesis, and recruits inflammatory cells in expressing cells as well as in neighboring (bystander) cells (reviewed in Ref. 16). The molecular mechanisms by which this powerful viral oncogene rewires the cell-signaling network are very complex. Not only is ORF74 highly constitutively active, but also its activity can be further potentiated by chemokines, such as chemokine (C–X–C motif) ligand 8/interleukin-8 (CXCL8/IL-8) and CXCL1/GROα.[17] Interestingly, chemokines CXCL10 and CXCL12 act as inverse agonists, and chemokines CXCL4, CXCL5, and CXCL7 act as neutral antagonists.[18–20] ORF74 activates numerous transcription factors, such as cAMP response element-binding protein, nuclear factor-kappaB, activator protein-1, and nuclear factor of activated T cells, resulting in the expression of a variety of angiogenic growth factors and proinflammatory chemokines and cytokines (e.g., vascular endothelial growth factor and CXCL8/IL-8).[21–23] Cell transformation caused by ORF74 is thus mediated partially by a paracrine mechanism.

Despite strong evidence for the essential role of ORF74 as a viral oncogene and potential novel target for the treatment of patients with Kaposi's sarcoma, no small-molecule allosteric modulators of this oncogene have been reported.

2.2. Epstein–Barr virus

Epstein–Barr virus (EBV), another representative of the γ-herpesvirus group, causes a lifelong latent infection in healthy individuals. Burkitt's lymphoma, Hodgkin's lymphoma, and nasopharyngeal carcinoma are the main diseases associated with EBV.[24] These diseases are particularly common in immunocompromised patients who cannot control the proliferation of EBV-infected B cells because of immune suppression.[25,26]

EBV encodes two vGPCRs—BILF1 and Epstein–Barr virus-induced receptor 2 (EBI2). BILF1 has low homology with the human chemokine (C–X–C motif) receptor 4. BILF1 is an orphan receptor, but due to its high constitutive activity, BILF1 signaling networks are well characterized.[4,8,27–30] It has been suggested that EBV may use BILF1 to control Gαi-activated pathways during viral lytic replication, thereby promoting disease progression.[28] In nude mice, BILF1 promoted tumor formation in 90% of the animals; the positive correlation between receptor activity and the ability to mediate cell transformation *in vitro* and tumor formation *in vivo* suggests that allosteric modulators which act as inverse agonists for BILF1 could inhibit cell transformation and be relevant therapeutic candidates.[27]

The second vGPCR of EBV, EBI2, is a highly constitutively active receptor that controls follicular B-cell migration and T-cell-dependent antibody production.[31–33] EBI2 has a low homology with lipid binding GPCRs.[34] In 2011, Liu *et al.*[35] identified oxysterols as endogenous ligands of this vGPCR. 7α,25-Dihydroxycholesterol (7α,25-OHC) is the most potent ligand and activator of EBI2 with a K_i of 0.45 nM. *In vitro* and *in vivo* studies showed (7α,25-OHC) can serve as a chemokine that directs EBI2-mediated migration of B cells.[35] Oxysterols are otherwise known to activate nuclear hormone receptors.[36,37] This discovery offers the opportunity to develop steroid-based synthetic derivatives to inhibit the constitutive activation of EBI2.

2.3. Human cytomegalovirus

The omnipresent human cytomegalovirus (HCMV), a member of β-herpesvirus group, causes a lifelong latent infection in healthy hosts. In patients with immature or suppressed immune systems (e.g., neonates, as well as AIDS, cancer, and transplant patients), HCMV can lead to severe and life-threatening disease. Studies of the effects of cytomegalovirus infection on cellular processes provide evidence that the virus contributes to the development of restenosis and atherosclerosis by increasing the migration of smooth muscle cells, inhibition of apoptosis, and augmentation of cellular proliferation.[38,39] The presence of the HCMV genome and antigens was confirmed in tumor cells (but not in adjacent normal tissue) of more than 90% of patients with malignancies like colon cancer, malignant glioma, prostate carcinoma, and breast cancer.[40–42] Also, HCMV infection in transplant patients contributes to transplant rejection.[43,44]

HCMV encodes four vGPCRs (US28, UL33, US27, and UL78) which demonstrate up to 33% sequence homology to human chemokine receptors.[1] US28 and UL33 are characterized by high constitutive activity; US27 and UL78 show no constitutive activity. US27, UL33, and UL78 are orphan receptors—no binding partners have been identified. US28 is the best characterized HCMV chemokine-like receptor which is known to bind to a broad spectrum of chemokines such as CCL2/MCP-1, CCL3/MIP-1α, CCL4/MIP-1β, CCL5/RANTES, and CX_3CL1/Fractalkine with sub-nanomolar affinity. These chemokines activate cell type- and ligand-specific US28-mediated signaling pathways.[45,46] The reprogramming of host cells by US28 leads to vascular smooth muscle cell migration and thus potentially accelerates atherosclerosis and promotes intestinal neoplasia in transgenic mice.[47–49] The view that US28 behaves as an oncogene and promotes tumorigenesis is not yet widely accepted.[45,47] UL33 is another homologue of human chemokine receptors and is characterized by high constitutive activity that may be used by HCMV to orchestrate multiple signaling networks within infected cells.[50] US28, UL33, and US27 are presumably also present in the viral membrane of HCMV.[51,52] The HCMV chemokine receptor homologue US27 is required for efficient spread by the extracellular route but not for direct cell-to-cell spread of HCMV.[53] The role of UL78 in the viral life cycle remains to be identified.

2.4. Other viruses expressing vGPCRs

Human herpesvirus 7 (HHV-7) belongs to the β-herpesvirus subfamily and infects children during infancy and then becomes latent. HHV-7 contains two genes (*U12* and *U51*) that encode putative homologues of cellular GPCRs.[54] *U12* and *U51* encode functional calcium-mobilizing receptors that bind CCL17/TARC, CCL19/ELC, CCL21/SLC, and CCL22/MDC.[55] Overall, these studies suggest that HHV-7 U51 is a positive regulator of virus replication *in vitro*, because it may promote membrane fusion and facilitates cell–cell spread of this highly cell-associated virus.[56]

3. ALLOSTERIC MODULATORS OF vGPCRs

Although a considerable amount of data demonstrates the important role of vGPCRs in viral dissemination and development of cancer, few efforts have been made to pharmacologically target vGPCRs. This section presents an overview of current molecular scaffolds that target US28 and EBI2. Allosteric modulators of vGPCRs described to date generally have only moderate affinity and limited selectivity. The development of highly potent and selective allosteric modulators of vGPCRs thus remains a challenge. Allosteric modulators offer considerable advantages over classical

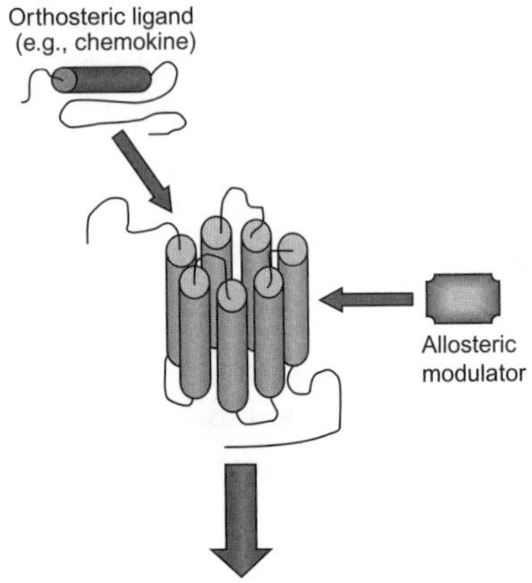

Figure 25.2 Allosteric modulation of vGPCRs. Allosteric modulators bind to a site topographically distinct from the orthosteric site on the receptor to alter either the affinity or efficacy of an endogenous ligand (chemokine) and thus shift the functional response. Allosteric modulators can also reduce the basal activity of the receptor in the absence of an endogenous ligand. (See Color Plate 25.2 in Color Plate Section.)

orthosteric ligands because they bind to a topographically distinct site from the orthosteric site (Fig. 25.2). Consequently, they alter GPCR conformation and interactive properties, both with respect to orthosteric ligands and downstream signaling partners in a positive or negative direction.[57,58] Additionally, allosteric binding sites are less conserved compared to orthosteric sites and thus offer the opportunity for better selectivity.

3.1. Allosteric modulators of US28

3.1.1 VUF2274 and its structural analogues

VUF2274 {5-(4-(4-chlorophenyl)-4-hydroxypiperidin-1-yl)-2,2-diphenylpentanenitrile} (**1**) was the first reported inhibitor of US28 receptor constitutive activity (Fig. 25.3).[59,60] This compound fully inhibited the constitutive activity of US28 in the phospholipase C (PLC) pathway with a potency of 3.5 μM. Structure–activity relationship studies showed that a 4-phenylpiperidine moiety is essential for affinity and activity of **1**.[60,61] This compound was initially reported as a CCR1 receptor antagonist,[62] which shares 33% homology with the US28 receptor.[59,60,62] Compound **1** does not inhibit constitutive ORF74- and R33-mediated accumulation of inositol triphosphate, but it interacts with adrenergic, dopamine, muscarinic, and serotonin receptors.[59,62,63] Despite the suboptimal selectivity profile,

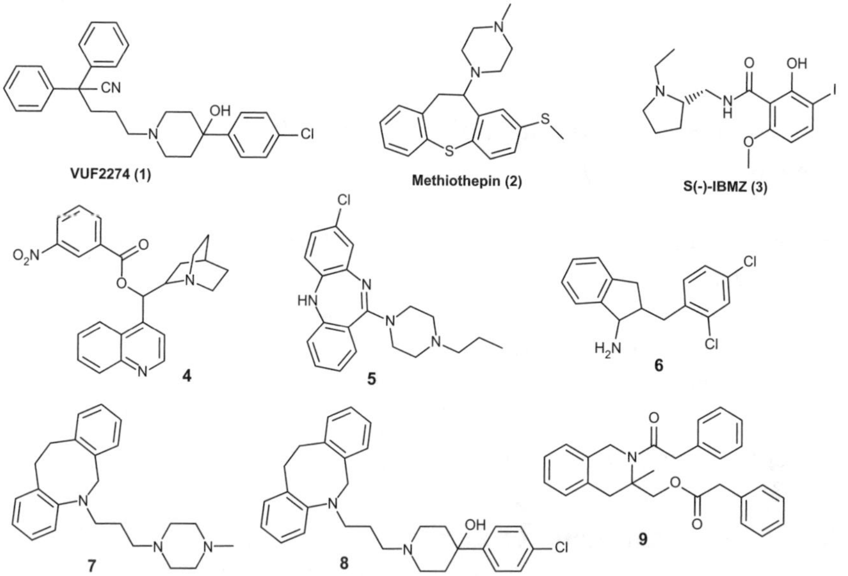

Figure 25.3 Allosteric modulators of the vGPCR of HCMV US28.

1 proved to be an interesting lead compound for the development of US28 inhibitors, leading to the discovery of novel compounds such as **8**.[59,64]

3.1.2 Dibenzothiepines, arylamines, and bicyclic compounds

Dihydrodibenzothiepines (e.g., methiothepin (**2**)), arylamines, such as *S*-iodobenzamide (**3**), and bicyclic compounds (e.g., cinchonidine derivatives, such as compound **4**) have been reported as modulators of US28 in the patent literature (Fig. 25.3).[65–68] Compounds **2** and **3** behave as agonists on US28.[65,67] For the remaining compounds, only the inhibition of fractalkine binding to US28 was reported, without any investigation of intrinsic activity. Continuous efforts to discover novel nonpeptidergic modulators that would inhibit constitutive activity of US28 resulted in a series of piperazinyldibenzodiazepine (e.g., **5**), cinchonidine (e.g., **4**), and indanylamine (e.g., **6**) derivatives (Fig. 25.3).[64] Although the successful modification of these scaffolds yielded a broad variety of structural analogues acting on US28, none of these modifications improved potency and efficacy. The only exception was structural hybridization of the tricyclic imipramine analogue **7** and the 4-phenylpiperidine moiety of **1**, which yielded a modulator **8** with negative efficacy comparable to **1** and increased antagonist potency against CCL5 binding (4.7 and 0.7 μM, respectively) (Fig. 25.3).[64]

3.1.3 *Isoquinoline and isoquinolinone derivatives*

The most recent molecular scaffolds reported to be efficacious inhibitors of US28 constitutive activity belong to the group of dihydrotetraisoquinolinone and tetrahydroisoquinoline **9** derivatives.[63] Compound **9** is a stronger partial agonist (a dose–response with maximal efficacy of −37% at 10 μM) in the reporter gene assay compared to **2** (−22%), although with an EC_{50} still in the micromolar range (3.4 μM).[63] These dihydroisoquinolinones demonstrate remarkable potency and efficacy on the dopamine D_{2L} receptor that underlies an obvious issue regarding the cross-reactivity with biogenic amine receptors similar to that reported for **1**.[62]

3.2. Allosteric modulator of EBI2

Rosenkilde *et al.* identified the first potent and efficacious inverse agonist, GSK682753A (**10**), of EBI2 (Fig. 25.4).[69] Inverse agonist **10** inhibits the constitutive activity of EBI2 in various functional assays with high potency (IC_{50}=2.6–53.6 nM) and efficacy (−75%), determined from a dose–response curve at a concentration of 10 μM. The selectivity of **10** was tested only on a limited set of constitutively active receptors, including GPR39, GPR17, MC1R, ORF74, and the ghrelin receptor, on which **10** had no inverse agonist effect. In the future, **10** will serve as a potent lead compound for the development of novel inverse agonists of vGPCRs and serve as a useful tool for further characterization of EBI2.

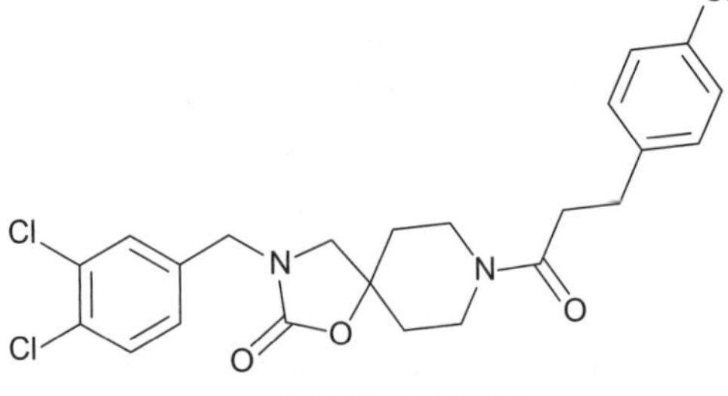

GSK682753A (10)

Figure 25.4 Allosteric modulator of the EBV vGPCR EBI2.

4. ALLOSTERIC MODULATORS OF vGPCRs AT WORK

Suboptimal potency and selectivity of the small number of currently available allosteric modulators of vGPCRs limits their use as chemical probes, but isolated examples reviewed herein demonstrate the utility of these allosteric modulators for the development of therapies directed toward the treatment of vGPCR-mediated pathologies.

4.1. Inhibition of US28-mediated HIV entry

In vitro experiments in cell lines coexpressing vGPCR US28 and CD4 showed that US28 serves as a cofactor for HIV-1 entry.[70] The US28 inverse agonist **1** at a concentration of 1 μM inhibited 60% of the US28-mediated HIV-1 entry in cells coexpressing US28 and CD4.[59]

4.2. Inhibition of US28-mediated signaling in HCMV-infected fibroblasts

The infection of human foreskin fibroblasts with HCMV induces a consistent increase in PLC activity[71] which is thought to be mediated by US28.[45,59,71] The US28 inverse agonist **1** inhibited US28-mediated PLC activation with an $IC_{50} \sim 0.8$ μM.[59]

4.3. Suppression of EBI2-mediated proliferation of murine B cells

Mice that overexpressed EBI2, specifically in B cells, were generated to mimic the expression pattern observed upon EBV infection.[69] EBI2 constitutive activity by **10** suppressed basal migration and antibody-induced proliferation of EBI2 expressing B cells *ex vivo* with an IC_{50} value 0.28 μM.[69]

5. CONCLUSIONS

vGPCRs are seven-transmembrane receptors with frequently striking sequence homology to host GPCRs. The reprogramming of cell signaling by vGPCRs often results in prosurvival and angiogenic effects, inflammation, transformation, proliferation, and increased viral replication that ultimately lead to lifelong infections and chronic diseases like atherosclerosis and cancer. Despite the fact that GPCRs are excellent drug targets, vGPCRs have received little attention. The development of potent and selective vGPCR allosteric modulators would have a significant impact on the deciphering of molecular mechanisms of negative vGPCR regulation and consequently on the development of therapies for various diseases caused by Herpesviruses.

The few studies reporting small weight inhibitors of vGPCRs indicate that selective targeting of these receptors represents a challenge due to suboptimal affinity and with it related selectivity issues. Alternatively, antibodies neutralizing or blocking vGPCRs could be used in proof-of-concept studies. After these issues are resolved and the drug candidates show favorable preclinical profile, these drugs could be used in clinical trials on the subset of patients carrying the desired biomarkers, but not primarily as a preventive therapy. This type of personalized medicine (the combination of vGPCR inhibitors and, e.g., currently available cancer treatments) is expected to improve treatment outcome.

REFERENCES

(1) Maussang, D.; Vischer, H.F.; Leurs, R.; Smit, M.J. *Mol. Pharmacol.* **2009**, *76*, 692.
(2) Couty, J.P.; Gershengorn, M.C. *Trends Pharmacol. Sci.* **2005**, *26*, 405.
(3) Alcami, A. *Nat. Rev. Immunol.* **2003**, *3*, 36.
(4) Sodhi, A.; Montaner, S.; Gutkind, J.S. *Nat. Rev. Mol. Cell Biol.* **2004**, *5*, 998.
(5) Nicholas, J. *J. Interferon Cytokine Res.* **2005**, *25*, 373.
(6) Overington, J.P.; Al-Lazikani, B.; Hopkins, A.L. *Nat. Rev. Drug Discov.* **2006**, *5*, 993.
(7) Rossi, D.; Zlotnik, A. *Annu. Rev. Immunol.* **2000**, *18*, 217.
(8) Slinger, E.; Langemeijer, E.; Siderius, M.; Vischer, H.F.; Smit, M.J. *Mol. Cell. Endocrinol.* **2011**, *331*, 179.
(9) Chang, Y.; Cesarman, E.; Pessin, M.S.; Lee, F.; Culpepper, J.; Knowles, D.M.; Moore, P.S. *Science* **1865**, *1994*, 266.
(10) Cesarman, E.; Chang, Y.; Moore, P.S.; Said, J.W.; Knowles, D.M. *N. Engl. J. Med.* **1995**, *332*, 1186.
(11) Dupin, N.; Fisher, C.; Kellam, P.; Ariad, S.; Tulliez, M.; Franck, N.; van Marck, E.; Salmon, D.; Gorin, I.; Escande, J.-P.; Weiss, R.A.; Alitalo, K.; Boshoff, C. *Proc. Natl. Acad. Sci. U.S.A.* **1999**, *96*, 4546.
(12) Lee, B.J.; Koszinowski, U.H.; Sarawar, S.R.; Adler, H. *J. Immunol.* **2003**, *170*, 243.
(13) Sandford, G.; Choi, Y.B.; Nicholas, J. *J. Virol.* **2009**, *83*, 13009.
(14) Guo, H.-G.; Sadowska, M.; Reid, W.; Tschachler, E.; Hayward, G.; Reitz, M. *J. Virol.* **2003**, *77*, 2631.
(15) Montaner, S.; Sodhi, A.; Ramsdell, A.K.; Martin, D.; Hu, J.; Sawai, E.T.; Gutkind, J.S. *Cancer Res.* **2006**, *66*, 168.
(16) Jham, B.C.; Montaner, S. *J. Cell. Biochem.* **2010**, *110*, 1.
(17) Gershengorn, M.C.; Geras-Raaka, E.; Varma, A.; Clark-Lewis, I. *J. Clin. Invest.* **1998**, *102*, 1469.
(18) Geras-Raaka, E.; Varma, A.; Clark-Lewis, I.; Gershengorn, M.C. *Biochem. Biophys. Res. Commun.* **1998**, *253*, 725.
(19) Geras-Raaka, E.; Varma, A.; Ho, H.; Clark-Lewis, I.; Gershengorn, M.C. *J. Exp. Med.* **1998**, *188*, 405.
(20) Rosenkilde, M.M.; Schwartz, T.W. *Mol. Pharmacol.* **2000**, *57*.
(21) Pati, S.; Cavrois, M.; Guo, H.-G.; Foulke, J.S., Jr.; Kim, J.; Feldman, R.A.; Reitz, M. *J. Virol.* **2001**, *75*, 8660.

(22) Sodhi, A.; Montaner, S.; Patel, V.; Zohar, M.; Bais, C.; Mesri, E.A.; Gutkind, J.S. *Cancer Res.* **2000**, *60*, 4873.
(23) Schwarz, M.; Murphy, P.M. *J. Immunol.* **2001**, *167*, 505.
(24) Hsu, J.L.; Glaser, S.L. *Crit. Rev. Oncol. Hematol.* **2000**, *34*, 27.
(25) Lucas, R.M.; Hughes, A.M.; Lay, M.L.J.; Ponsonby, A.L.; Dwyer, D.E.; Taylor, B.V.; Pender, M.P. *J. Neurol. Neurosurg. Psychiatry* **2011**, *82*, 1142.
(26) McManus, T.E.; Marley, A.M.; Baxter, N.; Christie, S.N.; Elborn, J.S.; O'Neill, H.J.; Coyle, P.V.; Kidney, J.C. *Eur. Respir. J.* **2008**, *31*, 1221.
(27) Lyngaa, R.; Norregaard, K.; Kristensen, M.; Kubale, V.; Rosenkilde, M.M.; Kledal, T.N. *Oncogene* **2010**, *29*, 4388.
(28) Paulsen, S.J.; Rosenkilde, M.M.; Eugen-Olsen, J.; Kledal, T.N. *J. Virol.* **2005**, *79*, 536.
(29) Nijmeijer, S.; Leurs, R.; Smit, M.J.; Vischer, H.F. *J. Biol. Chem.* **2010**, *285*, 29632.
(30) Zuo, J.; Quinn, L.L.; Tamblyn, J.; Thomas, W.A.; Feederle, R.; Delecluse, H.-J.; Hislop, A.D.; Rowe, M. *J. Virol.* **2011**, *85*, 1604.
(31) Gatto, D.; Wood, K.; Brink, R. *J. Immunol.* **2011**, *187*, 4621.
(32) Kelly, L.M.; Pereira, J.P.; Yi, T.; Xu, Y.; Cyster, J.G. *J. Immunol.* **2011**, *187*, 3026.
(33) Pereira, J.O.P.; Kelly, L.M.; Cyster, J.G. *Int. Immunol.* **2010**, *22*, 413.
(34) Joost, P.; Methner, A. *Genome Biol.* **2002**, *3 research0063.1.*
(35) Liu, C.; Yang, X.V.; Wu, J.; Kuei, C.; Mani, N.S.; Zhang, L.; Yu, J.; Sutton, S.W.; Qin, N.; Banie, H.; Karlsson, L.; Sun, S.; Lovenberg, T.W. *Nature* **2011**, *475*, 519.
(36) Willy, P.J.; Umesono, K.; Ong, E.S.; Evans, R.M.; Heyman, R.A.; Mangelsdorf, D.J. *Genes Dev.* **1995**, *9*, 1033.
(37) Jin, L.; Martynowski, D.; Zheng, S.; Wada, T.; Xie, W.; Li, Y. *Mol. Endocrinol.* **2010**, *24*, 923.
(38) Epstein, S.E.; Zhou, Y.F.; Zhu, J. *Am. Heart J.* **1999**, *138*, S476.
(39) Simanek, A.M.; Dowd, J.B.; Pawelec, G.; Melzer, D.; Dutta, A.; Aiello, A.E. *PLoS One* **2011**, *6*, e16103.
(40) Samanta, M.; Harkins, L.; Klemm, K.; Britt, W.J.; Cobbs, C.S. *J. Urol.* **2003**, *170*, 998.
(41) Cecilia, S.-N. *J. Clin. Virol.* **2008**, *41*, 218.
(42) Cobbs, C.; Soroceanu, L.; Denham, S.; Zhang, W.; Britt, W.; Pieper, R.; Kraus, M. *J. Neurooncol.* **2007**, *85*, 271.
(43) Grattan, M.T.; Moreno-Cabral, C.E.; Starnes, V.A.; Oyer, P.E.; Stinson, E.B.; Shumway, N.E. *JAMA* **1989**, *261*, 3561.
(44) Legendre, C.; Pascual, M. *Clin. Infect. Dis.* **2008**, *46*, 732.
(45) Vomaske, J.; Nelson, J.A.; Streblow, D.N. *Infect. Disord. Drug Targets* **2009**, *9*, 548.
(46) Boomker, J.M.; van Luyn, M.J.; The, T.H.; de Leij, L.F.; Harmsen, M.C. *Rev. Med. Virol.* **2005**, *15*, 269.
(47) Maussang, D.; Verzijl, D.; van Walsum, M.; Leurs, R.; Holl, J.; Pleskoff, O.; Michel, D.; van Dongen, G.A.M.S.; Smit, M.J. *Proc. Natl. Acad. Sci. U.S.A.* **2006**, *103*, 13068.
(48) Streblow, D.N.; Soderberg-Naucler, C.; Vieira, J.; Smith, P.; Wakabayashi, E.; Ruchti, F.; Mattison, K.; Altschuler, Y.; Nelson, J.A. *Cell* **1999**, *99*, 511.
(49) Bongers, G.; Maussang, D.; Muniz, L.R.; Noriega, V.M.; Fraile-Ramos, A.; Barker, N.; Marchesi, F.; Thirunarayanan, N.; Vischer, H.F.; Qin, L.; Mayer, L.; Harpaz, N.; Leurs, R.; Furtado, G.C.; Clevers, H.; Tortorella, D.; Smit, M.J.; Lira, S.A. *J. Clin. Invest.* **2010**, *120*, 3969.
(50) Casarosa, P.; Gruijthuijsen, Y.K.; Michel, D.; Beisser, P.S.; Holl, J.; Fitzsimons, C.P.; Verzijl, D.; Bruggeman, C.A.; Mertens, T.; Leurs, R.; Vink, C.; Smit, M.J. *J. Biol. Chem.* **2003**, *278*, 50010.
(51) Fraile-Ramos, A.; Pelchen-Matthews, A.; Kledal, T.N.; Browne, H.; Schwartz, T.W.; Marsh, M. *Traffic* **2002**, *3*, 218.
(52) Margulies, B.J.; Gibson, W. *Virus Res.* **2007**, *123*, 57.

(53) O'Connor, C.M.; Shenk, T. *J. Virol.* **2011**, *85*, 3700.
(54) Nakano, K.; Tadagaki, K.; Isegawa, Y.; Aye, M.M.; Zou, P.; Yamanishi, K. *J. Virol.* **2003**, 77, 8108.
(55) Tadagaki, K.; Nakano, K.; Yamanishi, K. *J. Virol.* **2005**, *79*, 7068.
(56) Zhen, Z.; Bradel-Tretheway, B.; Sumagin, S.; Bidlack, J.M.; Dewhurst, S. *J. Virol.* **2005**, *79*, 11914.
(57) Jeffrey Conn, P.; Christopoulos, A.; Lindsley, C.W. *Nat. Rev. Drug Discov.* **2009**, *8*, 41.
(58) Kenakin, T.; Miller, L.J. *Pharmacol. Rev.* **2010**, *62*, 265.
(59) Casarosa, P.; Menge, W.M.; Minisini, R.; Otto, C.; van Heteren, J.; Jongejan, A.; Timmerman, H.; Moepps, B.; Kirchhoff, F.; Mertens, T.; Smit, M.J.; Leurs, R. *J. Biol. Chem.* **2003**, *278*, 5172.
(60) Hulshof, J.W.; Casarosa, P.; Menge, W.M.; Kuusisto, L.M.; van der Goot, H.; Smit, M.J.; de Esch, I.J.; Leurs, R. *J. Med. Chem.* **2005**, *48*, 6461.
(61) Hulshof, J.W.; Vischer, H.F.; Verheij, M.H.; Fratantoni, S.A.; Smit, M.J.; de Esch, I.J.; Leurs, R. *Bioorg. Med. Chem.* **2006**, *14*, 7213.
(62) Hesselgesser, J.; Ng, H.P.; Liang, M.; Zheng, W.; May, K.; Bauman, J.G.; Monahan, S.; Islam, I.; Wei, G.P.; Ghannam, A.; Taub, D.D.; Rosser, M.; Snider, R.M.; Morrissey, M.M.; Perez, H.D.; Horuk, R. *J. Biol. Chem.* **1998**, *273*, 15687.
(63) Kralj, A.; Wetzel, A.; Mahmoudian, S.; Stamminger, T.; Tschammer, N.; Heinrich, M.R. *Bioorg. Med. Chem. Lett.* **2011**, *21*, 5446.
(64) Vischer, H.F.; Hulshof, J.W.; Hulscher, S.; Fratantoni, S.A.; Verheij, M.H.; Victorina, J.; Smit, M.J.; de Esch, I.J.; Leurs, R. *Bioorg. Med. Chem.* **2010**, *18*, 675.
(65) Schall, T. J.; McMaster, B. E.; Dairaghi, D. J. US Patent 2002/0127544 A1, **2002**.
(66) McMaster, B. E.; Schall, T. J.; Penfold, M.; Wright, J. J.; Dairaghi, D. J. US Patent 6,821,998 B2, **2004**.
(67) Schall, T. J.; McMaster, B. E.; Dairaghi, D. J. US Patent 2002/0193374 A1, **2002**.
(68) McMaster, B. E.; Schall, T. J.; Penfold, M.; Wright, J. J.; Dairaghi, D. J. US Patent 2003/0149055 A1, **2003**.
(69) Benned-Jensen, T.; Smethurst, C.; Holst, P.J.; Page, K.R.; Sauls, H.; Sivertsen, B.R.; Schwartz, T.W.; Blanchard, A.; Jepras, R.; Rosenkilde, M.M. *J. Biol. Chem.* **2011**, *286*, 29292.
(70) Pleskoff, O.; Treboute, C.; Alizon, M. *J. Virol.* **1998**, *72*, 6389.
(71) Minisini, R.; Tulone, C.; Lüske, A.; Michel, D.; Mertens, T.; Gierschik, P.; Moepps, B. *J. Virol.* **2003**, 77, 4489.

CHAPTER TWENTY-SIX

Recent Advances in Wnt/β-Catenin Pathway Small-Molecule Inhibitors

Daniel D. Holsworth*, Stefan Krauss†
*ODIN Therapeutics AS, Oslo, Norway
†SFI-CAST Biomedical Innovation Center, Unit for Cell Signaling, Oslo University Hospital, Oslo, Norway

Contents

ABBREVIATIONS

APC adenomatosis polyposis coli
AXIN2 axis inhibition protein 2
CBP cAMP response element-binding protein-binding protein
CK1-α cyclin-dependent kinase 1α
CK1ε cyclin-dependent kinase 1ε
CREB cyclic AMP response element-binding protein
DVL dishevelled

Annual Reports in Medicinal Chemistry, Volume 47
ISSN 0065-7743
http://dx.doi.org/10.1016/B978-0-12-396492-2.00026-6

GSK3β glycogen synthase kinase 3β
HGF hepatocyte growth factor
LEF lymphoid enhancer factor; transcription factor
LGR5 leucine-rich repeat-containing, G protein-coupled receptor 5
LRP5/6 LDL receptor-related protein 5/6
MM multiple myeloma
PARP poly (ADP-ribose) polymerase
PKA protein kinase A
PORC porcupine
RNF146 poly (ADP-ribose)-directed E3 ligase
SAR structure–activity relationship
STF super TOPFlash reporter containing TCF/LEF-binding site
TCF T cell factor; transcription factor
TNKS TRF-1-interacting ankyrin-related ADP-ribose polymerase
β-TrCP β-transducin repeat-containing protein

1. INTRODUCTION

Wnt/β-catenin signaling is a branch of a functional network that dates back to the first metazoans, and is involved in a broad range of biological systems, including stem cell biology, developmental biology, and adult organ systems. Simplified, the pathway may be described as Wnt protein binding to the cell surface LDL receptor-related protein 5/6 (LRP5/6)–Frizzled receptor complex and allowing β-catenin to travel to the nucleus where binding to T-cell factor/lymphoid enhancer factor (TCF/LEF) results in gene transcription. An overview of the Wnt pathway is shown in Fig. 26.1.

Specific organ systems that depend on Wnt/β-catenin signaling during their development and/or in their adult steady state include the cerebral cortex, hippocampus, eye, lens, spinal cord, limbs, bone, cartilage, somites, neural crest, skin, teeth, gut, lungs, heart, pancreas, liver, kidneys, mammary glands, the hematopoietic system, and the reproductive system. Deregulation of components of Wnt/β-catenin signaling is implied in a wide spectrum of diseases including degenerative diseases, metabolic diseases, and a number of cancers such as colon, breast, bladder, head and neck, nonsmall-cell lung, gastric, melanoma, prostate, leukemia, hepatocellular, pancreas adenocarcinoma, ovarian, and Wilms tumor.[1–4]

Serving several cellular functions, the key mediator of Wnt signaling, β-catenin, is found in a dynamic mode at multiple subcellular locations,

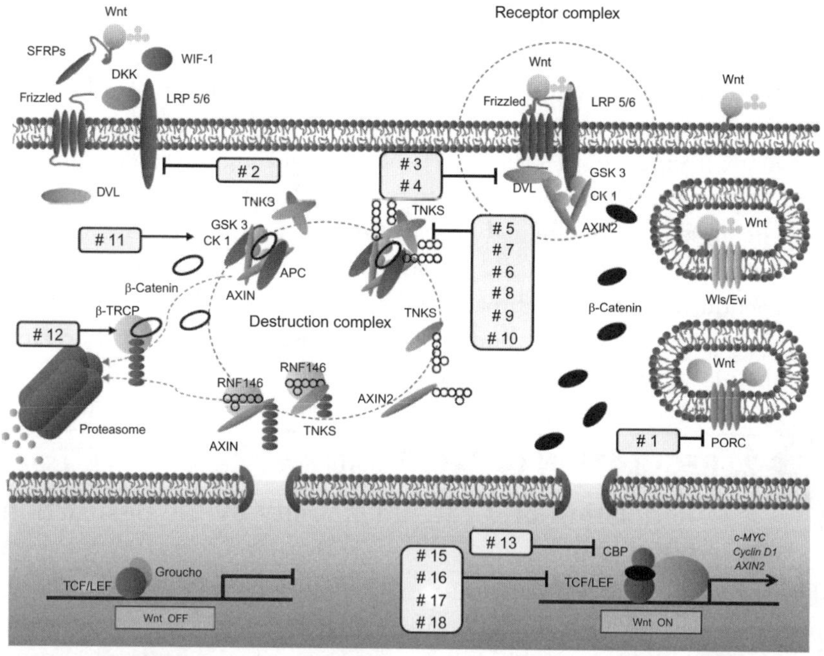

Figure 26.1 Model for Wnt signaling. Newly synthesized Wnt is palmitoylated by PORC and N-glycosylated (gray dots). Upon Wnt binding to the Frizzled/LRP5/6 receptor complex, β-catenin (blue circles) ceases to be degraded and enters the nucleus where it binds TCF/LEF. In the absence of active Wnt, β-catenin becomes phosphorylated (blue rings) in the destruction complex. Subsequently, it is ubiquitinated (green circles) and degraded in the proteasome. Destruction complex proteins are gradually poly-ADP-ribosylated (black rings) by TNKS. Poly-ADP-ribosylation of AXIN and TNKS leads to a destabilization of the destruction complex, followed by ubiquitination and degradation of AXIN and TNKS in the proteasome. Inhibitory substances are marked by a bar, while activating substances are marked by an arrow. The numbers (#) refer to the compounds in the text. (See Color Plate 26.1 in Color Plate Section.)

including junctions, where it contributes to the stabilization of cell–cell contacts, the cytoplasm where β-catenin thresholds are regulated by the destruction complex, and the nucleus where β-catenin is involved in transcriptional regulation and chromatin interactions. Wnt morphogens, cysteine-rich, secreted glycoproteins, are the central regulators of β-catenin thresholds. Through the LRP5/6–Frizzled receptor complex, Wnt morphogens regulate the location and activity of the destruction complex and, consequently, intracellular β-catenin levels. However, β-catenin thresholds are also influenced by multiple other factors, including hypoxia, hepatocyte growth factor, protein kinase A (PKA), and E-cadherin.

β-Catenin is a specialized member of the larger armadillo protein family that consists of three subfamilies: the p120 subfamily, the beta-subfamily (β-catenin and plakoglobin), and the more distant alpha subfamily.[5] The functional interplay between members of this protein family is not well understood, but an involvement of p120 in Wnt/β-catenin signaling was recently shown.[6] Further functional overlaps may exist, in particular, with plakoglobin.

The presence and stability of β-catenin in its various locations, as well as its shuffling through the cell, provide alternative intervention points for therapeutic reagents. The broad implications of Wnt/β-catenin signaling in development, the adult body, and in disease, render it a prime target for pharmacological research and development.

2. REGULATION OF β-CATENIN

2.1. Regulation of β-catenin stability by the destruction complex

The major gatekeeper for regulating β-catenin thresholds in the cell is the β-catenin destruction complex in the cytoplasm, a multiprotein complex containing several druggable biotargets. Although the precise molecular structure and composition of the destruction complex remains to be elucidated, the core of the destruction complex consists of axis inhibition protein 2 (AXIN2), adenomatous polyposis coli (APC), the priming kinase CK1-α, and the kinase glycogen synthase kinase 3β (GSK3β). The positively charged β-catenin associates with the destruction complex where cyclin-dependent kinase 1α (CK1-α) and GSK3β phosphorylate β-catenin at the N-terminal positions S45 and S33, and S37 and T41, respectively. The reduced positive charge on β-catenin causes its dissociation from the destruction complex. The β-transducin repeat-containing protein (β-TrCP), a part of the ubiquitin ligase complex, recognizes β-catenin, causing its ubiquitination, and β-catenin is subsequently degraded by the 26S proteasome. The stability of the destruction complex itself is regulated by tankyrase (TRF-1-interacting ankyrin-related ADP-ribose polymerase) and RNF146. Tankyrase (TNKS) belongs to the 22-member family of poly (ADP-ribose) polymerase (PARP) enzymes and plays a key role in the destabilization of the β-catenin destruction complex by adding negatively charged poly-ADP-ribose units to AXIN2 and to itself. TNKS exists in two highly homologous isoforms. Upon poly-ADP-ribosylation, TNKS 1/2 and AXIN2 presumably dissociate from the destruction complex and are recognized by the ubiquitin ligase RNF146, ubiquitinylated, and then degraded by the proteasome. The degree of poly-ADP-ribosylation at the destruction complex is positively regulated by the availability of NAD^+ and negatively regulated by poly (ADP-ribose) glycohydrolase. Taken together, the tankyrase/destruction complex is a highly dynamic protein complex that centrally regulates cytoplasmic β-catenin thresholds.[3,4,7–9]

2.2. Regulation of β-catenin by the Wnt receptor complex

In the secretory pathway, Wnt morphogens have to be modified by palmitoylation and glycosylation to become mature signaling proteins. Evidence suggests that the membrane-bound *O*-acyltransferase Porcupine (PORC) may be involved in palmitoylation and secretion of Wnt proteins, a process that has recently been established as drug sensitive. Upon binding of the Wnt morphogen, a Frizzled-LRP5/6 receptor complex forms the basic unit of the Wnt signalosome. The secreted proteins, Wnt inhibitor factor-1, Cerberus, and soluble Frizzled-related proteins can interfere with Wnt binding to the Frizzled-LRP5/6 receptor, while a Dickkopf/Kremen complex prevents the formation of the Wnt signalosome by recruiting LRP5/6 and thus preventing it from forming a complex with Frizzled.

Nineteen isoforms of the Wnt morphogen together with 10 isoforms of the Frizzled receptor in humans allow some functional diversity and differential cellular response. In the canonical branch of Wnt signaling, components of the destruction complex are drawn to the Wnt signalosome. This disables the destruction complex from phosphorylating β-catenin and leads to an elevation of cytoplasmic and nuclear β-catenin. At the cytoplasmic side of the Wnt signalosome, phosphorylated Dishevelled (DVL) binds to Frizzled and enables the positioning of AXIN2 to LRP5/6. Blocking DVL attenuates the Wnt signaling cascade leading to reduced β-catenin levels. DVL has recently been described as a drug biotarget.[7]

Alternatively, Wnt binding to the Frizzled-LRP5/6 signalosome induces phosphorylation of the armadillo protein p120 by CK1ε followed by its release to the nucleus.[6]

2.3. β-Catenin in the nucleus

In the nucleus, β-catenin appears to have multiple functions. The best understood role of β-catenin is its binding to members of the TCF/LEF zinc finger family of transcription factors to act as transcriptional activators. In contrast, TCF3 is predominantly a transcriptional repressor. The histone acetyltransferase cAMP response element-binding protein-binding protein (CBP) attenuates the complex and acts as a context-dependent β-catenin-binding transcriptional regulator.[9] Furthermore, p120 binding to the zinc finger transcription factor Kaiso releases its inhibition of the β-catenin–TCF4 transcriptional complex. Both β-catenin/TCF binding and β-catenin/CBP binding have been used as interference points for exploratory drugs.[7]

Also, β-catenin shuttling to and from the nucleus is regulated. A picture emerges where stability and nuclear uptake of β-catenin can be enhanced by the context-dependent C-terminal phosphorylation of β-catenin at S675 by PKA, while export of β-catenin from the nucleus is GSK3β dependent.[10] Both kinases are well-explored drug targets.

3. SELECTIVE SMALL-MOLECULE ANTAGONISTS OF WNT/β-CATENIN SIGNALING

Wnt/β-catenin signaling in tumorigenesis has provided a prime rationale for mapping druggable intervention points in the pathway. While an inhibition or normalization of the pathway is predominantly desired in the cancer arena, a controlled attenuation or increase of Wnt/β-catenin signaling may also be sought for certain aspects of regenerative medicine.

3.1. PORC inhibitor

PORC is involved in the palmitoylation and secretion of Wnt proteins. A PORC inhibitor would reduce secretion and maturation of the Wnt morphogen, thereby inhibiting paracrine Wnt signaling.

Compound **1** inhibits the Wnt pathway (IC_{50}: 27 nM, L-Wnt STF cells) by competitively inhibiting the function of PORC. Inhibition of LRP6 and DVL phosphorylation and cytoplasmic β-catenin accumulation was also observed upon exposure of **1** at 10 μM to L-Wnt super TOPFlash reporter containing TCF/LEF-binding site (STF) cells.[11]

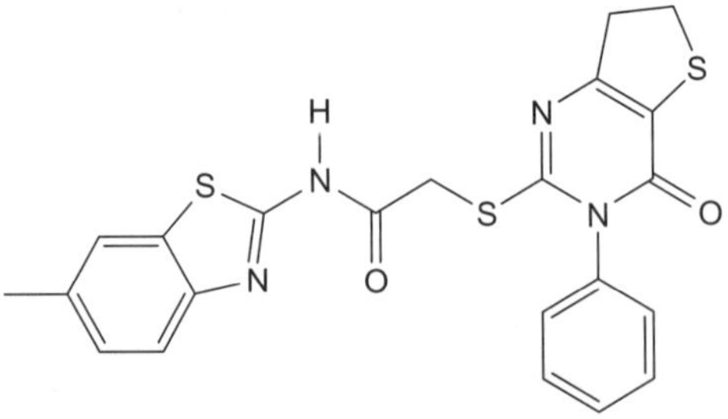

1

3.2. LRP6 inhibitor

The transmembrane proteins LRP5/6 and Frizzled form the core of the Wnt signalosome. An induced degradation of LRP5/6 would sever the cell's responsiveness to an incoming Wnt signal.

Niclosamide (**2**) has been shown to reduce cancer cell growth by inducing LRP6 degradation.[12] Niclosamide was also shown to suppress LRP6 expression and phosphorylation in HEK293 human embryonic kidney cells.

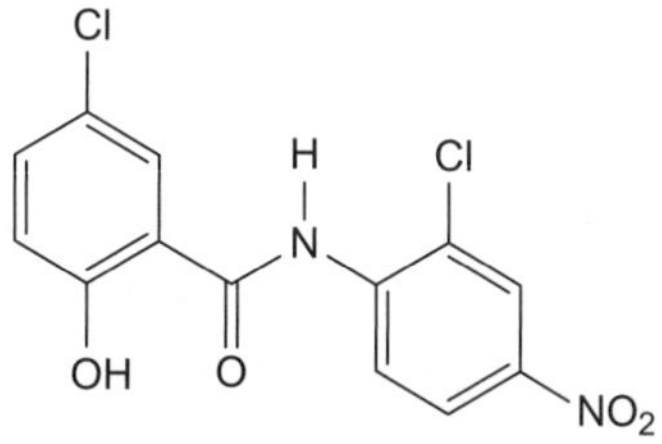

2

Niclosamide is also an inhibitor of multiple other pathways (NF-κB, Wnt, NOTCH, ROS, mTORC1, STAT3), most of which are implicated in cancer metastasis. Compound **2** inhibited cell growth (GI_{50}s) in human prostate (PC-3, DU145) and breast cancer (MDA-MB-231, T-47D) cell lines at 0.7–0.3 μM. It also inhibited S100A4 protein expression, cell proliferation, migration, and invasion in colon cancer cell lines.[13]

3.3. DVL inhibitors

DVL interacts with the membrane-bound Wnt Frizzled receptor. A DVL inhibitor would, among other things, reduce the interplay between the Wnt signalosome and the destruction complex and thus inhibit canonical Wnt signaling.

Sulindac (**3**), a nonsteroidal anti-inflammatory drug, was shown to block the PDZ domain of DVL (IC_{50}: 10.7 μM) and cause decreased nuclear accumulation of β-catenin and reduced expression of the metastatic mediator protein S100A4 in a human colon cancer xenograft model.[14]

Compound **4** exhibited a K_d of 10.6 μM toward the DVL PDZ domain in an assay system using fluorescence anisotropy.[15] At an IC_{50} of 12.5 μM, **4** reduced cell growth and β-catenin levels in the PC-3 prostate cell line.

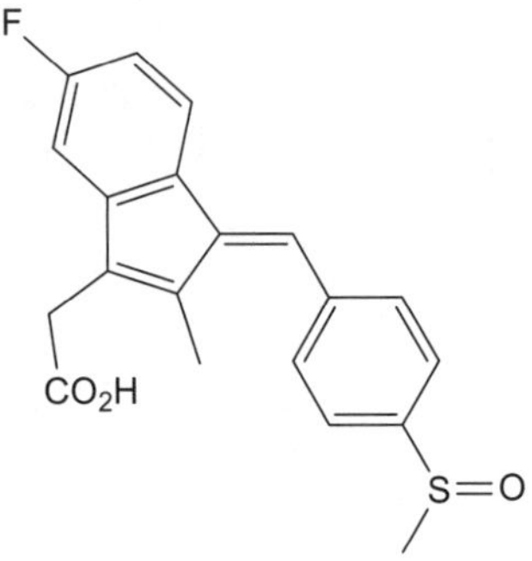

3

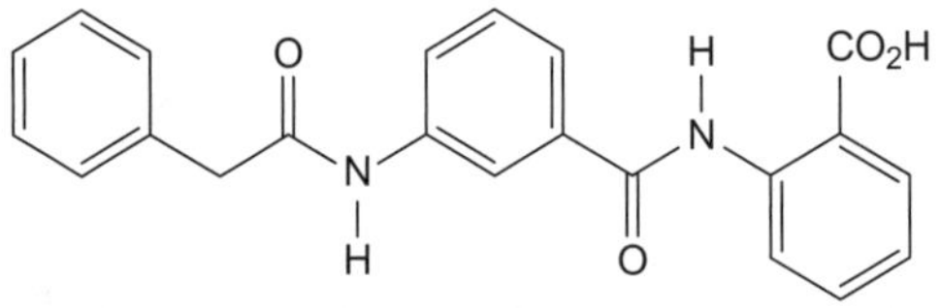

4

3.4. Tankyrase inhibitors

Tankyrase (TNKS) 1 and 2 have multiple cellular functions including marking destruction complex proteins by poly-ADP-ribosylation for ubiquitination and degradation. Since TNKS 1/2 are druggable modulators of Wnt/β-catenin signaling, they have recently received substantial attention. However, obtaining selectivity of small molecules between the PARP family members and the TNKS isoforms can be been challenging. Blocking the PARP domain of TNKS 1/2 can lead to a context-dependent inhibition or normalization of Wnt/β-catenin signaling.

The first small-molecule inhibitors of TNKS 1/2 were disclosed in 2009 (**5** and **6**). IWR-1 (**5**) exhibited good potency against TNKS 1/2 with IC_{50}s of 131 and 56 nM, respectively, while not affecting PARP-1 and 2.[11] IWR-1 inhibited Wnt-stimulated transcription activity (IC_{50}: 180 nM) and increased AXIN2 and phosphorylated β-catenin levels in DLD-1 colorectal carcinoma cells at 1 μM.

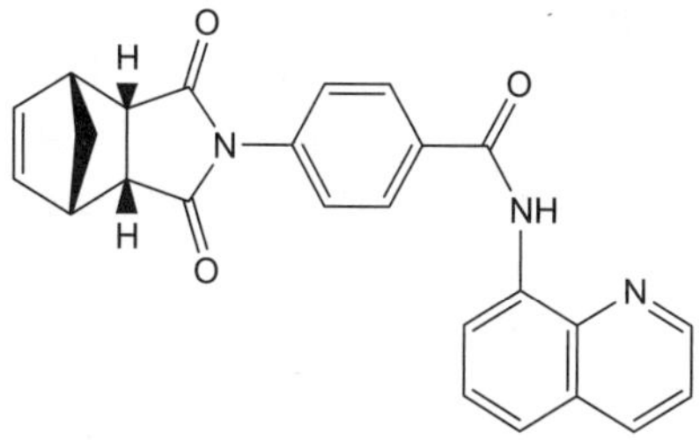

5

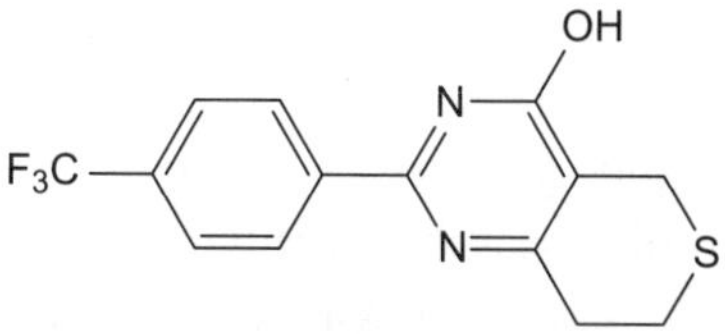

6

IWR-1 demonstrated *in vivo* activity by inhibiting zebra fish tail fin regeneration at 10 μM. IWR-1 suffers from instability in mouse liver microsomes ($T_{1/2} \sim 20$ min). Early structure–activity relationship (SAR) works to find IWR-1 analogs that exhibited improved potency and stability were met with modest success.[16] Compound **5** was obtained from a high-content screen searching for stimulators of cardiomyogenesis.[17] Optimization of **5** led to **7**, which exhibited an IC_{50} of 4 nM against the Wnt pathway in HEK293T cells, and demonstrated 45% greater efficacy in inducing cardiogenesis than **5**. Reduction of the double bond of the bicyclic ring and incorporation of a *trans*-cyclohexyl ring in place of the phenyl ring spacer of **7** led to a series that exhibited robust SAR.

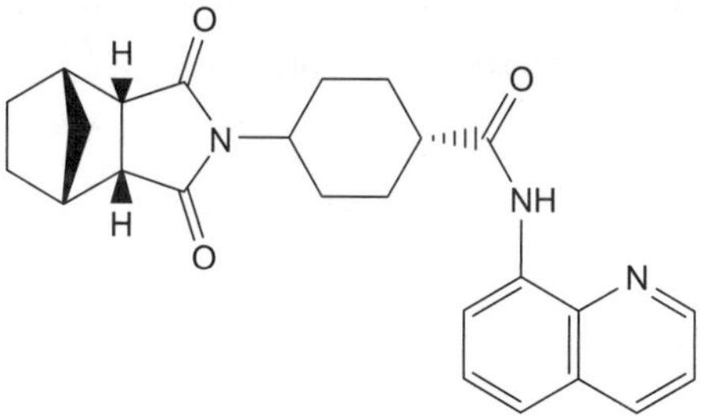

7

A recent X-ray crystal structure of **5** complexed with TNKS 2[18] illustrates that IWR-1 binds to the adenosine pocket of the PARP domain, which may account for its specificity to TNKS 1/2.

XAV939 (**6**) is a potent TNKS inhibitor (IC_{50}s: TNKS1: 11 nM, TNKS2: 4 nM (HEK293 STF cells)) that, in addition, inhibits PARP1 and 2 (IC_{50}: 2200 and 114 nM, respectively).[19] XAV939 increased AXIN2 and TNKS 1/2 protein levels and promoted the degradation of β-catenin in SW480 cells at 1 μM. XAV939 has been shown to reduce the growth of β-catenin-dependent DLD-1 colon carcinoma cells by 95% at 3.3 μM. An X-ray cocrystal structure of XAV939 bound to TNKS 2 showed that **6** binds in the nicotinamide pocket of the TNKS 2 PARP domain.[20,21] Along with **6**'s anticancer properties, it has also been shown to accelerate repair of oligodendrocyte progenitor cells from the brain and spinal cord after hypoxic and demyelinating injury.[22]

Recently, a structural analysis of PARP and TNKS inhibitors was conducted[21] to provide insight into the selectivity of TNKS inhibitors as compared to PARP 1–4 inhibitors. It was shown that the binding site near the NAD^+ pocket of PARPs is lined with hydrophilic residues, whereas in TNKS2, the residues are generally hydrophobic. Furthermore, TNKS2 has a more elongated binding site than PARPs 1–4, providing rationale to design TNKS-specific inhibitors.

The 1,2,4-triazole-containing compound (**8**) was identified as a selective Wnt pathway inhibitor (IC_{50}: 0.79 μM).[23] The biological target of **8** has recently been described as TNKS.[9] Compound **8** specifically inhibited induced Wnt signaling in a *Xenopus laevis* axis duplication assay at 0.8 pM, and was effective in stabilizing AXIN2 and reducing the active form of β-catenin in SW480 colon carcinoma cells. In SW480 cells, **8** reduced growth (GI_{50}) at 5 μM. Although **8** was found to be extremely labile to liver microsomes (human and mouse: 3 min), it reduced tumor growth of SW480 cells in a CB17/SCID xenograft model when dosed orally QD at 150 mg/kg for 21 days. It was also demonstrated that **8** inhibited tumor formation and growth in the small intestine and colon of APC^{min} mice by 48% when dosed at 150 mg/kg.

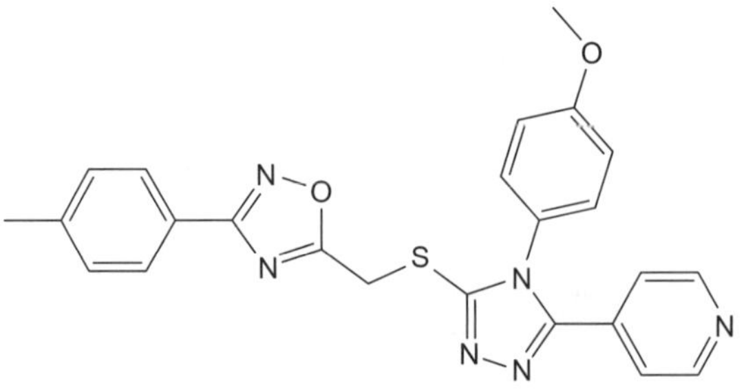

8

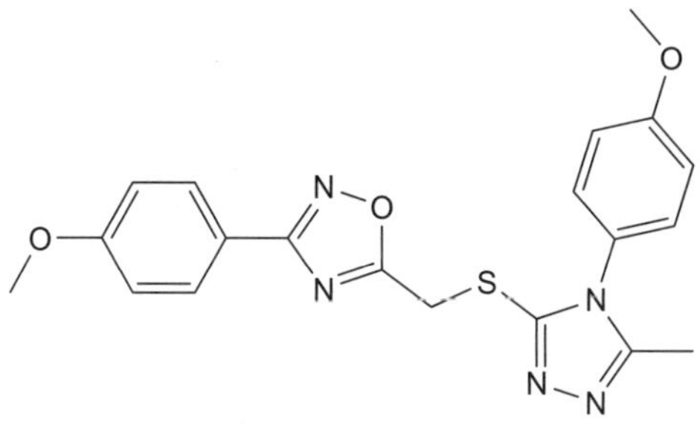

9

Compound **8** was independently identified and found to be a modest TNKS inhibitor (TNKS1, IC_{50}: 2.55 μM; TNKS2, IC_{50}: 0.65 μM (HEK293 STF cells)).[24] Early SAR development of **8** resulted in **9**. Compound **9** has an IC_{50} of 33 nM against TNKS2 and IC_{50}s greater than 19 μM for PARP1 and 2. By installing a methyl group in place of the pyridyl group of **8**, diminished P450 isozyme inhibition was observed. X-ray cocrystal structure of **9** in TNKS1 showed **9** bound in the adenosine and diphosphate linker portion of the NAD^+ donor site. Additional interactions of **9** in the hydrophobic section surrounding the diphosphate linker portion were also observed.

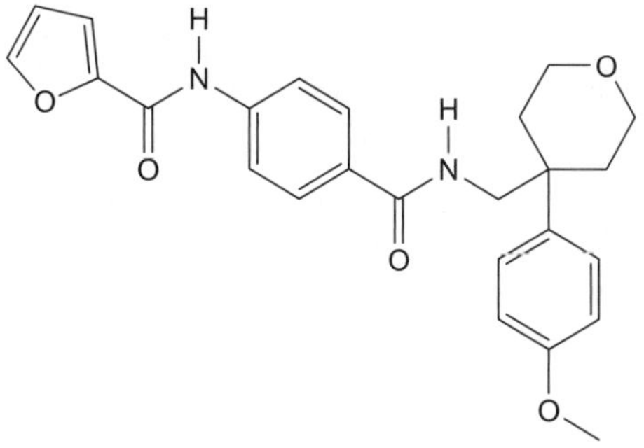

10

Compound **10** was shown to be a potent inhibitor of the Wnt pathway (IC_{50} in HEK293 reporter cells: 0.470 μM) and a selective inhibitor of TNKS (biochemical TNKS1 inhibition, IC_{50}: 1.9 μM; biochemical TNKS2 inhibition, IC_{50}: 0.83 μM) as compared to PARP1 ($IC_{50} > 20$ μM).[9] Similar to compound **6**, compound **10** stabilized AXIN2 protein levels; however, in contrast, **10** reduced TNKS protein levels in SW480 colorectal carcinoma cells. Compound **10** when exposed to SW480 cells displayed a general reduction of total β-catenin at 1 and 5 μM. It also reduced the expression of the Wnt target genes AXIN2, SP5, and NκD1 at 10 μM in SW480 and DLD1 colorectal cell lines. Compound **10** was administered orally at

100 mg/kg to $APC^{cko/cko}$ Lgr5-CreERT2^{+} mice and found to reduce mean total tumor area in the small intestine.

3.5. Casein kinase 1-α activity enhancer

The priming kinase CK1-α is part of the destruction complex, which phosphorylates β-catenin at serine 45. A drug that induces increased CK1-α/γ activity would regulate negatively the stability of β-catenin in the cytoplasm.

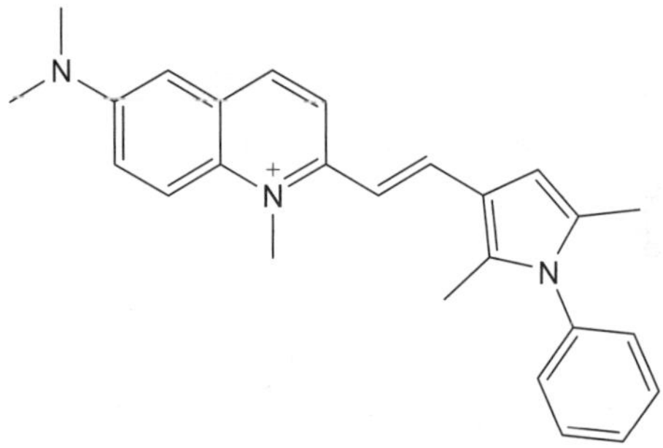

11

Compound **11** is an FDA-approved anthelmintic compound that potently (EC_{50}: 10 nM) inhibits Wnt gene transcription in HEK293 cells.[25] **11** binds to all casein kinase 1 family members and selectively enhances CK1-α activity, thus lowering cytoplasmic β-catenin levels. Furthermore, **11** promoted the degradation of the Pygopus PHD-finger protein within the nucleus and subsequently reduced Wnt-directed transcription. Compound **11** displayed toxicity when dosed to mice at 200 nM IV or IP, but at lower doses exhibited wound repair and postmyocardial infarction remodeling potential in mice.[26]

3.6. Wnt/β-catenin inhibitors enhancing β-TrCP/β-catenin interaction

A compound with the general structure **12** has been reported to decrease c-Myc, cyclin D1, and survivin expression in a multiple myeloma (MM) mouse model, and also prolonged survival of mice with MM.[27] **12** inhibited the proliferation of MM cells in a time- and dose-dependent manner (with IC_{50}s ranging from 11 to 82 nM[28]) by enhancing the β-TrCP/β-catenin interaction. Ubiquitination of cellular β-catenin was shown to be increased, and nuclear β-catenin levels reduced.

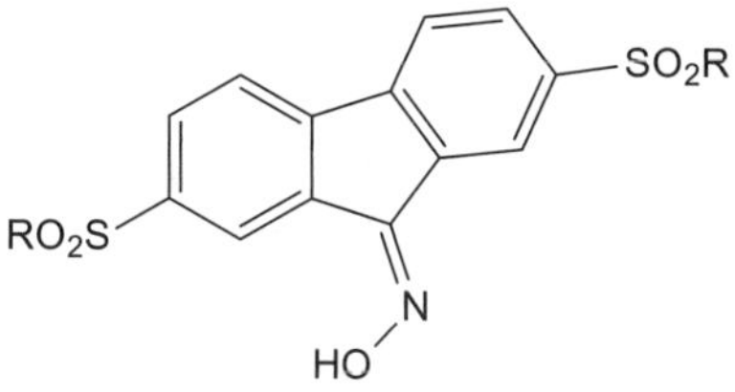

12

3.7. Inhibitors of β-catenin/TCF-LEF and β-catenin/CBP binding

Interference between β-catenin and its nuclear targets has long been a prominent target for drug discovery. Both binding between β-catenin and members of the TCF/LEF family and binding between β-catenin and CBP can be inhibited by small molecules. Such an inhibition would not lower β-catenin thresholds in the cytoplasm and nucleus, but would alter the transcription of Wnt/β-catenin downstream genes such as c-Myc, cyclin D1, and survivin.

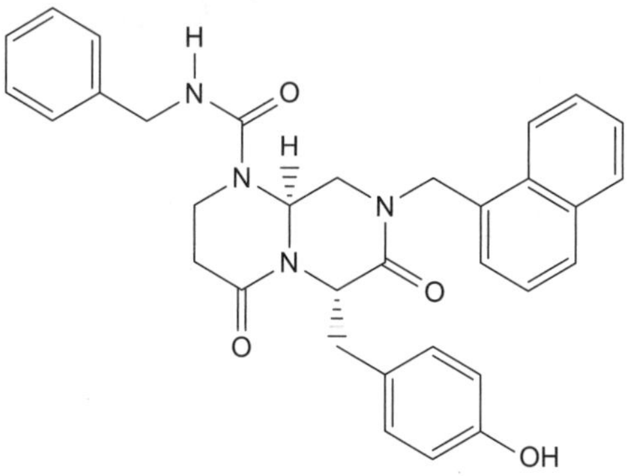

13

Compound **13** interacts with CBP, a context-dependent coactivator for Wnt/β-catenin transcription.[9,29] Compound 13 is a modest inhibitor (IC_{50}: 3 μM (SW480 TOPFlash cells)) of the Wnt/β-catenin signaling pathway and selective for CBP, but not its close homolog p300. β-Catenin can switch from interacting with CBP to interacting with p300, which causes cells to become less differentiated and behave more like stem cells. Compound **13** has been claimed to eliminate cancer stem cells without affecting normal somatic stem cells.[30] Recently, a nondisclosed homolog of **13** that also inhibits the β-catenin/CBP interaction (IC_{50}: 200 nM) has entered phase I trials for patients with solid tumors.[31]

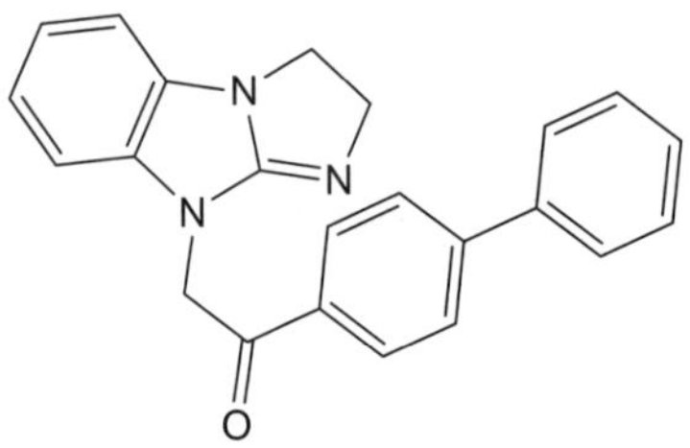

14

A cell-based, small-molecule screen for regulators of TCF-dependent transcription yielded **14**.[32] Compound **14** blocked nuclear and cytosolic β-catenin accumulation in mouse L-3 cells when incubated with a GSK3 inhibitor. Additional mechanistic studies suggest that **14** interacts through multiple molecular targets.

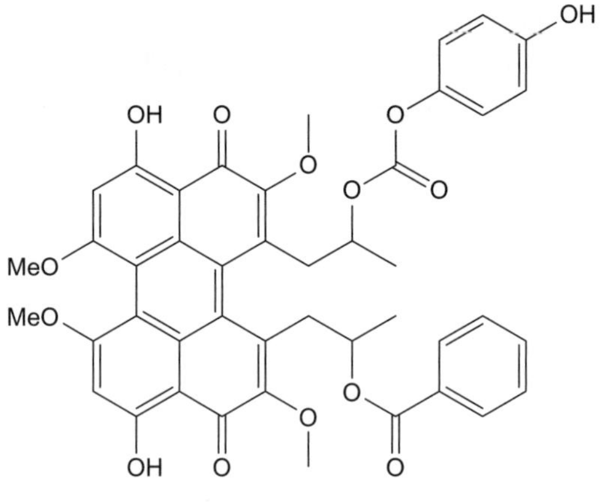

15

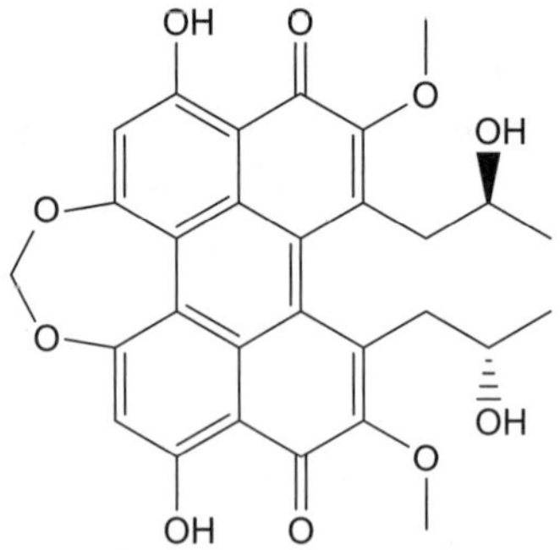

16

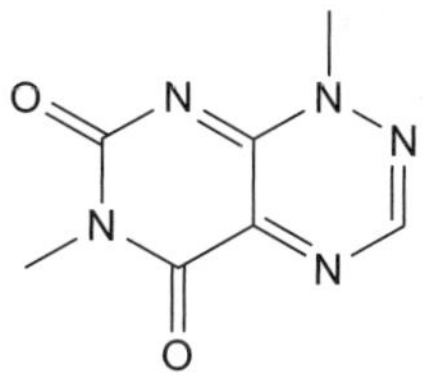

17

Compounds **15–17** inhibit the β-catenin/TCF/LEF interaction and transcriptional activity.[33] All three compounds were found to be cytotoxic to hepatoma cells (IC_{50}s 0.26–0.98 μM) but not to normal hepatocytes. Compound **15** was most active against the hepatoma cell lines (Hep40, Huh7, HepG2) tested. Compounds **15** and **16** were also efficient in selectively killing chronic lymphocytic leukemia (CLL) cells (LC_{50}s: 0.7–0.9 μM) over B cells,[34] and demonstrated *in vivo* efficacy in CLL mouse model when dosed at 25 mg/kg/day for 12 days.

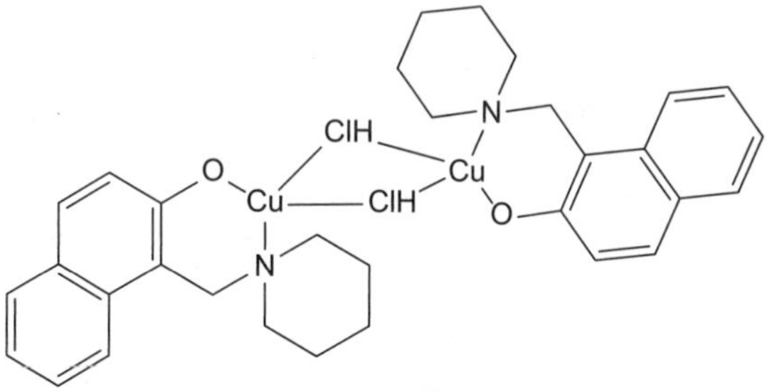

18

Compound **18** was obtained from a virtual screen of agonist putative binding sites of TCF4 with β-catenin.[35] **18** displayed competitive binding to GST-TCF4 with an IC_{50} of 5 μM. **18** also decreased HCT116 colon cancer cell viability (IC_{50}: 15 μM) and at 5 μM blocked 80% of the colony-forming capabilities of HCT116 cells as well as downstream target genes c-Myc and cyclin D1.

4. CONCLUSIONS

The search for viable inhibitors at multiple points along the Wnt/β-catenin signaling pathway has received considerable attention from pharmaceutical and academic groups. Drugs and druggable biotargets have been identified that inhibit Wnt/β-catenin signaling at the Wnt secretory pathway, at the Wnt signalosome, at the destruction complex, and at the nuclear β-catenin/TCF and β-catenin/CBP interface. Identification of additional chemotypes that possess drug-like properties, and mapping their precise biotarget, will be needed to fully validate druggable interference points within the Wnt/β-catenin complex pathway. Validation of Wnt/β-catenin inhibitors as viable anticancer reagents *in vitro*, in animal models, and in clinical trials is still at an early stage.

REFERENCES

(1) Tanaka, S.S.; Kojima, Y.; Yamaguchi, Y.L.; Nishinakamura, R.; Tam, P.P. *Dev. Growth Differ.* **2011**, *53*, 843.
(2) Grigoryan, T.; Wend, P.; Klaus, A.; Birchmeier, W. *Genes Dev.* **2008**, *22*, 2308.
(3) Camilli, T.C.; Weeraratna, A.T. *Biochem. Pharmacol.* **2010**, *80*, 705.
(4) Polakis, P. *Cold Spring Harb. Perspect. Biol.* **2012**, http://dx.doi.org/10.1101/cshperspect.a008052.
(5) Zhao, Z.M.; Reynolds, A.B.; Gaucher, E.A. *BMC Evol. Biol.* **2011**, *11*, 198.
(6) Casagolda, D.; Del Valle-Pérez, B.; Valls, G.; Lugilde, E.; Vinyoles, M.; Casado-Vela, J.; Solanas, G.; Batlle, E.; Reynolds, A.B.; Casal, J.I.; de Herreros, A.G.; Duñach, M.A. *J. Cell Sci.* **2010**, *123*, 2621.
(7) MacDonald, B.T.; Tamai, K.; He, X. *Dev. Cell* **2009**, *17*, 9.
(8) Callow, M.G.; Tran, H.; Phu, L.; Lau, T.; Lee, J.; Sandoval, W.N.; Liu, P.S.; Bheddah, S.; Tao, J.; Lill, J.R.; Hongo, J.A.; Davis, D.; Kirkpatrick, D.S.; Polakis, P.; Costa, M. *PLoS One* **2011**, *6*, e22595.
(9) Waaler, J.; Machon, O.; Tumova, L.; Dinh, H.; Korinek, V.; Wilson, S.R.; Paulsen, J.E.; Pedersen, N.M.; Eide, T.J.; Machonova, O.; Gardl, D.; Von Kries, J.P.; Krauss, S. *Cancer Res.* **2012**, *72*, 2695.
(10) Li, J.; Sutter, C.; Parker, D.S.; Blauwkamp, T.; Fang, M.; Cadigan, K.M. *EMBO J.* **2007**, *26*, 2284.
(11) Chen, B.; Dodge, M.E.; Tang, W.; Lu, J.; Ma, Z.; Fan, C.-W.; Wei, S.; Hao, W.; Kilgore, J.; Williams, N.S.; Roth, M.G.; Amatruda, J.F.; Chen, C.; Lum, L. *Nat. Chem. Biol.* **2009**, *5*, 100.
(12) Lu, W.; Lin, C.; Roberts, M.J.; Waud, W.R.; Piazza, G.A.; Li, Y. *PLoS One* **2011**, *6*, e29290.
(13) Sack, U.; Walther, W.; Scudiero, D.; Selby, M.; Kobelt, D.; Lemm, M.; Fichtner, I.; Schlag, P.M.; Shoemaker, R.H.; Stein, U. *J. Natl. Cancer Inst.* **2011**, *103*, 1018.
(14) Stein, U.; Arlt, F.; Smith, J.; Sack, U.; Hermann, P.; Walther, W.; Lemm, M.; Fichtner, I.; Shoemaker, R.H.; Schlag, P.M. *Neoplasia* **2011**, *13*, 131.
(15) Gandy, D.; Shan, J.; Zhang, X.; Rao, S.; Akunuru, S.; Li, H.; Zhang, Y.; Alpatov, I.; Zhang, X.A.; Lang, R.A.; Shi, D.-L.; Zheng, J.J. *J. Biol. Chem.* **2009**, *24*, 16256.
(16) Lu, J.; Ma, Z.; Hsieh, J.-C.; Fan, C.-W.; Chen, B.; Longgood, J.C.; Williams, N.S.; Amatruda, J.F.; Lum, L.; Chen, C. *Bioorg. Med. Chem. Lett.* **2009**, *19*, 3825.
(17) Lanier, M.; Schade, D.; Willems, E.; Tsuda, M.; Spiering, S.; Kalisiak, J.; Mercola, M.; Cashman, J.R. *J. Med. Chem.* **2012**, *55*, 697.
(18) Narwal, M.; Venkannagari, H.; Lehtio, L. *J. Med. Chem.* **2012**, *55*, 1360.
(19) Huang, S.-M.A.; Mishina, Y.M.; Liu, S.; Cheung, A.; Stegmeier, F.; Michaud, G.A.; Charlat, O.; Wiellette, E.; Zhang, Y.; Wiessner, S.; Hild, M.; Shi, X.; Wilson, C.J.; Mickanin, C.; Myer, B.; Fazal, A.; Tomlinson, R.; Serluca, F.; Shao, W.; Cheng, H.; Schultz, M.; Rau, C.; Schirle, M.; Schlegl, J.; Ghidelli, S.; Fawell, S.; Lu, C.; Curtis, D.; Kirschner, M.W.; Lengauer, C.; Finan, P.M.; Tallarico, J.A.; Bouwmeester, T.; Porter, J.A.; Bauer, A.; Cong, F. *Nature* **2009**, *461*, 614.
(20) Karlberg, T.; Markova, N.; Johansson, I.; Hammarstrom, M.; Schutz, P.; Weigelt, J.; Schuler, H. *J. Med. Chem.* **2010**, *53*, 5352.

(21) Wahlberg, E.; Karlberg, T.; Kouznetsova, E.; Markova, N.; Macchiarulo, A.; Thorsell, A.-G.; Pol, E.; Frostell, A.; Ekbald, T.; Oncu, D.; Kull, B.; Robertson, G.M.; Pelliciari, R.; Schuller, H.; Weigelt, J. *Nat. Biotechnol.* **2012**, *30*, 283.
(22) Fancy, S.P.; Harrington, E.P.; Yuen, T.J.; Silbereis, J.C.; Zhao, C.; Baranzini, S.E.; Bruce, C.C.; Otero, J.J.; Huang, E.J.; Nusse, R.; Franklin, R.J.; Rowitch, D.H. *Nat. Neurosci.* **2011**, *14*, 1009.
(23) Waaler, J.; Machon, O.; Von Kries, J.P.; Wilson, S.R.; Lundenes, E.; Wedlich, D.; Gradl, D.; Paulson, J.E.; Machonova, O.; Dembinski, J.L.; Dinh, H.; Krauss, S. *Cancer Res.* **2011**, *71*, 197.
(24) Shultz, M.D.; Kirby, C.A.; Stams, T.; Chin, D.N.; Blank, J.; Charlat, O.; Cheng, H.; Cheung, A.; Cong, F.; Feng, Y.; Fortin, P.D.; Hood, T.; Tyagi, V.; Xu, M.; Zhang, B.; Shao, W. *J. Med. Chem.* **2012**, *55*, 1127.
(25) Thorne, C.A.; Hanson, A.J.; Schneider, J.; Tahinci, E.; Orton, D.; Cselenyi, C.S.; Jernigan, K.K.; Meyers, K.C.; Hang, B.I.; Waterson, A.G.; Kim, K.; Melancon, B.; Ghidu, V.P.; Sulikowski, G.A.; LaFleur, B.; Salic, A.; Lee, L.A.; Miller, D.M.; Lee, E. *Nat. Chem. Biol.* **2010**, *6*, 829.
(26) Saraswati, S.; Alfaro, M.P.; Thorne, C.A.; Atkinson, J.; Lee, E.; Young, P.P. *PLoS One* **2010**, *5*, e15521.
(27) Ashihara, H.Y.; Strovel, J.W.; Nakagawa, Y.; Kuroda, J.; Nagao, R.; Yokota, A.; Takeuchi, M.; Hayashi, Y.; Shimazaki, C.; Taniwaki, M.; Strand, K.; Padia, J.; Hirai, H.; Kimura, S.; Maekawa, T. *Blood Cancer J.* **2011**, *1*, e43.
(28) Hisayuki, Y.; Ashihara, E.; Nagao, R.; Kimura, S.; Hirai, H.; Strovel, J. W.; Padia, J.; Cholody, W. M.; Maekawa, T. *Abstract 2866*, 51st ASH Annual Meeting, New Orleans, LA, December **2009**.
(29) Emami, K.H.; Nguyen, C.; Ma, H.; Kim, D.H.; Jeong, K.W.; Eguchi, M.; Moon, R.T.; Teo, J.-L.; Oh, S.W.; Kim, H.Y.; Moon, S.W.; Ha, J.R.; Kahn, M. *Proc. Natl. Acad. Sci. U.S.A.* **2004**, *101*, 12682.
(30) Kahn, M. *Am. Soc. Clin. Oncol.* **2011**, 435.
(31) Cha, J. Y.; Jung, J. -E.; Lee, K. -H.; Briaud, I.; Tenzin, F.; Pyon, Y.; Lee, D.; Chung, J. U.; Lee, J. H.; Oh, S. -W.; Jung, K. Y.; Pai, J. K.; Emami, K. *Abstract 3038*, 52nd ASH Annual Meeting, Orlando, FL, December **2010**.
(32) Ewan, K.; Pajak, B.; Stubbs, M.; Todd, H.; Barbeau, O.; Quevedo, C.; Botfield, H.; Young, R.; Ruddle, R.; Samuel, L.; Battersby, A.; Raynaud, F.; Allen, N.; Wilson, S.; Latinkic, B.; Workman, P.; McDonald, E.; Blagg, J.; Aherne, W.; Dale, T. *Cancer Res.* **2010**, *70*, 5963.
(33) Wei, W.; Chua, M.-S.; Grepper, S.; So, S. *Int. J. Cancer* **2009**, *126*, 2426.
(34) Gandhirajan, R.K.; Staib, P.A.; Minke, K.; Gehrke, I.; Plickert, G.; Schlosser, A.; Schmitt, E.K.; Hallek, M.; Kreuzer, K.-A. *Neoplasia* **2010**, *12*, 326.
(35) Tian, W.; Han, X.; Yan, M.; Xu, Y.; Duggineni, S.; Lin, N.; Luo, G.; Li, Y.M.; Han, X.; Huang, X.; An, J. *Biochemistry* **2012**, *51*, 724.

PART 7

Topics in Drug Design and Discovery

Editor: Peter R. Bernstein
PhaRmaB LLC, Rose Valley, Pennsylvania

CHAPTER TWENTY-SEVEN

Targeted Covalent Enzyme Inhibitors

Mark C. Noe, Adam M. Gilbert
Department of Worldwide Medicinal Chemistry, Pfizer Worldwide Research and Development, Groton, Connecticut, USA

Contents

1. INTRODUCTION

Small molecule drugs exert their pharmacological effects through binding to biomolecular targets, thereby modulating their activity (in the case of enzymes) or downstream signaling (in the case of receptors) with biological consequences that are relevant to the disease state being treated. The majority of small molecule drugs bind to their targets through noncovalent interactions (hydrogen bonds and hydrophobic interactions) or through reversible metal–ligand interactions. For these agents, PK/PD relationships are often understood in terms of equilibrium binding affinities (K_d) for their primary target(s) relative to steady-state free plasma concentrations. This seemingly uncomplicated scenario offers many advantages to the medicinal chemist: target potency and selectivity are determined through simple biological assays at equilibrium, reversibility of drug–target interactions allows tailoring the PD response

Annual Reports in Medicinal Chemistry, Volume 47
ISSN 0065-7743
http://dx.doi.org/10.1016/B978-0-12-396492-2.00027-8

through modifying the PK profile, and the risk of idiosyncratic toxicities is reduced because the target protein is not modified by the drug (excepting, of course, cases of reactive metabolite formation). However, compounds that covalently modify their drug targets are well represented in the pharmacopeia. At present, 39 drugs approved by the FDA bind covalently to their targets, and approximately one-third of all enzyme drug targets have at least one example of an approved covalent drug.[1,2] Many of these compounds were discovered accidentally, noteworthy examples being aspirin, clopidogrel, and omeprazole.[1] Indeed, the formation of a covalent bond between a small molecule drug and its target protein has been largely avoided as a design strategy due to risks associated with immunogenic responses to covalently modified proteins.

Selective covalent modification of a biomolecular target can offer several important advantages in drug discovery—principally through establishing nonequilibrium binding kinetics.[3] For example, it is extremely difficult to target protein–protein interactions with noncovalent small molecule inhibitors due to the expansive binding surface over which the protein–protein interaction occurs.[4] Likewise, targeting the ATP binding site of a protein kinase can be problematic with a noncovalent inhibitor due to high intracellular ATP concentrations that often exceed the enzyme's K_m for ATP by several orders of magnitude. Each of these limitations can theoretically be overcome through the nonequilibrium condition established by targeted covalent modification. Indeed, for the case of kinases, this advantage has been taken further to address the epidermal growth factor receptor (EGFR) kinase T790M resistance mechanism in tumor cells, where this single mutation causes a significant enhancement in ATP affinity and is successfully surmounted with covalent EGFR inhibitors binding at the ATP site.[5–7] Likewise, targeting a uniquely positioned nucleophilic residue within a target's active site offers the opportunity to achieve selectivity for otherwise highly homologous proteins. The case of covalent kinase inhibitors has been extensively studied, and recently reported kinase cysteinome bioinformatics and chemical proteomic methods to characterize reactive cysteines within an enzyme offer useful tools for developing cysteine-targeting strategies for these enzymes.[7–9] Finally, the nonequilibrium condition established through targeted covalent binding can enable significantly lower therapeutic exposures because for protein targets that regenerate slowly, the pharmacodynamic half-life of the drug is much longer than its pharmacokinetic half-life.[10]

Equation (27.1) depicts the differences between covalent and noncovalent enzyme inhibition and establishes the need for unique approaches to study and optimize targeted covalent enzyme inhibitors.

In the case of reversible enzyme inhibition, the K_i for the enzyme (and occasionally k_{on} and k_{off} in the case of slow kinetics to reach equilibrium) is the relevant parameter for drug affinity optimization. For covalent inhibitors, it is the ratio of k_{inact} (the rate constant for covalent modification) to K_i (the equilibrium constant for prereaction association between enzyme and drug), or k_{inact}/K_i, that is important. This value is equivalent to the second order rate constant for bimolecular reaction of the enzyme and its inhibitor. The measurement of k_{inact}/K_i can be quite labor intensive and involves preincubating enzyme with inhibitor, followed by introducing substrate and measuring the percentage of remaining enzyme activity. Preincubation time is varied to produce k_{app} (the apparent rate constant for enzyme inactivation) for a given inhibitor concentration. Varying the inhibitor concentration enables generating linear plots according to the equation: $1/k_{app} = 1/k_{inact} + K_i/(k_{inact} \times [I])$. If $[I] \ll K_i$, then this equation simplifies to $k_{app}/[I] = k_{inact}/K_i$.[11] Therefore, plotting k_{app} as a function of [I] generates linear plots with a slope of k_{inact}/K_i. Because these experiments are very labor intensive, many programs attempt to compare compounds with IC_{50} values determined by varying inhibitor concentration at a single set preincubation time. This methodology is inadequate for generating SAR data because the IC_{50} values are dependent on preincubation time (which can vary between experiments), and information on the fundamental kinetic parameters for enzyme inactivation is not produced.[1] Newer methods for determining k_{inact} and K_i, such as using fluorescence spectroscopy to monitor changes in inhibitor fluorescence upon covalent binding to the target enzyme or deriving these parameters directly from time-dependent IC_{50} values, offer streamlined alternatives that provide high-quality experimental data.[12,13]

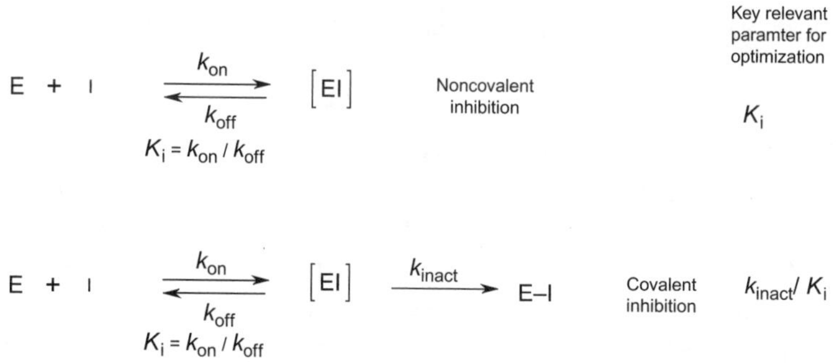

[27.1]

Kinetic and thermodynamic parameters describing covalent and noncovalent inhibition.

The question of broad target selectivity is an important consideration in programs utilizing covalent target modification strategies. This issue arises due to intrinsic reactivity of the drug molecule and the presence of surface nucleophiles on a number of proteins (such as albumin). The advent of mass spectrometry-enabled proteomics provides an important tool for determining selectivity of covalent target modification. Specially designed activity-based probes can be used to understand the functional activity of an entire enzyme family (for instance kinases or serine proteases, for which general probes have been developed).[14] When coupled to SILAC (stable isotope labeling with amino acids in cell culture) and click chemistry (to enable capture and enrichment of modified proteins, often using a biotinylated handle), these probes enable selectivity determination across relevant subsets of the cellular proteome. Likewise, it is possible to design clickable probes based on the covalent small molecule pharmacophore that enable capture and enrichment of all target proteins that covalently bind to the probe within the cell. Examples of programs where proteomic methods have been used to interrogate covalent enzyme inhibitor selectivity are presented in the case studies reviewed herein.

As mentioned earlier, the primary motivation for avoiding covalent inhibition approaches is the issue of immunotoxicity resulting from covalent protein modification that produces a hapten recognized as foreign by the immune system.[15] Wholesale strategies for avoiding intrinsically reactive functional groups (or seemingly innocuous groups that form reactive metabolites), coupled with reducing lipophilicity, have emerged as essential medicinal chemistry design principles for producing safer drug candidates.[16,17] These principles are augmented by extensive reactivity– and clinical exposure–toxicity relationship studies for drugs that produce reactive metabolites, which have resulted in dosage guidelines for compounds producing electrophilic species and zone classification systems based on dose and covalent binding to assess idiosyncratic toxicity risk.[18,19] However, it is important not to confuse the issue of toxicity from highly reactive species (epoxides, highly electrophilic Michael acceptors, etc.) produced by oxidative metabolism of compounds present at high concentrations in the liver with the safety risk of targeted covalent modification from mildly electrophilic functionality that is only reactive in the context of its protein target binding pocket.[1] The development of predictive immunotoxicity assays and more extensive clinical studies with targeted covalent inhibitors will better inform medicinal chemistry strategies pursuing targeted covalent inhibition.

The topic of covalent enzyme inhibition as a medicinal chemistry strategy has been reviewed several times over the past decade—often focusing on specific enzyme families or covalent modalities.[1,20] The purpose of this report is to broadly detail targeted covalent inhibitor approaches across a variety of enzyme families which have been published over the past 5 years. Covalent modification of essential enzyme cofactors as a design strategy and the development of covalent probes to enable chemical biology studies are beyond the scope of this review. This review will be divided into several sections based on reactivity of the electrophilic species (producing functionally irreversible or reversible covalent association with the target protein), and type of nucleophile with which the electrophile is known to associate in its biological context (sulfur, oxygen, or nitrogen).

2. FUNCTIONALLY REVERSIBLE COVALENT ENZYME INHIBITORS

This concept refers to compounds that covalently bind their target and subsequently dissociate from it with a rate that is faster than the physiological degradation rate of the protein. In principle, this approach offers improved potency relative to noncovalent inhibitors while minimizing production of long-lived covalently modified proteins or protein fragments that could trigger immunological responses. It could, therefore, be a preferred strategy in cases where target resynthesis rate is rapid, necessitating higher sustained plasma concentrations of the covalent inhibitor to produce prolonged pharmacodynamic coverage of the target.

2.1. Oxygen-targeting electrophiles

Serine hydrolases and proteases represent a diverse set of targets for drug discovery because they are implicated in many different disease states, including thrombosis, infection, neurological signaling, cancer, and inflammation. While there are many different approaches to inhibiting these enzymes, covalent modification of the active site serine is attractive due to its role as a nucleophile in enzyme catalysis. This strategy offers the possibility of designing tight binding, reversible inhibitors, although significant challenges are associated with achieving enzyme selectivity due to the homology and breadth of this enzyme family.[21] This approach is precedented by several marketed inhibitors of DPP4, HCV protease, human neutrophil elastase, gastric lipases, and acetylcholinesterase—the reader is referred to a comprehensive recent review for further information on these targets.[21]

Hepatitis C virus NS3/4a protease catalyzes the initial cleavage of the single HCV-encoded polypeptide into its component functional proteins and is essential for viral replication. Telaprevir, **1**, and boceprevir, **2**, are the only two marketed inhibitors of this protease and represent promising treatments for hepatitis C infection. These compounds were derived from peptidomimetic approaches studying peptide leads of varying length from P6′ to P5 of the natural substrate. Drug discovery efforts for **1** and **2** targeted the alpha keto amide serine trap based on its precedence in serine protease inhibition, knowledge of the NS5a/5b substrate, and product inhibition of the NS3/4a protease. The NS3a/4a cleavage sequences and structures of these inhibitors with each residue mapped onto the NS3a/4a binding site are shown below.

HCV NS3-4a substrate cleavage site sequences

P P′

DLEVVT-STWV
DLEVMT-STWV
DLEVTT-STWV
DLEIMT-SSWV

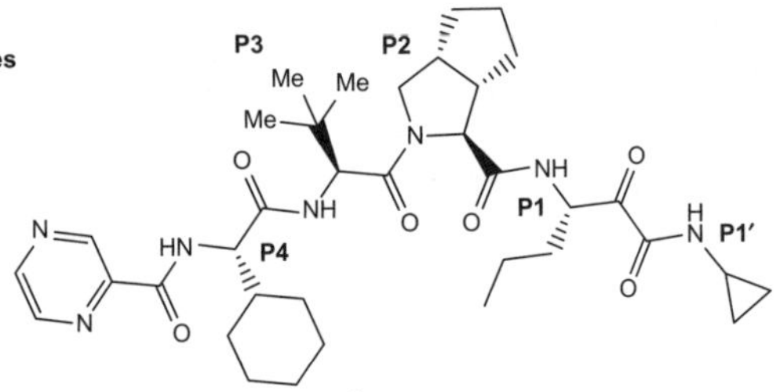

1

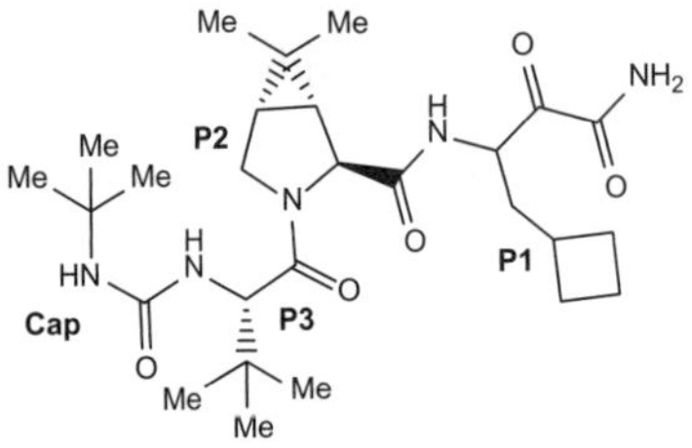

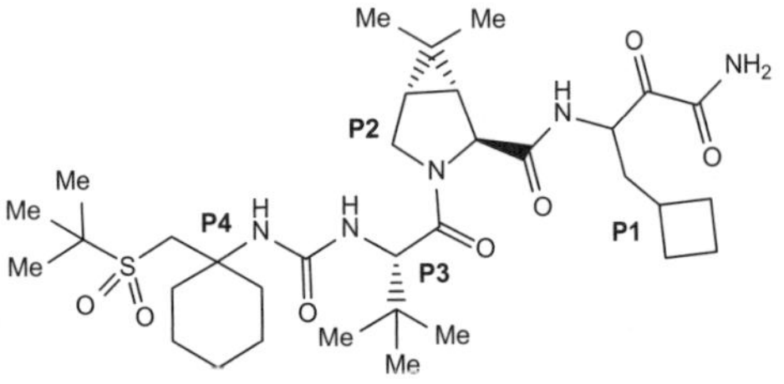

2 **3**

The potency and selectivity of these inhibitors are strongly influenced by structural features of their amino acid side chains, and iterative optimization at each site was required to achieve the proper balance of enzyme potency, activity in a cellular HCV replicon assay and suitable oral pharmacokinetics.[22–24] The S1 and S3 sites (binding regions for the P1 and P3 substituents) are shallow hydrophobic pockets accommodating small linear or branched aliphatic side chains. In the case of boceprevir, small cycloaliphatic substituents at P1 offered significantly enhanced selectivity versus other serine proteases (human neutrophil elastase). The P2 amino acid has a strong influence on enzyme affinity and cellular replicon potency, with substituted proline derivatives providing substantial increases in activity. This increased activity is attributed to conformational constraints associated with tying back what would otherwise be the isoleucine side chain present at the P2 position of the substrate, but removal of the additional hydrogen bond donor of the backbone amide in acyclic derivatives could also have positive impact on cellular penetration. The P4 (**1**) or P3 capping group (**2**) projects into a small hydrophobic pocket that is important for influencing potency and selectivity.

The binding mechanism for compound **1** involves an initial transient collision complex, followed by rearrangement within the binding site to produce the reversible covalent complex.[25] The dissociation half-life of this covalent complex is 58 min, with a steady-state inhibition constant (K_i*) of 7 nM.[24] Compound **1** demonstrated strong selectivity against other serine proteases, including kallikrein, thrombin, plasmin, and Factor Xa (no inhibition at 10 μM).[25] Compound **2** has a similar binding constant of 14 nM for NS3A, but a much longer dissociation half-life (20 h), and is 2200-fold selective versus human neutrophil elastase as a representative off-target serine protease.[23,26] Narlaprevir, **3**, is a structurally related compound that provides greater intrinsic activity in enzyme and replicon assays.[27,28] As with most anti-infective agents, daily dosage is high for these inhibitors (2.2–2.4 g/day for 12–44 weeks).

Aldehydes and activated ketones are also well-precedented electrophiles for serine hydrolases. Alpha-ketoheterocycles were shown to be highly potent, competitive, reversible inhibitors of fatty acid amide hydrolase (FAAH), a potential target for pain which regulates signaling through degrading fatty acid amides such as anandamide and oleamide at their site of action. Their mechanism of inhibition involves attack of the active site serine on the heteroaryl ketone, producing a covalent hemiketal complex that is stabilized by an intramolecular hydrogen bond of the resulting hydroxyl group to the heterocycle. The importance of this hydrogen bond is illustrated by the significant potency difference between **7** (4.7 nM) and **8** (22 μM).[29] The potency of these inhibitors is also strongly influenced by the electron withdrawing character of the heterocycle C5 substituents, as demonstrated by a well-defined correlation between potency ($-\log K_i$) and Hammett σ_p constant for the substituent ($\rho = 3.01$; $R^2 = 0.91$).[30] This pronounced substituent effect is attributed to strengthening the covalent bond between the hemiketal carbon and the active site serine (Ser241) of

FAAH. Inhibitors **4**, **5**, and **6** demonstrate exquisite potency ($K_i = 400–800$ pM). The C2 substituent mimics the hydrocarbon side chain of the fatty acid amide substrate, and the optimum length of the linker alkyl group was demonstrated to be six carbons. Conformationally, constraining the C2 lipophilic side chain by replacing some of the linker group methylenes with a terminal biphenyl or biphenyl ether results in exquisitely potent and selective FAAH inhibitors.[31] Extensive SAR studies were conducted at the central heterocycle,[29] C2 side chain,[31,32] and C5 substituent.[33] Compound **7** is a potent ($IC_{50} = 4.7$ nM) FAAH inhibitor with $>300\times$ selectivity over any other serine hydrolase,[34] increases anandamide levels *in vivo*, and demonstrates potent analgesic activity in models of inflammatory and neuropathic pain.[35,36]

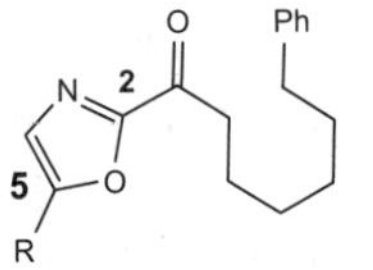

4: R=H
5: R=CO_2Me
6: R=CF_3

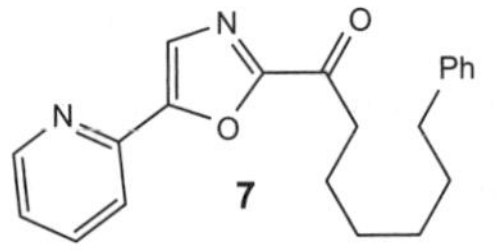

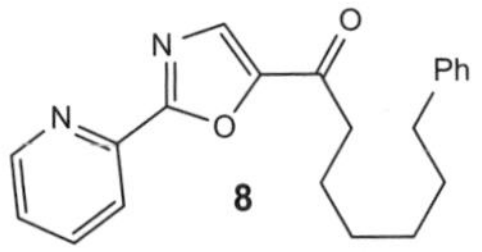

Peptidomimetic aldehydes and nitriles have also been used as electrophiles for prolyl oligopeptidase (POP), a cytosolic serine endopeptidase that has been pursued as a target for neurodegenerative and psychiatric diseases due to its role in neuropeptide maturation and degradation. Electrophiles are commonly appended to the alpha position of a pyrrolidine or piperidine, reflecting catalytic activity of the enzyme at the C-terminus of a proline peptide linkage. Compounds **9** and **10** are potent inhibitors of POP (Compound **10**, $K_i = 4.3$ nM), and crystallographic studies indicate that Ser554 forms a hemiacetal with the inhibitor aldehyde, the oxygen of which occupies the oxyanion hole formed by the active site Tyr473 and backbone NH of Asn555.[37] Pyrrolidine carbonitriles are also potent POP inhibitors, presumably forming a Pinner adduct with the active site serine. Compounds **11** ($IC_{50} = 20$ nM) and **12** ($IC_{50} = 3$ nM) are potent inhibitors of POP activity in cell extracts, whereas the conformationally constrained inhibitor **13** is considerably less active ($IC_{50} = 500$ nM).[38] These compounds demonstrated strong selectivity versus DPP IV activity in cells and cell extracts (Compound **13**, $IC_{50} > 100$ μM).

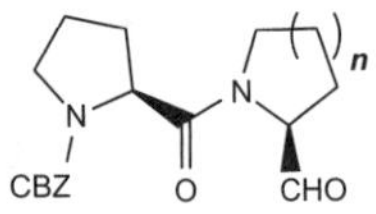

9: *n*=1
10: *n*=2

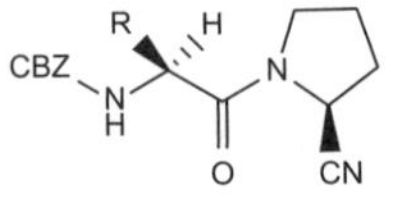

11: R=H
12: R=Me

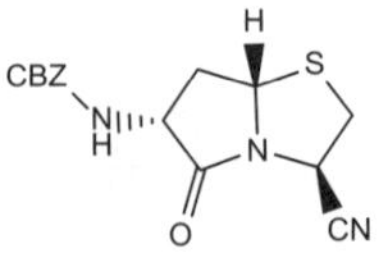

13

The aforementioned examples of reversible covalent inhibitors focus on slow dissociation of the covalent complex to release the original inhibitor, effectively setting up an equilibrium between target and inhibitor. There are also several well-known examples of compounds that acylate their target enzyme, wherein enzyme function can be restored through hydrolysis of the acylated enzyme. Some of the best known examples are β-lactam inhibitors of serine β-lactamases. Because this topic has been recently reviewed, this section will focus on non-β-lactam inhibitors of these enzymes.[39] The most advanced non-β-lactam inhibitor of these enzymes is NXL-104 **14**, which contains an anti-Bredt urea that acylates the nucleophilic serine of β-lactamases, releasing the nitrogen atom bearing the sulfate moiety to produce a carbamate with considerably greater stability than the serine esters formed by reaction of the enzyme with β-lactam inhibitors.[40] Against a variety of Class A and Class C serine β-lactamases, compound **14** demonstrated an IC_{50} between 5 and 170 nM after a 5-min preincubation. This compound is presently in Phase III clinical trials in combination with ceftazidime or ceftaroline as a broad spectrum antibacterial therapy for nosocomial infections.

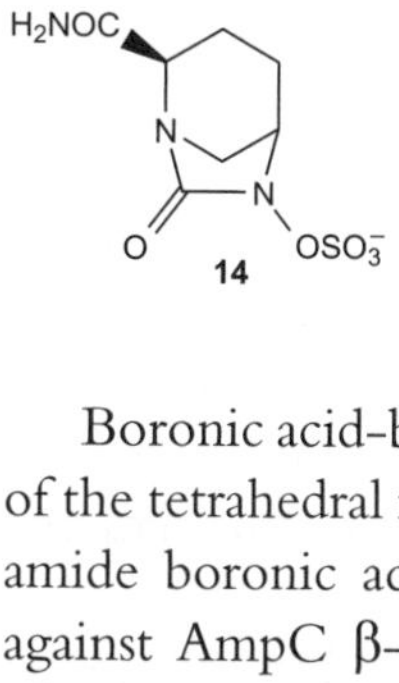

14

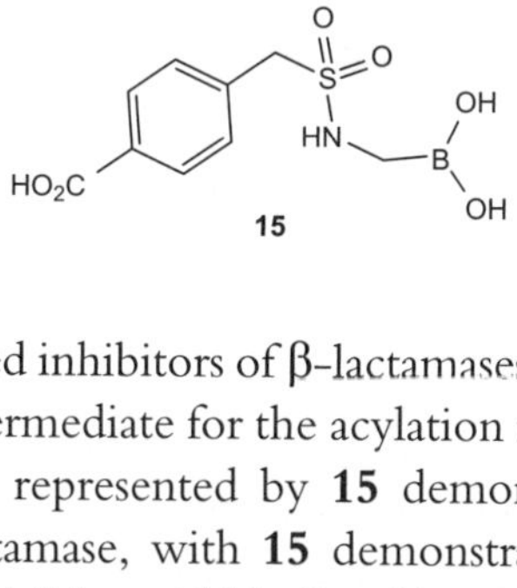

15

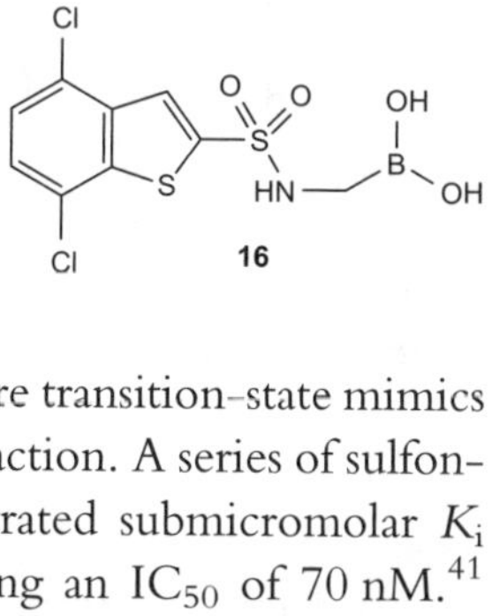

16

Boronic acid-based inhibitors of β-lactamases are transition-state mimics of the tetrahedral intermediate for the acylation reaction. A series of sulfonamide boronic acids represented by **15** demonstrated submicromolar K_i against AmpC β-lactamase, with **15** demonstrating an IC_{50} of 70 nM.[41] These compounds exhibit rapid binding kinetics, as evidenced by the lack of preincubation effects on enzyme potency. Crystallographic analysis of the complex between AmpC and these inhibitors reveals the covalent association of the active site serine Ser64 with the boronic acid to form the tetrahedral transition-state analogue with one oxygen of the boronate projecting into the oxyanion hole formed by backbone NH groups for Ser64 and Ala318. The other boronate oxygen interacts with conserved Tyr150 and an ordered water molecule known to participate in deacylation reactions of substrates. The related boronic acid **16** demonstrates activity across a broad array of β-lactamases from the TEM, AmpC, SHV, P99, and OXA classes, with IC_{50} values between 0.6 and 5.6 μM.[42]

Activated oxime esters were shown to be reversible covalent inhibitors of retinoblastoma binding protein 9, a serine hydrolase of unknown function that confers resistance to antiproliferative signaling of TGFβ and may be implicated in carcinogenesis.[43] Acylated thiazole oximes such as **17** were identified using fluorescent-labeled activity-based proteomic profile screening (fluopol-ABPP), wherein compounds are assessed for their ability to impede covalent labeling of the target protein by a fluorescent probe. These activated oxime esters acylate Ser75, forming a serine ester and releasing the oxime as a leaving group; enzyme function is regenerated upon hydrolysis of the serine ester. Compound **17** demonstrated an IC_{50} of 1.9 μM in this assay and showed no significant inhibition of other serine hydrolases across the 293T soluble proteome of transformed HEK cells.

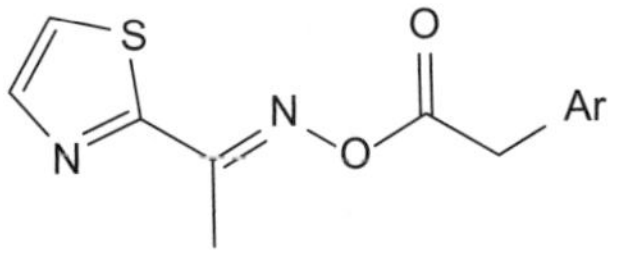

17: Ar=(4-Cl)Phenyl

2.2. Sulfur-targeting electrophiles

Several classes of reversible sulfur-targeting electrophiles have been reported, including the well-known peptidomimetic aldehydes and nitriles that are reversible covalent inhibitors of cysteine proteases, such as cathepsin S and cathepsin C, forming thiohemiacetals or thioimidates with their active site cysteine residue. Because these inhibitor classes have been recently reviewed, this section will focus on less exemplified electrophiles for reversible covalent modification of cysteine.[44,45]

Alkyl amidines form reversible covalent bonds with the active site cysteine residue of dimethylarginine dimethylaminohydrolase-1 (DDAH-1), which mediates the hydrolysis of N^{ω},N^{ω}-dimethyl-L-arginine, an endogenous inhibitor of nitric oxide synthetase.[46] Upon covalent bond formation with Cys274, the amidine nitrogen atoms participate in hydrogen bonding and ion pair interactions with the side chains of Asp79 and Glu78. Isothermal titration calorimetry was used to determine the potency of **18** against the wild-type enzyme (7 μM) and the C274S mutant (28 μM), which was shown crystallographically to be incapable of forming a covalent bond to the inhibitor's amidine group. The modest potency difference for these two orthologues suggests that the covalent bond is weak and does not contribute much to the overall binding affinity of the inhibitor.

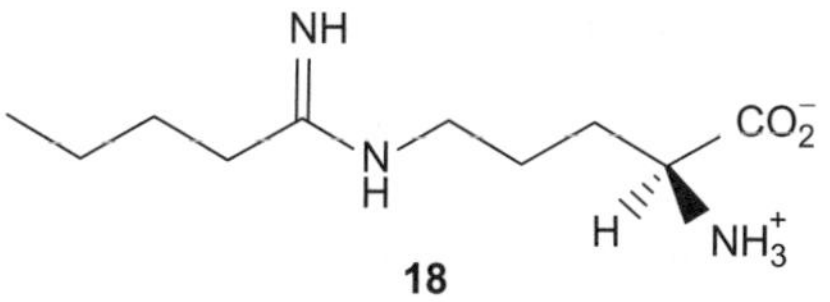

18

Benzylidene rhodanine derivatives were shown to be reversible covalent inhibitors of Hepatitis C NS5b RNA polymerase, an essential enzyme involved in transcription of viral RNA. These compounds, exemplified by **19**, covalently modify Cys366 through Michael reaction at the benzylidene group of the inhibitor.[47] This Michael reaction gets assistance from activation of the rhodanine ketone, which accepts a hydrogen bond from Ser367. Enzyme kinetic studies demonstrate noncompetitive behavior with nucleotide substrates, consistent with location of Cys366 outside the active site of the enzyme. These inhibitors impede growth of the RNA daughter strand by blocking the channel that the RNA uses to exit the enzyme. Reversibility was demonstrated through preincubation and dilution experiments, and compound **19** was the most potent inhibitor identified (IC_{50} = 200 nM). This compound demonstrated strong selectivity for NS5b RNA polymerase over cathepsin B; calpain; caspases 1, 3, 6, 7, and 8, and aldose reductase. This compound demonstrates low clearance (0.13 mL/min/kg) and volume of distribution (0.09 L/kg) in rat with a half-life of 9.2 h.[47]

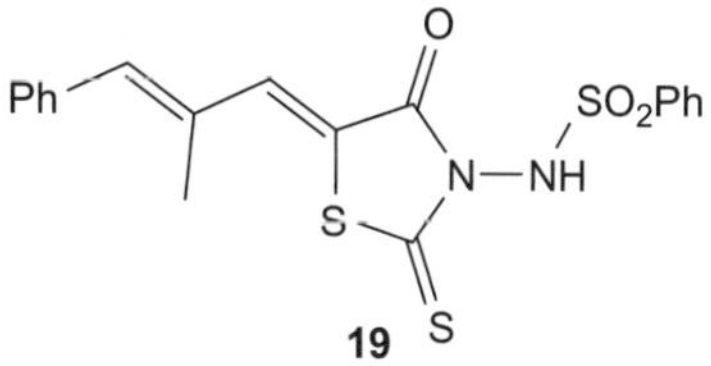

19

3. FUNCTIONALLY IRREVERSIBLE COVALENT ENZYME INHIBITORS

This concept refers to compounds that covalently bind their target and dissociate from it with a rate that is slower than the physiological degradation rate of the target. In cases where the target resynthesis rate is slow relative to the pharmacokinetic half-life of the compound, it offers the possibility for significantly prolonged pharmacodynamic effects.[20] However, the potential to produce immunogenic fragments of covalently modified proteins requires increased attention to selectivity and minimizing dose relative to functionally reversible targeted covalent inhibitors.

3.1. Sulfur-targeting electrophiles

A number of designed covalent inhibitors take advantage of the facile reaction between a cysteine residue in the target active site and a Michael acceptor. Covalent epidermal growth factor inhibitors are among the best-studied cases. The EGFRs (Her-1 and Her-2) are members of ErbB family of receptor tyrosine kinases. Uncontrolled activation of EGF receptors by overexpression of the receptors or of the ligands that activate EGFRs is implicated in cancers of the head, lung, breast, bladder, ovary, and kidneys. A number of first-line, efficacious, reversible EGFR inhibitors have been approved: Iressa, Tarceva, and lapatinib; however, most patients develop acquired resistance to these drugs. Covalent inhibitors with greater biochemical efficiency were developed to overcome this resistance. One of the first agents which showed clinical efficacy was HKI-272 or neratinib **20**. This compound shows good potency against EGFR (IC_{50}: 92 nM) and Her-2 (IC_{50}: 59 nM) and improved efficacy compared to EKB-569, a previously developed covalent EGFR inhibitor in Her-2-dependent tumor models.[48] The dimethylaminomethylene moiety attached to the acrylamide tail not only gives improved solubility but also presumably acts as an intramolecular base catalyzing the addition of Cys773 (EGFR) or Cys805 (Her-2) to the β-carbon of the acrylamide (structure **21**).[7] Phase I[49] and Phase II advanced non-small-cell lung cancer[50] and Erb-2-positive breast cancer[51] clinical studies have been published, with excellent efficacy seen in the breast cancer studies. In both trials, doses of 240 mg/day were used over several months with ~30% patients experiencing grades 3–4 diarrhea as the most common adverse event.

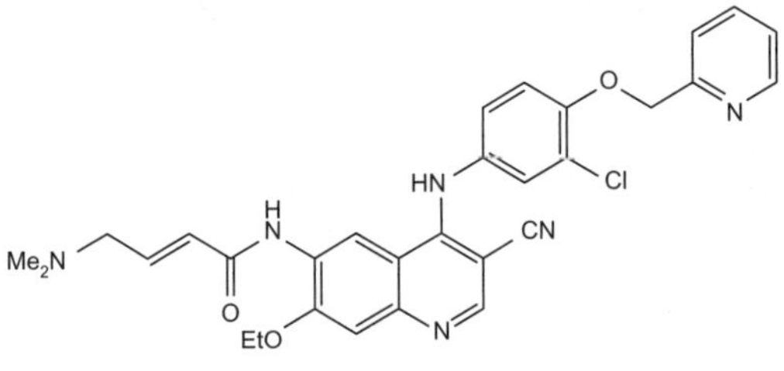

20
EGFR IC_{50}: 92 nM
Her-2 IC_{50}: 59 nM

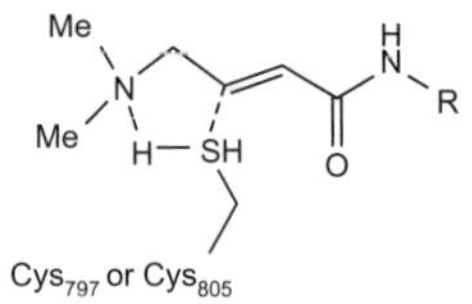

21

BIBW 2992 (afatinib **22**) is also an irreversible covalent EGFR inhibitor that is currently undergoing clinical assessment. Compound **22** shows excellent potency against EGFR (IC_{50}: 0.5 nM) and Her-2 (IC_{50}: 14 nM). It also shows excellent potency against two EGFR mutants, $EGFR^{L858R}$

(IC_{50}: 0.4 nM) and $EGFR^{L858R/T790M}$ (IC_{50}: 10 nM), the double mutant that is the cause for more than 50% of all acquired resistance to first-line reversible EGFR inhibitors.[52] The double L858R and T790M gatekeeper mutation provides resistance by producing a more active kinase (L858R) and by increasing ATP affinity (T790M).[5] Thus, covalent EGFR inhibitors can potentially overcome T790M resistance since, once covalently bound, they are no longer in a competitive, reversible equilibrium with ATP. Phase I trials of **22** showed good tolerability with maximum tolerated doses (MTDs) of 40–50 mg/day.[53] Slightly higher MTDs can be obtained using a 2–3 weeks on/1–2 weeks off dosing schedule. Two Phase II trials of **22** in NSCLC show positive results in EGFR inhibitor-pretreated patients with positive effects shown on progression-free survival and objective response rates, although these results show no advantage over currently marketed tyrosine kinase inhibitors. Currently, several Phase III trials are being conducted exploring **22** in EGFR-mutated NSCLC in comparison to cisplatin–pemetrexed therapy.

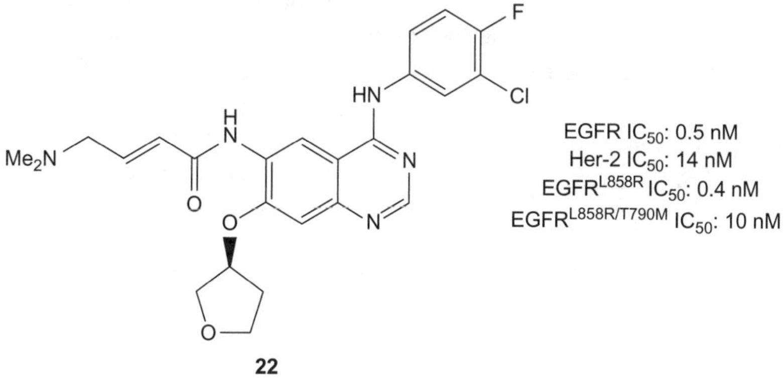

22

Another member of this class of covalent EGFR inhibitors is PF-00299804 or dacomitinib **23**. Compound **23** shows excellent potency across many EGFR family members: EGFR (IC_{50}: 6.0 nM), Her-2 or ERBB2 (IC_{50}: 46 nM), and Her-4 or ERBB4 (74 nM). Additionally, **23** shows efficacy in cell lines containing the L858R (H3255: IC_{50}: 0.007 μM) and the L858R/T790M (H3255 GR: IC_{50}: 0.119 μM) mutations which render the reversible EGFR inhibitor gefitinib **24** inactive.[54] Dacomitinib also shows excellent efficacy in SKOV3 and A431 human tumor xenograft models with MEDs of 15 and 11 mg/kg, excellent tumor drug exposure in these xenograft models, moderate clearance (24–49 mL/min/kg), and excellent oral bioavailability (56–80%) across multiple species.[55] Phase I clinical studies with **23** showed a MTD of 45 mg and a comparable safety profile to other EGFR inhibitors.[56] Compound **23** demonstrated proof of concept in patients with NSCLC and is currently entering Phase III studies.

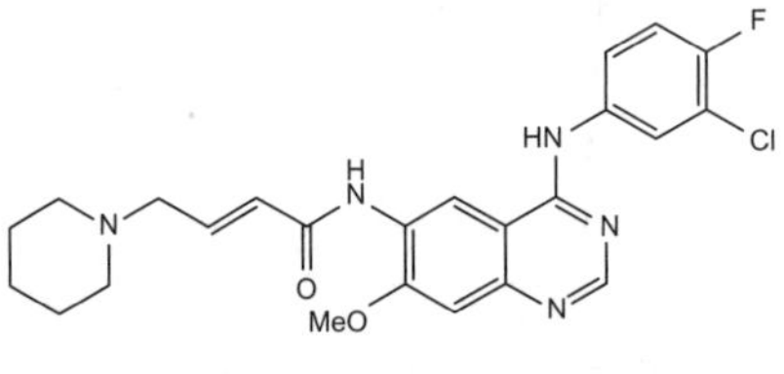

23

EGFR IC_{50}: 46 nM
Her-2 IC_{50}: 74 nM

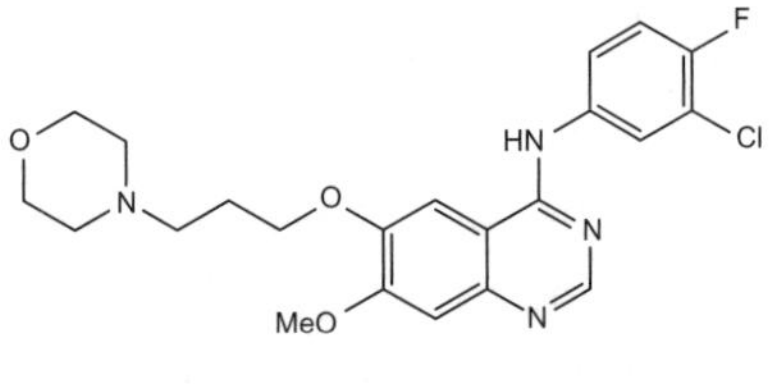

24

In comparison to the acrylamide Michael acceptor used in the compounds described above, electron-deficient acetylenes have also been used to target the active site cysteine residues in EGFR enzymes. 6-Ethynylthieno[3,2-*d*]- and 6-ethynylthieno[2,3-*d*]pyrimidin-4-anilines such as **25** have been shown to have good EGFR, Her-2, and Her-4 potency and form covalent adducts with their target enzymes. The basic amine on the alkyne terminus is believed to act as a general base to deprotonate the active site cysteine, aiding nucleophilic addition to the alkyne. This compound has potent antitumor activity in BT474 tumor xenograft models at doses from 10 to 100 mg/kg and shows superior efficacy to the reversible EGFR inhibitor lapatinib.[57]

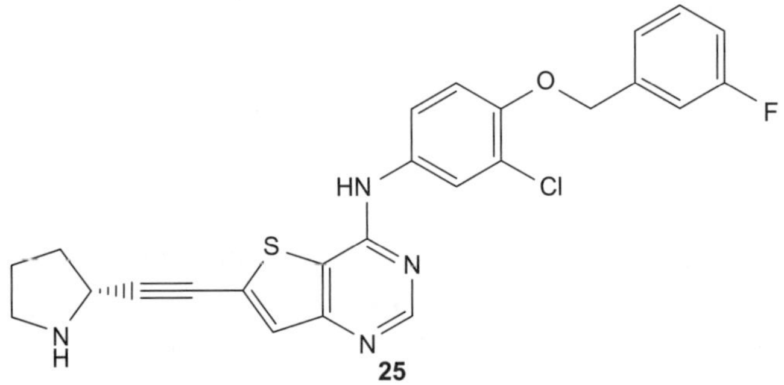

25

EGFR IC_{50}: 7 nM
Her-2 IC_{50}: 13 nM
Her-4 IC_{50}: 66 nM

Bruton's tyrosine kinase (BTK) is a member of Tec family kinases and is a key component in B-cell receptor signaling. Genetic loss of function[58] and siRNA studies[59] have implicated BTK as a mediator of inflammatory signaling, making it a suitable target to treat rheumatoid arthritis. One successful strategy was to take a reversible BTK inhibitor with poor selectivity over LCK and LYN and incorporate a piperidine containing an electrophilic acrylamide to specifically target Cys481, thereby producing greater kinase selectivity. The resulting compound, ibrutinib, **26** is a potent covalent irreversible BTK inhibitor (IC_{50}: 0.72 nM) with ~130-fold selectivity versus LCK and ~19-fold selectivity versus LYN.[60] Subsequent preclinical studies have shown **26** to be effective in diseases involving activation of the B-cell antigen receptor pathway,[61] especially chronic lymphocytic leukemia (CLL).[62] This compound has also been used as a tool to show that BTK is specifically required for IgE activation of human basophils compared to other basophil stimulators such as FMLP, C5a and IL-3.[63] A chemical probe was developed wherein **26** was linked to a BODIPY tag, enabling target occupancy to be determined by displacement of the probe with **26**.[64] This fluorescent probe demonstrates specific labeling of BTK in DOHH2 cells and mouse splenocytes, with no protein labeling detected in Jurkat cells, which do not express BTK protein. Compound **26** demonstrates complete target occupancy in mouse splenocytes for 12 h when given orally at 50 mg/kg. Compound **26** is currently in Phase II clinical trials for CLL.[65]

Another BTK inhibitor that has advanced to Phase II clinical studies is AVL-292.[66] This compound demonstrates strong potency against BTK in enzyme (IC_{50} = 0.5 nM) and Ramos cell-based (EC_{50} = 8 nM) assays, with excellent selectivity over Src-family kinases and B-cell signaling components (IC_{50} > 700 nM) across a family of 9 kinases. A chemical probe using a biotin tag for protein captured demonstrated 50% target occupancy in cell lysates at 5.9 nM concentration, in agreement with the Ramos cell assay results. In Phase I clinical studies, a single 2 mg/kg oral dose produced nearly complete target occupancy after 2 h that was sustained through the 8-h measurement period. In multidose studies, complete and sustained target occupancy was attained at the 250 mg QD dose level with good toleration. While the structure of AVL-292 has not been specifically disclosed, a single compound patent from Avila reports advanced physical form characterization for compound **27**, which is likely an advanced candidate compound, if not AVL-292 itself.[67,68]

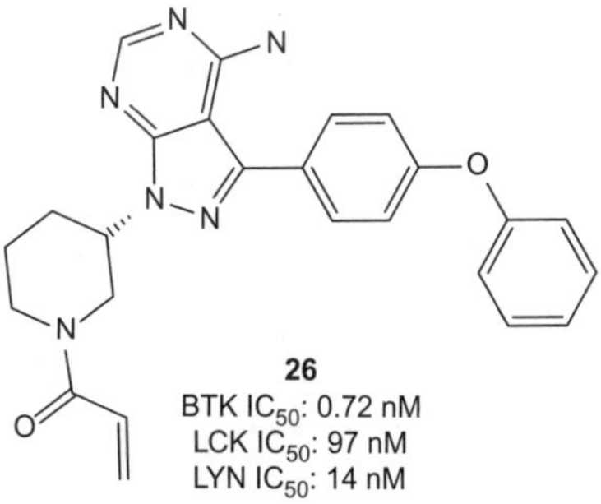

26
BTK IC_{50}: 0.72 nM
LCK IC_{50}: 97 nM
LYN IC_{50}: 14 nM

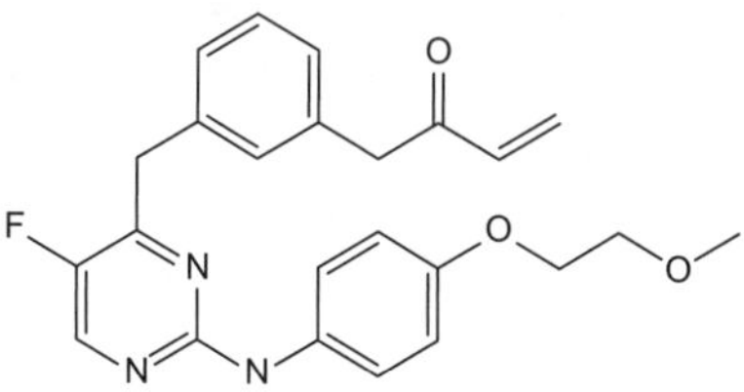

27
BTK IC_{50}: <10 nM
ITK IC_{50}: 10–100 nM

Nek2 is a serine/threonine kinase thought to play a role in bipolar spindle assembly driven by the microtubule motor protein Eg5 and is known to be overexpressed in breast tumor and diffuse large B-cell cancers. Nek2 possess a cysteine residue in its active site (Cys22) which is a suitable target for covalent inhibitors. Compound **28** (JH295) inhibits Nek2 (IC_{50}: 770 nM) and shows good selectivity over Cdk1, a key selectivity target, as Cdk1 inhibitors block mitotic entry and trigger mitotic exit. Incubation of 10 equivalents of **28** and 1 equivalent of Nek2 produces a single covalent adduct as seen by protein mass spectroscopy, and **28** was shown to inhibit cellular mitosis A549 cells after 95% of Nek2 activity was inhibited.[69]

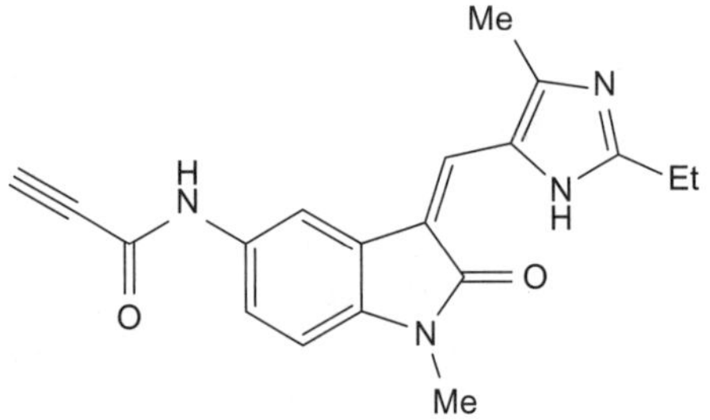

28

Fibroblast growth factor receptor (FGFR) kinases are a family of tyrosine kinases that play critical roles in normal development, wound healing, as well as tumor formation and progression. FGFR gain of function mutations and FGFR overexpression have been identified in bladder, gastric, colorectal, and additional cancers. Recent studies have identified **29** (FIIN-1) which effectively targets Cys486 of FGFR1 with an acrylamide electrophile. Compound **29** shows excellent potency for FGFR1, 2, and 3 (IC_{50}s: 9.2, 6.2, 11.9 nM) and only binds two other kinases with a K_d below 100 nM (Blk and Flt1) in a 402 kinase selectivity panel. This compound was confirmed to block activation of FGFR1 in MCF10A cells, and its binding to Cys486 was confirmed by incubating biotin labeled **29** in MCF10A cell lines expressing WT iFGFR1 and a C486S FGFR1 mutant, followed by immunoprecipitation, where binding to WT iFGFR1 was only seen. Compound **29**, however, shows weak potency against V561M FGFR1, a gatekeeper mutation that has previously been shown to reduce the FGFR1 potency of PD173074.[70]

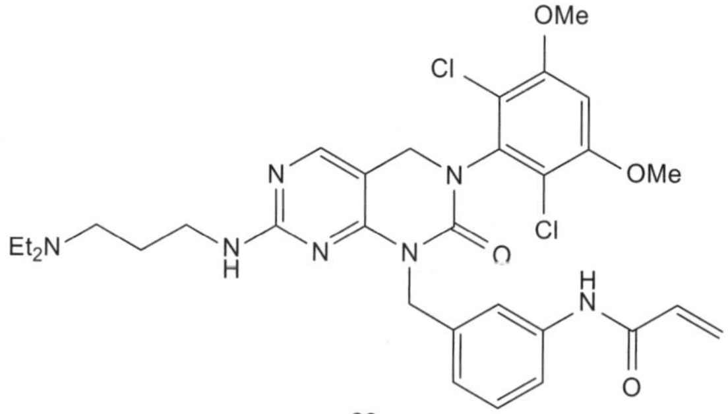

29

FGFR1-3 IC_{50}s: 6.2–11.9 nM
FGFR4 IC_{50}: 189 nM

Cysteine trapping with acrylamides has also been used as an effective strategy to design HCV NS3/4A viral protease inhibitors. Compound **30** shows excellent HCVP inhibition (IC_{50}: 2 nM) but poor activity against a C159S HCVP mutant (IC_{50}: 1782 nM), indicating Cys159 to be the target of the acrylamide electrophile. The authors were also able to develop a biotinylated analog of **30** which was used as an occupancy biomarker in Huh-7 replicon cells to show that practically complete occupancy is needed to arrest replicon activity. An HCV NS3/4A turnover rate of 8–24 h could also be determined using this approach.[71]

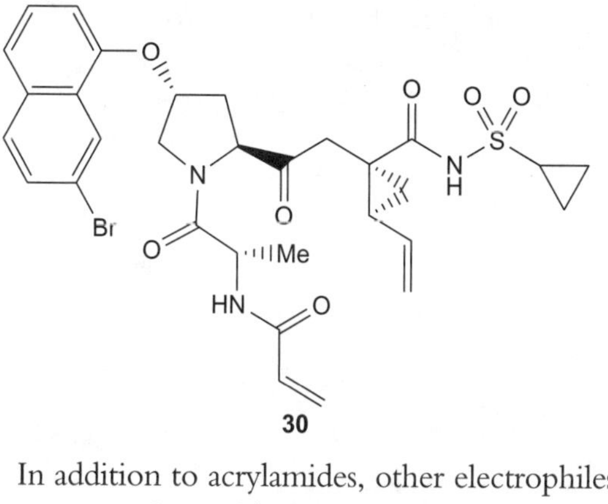

30

HCVP IC_{50}: 2 nM
C159S HCVP IC_{50}: 1782 nM

In addition to acrylamides, other electrophiles have been effectively used to target cysteine residues to produce covalent inhibitors. Glycogen synthase kinase 3 (GSK3) inhibitors can be switched from reversible to irreversible binders by changing the methyl ketone moiety that binds near Cys199 to the corresponding α-chloro or bromo ketones, which suffer nucleophilic attack. Compounds **31** and **32** are two representative inhibitors that show good to excellent GSK3 potency (**31** IC_{50}: 580 nM; **32** IC_{50}: 5 nM). Covalency was confirmed by dependence of GSK3 inhibition on preincubation time, and covalent binding of **32** was confirmed biophysically by MALDI–TOF analyses.[72]

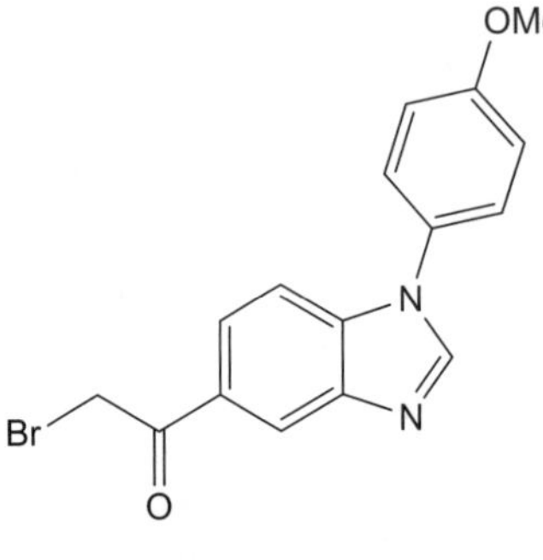

31
GSK3 IC_{50}: 580 nM

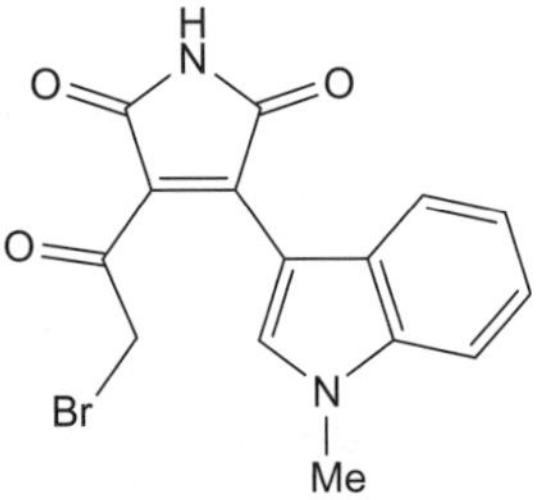

32
GSK3 IC_{50}: 5 nM

O-Acyl hydroxamates have been used to prepare aza-peptide covalent inhibitors of papain family cysteine proteases such as cathepsin B. Compound **33**, which contains an acyl-*N*-((4-methoxybenzoyl)oxy)hydrazinecarboxamide unit, shows potent inhibition of cathepsin B and cathepsin Z (cathepsin B IC_{50}: 13 nM, cathepsin Z IC_{50}: 170 nM) with very fast rates of enzyme association (cathepsin B k_{on}: 75,000 M^{-1} s^{-1}). The corresponding biotinylated analog of **34** showed strong binding to cathepsin B in labeling experiments and survived reducing conditions (pH 6.8, 150 mm DTT, 100 °C, 2 min) prior to SDS-PAGE gel loading.[73] The *p*-methoxybenzyl group creates an electron-deficient nitrogen on the hydrazinecarboxamide, facilitating attack of a nucleophilic cysteine leading to a *N*-mercaptohydrazinecarboxamide covalent structure.

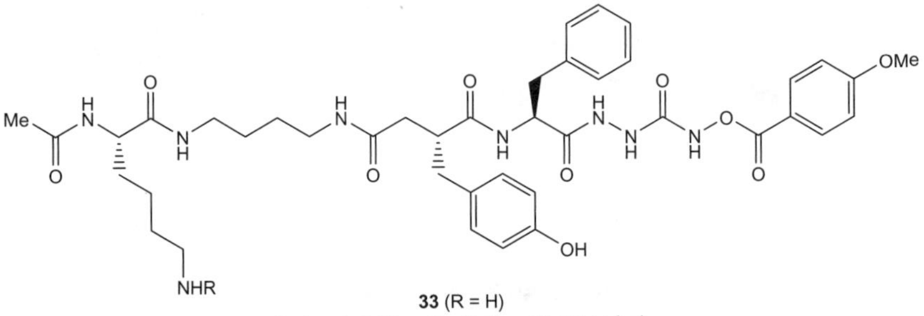

33 (R = H)
Cathepsin B IC_{50}: 13 nM; K_{ass}: 75,000 $M^{-1}s^{-1}$
Cathepsin L K_{ass}: 65,000 $M^{-1}s^{-1}$
Cathepsin Z IC_{50}: 170 nM
34 (R = biotin)

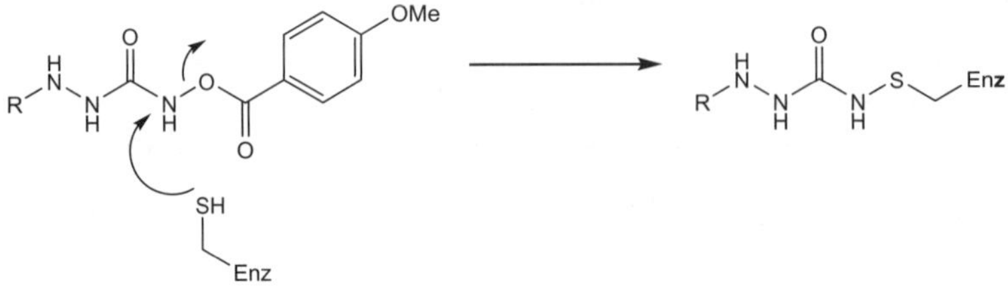

Tissue transglutaminase 2 (TG2), a Ca^{+2} enzyme responsible for making intermolecular peptide cross-links between the γ-carboxamide of a glutamine and an ε-amino group of a lysine, has been shown to be important in mitochondrial energy function.[74] Overactivity of TG2 has been closely associated with Celiac and Huntington's diseases. Covalent inhibitors of TG2 have been recently developed by targeting Cys277 using unsubstituted acrylamides **35** (TG2 IC_{50}: 10 nM) and **36** (TG2 IC_{50}: 120 nM). In addition to unsubstituted acrylamides, other electrophiles were incorporated that led to active compounds (substituted arcrylamides, oxirane-2-carboxamide, nitriles, ynamide, diazoketone). These analogs showed time-dependent TG2 inhibition and modest second order inhibition rate constants were determined (**36** k_{inact}/K_i: 1087 $\mu M^{-1}\ s^{-1}$). A log–log plot of k_{inact}/K_i versus IC_{50} (30 min incubation) shows a high correlation of these measures. Compound **35** shows excellent selectivity against TG1 and TG3 with modest selectivity shown over FXIIIa. *In vitro* stability/metabolism studies in mice shows moderate plasma stability ($T_{1/2}$: 248 min) and high mouse liver microsome intrinsic clearance (3203 mL/min/kg).[75]

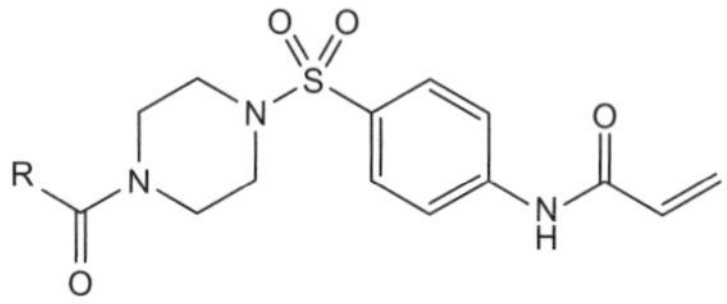

R = 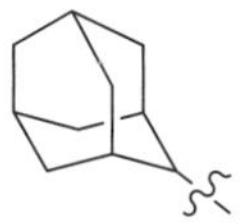

35
TG2 IC_{50}: 10 nM
TG1 IC_{50}: 3400 nM
TG3 IC_{50}: 80000 nM
FXIIIa IC_{50}: 180 nM
mouseplasma $T_{1/2}$: 248 min
mLM Clint: 3203 ml/min/kg

R = 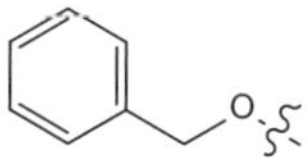

36
TG2 IC_{50}: 120 nM
TG2 k_{inact}/K_i: 1087 $\mu M^{-1}s^{-1}$
TG1 IC_{50}: 3700 nM
TG3 IC_{50}: >80,000 nM
FXIIIa IC_{50}: 420 nM

3.2. Oxygen-targeting electrophiles

Harder electrophiles (e.g., ureas, carbamates, epoxides) enable covalent modification of hydroxylated amino acids (e.g., serine) in a number of targets. FAAH, a serine hydrolase that hydrolyzes the endogenous endocannabinoid anandamide, can be inhibited with excellent potency (hFAAH k_{inact}/K_i: 40,300 $M^{-1}\ s^{-1}$: IC_{50}: 7.2 nM) by urea **37**, a covalent inhibitor that targets Ser241 in the active site of FAAH.[76] Compound **37** shows exquisite selectivity for FAAH as assessed by competitive activity-based protein profiling (ABPP) in human and mouse brain membranes and soluble liver proteomes using a rhodamine-tagged fluorophosphonate (FP) ABPP probe.[77] Even at 100 μM, **37** selectivity blocked labeling of only

FAAH in soluble liver lysates, whereas first-generation inhibitor URB597 (**38**) blocked labeling multiple proteins. Compound **37** shows excellent pharmacokinetic properties in both rat and dog and demonstrates statistically significant activity in a CFA model of inflammatory pain after oral administration at 0.1 mg/kg. Related covalent FAAH inhibitors have also been disclosed.[78,79] Compound **37** is currently in clinical trials for pain and other indications.

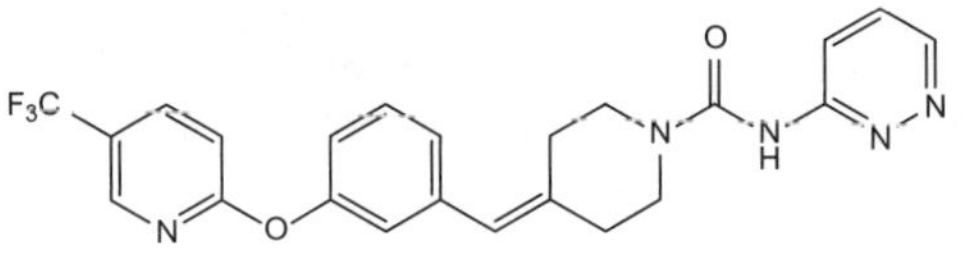

37

hFAAH k_{inact}/K_i: 40,300 $M^{-1}s^{-1}$

hFAAH IC_{50}: 7.2 nM

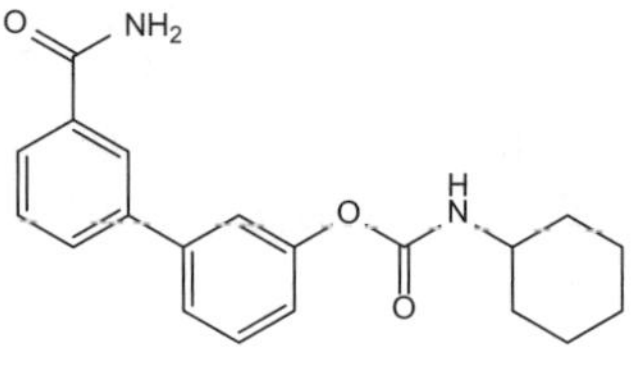

38

Monoacylglycerol lipase (MGL) is a serine hydrolase related to FAAH which hydrolyses 2-arachidonoylglycerol, an endocannabinoid-like anandamide. Compound **39** (LY 2183240) is an electrophilic urea that inhibits hMGL by targeting Ser129 with $IC_{50}=20$ nM. Tryptic digest of the compound **39**-inhibited hMGL followed by MALDI-TOF MS analysis clearly shows the incorporation of a dimethylamido (-C(O)NMe_2) fragment as a result of serine acylation.[80] Compound **39** also shows potent hFAAH inhibition (IC_{50}: 37 nM)[81] and displays nociceptive effects in mouse models of inflammatory pain.[82]

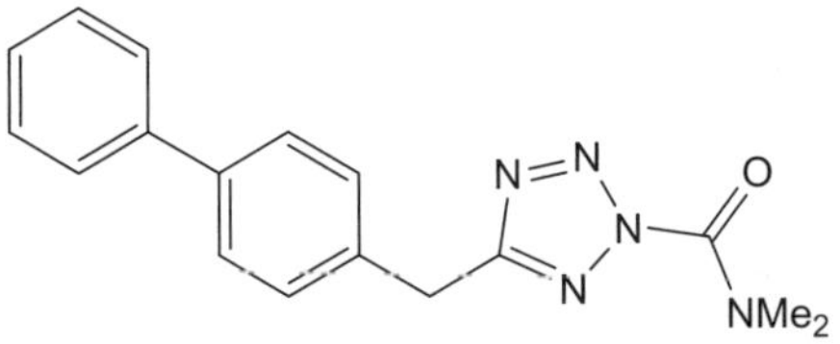

hMGL IC_{50}: 20, 54 nM

hFAAH IC_{50}: 37 nM

39

The serine hydrolase protein phosphatase methylesterase-1 (PME-1) demethylates protein phosphatase 2A, potentially leading to cancer and neurodegenerative disease. Potent covalent inhibitors of PME-1 have been identified, **40** (IC_{50}: 10 nM)[83] and **41** (IC_{50}: 600 nM),[84] using fluorescence polarization-activity-based protein profiling (fluopol-ABPP). Both **40** and **41** show excellent serine hydrolase selectivity via competitive ABPP using FP, chloroacetamide, and sulfonate ester probes. Compound **41** is believed to inhibit PME-1 via a covalent irreversible mechanism, since recovery of enzyme activity was not observed after gel filtration chromatography. The PME-1 adduct with compound **40** was isolated, but not with compound **41**.

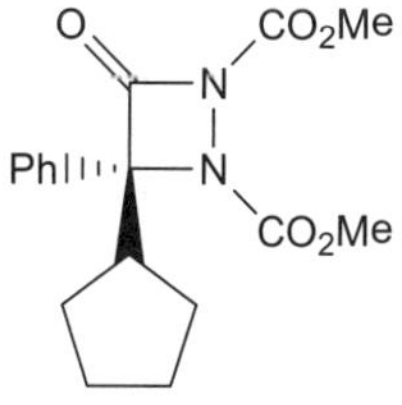

40
IC_{50}: 10 nM

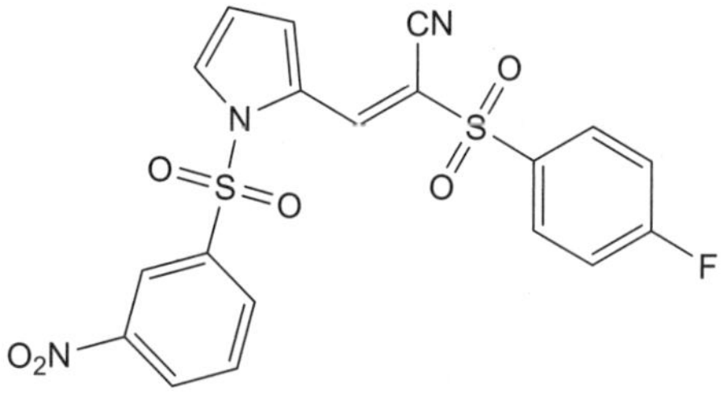

41
IC_{50}: 600 nM

Carfilzomib (**42**) is a potent inhibitor of the human 20S proteasome (k_{inact}/K_i: 34,000 M^{-1} s^{-1}) and shows more than 300-fold selectivity over other proteasome catalytic activities.[85] Inhibition of the 20 S Proteasome is a clinically validated approach for the treatment of hematological cancers (leukemia, lymphoma, and myeloma) as evidenced by bortezomib (PS-341, Velcade™), a proteasome inhibitor that shows efficacy in multiple myeloma and non-Hodgkin's lymphoma. Compound **42** inhibits the 20S proteasome by irreversibly binding to an active site serine. The compound induces apoptosis and growth arrest in hematologic (RPMI 8226 and HS-Sultan) and solid tumor (HT-29) cell lines and shows efficacy in HT-29, RL, and HS-Sultan human tumor xenograft models.[86] Compound **42** has progressed rapidly through the clinic and showed positive effects in Phase II trials of multiple myeloma.

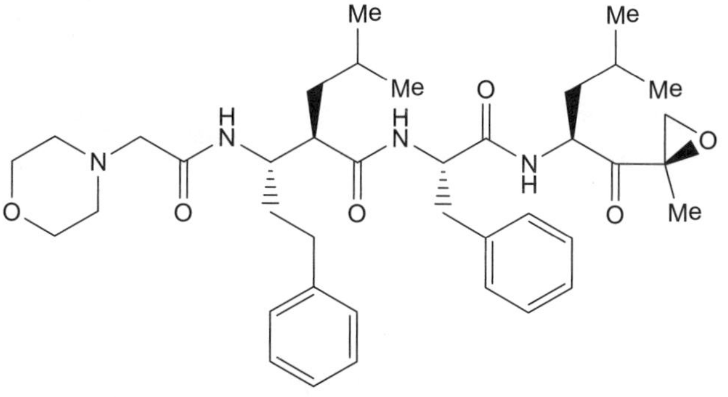

42
h20S proteosome k_{inact}/K_i: 34,000 $M^{-1}s^{-1}$
h20S proteosome 1 h IC_{50}: 6 nM

3.3. Nitrogen-targeting electrophiles

Beloranib (**43** ZGN-433) is a covalent inhibitor of methionine aminopeptidase 2 (METAP2). Originally developed as an angiogenesis inhibitor,[87] it was subsequently discovered to have antiobesity effects, and it is currently in early clinical trials. An active site histidine in METAP2 is believed to react with one of the epoxides of **43** leading to covalent enzyme inactivation.

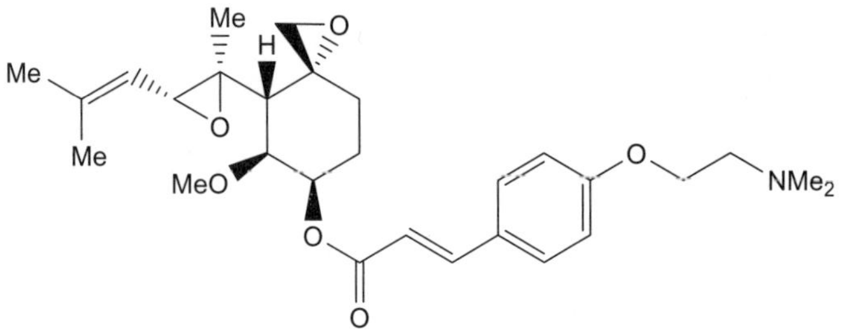

43

Macrophage migration inhibitory factor (MIF) plays a key role in the pathology of cancer and inflammatory diseases. While the mechanism of how MIF exerts its biological effects is not clear, it has tautomerase activity that is likely linked to its cytokine activity. It was recently been disclosed that irreversible inhibition of MIF can be achieved using 7-((arylamino)(pyridin-2-yl)methyl)quinolin-8-ols such as **44**. Compound **44** reacts with MIF tautomerase by making a covalent bond with Pro-1 as shown by X-ray crystallographic analysis and time-dependent inhibitory potency profile.[88]

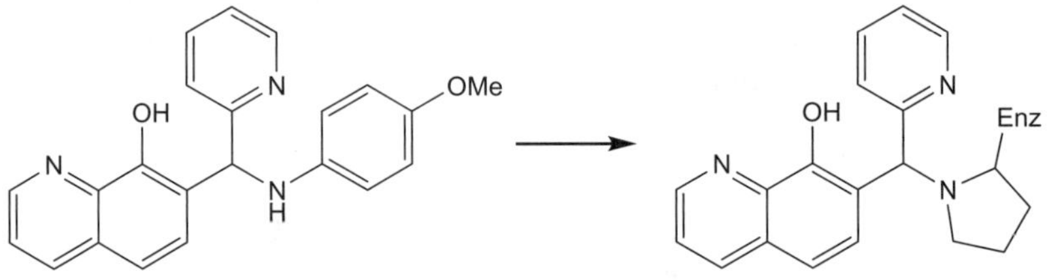

44

4. CONCLUSIONS

The past 5 years have witnessed a resurgence of interest in covalent enzyme inhibitors due to a renewed appreciation for their prevalence in the human pharmacopeia and the recognition that more challenging biological targets will require unconventional drug discovery strategies to produce clinical candidates. While covalent inhibitor approaches carry increased safety risk due to the presence of a reactive moiety, these concerns can be mitigated by appropriately targeting the inhibitor through modulating its electrophilicity, emphasizing noncovalent binding interactions to improve selectivity and minimizing dose with careful attention to pharmacokinetic properties and potency. Recent advances in protein mass spectrometry and chemical biology offer the possibility of quantifying selectivity against entire protein families in a cellular context, providing additional information to gauge safety risks. Many of these methods have yet to be applied systematically to targeted covalent inhibitor programs, offering significant opportunity for the medicinal chemist to devise new strategies for monitoring target occupancy and off-target effects for irreversible or slowly dissociating compounds. Indeed, the number of irreversible covalent inhibitor programs cited in this review where primary pharmacologic activity was judged based only on IC_{50} values demonstrates the broader need for more rigor in designing even basic screening strategies for covalent inhibitor approaches. The diversity of electrophilic moieties utilized for reversible and irreversible enzyme inhibition for programs exemplified herein illustrates the breadth of opportunities already available for generating targeted covalent inhibitors. However, there is significant room for elevating design sophistication through understanding electrophile reactivity under physiological conditions and in the context of their target protein binding pocket properties, which would enable selecting the optimum electrophile for targeted covalent inhibition. Further advances in this field promise to make covalent inhibition a more prominent tool for the medicinal chemist to approach contemporary targets that pose significant challenges for traditional noncovalent strategies.

REFERENCES

(1) Singh, J.; Petter, R.C.; Baillie, T.A.; Whitty, A. *Nat. Rev. Drug Discov.* **2011**, *10*, 307.
(2) Robertson, J.G. *Biochemistry* **2005**, *44*, 5561.
(3) Swinney, D.C. *Curr. Opin. Drug Discov. Devel.* **2009**, *12*, 31.
(4) Wells, J.A.; McClendon, C.L. *Nature* **2007**, *450*, 1001.
(5) Yun, C.-H.; Mengwasser, K.E.; Toms, A.V.; Woo, M.S.; Greulich, H.; Wong, K.-K.; Meyerson, M.; Eck, M.J. *Proc. Natl. Acad. Sci. U.S.A.* **2008**, *105*, 2070.
(6) Bonanno, L.; Jirillo, A.; Favaretto, A. *Curr. Drug Targets* **2011**, *12*, 922.
(7) Singh, J.; Petter, R.C.; Kluge, A.F. *Curr. Opin. Chem. Biol.* **2010**, *14*, 475.
(8) Garuti, L.; Roberti, M.; Bottegoni, G. *Curr. Med. Chem.* **2011**, *18*, 2981.
(9) Leproult, E.; Barluenga, S.; Moras, D.; Wurtz, J.-M.; Winssinger, N. *J. Med. Chem.* **2011**, *54*, 1347.
(10) Smith, A.J.T.; Zhang, X.; Leach, A.G.; Houk, K.N. *J. Med. Chem.* **2009**, *52*, 225.
(11) Kitz, R.; Wilson, I.B. *J. Biol. Chem.* **1962**, *237*, 3245.
(12) Kluter, S.; Simard, J.R.; Rode, H.B.; Grutter, C.; Pawar, V.; Raaijmakers, H.C.A.; Barf, T.A.; Rabiller, M.; van Otterlo Willem, A.L.; Rauh, D. *Chembiochem* **2010**, *11*, 2557.

(13) Krippendorff, B.-F.; Neuhaus, R.; Lienau, P.; Reichel, A.; Huisinga, W. *J. Biomol. Screen.* **2009**, *14*, 913.
(14) Johnson, D.S.; Weerapana, E.; Cravatt, B.F. *Future Med. Chem.* **2010**, *2*, 949.
(15) Naisbitt, D.J.; Gordon, S.F.; Pirmohamed, M.; Park, B.K. *Drug Saf.* **2000**, *23*, 483.
(16) Stepan, A.F.; Walker, D.P.; Bauman, J.; Price, D.A.; Baillie, T.A.; Kalgutkar, A.S.; Aleo, M.D. *Chem. Res. Toxicol.* **2011**, *24*, 1345.
(17) Hughes, J.D.; Blagg, J.; Price, D.A.; Bailey, S.; DeCrescenzo, G.A.; Devraj, R.V.; Ellsworth, E.; Fobian, Y.M.; Gibbs, M.E.; Gilles, R.W.; Greene, N.; Huang, E.; Krieger-Burke, T.; Loesel, J.; Wager, T.; Whiteley, L.; Zhang, Y. *Bioorg. Med. Chem. Lett.* **2008**, *18*, 4872.
(18) Park, B.K.; Boobis, A.; Clarke, S.; Goldring, C.E.P.; Jones, D.; Kenna, J.G.; Lambert, C.; Laverty, H.G.; Naisbitt, D.J.; Nelson, S.; Nicoll-Griffith, D.A.; Obach, R.S.; Routledge, P.; Smith, D.A.; Tweedie, D.J.; Vermeulen, N.; Williams, D.P.; Wilson, I.D.; Baillie, T.A. *Nat. Rev. Drug Discov.* **2011**, *10*, 292.
(19) Nakayama, S.; Atsumi, R.; Takakusa, H.; Kobayashi, Y.; Kurihara, A.; Nagai, Y.; Nakai, D.; Okazaki, O. *Drug Metab. Dispos.* **2009**, *37*, 1970.
(20) Potashman, M.H.; Duggan, M.E. *J. Med. Chem.* **2009**, *52*, 1231.
(21) Bachovchin, D.A.; Cravatt, B.F. *Nat. Rev. Drug Discov.* **2012**, *11*, 52.
(22) Prongay, A.J.; Guo, Z.; Yao, N.; Pichardo, J.; Fischmann, T.; Strickland, C.; Myers, J., Jr.; Weber, P.C.; Beyer, B.M.; Ingram, R.; Hong, Z.; Prosise, W.W.; Ramanathan, L.; Taremi, S.S.; Yarosh-Tomaine, T.; Zhang, R.; Senior, M.; Yang, R.-S.; Malcolm, B.; Arasappan, A.; Bennett, F.; Bogen, S.L.; Chen, K.; Jao, E.; Liu, Y.-T.; Lovey, R.G.; Saksena, A.K.; Venkatraman, S.; Girijavallabhan, V.; Njoroge, F.G.; Madison, V. *J. Med. Chem.* **2007**, *50*, 2310.
(23) Venkatraman, S.; Bogen, S.L.; Arasappan, A.; Bennett, F.; Chen, K.; Jao, E.; Liu, Y.-T.; Lovey, R.; Hendrata, S.; Huang, Y.; Pan, W.; Parekh, T.; Pinto, P.; Popov, V.; Pike, R.; Ruan, S.; Santhanam, B.; Vibulbhan, B.; Wu, W.; Yang, W.; Kong, J.; Liang, X.; Wong, J.; Liu, R.; Butkiewicz, N.; Chase, R.; Hart, A.; Agrawal, S.; Ingravallo, P.; Pichardo, J.; Kong, R.; Baroudy, B.; Malcolm, B.; Guo, Z.; Prongay, A.; Madison, V.; Broske, L.; Cui, X.; Cheng, K.-C.; Hsieh, T.Y.; Brisson, J.-M.; Prelusky, D.; Korfmacher, W.; White, R.; Bogdanowich-Knipp, S.; Pavlovsky, A.; Bradley, P.; Saksena, A.K.; Ganguly, A.; Piwinski, J.; Girijavallabhan, V.; Njoroge, F.G. *J. Med. Chem.* **2006**, *49*, 6074.
(24) Lin, C.; Kwong, A.D.; Perni, R.B. *Infect. Disord. Drug Targets* **2006**, *6*, 3.
(25) Perni, R.B.; Almquist, S.J.; Byrn, R.A.; Chandorkar, G.; Chaturvedi, P.R.; Courtney, L.F.; Decker, C.J.; Dinehart, K.; Gates, C.A.; Harbeson, S.L.; Heiser, A.; Kalkeri, G.; Kolaczkowski, E.; Lin, K.; Luong, Y.-P.; Rao, B.G.; Taylor, W.P.; Thomson, J.A.; Tung, R.D.; Wei, Y.; Kwong, A.D.; Lin, C. *Antimicrob. Agents Chemother.* **2006**, *50*, 899.
(26) Malcolm, B.A.; Liu, R.; Lahser, F.; Agrawal, S.; Belanger, B.; Butkiewicz, N.; Chase, R.; Gheyas, F.; Hart, A.; Hesk, D.; Ingravallo, P.; Jiang, C.; Kong, R.; Lu, J.; Pichardo, J.; Prongay, A.; Skelton, A.; Tong, X.; Venkatraman, S.; Xia, E.; Girijavallabhan, V.; Njoroge, F.G. *Antimicrob. Agents Chemother.* **2006**, *50*, 1013.
(27) Tong, X.; Arasappan, A.; Bennett, F.; Chase, R.; Feld, B.; Guo, Z.; Hart, A.; Madison, V.; Malcolm, B.; Pichardo, J.; Prongay, A.; Ralston, R.; Skelton, A.; Xia, E.; Zhang, R.; Njoroge, F.G. *Antimicrob. Agents Chemother.* **2010**, *54*, 2365.
(28) Bennett, F.; Huang, Y.; Hendrata, S.; Lovey, R.; Bogen, S.L.; Pan, W.; Guo, Z.; Prongay, A.; Chen, K.X.; Arasappan, A.; Venkatraman, S.; Velazquez, F.; Nair, L.; Sannigrahi, M.; Tong, X.; Pichardo, J.; Cheng, K.-C.; Girijavallabhan, V.M.; Saksena, A.K.; Njoroge, F.G. *Bioorg. Med. Chem. Lett.* **2010**, *20*, 2617.
(29) Garfunkle, J.; Ezzili, C.; Rayl, T.J.; Hochstatter, D.G.; Hwang, I.B.; Dale, L. *J. Med. Chem.* **2008**, *51*, 4392.

(30) Romero, F.A.; Inkyu, H.; Boger, D.L. *J. Am. Chem. Soc.* **2006**, *128*, 14004.
(31) Kimball, F.S.; Romero, F.A.; Ezzili, C.; Garfunkle, J.; Rayl, T.J.; Hochstatter, D.G.; Hwang, I.; Boger, D.L. *J. Med. Chem.* **2008**, *51*, 937.
(32) Hardouin, C.; Kelso, M.J.; Romero, F.A.; Rayl, T.J.; Leung, D.; Hwang, I.; Cravatt, B.F.; Boger, D.L. *J. Med. Chem.* **2007**, *50*, 3359.
(33) Romero, F.A.; Du, W.; Hwang, I.; Rayl, T.J.; Kimball, F.S.; Leung, D.; Hoover, H.S.; Apodaca, R.L.; Breitenbucher, J.G.; Cravatt, B.F.; Boger, D.L. *J. Med. Chem.* **2007**, *50*, 1058.
(34) Boger, D.L.; Miyauchi, H.; Du, W.; Hardouin, C.; Fecik, R.A.; Cheng, H.; Hwang, I.; Hedrick, M.P.; Leung, D.; Acevedo, O.; Guimaraes, C.R.W.; Jorgensen, W.L.; Cravatt, B.F. *J. Med. Chem.* **2005**, *48*, 1849.
(35) Lichtman, A.H.; Leung, D.; Shelton, C.C.; Saghatelian, A.; Hardouin, C.; Boger, D.L.; Cravatt, B.F. *J. Pharmacol. Exp. Ther.* **2004**, *311*, 441.
(36) Chang, L.; Luo, L.; Palmer, J.A.; Sutton, S.; Wilson, S.J.; Barbier, A.J.; Breitenbucher, J.G.; Chaplan, S.R.; Webb, M. *Br. J. Pharmacol.* **2006**, *148*, 102.
(37) Racys, D.T.; Rea, D.; Fueloep, V.; Wills, M. *Bioorg. Med. Chem.* **2010**, *18*, 4775.
(38) Lawandi, J.; Toumieux, S.; Seyer, V.; Campbell, P.; Thielges, S.; Juillerat-Jeanneret, L.; Moitessier, N. *J. Med. Chem.* **2009**, *52*, 6672.
(39) Mansour, T.S.; Bradford, P.A.; Venkatesan, A.M. *Annu. Rep. Med. Chem.* **2008**, *43*, 247.
(40) Stachyra, T.; Pechereau, M.-C.; Bruneau, J.-M.; Claudon, M.; Frere, J.-M.; Miossec, C.; Coleman, K.; Black, M.T. *Antimicrob. Agents Chemother.* **2010**, *54*, 5132.
(41) Eidam, O.; Romagnoli, C.; Caselli, E.; Babaoglu, K.; Pohlhaus, D.T.; Karpiak, J.; Bonnet, R.; Shoichet, B.K.; Prati, F. *J. Med. Chem.* **2010**, *53*, 7852.
(42) Tan, Q.; Ogawa, A.M.; Painter, R.E.; Park, Y.-W.; Young, K.; DiNinno, F.P. *Bioorg. Med. Chem. Lett.* **2010**, *20*, 2622.
(43) Bachovchin, D.A.; Wolfe, M.R.; Masuda, K.; Brown, S.J.; Spicer, T.P.; Fernandez-Vega, V.; Chase, P.; Hodder, P.S.; Rosen, H.; Cravatt, B.F. *Bioorg. Med. Chem. Lett.* **2010**, *20*, 2254.
(44) Laine, D.I.; Busch-Petersen, J. *Expert Opin. Ther. Pat.* **2010**, *20*, 497.
(45) Wiener, J.J.M.; Sun, S.; Thurmond, R.L. *Curr. Top. Med. Chem.* **2010**, *10*, 717.
(46) Lluis, M.; Wang, Y.; Monzingo, A.F.; Fast, W.; Robertus, J.D. *ChemMedChem* **2011**, *6*, 81.
(47) Powers, J.P.; Piper, D.E.; Li, Y.; Mayorga, V.; Anzola, J.; Chen, J.M.; Jaen, J.C.; Lee, G.; Liu, J.; Peterson, M.G.; Tonn, G.R.; Ye, Q.; Walker, N.P.C.; Wang, Z. *J. Med. Chem.* **2006**, *49*, 1034.
(48) Tsou, H.-R.; Overbeek-Klumpers, E.G.; Hallett, W.A.; Reich, M.F.; Floyd, M.B.; Johnson, B.D.; Michalak, R.S.; Nilakantan, R.; Discafani, C.; Golas, J.; Rabindran, S.K.; Shen, R.; Shi, X.; Wang, Y.-F.; Upeslacis, J.; Wissner, A. *J. Med. Chem.* **2005**, *48*, 1107.
(49) Wong, K.-K.; Fracasso, P.M.; Bukowski, R.M.; Lynch, T.J.; Munster, P.N.; Shapiro, G.I.; Jaenne, P.A.; Eder, J.P.; Naughton, M.J.; Ellis, M.J.; Jones, S.F.; Mekhail, T.; Zacharchuk, C.; Vermette, J.; Abbas, R.; Quinn, S.; Powell, C.; Burris, H.A. *Clin. Cancer Res.* **2009**, *15*, 2552.
(50) Sequist, L.V.; Besse, B.; Lynch, T.J.; Miller, V.A.; Wong, K.K.; Gitlitz, B.; Eaton, K.; Zacharchuk, C.; Freyman, A.; Powell, C.; Ananthakrishnan, R.; Quinn, S.; Soria, J.-C. *J. Clin. Oncol.* **2010**, *28*, 3076.
(51) Burstein, H.J.; Sun, Y.; Dirix, L.Y.; Jiang, Z.; Paridaens, R.; Tan, A.R.; Awada, A.; Ranade, A.; Jiao, S.; Schwartz, G.; Abbas, R.; Powell, C.; Turnbull, K.; Vermette, J.; Zacharchuk, C.; Badwe, R. *J. Clin. Oncol.* **2010**, *28*, 1301.
(52) Li, D.; Ambrogio, L.; Shimamura, T.; Kubo, S.; Takahashi, M.; Chirieac, L.R.; Padera, R.F.; Shapiro, G.I.; Baum, A.; Himmelsbach, F.; Rettig, W.J.; Meyerson, M.; Solca, F.; Greulich, H.; Wong, K.K. *Oncogene* **2008**, *27*, 4702.

(53) Subramaniam, D.S.; Hwang, J. *Expert Opin. Investig. Drugs* **2011**, *20*, 415.
(54) Engelman, J.A.; Zejnullahu, K.; Gale, C.-M.; Lifshits, E.; Gonzales, A.J.; Shimamura, T.; Zhao, F.; Vincent, P.W.; Naumov, G.N.; Bradner, J.E.; Althaus, I.W.; Gandhi, L.; Shapiro, G.I.; Nelson, J.M.; Heymach, J.V.; Meyerson, M.; Wong, K.-K.; Jänne, P.A. *Cancer Res.* **2007**, *67*, 11924.
(55) Gonzales, A.J.; Hook, K.E.; Althaus, I.W.; Ellis, P.A.; Trachet, E.; Delaney, A.M.; Harvey, P.J.; Ellis, T.A.; Amato, D.M.; Nelson, J.M.; Fry, D.W.; Zhu, T.; Loi, C.-M.; Fakhoury, S.A.; Schlosser, K.M.; Sexton, K.E.; Winters, R.T.; Reed, J.E.; Bridges, A.J.; Lettiere, D.J.; Baker, D.A.; Yang, J.; Lee, H.T.; Tecle, H.; Vincent, P.W. *Mol. Cancer Ther.* **2008**, *7*, 1880.
(56) Janne, P.A.; Boss, D.S.; Camidge, D.R.; Britten, C.D.; Engelman, J.A.; Garon, E.B.; Guo, F.; Wong, S.; Liang, J.; Letrent, S.; Millham, R.; Taylor, I.; Eckhardt, S.G.; Schellens, J.H.M. *Clin. Cancer Res.* **2011**, *17*, 1131.
(57) Wood, E.R.; Shewchuk, L.M.; Ellis, B.; Brignola, P.; Brashear, R.L.; Caferro, T.R.; Dickerson, S.H.; Dickson, H.D.; Donaldson, K.H.; Gaul, M.; Griffin, R.J.; Hassell, A.M.; Keith, B.; Mullin, R.; Petrov, K.G.; Reno, M.J.; Rusnak, D.W.; Tadepalli, S.M.; Ulrich, J.C.; Wagner, C.D.; Vanderwall, D.E.; Waterson, A.G.; Williams, J.D.; White, W.L.; Uehling, D.E. *Proc. Natl. Acad. Sci. U.S.A.* **2008**, *105*, 2773.
(58) Petro, J.B.; Rahman, S.M.J.; Ballard, D.W.; Khan, W.N. *J. Exp. Med.* **2000**, *191*, 1745.
(59) Heinonen, J.E.; Smith, C.I.E.; Nore, B.F. *FEBS Lett.* **2002**, *527*, 274.
(60) Pan, Z.; Scheerens, H.; Li, S.-J.; Schultz, B.E.; Sprengeler, P.A.; Burrill, L.C.; Mendonca, R.V.; Sweeney, M.D.; Scott, K.C.K.; Grothaus, P.G.; Jeffery, D.A.; Spoerke, J.M.; Honigberg, L.A.; Young, P.R.; Dalrymple, S.A.; Palmer, J.T. *ChemMedChem* **2007**, *2*, 58.
(61) Honigberg, L.A.; Smith, A.M.; Sirisawad, M.; Verner, E.; Loury, D.; Chang, B.; Li, S.; Pan, Z.; Thamm, D.H.; Miller, R.A.; Buggy, J.J. *Proc. Natl. Acad. Sci. U.S.A.* **2010**, *107*, 13075.
(62) Herman, S.E.M.; Gordon, A.L.; Hertlein, E.; Ramanunni, A.; Zhang, X.; Jaglowski, S.; Flynn, J.; Jones, J.; Blum, K.A.; Buggy, J.J.; Hamdy, A.; Johnson, A.J.; Byrd, J.C. *Blood* **2011**, *117*, 6287.
(63) MacGlashan, D., Jr.; Honigberg, L.A.; Smith, A.; Buggy, J.; Schroeder, J.T. *Int. Immunopharmacol.* **2011**, *11*, 475.
(64) Honigberg, L.A.; Smith, A.M.; Chen, J.; Thiemann, P.; Verner, E. Poster Presented at the American Society for Hematology Annual Meeting, Atlanta, GA, 2007.
(65) Advani, R.; Sharman, J.; Smith, S.; Pollyea, D.; Boyd, T.; Grant, B.; Kolibaba, K.; Buggy, J.; Hamdy, A.; Fowler, N. Oral Presentation at the American Society for Clinical Oncology Meeting, Chicago, Il, **2010**.
(66) Evans, E.; Tester, R.; Aslanian, S.; Mazdiyasni, H.; Ponader, S.; Tesar, B.; Chaturverdi, P.; Nacht, M.; Stiede, K.; Witowski, S.; Lounsbury, H.; Silver, B.; Burger, J.; Mahadevan, D.; Sharman, J.; Harb, W.; Petter, R.; Singh, J.; Westlin, W. Poster Presented at the ASH Annual Meeting, San Diego, CA, **2011**.
(67) Singh, J.; Petter, R.; Tester, R.; Kluge, A. F.; Mazdiyasni, H.; Westlin, W.; Niu, D.; Qiao, L. US Patent US 2010/0249092, **2010**.
(68) Witowski, S.; Westlin, W.; Tester, R. Patent Application WO 2012/021444A1, **2012**.
(69) Henise, J.C.; Taunton, J. *J. Med. Chem.* **2011**, *54*, 4133.
(70) Zhou, W.; Hur, W.; McDermott, U.; Dutt, A.; Xian, W.; Ficarro, S.B.; Zhang, J.; Sharma, S.V.; Brugge, J.; Meyerson, M.; Settleman, J.; Gray, N.S. *Chem. Biol.* **2010**, *17*, 285.
(71) Hagel, M.; Niu, D.; St Martin, T.; Sheets, M.P.; Qiao, L.; Bernard, H.; Karp, R.M.; Zhu, Z.; Labenski, M.T.; Chaturvedi, P.; Nacht, M.; Westlin, W.F.; Petter, R.C.; Singh, J. *Nat. Chem. Biol.* **2011**, *7*, 22.

(72) Perez, D.I.; Palomo, V.; Pérez, C.N.; Gil, C.; Dans, P.D.; Luque, F.J.; Conde, S.; Martinez, A. *J. Med. Chem.* **2011**, *54*, 4042.
(73) Verhelst, S.H.L.; Witte, M.D.; Arastu-Kapur, S.; Fonovic, M.; Bogyo, M. *Chembiochem* **2006**, 7, 943.
(74) Greenberg, C.S.; Birckbichler, P.J.; Rice, R.H. *FASEB J.* **1991**, *5*, 3071.
(75) Prime, M.E.; Andersen, O.A.; Barker, J.J.; Brooks, M.A.; Cheng, R.K.Y.; Toogood-Johnson, I.; Courtney, S.M.; Brookfield, F.A.; Yarnold, C.J.; Marston, R.W.; Johnson, P.D.; Johnsen, S.F.; Palfrey, J.J.; Vaidya, D.; Erfan, S.; Ichihara, O.; Felicetti, B.; Palan, S.; Pedret-Dunn, A.; Schaertl, S.; Sternberger, I.; Ebneth, A.; Scheel, A.; Winkler, D.; Toledo-Sherman, L.; Beconi, M.; Macdonald, D.; Muñoz-Sanjuan, I.; Dominguez, C.; Wityak, J. *J. Med. Chem.* **2012**, *55*, 1021.
(76) Johnson, D.S.; Stiff, C.; Lazerwith, S.E.; Kesten, S.R.; Fay, L.K.; Morris, M.; Beidler, D.; Liimatta, M.B.; Smith, S.E.; Dudley, D.T.; Sadagopan, N.; Bhattachar, S.N.; Kesten, S.J.; Nomanbhoy, T.K.; Cravatt, B.F.; Ahn, K. *ACS Med. Chem. Lett.* **2011**, *2*, 91.
(77) Liu, Y.; Patricelli, M.P.; Cravatt, B.F. *Proc. Natl. Acad. Sci. U.S.A.* **1999**, *96*, 14694.
(78) Meyers, M.J.; Long, S.A.; Pelc, M.J.; Wang, J.L.; Bowen, S.J.; Schweitzer, B.A.; Wilcox, M.V.; McDonald, J.; Smith, S.E.; Foltin, S.; Rumsey, J.; Yang, Y.-S.; Walker, M.C.; Kamtekar, S.; Beidler, D.; Thorarensen, A. *Bioorg. Med. Chem. Lett.* **2011**, *21*, 6545.
(79) Meyers, M.J.; Long, S.A.; Pelc, M.J.; Wang, J.L.; Bowen, S.J.; Walker, M.C.; Schweitzer, B.A.; Madsen, H.M.; Tenbrink, R.E.; McDonald, J.; Smith, S.E.; Foltin, S.; Beidler, D.; Thorarensen, A. *Bioorg. Med. Chem. Lett.* **2011**, *21*, 6538.
(80) Zvonok, N.; Pandarinathan, L.; Williams, J.; Johnston, M.; Karageorgos, I.; Janero, D.R.; Krishnan, S.C.; Makriyannis, A. *Chem. Biol.* **2008**, *15*, 854.
(81) Matuszak, N.; Muccioli, G.G.; Labar, G.; Lambert, D.M. *J. Med. Chem.* **2009**, *52*, 7410.
(82) Maione, S.; Morera, E.; Marabese, I.; Ligresti, A.; Luongo, L.; Ortar, G.; Di, M.V. *Br. J. Pharmacol.* **2008**, *155*, 775.
(83) Bachovchin, D.A.; Mohr, J.T.; Speers, A.E.; Wang, C.; Berlin, J.M.; Spicer, T.P.; Fernandez-Vega, V.; Chase, P.; Hodder, P.S.; Schürer, S.C.; Nomura, D.K.; Rosen, H.; Fu, G.C.; Cravatt, B.F. *Proc. Natl. Acad. Sci. U.S.A.* **2011**, *108*, 6811.
(84) Bachovchin, D.A.; Zuhl, A.M.; Speers, A.E.; Wolfe, M.R.; Weerapana, E.; Brown, S.J.; Rosen, H.; Cravatt, B.F. *J. Med. Chem.* **2011**, *54*, 5229.
(85) Davies, S.; Pandian, R.; Bolos, J.; Castaner, R. *Drugs Future* **2009**, *34*, 708.
(86) Demo, S.D.; Kirk, C.J.; Aujay, M.A.; Buchholz, T.J.; Dajee, M.; Ho, M.N.; Jiang, J.; Laidig, G.J.; Lewis, E.R.; Parlati, F.; Shenk, K.D.; Smyth, M.S.; Sun, C.M.; Vallone, M.K.; Woo, T.M.; Molineaux, C.J.; Bennett, M.K. *Cancer Res.* **2007**, *67*, 6383.
(87) Kim, J.-H.; Lee, S.-K.; Ki, M.-H.; Choi, W.-K.; Ahn, S.-K.; Shin, H.-J.; Hong, C.I. *Int. J. Pharm.* **2004**, *272*, 79.
(88) McLean, L.R.; Zhang, Y.; Li, H.; Li, Z.; Lukasczyk, U.; Choi, Y.-M.; Han, Z.; Prisco, J.; Fordham, J.; Tsay, J.T.; Reiling, S.; Vaz, R.J.; Li, Y. *Bioorg. Med. Chem. Lett.* **2009**, *19*, 6717.

CHAPTER TWENTY-EIGHT

Drug Design Strategies for GPCR Allosteric Modulators

P. Jeffrey Conn*, Scott D. Kuduk†, Darío Doller‡

*Department of Pharmacology, Vanderbilt Center for Neuroscience Drug Discovery, Vanderbilt University, Nashville, Tennessee, USA

†Department of Medicinal Chemistry, Merck Research Laboratories, West Point, Pennsylvania, USA

‡Discovery Chemistry & DMPK, Lundbeck Research DK, Valby, Denmark and Neuroinflammation Disease Biology Unit, Lundbeck Research USA, Paramus, New Jersey, USA

Contents

Annual Reports in Medicinal Chemistry, Volume 47
ISSN 0065-7743
http://dx.doi.org/10.1016/B978-0-12-396492-2.00028-X

ABBREVIATIONS

Ago-PAM allosteric ligand acting as agonist in the absence of orthosteric ligand and also potentiating the effect of an agonist
GPCR G-protein-coupled receptor
NAM negative allosteric modulator
PAM positive allosteric modulator
Pure PAM positive allosteric modulator not showing functional activity in the absence of an orthosteric ligand
SAM silent allosteric modulator

1. INTRODUCTION

1.1. The relevance of allosteric receptor modulation in today's therapeutic drug research

A large fraction of marketed prescription drugs are ligands acting at cell membrane G-protein-coupled receptors (GPCRs), making this class of proteins one of the two most fruitful therapeutic target families, together with enzymes.[1,2] Recently, terms such as "nondruggable GPCRs" have been coined to describe a unique group of GPCRs having strong biological hypothesis validation, but for which medicinal chemistry efforts based on orthosteric ligands (those binding at the same site than the endogenous ligand) have failed to produce marketed drugs (or even clinical development candidates).[3] These receptors remain attractive as biological targets for new drugs, but a novel medicinal chemistry strategy must be developed to modulate their biological effects.

Recently, a new "wave" of drug discovery research has emerged to fill this gap, purposely aiming to modulate target receptors by the use of allosteric ligands, which interact with the receptor at a binding site topographically distinct from the endogenous ligand. By binding at this new receptor region, challenging chemical space may be avoided. Moreover, enhanced subtype selectivity profiles may be obtained compared with that of an orthosteric agent binding to a highly conserved site, potentially leading to improved safety and pharmacology profiles. Furthermore, the lack of desensitization arising from receptor overstimulation under constant exposure to an agonist, and preservation of the temporal and local patterns of physiological activity of the endogenous ligand are additional appealing attributes of allosteric modulators.[3–6]

As research in allosteric modulator drug design intensifies, some empirical observations are emerging. The objective of this review is to briefly discuss these concepts and present a number of medicinal chemistry strategies to address current challenges in a productive manner from the drug discovery standpoint.

1.2. The medicinal chemist's jargon of allosterism: NAM, PAM, SAM, allosteric agonists, ago-PAMs, bitopic ligands, probe dependency

The binding interactions between a given allosteric ligand and a GPCR can generate a number of different conformational states, both on their own and when an orthosteric ligand is bound. The consequence of this added complexity results in pharmacological texture that is ligand-, receptor-, and cell dependent.[4] The terminology used to describe such a plethora of possibilities has recently been discussed in a number of excellent reviews.[5–7] The reader's familiarity with these terms is assumed.

1.3. What is so unique about designing allosteric modulators?

Benzodiazepines such as Librium, **1**, acting as $GABA_A$ receptor modulators, have inadvertently become the earliest clinical success story of allosteric modulation. Several drug candidates acting through allosteric mechanisms are currently undergoing clinical trials. The phenomenon appears quite broad in scope, as allosteric modulators have been reported for family A, B, and C GPCRs, as well as ion channels, kinases, and phospholipases.[5]

In retrospect, a number of "noncompetitive receptor antagonist" projects (e.g., CRF-1, NK_1, CGRP, GnRh receptors) had actually been toiling on negative allosteric modulators (NAMs) instead.[8] Thus, this is not necessarily "uncharted territory" for the medicinal chemist working to design drugs. However, it should be noted that thorough differentiation between antagonists and NAMs is not trivial or even currently possible (e.g., with orphan receptors of unknown endogenous agonist).[9]

Recently, Cinacalcet, **2**, a calcium-sensing receptor (CaSR) positive allosteric modulator (PAM), and Maraviroc, **3**, a CCR5 NAM, have reached the market. Besides the obvious merit of helping patients, these two drugs highlight the potential of allosteric ligands to accomplish unique biochemical tasks. Cinacalcet potentiates the effects of an inorganic agonist (Ca^{2+}) on a class C GPCR (CaSR) through a stimulus-biased mechanism (see Section 3.3).[10] Maraviroc precludes the functional outcome derived from interactions between small proteins called chemokines and the surface subunit of the HIV-1 envelope glycoprotein (gp120) signaling through CCR5, one of the main chemokine receptors involved in the HIV entry process. Thus, **3** competes favorably with other mechanisms of action such as chemically modified chemokines and monoclonal antibodies.[11]

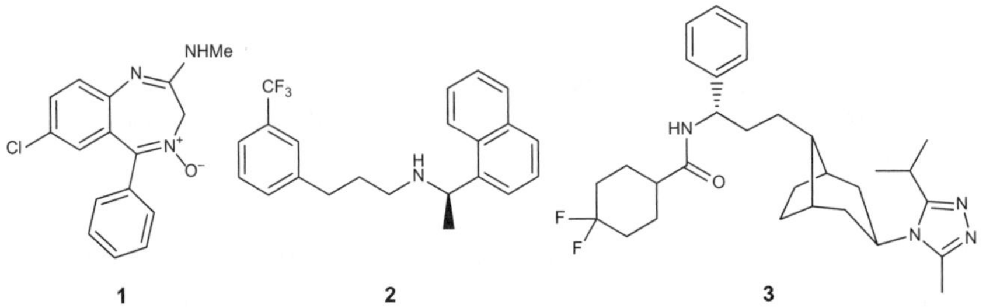

1 2 3

1.4. How do allosteric ligands exert their function?

For decades, medicinal chemists have learned to design drugs that bind at orthosteric sites using classical concepts such as "lock and key," where the rigidity of the binding site is a fundamental premise. Such rigidity supports the use of linear drug design strategies, where different structural regions can be optimized independently, and then the final drug candidate emerges from a combination of optimized fragments.

Affinity is a key optimization parameter and remains so in allosteric modulator drug design. However, two allosteric drugs may bind at the same receptor with the same affinity, yet provide different responses not only from the functional perspective (e.g., positive or negative modulation) but also by differing in the specific signal transduction pathway being engaged. This functional selectivity has been rationalized by considering protein dynamics concepts: the flexible nature of the protein receptor backbone and the changes in free energy landscape arising upon binding.[12] The existence of multiple distinct allosteric binding sites on the same target would also be consistent with experimental observations, and some evidence indicates this may be possible.[13] Thus, establishing structure–functional outcome relationships often leads to challenges in optimization strategies, adding unpredictability and complexity to drug discovery efforts.

2. STRUCTURE–ACTIVITY RELATIONSHIPS OF ALLOSTERIC LIGANDS

2.1. High-throughput screening of allosteric modulators: "Flat SAR"

From a tactical viewpoint, the process to discover an optimized drug candidate from an initial screening hit is divided in stages. After each stage, the qualities of the chemical entities are becoming closer to those desired for clinical success. As a consequence of progress in parallel synthesis and high-throughput screening technologies, screening campaigns based on functional readouts may produce a large number of confirmed hits, depending on library size and biological activity cutoffs. For example, a recent mGlu3 receptor "triple add"-based screening on a collection containing 829,000 compounds delivered 23 agonist hits, 121 PAM hits, and 866 antagonist hits after concentration–response curves and selectivity counter-screens.[14]

Before starting a lead identification or optimization program, a prioritization exercise should be undertaken, as not every hit will correspond to a series with chemically tractable SAR.[15] The specific hit prioritization strategy followed depends on the library composition (singletons and close analogs), and it includes considerations of favorable ligand metrics.[16] However, early efforts developing SAR from HTS allosteric modulator hits revealed that the phenomenon known as "*flat SAR*" appears considerably more widespread than with orthosteric ligands. This is the characteristic shown by some (but not all) hits, whereby very small changes of structure lead to inactive compounds, or point toward very shallow and narrow SAR patterns. Flat SAR is seen often with PAMs.[17] Thus, chemical tractability of allosteric modulators cannot be taken for granted. Unfortunately, physicochemical attributes and good ligand metrics do not foretell of allosteric modulator SAR tractability. Thus, the goal for hit prioritization of allosteric modulators is to select those chemotypes with *both* SAR tractability and encouraging ligand efficiency/physicochemical property attributes, therefore adding value to drug discovery programs through the discovery of such truly optimizable allosteric ligand chemical series.

2.2. Functional switches

Alternations of agonist/antagonist activity with small structural changes in a given chemotype are not unknown for orthosteric ligands.[18,19] However, it appears as though allosteric modulators show this phenomenon with a relatively high frequency. In addition, patterns of structural factors governing these switches can be subtle.[20]

The mGlu5 receptor has gained notoriety (and made drug discovery efforts more challenging) for the apparently unpredictable way in which these switches occur, in particular, with analogs derived from MPEP (**4**), an early tool compound. However, when close analogs are designed and their biological activity analyzed systematically, some trends can be picked up. For example, two independent reports on a number of *N*-substituted 2-(arylethynyl)-7,8-dihydro-1,6-naphthyridin-5(6*H*)-one analogs (**5**, **6**;

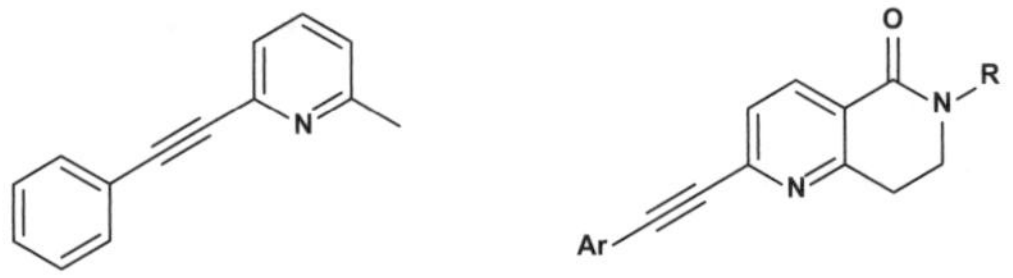

4

5 (Ar=Ph)
6 (Ar=3-F-Ph)

Compound	R	mGlu5 EC_{50} or IC_{50} (nM)	% Glu Max	Functionality
6a	H	290	72	PAM
6b	CH_3	170	34	NAM
6c	*t*-Bu	54	40	PAM
6d	*n*-Bu	56	46	PAM
6e	$CH_2(CH_2)_3$	660	34	NAM
6f	$CH_2(CH_2)_4$	130	53	PAM

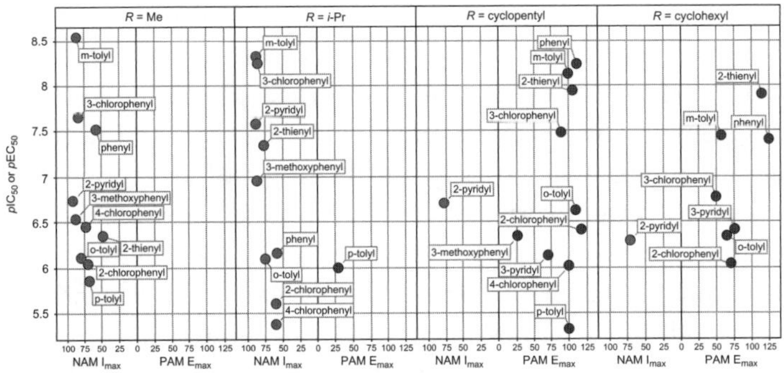

Figure 28.1 Switches in functional potency and efficacy for a series of *N*-substituted 2-(arylethynyl)-7,8-dihydro-1,6-naphthyridin-5(6H)-one analogs (**5**), Ar as specified in the table. (See Color Plate 28.1 in Color Plate Section.)

Fig. 28.1) showed that *N*-methyl or *N*-isopropyl substituents delivered predominantly NAMs, while increasing the *N*-substituent to cyclopentyl or cyclohexyl primarily yielded PAMs.[21,22]

It is important to note that not all chemical series binding to an allosteric site of a given receptor will present switches to the same degree. From a strategic viewpoint, the process of lead optimization is greatly facilitated when a chemotype is chosen where functional switches are nonexistent or present themselves only sporadically.

The untoward possibility exists that metabolites derived from allosteric modulator drugs may have different functional activity than the "parent." While specific examples of metabolic functional switches have not been published, it is clear from the literature that many of the changes that lead to functional switches could occur through hepatic metabolism. Again, this is not unique to allosteric site ligands, but the higher frequency with which these functional switches are seen in *in vitro* tests with these ligands may lead to an increased risk of misinterpreting observations from *in vivo* pharmacological studies, or when establishing pharmacokinetic/pharmacodynamic relationships if such metabolites remain in circulation. Therefore, conducting a comprehensive metabolite identification study may be necessary to support conclusions from *in vivo* tests.

2.3. Can functional switches be "designed"?

The challenge represented by switches of functional activity between positive and negative allosterism within a chemotype can be transformed into an opportunity to design compounds with specific properties. For example, the synthesis of dimethyl derivatives of [5.6.5] spiro bicyclic lactam Pro-Leu-Gly-NH_2 peptidomimetics **7** and **8** was carried out to test the hypothesis that by placing geminal methyl groups on the β-methylene carbon of the thiazolidine ring steric bulk would be introduced into the topological space that the β-methylene carbon is believed to occupy when binding to the dopamine D_2 receptor. As a result of this modification, a PAM was converted into a NAM.[23]

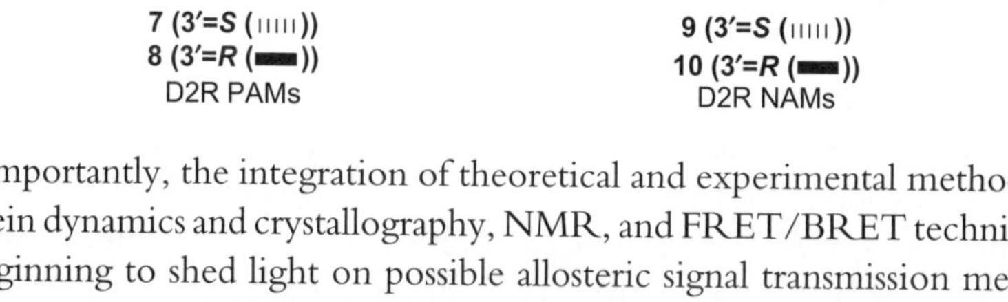

7 (3′=*S* (IIIII))
8 (3′=*R* (▬))
D2R PAMs

9 (3′=*S* (IIIII))
10 (3′=*R* (▬))
D2R NAMs

Importantly, the integration of theoretical and experimental methods in protein dynamics and crystallography, NMR, and FRET/BRET techniques is beginning to shed light on possible allosteric signal transmission mechanisms, aiming to facilitate the design of much improved allosteric drugs.[24] A strategy was explored using a FRET-based binding assay to identify compound **11** as an mGlu2 receptor silent allosteric modulator (SAM). Although **11** was relatively weak ($K_i = 6.6$ μM), higher-affinity close analogs with mGlu2 NAM functionality were quickly obtained.[25] While this exercise provided "proof of concept" that SAMs may be stepping stones in programs aiming for identification of allosteric ligands using novel technologies, the *predictive* use of such strategy is far from being realized.

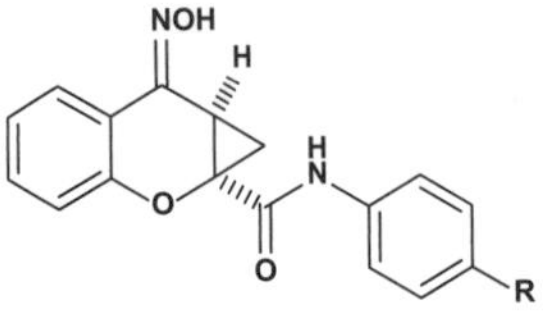

Compound	*R*	FRET K_i (μM)	NAM IC_{50} (μM)
11	F	6.6	Not active up to 100 μM
12	Cl	1.0	0.8
13	Me	0.6	1.5
14	OMe	0.8	1.0

3. FUNCTIONAL SELECTIVITY

Allosteric modulators, even those with very closely related chemical structures, can vary in their capacity to activate or inhibit functional responses in cells upon binding to a GPCR. This may originate from a number of different mechanisms.

3.1. Probe dependency

Theoretical support and experimental evidence exist, demonstrating that the specific nature of the effect of an allosteric modulator at a given receptor may vary with different orthosteric ligands.[26] Because of probe dependency, it is important to use a physiologically relevant orthosteric ligand (typically an agonist) during screening. This may be an issue when dealing with endogenous ligands which are unstable (e.g., acetylcholine); are intrinsic components of cellular systems, thus making their exact concentration variable (e.g., GABA or glutamate); have weak affinities for the orthosteric site, requiring prohibitively high concentrations for *in vitro* testing due to cell toxicity (e.g., glutamate at the mGlu7 receptor); or when more than one endogenous agonist exist (e.g., GLP-1(7-36)NH_2 and oxyntomodulin, endogenous GLP-1 receptor agonists[27]; L-serine-O-phosphate as endogenous agonist at mGlu4 receptor[28]).

In some cases, a stable (e.g., the cholinergic agonist pilocarpine) or a potent nonendogenous agonist (e.g., synthetic group 3 mGlu receptor agonist L-AP4) may be used, provided some measures are taken to mitigate the risk introduced. If more than one endogenous agonist exist, occasional testing against all known modulators may be advisable. If so, early characterization in appropriate *in vitro* and *in vivo* screens is recommended.

3.2. Receptor hetero- or homo-oligomerization

Adding complexity to this picture, functional effects derived from GPCR activation may be mediated by receptor hetero- or homo-oligomerization. For example, dopamine, serotonin, and glutamate play a role in the pathophysiology of schizophrenia. In the brain, a functional cross talk between the serotonin receptor 5-HT_{2A} and the mGlu2 receptor has been demonstrated[29]; however, its biological significance has been challenged.[30]

3.3. Biased signaling

Upon ligand binding, GPCRs generate second messengers (e.g., cAMP, calcium, phosphoinositides) by acting through different signaling pathways besides heterotrimeric G-proteins, such as β-arrestins and GPCR kinases.[31] If the endogenous orthosteric ligand acts by stabilizing a subset of receptor conformations considered "unbiased," allosteric ligands might stabilize a different receptor conformation subset, leading to signaling through different pathways. Thus, GPCR potentiation by ligands does not always lead to uniform activation of all potential signaling pathways mediated by a given receptor. When comparing structurally different ligands, some of these may be biased toward producing subsets of receptor behaviors.[32]

This introduces a challenge, as well as an opportunity, to design compounds aiming specifically at a biological pathway inducing therapeutically meaningful agonist bias.[33] The first evidence that an allosteric modulator used in clinical practice (Cinacalcet) exhibits stimulus bias was recently reported.[10] Previously, GLP-1 receptor allosteric ligands **15** and **16** were found to differentially modulate endogenous and exogenous peptide responses in a pathway-selective manner.[27] The detection of such stimulus bias at a GPCR requires investigation across multiple signaling pathways and the development of methods to quantify the effects of allosteric ligands on orthosteric ligand affinity and cooperativity at each pathway. Evaluating compounds in the appropriate tissue for a given therapeutic indication will mitigate this risk.

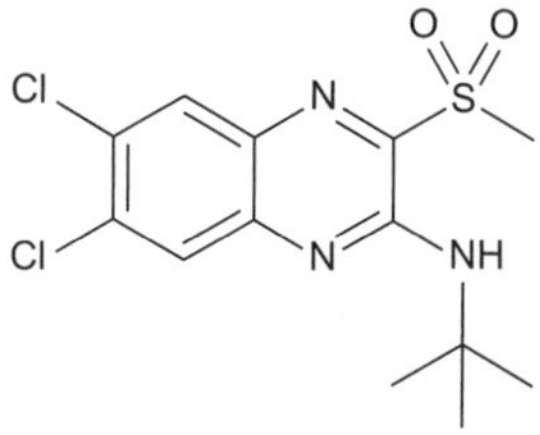

15

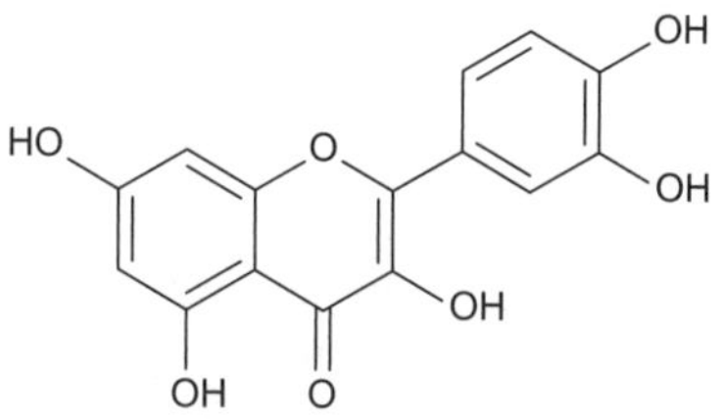

16

4. WHAT TO OPTIMIZE: IC_{50}? EC_{50}? E_{MAX}? FOLD-SHIFT? LOG($\alpha\beta$)?

While allosteric modulation has been reported for a variety of receptor classes, the compound optimization strategy for allosteric ligands of different biological targets requires careful consideration of compound-dependent attributes, as well as the specific biological modulation aimed for. In other words, two allosteric modulator programs at two different targets within the same family (i.e., mGlu2 or mGlu3 receptors) may optimize compounds toward a different set of allosteric ligand properties.

4.1. "Pure PAMs"

Quantification of pharmacological effects of allosteric ligands is more complex than for orthosteric ligands.[34] Generally, concentration–response curves are compared using *in vitro* tests, measuring the orthosteric agonist functional response in the presence of increasing concentrations of allosteric modulator. Excellent discussions of these experiments and the mathematical models used have recently been published.[5,6,35] In simple systems, using EC_{50}, E_{MAX}, and orthosteric agonist fold-shift values (for PAMs) or IC_{50} values (for NAMs) may provide guidance to select improved compounds. Dealing with more complex systems requires a combination of parameters related to the agonist and the allosteric ligand affinities (K_A and K_B, respectively), their capacity to exhibit agonism when acting alone (τ_A and τ_B, respectively), and two cooperativity factors quantifying the effect of the allosteric ligand on the affinity of the orthosteric ligand (α) or the effect of the modulator on the efficacy of the orthosteric ligand (β, Fig. 28.2).

4.2. Are compounds with ago-PAM activity and pure PAMs truly different?

When developing SAR for PAMs, an alternation is sometimes seen between positive allosteric modulation not showing functional activity in the absence of an orthosteric ligand (pure PAM), and allosteric ligand acting as agonist in the absence of orthosteric ligand and also potentiating the effect of an agonist (ago-PAM) activity upon minor structural changes within the same chemotype. Generally, the agonist EC_{50} values are above those for modulation. This may hinder *in vitro* determinations based on the "triple add" protocol, as the agonist activity causes desensitization and attenuates the functional response.

Mechanistically, true PAMs require a certain endogenous agonist tone to exert their effect, whereas ago-PAMs might act independently of the agonist tone. The translation of these *in vitro* effects to functional electrophysiology and behavioral readouts was recently explored by comparing optimized

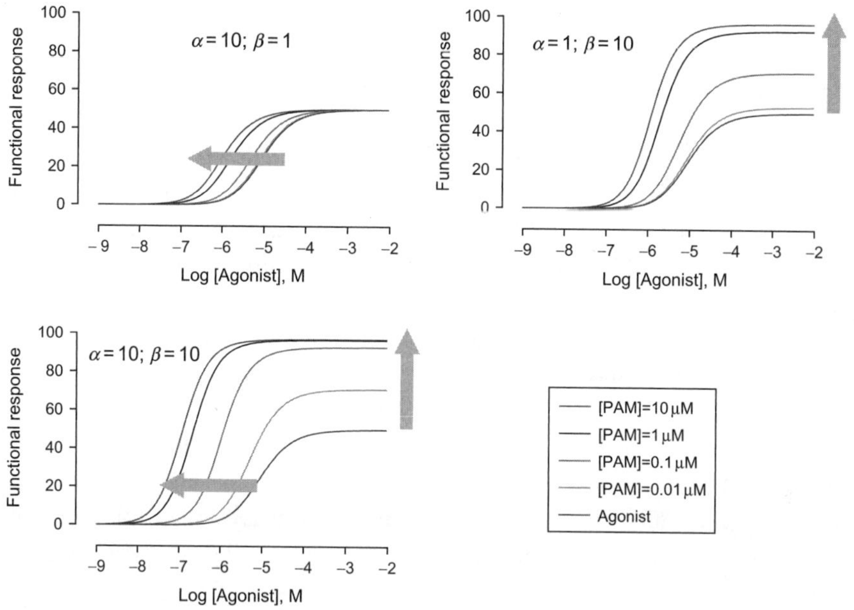

Figure 28.2 Computer simulation of curves showing the impact of allosteric modulation on an agonist functional response for different combinations of α and β. Calculations based on Leach equation [6], using the following parameters: $pK_A = 5$; $pK_B = 7$; $E_m = 100$; $\tau_A = 1$; $\tau_B = 0$; $n_H = 1.5$. (See Color Plate 28.2 in Color Plate Section.)

mGlu5 receptor ago-PAMs such as VU0360172 (**17**) with maximal intrinsic agonist activity in cell lines, along with close structural analogs acting as pure PAMs such as VU0361747 (**18**), devoid of intrinsic agonist activity.[36]

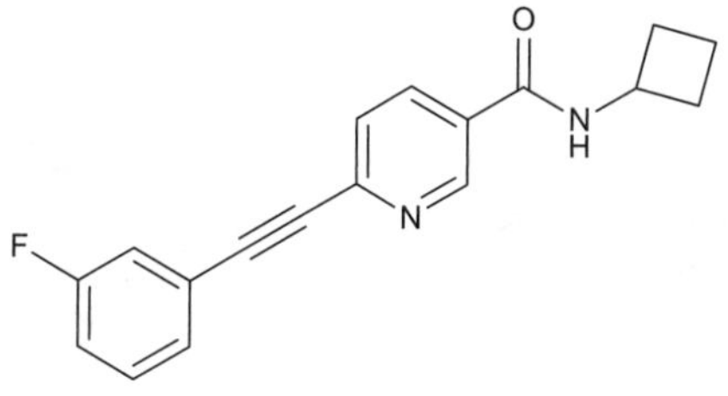

17

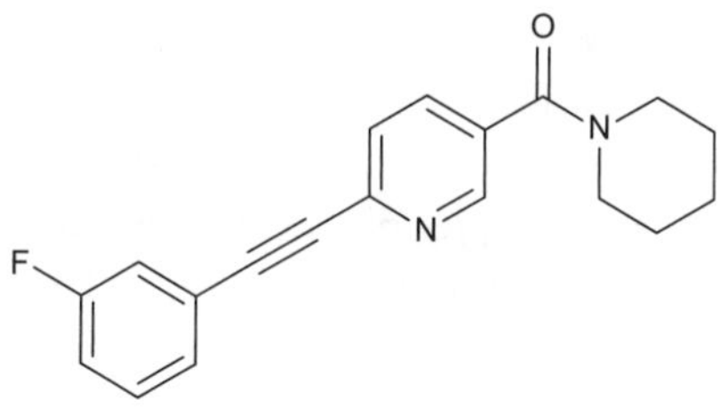

18

Testing these compounds with cell lines engineered to have different levels of mGlu5 receptor expression, ago-PAM activity was only seen with high expression levels. All compounds examined in this study behaved as pure PAMs in native systems studied, regardless of whether they exhibited agonist activity in overexpressing cell lines. Pure PAMs and ago-PAMs showed identical efficacy in reversing amphetamine-induced hyperlocomotor activity, a preclinical model of potential antipsychotic effects.

These studies suggest that the presence of ago-PAM activity observed in overexpressing cell lines may not represent functionally relevant agonist activity in native systems. However, this does not rule out the possibility that functionally relevant ago-PAM activity can be observed. The studies with **17** and **18** were followed up with efforts to optimize compounds with authentic ago-PAM activity using a cell line expressing low levels of mGlu5 receptors. This effort yielded compounds that showed robust agonist activity in astrocytes and hippocampal neurons. Thus, it is possible to develop true ago-PAMs that behave very differently from pure PAMs or compounds that only show ago-PAM activity in overexpressing cell lines.[37] Another example of how the level of receptor expression in cell lines used during *in vitro* screens affects the functional pharmacology observed for different physiological and pharmacological GLP-1 receptor ligands was recently reported.[38]

Again, the key message is that there is no one-size-fits-all strategy. From target validation through lead optimization, judgment should be used based on the specific nature of the target and the biological modulation desired for a given pharmacological action.

5. DO *IN VITRO* PROFILES TRANSLATE TO *IN VIVO* PHARMACOLOGY?

While secondary screens are important elements of drug discovery research in an allosteric modulator program, they are not always capable of explaining disconnects that arise between *in vitro* and *in vivo* pharmacology. As an example, piperazine **19**[39] and 4-cyanopiperidine **20**[40] were recently reported as highly M_1-selective PAMs. Both of these PAMs exhibited good pharmacokinetic properties, sufficient brain penetration, and were efficacious in a mouse contextual fear-conditioning model of episodic memory, a task that requires the hippocampus, where the M_1 receptor is densely expressed. During the course of the SAR work on PAMs **19** and **20**, closely related piperidines **21** and **22** were characterized and evaluated in the aforementioned mouse contextual fear-conditioning model. However, **21**or **22** did not exhibit any activity in this assay, despite high plasma and CSF exposures and excellent FLIPR potency at the mouse receptor.[41]

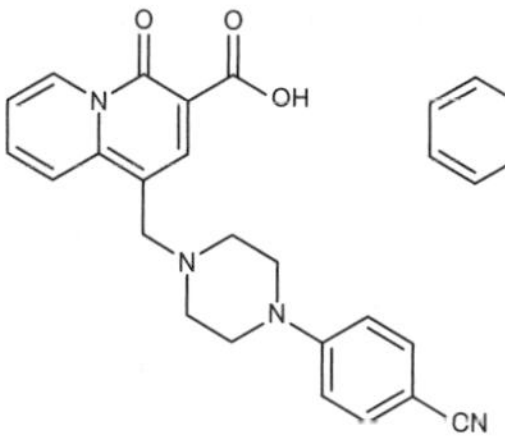
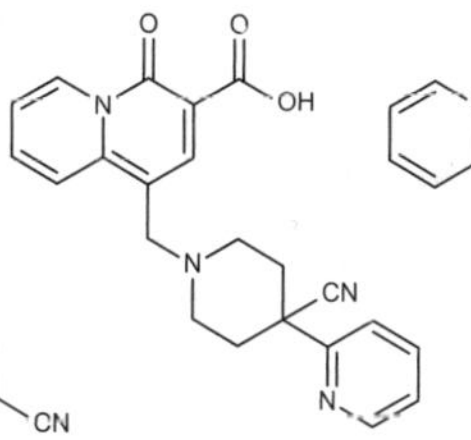
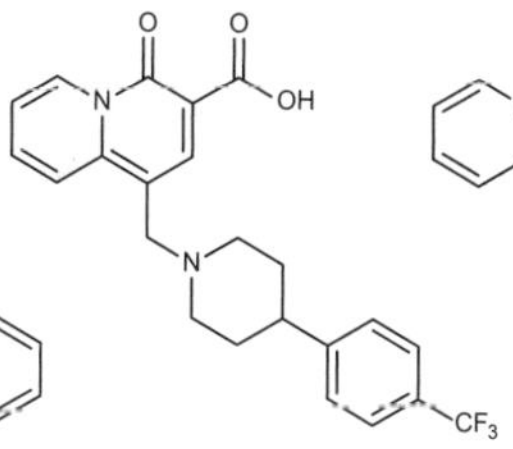
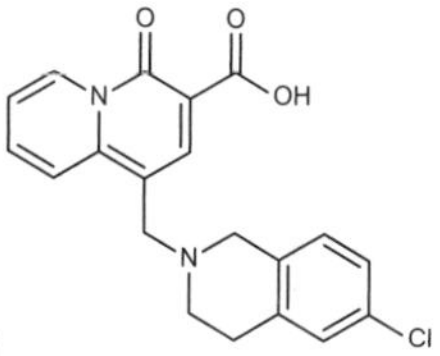

19 **20** **21** **22**

The exact reason for the *in vitro–in vivo* disconnect is not obvious given the very subtle molecular modification between a piperidine or piperazine/4-cyanopiperidine. In terms of their overall profiles, they were similar or more potent than **19** and **20** in a Gαs-coupled Ca^{2+} FLIPR assay and also exhibited similar pharmacological behavior in their ability to sensitize the response to varying doses of acetylcholine in the presence of a fixed concentration of PAM. Last, **21** and **22** were also evaluated for activity in the β-arrestin pathway, which recruits different signaling proteins than the Gαs-coupled Ca^{2+} release measured by the FLIPR assay, in the presence and absence of an EC_{15} concentration of ACh. Both PAMs behaved comparably in both FLIPR and β-arrestin assays relative to **19** and **20**, suggesting that possible different effects of these compounds on these specific signaling pathways are not responsible for the lack of *in vivo* activity. Thus, in addition to keeping a keen eye toward primary and secondary screens, one should be aware of potential *in vitro–in vivo* divide that may occur during optimization.

6. ASSESSING ALLOSTERIC SITE OCCUPANCY THROUGH RADIOLIGANDS AND PET AGENTS

As allosteric modulators represent a novel strategy of producing drug action, it is imperative to count on tools to directly measure receptor engagement, thus enhancing survival of new molecular entities through the discovery and development lifecycle.[42] Different allosteric modulators can act not only at multiple distinct, overlapping sites but also at non-overlapping sites. For compounds acting in a fully competitive manner at a defined site, PET occupancy has a predictable relationship with efficacy. Thus, a critical issue in developing a successful allosteric PET ligand is to confirm that it acts in a fully competitive manner at a defined site. Indeed, PET-based occupancy using $[^{18}F]$-FPEB (**23**) correlates with efficacy.[37]

In spite of challenges encountered with the design of allosteric modulators, a number of allosteric PET ligands have reached clinical testing and proof-of-concept studies and are becoming key translational tools to determine receptor occupancy. Indeed, allosteric PET agents such as [^{18}F]-FPEB (**23**), [^{11}C]-ABP688 (**24**), and [^{18}F]-SP203B (**25**) have already been successfully used in the clinic to assess receptor occupancy at the mGlu5 receptor.[43]

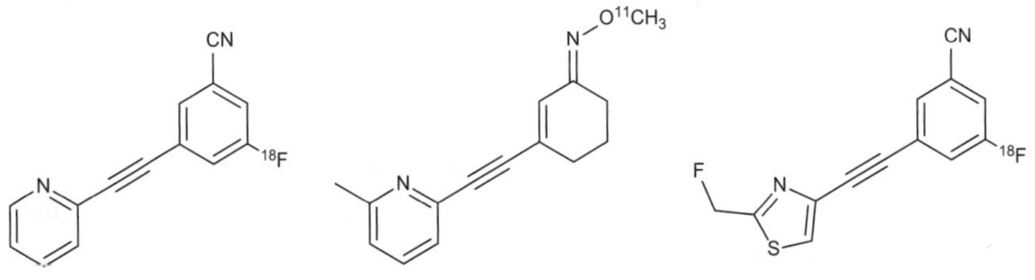

23 **24** **25**

7. MEDICINAL CHEMISTRY STRATEGIES FOR ALLOSTERIC LIGANDS

a. Attention during hit selection is vital. Do not take SAR for granted, and do explore chemical tractability, ideally using libraries. Physicochemical properties of initial hits are important, but are not the sole predictor of chemical tractability, so one should not prioritize hits based only on ligand metrics.

b. Consider stimulus bias and functional switches in your hit selection. Thoroughly profile hits in binding and functional GPCR screens. Preferably, work with chemotypes not showing the PAM⇆NAM switch. It is generally possible to "lock" the functional switch to a given chemotype.

c. Do invest in secondary screens early on. For example, if the primary screen is a calcium mobilization assay, a GTPγ^{32}S-binding assay would be helpful to confirm the functional activity of the different series under investigation. In addition, *in vitro* screens in native tissue (e.g., brain slice electrophysiology) will go a long way to support the biological activity of the compounds under study.

d. After every few rounds of SAR iteration, confirm that the pharmacology remains unchanged. Compounds with very similar profiles *in vitro*, and for DMPK *in vivo*, may behave significantly differently during studies using *in vivo* models.

e. For *in vivo* imaging work and *ex vivo* receptor occupancy studies, it is preferred to work with a radioligand in the same chemotype. Candidate radioligand compounds should be thoroughly characterized to support the competitive nature of their interactions with the drug candidate compounds.

f. SAR at these targets tends to be nonlinear. The "rank order" of substituents in a given series may not translate identically to a different (yet very closely related) chemotype leading to cases of "nonadditive" SAR. Whenever possible, select "islands" of substituents leading to good compounds, and develop SAR using libraries, even with very similar cores.

g. When interpreting results from *in vivo* or metabolism-enabled *in vitro* tests, consider the potential presence of metabolites, which may have opposite pharmacology and confound biological readouts.

8. CONCLUSIONS

GPCR allosteric ligand design is at a stage where scholarship is still developing, and descriptive reports far exceed detailed mechanistic and structural understanding of this phenomenon.[44] Today, designing allosteric ligands presents a number of challenges beyond those typically faced with their orthosteric counterparts. In addition to the usual physicochemical, pharmacokinetic and safety/toxicology optimization, and translation into the clinical setting, knowledge of how chemical structure impacts attributes such as differential modulation of ligand affinity and efficacy, probe dependency, and functional selectivity is currently developing.

At the same time, allosteric ligands hold the promise of improved molecular selectivity, tissue selectivity, and pathways specificity, all of which may contribute to realizing the vision of personalized medicine. Naturally, this requires profound knowledge of very complex biological systems. Thus, medicinal chemists have before us a unique opportunity to add value to drug discovery programs by embracing collaborations in disciplines laying at the interface between chemistry, *in vitro* pharmacology, protein dynamics, and molecular biology.

REFERENCES

(1) Jacoby, E.; Bouhelal, R.; Gerspacher, M.; Seuwen, K.A. *ChemMedChem* **2006**, *1*, 761.
(2) Hopkins, A.L.; Groom, C.R. *Nat. Rev. Drug Discov.* **2002**, *1*, 727.
(3) Ballesteros, J.; Ransom, J. *Drug Discov. Today Ther. Strateg.* **2006**, *3*, 445.
(4) May, L.T.; Leach, K.; Sexton, P.M.; Christopoulos, A. *Annu. Rev. Pharmacol. Toxicol.* **2007**, *47*, 1.
(5) Melancon, B.J.; Hopkins, C.R.; Wood, M.R.; Emmitte, K.A.; Niswender, C.M.; Christopoulos, A.; Conn, P.J.; Lindsley, C.W. *J. Med. Chem.* **2012**, *55*, 1445.
(6) Valant, C.; Lane, J.R.; Sexton, P.M.; Christopoulos, A. *Annu. Rev. Pharmacol. Toxicol.* **2012**, *52*, 153.

(7) Kenakin, T.P. *Br. J. Pharmacol.* **2012**, *165*, 1659.
(8) Gregory, K.J.; Sexton, P.M.; Christopoulos, A. *Curr. Protoc. Pharmacol.* **2010**, *51*, 1. http://dx.doi.org/10.1002/0471141755.ph0121s51 Chapter 1: Unit 1.21.
(9) Leach, K.; Sexton, P.M.; Christopoulos, A. *Curr. Protoc. Pharmacol.* **2011**, *52*, 1. http://dx.doi.org/10.1002/0471141755.ph0122s52 Chapter 1: Unit 1.22.
(10) Davey, A.E.; Leach, K.; Valant, C.; Conigrave, A.D.; Sexton, P.M.; Christopoulos, A. *Endocrinology* **2012**, *153*, 1232.
(11) Garcia-Perez, J.; Rueda, P.; Alcami, J.; Rognan, D.; Arenzana-Seisdedos, F.; Lagane, B; Kellenberger, E. *J. Biol. Chem.* **2011**, *286*, 33409.
(12) Nussinov, R.; Tsai, C.-J. *Curr. Pharm. Des.* **2012**, *18*, 1311.
(13) Sheffler, D.J.; Gregory, K.J.; Rook, J.M.; Conn, P.J. *Adv. Pharmacol.* **2011**, *62*, 37.
(14) Pratt, S.D.; Mezler, M.; Geneste, H.; Bakker, M.H.M.; Hajduk, P.J.; Gopalakrishnan, S.M. *Comb. Chem. High Throughput Screen.* **2011**, *14*, 631.
(15) Rydzewski, R.M. *Real World Drug Discovery*, 1st ed.; Elsevier: Amsterdam, 2008; p. 220.
(16) Tarcsay, A.; Nyíri, K.; Keserű, G.M. *J. Med. Chem.* **2012**, *55*, 1252.
(17) Hopkins, C.R.; Lindsley, C.W.; Niswender, C.M. *Future Med. Chem.* **2009**, *1*, 501.
(18) For a recent example of a switch of a non-peptide AT2 receptor agonist to structurally related antagonists see: Murugaiah, A.M.S.; Wu, X.; Wallinder, C.; Mahalingam, A.K.; Wan, Y.; Skoüld, C.; Botros, M.; Guimond, M.-O.; Joshi, A.; Nyberg, F.; Gallo-Payet, N.; Hallberg, A.; Alterman, M. *Bioorg. Med. Chem. Lett.* **2010**, *18*, 4570.
(19) For an early review on agonist/antagonist switches see: Sugg, E.E. *Annu. Rep. Med. Chem.* **1997**, *32*, 277.
(20) Wood, M.R.; Hopkins, C.R.; Brogan, J.T.; Conn, P.J.; Lindsley, C.W. *Biochemistry* **2011**, *50*, 2403.
(21) Sams, A.G.; Mikkelsen, G.K.; Brodbeck, R.M.; Pu, X.; Ritzén, A. *Bioorg. Med. Chem. Lett.* **2011**, *21*, 3407.
(22) Williams, R.; Manka, J.T.; Rodriguez, A.L.; Vinson, P.N.; Niswender, C.M.; Weaver, C.D.; Jones, C.K.; Conn, P.J.; Lindsley, C.W.; Stauffer, S.R. *Bioorg. Med. Chem. Lett.* **2011**, *21*, 1350.
(23) Bhagwanth, S.; Mishra, S.; Daya, R.; Mah, J.; Mishra, R.K.; Johnson, R.L. *ACS Chem. Neurosci.* **2012**, *3*, 274. http://dx.doi.org/10.1021/cn200096u.
(24) Trzaskowski, B.; Latek, D.; Yuan, S.; Ghoshdastider, U.; Debinski, A.; Filipek, S. *Curr. Med. Chem.* **2012**, *19*, 1090.
(25) Schann, S.; Mayer, S.; Franchet, C.; Frauli, M.; Steinberg, E.; Thomas, M.; Baron, L.; Neuville, P. *J. Med. Chem.* **2010**, *53*, 8775.
(26) Valant, C.; Felder, C.C.; Sexton, P.M.; Christopoulos, A. *Mol. Pharmacol.* **2012**, *81*, 41.
(27) Koole, C.; Wootten, D.; Simms, J.; Valant, C.; Sridhar, R.; Woodman, O.L.; Miller, L.J.; Summers, R.J.; Christopoulos, A.; Sexton, P.M. *Mol. Pharmacol.* **2010**, *78*, 456.
(28) Antflick, J.E.; Vetiska, S.; Baizer, J.S.; Yao, Y.; Baker, G.B.; Hampson, D.R. *Brain Res.* **2009**, *1300*, 1.
(29) Fribourg, M.; Moreno, J.L.; Holloway, T.; Provasi, D.; Baki, L.; Mahajan, R.; Park, G.; Adney, S.K.; Hatcher, C.; Eltit, J.M.; Ruta, J.D.; Albizu, L.; Li, Z.; Umali, A.; Shim, J.; Fabiato, A.; MacKerell, A.D., Jr.; Brezina, V.; Sealfon, S.C.; Filizola, M.; González-Maeso, J.; Logothetis, D.E. *Cell* **2011**, *147*, 1011.
(30) Delille, H.K.; Becker, J.M.; Burkhardt, S.; Bleher, B.; Terstappen, G.C.; Schmidt, M.; Meyer, A.H.; Unger, L.; Marek, G.J.; Mezler, M. *Neuropharmacology* **2012**, *62*, 2184. http://dx.doi.org/10.1016/j.neuropharm.2012.01.010.
(31) Reiter, E.; Ahn, S.; Shukla, A.K.; Lefkowitz, R.J. *Annu. Rev. Pharmacol. Toxicol.* **2012**, *52*, 179.
(32) Digby, G.J.; Conn, P.J.; Lindsley, C.W. *Curr. Opin. Drug Discov. Devel.* **2010**, *13*, 587.

(33) Kenakin, T.; Watson, C.; Muniz-Medina, V.; Christopoulos, A.; Novick, S. *ACS Chem. Neurosci.* **2012**, *3*, 193. http://dx.doi.org/10.1021/cn200111m.
(34) Kenakin, T.P. *ACS Chem. Biol.* **2009**, *4*, 249.
(35) Langmead, C.J. *Methods Mol. Biol.* **2011**, *746*, 195.
(36) Noetzel, M.J.; Rook, J.M.; Vinson, P.N.; Cho, H.; Days, E.; Zhou, Y.; Rodriguez, A.L.; Lavreysen, H.; Stauffer, S.R.; Niswender, C.M.; Xiang, Z.; Daniels, J.S.; Jones, C.K.; Lindsley, C.W.; Weaver, C.D.; Conn, P.J. *Mol. Pharmacol.* **2011**, *81*, 120.
(37) Conn, P.J.; Stauffer, S.R.; Zhou, S.; Manka, J.; Williams, R.; Noetzel, M.J.; Gregory, K.J.; Vinson, P.; Niswender, C.M.; Jones, C.K.; Steckler, T.; MacDonald, G.; Lindsley, C.W. 7th International Meeting in Metabotropic Glutamate receptors. Taormina, Italy, October 2–7, **2011**.
(38) Knudsen, L.B.; Hastrup, S.; Underwood, C.R.; Wulff, B.S.; Fleckner, J. *Regul. Pept.* **2012**, *175*, 21. http://dx.doi.org/10.1016/j.regpep. 2011.12.006.
(39) Kuduk, S.D.; Chang, R.K.; Di Marco, C.N.; Ray, W.J.; Ma, L.; Wittmann, M.; Seager, M.A.; Koeplinger, K.A.; Thompson, C.D.; Hartman, G.D.; Bilodeau, M.T. *ACS Med. Chem. Lett.* **2010**, *1*, 263.
(40) Kuduk, S.D.; Chang, R.K.; Di Marco, C.N.; Pitts, D.R.; Greshock, T.J.; Ma, L.; Wittmann, M.; Seager, M.; Koeplinger, K.A.; Thompson, C.D.; Hartman, G.D.; Bilodeau, M.T.; Ray, W.J. *J. Med. Chem.* **2011**, *13*, 4773.
(41) Kuduk, S.D.; Chang, R.K.; Di Marco, C.N.; Ray, W.J.; Ma, L.; Wittmann, M.; Seager, M.; Koeplinger, K.A.; Thompson, C.D.; Hartman, G.D.; Bilodeau, M.T. *Bioorg. Med. Chem. Lett.* **2011**, *21*, 1710.
(42) Morgan, P.; Van Der Graaf, P.H.; Arrowsmith, J.; Feltner, D.E.; Drummond, K.S.; Wegner, C.D.; Street, S.D.A. *Drug Discov. Today* **2012**, *17*, 419. http://dx.doi.org/10.1016/j.drudis.2011.12.020.
(43) Mu, L.; Schubiger, P.A.; Ametamey, S.M. *Curr. Top. Med. Chem.* **2010**, *10*, 1558.
(44) Canals, M.; Sexton, P.M.; Christopoulos, A. *Trends Biochem. Sci.* **2011**, *36*, 663. http://dx.doi.org/10.1016/j.tibs.2011.08.005.

CHAPTER TWENTY-NINE

Progress in the Development of Non-ATP-Competitive Protein Kinase Inhibitors for Oncology

Campbell McInnes
Pharmaceutical and Biomedical Sciences, South Carolina College of Pharmacy, University of South Carolina, Columbia, South Carolina, USA

Contents

1. INTRODUCTION

Alternative approaches for inhibitor development in targeting sites other than the ATP cleft[1] are increasingly being pursued in the search for new therapeutics targeting protein kinases.[2,3] While approved kinase inhibitor drugs offer benefit in cancer treatment, further advances are required to avoid off-target-related toxicities, effect tumor selective cell killing, and improve survival rates in oncology patients. In addition, lack of target selectivity may be a significant issue in chronic diseases where drugs are administered over a long period. Protein–protein interactions involved in kinase regulation and substrate recognition offer high potential for selectivity and avoid decreased efficacy as a result of competition with high intracellular ATP concentrations. These interfaces, however, are more challenging for inhibitor discovery. We discuss several examples where regulatory and substrate-binding sites have been successfully blocked and therefore offer potential for further drug discovery and development.

Annual Reports in Medicinal Chemistry, Volume 47
ISSN 0065-7743
http://dx.doi.org/10.1016/B978-0-12-396492-2.00029-1

We also discuss the latest developments in methodology for drug discovery that can be applied to the identification of protein–protein interaction inhibitors, including the validation of the REPLACE strategy which has been successfully applied to two inhibitors of protein kinase oncology targets, CDK2/cyclin A and PLK1.[4]

2. INHIBITION OF CYCLIN-DEPENDENT KINASES THROUGH THE CYCLIN GROOVE

The cyclin groove of the cell cycle CDKs is involved in recruitment of substrates prior to phosphorylation and thus provides a target for inhibition of kinase activity in antitumor drug discovery. Certain CDK substrates including the Rb and E2F proteins must undergo cyclin binding before phosphorylation, and therefore, inhibitors of the cyclin groove also block substrate-specific kinase activity. In contrast to CDK inhibitors targeting the ATP-binding site which have generally not fared well in clinical trials, this non-ATP-competitive approach can be used to generate highly selective and cell cycle-specific CDK inhibitors that have reduced inhibition of transcription mediated through CDK7 and 9.[2,5]

CDK2/cyclin A,E and CDK4/cyclin D1 are responsible for regulation of G1 and S phase of the cell cycle and phosphorylate mainly substrates containing the cyclin-binding motif (CBM).[6–8] The cyclin groove interacts with and recruits specific substrates prior to phosphorylation, and endogenous CDK inhibitors such as $p21^{WAF1}$ and $p27^{KIP1}$ block this site in addition to the ATP cleft (Fig. 29.1). Inactivation of CDK inhibitory proteins (i.e., through mutation) can be a transforming event and leads to cells bypassing the G1 checkpoint. CBM peptides that bind to the cyclin groove have been demonstrated to act as potent kinase inhibitors through blocking substrate recruitment and when administered in cell-permeable form have significant antitumor efficacy. Lack of specificity for cell cycle versus transcriptional CDKs is a major potential drawback of ATP-competitive CDK inhibitors, and toxicities arising from inhibition of CDKs 7, 8, and 9 may be the cause of undesirable effects observed with clinically investigated inhibitors. Since only A-, D-, and E-type cyclins possess a functional cyclin groove, it is possible to generate cell cycle-specific CDK inhibitors through non-ATP-competitive approaches.

Extensive structure–activity relationship data on the molecular determinants and binding groups have been obtained for the CBM, and crystallographic analysis of cyclin groove inhibitor (CGI) peptide–cyclin complexes

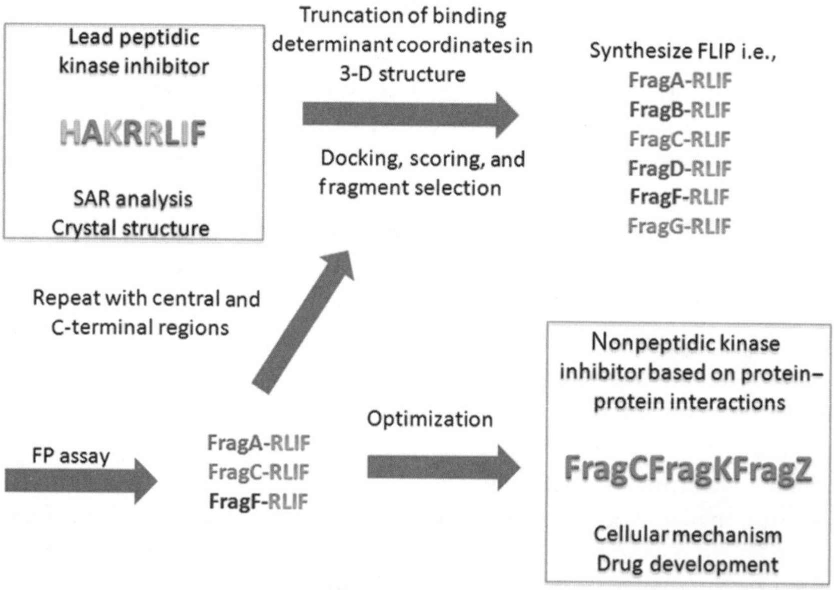

Figure 29.1 Overview of the REPLACE strategy for generation of more drug-like protein–protein interaction inhibitors. (See Color Plate 29.1 in Color Plate Section.)

shows that it binds in an extended conformation and that three subsites are important. These are (a) a primary hydrophobic pocket interacting with two lipophilic peptide side chains (typically a Leu and a Phe), (b) a secondary hydrophobic pocket contacting an Ala or Val side chain, and (c) a bridging site providing complementarity with basic residues of the peptide.

In efforts to develop more drug-like CGIs, peptidomimetic approaches have been applied.[9–12] Recently, the SAR of the p21 CBM was explored. After combination of individual substitutions shown to result in potency increases, the largely non-natural amino acid containing sequence, AAURSLNpfF was identified which inhibited CDK2/cyclin A with low nanomolar activity.[12] Truncation studies of the p21 sequence generated pentapeptide CGIs that were scaffolds for further drug development. In this context, 4-fluoro substitution of the C-terminal phenylalanine residue led to enhancement of binding affinity. A 3-chloro substituent on the Phe side chain of the peptide PVKRRLFG[13] also results in similar potency gains.

Another approach to developing compounds that are more pharmaceutically appropriate involved a rigidification strategy through the synthesis of peptides incorporating a cyclic restraint. Cyclization of a lysine side chain to the C-terminal glycine and thus mimicking an intermolecular hydrogen bond observed in the p27/cyclin A crystal structure (1JSU) led to a decrease in the entropic cost of binding thereby improving *in vitro* potency. Crystal structure of these peptides with cyclin A revealed their binding mode and how this information might be exploited in future design efforts.[14]

In order to develop a general method that could be applied to protein–protein interactions, the REPLACE strategy was conceived and exemplified against CDK2/cyclin A as a target (Fig. 29.1). This method utilizes computational and synthetic chemistry so as to identify fragment alternatives for individual binding determinants.[4] In the case where SAR has previously been established, the peptide is then truncated from the N- or C-terminus in the binding site in the crystal structure. Small drug-like fragments (containing requisite functionality for conjugation to the truncated peptide) are then docked into the cavity previously containing the binding determinant. These fragments are then selected for addition to the truncated peptide sequence after prioritization through scoring functions. In the proof of concept for this approach, a library of 74 fragment ligated inhibitory peptides (FLIPs) was synthesized after *in silico* prediction and tested in a fluorescence polarization-binding assay for activity against CDK2/cyclin A. As a result, 19 hits were obtained with the most active compounds binding with higher affinity relative to the intact pentapeptide.

The discovery of fragment alternatives for the critical N-terminal arginine-binding determinant (p21 peptide HAKRRLIF, structure A, Fig. 29.2) therefore resulted in validation of REPLACE as a strategy. The most successful N-terminal capping group obtained was a phenyltriazole scaffold contacting the secondary hydrophobic subsite occupied by Ala2. Further application of REPLACE to the C-terminal phenylalanine residue (again highly sensitive to replacement) led to the identification of bis-aryl ether capping groups that appropriately mimicked the aromatic side chain in contacting the larger hydrophobic pocket.[4] Crystallographic studies of the FLIPs provided insights for structure-guided design. Overall, the REPLACE strategy was demonstrated to be a viable approach for conversion of peptide inhibitors into more stable and drug-like compounds, and proof of concept was obtained for non-ATP-competitive CDK2 inhibitors through the cyclin groove.

The REPLACE method has certain advantages over conventional approaches of fragment-based design. Potential capping groups are tested while conjugated to truncated peptide sequences, and therefore a successful FLIP recovers binding of the intact native peptide. The peptidic portion acts as an anchor and circumvents the requirement for a highly sensitive detection method in contrast to conventional fragment-based discovery (FBD)

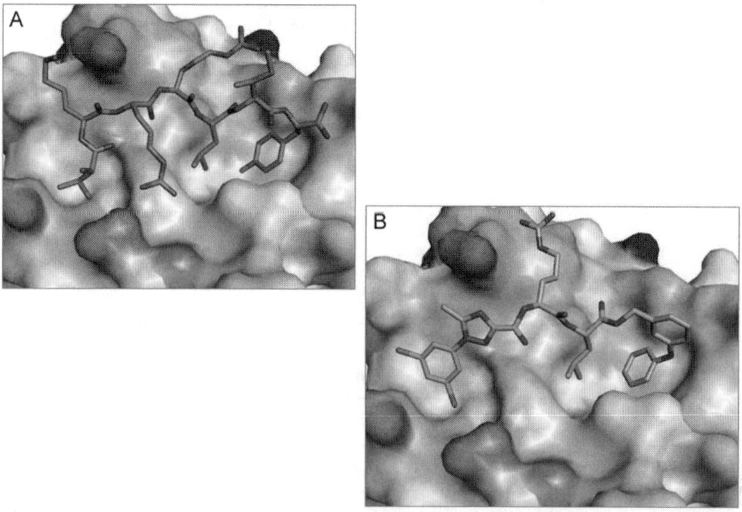

Figure 29.2 REPLACE-mediated conversion of the cyclin groove inhibitor HAKRRLIF (A) into an N- and C-terminally capped dipeptide (B). (See Color Plate 29.2 in Color Plate Section.)

which necessitates detection of millimolar binding affinity.[15] An additional requirement of FBD guided by X-ray crystallography is the use of fragments with high levels of solubility. The REPLACE strategy evaluates fragments as peptide hybrids thus providing a solubility handle for the inhibitor. Optimization of the identified fragments is completed while in FLIP context therefore further avoiding the need for expensive and difficult methods for detecting binding.

REPLACE optimization has been pursued through a library of heterocyclic capping groups based on the phenyltriazole system previously identified,[4] which explored the SAR for substitution of the phenyl ring and replacement of the triazole with other 5-membered heterocycles.[16] FLIPs capped with pyrazole, furan, pyrrole, and imidazole rings determined that the phenyl 1,2,4-triazole Ncaps were the most active of the group (Table 29.1). Although weaker in affinity, the pyrazole scaffold provided important SAR information regarding the secondary pocket and specifically that the 3-chloro phenyl analog (**5764**) was shown to be the most potent in the CDK2/cyclin A context. The 4-chlorophenyl derivative in the phenyltriazole context (**5774**) was shown to be the first peptide–small molecule hybrid identified with significant potency on CDK4/cyclin D1 and had increased potency relative to the RRLIF pentapeptide. The greater activity of the phenyltriazole system can be attributed to the increased

Table 29.1 Structure activity of phenylheterocyclic N-terminal partial ligand alternatives

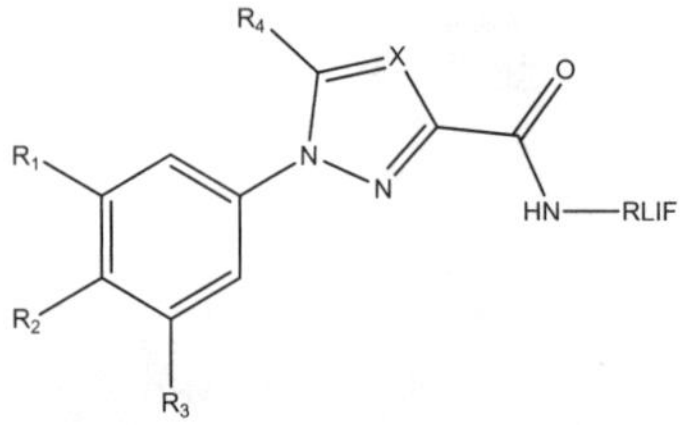

	SCCP ID	R1	R2	R3	R4	X	CDK2/cyclin A IC_{50} (μM)	CDK4/cyclin D1 IC_{50} (μM)
Triazole	5843	H	H	H	H	N	16.2±3	48.7
	5773	Cl	H	Cl	CH_3	N	4±0.6	27
	5774	H	Cl	H	CH_3	N	11.5±3.3	11.3
Pyrazole	5762	H	H	H	CH_3	C	40.3±6.5	53.8
	5763	Cl	H	Cl	CH_3	C	21.8±13.7	100–180
	5764	Cl	H	H	CH_3	C	11.9±2.0	45
	5771	F	H	H	CH_3	C	29.6±12.2	69.6
	5765	H	Cl	H	CH_3	C	33.7±8.1	49

H-bond strength relative to the other heterocycles and a potential bridging water molecule, which is not present in the phenylpyrazole complex.

Validation of fragment alternatives identified as individual capping groups when combined into a single inhibitor is a key aspect of REPLACE since it is possible that conformational changes result after modification of peptide determinants. Optimized N-capping and C-capping groups were combined into individual molecules (structure B, Fig. 29.2) in order to accomplish this and revealed that 3-phenoxybenzylamine had decreased activity when combined with the 3,5-DCPT-Arg-Leu, relative to the peptide context. It was found, however, that addition of halogens onto the phenoxy substituent contacting the primary hydrophobic pocket resulted in increased activity.[17] Either a 3-fluoro or a 4-fluoro substituted bis-aryl ether had enhanced potency, thus replicating the corresponding increase observed in the peptide context.[12] These data indicate that potency lost by combining partial ligand alternatives can be recovered through subsequent optimization.

Another approach to CDK2/cyclin A inhibition through the cyclin groove included combination of fragment-like structures with peptidomimetic design and synthesis. The octapeptide, PVKRRLFG was employed as the starting point, and consistent with cyclin groove interactions, Arg, Leu, and Phe were delineated as critical determinants.[13] The identification of a smaller tetrapeptide lead was optimized through design, and further rigidification of its structure resulted in aminothiazole capping group variants replacing the critical arginine while recovering binding lost through truncation.[13] An optimized CGI was generated with 500-fold increase in activity through reduction of conformational freedom at the C-terminal group using a *trans*-2-arylcyclohexyl combined with the aminothiazole Ncap (compound **1**, Fig. 29.3, 0.021 μM). A proline mimetic was used to replace the second arginine of PVKRRLFG and resulted in peptidomimetic inhibitor with greater drug-likeness as a trade-off for slightly decreased potency (compound **2**, Fig. 29.3, 0.04 μM). In summary, this work led to the generation of a potent noncharged CDK2/cyclin A inhibitor containing fewer rotatable bonds and therefore making it considerably more drug-like.

Although highly potent peptide and small-molecule inhibitors of CDK2/cyclin A substrate recruitment have been generated, much less attention has been focused on the cyclin groove of the CDK4/cyclin D1, a validated anticancer drug target. Structures of the cyclin groove of cyclin D in complex with CGIs, and *in vitro* binding of peptide analogs were examined in detail in order to determine the basis for peptide affinity and selectivity.[18] Cyclin groove comparison of cyclin A2 and D1 resulting from these studies determined that residue substitutions result in significant differences in the acidic region contacting the critical arginine and in the volume of the

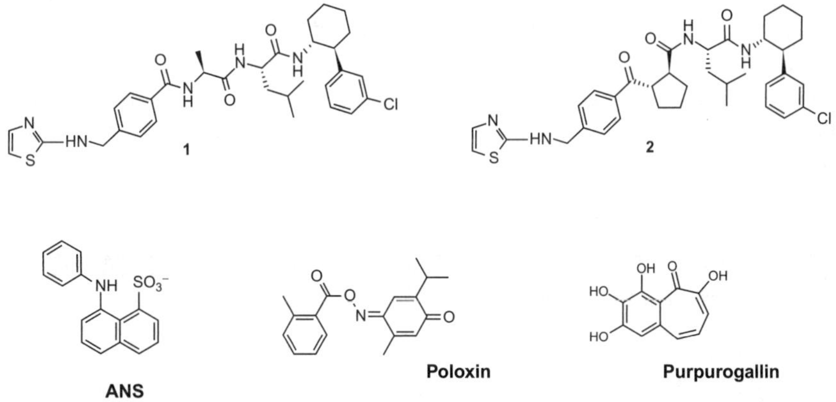

Figure 29.3 Chemical structures of non-ATP-competitive CDK and PLK inhibitors.

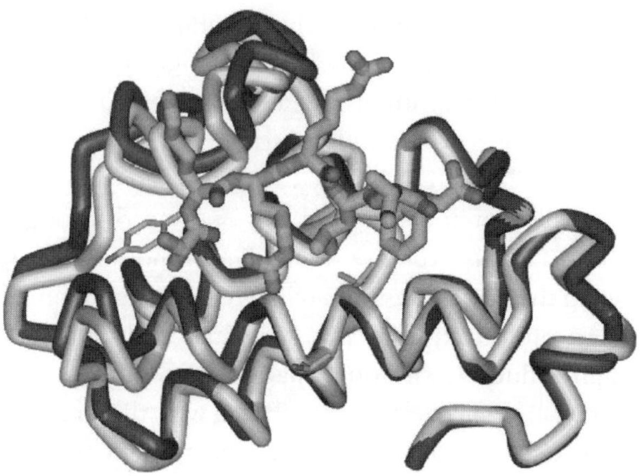

Figure 29.4 The cyclin-binding grooves of cyclin A (blue ribbon) and cyclin D1 (yellow ribbon) are compared through overlay. Unique residues of cyclin D1 shown in orange are Tyr127, Thr62, and Val60 (left to right). (See Color Plate 29.4 in Color Plate Section.)

primary and secondary lipophilic pockets (Fig. 29.4). Structural differences were then further investigated through incorporation of non-natural phenylalanine replacements into p21 and p107 octapeptide contexts. Of special note, a 3-thienylalanine substitution was effective and in addition, cyclohexylalanine led to increased binding for cyclin D1 relative to cyclin A2. Subsequently, the interactions of the CDKI, $p27^{KIP1}$, were modeled with cyclin D1 and provided significant insights into the endogenous inhibition of CDK4. Unique features of the CBG of cyclin D1 were observed including an extension to the primary pocket which can potentially be exploited in the design of non-ATP-competitive CDK inhibitors.

As a whole, these studies validate the cyclin grooves of CDK2/cyclin A and CDK4/cyclin D as druggable sites in order to create potent and cell cycle-specific non-ATP-competitive inhibitors and generate considerable insights into how these compounds can be modified through structure-guided design in order to develop chemical biology probes and potential therapeutics based on selective inhibition of CDK4/cyclin D activity.

3. ALTERNATIVE STRATEGIES IN THE DEVELOPMENT OF NON-ATP-COMPETITIVE CDK INHIBITORS

Another strategy for the development of antitumor therapeutics based on CDK inhibition and beyond targeting substrate recruitment sites is to block the association of the cyclin positive regulatory and catalytic subunits and therefore prevent formation of the activated CDK–cyclin complex.

This should theoretically deliver similar results as can be achieved with inhibition of the ATP-binding site and the cyclin groove. Subunits that form protein complexes generally bind too strongly to be competitively dissociated. CDKs and cyclins belong to a class of proteins that exist in both monomeric and heterodimeric form in response to their environment and thus have relatively high K_d values (1 μM to 1 nM).[19] The CDK2–cyclin A complex has a dissociation constant of approximately 50 nM[20] and therefore is in a feasible affinity range for complex dissociation induced by small molecules. Toward this end, an all d-amino acid hexapeptide (NBI1) has been identified that interferes with the formation of the CDK2–cyclin A complex. Inhibition of kinase activity was demonstrated and direct binding to cyclin A was revealed using surface plasmon resonance.[21] In addition, a cell-permeable derivative of NBI1 resulted in apoptotic cell death and antiproliferative effects in tumor cell lines. This study demonstrates the proof of concept that protein–protein interactions regulating CDK activation can be successfully targeted and may be useful in the generation of non-ATP-competitive cell cycle-based cancer therapeutics.

Further studies on the disruption of the CDK–cyclin interaction were undertaken through probing the binding of the fluorophore, 8-anilino-1-naphthalene sulfonate (ANS; Fig. 29.3) with CDK2 and cyclin A using fluorescence spectroscopy and protein crystallography.[22] Results showed that ANS binds to CDK2 with a K_d of 37 μM. Additional experiments revealed that cyclin A was able to displace the fluorescent dye from CDK2 with an EC_{50} value of 0.6 μM. Insights into this novel observation were provided through cocrystallization of monomeric CDK2 with ANS. Other complexes with ATP-competitive inhibitors determined that two ANS molecules bound in a significant cleft extending from the C-helix and were positioned beside each other. Large conformational changes in the catalytic subunit are induced by the binding of ANS and thereby allosterically blocking the binding of CDK2 to cyclin A. The ANS binding pocket is therefore a potentially druggable binding site for inhibitors acting on CDK2/cyclin A through a novel allosteric mechanism.

4. POLO-BOX DOMAIN INHIBITORS OF POLO-LIKE KINASES

The two domains of PLKs include the kinase catalytic region and the polo-box domain (PBD) with the latter being required for subcellular localization and many PLK functions.[23,24] The role of the PBD in mitosis and potential for oncology have been demonstrated through administration of a permeabilized PB1 fragment to tumor cells which resulted in mitotic arrest, misaligned chromosomes, and multiple centrosomes.[25] Competition between the PB fragment and endogenous PLK1 for binding to substrates is the probable cause for the observed effects.[26–28] A phosphopeptide consensus sequence has been demonstrated to interact with the region containing the two PBs (residues 326–603), and further investigation determined that a number of important mitotic proteins, including CDC25C, contain this motif and avidly bind to the PBD of PLK1.

Several recent crystallographic structures of the human PLK1 PBD in complex with phosphopeptide inhibitors have provided insights into the structural basis for peptide binding which occurs in a shallow groove at the interface of the two PBs. These PBD ligands interact through electrostatic (Gln^2, Ser^3, and Thr^4) and van der Waals contacts (Met^1, Leu^6) and potentially through bridging water molecules.[29–32] Within the PBD, the two PBs were shown to have an almost identical fold despite having relatively low sequence identity.

Insights into PBD inhibitor binding obtained from these crystal structures, in addition to the biological relevance of the PBD, suggest these are druggable interfaces that can be exploited in design of non-ATP-competitive inhibitors (Fig. 29.5). Further rationale for pursing the PBD as a drug target is provided through precedence for similar interfaces found in adaptor proteins that bind to tyrosine kinases.[33] For example, Src-homology (SH2 and SH3) domains recognize phosphopeptide motifs such as pTyr motifs.[34,35]

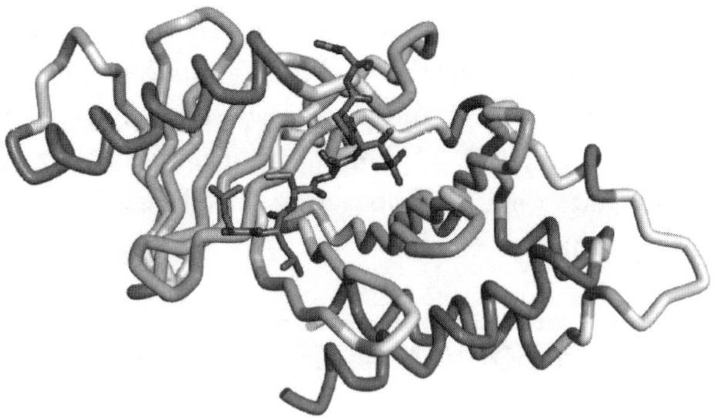

Figure 29.5 Ribbon representation of the polo-box domain showing the CDC25c PBD crystal structure (3BZI) with the peptide bound at the intersection of the two polo boxes. The Leu-Leu-Cys tripeptide replaced by fragment alternatives is on the left side of the figure. (See Color Plate 29.5 in Color Plate Section.)

Additional impetus for developing inhibitors of the PBD has come to light in recent studies. Evidence has emerged that PLK3 acts directly or indirectly as a tumor suppressor in that it is activated as a result of DNA damage and its inhibition stimulates proliferation.[36,37] These findings have been coupled with recent data showing that minimized phosphopeptides bind potently and selectively to the PLK1 PBD and not to the PLK2 and 3 phosphopeptide-binding site.[38–42] In contrast to this, many of the reported ATP-competitive PLK inhibitors, although selective for PLKs, bind individual isoforms with similar affinity. As a result of the roles of PLK3 in countering PLK1, isotype selective compounds will be important. Hence, drugs specifically targeting the PLK1 would obviate effects of blocking PLK3 since this may contribute to tumorigenesis in normal cells.

In attempts to discover small-molecule PBD inhibitory compounds, Reindl and colleagues[43] have applied high-throughput screening (HTS) approaches to the PBD although with limited success. A library of 22,461 small molecules was screened in a fluorescence polarization competitive-binding assay with poloxin (Fig. 29.3, IC_{50} of 4.8 ± 1.3 μM), thymoquinone (apparent IC_{50} of 1.14 ± 0.04 μM), and poloxipan[43,44] being identified. Their low activity, limited scope for optimization, and redox potential of these series suggest that better compounds are needed. Although purpurogallin[45] (Fig. 29.3, $IC_{50} = 500$ nM), another compound identified by HTS methods, has improved activity toward the PBD of PLK1, it would require substantial modification in order to improve drug-like properties. The modest success of HTS attempts in the search for PBD inhibitory compounds leads to the conclusion that alternate drug discovery strategies would be expedient with this target.

Further to this, REPLACE, which has been described above and validated for CDK2/cyclin A as a drug target, was applied to the PBD to discover small-molecule alternatives and generate non-ATP-competitive inhibitors that are selective for PLK1. The basis for the application of REPLACE in this context was the CDC25c substrate peptide, LLCS[pT]PNGL which is primed by CDK1 phosphorylation, resulting in high affinity binding to the PBD. Prior to efforts using REPLACE, peptide analogs were synthesized and tested in order to delineate the structure–activity relationship of the CDC25c sequence and to further describe the key binding determinants. Further peptide derivatives were explored based on modification of the sequences from PBIP1 (PLHS[pT]AI) proteins which form complexes with the PLK1 PBD.[38] In the first instance, the activity requirement for the phosphothreonine residue was investigated in both peptide contexts. Results demonstrated that isosteric Glu replacements bound with weak affinity, thereby providing information that is useful for inhibitor design. An unexpected and serendipitous result was obtained from this study in the observation that acetylation of the N-terminus is crucial for truncated peptides to bind with high affinity to the PBD and thus provided extended SAR data for PBIP1 peptides previously reported.[38] Further examination of the complex structures for the PBD/PBIP1 peptides revealed that the structural basis for this contribution derives from the close proximity of an arginine residue. In the nonacetylated peptide, the basic guanidinium side chain would repel the positively charged N-terminus compared to the acetylated context where H-bonds to the amide are apparent from the crystal structure. Further SAR from alanine replacement of Leu2 of the PBIP1 sequence indicated a threefold loss in binding. Prior to application of REPLACE, it was established that the hydrophobic tripeptide (LLC to generate S[pT]PNGL) of the Cdc25C sequence is a critical determinant as shown through inactivity of the truncated peptide.

Since the peptide SAR studies generated additional information that would be exploited in the design of nonpeptidic inhibitory compounds, the REPLACE strategy was applied to the N-terminal tripeptide using available crystal structures of peptides bound to the PBD. In the first iterations of REPLACE, the three N-terminal residues were excised from the 3-D structure and subsequently 1800 drug-like carboxylate-containing fragments were docked into the vacated pocket of the receptor. While weak partial ligand alternatives were identified initially, subsequent iterations of REPLACE resulted in derivatized benzoic acid-capped peptides (Table 29.2). As shown in this table, with iterations and optimization of REPLACE, compounds progressed in activity from 320 (1G2-S[pT]PNGL) to less than 10 μM (3G2-S[pT]PNGL). This simple scaffold mimics the interactions of the Leu-Leu-Cys tripeptide within one log of the activity of the endogenous Cdc25C peptide and almost equivalent in potency to the truncated PBIP1 peptides, however, with approximately 1/3 the molecular weight. FLIPs displayed a clear structure–activity relationship and activity below 10 μM. A significant aspect of this series is that it provides sufficient potential diversity for optimization through exploiting interactions observed in the peptide complex structures and SAR data provided in the peptide context. In addition, testing of these compounds in a competitive-binding assay for PLK3 determined that PLK1 selectivity was retained. Further optimization through REPLACE, for example, incorporation of an H-bond

Table 29.2 PLK PBD *in vitro* binding and cellular activity for FLIP compounds (Ncap-S[pT]PNGL)

Abbreviation	N-capping group	PLK1 PBD FP IC_{50} (μM)	PLK3 PBD FP IC_{50} (μM)	Apoptosis at 24 h (30 μM)	Aberrant mitoses at 24 h (30 μM)
1G2-S[pT]PNGL	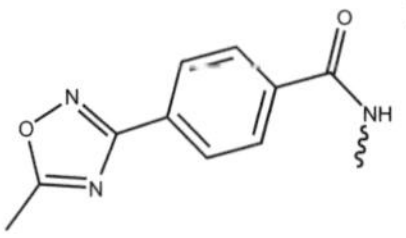	320	ND	ND	ND
2G1-S[pT]PNGL	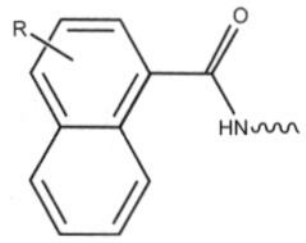	99	ND	ND	ND
3G1-S[pT]PNGL	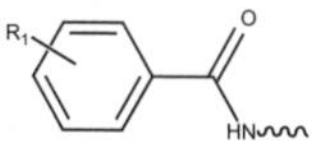	16.5	>600	36.2%	47.9%
3G2-S[pT]PNGL	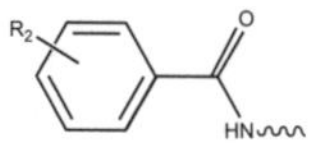	8.6	>600	55.5%	53.8%
Ac-PLH S[pT]A	–	2	>600	34.0%	40.3%

acceptor group to the benzoic acid scaffold in the manner of the critical acetyl group described above could provide enhanced binding affinity. Recent studies of peptide analogs resulted in the serendipitous discovery of a hydrophobic channel that is exploited by lipophilic substituents incorporated on a histidine residue of PBIP sequence.[46] These and more recent compounds resulted in high affinity binding to the PBD.[47] The crystal structures obtained from these experiments suggest that appropriate modification of PBD peptide capping groups would result in similar activity increases, and resulting affinity gains should allow truncation and replacement of the C-terminus through capping groups which mimic the phosphothreonine and C-terminal residues.

As a prelude to the further identification and characterization of nonpeptidic PBD inhibitors, FLIPs and peptides were delivered into tumor cells to determine if phenotypes observed consistent with PLK1 inhibition and specifically through the PBD were obtained.[48] Intracellular administration was achieved through the use of the QQ transfection reagent, and after treatment of HELA cells, a significant reduction of PLK1 localization to centrosomes was observed with both capped and peptidic compounds. Further investigation determined that aberrant mitoses as visualized by mono and multipolar spindles, abnormal chromosome alignment during metaphase (Table 29.2, 3G1 and 3G2-S[pT]PNGL), and cell death through apoptosis occurred in a dose-dependent fashion. Cellular data obtained for third-generation FLIPs indicate that, in spite of lower affinity, peptide–small molecule hybrids have improved antitumor activity relative to purely peptidic inhibitors (Table 29.2, 3G1 and 3G2-S[pT]PNGL). It is possible that the N-terminal capping group increases cellular half-life through improvements in compound stability toward proteolytic enzymes.[49] As a whole, the data acquired for FLIPs discovered through REPLACE provide additional validation that it is an effective method for producing more drug-like and less peptidic inhibitors of the PBD of PLK1.

5. CONCLUSIONS

Although protein–protein interactions represent potential drug targets in the identification of selective kinase inhibitors, improvements in methodology are required. Two case studies of successful application of the REPLACE strategy are presented in the discovery of nonpeptidic and non-ATP-competitive protein–protein interaction inhibitors of validated oncology targets. More drug-like inhibitors of cyclin-dependent kinase substrate recruitment have been identified through replacement of N- and C-terminal determinants of a potent octapeptide. The PBD provides the opportunity to develop selective PLK1 inhibitors through its phosphopeptide-binding site, and progress is described in the identification of FLIPs with preliminary antitumor activity. In summary, REPLACE has been exemplified against two protein kinase targets and is a viable strategy for the conversion of peptidic inhibitors of protein–protein interactions into more drug-like molecules. In addition, REPLACE has significant advantages over other fragment-based design methods in that it obviates the need for highly soluble capping groups and the availability of sensitive binding detection.

REFERENCES

(1) McInnes, C.; Fischer, P.M. *Curr. Pharm. Design* **2005**, *11*, 1845.
(2) Kirkland, L.O.; McInnes, C. *Biochem. Pharmacol.* **2009**, 77, 1561.
(3) Garuti, L.; Roberti, M.; Bottegoni, G. *Curr. Med. Chem.* **2010**, *17*, 2804.

(4) Andrews, M.J.; Kontopidis, G.; McInnes, C.; Plater, A.; Innes, L.; Cowan, A.; Fischer, P.M. *Chembiochem* **1909**, *2006*, 7.
(5) Wang, S.; Griffiths, G.; Midgley, C.A.; Barnett, A.L.; Cooper, M.; Grabarek, J.; Fischer, P.M. *Chem. Biol.* **2010**, *17*, 1111.
(6) McInnes, C.; Andrews, M.J.; Zheleva, D.I.; Lane, D.P.; Fischer, P.M. *Curr. Med. Chem. Anticancer Agents* **2003**, *3*, 57.
(7) Zheleva, D.I.; McInnes, C.; Gavine, A.L.; Zhelev, N.Z.; Fischer, P.M.; Lane, D.P. *J. Pept. Res.* **2002**, *60*, 257.
(8) Chen, Y.N.; Sharma, S.K.; Ramsey, T.M.; Jiang, L.; Martin, M.S.; Baker, K.; Adams, P.D.; Bair, K.W.; Kaelin, W.G. *Proc. Natl. Acad. Sci. U.S.A.* **1999**, *96*, 4325.
(9) Atkinson, G.E.; Cowan, A.; McInnes, C.; Zheleva, D.I.; Fischer, P.M.; Chan, W.C. *Bioorg. Med. Chem. Lett.* **2002**, *12*, 2501.
(10) Mendoza, N.; Fong, S.; Marsters, J.; Koeppen, H.; Schwall, R.; Wickramasinghe, D. *Cancer Res.* **2003**, *63*, 1020.
(11) Lowe, E.D.; Tews, I.; Cheng, K.Y.; Brown, N.R.; Gul, S.; Noble, M.E.; Gamblin, S.J.; Johnson, L.N. *Biochemistry* **2002**, *41*, 15625.
(12) Kontopidis, G.; Andrews, M.J.; McInnes, C.; Cowan, A.; Powers, H.; Innes, L.; Fischer, P.M. *Structure* **2003**, *11*, 1537.
(13) Castanedo, G.; Clark, K.; Wang, S.; Tsui, V.; Wong, M.; Nicholas, J.; Wickramasinghe, D.; Marsters, J.C.; Sutherlin, D. *Bioorg. Med. Chem. Lett.* **2006**, *16*, 1716.
(14) Andrews, M.J.I.; McInnes, C.; Kontopidis, G.; Innes, L.; Cowan, A.; Plater, A.; Fischer, P.M. *Org. Biomol. Chem.* **2004**, *2*, 2735.
(15) Carr, R.A.; Congreve, M.; Murray, C.W.; Rees, D.C. *Drug Discov. Today* **2005**, *10*, 987.
(16) Nandha Premnath, P.; Bolger, J. K.; Liu, S.; McInnes, C. Proceedings of the 102nd Annual Meeting of the American Association for Cancer Research, **2011**; Abstract nr 1351A.
(17) Bolger, J. K.; Nandha Premnath, P.; Liu, S.; McInnes, C. *Abstracts of Papers*, 242nd ACS National Meeting & Exposition, Denver, CO, **2011**, MEDI-67.
(18) Liu, S.; Bolger, J.K.; Kirkland, L.O.; Premnath, P.N.; McInnes, C. *ACS Chem. Biol.* **2010**, *5*, 1169.
(19) Larsen, T.A.; Olson, A.J.; Goodsell, D.S. *Structure* **1998**, *6*, 421.
(20) Heitz, F.; Morris, M.C.; Fesquet, D.; Cavadore, J.C.; Dore, M.; Divita, G. *Biochemistry* **1997**, *36*, 4995.
(21) Canela, N.; Orzáez, M.; Fucho, R.; Mateo, F.; Gutierrez, R.; Pineda-Lucena, A.; Bachs, O.; Pérez-Payá, E. *J. Biol. Chem.* **2006**, *281*, 35942.
(22) Betzi, S.; Alam, R.; Martin, M.; Lubbers, D.J.; Han, H.; Jakkaraj, S.R.; Georg, G.I.; Schönbrunn, E. *ACS Chem. Biol.* **2011**, *6*, 492.
(23) Lee, K.S.; Grenfell, T.Z.; Yarm, F.R.; Erikson, R.L. *Proc. Natl. Acad. Sci. U.S.A.* **1998**, *95*, 9301.
(24) Lee, K.S.; Song, S.; Erikson, R.L. *Proc. Natl. Acad. Sci. U.S.A.* **1999**, *96*, 14360.
(25) Yuan, J.; Kramer, A.; Eckerdt, F.; Kaufmann, M.; Strebhardt, K. *Cancer Res.* **2002**, *62*, 4186.
(26) May, K.M.; Reynolds, N.; Cullen, C.F.; Yanagida, M.; Ohkura, H. *J. Cell. Biol.* **2002**, *156*, 23.
(27) Nakajima, H.; Toyoshima-Morimoto, F.; Taniguchi, E.; Nishida, E. *J. Biol. Chem.* **2003**, *278*, 25277.
(28) Neef, R.; Preisinger, C.; Sutcliffe, J.; Kopajtich, R.; Nigg, E.A.; Mayer, T.U.; Barr, F.A. *J. Cell. Biol.* **2003**, *162*, 863.
(29) Cheng, K.Y.; Lowe, E.D.; Sinclair, J.; Nigg, E.A.; Johnson, L.N. *EMBO J.* **2003**, *22*, 5757.

(30) Elia, A.E.H.; Rellos, P.; Haire, L.F.; Chao, J.W.; Ivins, F.J.; Hoepker, K.; Mohammad, D.; Cantley, L.C.; Smerdon, S.J.; Yaffe, M.B. *Cell* **2003**, *115*, 83.
(31) Leung, G.C.; Hudson, J.W.; Kozarova, A.; Davidson, A.; Dennis, J.W.; Sicheri, F. *Nat. Struct. Biol.* **2002**, *9*, 719.
(32) Garcia-Alvarez, B.; de Carcer, G.; Ibanez, S.; Bragado-Nilsson, E.; Montoya, G. *Proc. Natl. Acad. Sci. U.S.A.* **2007**, *104*, 3107.
(33) Pawson, T.; Scott, J.D. *Science* **1997**, *278*, 2075.
(34) Sawyer, T.K.; Bohacek, R.S.; Dalgarno, D.C.; Eyermann, C.J.; Kawahata, N.; Metcalf, C.A.; Shakespeare, W.C.; Sundaramoorthi, R.; Wang, Y.; Yang, M.G. *Mini-Rev. Med. Chem.* **2002**, *2*, 475.
(35) Fischer, P.M. *Curr. Med. Chem.* **2004**, *11*, 1563.
(36) Bahassi, E.M.; Conn, C.W.; Myer, D.L.; Hennigan, R.F.; McGowan, C.H.; Sanchez, Y.; *et al. Oncogene* **2002**, *21*, 6633.
(37) Xie, S.; Wu, H.; Wang, Q.; Cogswell, J.P.; Husain, I.; Conn, C.; Stambrook, P.; Jhanwar-Uniyal, M.; Dai, W. *J. Biol. Chem.* **2001**, *276*, 43305.
(38) Yun, S.M.; Moulaei, T.; Lim, D.; Bang, J.K.; Park, J.E.; Shenoy, S.R.; Liu, F.; Kang, Y.H.; Liao, C.; Soung, N.K.; Lee, S.; Yoon, D.Y.; Lim, Y.; Lee, D.H.; Otaka, A.; Appella, E.; McMahon, J.B.; Nicklaus, M.C.; Burke, T.R.; Yaffe, M.B.; Wlodawer, A.; Lee, K.S. *Nat. Struct. Mol. Biol.* **2009**, *16*, 876.
(39) Lee, H.J.; Hwang, H.I.; Jang, Y.J. *Cell Cycle* **2010**, *9*, 2389–2398.
(40) van Vugt, V.A.; Medema, R.H. *Cell Cycle* **2004**, *3*, 1383.
(41) Smits, V.A.; Klompmaker, R.; Arnaud, L.; Rijksen, G.; Nigg, E.A.; Medema, R.H. *Nat. Cell. Biol.* **2000**, *2*, 672.
(42) van Vugt, M.A.; Smits, V.A.; Klompmaker, R.; Medema, R.H. *J. Biol. Chem.* **2001**, *276*, 41656.
(43) Reindl, W.; Yuan, J.; Kramer, A.; Strebhardt, K.; Berg, T. *Chem. Biol.* **2008**, *15*, 459.
(44) Reindl, W.; Yuan, J.; Kramer, A.; Strebhardt, K.; Berg, T. *Chembiochem* **2009**, *10*, 1145.
(45) Watanabe, N.; Sekine, T.; Takagi, M.; Iwasaki, J.; Imamoto, N.; Kawasaki, H.; Osada, H. *J. Biol. Chem.* **2009**, *284*, 2344.
(46) Liu, F.; Park, J.E.; Qian, W.J.; Lim, D.; Graber, M.; Berg, T.; Yaffe, M.B.; Lee, K.S.; Burke, T.R. *Nat. Chem. Biol.* **2011**, *7*, 595.
(47) Liu, F.; Park, J.E.; Qian, W.J.; Lim, D.; Scharow, A.; Berg, T.; Yaffe, M.B.; Lee, K.S.; Burke, T.R. *ACS Chem. Biol.* **2012**, *7*, 805–810.
(48) Li, Q.; Huang, Y.; Xiao, N.; Murray, V.; Chen, J.; Wang, J. *Methods Cell. Biol.* **2008**, *90*, 287.
(49) McInnes, C.; Estes, K.; Baxter, M.; Yang, Z.; Boshra Farag, D.; Johnson, P.; Lazo, J.; Wyatt, M.D. *Mol. Cancer Ther.* **2012**, *11*, 1683–1692.

PART 8

Trends

Editor: Joanne Bronson
Bristol-Myers Squibb, Wallingford, Connecticut

CHAPTER THIRTY

New Chemical Entities Entering Phase III Trials in 2011

Gregory T. Notte
Gilead Sciences Inc., San Mateo, California, USA

Contents

1. SELECTION CRITERIA

- The Phase III clinical trial must have been registered with ClinicalTrials.gov in 2011 (or before) and scheduled to begin in 2011.
- This list was compiled using publically available information.[1] It was intended to give an overview of small molecule chemical matter entering Phase III and may not be all-inclusive.
- The chemical structure must be available. References describing the medicinal chemistry discovery effort were included if available. Otherwise, a reference was chosen which describes clinical efficacy or pharmacokinetics. Patent applications were cited only in the absence of alternative references.
- It must be the first time that this compound has reached Phase III for any indication as a single agent or in combination.
- The compound must be synthetic in origin. The following classes of drugs were not included: biologics, inorganic or organometallic compounds,

Annual Reports in Medicinal Chemistry, Volume 47
ISSN 0065-7743
http://dx.doi.org/10.1016/B978-0-12-396492-2.00030-8

ssRNA, endogenous substances, radiopharmaceuticals, natural polypeptides, or herbal extracts.
- New formulations or single enantiomers of a previously approved drug were not included; novel prodrugs were included.
- Compounds meeting the criteria are shown as the free base except those containing quaternary nitrogens.

2. FACTS AND FIGURES

- In 2011, there were 1537 Phase III trials registered at ClinicalTrials.gov that were classified as having a "drug intervention."
- Of the registered trials, 42 molecules (2.7%) met the selection criteria.
- For the molecules contained herein[2]:
  - Average molecular weight = 542 (rangc = 228–1917)
  - Average cLog P = 3.6 (range = −6.2 to 7.3)
- The top three indications from all trials were as follows:
  - Type 1 and 2 diabetes (94 trials)
  - Hepatitis C (40 trials)
  - Chronic obstructive pulmonary disorder (37 trials)

3. NCE LIST

1. Acolbifene (EM-652)

Sponsor	Endoceutics
MW/cLog *P*	457.56/6.6
CAS#	182167-02-8
Start/end date	October 2011–May 2012
Indication	Vasomotor symptoms
Mechanism of action (MOA)	Selective estrogen receptor modulator[3]
ClinicalTrials.gov Identifier	NCT01452373

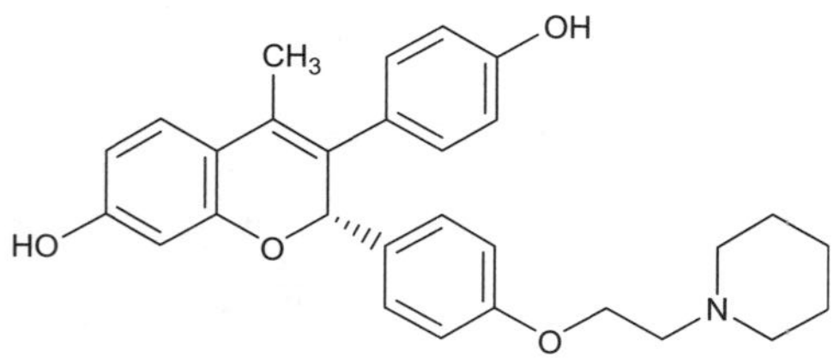

2. Alisporivir (DEB-025)

Sponsor	Novartis
MW/cLog *P*	1216.64/3.8
CAS#	254435-95-5
Start/end date	March 2011–April 2013
Indication	HCV infection
MOA	Cyclophilin inhibitor[4]
ClinicalTrials.gov Identifier	NCT01318694

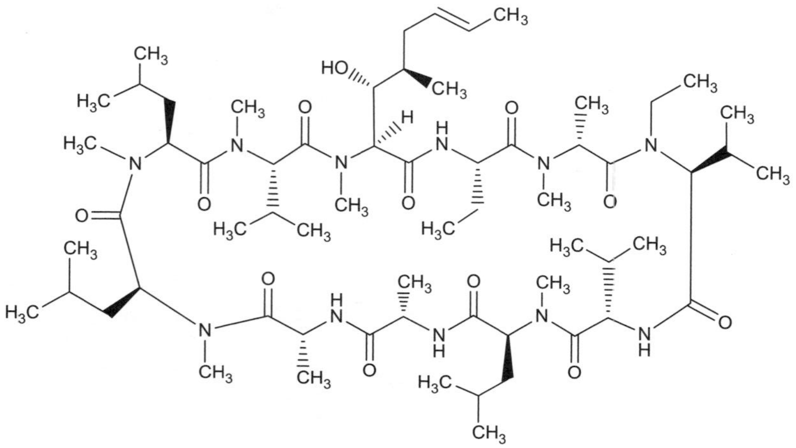

3. Amitifadine (EB-1010)

Sponsor	Euthymics Bioscience
MW/cLog *P*	228.12/2.4
CAS#	410074-73-6
Start/end date	February 2011–March 2012
Indication	Major depression
MOA	Triple reuptake inhibitor of the dopamine, norepinephrine, and serotonin transporters[5]
ClinicalTrials.gov Identifier	NCT01318434

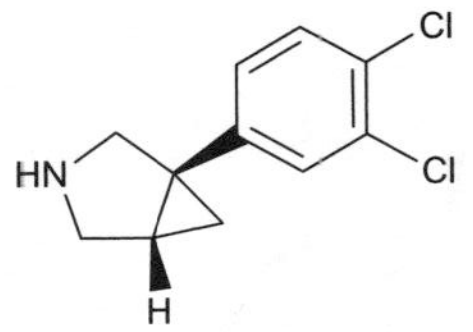

4. Anamorelin (RC-1291)

Sponsor	Helsinn Therapeutics
MW/cLog *P*	564.7/2.8
CAS#	249921-19-5
Start/end date	July 2011–July 2013
Indication	Non-small cell lung cancer cachexia
MOA	Growth hormone secretagogue receptor agonist[6]
ClinicalTrials.gov Identifier	NCT01387269

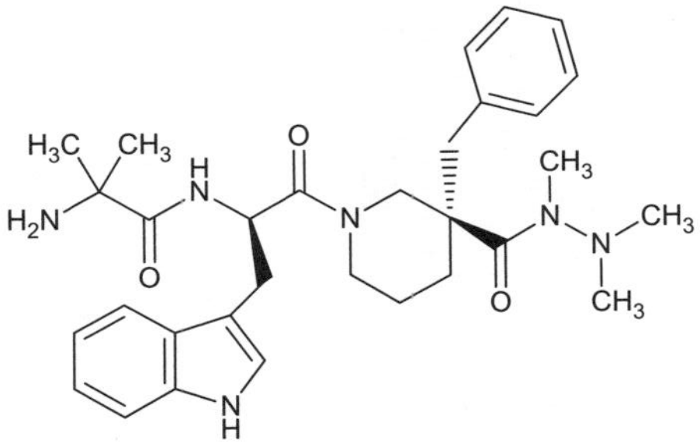

5. Arbaclofen placarbil (XP-19986)

Sponsor	Xenoport
MW/cLog *P*	399.87/4.3
CAS#	847353-30-4
Start/end date	May 2011–August 2013
Indication	Multiple sclerosis
MOA	GABA(B) agonist (prodrug of R-baclofen)[7]
ClinicalTrials.gov Identifier	NCT01359566

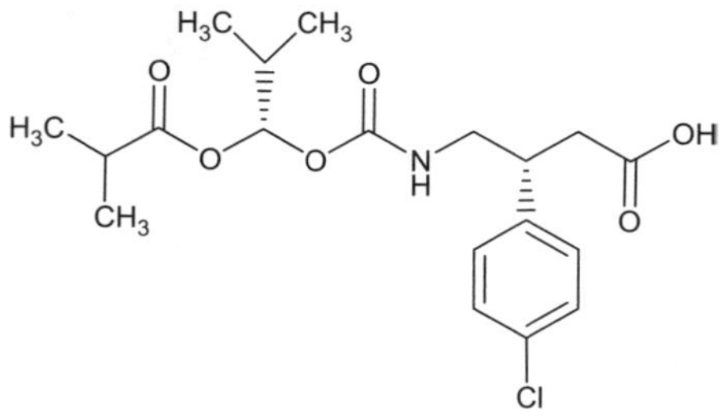

6. Avatrombopag (E-5501)

Sponsor	Eisai
MW/cLog *P*	649.65/4.1
CAS#	570406-98-3
Start/end date	December 2011–December 2012
Indication	Idiopathic thrombocytopenic purpura
MOA	Thrombopoietin receptor (c-Mpl) agonist[8]
ClinicalTrials.gov Identifier	NCT01433978

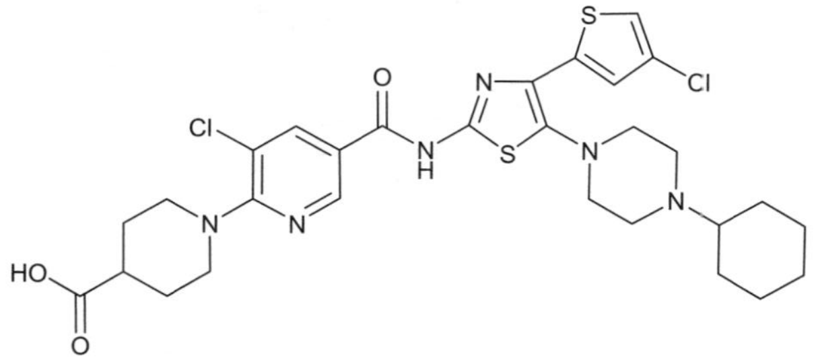

7. *Bardoxolone Methyl (RTA-402)*

Sponsor	Abbott
MW/cLog *P*	505.69/6.1
CAS#	218600-53-4
Start/end date	June 2011–June 2013
Indication	Chronic kidney disease with type II diabetes mellitus
MOA	Nrf2 inducer[9]
ClinicalTrials.gov Identifier	NCT01351675

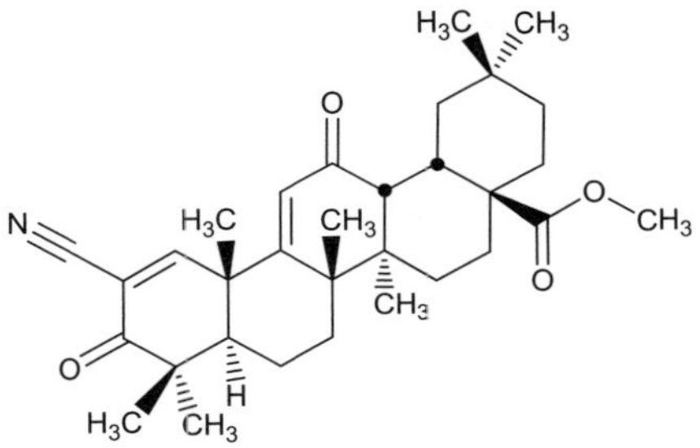

8. *Bedaquiline (TMC-207)*

Sponsor	Tibotec
MW/cLog *P*	555.50/7.3
CAS#	843663-66-1
Start/end date	November 2011–October 2014
Indication	Tuberculosis
MOA	ATP synthase inhibitor[10]
ClinicalTrials.gov Identifier	NCT01464762

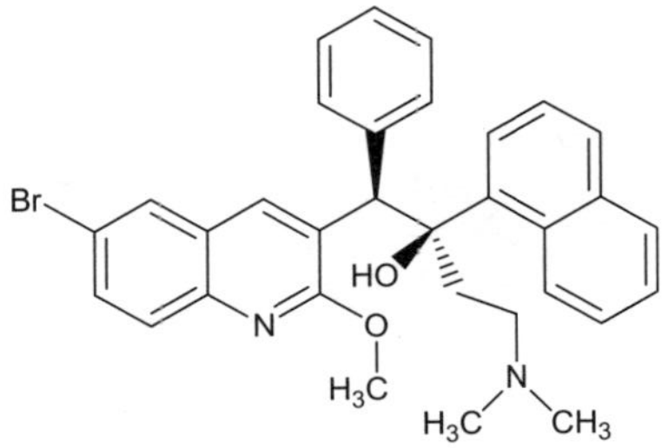

9. BI-201335

Sponsor	Boehringer Ingelheim
MW/cLog *P*	869.82/6.0
CAS#	801283-95-4
Start/end date	April 2011–February 2014
Indication	HCV infection
MOA	NS3/4A protease inhibitor[11]
ClinicalTrials.gov Identifier	NCT01343888

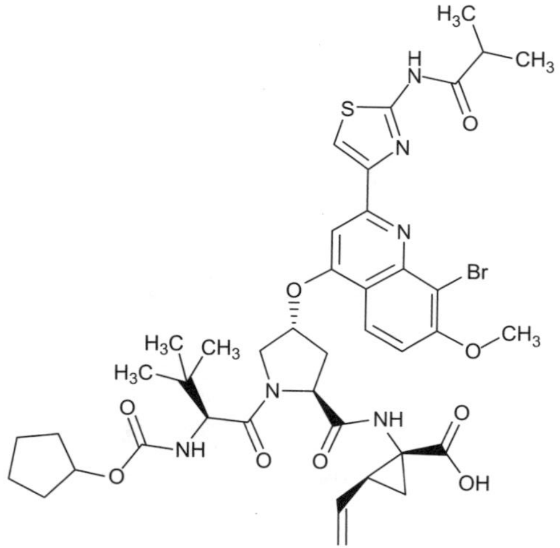

10. Brexpiprazole (OPC-34712)

Sponsor	Otsuka Pharmaceutical
MW/cLog *P*	433.57/4.6
CAS#	913611-97-9
Start/end date	June 2011–May 2013
Indication	Major depressive disorder
MOA	Dopamine D2 receptor partial agonist[12]
ClinicalTrials.gov Identifier	NCT01360632

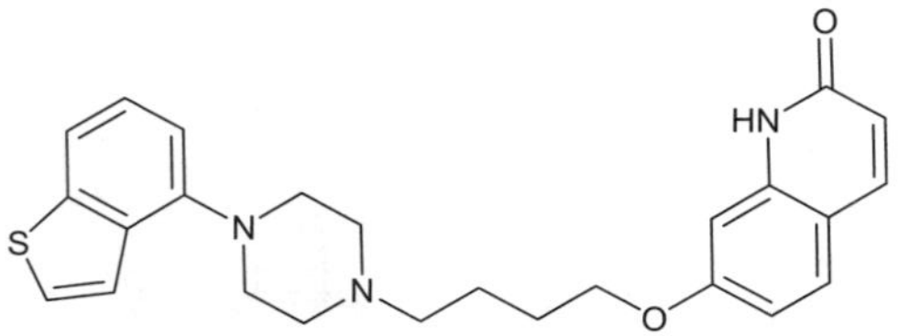

11. Ceftolozane sulfate (CXA-201)

Sponsor	Cubist
MW/cLog *P*	764.77/−6.2
CAS#	936111-69-2
Start/end date	June 2011–December 2012
Indication	Complicated urinary tract infection (formulated in combination with tazobactam)
MOA	Cell wall biosynthesis inhibitor[13]
ClinicalTrials.gov Identifier	NCT01345929

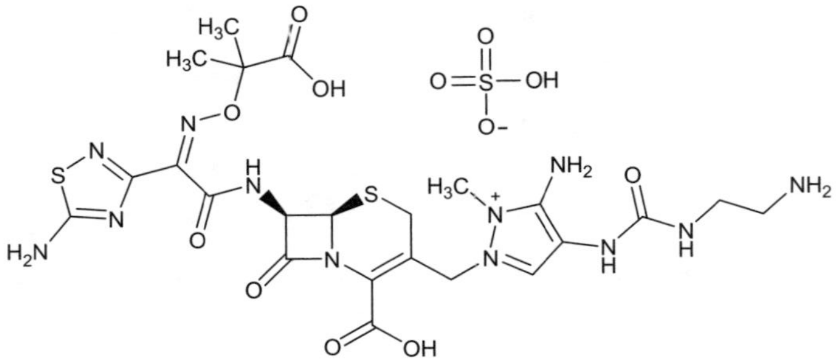

12. CF-101

Sponsor	Can-Fite Biopharma
MW/cLog *P*	510.29/0.48
CAS#	152918-18-8
Start/end date	July 2011–August 2012
Indication	Dry eye
MOA	Adenosine A3 agonist[14]
ClinicalTrials.gov Identifier	NCT01235234

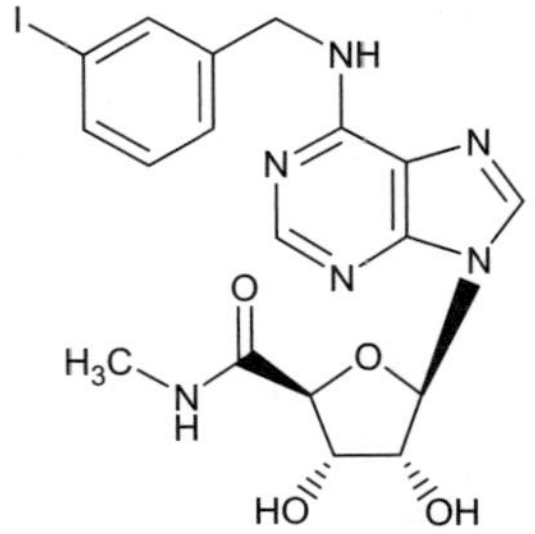

13. Dabrafenib (GSK-2118436)

Sponsor	GlaxoSmithKline
MW/cLog *P*	519.56/4.6
CAS#	1195765-45-7
Start/end date	January 2011–June 2012
Indication	BRAF mutation positive advanced or metastatic melanoma
MOA	Raf kinase B inhibitor[15]
ClinicalTrials.gov Identifier	NCT01227889

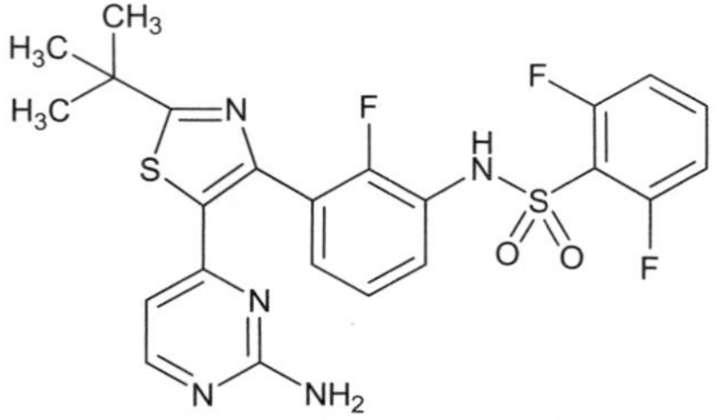

14. Daclatasvir (BMS-790052)

Sponsor	Bristol-Myers Squibb
MW/cLog *P*	738.88/5.4
CAS#	1009119-64-5
Start/end date	September 2011–October 2013
Indication	HCV infection
MOA	HCV NS5A inhibitor[16]
ClinicalTrials.gov Identifier	NCT01389323

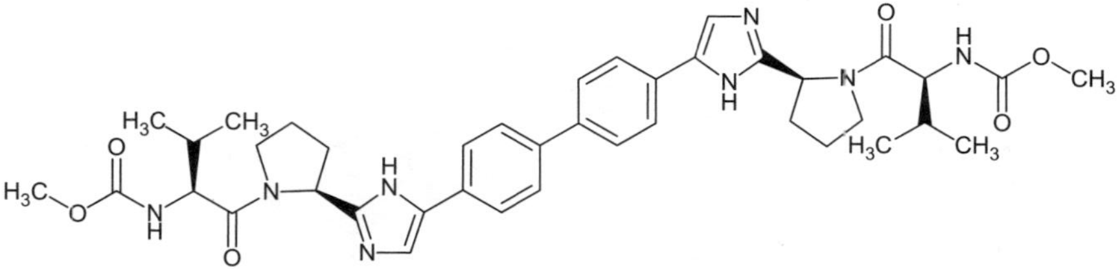

15. Delamanid (OPC-67683)

Sponsor	Otsuka Pharmaceutical
MW/cLog *P*	534.48/5.3
CAS#	681492-22-8
Start/end date	September 2011–August 2013
Indication	Multidrug-resistant tuberculosis
MOA	Cell wall biosynthesis inhibitor[17]
ClinicalTrials.gov Identifier	NCT01424670

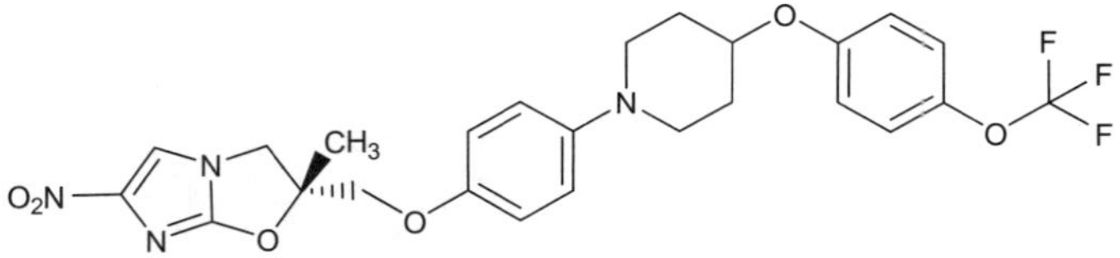

16. Dovitinib (TKI-258)

Sponsor	Novartis
MW/cLog *P*	392.43/2.6
CAS#	405169-16-6
Start/end date	January 2011–January 2014
Indication	Metastatic renal cell carcinoma
MOA	Angiogenesis inhibitor (multiple kinase targets)[18]
ClinicalTrials.gov Identifier	NCT01223027

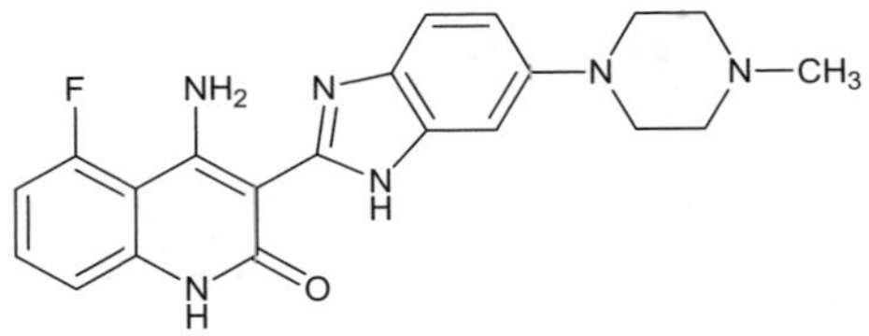

17. EC-145

Sponsor	Endocyte
MW/cLog *P*	1917.04/2.1
CAS#	742092-03-1
Start/end date	April 2011–June 2013
Indication	Ovarian cancer
MOA	Folate receptor antagonist[19]
ClinicalTrials.gov Identifier	NCT01170650

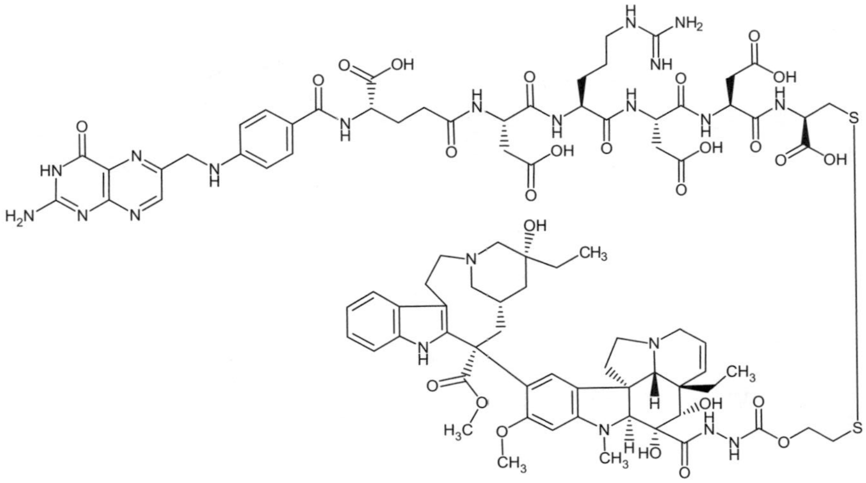

18. Enobosarm (Ostarine)

Sponsor	GTx
MW/cLog *P*	389.33/3.4
CAS#	841205-47-8
Start/end date	July 2011–January 2012
Indication	Muscle wasting
MOA	Selective androgen receptor modulators (SARM)[20]
ClinicalTrials.gov Identifier	NCT01355484

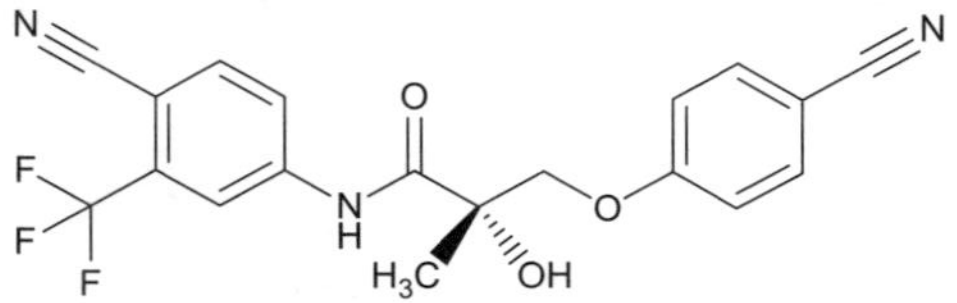

19. Eprotirome (KB-2115)

Sponsor	Karo Bio
MW/cLog *P*	487.14/5.1
CAS#	355129-15-6
Start/end date	July 2011–discontinued
Indication	Heterozygous familial hypercholesterolemia
MOA	Thyroid hormone receptor beta agonist[21]
ClinicalTrials.gov Identifier	NCT01410383

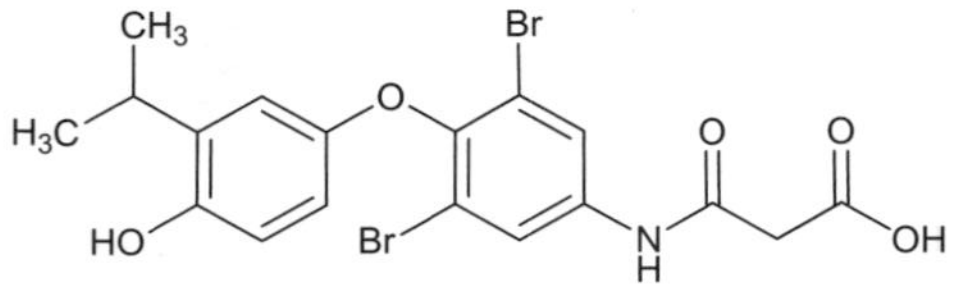

20. Ganetespib (STA-9090)

Sponsor	Synta
MW/cLog *P*	364.40/3.7
CAS#	888216-25-9
Start/end date	May 2011–July 2012
Indication	Non-small cell lung cancer
MOA	Heat-shock protein 90 (hsp90) inhibitor[22]
ClinicalTrials.gov Identifier	NCT01348126

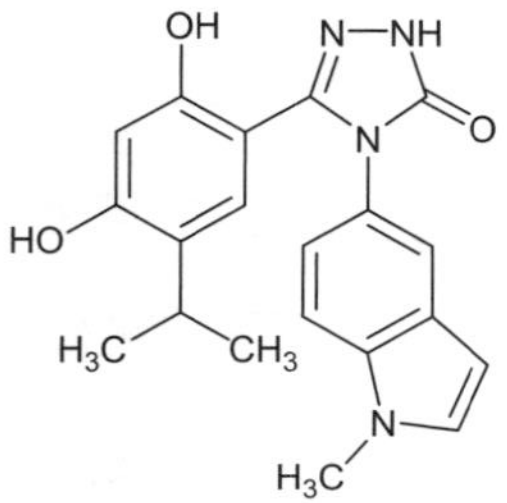

21. GS-7977 (PSI-7977)

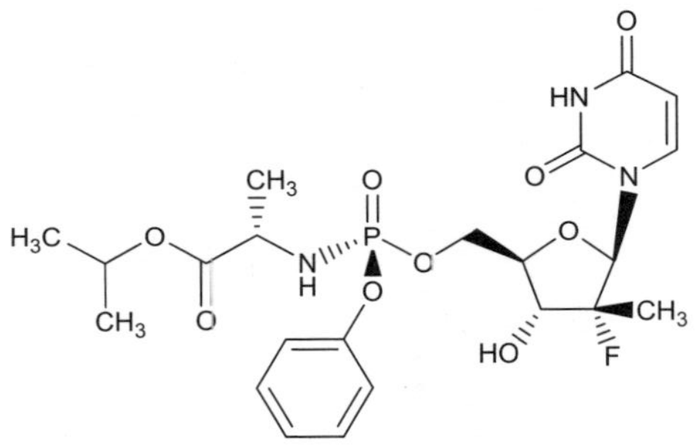

Sponsor	Gilead
MW/cLog *P*	529.45/0.84
CAS#	1190307-88-0
Start/end date	December 2011–January 2013
Indication	HCV infection
MOA	HCV NS5B polymerase inhibitor (phosphoramidate prodrug)[23]
ClinicalTrials.gov Identifier	NCT01542788

22. Iferanserin (VEN-309)

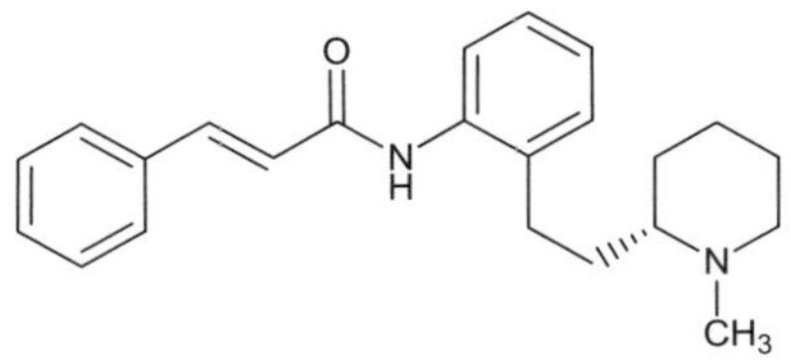

Sponsor	Ventrus Biosciences
MW/cLog *P*	348.48/4.5
CAS#	58754-46-4
Start/end date	July 2011–April 2012
Indication	Internal hemorrhoids
MOA	5-Hydroxytryptamine 2A receptor[24]
ClinicalTrials.gov Identifier	NCT01355874

23. Lenvatinib (E-7080)

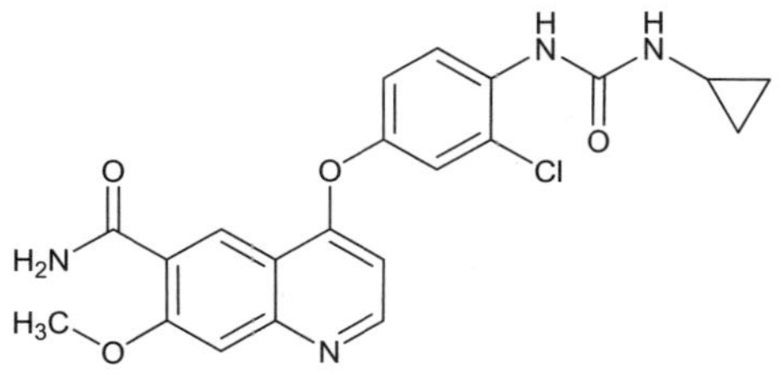

Sponsor	Eisai/Quintiles
MW/cLog *P*	426.85/3.4
CAS#	417716-92-8
Start/end date	March 2011–July 2013
Indication	Thyroid cancer
MOA	KDR/Flk-1 tyrosine kinase inhibitor[25]
ClinicalTrials.gov Identifier	NCT01321554

24. Lesinurad (RDEA-594)

Sponsor	Ardea Biosciences
MW/cLog *P*	404.28/4.0
CAS#	878672-00-5
Start/end date	December 2011–January 2013
Indication	Gout
MOA	Urate transporter 1 inhibitor[26]
ClinicalTrials.gov Identifier	NCT01493531

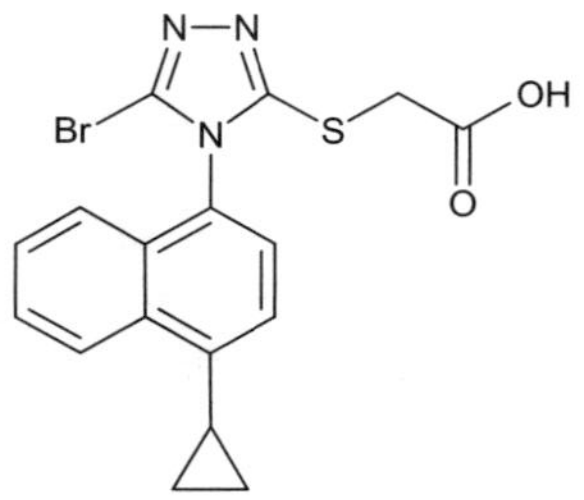

25. Naloxegol (NKTR-118)

Sponsor	AstraZeneca
MW/cLog *P*	651.78/0.44
CAS#	854601-70-0
Start/end date	March 2011–August 2012
Indication	Opioid-induced constipation
MOA	Peripheral opioid antagonist (pegylated naloxol)[27]
ClinicalTrials.gov Identifier	NCT01309841

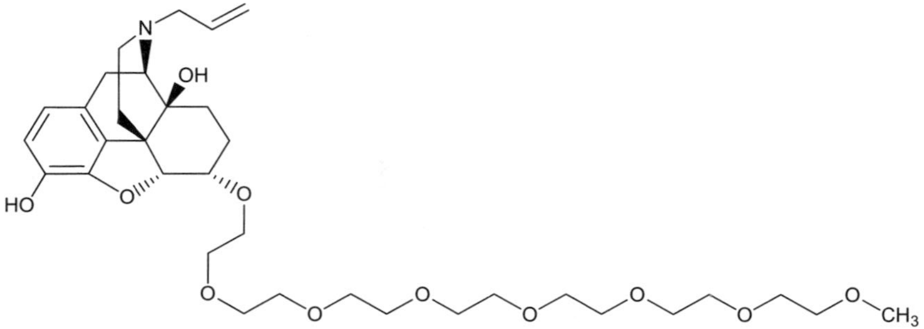

26. Netupitant (Ro-67-3189)

Sponsor	Helsinn
MW/cLog *P*	578.59/6.5
CAS#	290297-26-6
Start/end date	April 2011–July 2012
Indication	Chemotherapy-induced nausea
MOA	Tachykinin NK-1 antagonist[28]
ClinicalTrials.gov Identifier	NCT01339260

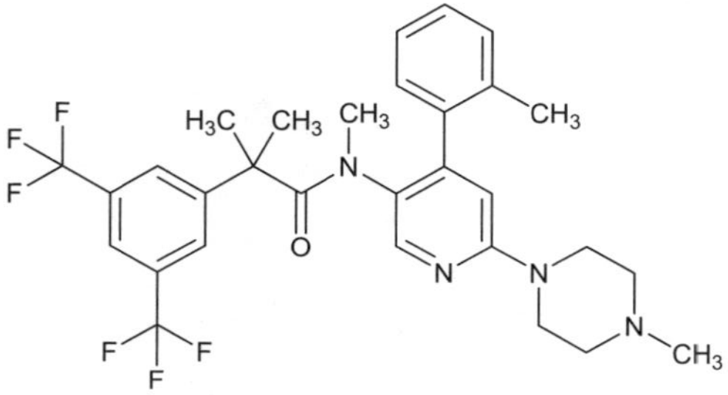

27. Opicapone (BIA-9-1067)

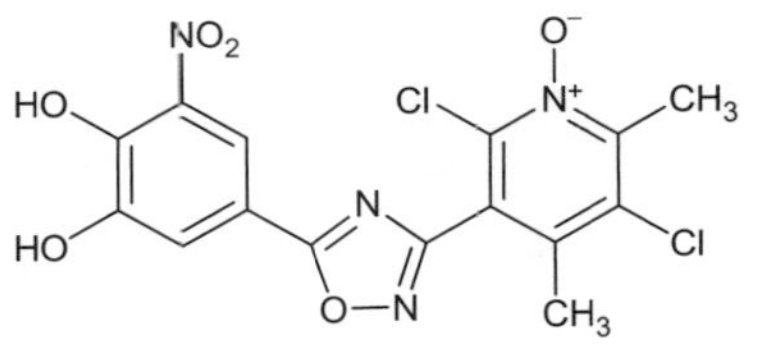

Sponsor	Bial
MW/cLog *P*	413.17/3.0
CAS#	923287-50-7
Start/end date	March 2011–October 2012
Indication	Parkinson's disease
MOA	Catechol-*O*-methyltransferase (COMT) inhibitor[29]
ClinicalTrials.gov Identifier	NCT01227655

28. Ozenoxacin (T-3912)

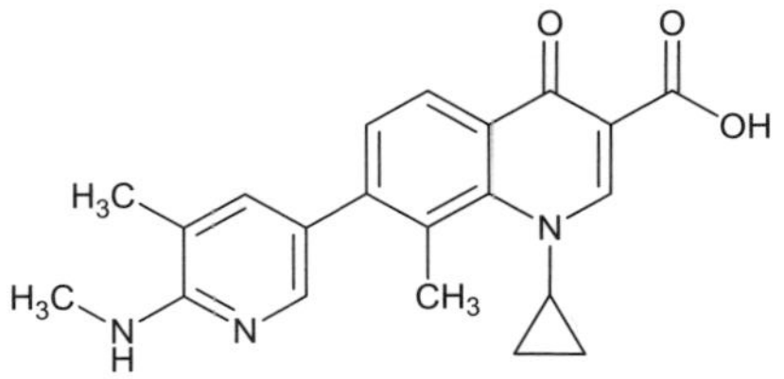

Sponsor	Ferrer
MW/cLog *P*	363.41/3.4
CAS#	245765-41-7
Start/end date	October 2011–July 2012
Indication	Impetigo (skin infection)
MOA	DNA topoisomerase IV inhibitor[30]
ClinicalTrials.gov Identifier	NCT01397461

29. Perphenazine 4-aminobutyrate (BL-1020)

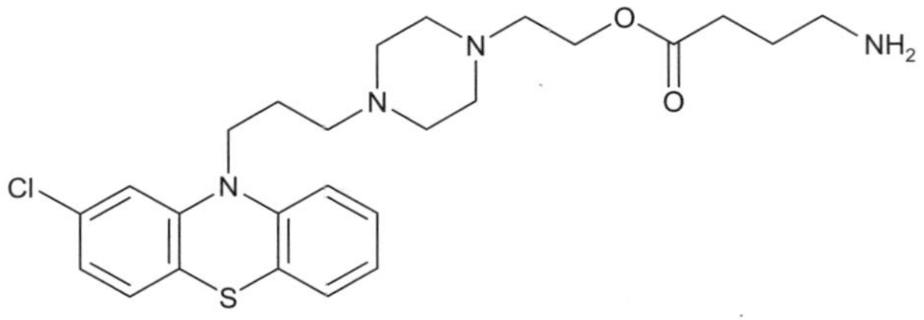

Sponsor	BioLineRx
MW/cLog *P*	489.08/4.5
CAS#	751477-01-7
Start/end date	May 2011–1H, 2013
Indication	Schizophrenia
MOA	Dopamine receptor antagonist and GABA receptor agonist[31]
ClinicalTrials.gov Identifier	NCT01363349

30. Sapacitabine (CS-682)

Sponsor	Cyclacel
MW/cLog *P*	490.64/5.6
CAS#	151823-14-2
Start/end date	January 2011–September 2013
Indication	Acute myeloid leukemia
MOA	DNA synthesis inhibitor[32]
ClinicalTrials. gov Identifier	NCT01303796

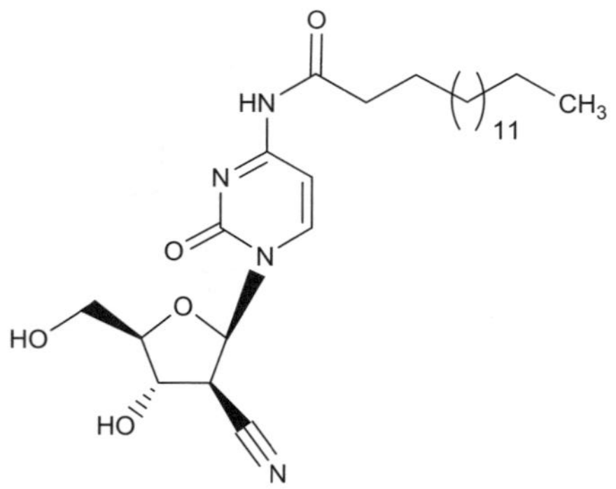

31. SAR-302503 (TG-101348)

Sponsor	Sanofi
MW/cLog *P*	524.68/6.6
CAS#	936091-26-8
Start/end date	October 2011–February 2013
Indication	Hematopoietic neoplasm
MOA	Janus kinase 2 (Jak2) inhibitor[33]
ClinicalTrials. gov Identifier	NCT01437787

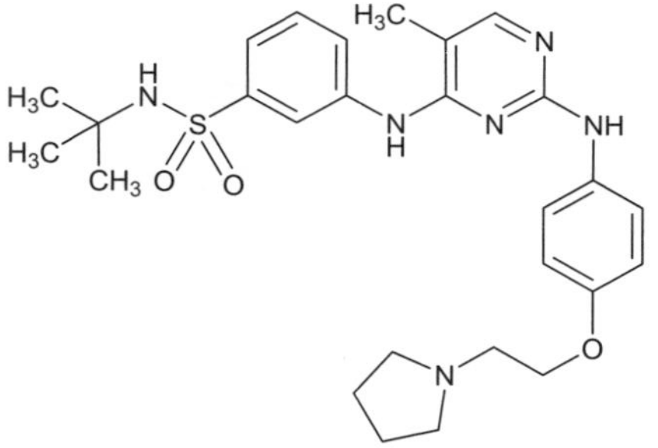

32. Setipiprant (ACT-129968)

Sponsor	Actelion
MW/cLog *P*	402.42/4.3
CAS#	866460-33-5
Start/end date	December 2011–March 2012
Indication	Seasonal allergic rhinitis
MOA	CRTH2 antagonist[34]
ClinicalTrials. gov Identifier	NCT01484119

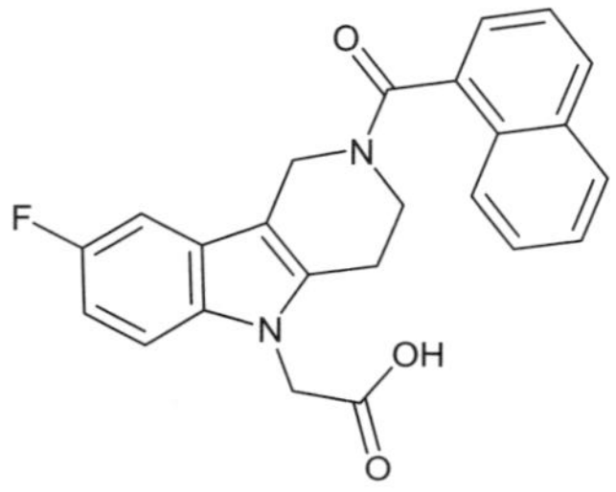

33. TAK-438

Sponsor	Takeda
MW/cLog *P*	345.39/2.5
CAS#	881681-00-1
Start/end date	September 2011–June 2013
Indication	Gastric and duodenal ulcers
MOA	Potassium-competitive acid blockers via H^+/K^+ ATPase inhibition[35]
ClinicalTrials.gov Identifier	NCT01452763

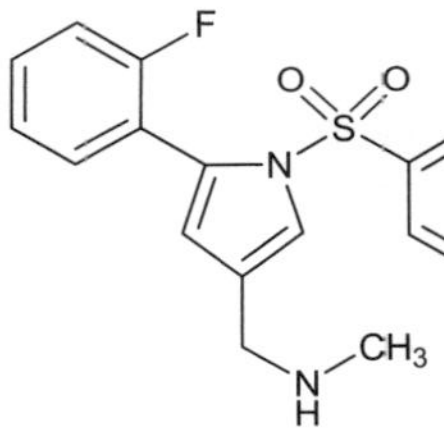

34. TAK-875

Sponsor	Takeda
MW/cLog *P*	524.63/4.7
CAS#	1000413-72-8
Start/end date	November 2011–July 2013
Indication	Type 2 diabetes mellitus
MOA	Free fatty acid receptor 1 agonist[36]
ClinicalTrials.gov Identifier	NCT01456195

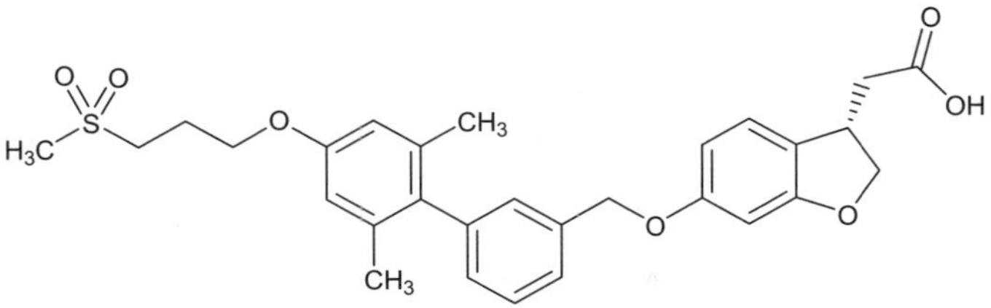

35. Tasquinimod (ABR-215050)

Sponsor	Active Biotech
MW/cLog *P*	406.36/2.5
CAS#	254964-60-8
Start/end date	March 2011–January 2016
Indication	Prostate cancer
MOA	Vascular endothelial growth factor (VEGF) inhibitor[37]
ClinicalTrials.gov Identifier	NCT01234311

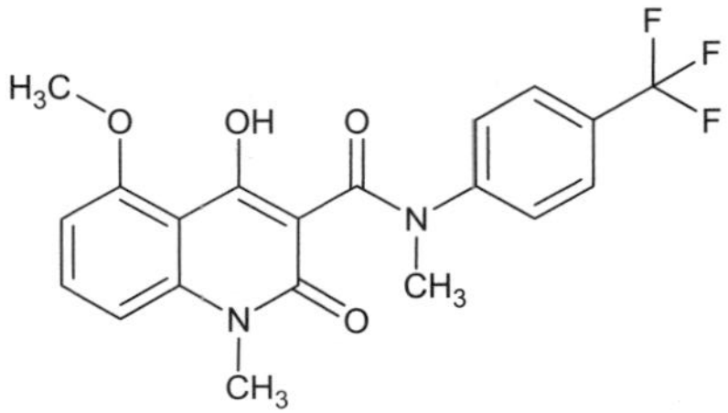

36. TH-302

Sponsor	Threshold
MW/cLog *P*	449.04/1.0
CAS#	918633-87-1
Start/end date	September 2011–June 2014
Indication	Soft-tissue sarcoma
MOA	Hypoxia-activated DNA synthesis inhibitor[38]
ClinicalTrials.gov Identifier	NCT01440088

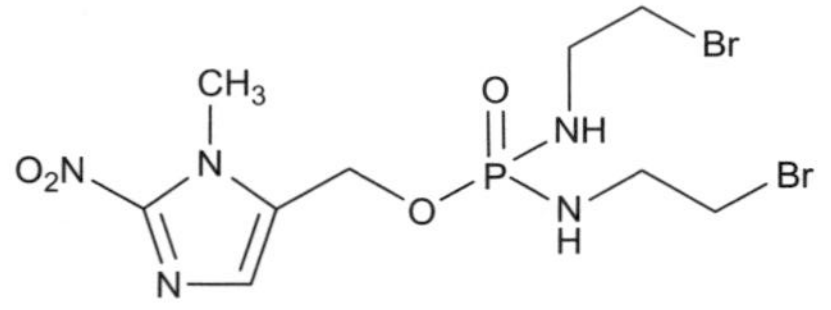

37. Tozadenant (SYN-115)

Sponsor	Biotie Therapeutics
MW/cLog *P*	406.50/1.5
CAS#	870070-55-6
Start/end date	March 2011–June 2013
Indication	Parkinson's disease
MOA	Adenosine A_{2A} receptor antagonist[39]
ClinicalTrials.gov Identifier	NCT01283594

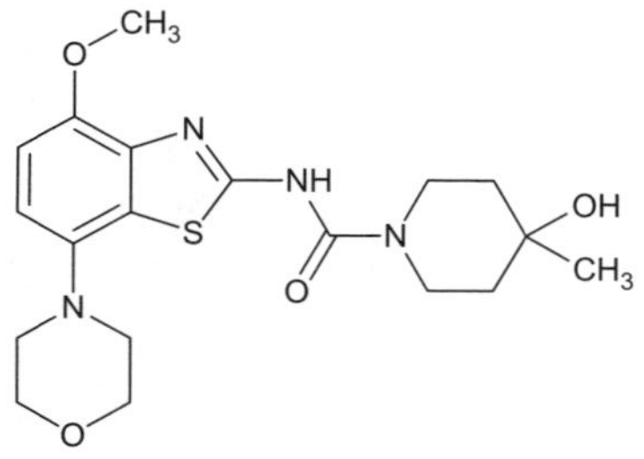

38. Trelagliptin (SYR-472)

Sponsor	Takeda
MW/cLog *P*	357.38/1.1
CAS#	865759-25-7
Start/end date	September 2011–July 2013
Indication:	Type 2 diabetes mellitus
MOA	Dipeptidyl peptidase IV inhibitor[40]
ClinicalTrials.gov Identifier	NCT01431807

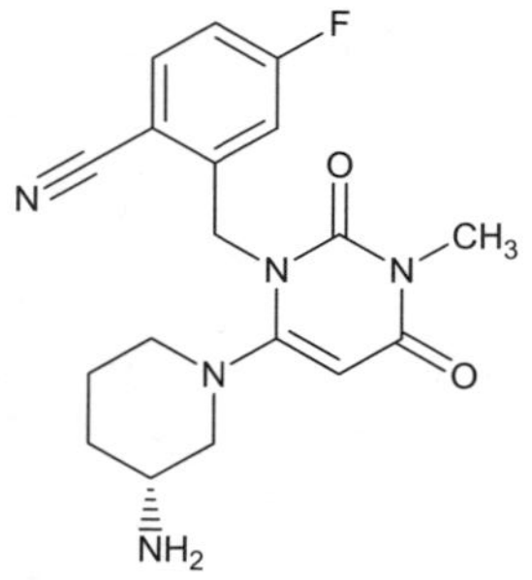

39. Ulimorelin (TZP-101)

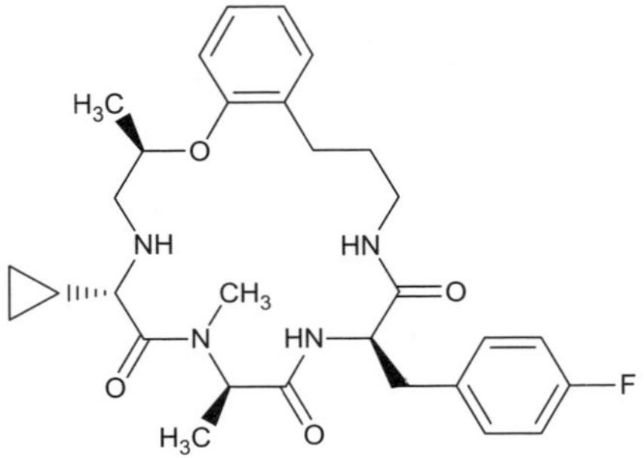

Sponsor	Tranzyme
MW/cLog *P*	538.65/5.3
CAS#	842131-33-3
Start/end date	January 2011–January 2012
Indication	Postoperative ileus
MOA	Ghrelin receptor agonist[41]
ClinicalTrials. gov Identifier	NCT01285570

40. Umeclidinium Bromide (GSK-573719)

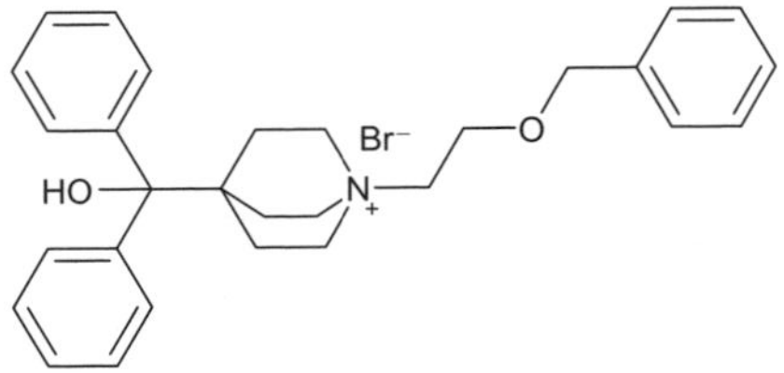

Sponsor	GlaxoSmithKline
MW/cLog *P*	508.49/1.6
CAS#	869113-09-7
Start/end date	January 2011–July 2012
Indication	Chronic obstructive pulmonary disorder
MOA	Muscarinic M3 antagonist[42]
ClinicalTrials. gov Identifier	NCT01316887

41. Vaniprevir (MK-7009)

Sponsor	Merck
MW/cLog *P*	757.94/3.6
CAS#	923590-37-8
Start/end date	June 2011–September 2013
Indication	HCV infection
MOA	NS3/4A protease inhibitor[43]
ClinicalTrials.gov Identifier	NCT01370642

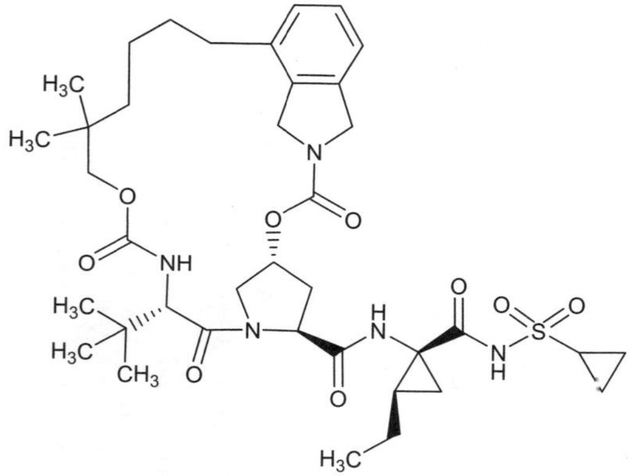

42. Zicronapine (Lu-31-130)

Sponsor	Lundbeck
MW/cLog *P*	354.92/6.1
CAS#	170381-16-5
Start/end date	April 2011–August 2012
Indication	Schizophrenia
MOA	D1/D2 and 5-HT2 antagonist[44]
ClinicalTrials.gov Identifier	NCT01295372

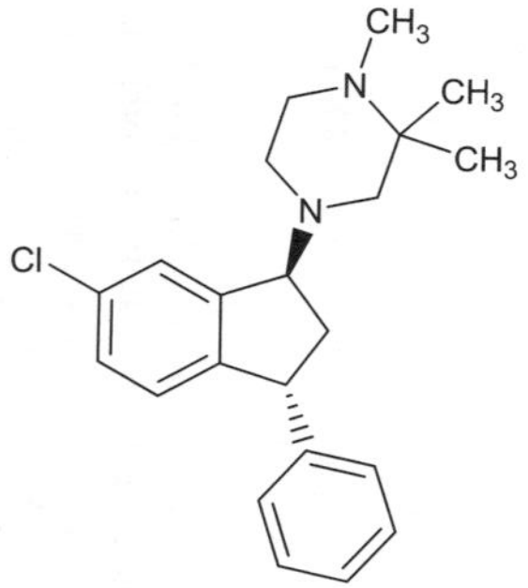

REFERENCES

(1) The following two websites were used extensively in compiling the information contained herein: http://www.clinicaltrials.gov/http://www.ama-assn.org/ama/pub/physician-resources/medical-science/united-states-adopted-names-council.page. Any additional information could be obtained via a thorough web-search or from the Sponsor's website.

(2) Calculated using ChemDraw 11 or ACD labs software if ChemDraw was unable to produce a value.

(3) Labrie, F.; Labrie, C.; Bélanger, A.; Simard, J.; Giguère, V.; Tremblay, A.; Tremblay, G.; *Steroid, J. Biochem. Mol. Biol.* **2001**, *79*, 213.

(4) Watashi, K. *Curr. Opin. Investig. Drugs* **2010**, *11*, 213.

(5) Epstein, J.W.; Brabander, H.J.; Fanshawe, W.J.; Hofmann, C.M.; McKenzie, T.C.; Safir, S.R.; Osterberg, A.C.; Cosulich, D.B.; Lovell, F.M. *J. Med. Chem.* **1981**, *24*, 481.

(6) Paul, B.J.; Littler, B.J.; Jos, F.; Vogt, P.F. *Org. Proc. Res. Dev.* **2006**, *10*, 339.

(7) Xu, F.; Peng, G.; Phan, T.; Dilip, U.; Chen, J.L.; Chernov-Rogan, T.; Zhang, X.; Grindstaff, K.; Annamalai, T.; Koller, K.; Gallop, M.A.; Wustrow, D.J. *Bioorg. Med. Chem. Lett.* **2011**, *21*, 6582.

(8) Fukushima-Shintani, M.; Suzuki, K.; Iwatsuki, Y.; Abe, M.; Sugasawa, K.; Hirayama, F.; Kawasaki, T.; Nakahata, T. *Eur. J. Haematol.* **2009**, *82*, 247.

(9) Honda, T.; Rounds, B.V.; Bore, L.; Finlay, H.J.; Favaloro, F.G., Jr.; Suh, N.; Wang, Y.; Sporn, M.B.; Gribble, G.W. *J. Med. Chem.* **2000**, *43*, 4233.

(10) Andries, K.; Verhasselt, P.; Guillemont, J.; Göhlmann, H.W.H.; Neefs, J.; Winkler, H.; Van Gestel, J.; Timmerman, P.; Zhu, M.; Lee, E.; Williams, P.; de Chaffoy, D.; Huitric, E.; Hoffner, S.; Cambau, E.; Truffot-Pernot, C.; Lounis, N.; Jarlier, V. *Science* **2005**, *307*, 223.

(11) Llinás-Brunet, M.; Bailey, M.D.; Goudreau, N.; Bhardwaj, P.K.; Bordeleau, J.; Bös, M.; Bousquet, Y.; Cordingley, M.G.; Duan, J.; Forgione, P.; Garneau, M.; Ghiro, E.; Gorys, V.; Goulet, S.; Halmos, T.; Kawai, S.H.; Naud, J.; Poupart, M.; White, P.W. *J. Med. Chem.* **2010**, *53*, 6466.

(12) Yamashita, H.; Matsubara, J.; Oshima, K.; Kuroda, H.; Ito, N.; Miyamura, S.; Shimizu, S.; Tanaka, T.; Oshiro, Y.; Shimada, J.; Maeda, K.; Tadori, Y.; Amada, N.; Akazawa, H.; Yamashita, J.; Mori, A.; Uwahodo, Y.; Masumoto, T.; Kikuchi, T.; Hashimoto, K.; Takahashi, H. Patent Application WO 2006/112464-A1, **2006**.

(13) Toda, A.; Ohki, H.; Yamanaka, T.; Murano, K.; Okuda, S.; Kawabata, K.; Hatano, K.; Matsuda, K.; Misumi, K.; Itoh, K.; Satoh, K.; Inoue, S. *Bioorg. Med. Chem. Lett.* **2008**, *18*, 4849.

(14) Baraldi, P.G.; Cacciari, B.; Pinea de las Infantas, M.J.; Romagnoli, R.; Spalluto, G.; Volpini, R.; Costanzi, S.; Vittori, S.; Cristalli, G.; Melman, N.; Park, K.; Ji, X.; Jacobson, K.A. *J. Med. Chem.* **1998**, *41*, 3174.

(15) Zambona, A.; Niculescu-Duvaza, I.; Niculescu-Duvaza, D.; Maraisb, R.; Springer, C.J. *Bioorg. Med. Chem. Lett.* **2012**, *22*, 789.

(16) Romine, J.L.; St. Laurent, D.R.; Leet, J.E.; Martin, S.W.; Serrano-Wu, M.H.; Yang, F.; Gao, M.; O'Boyle II, D.R.; Lemm, J.A.; Sun, J.; Nower, P.T.; Huang, X.; Deshpande, M.S.; Meanwell, N.A.; Snyder, L.B. *ACS Med. Chem. Lett.* **2011**, *2*, 224.

(17) Sasaki, H.; Haraguchi, Y.; Itotani, M.; Kuroda, H.; Hashizume, H.; Tomishige, T.; Kawasaki, M.; Matsumoto, M.; Komatsu, M.; Tsubouchi, H. *J. Med. Chem.* **2006**, *49*, 7854.

(18) Renhowe, P.A.; Pecchi, S.; Shafer, C.M.; Machajewski, T.D.; Jazan, E.M.; Taylor, C.; Antonios-McCrea, W.; McBride, C.M.; Frazier, K.; Wiesmann, M.; Lapointe, G.R.; Feucht, P.H.; Warne, R.L.; Heise, C.C.; Menezes, D.; Aardalen, K.; Ye, H.; He, M.;

Le, V.; Vora, J.; Jansen, J.M.; Wernette-Hammond, M.E.; Harris, A.L. *J. Med. Chem.* **2009**, *52*, 278.

(19) Vlahov, I.R.; Krishna, H.; Santhapuram, R.; Kleindl, P.J.; Howard, S.J.; Stanford, K.M.; Leamon, C.P. *Bioorg. Med. Chem. Lett.* **2006**, *16*, 5093.

(20) Duke III, C.B.; Jones, A.; Bohl, C.E.; Dalton, J.T.; Miller, D.D. *J. Med. Chem.* **2011**, *54*, 3973.

(21) Ladenson, P.W.; Kristensen, J.D.; Ridgway, E.C.; Olsson, A.G.; Carlsson, B.; Klein, I.; Baxter, J.D.; Angelin, B. *N. Engl. J. Med.* **2010**, *362*, 906.

(22) Proia, D.A.; Foley, K.P.; Korbut, T.; Sang, J.; Smith, D.; Bates, R.C.; Liu, Y.; Rosenberg, A.F.; Zhou, D.; Koya, K.; Barsoum, J.; Blackman, R.K. *PLoS One* **2011**, *6*, e18552.

(23) Sofia, M.J.; Bao, D.; Chang, W.; Du, J.; Nagarathnam, D.; Rachakonda, S.; Reddy, P.G.; Ross, B.S.; Wang, P.; Zhang, H.; Bansal, S.; Espiritu, C.; Keilman, M.; Lam, A.M.; Steuer, H.M.M.; Niu, C.; Otto, M.J.; Furman, P.A. *J. Med. Chem.* **2010**, *53*, 7202.

(24) Herold, A.; Dietrich, J.; Aitchison, R. *Clin. Ther.* **2012**, *34*, 329.

(25) Matsui, J.; Funahashi, Y.; Uenaka, T.; Watanabe, T.; Tsuruoka, A.; Asada, M. *Clin. Cancer Res.* **2008**, *14*, 5459.

(26) Pema, K.M. *Drugs Future* **2011**, *36*, 875.

(27) Diego, L.; Atayee, R.; Helmons, P.; Hsiao, G.; von Gunten, C.F. *Expert Opin. Investig. Drugs* **2011**, *20*, 1047.

(28) Hoffmann, T.; Boes, M.; Stadler, H.; Schnider, P.; Hunkeler, W.; Godel, T.; Galley, G.; Ballard, T.M.; Higgins, G.A.; Poli, S.M.; Sleight, A.J. *Bioorg. Med. Chem. Lett.* **2006**, *16*, 1362.

(29) Kiss, L.E.; Ferreira, H.S.; Torrão, L.; Bonifácio, M.J.; Palma, P.N.; Soares-da-Silva, P.; Learmonth, D.A. *J. Med. Chem.* **2010**, *53*, 3396.

(30) Yamakawa, T.; Nishimura, S. *J. Control. Release* **2003**, *86*, 101.

(31) Nudelman, A.; Gil-Ad, I.; Shpaisman, N.; Terasenko, I.; Ron, H.; Savitsky, K.; Geffen, Y.; Weizman, A.; Rephaeli, A. *J. Med. Chem.* **2008**, *51*, 2858.

(32) Kantarjian, H.; Garcia-Manero, G.; O'Brien, S.; Faderl, S.; Ravandi, F.; Westwood, R.; Green, S.R.; Chiao, J.H.; Boone, P.A.; Cortes, J.; Plunkett, W. *J. Clin. Oncol.* **2010**, *28*, 285.

(33) Pardanani, A.; Gotlib, J.R.; Jamieson, C.; Cortes, J.E.; Talpaz, M.; Stone, R.M.; Silverman, M.H.; Gilliland, D.G.; Shorr, J.; Tefferi, A. *J. Clin. Oncol.* **2011**, *29*, 789.

(34) Pettipher, R.; Whittaker, M. *J. Med. Chem.* **2012**, *55*, 2915.

(35) Shin, J.M.; Inatomi, N.; Munson, K.; Strugatsky, D.; Tokhtaeva, E.; Vagin, O.; Sachs, G. *J. Pharmacol. Exp. Ther.* **2011**, *339*, 412.

(36) Negoro, N.; Sasaki, S.; Mikami, S.; Ito, M.; Suzuki, M.; Tsujihata, Y.; Ito, R.; Harada, A.; Takeuchi, K.; Suzuki, N.; Miyazaki, J.; Santou, T.; Odani, T.; Kanzaki, N.; Funami, M.; Tanaka, T.; Kogame, A.; Matsunaga, S.; Yasuma, T.; Momose, Y. *ACS Med. Chem. Lett.* **2010**, *1*, 290.

(37) Isaacs, J.T.; Pili, R.; Qian, D.Z.; Dalrymple, S.L.; Garrison, J.B.; Kyprianou, N.; Björk, A.; Olsson, A.; Leanderson, T. *Prostate* **2006**, *66*, 1768.

(38) Duan, J.X.; Jiao, H.; Kaizerman, J.; Stanton, T.; Evans, J.W.; Lan, L.; Lorente, G.; Banica, M.; Jung, D.; Wang, J.; Ma, H.; Li, X.; Yang, Z.; Hoffman, R.M.; Ammons, W.S.; Hart, C.P.; Matteucci, M. *J. Med. Chem.* **2008**, *51*, 2412.

(39) Black, K.J.; Koller, J.M.; Campbell, M.C.; Gusnard, D.A.; Bandak, S.I. *J. Neurosci.* **2010**, *30*, 16284.

(40) Zhang, Z.; Wallace, M.B.; Feng, J.; Stafford, J.A.; Skene, R.J.; Shi, L.; Lee, B.; Aertgeerts, K.; Jennings, A.; Xu, R.; Kassel, D.B.; Kaldor, S.W.; Navre, M.; Webb, D.R.; Gwaltney, S.L. *J. Med. Chem.* **2011**, *54*, 510.

(41) Hoveyda, H.R.; Marsault, E.; Gagnon, R.; Mathieu, A.P.; Vezina, M.; Landry, A.; Wang, Z.; Benakli, K.; Beaubien, S.; Saint-Louis, C.; Brassard, M.; Pinault, J.; Ouellet, L.; Bhat, S.; Ramaseshan, M.; Peng, X.; Foucher, L.; Beauchemin, S.; Bherer, P.; Veber, D.F.; Peterson, M.L.; Fraser, G.L. *J. Med. Chem.* **2011**, *54*, 8305.
(42) Lainé, D.I.; McCleland, B.; Thomas, S.; Neipp, C.; Underwood, B.; Dufour, J.; Widdowson, K.L.; Palovich, M.R.; Blaney, F.E.; Foley, J.J.; Webb, E.F.; Luttmann, M.A.; Burman, M.; Belmonte, K.; Salmon, M. *J. Med. Chem.* **2009**, *52*, 2493.
(43) McCauley, J.A.; McIntyre, C.J.; Rudd, M.T.; Nguyen, K.T.; Romano, J.J.; Butcher, J.W.; Gilbert, K.F.; Bush, K.J.; Holloway, M.K.; Swestock, J.; Wan, B.; Carroll, S.S.; DiMuzio, J.M.; Graham, D.J.; Ludmerer, S.W.; Mao, S.; Stahlhut, M.W.; Fandozzi, C.M.; Trainor, N.; Olsen, D.B.; Vacca, J.P.; Liverton, N.J. *J. Med. Chem.* **2010**, *53*, 2443.
(44) Bogeso, K.P.; Arnt, J.; Frederiksen, K.; Hansen, H.O.; Hyttel, J.; Pedersen, H. *J. Med. Chem.* **1995**, *38*, 4380.

CHAPTER THIRTY-ONE

To Market, To Market—2011

Joanne Bronson*, Murali Dhar†, William Ewing‡, Nils Lonberg§
*Bristol-Myers Squibb Company Wallingford, Connecticut, USA
†Bristol-Myers Squibb Company Princeton, New Jersey, USA
‡Bristol-Myers Squibb Company Pennington, New Jersey, USA
§Bristol-Myers Squibb Company Milpitas, California, USA

Contents

Annual Reports in Medicinal Chemistry, Volume 47
ISSN 0065-7743
http://dx.doi.org/10.1016/B978-0-12-396492-2.00031-X

OVERVIEW

This year's To-Market-To-Market chapter provides summaries for 26 new molecular entities (NMEs) that received first time approval world-wide in 2011.[1] Nineteen of these NMEs received their first approval in the United States. Three of the remaining first-time NMEs are from the European Union (EU), two are from Japan, one is from China, and one is from South Korea. It is noteworthy that the U.S. Food and Drug Administration (FDA) approved a total of 30 NMEs in 2011, a considerable increase from 21 approvals in 2010 and the highest number of approvals since 2004.[2] This total includes the 19 U.S.-approved NMEs covered in this chapter and 6 NMEs previously approved outside the United States, with the remainder being agents such as vaccines and imaging agents that are outside the scope of this chapter. Twenty one of the NMEs covered in the 2011 To-Market-To-Market chapter are small molecules and five are biologic agents. Anticancer agents topped the list in 2011 with seven first-time NMEs in total, of which five are small molecules and two are biologic agents. In the infectious disease area, three antiviral agents and one antibacterial agent were approved. There were three first-time NME drug approvals each for central nervous system, cardiovascular, and endocrine diseases. There were also three approvals in the immunology therapeutic area, with one small molecule and two biologic agents. The remaining NMEs include a biologic agent for macular degeneration, a small molecule for the rare disease transthyretin familial amyloidosis, and a small molecule for urinary incontinence. The following overview is organized by therapeutic area, with drugs covered in this year's chapter described first, followed by additional approvals that are of interest but not covered in detail.

Four of the five small molecules approved in the anticancer area are kinase inhibitors. Crizotinib (Xalkori®) is an ATP competitive, dual inhibitor of tyrosine kinases c-MET (mesenchymal–epithelial transition factor) kinase and anaplastic lymphoma kinase (ALK). The U.S. FDA approved crizotinib for the treatment of non-small-cell lung cancer at 250 mg administered orally twice daily. Ruxolitinib (Jakafi®) is the first approved oral inhibitor of JAK1 and JAK2 for the treatment of patients with intermediate or high-risk myelofibrosis. Ruxolitinib was approved in the U.S. at a recommended dose of 20 mg administered orally twice daily. Vandetanib (Caprelsa®) is an oral tyrosine kinase inhibitor that was approved by the U.S. FDA at 300 mg once daily for the treatment of symptomatic or progressive medullary thyroid cancer (MTC) in adult patients with inoperable advanced or metastatic disease. Vemurafenib (Zelboraf®) is an inhibitor of B-raf kinase that is specifically recommended for the treatment of patients with the $BRAF^{V600E}$ mutation. Along with the human monoclonal antibody ipilimumab, vemurafenib is the second approved therapy in 2011 for the treatment of metastatic melanoma. The U.S. FDA approved vemurafenib at 960 mg administered orally twice daily for the treatment of patients with metastatic melanoma, as detected by the cobas 4800 BRAF V600 mutation test. Abiraterone acetate (Zytiga®) inhibits CYP17, a key enzyme in androgen biosynthesis, and is the third approved therapy in the past 2 years for the treatment of metastatic castration-resistant prostate cancer (mCRPC). The U.S. FDA approved abiraterone acetate for the treatment of mCRPC at 1000 mg once daily orally in combination with prednisone. Abiraterone acetate joins two therapies approved for prostate cancer in 2010, sipuleucel-T and cabazitaxel. The two biologic agents approved for cancer therapy are brentuximab (Adcetris™) and ipilimumab (Yervoy®). Brentuximab verdotin is an antibody–drug conjugate that binds to CD30 positive tumor cells and delivers the microtubule inhibitor monomethyl auristatin E (MMAE). It was approved as an intravenous infusion by the U.S. FDA for treatment of last-line Hodgkin's lymphoma (HL) and systemic anaplastic large cell leukemia (ALCL). Ipilimumab is an immunostimulatory antibody that was approved by the U.S. FDA as an intravenous infusion for treatment of unresectable or metastatic melanoma. The antibody augments host immune responses by blocking negative signaling mediated by the T cell surface molecule CTLA-4. The U.S. FDA also approved the orphan drug Erwinaze (asparaginase *Erwinia chrysanthemi*) as a component of a multiagent regimen for the treatment of patients with acute lymphoblastic leukemia who have developed hypersensitivity to the *E. coli*-derived enzyme.[3] Erwinaze is an enzyme that breaks down asparagine, which is essential for cell survival. Leukemia cells cannot produce this amino acid, whereas normal cells are able to make enough asparagine to survive.

In the infectious diseases area, boceprevir (Victrelis®) and telaprevir (Incivek®) were the first hepatitis C virus (HCV) NS3-4A protease inhibitors approved for the treatment of HCV infection. Boceprevir is a peptidomimetic HCV NS3-4A protease inhibitor that binds covalently, but reversibly, to the active site serine. The approved dose of boceprevir is 800 mg given orally three times daily in combination with peginterferon alfa plus ribavirin. The peptidomimetic telaprevir likewise acts as a covalent, reversible, inhibitor of HCV protease. The recommended dosage of telaprevir (Incivek) is 750 mg taken orally three times a day with food in combination with peginterferon alfa plus ribavirin. For treatment of human immunodeficiency virus (HIV) infection, the non-nucleoside reverse transcriptase (RT) inhibitor rilpivirine (Edurant®) was approved as a 25-mg oral tablet to be taken once daily with a meal in combination with other antiretroviral agents. The U.S. FDA also approved the specific three-drug combination Complera®, a fixed dose combination of rilpivirine (25 mg) with emtricitabine (200 mg) and tenofovir (300 mg) to be given once daily taken orally with meals.[4] The macrolide natural product fidaxomicin (Dificid®) was approved by the U.S. FDA as an oral agent for the treatment of *Clostridium difficile*-associated diarrhea. The recommended dose is 200 mg twice daily for a treatment course of 10 days. There is a lower rate of recurrence of *C. difficile* infections (CDI) with fidaxomicin compared with other agents.

In the cardiovascular area, first-time approvals were achieved in 2011 for apixaban (Eliquis®) and edoxaban (Lixiana®), two antithrombotic drugs that act as direct inhibitors of Factor Xa (FXa), an enzyme with a central role in producing thrombin by both the intrinsic and extrinsic pathways of blood coagulation. The first approved FXa inhibitor was rivaroxaban (Xarelto®), which received EU approval in 2008 and, more recently, was approved in the United States in 2011 with a recommended dose of 10 mg once daily to reduce the risk of blood clots, deep vein thrombosis (DVT), and pulmonary embolism (PE) following knee or hip replacement surgery. Apixaban was approved in the EU for prevention of venous thromboembolic events (VTE) in adult patients who have undergone elective hip or knee replacement surgery. The recommended dose is 2.5 mg given orally twice daily. Edoxaban was approved in Japan at doses of 15 and 30 mg once daily for the prevention of venous thromboembolism after major orthopedic surgery. For lowering blood pressure in patients with hypertension, azilsartan medoxomil (Edarbi®), a potent and selective angiotensin II receptor antagonist, was approved by the U.S. FDA at 40 and 80 mg once daily. In late 2011, the U.S. FDA approved Edarbyclor™, a fixed dose combination of azilsartan medoxomil and the diuretic chlorthalidone in a once-daily, single tablet for patients with hypertension.[5] The recommended starting dose of Edarbyclor is 40 mg azilsartan medoxomil combined with 12.5 mg chlorthalidone; the maximum dose is a 40/25-mg azilsartan medoxomil/chlorthalidone combination.

In the endocrine disease therapeutic area, there were approvals for treatment of diabetes, erectile dysfunction (ED), and osteoporosis. Linagliptin (Tradjenta®) is an inhibitor of dipeptidyl peptidase-4 (DPP-4) that was approved by the U.S. FDA to treat Type 2 diabetes along with diet and exercise. The recommended dose of linagliptin is 5 mg given orally once daily. Tradjenta® is the fifth marketed inhibitor of DPP-4. In 2011, the U.S. FDA approved Juvisync®, the first fixed-dose combination of a DPP-4 inhibitor (sitagliptin) with a cholesterol-lowering agent (simvastatin) for use in adults who need both agents.[6] Tablets containing different doses of sitagliptin and simvastatin in fixed-dose combination were developed to meet the different needs of individual patients. The type 5 phosphodiesterase (PDE5) inhibitor avanafil (Zepeed) was approved in South Korea for the treatment of ED at doses of 100 and 200 mg. Avanafil is reported to have a fast onset of action and to be the most selective PDE5 inhibitor on the market. The vitamin D analogue eldecalcitol (Edirol®) was approved in Japan for once daily oral treatment of osteoporosis at 0.5 and 0.75 μg doses. Eldecalcitol binds more potently to the vitamin D receptor than calcitrol, the active form of vitamin D.

Approvals in the central nervous system therapeutic area cover a range of disorders, including epilepsy, depression, and restless legs syndrome (RLS). Retigabine was approved in early 2011 in the EU (Trobalt) and in mid-2011 in the U.S. (Ezogabine/Potiga™) for the adjunctive treatment of partial-onset seizures in adults who have epilepsy. Retigabine acts as a selective positive allosteric modulator of KCNQ2-5 potassium channels, which are key regulators of neuronal excitability. The recommended dose of retigabine is 300 mg/day tid increasing by 150 mg/day tid in 1-week intervals up to 600 mg/day tid to 1200 mg/day tid depending on tolerability. Vilazodone (Viibryd®), a novel antidepressant agent that combines potent serotonin reuptake inhibition with partial agonist functional activity at 5-HT1A receptors, was approved in the U.S. for the treatment of major depressive disorder (MDD). The recommended dose of vilazodone is 40 mg once daily following titration from 10 mg for 7 days to 20 mg for 7 days. Gabapentin enacarbil (Horizant®), a novel prodrug of gabapentin, was approved in the U.S. for the treatment of moderate-to-severe RLS in adults. The recommended dose is 600 mg taken orally once daily taken with food at about 5 p.m. Although gabapentin has shown evidence of efficacy in RLS, its mechanism of action remains unclear.

In the immunology area, first-time NMEs were approved for organ transplant rejection, lupus, and rheumatoid arthritis (RA). Belatacept (Nulojix®), a recombinant fusion protein that blocks CD28-mediated T-cell costimulation, is a selective immunosuppressive that was approved by the U.S. FDA for prophylactic prevention of rejection in kidney transplant recipients. Belatacept is formulated as a lyophilized powder for intravenous infusion and is dosed at 10 mg/kg during an initial phase and 5 mg/kg during a maintenance phase. Belimumab (Benlysta®), a monoclonal antibody that binds to human B lymphocyte stimulator protein, became the first new drug approved by the U.S. FDA in over 50 years for treating the autoimmune disease systemic lupus erythematosus (SLE). The soluble factor neutralized by belimumab stimulates cell proliferation and antibody secretion in B lineage cells. Belimumab is formulated as a lyophilizate for intravenous infusion, and is administered as a 1-h infusion at 10 mg/kg every 2 weeks for the first three doses, followed by a maintenance regimen of 10 mg/kg every 4 weeks. Iguratimod (Iremod) was approved in China as an oral, disease modifying antirheumatic drug (DMARD) for treatment of RA, taken orally at 25 mg twice daily. Iguratimod inhibits PGE_2 production and COX-2 m-RNA expression in cultured fibroblasts.

Additional approvals in 2011 include drugs for macular degeneration, a rare amyloidosis disease and urinary incontinence. The recombinant fusion protein aflibercept (Eylea®), administered as an intravitreal injection, was approved in the U.S. for treatment of patients with neovascular (wet) age-related macular degeneration (AMD). Aflibercept acts as a soluble decoy receptor that binds all VEGF isoforms more tightly than their native receptors, thereby diverting VEGF from its normal function. The recommended dose for aflibercept is 2 mg administered by intravitreal injection every 4 weeks for the first 3 months, and then every 8 weeks thereafter. Tafamidis meglumine (Vyndaqel®) was approved in the EU for the treatment of Transthyretin Familial Amyloid Polyneuropathy (TTR-FAP) in adult patients with stage 1 symptomatic polyneuropathy. It is the first approved medicine for TTR-FAP, which is a rare, progressive, and fatal disorder characterized by neuropathy, cardiomyopathy, renal failure, and blindness. Tafamidis has a novel mechanism of action in that it stabilizes both the wild type and mutant forms of the transthyretin tetramer and prevents tetramer dissociation by noncooperatively binding to the two thyroxine binding sites. The recommended dose of tafamidis is 20 mg orally once daily. Mirabegron (Betanis®) was approved in Japan for the treatment of urgency, urinary frequency, and urinary urge urinary incontinence associated with overactive bladder (OAB). Mirabegron activates the β3-adrenoceptor (β3-AR) receptor, which is located on the bladder and involved in bladder smooth muscle relaxation. The recommended dose of mirabegron is 50 mg once daily.

1. ABIRATERONE ACETATE (ANTICANCER)[7–14]

Class:	Cytochrome P450 17A inhibitor
Country of origin:	United Kingdom
Originator:	Institute of Cancer Research, London
First introduction:	United States
Introduced by:	Janssen Biotech Inc.
Trade name:	Zytiga®
CAS registry no:	154229-18-2
Molecular weight:	391.55

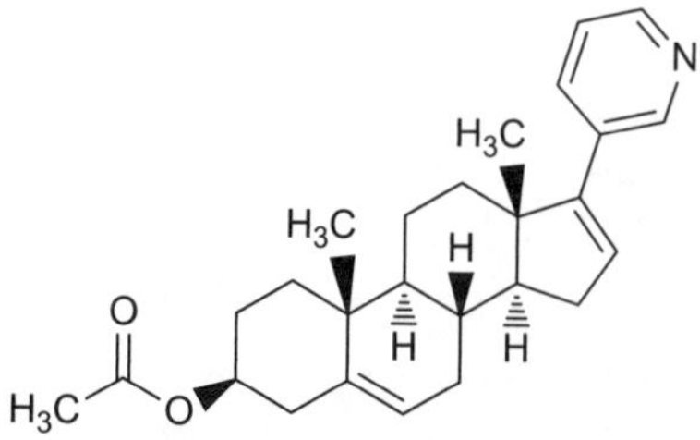

In April 2011, the United States FDA approved abiraterone acetate (CB7630) in combination with the steroid prednisone for the treatment of metastatic castration-resistant prostate cancer (mCRPC) for patients who were previously treated with a docetaxel containing regimen for late-stage disease. Prostate cancer is the leading cause of cancer death in American men: 2010 statistics from the U.S. National Cancer Institute indicate that ~ 220,000 men were diagnosed with prostate cancer.[7] Chemotherapeutic options include docetaxel-based therapy with a median survival time of ~ 18 months. Surgical castration by androgen deprivation therapy is the cornerstone treatment for advanced prostate cancer. Although tumor growth is initially blocked, patients eventually develop disease progression resulting in a lethal drug resistant stage called castration-resistant prostate cancer (CRPC). However, many patients with CRPC respond to administration of secondary hormonal manipulations suggesting tumors are still dependent on the androgen receptor (AR) for signaling. Abiraterone acetate affects prostate, testicular, and adrenal androgens by irreversibly inhibiting both the lyase and hydroxylase activity of cytochrome P450 17A (CYP17) signaling pathways ($IC_{50's}$ of 2.9 and 4 nM, respectively) thereby decreasing testosterone levels.[8,9] The 16,17-double bond of the steroid scaffold of abiraterone acetate imparts irreversible binding to CYP17, while the 3-pyridyl substitution confers specificity in binding to CYP17.[10] Abiraterone was developed as a prodrug (abiraterone acetate) to improve its bioavailability—the acetate moiety is rapidly cleaved *in vivo* leading to the generation of abiraterone. The key step in a reported large-scale synthesis of abiraterone acetate[11] involves a palladium-catalyzed cross coupling reaction between diethyl(3-pyridy1)borane and a vinyl iodide derived from dehydroepiandrosterone. Preclinical *in vivo* studies in mice indicated that abiraterone acetate reduced plasma testosterone to less than 0.1 nmol/L despite a three- to fourfold increase in the plasma levels of luteinizing hormone. There were no significant adrenal weight changes suggesting no significant inhibition of corticosterone production or a compensatory increase in the adrenocorticotropic hormone.[12]

In Phase I clinical trials, it was shown that abiraterone acetate is rapidly absorbed and deacetylated in the liver to abiraterone. There was a nonlinear increase in AUC and C_{max} with increasing doses. The mean terminal half-life was 10.3 h and clearance in fed state was 307 L/h. Coadministration of abiraterone acetate with food led to significant increases in both C_{max} and AUC. Abiraterone acetate is highly protein bound (>99%) and has a steady-state Vss (mean $\pm$ SD) of 19,669 $\pm$ 13,358 L. The safety and efficacy of abiraterone acetate was assessed in randomized, placebo-controlled multicenter Phase III trials in patients with mCRPC who had received prior chemotherapy containing docetaxel. A total of 1195 patients were randomized 2:1 to receive either abiraterone acetate (1000 mg once daily, oral dosing) in combination with prednisone (5 mg, twice daily, oral dosing) or placebo (once daily) plus prednisone (5 mg, twice daily, oral dosing). The median overall survival (OS) (primary end point) was 15.8 months for patients in the abiraterone acetate treatment group compared to 11.2 months for patients in the placebo group. Interim analysis of prostate-specific antigen (PSA) progression (secondary end point) clearly showed the superiority of abiraterone acetate compared to placebo (10.2 months compared to 6.6 months). The PSA response rate was also high: 29% compared to 6% for placebo.[13] Most common serious adverse events for abiraterone acetate versus placebo included fluid retention (30.5% vs. 22.3%), hypokalemia (17.1% vs. 8.4%), hypertension (9.7% vs. 7.9%), hepatic transaminase abnormalities (10.4% vs. 8.1%), and cardiac abnormalities (13.3% vs. 10.4%).[14] The recommended dose of abiraterone acetate is 1000 mg administered orally once daily in combination with prednisone 5 mg administered orally twice daily.

2. AFLIBERCEPT (OPHTHALMOLOGIC, MACULAR DEGENERATION)[15–22]

Class:	Recombinant fusion protein
Country of origin:	United States
Originator:	Regeneron Pharmaceuticals
First introduction:	United States
Introduced by:	Regeneron Pharmaceuticals
Trade name:	Eylea®
Expression system:	Recombinant Chinese hamster ovary CHO-cell line
CAS registry no:	862111-32-8
Molecular weight:	~115 kDa

In November 2011, the U.S. FDA approved the recombinant fusion protein aflibercept, administered as an intravitreal injection, for the treatment of patients with neovascular (wet) age-related macular degeneration (AMD).[15] Macular degeneration results from damage to the macula, the part of the eye responsible for sharp, detailed central vision.[16] AMD, which is the most common cause of vision loss in people over 60, is classified as nonneovascular (dry) or neovascular (wet). All cases of AMD begin as the dry form, but 10% progress to the wet form. Wet AMD results from abnormal growth of blood vessels in the retina. Damage due to leakage of blood and fluid can occur rapidly and lead to scarring of the retina and severe central vision loss. About 200,000 new cases of wet AMD are diagnosed each year in North America; the U.S. National Eye Institute estimates that the prevalence of advanced AMD will grow to nearly 3 million by 2020.[16] One approach for treatment of AMD involves inhibition of vascular endothelial growth factor (VEGF), which plays a key role in ocular neovascularization (angiogenesis).[17] Previously approved anti-VEGF agents developed for ophthalmic use are pegaptanib, an aptamer targeting the $VEGF_{165}$ isoform, and ranibizumab, a recombinant VEGF-specific antibody fragment. In addition, the anticancer agent bevacizumab (Avastin), a monoclonal VEGF-specific antibody, has been used off-label to treat wet AMD. Aflibercept differs from these agents in that it acts as a soluble decoy receptor that binds all VEGF isoforms more tightly than their native receptors, thereby diverting VEGF from its normal function.[18] Aflibercept consists of a fusion of the second Ig domain of human VEGF receptor 1 and the third Ig domain of human VEGF receptor 2 fused to the constant region of (Fc) of human immunoglobulin G1. Aflibercept is a dimeric glycoprotein with a protein molecular weight of 97 kDa and glycoside molecular weight of 18 kDa. It is produced in recombinant Chinese hamster ovary (CHO) cells that overexpress the fusion protein. Aflibercept has subpicomolar affinity for VEGF-A (K_D=0.66 pM for VEGF-A_{165} and 0.19 pM for VEGF-A_{121}), the major driver of pathological angiogenesis and vascular leak in wet AMD.[19] Ranibizumab and bevacizumab bind VEGF-A_{165} with lower affinity (K_D=20.6 and 35.1 pM, respectively). In addition, the association rate for aflibercept binding to VEGF-A was orders of magnitude faster than for ranibizumab and bevacizumab.

Following intravitreal administration of aflibercept to patients with wet AMD, the C_{max} of free aflibercept in plasma was 0.02 μg/mL, with a T_{max} of 1–3 days.[20] Aflibercept was undetectable in the plasma after 2 weeks. The safety and efficacy of aflibercept in patients with wet AMD were evaluated in two 52-week randomized, double-masked, active-controlled studies (VIEW-1 and VIEW-2 trials).[20,21] The primary endpoint was the proportion of patients who maintained vision, as defined by losing fewer than 15 letters on the best corrected visual acuity (BVCA) scale. A secondary endpoint was the change in BVCA from baseline as measured by a letter score. Aflibercept was dosed intravitreally in one of three regimens: 2 mg every 4 weeks followed by 2 mg every 8 weeks, 2 mg every 4 weeks, or 0.5 mg every 4 weeks. The active control ranibizumab was given at 0.5 mg every 4 weeks. Efficacy results for either 2 mg dose regimen of aflibercept were equivalent to ranibizumab in both the primary and secondary endpoints. The proportions of patients that maintained visual acuity were 94% in VIEW-1 and 95% in VIEW2 for aflibercept at 2 mg every 4 weeks followed by 2 mg every 8 weeks; the same response rate was seen with ranibizumab at 0.5 mg every week. The most common adverse events (≥5%) for aflibercept were similar to those for ranibizumab and included conjunctival hemorrhage (25%), eye pain (9%), cataract (7%), vitreous detachment (6%), vitreous floaters (6%), and increased ocular pressure (5%). The incidence of immunoreactivity to aflibercept was similar before and after treatment for 52 weeks (1–3%). The recommended dose for aflibercept is 2 mg administered by intravitreal injection every 4 weeks for the first 3 months, and then every 8 weeks thereafter. Aflibercept is also undergoing extensive evaluation as an anticancer agent.[22]

3. APIXABAN (ANTITHROMBOTIC)[23–37]

Class:	Factor Xa inhibitor
Country of origin:	United States
Originator:	Bristol Myers Squibb Company
First introduction:	European Union
Introduced by:	Bristol-Myers Squibb Company and Pfizer Inc.
Trade name:	Eliquis®
CAS registry no:	503612-47-3
Molecular weight:	459.5

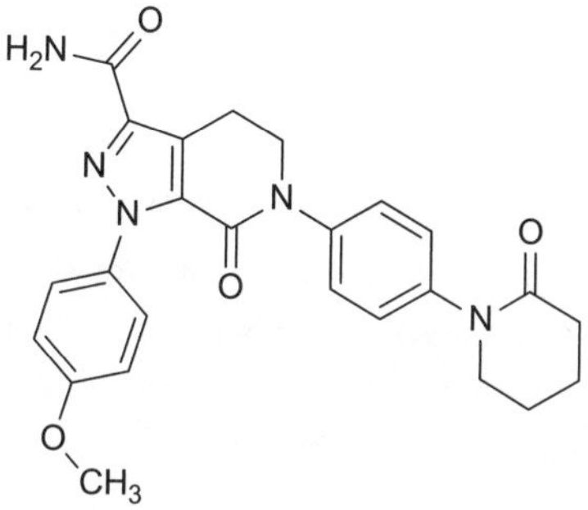

Eliquis® (apixaban), a direct inhibitor of factor Xa (FXa), was approved by the European Commission on May 18, 2011 for prevention of venous thromboembolic events (VTE) in adult patients who have undergone elective hip or knee replacement surgery. VTE, a condition characterized by clot formation in a vein, is a potentially lethal and debilitating cardiovascular condition. VTE is the third most common cardiovascular disease after stroke and acute coronary syndrome (ACS) and can result from hereditary conditions, acquired risk factors, and surgery. Both deep vein thrombosis (DVT) and pulmonary embolism (PE) are disease conditions of VTE. DVT describes clot formation in the veins of the legs and pelvis. PE results when a dislodged clot migrates to the veins in the lungs. Hip or knee replacement surgery greatly increases the risk of DVT by 40–60% with the risk of VTE persisting 3 months post surgery.[23–25] Anticoagulant prophylaxis after surgery reduces the risk of VTE, with studies showing the duration of prophylaxis treatment as an important factor to reducing the incidence and severity of DVT.[25,26] Warfarin, low-molecular-weight heparins (LMWH), and fondaparinux are approved agents used for the prophylaxis of VTE, but these drugs have some limitations. Warfarin has a narrow therapeutic window resulting in the need for frequent monitoring and dose adjustments and requires careful diet control. LMWHs require injection with the timing of administration being important to reduce surgical bleeding complications.[27] FXa is a blood coagulation cascade serine protease that converts prothrombin to thrombin, setting in motion the formation of fibrin which is then cross-linked to form a fibrin rich clot. After vascular injury, the extrinsic pathway of blood coagulation is initiated by tissue factor binding to FVII to generate FVIIa. The resultant tissue factor–FVIIa complex cleaves a single amino acid from the heavy chain of FX to produce FXa. The assembly of FVa, FXa, and calcium ions on a phospholipid membrane (the tenase complex) then converts prothrombin to thrombin. This initially generated thrombin amplifies the coagulation cascade by converting FV to FVa, FXIII to FXIIIa, and FXI to FXIa. The activation of FXIa initiates the intrinsic pathway which leads to increased concentration of FXa, thrombin, and FIXa. The resulting increase in thrombin production accelerates the formation of fibrin.[28–31] With the central role of FXa in producing thrombin by both the intrinsic and extrinsic pathways of blood coagulation established and with the favorable results from pharmacological studies using selective peptide inhibitors isolated from bats, ticks, and snakes, efforts to identify drug inhibitors of FXa were initiated.[32] The discovery of apixaban was the culmination of a succession of novel and innovative medicinal chemistry discoveries starting with the identification of nonpeptide leads, rational drug design using computer-aided and X-ray crystallographic information, and the building of drug-like properties through the systematic replacement of basic groups with neutral moieties. One remarkable finding was the discovery of exquisitely potent picomolar inhibitors with high selectivity over other serine proteases in the blood coagulation cascade.[33] Apixaban arose from modifications to razaxaban by constraining a pyrazole amide to form a bicyclic pyrazolo-pyridinone scaffold. Optimization of the P1 group resulted in the identification of the nonbasic methoxy phenyl group, while a P4 piperidinone improved the balance of potency and pharmacokinetics with low Vdss. The synthesis of apixaban begins with the generation of a hydrazone of 4-methoxyaniline which is then used in a 3+2 cycloaddition with a dihydropiperidinone to form a bicyclic pyrazolo-pyridinone scaffold. The distal piperidinone group is installed using an Ullmann coupling reaction followed by aminolysis of an ethyl ester on the pyrazole ring to complete the synthesis of apixaban.

Apixaban inhibits human FXa with a $K_i = 80$ pM. After oral administration in a dog PK study, apixaban was shown to be 58% bioavailable with a slow clearance rate (0.02 L/Kg/h), low volume of distribution (0.2 L/kg), and a half-life of 5.8 h.[33] In the rabbit arteriovenous (A/V) shunt model, apixaban inhibited thrombus formation in a dose-dependent manner with an IC_{50} value of 329 nM.[34]

Oral administration of apixaban in a single-dose study in male healthy volunteers showed rapid absorption reaching T_{max} in 1 h (0.50–2.00) with an average terminal half-life of 12.7 h $\pm$ 8.55.[35] A study of the metabolism and drug disposition of [^{14}C]-apixaban showed excretion in the feces (56% of recovered dose) and urine (24.5–28.8% of recovered dose). The parent drug is the major component in plasma, urine, and feces. The major metabolite of apixaban, the sulfate of *O*-desmethyl-apixaban, is inactive. Apixaban has been studied in nine Phase III studies assessing its role in preventing VTE in patients undergoing total knee replacement (TKR) and total hip replacement (THR). Apixaban has also been studied in the secondary prevention of cardiovascular events in ACS and in the prevention of stroke in patients with atrial fibrillation. In the apixaban versus acetylsalicylic acid to prevent strokes trial, 5600 patients with atrial fibrillation who were not eligible for warfarin were given either 5 mg of apixaban bid or aspirin (81 or 324 mg) for 36 months or until end of study. The trial was stopped early due to the significant benefit found in the apixaban dosing group. Apixaban was found to reduce the risk of stroke or systemic embolic events 54%, was superior to aspirin, and showed no increased risk of major bleeding. In the 18,000 patient Apixaban for the Prevention of Stroke in Subjects with Atrial Fibrillation (ARISTOTLE) trial, apixaban 5 mg twice daily was compared to warfarin in patients with atrial fibrillation. In the ARISTOTLE trial, apixaban showed superiority to warfarin in preventing stroke or systemic embolism. In comparison to warfarin, apixaban was found to cause less bleeding and resulted in lower mortality.[36] Four clinical trials have been conducted in patients undergoing major orthopedic surgery. The Phase II APROPOS (apixaban prophylaxis in patients undergoing total knee replacement) study compared apixaban (qd and bid doses) to warfarin and enoxaparin. In this study, apixaban dose dependently decreased the incidence of VTE. The Phase III studies ADVANCE-1 and ADVANCE-2 assessed apixaban in TKR patients, while the ADVANCE-3 study assessed apixaban in THR patients. The ADVANCE trials all used an oral dose of 2.5 mg bid apixaban. In the ADVANCE-2 trial, apixaban was shown to be superior to enoxaparin in the composite end-point of DVT, PE, and all-cause mortality, with the major bleeding rates being comparable between the two drugs. ADVANCE-3 also demonstrated apixaban to be superior to enoxaparin with similar bleeding risk.[27,37] These studies formed the basis for the approval by the European Commission of Eliquis® 2.5 mg (apixaban) on May 18, 2011 for prevention of venous thromboembolism after elective hip or knee replacement surgery. Eliquis® is an oral, direct inhibitor of FXa and is codeveloped by Bristol-Myers Squibb and Company and Pfizer Inc.

4. AVANAFIL (MALE SEXUAL DYSFUNCTION)[38–45]

Class:	PDE5 inhibitor
Country of origin:	Japan
Originator:	Mitsubishi Tanabe Pharma Corporation
First introduction:	South Korea
Introduced by:	JW Choongwae Pharmaceutical Corp, Vivus, Inc., Mitsubishi Pharma Corporation
Trade name:	Zepeed
CAS registry no:	330784-47-9
Molecular weight:	483.95

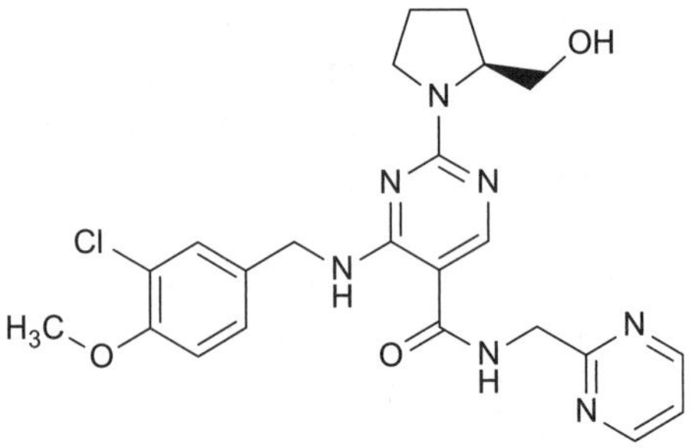

Avanafil (Zepeed) was approved by the Korean Health Ministry for the treatment of erectile dysfunction (ED) in August 2011.[38] Avanafil is a highly selective type 5 phosphodiesterase (PDE5) inhibitor. Other PDE5 inhibitors on the market are Pfizer's Viagra® (sildenafil), Eli Lilly & Co.'s Cialis® (tadalafil), Bayer's Levitra® (vardenafil), and Dong-A Pharmaceutical Co.'s Zydena® (udenafil). PDE5 inhibitors have revolutionized the treatment of ED by providing an efficacious, convenient, and safe treatment option to patients. The physiological event for stimulation and maintenance of an erection is governed by vascular pressure changes in the corpora caverona through the nitric oxide (NO)/cyclic guanosine monophosphate (cGMP) pathway. Increase in the release of NO causes the relaxation of smooth muscle cells which increases blood flow in the penis.[39] Inhibition of PDE5 prevents the cleavage of cGMP thus prolonging NO production. Of the five PDE5 inhibitors on the market, there are differences with regard to selectivity. Sildenafil, vardenafil, and avanafil show some PDE6 activity, while tadalafil shows some inhibition of PDE11. PDE6 is located in retinal photoreceptors and a side effect of PDE6 inhibition seen in some patients is color vision abnormalities. PDE11 is located in the testes and prostrate, but no side effects have been ascribed to inhibition of this enzyme. Avanafil is reported to be the most selective PDE5 inhibitor on the market. The onset of T_{max} and half-life also varies among the marketed PDE5 inhibitors. Sildenafil has a T_{max} at 1 h and a half-life of 3–5 h. Vardenafil is somewhat similar with a T_{max} of 0.6 h and a half-life of 4–6 h. Tadalafil has the longest half-life among the marketed drugs with a half-life of 17 h. Avanafil has a fast onset of action reaching T_{max} in 0.6 h with a half-life of 1.2 h.[40] A synthesis of avanafil (TA-1790) is described in the patent literature.[41] The pharmacokinetic profile of avanafil was studied in two trials, a single dose (200 mg) study and a repeat bid dosing study over 7 days. From these studies, the T_{max} for avanafil was found to be 0.5–0.7 h, with a mean half-life of 1.1–1.2 h. Metabolism studies showed that avanafil is metabolized by CYPs with no metabolite being more potent or having a longer half-life than avanafil.[42,43] The main elimination route of avanafil is through the bile and feces. Avanafil was also found to be reabsorbed through enterohepatic recirculation. In a 16-week Phase III study (390 patients), avanafil at 100 and 200 mg doses showed positive results on erectile function in men with both diabetes and ED. The effectiveness rate rose from 32% to 54% with the 100 mg dose, and from 42% to 63% with the 200 mg dose, versus an increase of 36% to 42% in the placebo group.[44] In another Phase III trial studying 646 subjects over 12 weeks at 50, 100, and 200 mg doses, all avanafil doses were shown to be more effective than placebo. Successful intercourse was demonstrated $>$6 h post dose in the avanafil dose group (59–83% success) versus placebo (25% success).[45] Avanafil (Zepeed) was approved by the Korean Health Ministry in August 2011 at doses of 100 and 200 mg for the treatment of ED and is being comarketed by JW Choongwae Pharmaceutical Corp, Vivus, Inc., and Mitsubishi Pharma Corporation.

5. AZILSARTAN MEDOXOMIL (ANTIHYPERTENSIVE)[46–52]

Class:	Angiotensin II receptor blocker
Country of origin:	United States
Originator:	Takeda
First introduction:	United States
Introduced by:	Takeda
Trade name:	Edarbi®
CAS registry no:	863031-21-4 (free acid) 863031-24-7 (salt)
Molecular weight:	568.53 (free acid) 607.63 (salt)

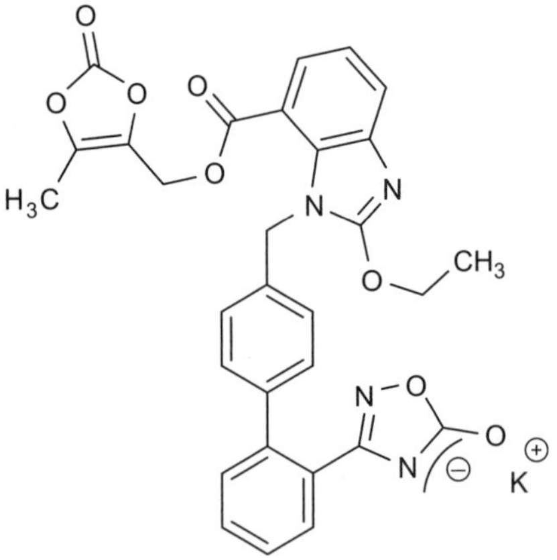

Azilsartan medoxomil (Edarbi®), an angiotensin II receptor antagonist, was approved by the U.S. FDA in February 2011 for the treatment of hypertension in adults. Hypertension is a major risk factor for cardiovascular disease and stroke. A report from the CDC on hypertension trends in the United States over 1999–2008 showed that the prevalence of hypertension remains at 30% of the population (74.5 million individuals).[46] Some of the treatment options available for patients are drugs that act on the renin–angiotensin system (RAS). The RAS modulates both blood pressure (BP) and fluid homeostasis. Stimulation of the angiotensin II type I receptor (AT_1), a G-protein coupled receptor (GPCR) located on the vasculature, by the endogenous octapeptide angiotensin II (Ang-II) causes vasoconstriction. Ang-II also interacts with AT_1 receptors in the kidney causing modulation of sodium and water reabsorption. The formation of Ang-II results from the cleavage of Ang-I by angiotensin converting enzyme, which, in turn, is formed as the cleavage product of angiotensinogen by renin. Antagonists of the AT_1 receptor limit the vasoconstrictive effects set in motion by Ang-II.[47] There are eight approved angiotensin receptor blockers (ARB) on the market. The discovery of azilsartan was the result of a medicinal chemistry effort to identify an ARB with a different carboxylic acid isostere than the ones found in the marketed ARBs. Several of the marketed ARBs use a tetrazole group as a carboxylic acid isostere. The medicinal chemistry approach that led to azilsartan involved the replacement of this commonly used tetrazole with a 5-oxo-1,2,4-oxadiazole group. Azilsartan can be synthesized by Suzuki coupling of *p*-tolyl boronic acid to 2-bromobenzonitrile, followed by bromination of the methyl group. The bromide is displaced to introduce a protected 2-ethoxy-1H-benzo[d]imidazole-7-carboxylate. The cyano group is converted to a hydroxylamidine, followed by reaction with an alkyl-chloroformate and intramolecular cyclization to form the 5-oxo-1,2,4-oxadiazole ring.[48] The acid is then deprotected and converted to a prodrug. The parent, azilsartan has been extensively characterized *in vitro* and compared with other marketed AT_1 antagonists olmesartan, valsartan, telmisartan, irbesartan, and candesartan. Azilsartan was found to be a potent ($IC_{50}=2.6$ nM), selective, inverse agonist of the AT_1 receptor. From washout experiments, azilsartan was found have slow dissociation from the receptor and thus is characterized as an insurmountable antagonist.[49] This feature is proposed to provide the long acting clinical efficacy that has been observed in Phase III trials.

In human studies, azilsartan medoxomil was found to be rapidly hydrolyzed to azilsartan. The pharmacokinetic profile from Phase I studies showed a 12-h half-life and 60% bioavailability. Azilsartan medoxomil is metabolized by CYP2C9. A Phase III trial with 1291 patients having a baseline 24-h mean systolic BP of 145 mm Hg evaluated azilsartan medoxomil at 40 and 80 mg doses in comparison to valsartan (320 mg) and olmesartan medoxomil (40 mg). The primary endpoints of this study were improvement in ambulatory BP and clinical BP as measured by a change from baseline in 24-h mean systolic BP. In this study, azilsartan medoxomil (80 mg) was found to be superior to valsartan (320 mg) and olmesartan (40 mg) in lowering 24-h systolic BP.[50,51] Another Phase III trial compared azilsartan medoxomil (20, 40, and 80 mg) to olmesartan medoxomil (40 mg) in patients (n=1275) with primary hypertension and baseline 24-h mean ambulatory systolic pressure between 130 and 170 mm Hg over 6 weeks of treatment. In this study, azilsartan medoxomil (80 mg) was found to be superior to olmesartan medoxomil (40 mg) by showing an additional 2.1 mm Hg drop in systolic BP. In both of these studies, azilsartan medoxomil was safe and well tolerated.[52] The results from the Phase III studies suggest that within the ARB class of drugs, azilsartan medoxomil could provide higher efficacy in lowering hypertension. Azilsartan medoxomil (Edarbi®) 40 and 80 mg once daily was approved by the U.S. FDA in February 2011 for the treatment of hypertension in adults. The 40 mg dose is recommended for patients who are being treated with high-dose diuretics to reduce salt in the body.

6. BELATACEPT (IMMUNOSUPRESSIVE)[53–61]

Class:	Recombinant fusion protein
Country of origin:	United States
Originator:	Bristol-Myers Squibb
First introduction:	United States
Introduced by:	Bristol-Myers Squibb
Molecular weight:	~90 kDa
Trade name:	Nulojix®
Expression system:	Rodent CHO-cell line
CAS registry no:	706808-37-9

Belatacept is an immunosuppressive that was approved by the U.S. FDA for prophylactic prevention of rejection in kidney transplant recipients. Approximately 16,000 kidney transplants are performed each year in the United States, and graft rejection is typically prevented through post-transplant induction of immunosuppression with antithymocyte globulin or interleukin 2 receptor antagonists (basilixumab), a calcineurin inhibitor (tacrolimus or cyclosporine A), and concomitant agents such as mycophenolate mofetil and corticosteroids.[53,54] Belatacept, which selectively blocks T cell costimulation, provides an alternative to the calcineurin inhibitors, which are associated with adverse effects on renal function as well as cardiovascular and metabolic parameters. T-cell activation by antigen-presenting cells (APCs) requires a costimulatory signal in addition to the signal mediated by interaction of the T-cell receptor and MHC-bound peptide antigen. Belatacept can block that second signal by binding to APC costimulatory surface proteins CD80 and CD86 to inhibit their interaction with the T-cell signaling molecule CD28. Belatacept is a recombinant protein comprising a hinge-region modified (to reduce Fc receptor binding) carboxyl-terminal human IgG1 Fc domain fused to a modified extracellular domain of human CTLA-4 that includes two amino acid substitutions in the ligand-binding site.[55] As a result of this modification, belatacept binds more tightly to CD80 and CD86 than the parent molecule abatacept (Orencia®, FDA approved for treatment of adult RA and juvenile idiopathic arthritis), from which its structure is derived. CTLA-4 is a T-cell negative signaling molecule that is structurally related to CD28 and shares its two ligands. While belatacept incorporates CTLA-4 sequences to bind CD80 and CD86, the immunosuppressive activity of the drug suggests that its principal mechanism of action is blockade of CD28-mediated T-cell activation.

Belatacept is formulated as a lyophilized powder for intravenous infusion. It is dosed at 10 mg/kg on the day of transplantation and again on days 5, 14, 28, 56, and 84. It is then dosed at 5 mg/kg every 4 weeks beginning 16 weeks after transplantation. At the 10 mg/kg dose, the peak concentration of the drug is 247 ± 68 μg/mL and the terminal half-life is 9.8 ± 3.2 days. At the 5 mg/kg dose, the peak concentration of the drug is 139 ± 28 μg/mL and the terminal half-life is 8.2 ± 2.4 days. Belatacept reactive antibodies were detected in approximately 8% of 372 patients tested prior to treatment, and in another 2% post treatment. Development of antidrug antibodies was not associated with altered clearance. Belatacept was tested in two randomized, controlled Phase III trials. One of these trials randomized 686 patients receiving a kidney transplant from living or standard criteria deceased donors (the BENEFIT study[56–59]), and the second randomized 578 patients receiving extended criteria donor kidneys (the BENEFIT-EXT study[58–61]). Patients receiving extended criteria donor kidneys have a lower probability of allograft survival over time due to the more marginal quality of these organs. For both studies, patients were randomized 1:1:1 to 3 arms. One cohort received cyclosporine

A, and the other two cohorts received belatacept under two different dosing regimens: the prescribed dosing regimen and a more intense regimen that included more frequent and higher doses of drug over the first 6 months. Overall efficacy outcomes were similar for the two different belatacept dosing regimens. Acute rejection episodes were reported more frequently in belatacept-treated patients than cyclosporine-treated patients. In the BENEFIT study, acute rejection episodes at 3 years were reported in 19.9% of patients receiving the recommended dose of belatacept compared to 10.4% of patients receiving cyclosporine A, while graft loss frequencies were 2.2% and 3.6%, respectively. Ninety-one percent of patients receiving the recommended dose of belatacept were alive with functioning grafts at 3 years, compared to 86.9% of the cyclosporine A-treated patients. In the BENEFIT-EXT study, acute rejection episodes at 3 years were reported in 24% of prescribed-dose belatacept-treated patients compared to 22.8% of cyclosporine A-treated patients, while graft loss frequencies were 12% and 12.5%, respectively; and 81.7% of prescribed-dose belatacept-treated patients were alive with functioning grafts at 3 years, compared to 77.7% of the cyclosporine A-treated patients. Overall kidney graft function, as assessed by glomerular filtration rate, was better at 3 years in patients receiving the recommended belatacept regimen (65.8 and 42.2 mL/min/1.73 m^2 for BENEFIT and BENEFIT-EXT, respectively) versus cyclosporine A-treated patients (44.4 and 31.5 mL/min/1.73 m^2 for BENEFIT and BENEFIT-EXT, respectively). As with all immunosuppressive therapies, malignancies and infections are a clinical concern. The principle safety concerns associated with belatacept are posttransplant lymphoproliferative disorder (PTLD), of which the predominant site of presentation was the central nervous system, and serious infections, including progressive multifocal leukoencephalopathy (PML). At 3 years of follow-up, 1.2% of patients receiving the recommended belatacept regimen had developed PTLD versus 0.2% of patients receiving cyclosporine A. Lack of immunity to Epstein–Barr virus was identified as the greatest risk factor for development of PTLD with belatacept, and therefore, use in such patients is contraindicated. Two cases of fatal PML have been reported in clinical studies in patients receiving the more intensive belatacept dosing regimen. No cases of PML have been reported in patients receiving the recommended dosing regimen.

7. BELIMUMAB (IMMUNOSUPPRESSIVE, LUPUS)[62–68]

Class:	Recombinant monoclonal antibody
Country of origin:	United States
Originator:	Human Genome Sciences
First introduction:	United States
Type:	Human IgG1λ, anti-BLys
Introduced by:	Human Genome Sciences
Molecular weight:	~147 kDa
Trade name:	Benlysta®
Expression system:	Mammalian cell line
CAS registry no:	356547-88-1

Belimumab was approved by the U.S. FDA in March 2011 for treatment of patients with active, autoantibody-positive, systemic lupus erythematosus (SLE). SLE is a serious, and sometimes fatal, autoimmune disease that can affect multiple organs. Prevalence estimates for SLE vary widely across populations. It was reported to be the underlying cause of over 1400 U.S. deaths in 1998.[62] Existing treatment options include broad-based immunosuppressive drugs such as corticosteroids, hydroxychloroquine, cyclophosphamide, azathioprine, mycophenolate mofetil, and methotrexate. Prior to belimumab, no new drugs have been approved for SLE in the past 50 years.[63] Belimumab is a human sequence monoclonal antibody that binds to human B lymphocyte stimulator protein (BLys; also known as BAFF, THANK, TALL-1, TNFSF13B, and zTNF4), which is overexpressed in SLE patients. BLys is expressed on the cell surface of monocytes, macrophages, and monocyte-derived dendritic cells and is cleaved to form a soluble cytokine that stimulates cell proliferation and antibody secretion in B lineage cells that express the BLys receptor BAFF-R (also known as BR-3). The precise function of two additional BLys receptors, TACI and BCMA, is less clear. Belimumab inhibits BLys interaction with these three receptors leading to measurable reductions in autoantibodies, such as anti-dsDNA, and specific circulating B cell lineage compartments including antibody producing plasma cells.[64] Belimumab was discovered by screening molecules from a phage display library of human single-chain Fv sequences for BLys binding and functional blocking, followed by optimization of the lead through the introduction of variant residues in V_H CDR3.[65]

Belimumab is formulated as a lyophilizate for intravenous infusion and is administered as a 1-h infusion at 10 mg/kg every 2 weeks for the first three doses, followed by a maintenance regimen of 10 mg/kg every 4 weeks. At this dose and schedule, the mean C_{max} was 313 μg/mL and the observed terminal half-life was 19.4 days. The single-dose C_{max} and terminal half-life at 10 mg/kg were reported to be 192.4 ± 34.9 μg/mL and 10.63 ± 2.89 days, respectively.[66] Antidrug antibodies were detected at a frequency of 0.7%; however, high drug levels could have interfered with the assay as the antidrug antibody frequency was higher (4.8%) in patients treated at a lower dose (1 mg/kg). Two placebo-controlled, randomized Phase III trials of belimumab have been reported. Both trials enrolled autoantibody-positive (antinuclear or anti-dsDNA) patients with active disease (SELENA-SLEDAI score ≥ 6) and included three different dosing arms, randomized 1:1:1, to receive placebo, 1 mg/kg, or 10 mg/kg belimumab on days 0, 14, and 28, and then every 28 days. Patients with severe active lupus nephritis or CNS disease were excluded. The two trials differed in length, with patients dosed either through week 52 (867 patients, BLISS-52 trial[67]) or through week 72 (819 patients, BLISS-76 trial[68]). The primary endpoint for both trials was the response rate at week 52 as measured by the systemic lupus erythematosus responder index, which is based on improvement in disease activity without worsening of the overall disorder or development of substantial disease activity in new organ systems. The response rates for the 10 mg/kg belimumab-treated cohorts in both trials were significantly greater than reported for the placebo-treated groups (58% vs. 44%, $p=0.0129$; and 43.2% vs. 33.5%, $p=0.017$). The response rate at 76 weeks was measured in the BLISS-76 trial and was reported to be 38.5% in the belimumab 10 mg/kg cohort compared to 32.4% ($p=0.13$) in the placebo group. The incidence of severe SLE flares over 76 weeks in the belimumab cohort was 20.5% compared to 26.5% ($p=0.13$) in the placebo group. In the BLISS-52 trial, the incidence of severe flares over 52 weeks was 14% in the belimumab 10 mg/kg cohort compared to 23% ($p=0.0055$) for the placebo group. For both trials, the rates of adverse events, laboratory abnormalities, and infections were similar for the belimumab 10 mg/kg and placebo groups. There were no malignancies in any of the cohorts for the BLISS-52 trial, and 7 in the BLISS-76 trial (2 and 1 in the belimumab 10 mg/kg and placebo groups respectively). Rates of serious infections (grade 3 or 4) for the belimumab 10 mg/kg groups in the two trials were 2% and 2.6%, compared with 3% and 4% for the placebo groups. The death rate for both groups was 1% in the BLISS-52 trial. There were no reported deaths in the placebo group in the BLISS-76 trial, and 1 death in the belimumab 10 mg/kg group.

8. BOCEPREVIR (ANTIVIRAL)[69–77]

Class:	Hepatitis C virus NS3-4A protease inhibitor
Country of origin:	United States
Originator:	Merck/Schering
First introduction:	United States
Introduced by:	Merck
Trade name:	Victrelis®
CAS registry no:	394730-60-0
Molecular weight:	519.68

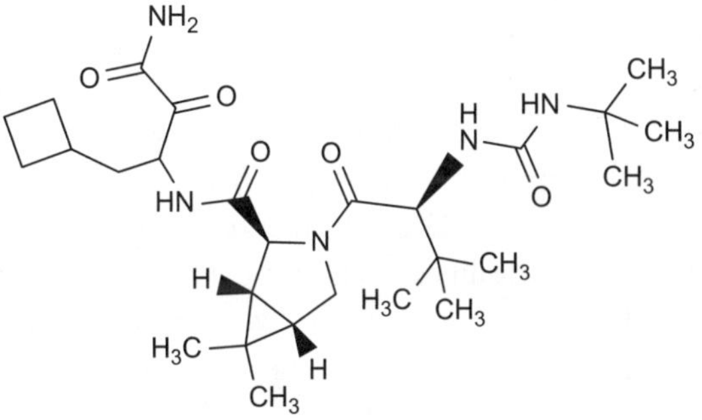

In May 2011, the U.S. FDA approved boceprevir (SCH-503034), to be given in combination with peginterferon alfa plus ribavirin, for the treatment of patients with chronic hepatitis C genotype 1 viral infection.[69] Boceprevir and telaprevir (*vide infra*) are the first hepatitis C virus (HCV) protease inhibitors to be approved for the treatment of HCV infection.[70] More than 170 million people are chronically infected with HCV worldwide, with >350,000 people dying each year from HCV-related liver diseases.[71] Chronic HCV infection causes inflammation of the liver and can lead to diminished liver function, liver failure, and liver cancer. While no HCV vaccine is available, HCV infection can be cured in some patients by treatment with a combination of peginterferon and ribavirin (PR). However, HCV genotype 1, which is the most prevalent subtype in the United States, Europe, and Japan, is the least responsive to PR therapy with rates of sustained virologic response (SVR) of less than 50%. Boceprevir is an inhibitor of HCV NS3-4A protease, an essential enzyme required by HCV for posttranslational processing of viral proteins into their mature forms. Boceprevir binds covalently, but reversibly, to the active site serine by addition of the hydroxyl group to the keto-amide functionality. Boceprevir inhibits HCV NS3-4A protease with a K_i of 14 nM. In cell culture, the EC_{50} of boceprevir was 200 nM for an HCV replicon constructed from genotype 1b. Boceprevir was two- to threefold less potent against HCV replicon from genotypes 1a, 2, and 3. The potency of boceprevir decreased threefold in the presence of human serum. Boceprevir was discovered through a series of systematic truncations and modifications of a keto-amide undecapeptide lead molecule.[72] Structure-based drug design and crystallography were used extensively in the design process, particularly in optimization of the P1 (cyclobutylalanine) and P2 (dimethylcyclopropylproline) moieties to impart potency and selectivity, and introduction of the keto-amide moiety to improve potency.[73] Boceprevir is synthesized by coupling of 3-amino-4-cyclobutyl-2-hydroxybutyramide or the related oxobutyramide with a cyclopropyl-pyrrolidine carboxylic acid intermediate. The pyrrolidine derivative can be prepared via cyclopropanation of a bicyclic lactam derivative or by conversion of 3,3-dimethylcyclopropane-1,2-dicarboxylic acid to the pyrrolidine in a multistep route.[74] Boceprevir is a 1:1 mixture of diastereomers at the readily epimerizable position α to the keto group. Boceprevir showed 26% oral bioavailability in rats and a liver/plasma ratio of ~30.

The pharmacokinetics of boceprevir were evaluated in a number of Phase I clinical studies at doses of 50–800 mg.[75] Pharmacokinetic profiles were similar between healthy, HCV-infected, and renally-impaired subjects. Boceprevir was rapidly absorbed with a mean T_{max} of 1–2.25 h and a mean terminal phase half-life of 7–15 h across the dosing range. Food enhanced absorption by up to 65% at the 800 mg dose. The C_{max} of boceprevir given at 800 mg three times a day alone was ~1900 ng/mL. The volume of distribution (V_d/F) was 772 L at steady state, which was achieved after 1 day of three times daily dosing. Boceprevir is 75% protein bound in human plasma. The major route of metabolism is reduction by aldoketoreductase to give an inactive ketone-reduced metabolite. Metabolism by CYP-mediated oxidation occurs to a lesser extent. Boceprevir itself is an inhibitor of CYP3A4 and P-glycoprotein. Dosing of ^{14}C-boceprevir showed that the compound is eliminated primarily by the liver, with 79% of radioactivity being recovered in feces and 9% in urine. Boceprevir was evaluated in two Phase III studies in 1500 adult patients who were previously untreated (SPRINT-2)[76] or who had inadequate response on previous PR therapy (RESPOND-2).[77] The key outcome measure was SVR as defined by undetectable plasma levels of HCV-RNA at follow-up week 24. In SPRINT-2, the addition of boceprevir to a PR regimen gave SVR rates of 63–66% for the boceprevir/PR arms compared with 38% SVR for PR alone. The results were similar with 24 and

44 weeks of boceprevir. In RESPOND-2, SVR rates were 59–66% for the boceprevir/PR arms compared with 23% SVR for PR alone. Serious adverse events occurred in ~10% of treated patients. Fatigue, headache, and nausea were the most common clinical adverse events in all treatment groups, with no significant differences between subjects receiving boceprevir and PR or PR alone. Dysgeusia (altered taste) was more than twice as common in boceprevir-treated patients. Anemia was also more common in boceprevir-treated patients than in control subjects, with 42% of boceprevir-treated patients receiving erythropoietin compared with 24% on PR alone. The approved dose of boceprevir (Victrelis®) is 800 mg (four 200 mg capsules) given orally three times daily along with the PR regimen.

9. BRENTUXIMAB VERDOTIN (ANTICANCER)[78–83]

Country of origin:	United States
Class:	Antibody–drug conjugate
Originator:	Seattle Genetics
First introduction:	United States
Type:	Chimeric IgG1κ (anti-CD30), tubulin inhibitor conjugate
Introduced by:	Seattle Genetics
Molecular weight:	~153 kDa
Trade name:	Adcetris™
Expression system (antibody component):	Rodent CHO-cell line
CAS registry no:	914088-09-8

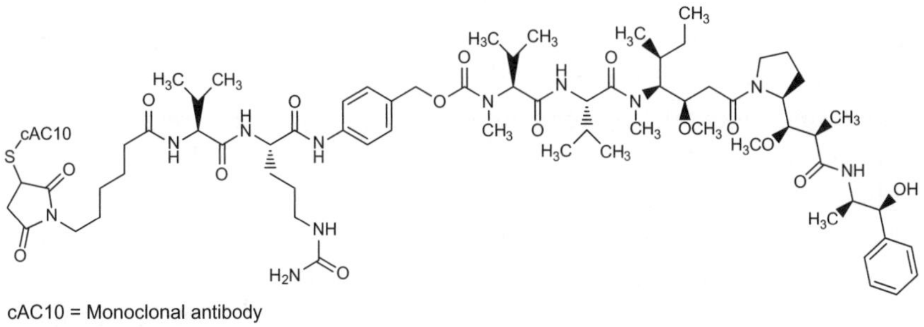

cAC10 = Monoclonal antibody

Brentuximab verdotin was approved by the U.S. FDA in August 2011 for treatment of Hodgkin's lymphoma (HL) in patients who have failed autologous stem cell transplant (ASCT) or ASCT ineligible patients who have failed at least two prior chemotherapy regimens, and for second line treatment of systemic anaplastic large cell leukemia (ALCL). The cell surface target of brentuximab verdotin is the lymphocyte activation marker CD30, a TNF receptor family protein that is abundantly expressed on both HL and ALCL cells. HL and ALCL are rare malignancies. HL is the more common of the two, with an estimated incidence of approximately 9000 new cases in the United States in 2012.[78,79] The majority of patients with either of these cancers achieve durable complete remissions in the front-line setting with conventional combination chemotherapy and radiation. However, a fraction of the patients relapse and are then treated with salvage chemotherapy regimens that can include consolidation with ASCT. Patients who fail this second-line treatment are considered incurable: there are estimated to be over 1000 HL deaths annually in the United States.[78,79] Brentuximab verdotin is a chimeric (mouse V region/human C region) CD30 binding monoclonal antibody (cAC10) that is conjugated via cysteine residues to a small molecule comprising a cysteine reactive dipeptide linker moiety and the microtubule polymerization inhibitor monomethyl auristatin E (MMAE).[80] The antibody component of the drug binds to CD30 expressing tumor cells, and the active cytotoxic component, MMAE, is released by proteolytic cleavage of the dipeptide linker moiety.

The drug is formulated as a lyophilizate for intravenous infusion. It is dosed at 1.8 mg/kg (for patients over 100 kg, the dose is calculated using 100 kg) given as a 30-min infusion every 3 weeks for up to 16 cycles, or until disease progression or unacceptable toxicity. In a Phase I study using this dosing regimen, the observed C_{max} for the intact antibody–drug conjugate was 32 μg/mL (coefficient of variation of the geometric mean = 29%), and the terminal half-life was 4.4 days (coefficient of variation of the geometric mean = 38%).[81] In two Phase II studies using the same dosing regimen, approximately 7% and 30% of the treated patients developed either persistent or transient antibody responses to the chimeric antibody portion of the drug, respectively. Approximately 60% of the antibody-positive samples tested comprised neutralizing antibodies. These two open-label, single-arm, Phase II studies, in HL and ALCL, formed the basis of brentuximab verdotin's regulatory approval.[82,83] The HL Phase II trial enrolled 102 relapsed or refractory, post-ASCT patients. The overall objective response rate was 73%, including 32% CR and 40% PR. The median duration of the responses was 6.7 months, with a 20.5 and 3.5 month median duration for the CRs and PRs, respectively. The ALCL

Phase II trial enrolled 58 relapsed or refractory patients, including 42 patients in the more difficult to treat ALK-negative category. The overall objective response rate was 86%, including 57% CR and 29% PR. The median duration of the responses was 12.6 months, with a 13.2 and 2.1 month median duration for the CRs and PRs, respectively. Pooling data from the 160 patients in these two trials, the most common serious adverse events (grades 3 or 4), were neutropenia (21%), thrombocytopenia (9%), anemia (7%), peripheral sensory (9%), and motor neuropathy (4%).

10. CRIZOTINIB (ANTICANCER)[84–89]

Class:	ALK/c-MET multitargeted receptor tyrosine kinase inhibitor
Country of origin:	United States
Originator:	Pfizer
First introduction:	United States
Introduced by:	Pfizer
Trade name:	Xalkori®
CAS registry no:	877399-52-5
Molecular weight:	450.34

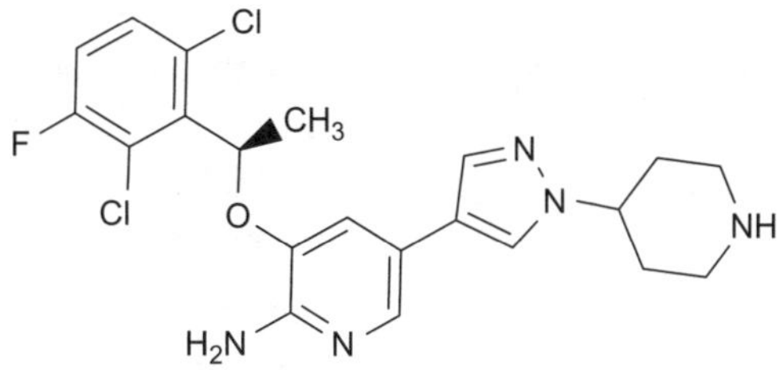

In August 2011, the United States FDA approved crizotinib (PF-02341066) for the treatment of anaplastic lymphoma kinase (ALK) rearranged non-small-cell lung cancer (NSCLC). Lung cancer is the second most common form of cancer in men (after prostate cancer) and in women (after breast cancer). The National Cancer Institute estimates ~220,000 new cases and ~157,000 deaths resulting from lung cancer in 2011.[84] NSCLC accounts for nearly 85% of lung cancer, with 15% as small-cell lung cancer. Chemotherapeutic options include platinum-based therapy with a median survival rate of less than a year. Other treatment options for NSCLC include the epidermal growth factor receptor (EGFR) tyrosine kinase inhibitors gefitinib and erlotinib for patients with EGFR mutations. Crizotinib is recommended for NSCLC that are driven by the ALK oncogene, which is activated by a breakage in chromosome 2 and a refusion of the two fragment genes in the opposite direction. Crizotinib is a dual ATP competitive inhibitor of tyrosine kinases c-MET (Mesenchymal-Epithelial Transition Factor) kinase (cellular $IC_{50}=8$ nM) and ALK (cellular $IC_{50}=20$ nM), both of which are important targets for cancer chemotherapy.[85] When crizotinib was tested for selectivity versus other kinases it was found to have enzyme $IC_{50's}$ within 100-fold multiples of c-MET for 13 of the 120 kinases tested. In cellular assays, crizotinib was found to inhibit RON (recepteur d'origine nantais) kinase with a 10-fold selectivity window over c-MET.[85] In preclinical studies, crizotinib regressed tumor growth within 15 days when administered to severe combined immunodeficient–beige mice bearing Karpas299 anaplastic large cell lymphoma tumor xenografts, at 100 mg/kg/day (oral dosing).[86] A process route to crizotinib has been described recently.[87] Key steps include chemoselective reduction of a nitropyridine group in the presence of halogens using sponge-nickel catalysts in methanol and a selective Suzuki reaction of an advanced 3-bromopyridine intermediate with a pyrazole-based pinacol boronate.

In an open-label, multicenter, Phase I trial involving 167 patients, peak plasma concentrations (C_{max}) were reached 4 h after oral administration of a single dose of 250 mg of crizotinib.[88] The mean absolute bioavailability was determined to be 43%. Increasing pH and a high-fat meal appeared to lower the bioavailability of crizotinib. The half-life of crizotinib was 43–51 h. Crizotinib is a substrate of P-gp and CYP 3A4 and a moderate inhibitor of CYP 3A4. The safety and efficacy of crizotinib (250 mg, BID, oral dosing) was assessed in two multicenter, single-arm studies labeled Study A and Study B.[89] The 136 patients in study A with ALK-positive NSCLC were identified using the Vysis ALK Break-Apart fluorescence *in situ* hybridization probe kit (Abbott Laboratories). The 119 ALK-positive patients in study B were identified using various local clinical trial assays. Except for 15 patients in Study B, all patients were on prior systemic therapy. The primary end point in both trials was overall response rate (ORR), defined as complete responses and partial responses. The ORR in study A was 50% with one complete response. In study B, the ORR was 61% with two complete responses. In a retrospective analysis on overall survival in patients with advanced NSCLC harboring ALK gene rearrangement, the 1-year survival was 77% for crizotinib-treated patients compared with 73% in historical controls, whereas the 2-year survival was 64% versus 33%, respectively. The most common serious adverse events ($>25\%$) for crizotinib were vision disorder, nausea, diarrhea, vomiting, edema, and constipation observed in both studies. The recommended dose of crizotinib is 250 mg administered orally twice daily for the treatment of patients with locally advanced or metastatic NSCLC, as detected by an FDA-approved diagnostic test designed to detect rearrangements of the ALK gene.

11. EDOXABAN (ANTITHROMBOTIC)[24–27,32,90–96]

Class:	Factor Xa inhibitor
Country of origin:	Japan
Originator:	Daiichi Pharmaceutical
First introduction:	Japan
Introduced by:	Daiichi Sankyo Company
Trade name:	Lixiana®
CAS registry no:	480448-29-1 (free base) 480449-71-6 (salt)
Molecular weight:	548.06 (free base) 738.28 (salt)

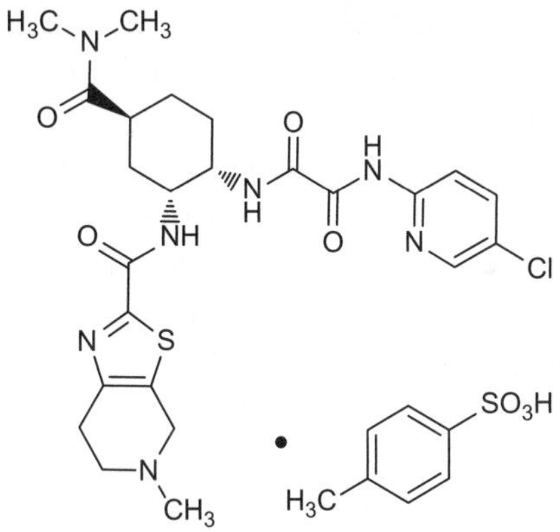

Edoxaban (Lixiana®), a direct inhibitor of Factor Xa (FXa) was approved by the Minister of Health, Labor, and Welfare in Japan in April 2011 for the prevention of venous thromboembolism (VTE) after major orthopedic surgery.[90] VTE, a condition characterized by clot formation in a vein, can result from hereditary conditions, acquired risk factors, and surgery. In the United States, VTE is the third most common cardiovascular disease after stroke and ACS. Both deep vein thrombosis (DVT), a condition describing clot formation in the veins of the legs and pelvis, and pulmonary embolism (PE), a condition that arises from a dislodged clot migrating to the veins of the lung, are disease conditions of VTE. The risk of DVT is increased by 40–60% in patients undergoing hip or knee replacement surgery with the risk of VTE persisting 3 months post surgery.[90,24,25] Anticoagulant prophylaxis after surgery reduces the risk of VTE, with studies showing the duration of prophylaxis treatment as an important factor in reducing the incidence and severity of DVT.[25,26] Drugs in other classes that are used for the postoperative treatment of VTE have some limitations. Warfarin has a narrow therapeutic window resulting in the need for frequent monitoring and dose adjustments and requires food restrictions and careful diet control. Low molecular weight heparins (LMWH) are intravenously administered drugs with the timing of administration important to reduce surgical bleeding complications.[27] The generation of FXa, a blood coagulation cascade serine protease, converts prothrombin to thrombin setting in motion clot formation through the generation of fibrin which when cross-linked forms fibrin rich clot. With central role of FXa in producing thrombin by both the intrinsic and extrinsic pathways of blood coagulation and the favorable results from pharmacological studies using selective peptide inhibitors, efforts to identify drug inhibitors of FXa were initiated.[32] The discovery of edoxaban resulted from the identification of a 1,2-*cis*-diaminocyclohexane template, the discovery of the novel 5-methyl-4,5,6,7-tetrahydrothiazolo[5,4-c]pyridine as an S4 binding element and systematic studies of novel and efficient P1 groups.[91–93] Edoxaban inhibits hFXa with a $K_i = 0.56$ nM. Edoxaban is a weak inhibitor of thrombin $K_i = 6000$ nM, has >10,000 fold selectivity over other serine proteases in the coagulation cascade, and demonstrates selectivity over trypsin and chymotrypsin. The synthesis of this class of FXa inhibitors begins with assembly of the diaminocyclohexane scaffold, *cis*-*t*-butyl ((1*R*,2*S*,5*S*)-2-amino-5-(dimethylcarbamoyl)cyclohexyl)carbamate. The free amine is coupled to the P1 group, 2-((5-chloropyridin-2-yl) amino)-2-oxoacetic acid. The Boc protecting group on the amine of the scaffold is removed and the resulting free amine is coupled to the P4 group, 5-methyl-4,5,6,7-tetrahydrothiazolo[5,4-*c*]pyridine-2-carboxylic acid. In monkey studies, when dosed at 1 mg/kg orally, edoxaban showed a rapid onset of action and reached a C_{max} at 4 h with concentrations detectable at 24 h and a bioavailability of 60%. In rats and rabbits, edoxaban produced a dose-dependent reduction in thrombus formation.[94]

In a single ascending dose study in healthy male subjects at doses of 10, 30, 90, 120, and 150 mg, edoxaban was well tolerated with no adverse effects and showed dose proportional increases in exposure. The half-life in the dose ranges of 30–150 mg were 8.56–10.7 h. The range for the mean renal clearance was 34.7–38.8%. Edoxaban is a substrate for P-gp. There was no effect of food on pharmacokinetics.[95] Edoxaban has been studied in Phase III clinical trials for the prophylactic treatment and prevention of

VTE and for the prevention of stroke in patients with atrial fibrillation. Several Phase III trials assessing edoxaban for the prevention of VTE in TKR and THR have been completed. In the STARS E-3 trial (761 Japanese patients), edoxaban was given at a dose of 30 mg once daily and compared to 20 mg bid of enoxaparin. In this study, edoxaban was shown to be superior to enoxaparin on the primary endpoint of reducing symptomatic PE, and symptomatic and asymptomatic DVT in patients that had undergone total knee surgery. The STARS J-5 trial studied 610 Japanese patients that had undergone hip arthroplasty and used the same dosing regimen as the STARS E-3 trial, with 30 mg once daily of edoxaban and 20 mg bid of enoxaparin. STARS J-5 also demonstrated that edoxaban was superior to enoxaparin in reducing PE and symptomatic and asymptomatic DVT.[96] Edoxaban (Lixiana®) doses of 15 and 30 mg once daily were approved by the Ministry of Health, Labor, and Welfare in Japan on April 22, 2011 for the prevention of VTE after major orthopedic surgery. Lixiana® is being developed and marketed by Daiichi Sankyo Company Limited.

12. ELDECALCITOL (OSTEOPOROSIS)[97–102]

Class:	Vitamin D3 derivative
Country of origin:	Japan
Originator:	Chugai Pharmaceutical/Roche
First introduction:	United States
Introduced by:	Taisho Pharmaceutical Holdings and Chugai Pharmaceutical
Trade name:	Edirol®
CAS registry no:	104121-92-8
Molecular weight:	490.71

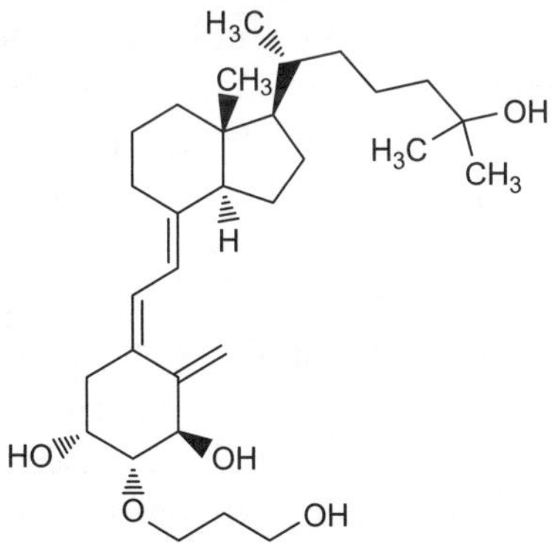

Eldecalcitol (Edirol®) was approved in January 2011 by the Japanese Ministry of Health, Labor, and Welfare for the treatment of osteoporosis. The prevalence of osteoporosis is increasing as the mean age of populations in Japan, Europe, United States, and several other countries increases. Japan, in particular, has one of the longest life expectancies in the world, with a mean age of 77 years for men and 84 years for women. In an aging population, the weakening of the bone matrix from osteoporosis increases the risk of severe fractures, such as spinal and hip fractures, thus creating a burden on quality of life and the health care system.[97] Some of the contributing risk factors are prior incidence of fracture, reduction of bone mineral density, and increasing age. Bone health is dependent on calcium homeostasis, which is maintained through the actions of vitamin D. The active metabolite of vitamin D, 1,25-dihydroxyvitamin D, exerts its action through interactions with the calcitriol receptor, a nuclear hormone receptor that is responsible for calcium absorption and bone formation and depletion.[98] Because of vitamin D's central role in the bone health, vitamin D and analogs of vitamin D have been used to treat patients diagnosed with osteoporosis.[99] Eldecalcitol is an analog of the active form of vitamin D, calcitriol, in which the lower cyclohexane ring contains a hydroxypropyl group. The synthesis of eldecalcitol involves the assembly of two units, a fully protected (3*S*,4*S*,5*R*)-oct-1-en-7-yne-3,4,5-triol and a fused bicyclic system, (*R*)-6-((1*R*,3a*R*,7a*R*,*E*)-4-(bromomethylene)-7a-methyloctahydro-1*H*-inden-1-yl)-2-methylheptan-2-ol, through a Diels-Alder reaction to give fully protected eldecalcitol. The hydroxyl groups are then deprotected to give the parent molecule.[100] Eldecalcitol binds to the vitamin D receptor 2.7-fold more potently than calcitriol, while only weakly inhibiting serum parathyroid hormone. Eldecalcitol in animal studies (ovariectomized rats) showed improvements in bone mass while lowering bone resorption, thus demonstrating its effectiveness against osteoporosis in preclinical models.

In healthy male volunteers, eldecalcitol at oral doses from 0.1 to 1.0 μg once daily showed rapid absorption and linear PK with a long elimination half-life of 8 h reaching C_{max} in 3.4 h. Unlike vitamin D, eldecalcitol is not a substrate for CYP3A4. Eldecalcitol has been studied in two Phase III trials comparing its effectiveness in treating osteoporosis versus alpha-calcitriol. Biomarkers of osteoporosis, serum bone-specific alkaline phosphatase, and serum osteocalcin were significantly reduced in patients receiving eldecalcitol orally 1.0 μg once daily when compared to either placebo or alpha-calcitriol. In another Phase III study, eldecalcitol at a 0.75-μg dose once daily was found to be superior to alpha-calcitriol in reducing the incidence of vertebrae fractions over 3 years. Eldecalcitol also was found to increase calcium levels and bone mineral density[101,102] over a 3-year period. Eldecalcitol (Edirol®) was approved in January 2011 by the Japanese Ministry of Health, Labour, and Welfare for the treatment of patients with osteoporosis (0.5 and 0.75 μg doses). Edirol® is being codeveloped by Chugai Pharmaceutical and Taisho Pharmaceutical Holdings.

13. FIDAXOMICIN (ANTIBACTERIAL)[103–108]

Class:	Bacterial RNA polymerase inhibitor
Country of origin:	United States
Originator:	Optimer Pharmaceuticals
First introduction:	United States
Introduced by:	Optimer Pharmaceuticals
Trade name:	Dificid®
CAS registry no:	873857-62-6
Molecular weight:	1058.04

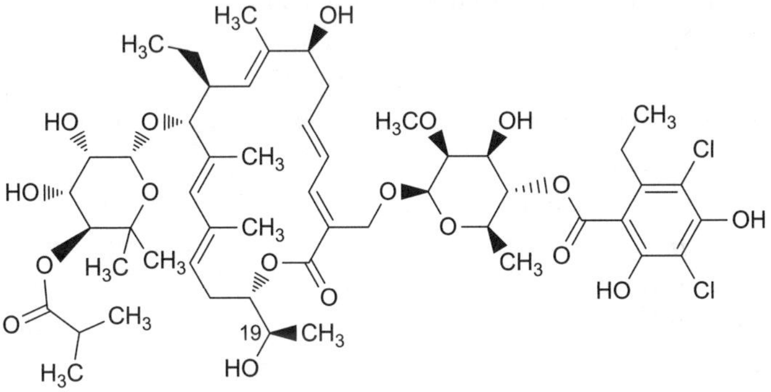

Fidaxomicin (OPT-80) was approved by the U.S. FDA in May 2011 for the treatment of *Clostridium difficile*-associated diarrhea (CDAD), joining metronidazole and vancomycin as drugs recommended for treatment of *C. difficile* infections (CDI).[103,104] *C. difficile* is a gram-positive, anaerobic bacterium that is responsible for a variety of gastrointestinal (GI) infections. It is the causative agent in up to 25% of all antibiotic-associated diarrhea and 50–75% of antibiotic-associated colitis; severe cases of CDAD can result in death. Up to 2% of all hospitalized patients and up to 4% of intensive care patients are estimated to be infected with this pathogen, which has surpassed methicillin-resistant *Staphylococcus aureus* as the most common cause of hospital-acquired infections. Metronidazole is the preferred treatment for mild to moderate CDI, but is less effective than vancomycin for severe infections, including those caused by a hypervirulent strain of *C. difficile.* Vancomycin is effective in these cases but is also associated with selection of resistant bacteria. Fidaxomicin, also known as lipiarmycin and tiacumicin, is an 18-membered macrolide natural product that was first reported in mid-1970s[105] and is produced by fermentation.[106] Fidaxomicin and its primary metabolite OP-1118, which results from hydrolysis of the isobutyryl ester, are narrow-spectrum antibacterial agents with activity against gram-positive aerobic and anaerobic organisms, but not against gram-negative organisms. Fidaxomicin and OP-1118 exert their antibacterial activity by inhibiting bacterial RNA polymerase, thereby inhibiting bacterial protein synthesis. The MIC_{90} (minimum inhibitory concentration to kill 90% of bacteria) for fidaxomicin against *C. difficile* is 0.125–0.25 μg/mL; OP-1118 is 4- to 16-fold less potent than the parent compound. Fidaxomicin has been reported to spare native intestinal flora such as *Bacteroides* spp. and as such, may prevent selection of drug-resistant bacteria. Fidaxomicin is bactericidal to *C. difficile* and has a low propensity for resistance development with no cross-resistance to existing antibiotics.

Fidaxomicin shows minimal systemic absorption following oral administration in preclinical studies and humans.[107] Following single- and multiple-dosing regimens in healthy adults and CDI patients, plasma concentrations of fidaxomicin are negligible while fecal concentrations reach very high levels. In a Phase IIa study, fecal concentrations reached 1433 μg/g in subjects treated with 400 mg/day for 10 days, a concentration that is >5000 times higher than the MIC_{90} for *C. difficile.* The metabolite OP-1118 is found in feces at a ~2:1 ratio of parent to metabolite. Fidaxomicin was evaluated in two Phase III randomized, multicenter, double-blinded trials designed to show noninferiority to vancomycin.[103] In a Phase III trial with 629 randomized patients having mild to severe CDI, 88.2% of patients treated with fidaxomicin (oral, 200 mg/day, twice a day, 10 days) achieved clinical cure compared with 85.8% of patients treated with vancomycin (oral, 125 mg/day, four times a day, 10 days). Patients receiving fidaxomicin had a statistically significant reduction in recurrence of CDI for all strains of *C. difficile* compared with vancomycin (15.4% vs. 25.3%). In a second Phase III trial, similar results were seen with similar efficacy and lower rates of recurrence for fidaxomicin compared with vancomycin. Fidaxomicin was reported to be well tolerated in clinical studies. The major adverse events were similar to those for vancomycin and included nausea (11%), vomiting (7%), abdominal pain (6%), and GI hemorrhage (4%). Fidaxomicin (Dificid®) is available in the United States as a 200-mg tablet, with a recommended dose of 200 mg twice daily with or without food, for a treatment course of 10 days. The lower rate of recurrence of CDI with fidaxomicin makes it an attractive option for the treatment of CDI.[108]

14. GABAPENTIN ENACARBIL (RESTLESS LEG SYNDROME)[109–117]

Class:	Gamma aminobutyric acid analogue
Country of origin:	United States
Originator:	Xenoport
First introduction:	United States
Introduced by:	GlaxoSmithKline
Trade name:	Horizant®
CAS registry no:	478296-72-9
Molecular weight:	329.39

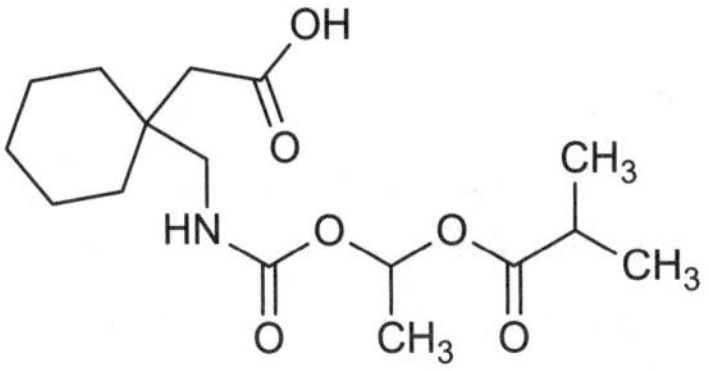

In April 2011, the U.S. FDA approved gabapentin enacarbil (XP-13512) for the treatment of moderate-to-severe Restless Legs Syndrome (RLS) in adults.[109] RLS is a neurological disorder that is characterized by an urge to move the legs, usually accompanied or caused by unpleasant sensations in the legs. Symptoms begin or worsen during periods of inactivity are relieved by movement (unlike leg cramps) and are typically worse in the evening. RLS results in significant sleep disturbances for patients and a reduction in daily function and quality of life. It is estimated that 2–3% of the United States population experience RLS symptoms that are severe enough to warrant pharmacological treatment. The underlying mechanism of RLS is not well understood. Two dopamine agonists, ropinirole and pramipexole, are currently approved in the U.S. for the treatment of RLS. While these agents provide relief for many RLS sufferers, they are not effective in all patients and can be associated with issues such as increased severity of symptoms and significant side effects. Gabapentin, which is approved in the United States for treatment of convulsions and for postherpetic neuralgia, has shown evidence of efficacy for RLS in clinical trials, although the mechanism of action is unclear. Gabapentin has variable and unpredictable bioavailability, which limits its utility as an oral agent. Gabapentin enacarbil is a novel prodrug of gabapentin that was designed to be recognized as a substrate for two high-capacity nutrient transports, monocarboxylate transporter type 1 and sodium-dependent multivitamin transporter, and to be efficiently cleaved after absorption to give gabapentin.[110] The separated enantiomers of gabapentin enacarbil have similar cleavage rates in human tissues. Preclinical studies showed that gabapentin enacarbil provides good systemic exposure of gabapentin in rats and monkeys.[111] Gabapentin enacarbil is prepared as a racemic mixture from gabapentin either by sequential coupling with 1-chloroethyl chloroformate in the presence of trimethylsilyl chloride and triethylamine followed by addition of isobutyric acid[110] or by direct coupling with an activated 1-(isobutyryloxy) ethyl carbonate.[112]

In healthy volunteers, gabapentin enacarbil was well absorbed and gave dose-proportional exposure of gabapentin with a $T_{\max}$ of 2.1–2.6 h and oral bioavailability based on urinary recovery of $>70\%$.[113] Oral bioavailability increased approximately twofold when taken with a meal. An extended release form gave a $T_{\max}$ of 7.3–9.8 h, with oral bioavailability of 82–86%. The efficacy of gabapentin enacarbil was demonstrated in two 12-week randomized, double-blinded, placebo-controlled clinical trials in adult RLS patient using endpoints based on the International Restless Legs Syndrome (IRLS) and Clinical Global Impression of Improvement (CGI-I) rating scales. In trial XP052, subjects with an initial IRLS score of ≥ 15 were treated with 1200 mg of gabapentin enacarbil or placebo, taken with food at 5 p.m. for 12 weeks.[114] Subjects who received gabapentin enacarbil showed a mean change in IRLS score of -13.2 and 76% responders based on the CGI-I scale, which was significantly improved compared to placebo controls (IRLS mean change of -8.8; 39% responders). The most commonly reported adverse events were somnolence (27% vs. 7% for placebo) and dizziness (20% vs. 5% for placebo). The XP053 12-week trial showed both 600 and 1200 mg of gabapentin enacarbil to significantly improve RLS symptoms and sleep disturbance compared with placebo.[115] Maintained improvements and long-term tolerability were demonstrated in a 9-month randomized, controlled study with 1200 mg of gabapentin enacarbil.[116] Gabapentin enacarbil (Horizant®) is supplied in an extended release tablet form, with the recommended dose of 600 mg once daily taken with food at about 5 p.m. Patients are warned not to drive until they have sufficient experience with Horizant® to assess whether it will impair their ability to drive.[117]

15. IGURATIMOD (ANTIARTHRITIC)[118–124]

Class:	Disease modifying antirheumatic drug
Country of origin:	Japan
Originator:	Toyama
First introduction:	China
Introduced by:	Simcere Pharmaceutical
Trade name:	Iremod
CAS registry no:	123663-49-0
Molecular weight:	374.37

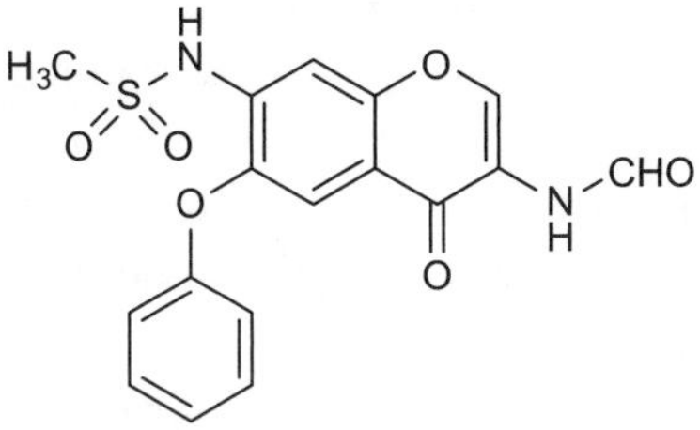

In August 2011, China's State FDA approved Simcere Pharmaceutical Group's new drug application for iguratimod (T-614), a disease modifying anti-rheumatic drug (DMARD) for the treatment of rheumatoid arthritis (RA). The World Health Organization estimates the prevalence of RA to be 0.3–1% world-wide. RA is more common in women and developed countries. According to statistics from the Center for Disease Control and Prevention, ~1.5 million adults in the United States had RA in 2007. Prevalence of RA in mainland China ranges from 0.2% to 0.37% which is similar to most Asian countries.[118] While the exact pathophysiology of RA is not known, genetic predisposition along with environmental and hormonal triggers contribute to the autoreactivity of the immune system. The pathogenesis of RA is multifactorial and includes synovial cell proliferation, fibrosis, pannus formation, and bone and cartilage erosion. The current paradigm for treating RA involves treatment with DMARD's such as methotrexate followed by injectable biologics such as etanercept, anakinra, tocilizumab, abatacept, rituximab, and others. Biochemical studies done at Toyama suggest that iguratimod inhibits PGE_2 production and COX-2 m-RNA expression in cultured fibroblasts suggesting a profile similar to selective COX-2 inhibitors.[119] Other mechanisms, such as suppression of NF-kB activation, have been proposed to explain the mechanism of action of iguratimod.[120] Preclinical *in vivo* studies indicated that iguratimod was effective in an established adjuvant-induced arthritis model ($ED_{40} = 3.6$ mg/kg) in rats and also efficacious in a type II collagen-induced arthritis model in DBA/1J mice at 30 mg and 100 mg/kg.[121] The key step in a four-step synthesis of iguratimod starting from readily available materials involves a pyrone-ring annulation with *N*,*N*-dimethylformamide dimethylacetal.[121]

In a controlled, randomized, double blind, parallel group Phase III study in 376 patients using American College of Rheumatology (ACR)20 response scores, iguratimod (25 mg for the first 4 weeks and 50 mg for the subsequent 24 weeks, oral dosing, daily) was superior to placebo (53.8% vs. 17.2%, $P < 0.001$) and was not inferior when compared to another DMARD salazosulfapyridine (63.1% vs. 57.7%).[122] A transient increase in hepatic enzyme levels, dermatological disorder (low frequency), abdominal pain, anemia, and symptoms related to gastrointestinal tract disorders were noted in this study. The long-term safety of iguratimod was assessed in a 52-week clinical study in 394 Japanese RA patients.[123] Increases in alanine aminotransferase and aspartate aminotransferase (19.4% and 18.3%, respectively) were noted in this study. ACR20 response rates were 46.9% at week 28 and 41.0% at week 52. A randomized, placebo-controlled, 24-week Phase II clinical study in 280 patients was conducted in China.[124] Ninety-five patients were assigned to the placebo group, while iguratimod was given to the remaining patients at daily oral doses of 50 mg ($n = 93$) or 25 mg ($n = 92$). After 24 weeks, the ACR20 scores for the 25 mg, 50 mg, and placebo groups were 39.13%, 61.29%, and 24.21%, respectively, while the ACR50 scores were 23.91%, 31.18%, and 7.37%, respectively. Adverse events noted in the 25 and 50 mg groups were upper abdominal discomfort, leucopenia, elevated serum alanine aminotransferase, skin rash, and/or pruritus. Simcere Pharmaceutical Group has conducted clinical trials comparing iguratimod with methotrexate in RA patients, but results are not available. Iguratimod is taken orally at 25 mg twice daily.

16. IPILIMUMAB (ANTICANCER)[78,125–131]

Class:	Recombinant monoclonal antibody
Country of origin:	United States
Originator:	Bristol-Myers Squibb
First introduction:	United States
Type:	Human IgG1κ, anti-CTLA-4
Introduced by:	Bristol-Myers Squibb
Molecular weight:	~148 kDa
Trade name:	Yervoy®
Expression system:	Rodent CHO-cell line
CAS registry no:	477202-00-9

Ipilimumab is a CTLA-4 blocking antibody that was approved by the U.S. FDA in March 2011 for the treatment of unresectable or metastatic melanoma. Metastatic melanoma has a long-term remission rate of less than 10%, with over 9000 estimated U.S. deaths in 2012.[78,125] Ipilimumab was discovered by immunization of transgenic mice comprising human immunoglobulin genes to generate a human CTLA-4-specific human sequence monoclonal antibody.[126] CTLA-4 is a membrane-bound T-cell protein that delivers a negative signal on engagement of its ligands CD80 or CD86. Therefore, ipilimumab has an indirect effect on tumors that is mediated through blockade of CTLA-4 negative signaling and augmentation of T-cell activity. It is one of a handful of immunostimulatory drugs (e.g., aldesleukin and interferon alfa-2b) now approved for cancer therapy. The drug is formulated as a solution for intravenous infusion. The recommended dosage is 3 mg/kg administered intravenously over 90 min every 3 weeks for a total of four doses. With this dosing regimen, ipilimumab C_{min} at steady state was observed to be 21.8 mcg/mL (±11.2), and the terminal half-life was 14.7 days. Blood samples from 1024 ipilimumab-treated patients were tested for the presence of antidrug antibodies. Antibodies were detected in 1.1% of the patients; however, they were not neutralizing, and none of these patients experienced infusion reactions. Because T-cell activation from treatment with ipilimumab can result in severe, and even fatal, immune-related reactions, the prescribing information includes specific warnings and precautions on the management of events such as enterocolitis, hepatitis, dermatitis, neuropathies, endocrinopathies, and ocular inflammatory diseases. These include recommendations on discontinuation of ipilimumab treatment and on the use of corticosteroids to treat these reactions. Corticosteroid treatment of immune-related reactions does not appear to abrogate the efficacy of ipilimumab.[127]

Ipilimumab has been tested in two Phase III trials: one enrolling previously treated metastatic melanoma patients and the second in previously untreated patients. The first of these was a 3-arm (ipilimumab plus peptide vaccine, ipilimumab alone, vaccine alone; randomized 3:1:1) trial that enrolled 676 stage III or IV metastatic melanoma patients, whose disease had progressed while on a prior therapy.[128] Median survival for the ipilimumab plus vaccine arm was 10 months. For the ipilimumab alone and the vaccine alone arms, median survival was 10.1 and 6.4 months, respectively. This was the first Phase III, randomized, controlled trial to report a survival advantage for a therapeutic agent in metastatic melanoma patients. Consistent with the immunostimulatory mechanism of action of the drug, immune-related adverse events, such as rash, pruritis, vitiligo, diarrhea, colitis, hypothyroidism, hypopituitarism, hypohysitis, and adrenal insufficiency, were reported in approximately 60% of the ipilimumab-treated patients. The percentage of patients experiencing grade 3 immune-related adverse events was 9.7% and 12.2% in the two ipilimumab arms and 3% in the vaccine alone arm. Grade 4 immune-related events occurred at 0.5% and 2.3% in the ipilimumab arms. Fourteen deaths (2.1%) were reported to be related to study drug: eight in the combination arm, four in the ipilimumab alone arm, and two in the vaccine alone arm. Immune-related adverse events were managed by close patient follow-up and the administration of high-dose systemic corticosteroids as necessary for grade 3 and 4 events. The median time to resolution of grade 2 or higher diarrhea (the most common immune-related adverse event) was 2.0 and 2.3 weeks in the ipilimumab treatment arms. Four patients with diarrhea or colitis were treated with infliximab. Because the peptide vaccine used in two of the trial arms had been optimized for a single human leukocyte antigen (HLA) subtype, enrollment in the trial was restricted to patients positive for that subtype (HLA-A*0201). However, in a retrospective analysis of data from four different Phase II trials, no HLA subtype-associated differences in survival or safety were observed.[129] Although the Phase III trial did not include a placebo arm, survival in the vaccine alone arm was consistent with historical data for other experimental agents in metastatic melanoma. A meta-analysis of data from 42 different Phase II trials of experimental agents involving 2100 metastatic melanoma patients between 1977 and 2005 found a 1-year survival rate 25.5%.[130] The vaccine alone arm of the ipilimumab Phase III showed 25.3% survival at 1 year, compared to 43.6% and 45.6% survival for the ipilimumab treatment arms. In the second Phase III clinical trial, a combination of ipilimumab therapy together with the cytotoxic drug dacarbazine was tested against dacarbazine alone in previously untreated metastatic melanoma patients.[131] The dose and schedule of ipilimumab in this trial was different from that used in the first Phase III (which was used to define the recommended dosage). Ipilimumab was given at 10 mg/kg instead of 3 mg/kg, and after four doses, some patients received additional doses. The 502 enrolled patients were randomized 1:1 to two treatment arms. Patients received 10 mg/kg ipilimumab, or a placebo, and 850 mg/m^2 dacarbazine on weeks 1, 4, 7, and 10, followed by 850 mg/m^2 dacarbazine every 3 weeks through week 22. Patients with stable disease or an objective response, and no dose-limiting toxicities, were subsequently dosed with 10 mg/kg ipilimumab, or placebo, every 12 weeks. Median survival was 11.2 months in the combination arm and 9.1 months in the dacarbazine alone arm. Survival rates at 1, 2, and 3 years were also higher in the ipilimumab-treated arm (47.3%, 28.5%, and 20.8% vs. 36.3%, 17.9%, and 12.2%). Grade 3 adverse events occurred in 40.1% of the ipilimumab patients, compared to 17.9% of the patients treated with dacarbazine alone. Rates of grade 4 events were 16.2% and 9.6% in the ipilimumab and dacarbazine alone arms, respectively. There were no reported drug-related deaths in the ipilimumab-treated patients. As in the previous Phase III trial, immune-related adverse events were the most common study drug-related events. However, the combination with dacarbazine appeared to result in a different spectrum of observed immune-related adverse events. In contrast to the previous trial, where diarrhea was the most commonly reported serious immune-related adverse event, elevated liver-function values were the most common immune-related adverse event when ipilimumab was combined with cytotoxic therapy. Grade 3 or 4 immune-related hepatitis was reported in 31.6% of the ipilimumab-treated patients compared to 2.4% of the patients in the dacarbazine alone arm.

17. LINAGLIPTIN (ANTIDIABETIC)[132–140]

Class:	Dipeptidyl peptidase-4 (DPP-4) inhibitor
Country of origin:	United States
Originator:	Boehringer Ingelheim
First introduction:	United States
Introduced by:	Boehringer Ingelheim and Eli Lilly and Company
Trade name:	Tradjenta®
CAS registry no:	668270-12-0
Molecular weight:	472.54

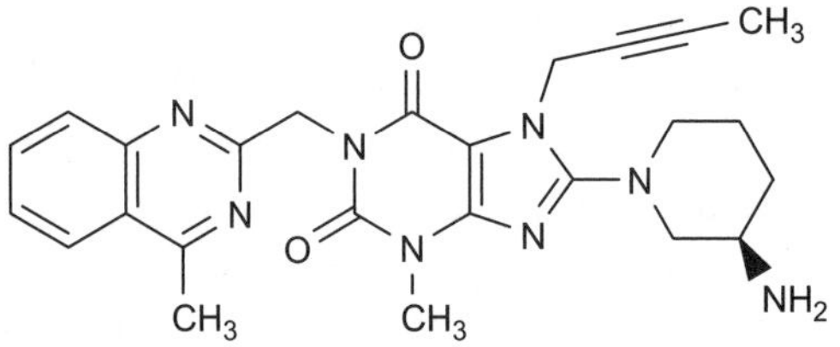

Linagliptin (trade names Tradjenta® and Trajetna®) is an inhibitor of dipeptidyl peptidase-4 (DPP-4) that was approved by the U.S. FDA in May 2011 for the treatment of Type 2 diabetes along with diet and exercise. Type 2 diabetes is a disease in which insulin resistance and beta-cell dysfunction lead to hyperglycemia. According to the American Diabetic Association, diabetes is the seventh leading cause of death and increases the risk of heart disease and stroke by two to four times. Macro- and microvascular complications result from the progression of the severity of diabetes. The prevalence of diabetes continues to increase world-wide with an estimated 370 million people projected to be affected by 2030. The current number of cases in the United States (8% of the population) is predicted to double by 2050. In the United States, the economic impact of diabetes was estimated at $176 billion in 2007, with $116 billion attributed medical expenditures.[132,133] The progression of diabetes is attributed to several factors. Patients in a hyperglycemic state exhibit increases in free fatty acids, cytokines, adipokines, and associated metabolites leading to the loss of beta-cell function and beta-cell mass in islets.[134] As islet function is lost, the severity of insulin resistance increases. The introduction of dipeptidyl-peptidase IV inhibitors (DPP-4 inhibitors) has brought a novel class of insulinotropic agents into the treatment options available to type 2 diabetic patients. Glucagon-like peptide-1 (GLP-1), an endogenous 30-amino acid peptide that plays a central role in glucose homeostasis, is inactivated by cleavage of the N-terminal dipeptide sequence (His-Ala) through the peptidase action of DPP-4. This inactivation occurs rapidly, with the half-life of circulating GLP-1 being <2 min. DPP-4 has several other substrates including another beneficial incretin peptide, gastric inhibitory peptide (GIP). Inhibitors of DPP-4 have been shown in man to increase GLP-1 and GIP levels two- to threefold. Because insulin secretion via the actions of GLP-1 occurs only in response to rising glucose levels, the risk of hypoglycemia is low, resulting in the wide acceptance of DPP-4 inhibitors into clinical practice. DPP-4 inhibitors are primarily once a day, weight neutral drugs with a favorable adverse-effect profile. As shown by animal studies, the class can decrease beta-cell apoptosis and increase beta-cell survival. In animal models, DPP-4 inhibitors increase the number of insulin positive beta-cells in islets. Islet insulin content is found to be increased, and glucose-stimulated insulin secretion in isolated islets is improved.[132–135] Linagliptin (BI-1356) has been described as a potent highly selective, slow-off rate and long acting inhibitor of DPP-4. Linagliptin arose from optimization efforts of xanthine-based DPP-4 inhibitors with the initial lead identified from an HTS campaign. After optimizing the activity of the initial micromolar lead, two issues that needed to be addressed were activity for hERG and muscarinic receptor M_1. Introduction of a butynyl group at the N7 position of the xanthine ring gave much reduced M_1 affinity with no measureable hERG activity. Linagliptin inhibits DPP-4 with an $IC_{50} = 1$ nM and is highly selective (>10,000-fold) against DPP-8 and DPP-9. Linagliptin shows no interactions with CYPs up to 50 μM. The described synthesis of linagliptin starts with 8-bromoxanthine, which is alkylated at the N-7 position to introduce the butyne group, followed by alkylation of the N-1 group to introduce the methyl-quinazoline group. Displacement of the bromide with (*R*)-Boc-3-amino-piperidine followed by deprotection gives linagliptin.[136] When administered to *db/db* mice orally, linagliptin dose dependently reduced glucose excursion from 0.1 mg/kg (15% inhibition) to 1 mg/kg (66% inhibition). Linagliptin was reported to have a longer *in vivo* duration of action on glucose tolerance when compared to other DPP-4 inhibitors.[137]

The initial Phase I PK studies in male type 2 diabetic patients studied doses of linagliptin at 1, 2.5, 5, and 10 mg once daily for 12 days. Linagliptin showed a less than dose proportional increase in exposure, a short accumulation half-life (8.6–23.9 h), and a rapid achievement of steady-state concentrations (2–5 days). Linagliptin showed a long terminal half-life of 113–131 h.[138] Linagliptin is eliminated primarily intact, with the primary route of excretion being fecal (84.7%) and renal excretion being a minor pathway (5.4%). With the low renal clearance, no dose adjustment is needed for patients with renal impairment. The most abundant metabolite, CD-1790, results from replacement of the

3-(*R*)-amino group of the piperidine by a 3-(*S*)-hydroxyl group. CD-1790 is formed at low levels (< 10% of parent drug concentrations)[139] via CYP3A4-mediated conversion of the amino group to a ketone, followed by reduction to the alcohol primarily by aldo-keto reductases. A comprehensive meta-analysis of the nine clinical trials of linagliptin as either monotherapy or in combination with metformin, sulfonylureas, or piaglitazone has been reported.[140] Linagliptin as monotherapy or in combination therapy gives a statistically significant reduction in fasting plasma glucose and HbA1C levels. Linagliptin was shown to be safe and well tolerated. Adverse events were minimal and not statistically different from placebo. In the eight Phase III trials, linagliptin did not increase cardiovascular events relative to placebo. The 6000 patient CAROLINA study, comparing the effects of 5 mg linagliptin to glimeride on cardiovascular events, is ongoing. Linagliptin (Tradjenta®), a highly selective, long acting DPP-4 inhibitor for the treatment of type 2 diabetes, received regulatory approval from the U.S. FDA in May 2011 as monotherapy at a 5-mg dose taken once daily and as combination therapy as an adjunct to diet and exercise to improve glycemic control in adults with type 2 diabetes mellitus. Tradjenta® is the fifth marketed inhibitor of DPP-4 and is comarketed by Boehringer Ingelheim and Eli Lilly and Company.

18. MIRABEGRON (URINARY TRACT/BLADDER DISORDERS)[141–144]

Class:	β_3-Adrenoceptor agonist
Country of origin:	Japan
Originator:	Astellas Pharma Inc.
First introduction:	Japan
Introduced by:	Astellas Pharma Inc.
Trade name:	Betanis®
CAS registry no:	223673-61-8 (free base)
	223672-18-2 (salt)
Molecular weight:	396.51 (free base)
	469.43 (salt)

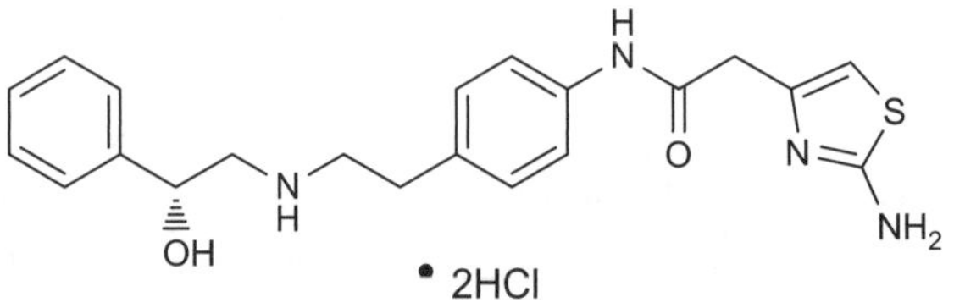

Betanis® (Mirabegron) was approved in July 2011 by the Japanese Ministry of Health, Labour, and Welfare for the treatment of urgency, urinary frequency, and urinary urge urinary incontinence associated with overactive bladder (OAB). A study to determine the incidence of OAB in Japan found 14% of men and 11% of women had the condition with prevalence increasing with age.[141] Mirabegron activates the β_3-adrenergic receptor (β3-AR), which is located on the bladder and involved in bladder smooth muscle relaxation. Mirabegron is synthesized by coupling 4-nitrophenethyl amine to (*R*)-2-hydroxy-2-phenylacetic acid. The resulting amide is reduced to an amine. The nitro group is then reduced and the resulting aniline is coupled to 2-(2-aminothiazol-4-yl) acetic acid to give mirabegron.[142] Mirabegron has an EC_{50} of 22 nM (intrinsic activity=0.8) for β3-AR with no detectable activity for β_1- and β_2-AR ($EC_{50} > 10{,}000$ nM). In an anesthetized rat rhythmic bladder contraction model in which bladder contractions are induced by saline, mirabegron at 3 mg/kg iv decreased the frequency of rhythmic bladder contraction without suppressing contraction amplitude. These data suggest that the activation of β3-AR increases bladder capacity without influencing the frequency of bladder contraction. In Phase I clinical studies, mirabegron was found to be metabolized by CYP3A4 and CYP2D6. Genetic polymorphism of CYP2D6 causes variability in the metabolism of mirabegron in humans. In patients characterized as poor metabolizers (PM), the amount of mirabegron excreted in the urine is higher (15.4%) versus patients that are extensive metabolizers (EM) where the amount in the urine is lower (11.7%). The terminal half-life in PM patients was 25 h and in the EM patients 23 h.[143] In a randomized, double-blinded Phase III study with 1976 patients comparing mirabegron 50 and 100 mg once daily to tolterodine extended-release (ER) 4 mg over 4 weeks in OAB patients, the frequency of incontinence per 24 h was reduced in both mirabegron dosing groups relative to the tolterodine-treated group. In another Phase III trial, mirabegron at 50 and 100 mg qd was studied over a 12-week period in patients with OAB. This 1328 patient study showed improvement in the reduction of incontinence episodes per 24 h versus placebo (−1.13, −1.47, and −1.63 for placebo, mirabegron 50 mg, and mirabegron 100 mg, respectively; $p < 0.05$).[144] Mirabegron, 50 mg once daily, was approved in July 2011 by the Japanese Ministry of Health, Labour, and Welfare for the treatment of urgency, urinary frequency, and urinary urge urinary incontinence associated with OAB. Betanis® is being developed by Astellas Pharma Inc.

19. RETIGABINE (ANTICONVULSANT)[145–156]

Class:	Potassium channel opener
Country of origin:	Germany
Originator:	ASTA Medica group
First introduction:	European Union
Introduced by:	Valeant/Glaxo Smith Kline
Trade name:	Trobalt (European Union), Potiga™ (United States)
CAS registry no:	150812-12-7
Molecular weight:	303.33

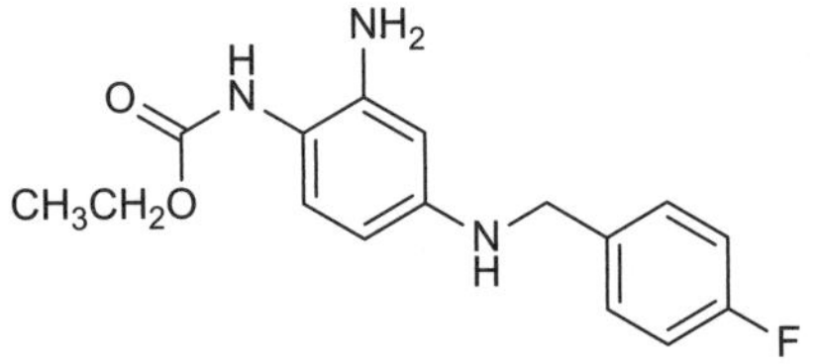

Retigabine was approved in March 2011 by the European Commission for the adjunctive treatment of partial-onset seizures in adults who have epilepsy; in June 2011, the U.S. FDA approved the same drug, known in the United States as ezogabine.[145–147] Epilepsy is a neurological disorder characterized by a predisposition to recurrent unprovoked seizures. Seizures are caused by abnormal electrical disturbances in the brain, with partial-onset seizures being localized to a limited area of the brain and generalized seizures affecting the whole brain. Epilepsy is estimated to affect more than 50 million people world-wide, with partial seizures occurring in ~60% of patients with epilepsy.[146] More than 20 antiepileptic drugs are available, however, 30–40% of patients continue to experience seizures on treatment with existing drugs.[148] Retigabine differs from all currently approved antiepileptic drugs in that it acts as a selective positive allosteric modulator (opener) of KCNQ2-5 potassium channels, which are key regulators of neuronal excitability.[149] The discovery of retigabine was based on modification of an analgesic agent flupirtine that had serendipitously shown potent anticonvulsant activity in animal models of epilepsy. Changing a central 2,3,6-triaminopyridine to a 1,2,4-triaminobenzene decreased analgesic activity while enhancing antiepileptic activity.[150] Retigabine (D-23129) was shown to be a broad spectrum anticonvulsant with oral activity in a variety of animal models. The mechanism of action of retigabine was discovered well after its *in vivo* activity was recognized.[151] Retigabine can be prepared by reductive amination of 2-ethoxycarbonylamino-5-(4-fluorobenzylamino)-nitrobenzene with 4-fluorobenzaldehyde, followed by hydrogenation in the presence of Raney nickel.[152]

The oral pharmacokinetic profile of retigabine was determined in healthy volunteers at doses of 25–600 mg and in patients at doses up to 1200 mg/day.[146] Retigabine is rapidly absorbed with T_{max} occurring between 0.5 and 2 h, and its absolute bioavailability relative to an intravenous dose is 60%. Retigabine undergoes extensive metabolism and is eliminated primarily by renal excretion as shown by administration of radiolabeled retigabine where 84% of the dose was eliminated in urine and 14% in feces. Parent drug was 36% of the dose in the urine, with 18% of the *N*-acetyl metabolite, and 24% of the *N*-glucuronide metabolite. Plasma protein binding of retigabine is ~80% and the volume of distribution is 2–3 L/kg at steady state. The plasma half-life of retigabine is 6–10 h, with an elimination half-life of 7–11 h. Coadministration of retigabine had no clinically significant effect on trough concentrations of various antiepileptic drugs. However, the clearance of retigabine was increased when dosed with carbamazepine or phenytoin. Retigabine is neither a substrate nor an inhibitor of CYP enzymes. In mutagenesis assays,[153] retigabine was negative in the *in vitro* Ames assay, the *in vitro* CHO *Hprt* gene mutation assay, and the *in vivo* mouse micronucleus assay. It was positive in the *in vitro* chromosomal aberration assay in human lymphocytes. The efficacy of retigabine was established in three randomized, double-blind, placebo-controlled studies in >1200 adult patients at oral doses of 600, 900, or 1200 mg/day tid. Enrolled patients had a diagnosis of localization-related epilepsy which was refractory to 1–3 antiepileptic drugs. The primary endpoint was change in seizure frequency from baseline; the secondary endpoint was responder rate defined as the percentage of patients with ≥50% reduction in total partial seizure frequency. In all three trials, patients had a forced titration period, wherein retigabine was initially given at 300 mg/day tid with an increase at 1-week intervals of 150 mg/day to the targeted dose, followed by a maintenance period of 8–12 weeks. In the first trial, decreases in seizures were 600 mg/day, −23%; 900 mg/day, −29%; 1200 mg/day, −35%; and placebo, −13%.[154] In the RESTORE-1 trial, 1200 mg/day tid of retigabine gave a 44% decrease in seizure frequency and a 44% responder rate (placebo: 17.5% decrease, 17.8% responder rate).[155] The RESTORE-2 trial showed significantly greater reductions in seizure frequency for retigabine at 600 mg/day (−28%) and 900 mg/day (−40%) compared with placebo (−16%).[156] Responder rates were also significantly improved. Across all trials, there were more study dropouts during the forced titration phase than during the maintenance phase. Adverse events noted in at least 10% of patients across trials were primarily related to the central nervous system and included somnolence, dizziness, fatigue, and confusion. The recommended dose of retigabine (Trobalt/Potiga™) is 300 mg/day tid increasing by 150 mg/day tid in 1-week intervals to 600 mg/day tid to 1200 mg/day tid depending on tolerability.

20. RILPIVIRINE (ANTIVIRAL)[4,157–165]

Class:	HIV nonnucleoside reverse transcriptase inhibitor
Country of origin:	Belgium
Originator:	Janssen
First introduction:	United States
Introduced by:	Tibotec (a subsidiary of Johnson and Johnson)
Trade name:	Edurant®
CAS registry no:	500287-72-9 (free base) 700361-47-3 (salt)
Molecular weight:	366.42 (free base) 402.88 (salt)

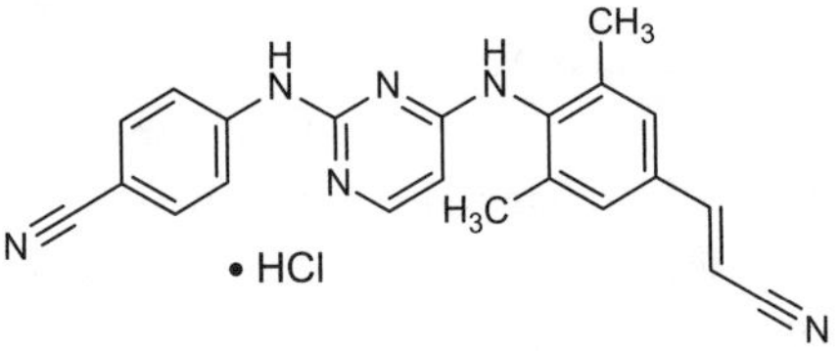

In May 2011, the U.S. FDA approved rilpivirine in combination with other antiretroviral agents for the treatment of human immunodeficiency virus (HIV) 1 infection in treatment-naïve adult patients. HIV infection results in destruction of host CD4+ T-cells, an essential part of the immune system, and can lead to acquired immune deficiency syndrome (AIDS), which is characterized by low T-cell count and the presence of opportunistic infections. In a joint UNAIDS/WHO report, it was estimated that in 2008, 33 million people worldwide were living with HIV infection and 2 million people died from AIDS-related deaths.[157] The U.S. CDC estimated in 2009 that 1.2 million people in the U.S. were living with HIV infection.[157] Drugs for treating HIV infection have been approved from several classes, including HIV reverse transcriptase (RT) inhibitors (nucleosides/nucleotides and nonnucleosides), HIV protease inhibitors, HIV fusion/entry inhibitors, HIV integrase inhibitors, and others.[158] Treatment of HIV infection was revolutionized in the mid-1990s with the recognition that combination therapy with multiple antiretroviral agents was highly effective in controlling HIV replication and inducing immune recovery. There remains a need for new HIV treatments that can be tolerated in long-term dosing with minimal side-effects, retained potency, and simplified dosing schedules. Rilpivirine is a member of the nonnucleoside reverse transcriptase inhibitor (NNRTI) class of anti-HIV agents. It is highly potent against a range of wild-type HIV strains (EC_{50} = 0.07–1.0 nM), ~10–20 times more potent than the NNRTI efavirenz (Sustiva®), and active against HIV strains resistant to other NNRTIs.[159,160] The discovery of rilpivirine was guided by molecular modeling and X-ray crystallography of HIV-1 RT complexed with inhibitors.[160,161] The synthesis of rilpivirine is accomplished by an efficient 6-step route in which the key step is coupling of 4-((4-chloropyrimidin-2-yl)amino)benzonitrile with (*E*)-3-(4-amino-3,5-dimethylphenyl)acrylonitrile.[159] In preclinical studies, the oral bioavailability of rilpivirine was 31% in dogs and 32% in rats.

When rilpivirine was given to HIV-seronegative male volunteers in a 7-day study with single daily doses ranging from 25 to 200 mg, T_{max} was 3–4 h, C_{max} was 2–3 times higher on day 7 than on day 1, and the terminal half-life was 34–55 h.[159,162,163] The C_{max} decreased by 45% when given under fasting conditions, indicating that rilpivirine should be given with food. Rilpivirine is highly protein bound (99.7%) and is a substrate but not an inhibitor of CYP3A4. Based on drug–drug interaction studies, rilpivirine should not be coadministered with drugs that cause CYP3A4 induction or gastric pH increase as these may lead to significant decreases in plasma concentrations. In studies with ^{14}C-labeled rilpivirine, 85% of radioactivity was recovered in feces and 6% in urine. Unchanged drug accounted for 25% of drug-related material in feces. Glutathione-dependent conjugative metabolism is the primary pathway of metabolism in hepatocytes from rodents and humans. The efficacy, safety, and tolerability of rilpivirine in comparison with efavirenz were assessed in two 48-week Phase III trials in patients not previously given antiretroviral therapy. A 25-mg dose of rilpivirine was selected based on having the best efficacy/side-effect profile in a 96-week Phase IIb trial extended to 192 weeks. In the ECHO trial,[164] rilpivirine (25 mg, once daily) and efavirenz (600 mg, once daily) were dosed in combination with tenofovir and emtricitabine; in the THRIVE[165] trial, rilpivirine was combined with an investigator-selected regimen of tenofovir/emtricitabine, zidovudine/lamivudine, or abacavir/lamivudine. Pooled results from the two trials showed that rilpivirine was equivalent to efavirenz in reducing viral load, with 83% of patients having undetectable virus levels in the rilpivirine combination regimen compared with 80% for the efavirenz combination regimen. Patients with higher initial viral load were more likely to not respond to rilpivirine than efavirenz; patients who failed rilpivirine therapy had more drug resistance than those who failed on efavirenz. Compared with efavirenz, rilpivirine showed a lower incidence of CNS effects, rash, and serum lipid disturbances. Rilpivirine (Edurant®) was approved as a 25-mg oral tablet to be taken once daily with a meal in combination with other antiretroviral drugs for the treatment of HIV1 infection. In August 2011, the FDA approved the specific three-drug combination Complera®, a fixed dose combination of rilpivirine (25 mg) with emtricitabine (200 mg), and tenofovir (300 mg) to be given once daily taken orally with meals.[4]

21. RUXOLITINIB (ANTICANCER)[166–171]

Class:	Janus kinase 1 and 2 inhibitor
Country of origin:	United States
Originator:	Incyte Corporation
First introduction:	United States
Introduced by:	Incyte Corporation
Trade name:	Jakafi®
CAS registry no:	941678-49-5 (free base)
	1092939-17-7 (salt)
Molecular weight:	306.37 (free base)
	404.36 (salt)

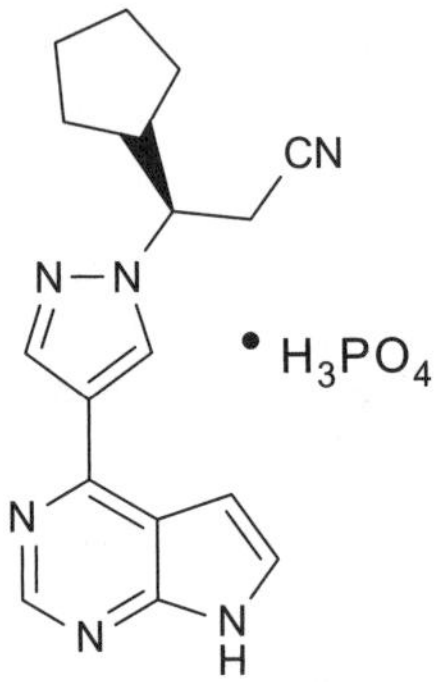

In November 2011, the U.S. FDA approved ruxolitinib (INCB018424) for the treatment of patients with intermediate or high-risk myelofibrosis. The annual incidence rate of primary myelofibrosis in European, Australian, and North American populations is estimated to range from 0.3 to 1.5 cases per 100,000 persons.[166] Treatment options for myelofibrosis include allogeneic hematopoietic stem cell transplantation (allo-HSCT) although this is more appropriate for younger, sufficiently healthy patients with high-risk myelofibrosis for whom a suitable donor is available. Other options include (a) treatment with hydroxyurea or low dose thalidomide and prednisone mainly to alleviate organomegaly and cytopenia (b) splenectomy, and (c) spleen irradiation. Myelofibrosis is associated with dysregulated Janus-associated kinases (JAKs) JAK1 and JAK2. Ruxolitinib is an ATP-competitive inhibitor of JAK1 and JAK2 ($IC_{50's}$ of 3.3 ± 1.2 nM and 2.8 ± 1.2 nM, respectively) and inhibition occurs regardless of the $JAK2^{V617F}$ mutational status. Ruxolitinib is a moderately potent inhibitor of the related JAK, TYK2 ($IC_{50} = 19 \pm 3.2$ nM) but is selective versus JAK3 ($IC_{50} = 428 \pm 243$ nM). It was also selective versus a panel of 26 other kinases at concentrations approximately 100-fold the IC_{50} of JAK1 and JAK2.[167] Inhibition of JAK1 and JAK2 downregulates the JAK-signal transducer and activator of transcription (STAT) pathway, inhibiting myeloproliferation, inducing apoptosis, and reducing numerous cytokine plasma levels. Consistent with this hypothesis, ruxolitinib inhibited IL-6 and thrombopoietin induced STAT-3 phosphorylation in an *in vitro* human whole blood assay with $IC_{50's}$ of 282 ± 54 nM and 281 ± 62 nM, respectively.[167] In a mouse model of myeloproliferative neoplasms, ruxolitinib dosed orally reduced enlargement of the spleen; eliminated neoplastic cells from the spleen, liver, and bone marrow; normalized histology of affected organs; and prolonged survival. In addition, significant suppression of IL-6 levels was seen and TNF-α levels were normalized.[167] An enantiospecific synthesis of ruxolitinib employing an organocatalytic asymmetric aza-Michael addition of pyrazoles to (*E*)-3-cyclopentylacrylaldehyde with diarylprolinol silyl ether as the catalyst has been described.[168]

The pharmacokinetics and pharmacodynamics of ruxolitinib were evaluated in single and multiple ascending dosing studies in healthy volunteers.[169] In the single ascending dose studies (5, 10, 25, 50, 100, and 200 mg), ruxolinitib was rapidly absorbed with a mean T_{max} of < 2 h and a mean terminal half-life of <5 h. Clearance was low (~ 20 L/h) and the apparent volume of distribution at the terminal phase was moderate (79–97 L). The apparent volume of distribution of ruxolitinib in myelofibrosis patients, at steady-state, was 53–65 L. Ruxolinitib is a substrate for CYP3A4 and is significantly protein bound ($\sim 97\%$). The safety and efficacy of ruxolinitib were assessed in two Phase III trials in patients with myelofibrosis. Study 1 (COMFORT-I)[170] was a double-blind, randomized, placebo-controlled study in 309 patients who were refractory to or were not candidates for available therapy. The ongoing Study 2 (COMFORT-II)[171] is an open-label, randomized study in 219 patients who were randomized 2:1 to ruxolinitib versus best available therapy. The primary efficacy endpoint was the proportion of patients achieving greater than or equal to a 35% reduction from baseline in spleen volume at week 24 (for Study 1) or week 48 (for Study 2) as measured by MRI or CT. In Study 1, efficacy analysis of primary endpoints indicated that the response rate was 41.9% among recipients of ruxolitinib at a starting dosage of 15 or 20 mg twice daily (depending on baseline platelet count) and then optimized during treatment, compared with 0.7% for placebo ($p < 0.0001$). In the ongoing Study 2, a response rate of 28.5% has been reported. The most common hematologic adverse reactions (incidence >20%) are thrombocytopenia and anemia. The most common nonhematologic adverse reactions (incidence >10%) are bruising, dizziness, and headache. The recommended dose of ruxolitinib is 20 mg administered orally twice daily.

22. TAFAMIDIS MEGLUMINE (NEURODEGENERATION)[172–176]

Class:	Transthyretin stabilizer
Country of origin:	United States
Originator:	Scripps Research Institute
First introduction:	European Union
Introduced by:	Pfizer
Trade name:	Vyndaqel®
CAS registry no:	951395-08-7 (salt) 594839-88-0 (free acid)
Molecular weight:	503.33 (salt) 308.12 (free acid)

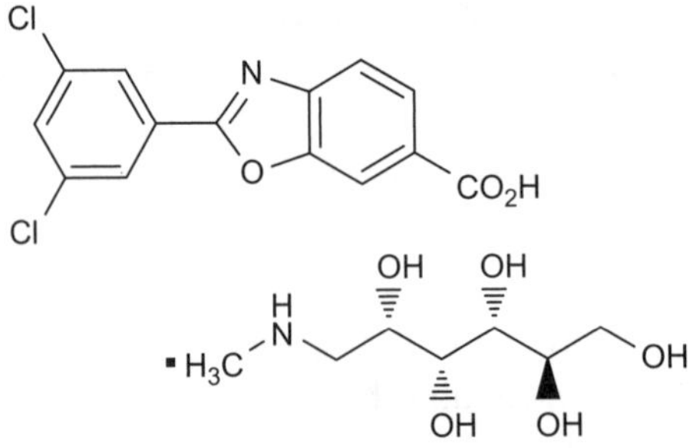

In November 2011, the European Commission approved tafamidis meglumine (Fx-1006A, PF-06291826) for the treatment of transthyretin familial amyloid polyneuropathy (TTR-FAP) in adult patients with stage 1 symptomatic polyneuropathy. TTR-FAP is a rare, progressive, and fatal disorder which presents itself phenotypically in the form of neuropathy, cardiomyopathy, renal failure, and blindness. TTR-FAP affects approximately 8000–10,000 patients worldwide.[172] Since transthyretin (TTR) is primarily synthesized in the liver, liver transplantation is an option for younger patients with the common Val30Met (V30M) mutation but is of little benefit for older patients or those with other mutations (*vide infra*). TTR is a 55-kDa homotetramer (each monomer is comprised of 127 amino acids) that transports thyroxine and retinol. The dissociation of the TTR tetramer into monomers is thought to be the rate-limiting step in the pathogenesis of TTR-FAP. The monomers, which are rich in beta sheet structures, undergo partial denaturation resulting in amyloid deposits. The hereditary form of the disease is caused by autosomal dominant mutations in the TTR gene. Tafamidis stabilizes both the wild type and mutant forms of TTR tetramer and prevents tetramer dissociation[173] by noncooperatively binding to the two thyroxine binding sites. Tafamidis is the first approved medicine for TTR-FAP. The K_d values for tafamidis for the two thyroxine binding sites on TTR, as determined by isothermal titration calorimetry, were 3 nM and 278 nM, respectively. In another *in vitro* study using wild type TTR, V30M mutant TTR, and V122I mutant TTR, it was shown that tafamidis inhibited fibril formation in a concentration-dependent manner reaching EC_{50} at a tafamidis:TTR stoichiometry of <1 (EC_{50} was in the range of 2.7–3.2 μM, corresponding to a tafamidis:TTR stoichiometry range of 0.75–0.9).[174] Tafamidis has been synthesized by coupling 4-amino-3-hydroxybenzoic acid with 3,5-dichlorobenzoyl chloride followed by dehydration using *p*-toluenesulfonic acid.[174] A recent publication describes the use of a nickel-catalyzed C-H arylation employing 1,3-dichloro-5-iodobenzene and a benzoxazole amide.[175]

In Phase I clinical trials in healthy volunteers, it was shown that tafamidis (20 mg, oral dosing) was rapidly absorbed with a T_{max} of 0.5 h and mean C_{max} of 1431 ng/mL. The mean $AUC_{0-\infty}$ derived from plasma concentration profile for tafamidis was 47,864.31 ng.h/ml, the mean apparent total clearance was 0.44 L/h, and the mean half-life of the terminal elimination phase was 54 h. Volume of distribution of the central compartment (V_c/F) and volume of distribution of the peripheral compartment (V_p/F) estimates were 0.48 L and 18.9 L, respectively.[176]

Tafamidis is highly protein bound (>99.5%). Following the completion of a pivotal Phase II/III trial, a 1-year open-label extension study was conducted to evaluate long-term safety and efficacy of tafamidis in which patients who completed the 18-month pivotal study were eligible to enroll. Of the 86 patients enrolled in the study, 45 were previously on tafamidis and 41 were previously on placebo. Patients treated with tafamidis for 30 months had less neurologic deterioration than patients who began tafamidis 18 months later (the placebo-tafamidis group), showing a 55.9% preservation of function as measured by the Neuropathy Impairment Score-Lower Limb (NIS-LL, primary endpoint). In addition, the secondary endpoints demonstrated that patients treated with tafamidis over 30 months showed preservation in large (66% preservation) and small nerve fiber function (45.5%). Despite having more severe disease (i.e., those patients initiating treatment 18 months later), initiation of tafamidis in patients previously on placebo resulted in slowing of disease progression.[176] Most common adverse drug reactions reported included diarrhea, upper abdominal pain, urinary tract infection, and vaginal infection. The recommended dose of tafamidis is 20 mg orally once daily.

23. TELAPREVIR (ANTIVIRAL)[69,70,177–183]

Class:	Hepatitis C virus NS3-4A Protease inhibitor
Country of origin:	United States
Originator:	Eli Lilly
First introduction:	United States
Introduced by:	Vertex Pharmaceuticals
Trade name:	Incivek®
CAS registry no:	402957-28-2
Molecular weight:	679.85

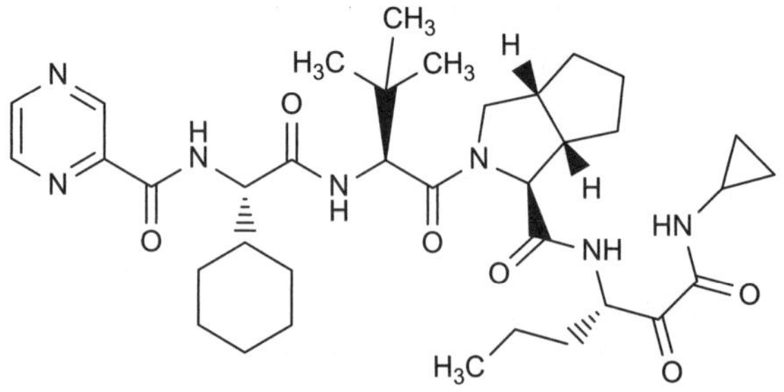

The hepatitis C virus (HCV) protease inhibitor telaprevir (VX-950, MP-424, LY-570310) was approved by the U.S. FDA in May 2011 for the treatment of genotype 1 chronic HCV infection in adult patients in combination with peginterferon alfa and ribavirin (PR).[69,177] Telaprevir and boceprevir (*vide supra*) are the first two HCV protease inhibitors to be approved for treatment of HCV infection. The World Health Organization estimates that 3% of the world population (170 million people) is infected with HCV and that 3–4 million new cases occur each year.[70] HCV is an RNA virus that causes acute and chronic liver disease. Infected individuals may remain asymptomatic for years, but >70% of people chronically infected with HCV progress to serious illness, including cirrhosis and hepatocellular carcinoma. Before the introduction of HCV protease inhibitors, the standard of care for treating HCV infection was a combination of PR. Some patients are cured of HCV infection with this regimen, as defined by a sustained virologic response (SVR) without detectable HCV RNA for 6 months after completion of therapy. However, with HCV genotype 1, which is the most prevalent subtype in the United States, Europe, and Japan, only 40–50% of patients achieve viral cure with PR therapy. In addition, the PR regimen has many adverse side effects, making it difficult for patients to tolerate the 48-week treatment duration. Telaprevir is a HCV NS3-4A protease inhibitor that exerts its antiviral effect by blocking the release of nonstructural viral proteins from a polyprotein precursor. Telaprevir is a potent inhibitor of the protease (IC_{50} = 10 nM) and is active in cell culture (HCV 1b replicon assay, EC_{50} = 354 nM).[177] Telaprevir was identified from efforts to truncate a decamer peptide inhibitor derived from the natural substrate NS5A-5B and was guided by structure-based design. The keto-amide group of telaprevir forms a covalent, reversible bond with the active site serine hydroxyl of the protease and compensates for the loss of affinity resulting from truncation of the peptide. Despite the presence of the reactive keto-amide group, telaprevir is >500-fold less potent against other serine proteases. Synthesis of the key octahydrocyclopenta[*c*]pyrrole-1-carboxylic acid fragment of telaprevir is achieved by α-deprotonation of Boc-protected 3-azabicyclo[3.3.0]nonane followed by reaction with CO_2 and resolution of the racemic acid.[178,179] Alternatively, deprotonation is carried out in the presence of a chiral amine to give the enantiomerically enriched acid.

Initial evaluation of telaprevir in healthy volunteers at doses up to 1250 mg every 8 h for 5 days showed the drug to be well tolerated.[180] Telaprevir was next evaluated in a Phase Ib study as a single agent in HCV-infected patients at doses of 450, 750, or 1250 mg, three times daily for 14 days. While a profound and rapid reduction in plasma HCV RNA was noted, viral breakthrough due to selection of variants with decreased sensitivity was observed in some patients, indicating that telaprevir should not be used as monotherapy. The 750-mg dose gave the most consistent antiviral activity and lowest viral breakthrough and was chosen as the clinical dose.[180] Exposures are higher in combination with PR than when given alone and exposure is 330% higher when given with a high fat meal. Telaprevir reaches T_{max} 4–5 h after dosing and has an elimination half-life of 4–5 h. The major metabolites are the (*R*)-isomer at the position α to the keto-amide (30 times less potent than the (*S*)-isomer), pyrazinoic acid, and the inactive-reduced ketoamide. Dosing of ^{14}C-telaprevir gave 90% recovery of radioactivity, with 82% in the feces and 1% in urine. Telaprevir is a substrate and inhibitor for CYP3A4 and a substrate for P-glycoprotein. It is 59–76% bound to plasma proteins. The efficacy of telaprevir in combination with PR was established in three Phase III clinical trials with the primary endpoint of SVR, defined as undetectable HCV RNA 24 weeks after the end of treatment. The ADVANCE trial was a randomized, double-blind, placebo-controlled study in treatment naïve patients.[181] SVR was 75% in patients given telaprevir/PR for 12 weeks followed by PR to week 24 or week 48 versus 44% SVR on PR alone. The patients with the shorter duration of PR follow-on therapy were those who had achieved an extended rapid viral response (eRVR), defined as undetectable HCV RNA at weeks 4 and 12. The SVR rate was 92% in this subset of subjects. The ILLUMINATE trial[182] was a randomized, open-label trial in treatment-naive subjects given PR plus 750 mg of telaprevir three times a day for 12 weeks followed by PR. The SVR rate in patients with eRVR was 92% in patients receiving 12 weeks of follow-up on PR versus 90% in patients receiving 36 weeks of PR follow-up, indicating noninferiority of the shorter follow-up duration. The REALIZE trial[183] enrolled patients who were previously treated with PR and either had poor responses or relapsed following treatment. SVR rates were significantly higher in the telaprevir/PR patients compared with PR alone: 86% versus 22% in relapsers, 59% versus 15% in partial responders, and 32% versus 5% in non-responders. The most common adverse events with telaprevir/PR combination therapy were rash, fatigue, pruritis, nausea, anemia, and diarrhea. The largest differences in adverse reactions compared with PR alone were with rash (56%), pruritis (47%), and anemia (36%). Adverse events were reversible after discontinuation of drug. The recommended dosage of telaprevir (Incivek®) is 750 mg taken three times a day with food, and in combination with peginterferon alfa and ribavirin.

24. VANDETANIB (ANTICANCER)[184–189]

Class:	EGFR family, VEGF, RET inhibitor
Country of origin:	United Kingdom
Originator:	Astra Zeneca
First introduction:	United States
Introduced by:	AstraZeneca
Trade name:	Caprelsa®
CAS registry no:	443913-73-3
Molecular weight:	475.36

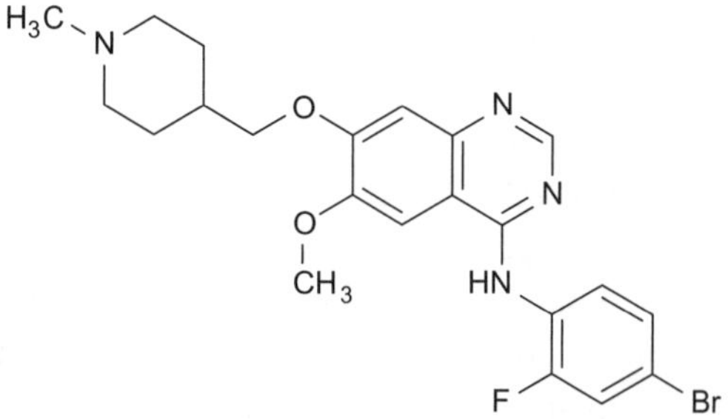

In April 2011, the U.S. FDA approved vandetanib (ZD6474) for the treatment of symptomatic or progressive medullary thyroid cancer (MTC) in adult patients with inoperable advanced or metastatic disease. According to statistics from the National Cancer Institute, there are 56,000 new cases of thyroid cancer each year in the United States.[184] Of the four types of thyroid cancer (papillary, follicular, medullary, and anaplastic), the incidence of MTC is 4%. The prognosis for MTC is poor. Surgery and radiotherapy are commonly employed to remove the primary tumor, but the treatment of metastatic disease remains a challenge with 5-year survival rates of ~40%. Twenty-five percent of MTC cases are primarily associated with the RET (REarranged during Transfection) oncogene. Vandetanib inhibits KDR/VEGFR2, VEGFR3, EGFR, and RET kinases with $IC_{50's}$ of 40, 110, 500, and < 100 nM, respectively.[185] In athymic mice bearing MTC tumors, a 14.5-fold reduction of tumor volume was observed after 45 days of treatment with vandetanib at 50 mg/kg/day. The decrease in tumor volume was accompanied by decreases in mitotic index (Ki67) and tumor angiogenesis in treated xenografts. Key steps in the synthesis of vandetanib include the displacement of the chlorine atom from 7-benzyloxy-4-chloro-6-methoxyquinazoline with 4-bromo-2-fluoroaniline under acidic conditions in a protic solvent and a Mitsunobu reaction of a *N*-protected piperidine alcohol with a phenol.[186,187]

In a study conducted on 231 patients with MTC, oral administration of vandetanib at 300 mg/day had a mean clearance of approximately 13.2 L/h, a mean volume of distribution of approximately 7450 L, a median T_{max} of 6 h, and a median plasma half-life of 19 days. The *in vitro* protein binding for vandetanib was ~90%.[188] The efficacy of vandetanib was assessed in a single, double-blind, placebo-controlled study involving 331 patients with unresectable locally advanced or metastatic MTC.[189] Of the 331 patients enrolled in the study, 231 were randomized to vandetanib (300 g) and 100 were randomized to placebo. Progression free survival (PFS), which was determined according to the Response Evaluation Criteria in Solid Tumors (RECIST), was used as the efficacy end point. Statistically significant improvements in PFS were observed in patients randomly assigned to vandetanib compared to placebo after a median follow-up period of 24 months. Patients with the most common activating gene mutation in MTCs, *RET* Met918Thr, had better response rates and more prolonged PFS than patients without this mutation. Most common adverse events for vandetanib versus placebo included diarrhea, rash, acne, nausea, hypertension, headache, fatigue, decreased appetite, abdominal pain, and QTc prolongation. The recommended dose of vandetanib is 300 mg administered orally once daily.

25. VEMURAFENIB (ANTICANCER)[190–195]

Class:	Kinase inhibitor
Country of origin:	United States
Originator:	Plexxikon
First introduction:	United States
Introduced by:	Hoffman-La Roche.
Trade name:	Zelboraf®
CAS registry no:	918504-65-1
Molecular weight:	489.9

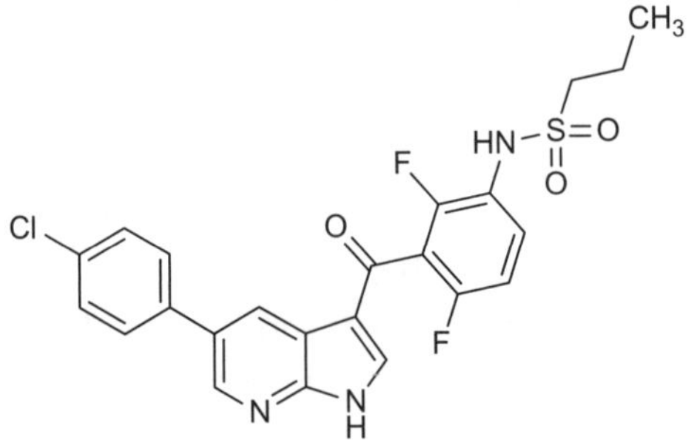

In August 2011, the United States FDA approved vemurafenib (PLX-4032, RO-5185426) for the treatment of patients with metastatic melanoma with the BRAF^{V600E} mutation. Statistics from the National Cancer Institute indicate ~76,250 new cases of melanoma will be diagnosed in 2012.[190] Immunotherapeutic options include treatment with IL-2 and more recently with ipilimumab. Chemotherapeutic options include treatment with dacarbazine. However, overall survival (OS) rates with either IL-2 or dacarbazine treatment are not high. Vemurafenib has been developed as a targeted therapy for patients with the BRAF gene mutation since oncogenic B-raf signaling is implicated in approximately 50% of melanomas. A mutation in the kinase domain at nucleotide 1799 of the BRAF gene results in the substitution of valine to glutamic acid leading to the constitutive activation of B-raf kinase. This results in excessive cell proliferation due to dysregulated downstream signaling and gene expression. Vemurafenib was identified based on an initial high-throughput screen followed by the extensive use of structure-based drug design.[191,192] Vemurafenib is a potent inhibitor of B-Raf^{V600E} kinase ($IC_{50}=13$ nM) compared to its potency against wild-type B-raf ($IC_{50}=160$ nM) and is fairly selective versus a panel of 200 kinases.[192] It does inhibit other kinases (RAF1, SRMS, ACK1, MAP4K5, and FGR) and mutant B-raf kinases (BRAF^{V600K}, BRAF^{V600D}, and BRAF^{V600R}) with enzyme $IC_{50's}$ of <100 nM.[193] Vemurafenib was tested against a panel of 17 melanoma cell lines and was a potent inhibitor in cell lines expressing the B-Raf^{V600E} mutant. In a murine LOX melanoma xenograft model, vemurafenib significantly inhibited tumor growth and induced tumor regression when dosed at 12.5, 25, and 75 mg/kg BID.[193] Efficacy was also seen in the A-375, COLO 829, and C8161 melanoma models. Key steps in the synthesis of vemurafenib[194] are a Friedel-Crafts reaction of 5-bromopyrrolo[2,3-*b*]pyridine with 2,6-difluoro-3-(propylsulfonamido)benzoyl chloride followed by Suzuki coupling with 4-chlorophenylboronic acid.

Pooled data from 458 patients with BRAF mutation-positive metastatic melanoma dosed with vemurafenib at 960 mg (BID) indicated a median T_{max} of ~3 h with a mean C_{max} and AUC_{0-12} of 62 ± 17 μg/mL and 601 ± 170 μg h/mL, respectively. Apparent volume of distribution and clearance are estimated to be 106 L and 31 L/day, respectively, and the half-life is estimated to be 57 h. The safety and efficacy of vemurafenib was assessed in a randomized, open label trial involving 675 patients.[195] Of these, 337 patients were assigned to vemurafenib, 960 mg orally twice daily, and the rest were assigned to dacarbazine at 1000 mg/m^2 administered intravenously once every 3 weeks. Efficacy measures included OS and investigator-assessed progression-free-survival (PFS). Interim analysis of the trial after 6 months indicated that the OS was 6.2 months for patients treated with vemurafenib and 4.5 months for patients treated with dacarbazine. The median PFS was 5.3 months for patients receiving vemurafenib and 1.6 months for patients receiving dacarbazine. In a second, single-arm trial involving 132 patients with $BRAF^{V600E}$ mutation and who had received at least one systemic therapy, the median duration of response was 6.5 months when vemurafenib was administered at 960 mg orally twice daily.[195] Most common adverse events (>30%) are arthralgia, rash, alopecia, fatigue, photosensitivity reaction, nausea, pruritus, and skin papilloma. Cutaneous squamous cell carcinoma occurred in ~24% of patients treated with vemurafenib. The recommended dose of vemurafenib is 960 mg administered orally twice daily for the treatment of patients with metastatic melanoma, as detected by the cobas 4800 BRAF V600 mutation Test (Roche Molecular Systems, Inc.).

26. VILAZODONE HYDROCHLORIDE (ANTIDEPRESSANT)[196–204]

Class:	Dual serotonin reuptake inhibitor and 5-HT1A partial agonist
Country of origin:	Germany
Originator:	Merck KGaA
First introduction:	United States
Introduced by:	Forest Laboratories
Trade name:	Viibryd®
CAS registry no:	163521-12-8 (free base)
	163521-08-2 (salt)
Molecular weight:	441.52 (free base)
	477.99 (salt)

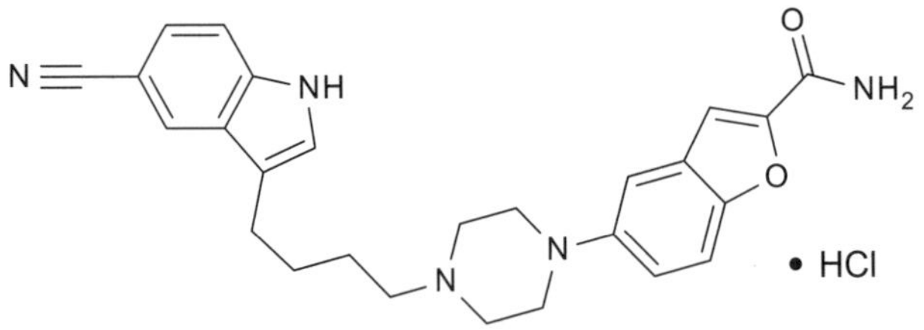

In January 2011, the U.S. FDA approved vilazodone for the treatment of major depressive disorder (MDD).[196,197] MDD is a serious medical condition that is characterized by persistent low mood, sadness, loss of interest in previously enjoyed activities, feelings of guilt or worthlessness, and thoughts of death or suicide. It is estimated that ~15% of the United

States population will experience MDD in their lifetime, with rates being higher in women than men. Treatments for MDD include antidepressant medications, psychotherapy, and electroconvulsive therapy. Among antidepressant treatments, the selective serotonin reuptake inhibitors (SSRIs) are the most commonly prescribed medications. Although these agents are generally safe and effective, fewer than 50% of all patients with depression show full remission with optimized SSRI treatment.[198] In addition, SSRIs must typically be dosed for 1–2 weeks before clinical benefit is seen. SSRIs act by blocking the serotonin (5-hydroytryptamine, 5-HT) transporter, which leads to increased levels of extracellular 5-HT and acute stimulation of presynaptic 5-HT1A autoreceptors with an initial reduction of 5-HT cell firing and 5-HT release. Repeated dosing of SSRIs is thought to lead to desensitization of 5-HT1A autoreceptors, thereby increasing 5-HT levels and augmenting serotonergic neurotransmission. Combination of SSRI activity with 5-HT1A partial agonism or antagonism is an approach to improving the onset of antidepressant efficacy by blunting the initial activation of the 5-HT1A receptor. Vilazodone is a novel antidepressant agent that combines potent serotonin reuptake inhibition ($IC_{50}=0.2$ nM) with high affinity for 5-HT1A receptors ($IC_{50}=0.5$ nM) and partial 5-HT1A receptor agonist functional activity. Vilazodone has good selectivity over other monoamine receptors and is efficacious in preclinical models of depression in rats and mice.[199] Vilazodone is an indolylbutylpiperazine derivative that has been prepared by coupling of an indolylbutyl chloride or tosylate with 5-(piperazin-1-yl)benzofuran-2-carboxamide or the corresponding ester.[200] The cyanoindole portion of vilazodone is important for conferring high affinity for both the serotonin transporter and the 5-HT1A receptor. Para-substitution on the phenyl group attached to the piperidine moiety reduces affinity for dopamine receptors, while the carboxamide group provides improved pharmacokinetic properties.[200]

In humans, vilazodone shows dose proportional increases in exposure at doses ranging from 5 to 80 mg.[196] Doses higher than 40 mg were poorly tolerated. Steady state is achieved after 3 days, with a plasma accumulation factor of 1.8. Vilazodone is 96–99% protein bound and has a high volume of distribution. Studies with radiolabeled vilazodone gave 85% recovery of radioactivity, with 65% in feces and 20% in urine. Only 3% was recovered as unchanged drug. Vilazodone is primarily metabolized by CYP3A4 to an inactive 6-hydroxylindole derivative.[201,202] When given with food, the C_{max} for vilazodone increases 147–160%, the AUC increases 64–85%, T_{max} is 4–5 h, and bioavailability is 72%. The efficacy of vilazodone was demonstrated in two Phase III short term (8-week), randomized, double-blind, placebo-controlled trials with ~900 subjects.[203] In these studies, vilazodone was given at a 10-mg dose for 7 days, a 20-mg dose for the next 7 days, and a 40-mg daily dose thereafter. The primary endpoint was the mean change in the Montgomery-Asberg Depression Rating Scale (MADRS), with the secondary endpoints including the change in the 17-item Hamilton Rating Score for Depression (HAM-D-17). After 8 weeks, there was a significant reduction from baseline in the MADRS (−12.9 for vilazodone treatment vs. −9.6 for placebo) and HAM-D-17 scores (−13.3 for vilazodone treatment vs. −10.3 for placebo). Data from a 52-week, open-label trial in ~600 subjects were used for evaluation of efficacy, safety, and tolerability.[204] Patients were titrated to a 40-mg dose and showed a significant improvement in MADRS total score (29.9 at baseline to 7.1 at week 52) and in other measures of efficacy. Common adverse events were diarrhea and nausea in both the 8- and 52-week trials and were considered mild to moderate. In the 8-week trials, there was a lower incidence of weight gain and sexual dysfunction compared with placebo. As with all approved antidepressant drugs in the United States, vilazodone has a black box warning describing the increased risk of suicidal thinking and behavior in children, adolescents, and young adults. The recommended dose of vilazodone is 40 mg once daily following titration from 10 mg for 7 days to 20 mg for 7 days. Vilazodone should be taken with food and the dose of vilazodone should be decreased to 20 mg if given in combination with a CYP3A4 inhibitor. Because there were no active comparators in clinical trials with vilazodone, differentiation from other antidepressants remains to be established.

Note: The authors are employees of Bristol-Myers Squibb and own stock in the company.

REFERENCES

(1) The collection of new therapeutic entities first launched in 2011 originated from the following sources: Prous Integrity Database; Thomson-Reuters Pipeline Database; The Pink Sheet; Drugs@FDA Website; FDA News Releases; IMS R&D Focus; Adis Business Intelligence R&D Insight; Pharmaprojects.
(2) http://www.fda.gov/Drugs/DevelopmentApprovalProcess/DrugInnovation/ucm285554.htm.
(3) http://www.fda.gov/NewsEvents/Newsroom/PressAnnouncements/ucm280525.htm.
(4) http://www.fda.gov/ForConsumers/ByAudience/ForPatientAdvocates/HIVandAIDSActivities/ucm267592.htm.
(5) http://www.accessdata.fda.gov/drugsatfda_docs/label/2011/202331lbl.pdf.
(6) http://www.merck.com/product/usa/pi_circulars/j/juvisync/juvisync_pi.pdf.
(7) http://seer.cancer.gov/statfacts/html/prost.html.

(8) O'Donnell, A.; Judson, I.; Dowsett, M.; Raynaud, F.; Dearnaley, D.; Mason, M.; Harland, S.; Robbins, A.; Halbert, G.; Nutley, B.; Jarman, M. *Br. J. Cancer* **2004**, *90*, 2317.
(9) Potter, G.A.; Barrie, S.E.; Jarman, M.; Rowlands, M.G. *J. Med. Chem.* **1995**, *38*, 2463.
(10) Jarman, M.; Barrie, S.E.; Llera, J.M. *J. Med. Chem.* **1998**, *41*, 5375.
(11) Potter, G.A.; Hardcastle, I.R.; Jarman, M. *Org. Prep. Proc. Int.* **1997**, *29*, 123.
(12) Barrie, S.E.; Potter, G.A.; Goddard, P.M.; Haynes, B.P.; Dowsett, M.; Jarman, M. *J. Steroid Biochem. Mol. Biol.* **1994**, *50*, 267.
(13) de Bono, J.S.; Logothetis, C.J.; Molina, A.; Fizazi, K.; North, S.; Chu, L.; Chi, K.N.; Jones, R.J.; Goodman, O.B., Jr.; Saad, F.; Staffurth, J.N.; Mainwaring, P.; Harland, S.; Flaig, T.W.; Hutson, T.E.; Cheng, T.; Patterson, H.; Hainsworth, J.D.; Ryan, C.J.; Sternberg, C.N.; Ellard, S.L.; Fléchon, A.; Saleh, M.; Scholz, M.; Efstathiou, E.; Zivi, A.; Bianchini, D.; Loriot, Y.; Chieffo, N.; Kheoh, T.; Haqq, C.M.; Scher, H.I. *N. Engl. J. Med.* **2011**, *364*, 1995.
(14) Sonpavde, G.; Attard, G.; Bellmunt, J.; Mason, M.D.; Malavaud, B.; Tombal, B.; Sternberg, C.N. *Eur. Urol.* **2011**, *60*, 270.
(15) Stewart, M.W.; Grippon, S.; Kirkpatrick, P. *Nat. Rev. Drug Discov.* **2012**, *11*, 269.
(16) http://www.nei.nih.gov/health/maculardegen/armd_facts.asp.
(17) Stewart, M.W. *Mayo Clin. Proc.* **2012**, *87*, 77.
(18) Holash, J.; Davis, S.; Papadopoulos, N.; Croll, S.D.; Ho, L.; Russell, M.; Boland, P.; Leidich, R.; Hylton, D.; Burova, E.; Ioffe, E.; Huang, T.; Radziejewski, C.; Bailey, K.; Fandl, J.P.; Daly, T.; Wiegand, S.J.; Yancopoulos, G.D.; Rudge, J.S. *Proc. Nat. Acad. Sci. U.S.A.* **2002**, *99*, 11393.
(19) Papadopoulos, N.; Martin, J.; Ruan, Q.; Rafique, A.; Rosconi, M.P.; Shi, E.; Pyles, E.A.; Yancopoulos, G.D.; Stahl, N.; Weigand, S.J. *Angiogenesis* **2012**, *15*, 171.
(20) http://www.accessdata.fda.gov/drugsatfda_docs/label/2011/125387lbl.pdf.
(21) Ohr, M.; Kaiser, P.K. *Expert Opin. Pharmacother.* **2012**, *13*, 585.
(22) Gaya, A.; Tse, V. *Cancer Treat. Rev.* **2012**, *38*, 484.
(23) Kim, E.S.H.; Bartholomew, J.R. Cleveland Clinic, August 1, 2010. Available at www.clevelandclinicmeded.com/medicalpubs/diseasemanagement/cardiology/venous-thromboembolism/.
(24) White, R.H. *Circulation* **2003**, *107*, 1.
(25) Quinlan, D.J.; Eikelboom, J.W.; Dahl, O.E.; Eriksson, B.I.; Sidhu, P.S.; Hirsh, J. *J. Thromb. Haemost.* **2007**, *5*, 1438.
(26) Kearon, C. *Chest* **2003**, *124*, 386S.
(27) Martin, M.T.; Nutescu, E.A. *Curr. Med. Res. Opin.* **2011**, *27*, 2123.
(28) Mann, K.G.; Nesheim, M.E.; Church, W.R.; Haley, P.; Krishnaswamy, S. *Blood* **1990**, *76*, 1.
(29) Davie, E.W. *Thromb. Haemost.* **1995**, *74*, 1.
(30) Davie, E.W. *J. Biol. Chem.* **2003**, *278*, 50819.
(31) Graham, J.B. *J. Thromb. Haemost.* **2003**, *1*, 871.
(32) Keiser, B. *Drugs Fut.* **1998**, *23*, 423.
(33) Pinto, D.J.P.; Orwat, M.J.; Koch, S.; Rossi, K.A.; Alexander, R.S.; Smallwood, A.; Wong, P.C.; Rendina, A.R.; Luettgen, J.M.; Knabb, R.M.; He, K.; Xin, B.; Wexler, R.R.; Lam, P.Y.S. *J. Med. Chem.* **2007**, *50*, 5339.
(34) Wong, P.C.; Crain, E.J.; Xin, B.; Wexler, R.R.; Lam, P.Y.S.; Pinto, D.J.; Luettgen, J.M.; Knabb, R.M. *J. Thromb. Haemost.* **2008**, *6*, 820.
(35) Raghavan, N.; Frost, C.E.; Yu, Z.; He, K.; Zhang, H.; Humphreys, W.G.; Pinto, D.; Chen, S.; Bonacorsi, S.; Wong, P.C.; Zhang, D. *Drug Metab. Dispos.* **2009**, *37*, 74.
(36) Granger, C.B.; Alexander, J.H.; McMurray, J.J.; Lopes, R.D.; Hylek, E.M.; Hanna, M.; Al-Khalidi, H.R.; Ansell, J.; Atar, D.; Avezum, A.; Bahit, M.C.;

Diaz, R.; Easton, J.D.; Ezekowitz, J.A.; Flaker, G.; Garcia, D.; Geraldes, M.; Gersh, B.J.; Golitsyn, S.; Goto, S.; Hermosillo, A.G.; Hohnloser, S.H.; Horowitz, J.; Mohan, P.; Jansky, P.; Lewis, B.S.; Lopez-Sendon, J.L.; Pais, P.; Parkhomenko, A.; Verheugt, F.W.; Zhu, J.; Wallentin, L. *N. Engl. J. Med.* **2011**, *365*, 981.
(37) Prom, R.; Spinler, S.A. *Ann. Pharmacother.* **2011**, *45*, 1262.
(38) PharmAsia News Oct 21st, 2011 http://www.elsevierbi.com/publications/pharmasia-news?issue=Oct-21-2011.
(39) Alwaal, A.; Al-Mannie, R.; Carrier, S. *Drug Des. Devel. Ther.* **2011**, *5*, 435.
(40) Limin, M.; Johnsen, N.; Hellstrom, W.J.G. *Expert Opin. Invest. Drugs* **2010**, *19*, 1427.
(41) Yamada, K.; Matsuki, K.; Omori, K.; Kikkawa, K. Patent Application US 2004/0142930, **2004**.
(42) Peterson, C.; Swearingen, D. *J. Sex. Med.* **2006**, *3*, 253.
(43) Jung, J.; Choi, S.; Cho, S.H.; Ghim, J.-L.; Hwang, A.; Kim, U.; Kim, B.S.; Koguchi, A.; Miyoshi, S.; Okabe, H.; Bae, K.-S.; Lim, H.-S. *Clin. Ther.* **2010**, *32*, 1178.
(44) VIVUS Announces Positive Results From Avanafil Phase 3 Study in Diabetics Presented at the 47th EASD Annual Meeting. 13 September **2011**. Available from: http://ir.vivus.com/releasedetail.cfm?ReleaseID=604819.
(45) Goldstein, I.; McCullough, A.R.; Jones, L.A.; Hellstrom, W.J.; Bowden, C.H.; DiDonato, K.; Trask, B.; Day, W.W. *J. Sex. Med.* **2012**, *9*, 1122.
(46) Yoon, S.S.; Ostchega, Y.; Louis, T. NCHS Data Brief, **2010**, *48*. http://www.cdc.gov/nchs/data/databriefs/db48.pdf.
(47) Lavoie, J.L.; Sigmund, C.D. *Endocrinology* **2003**, *144*, 2179.
(48) Kohara, Y.; Kubo, K.; Imamiya, E.; Wada, T.; Inada, Y.; Naka, T. *J. Med. Chem.* **1996**, *39*, 5228.
(49) Ojima, M.; Igata, H.; Tanaka, M.; Sakamoto, H.; Kuroita, T.; Kohara, Y.; Kubo, K.; Fuse, H.; Imura, Y.; Kusumoto, K.; Nagaya, H. *J. Pharmacol. Exp. Ther.* **2011**, *336*, 801.
(50) *Azilsartan Medoxomil (TAK 491) Investigator's Brochure.* 3rd ed. Deerfield, IL: Takeda Global Research & Development Center, Inc; 2007.
(51) White, W.B.; Weber, M.A.; Sica, D.; Bakris, G.L.; Perez, A.; Cao, C.; Kupfer, S. *Hypertension* **2011**, *57*, 413.
(52) Bakris, G.L.; Sica, D.; Weber, M.; White, W.B.; Roberts, A.; Perez, A.; Cao, C.; Kupfer, S. *J. Clin. Hypertens.* **2011**, *13*, 81.
(53) Knoll, G. *Drugs* **2008**, *68*(Suppl. 1), 3.
(54) Bia, M.; Adey, D.B.; Bloom, R.D.; Chan, L.; Kulkarni, S.; Tomlanovich, S. *Am. J. Kidney Dis.* **2010**, *56*, 189.
(55) Larsen, C.P.; Pearson, T.C.; Adams, A.B.; Tso, P.; Shirasugi, N.; Strobert, E.; Anderson, D.; Cowan, S.; Price, K.; Naemura, J.; Emswiler, J.; Greene, J.; Turk, L.A.; Bajorath, J.; Townsend, R.; Hagerty, D.; Linsley, P.S.; Peach, R.J. *Am. J. Transplant.* **2005**, *5*, 443.
(56) Vincenti, F.; Charpentier, B.; Vanrenterghem, Y.; Rostaing, L.; Bresnahan, B.; Darji, P.; Massari, P.; Mondragon-Ramirez, G.A.; Agarwal, M.; Di Russo, G.; Lin, C.S.; Garg, P.; Larsen, C.P. *Am. J. Transplant.* **2010**, *10*, 535.
(57) Vincenti, F.; Larsen, C.P.; Alberu, J.; Bresnahan, B.; Garcia, V.D.; Kothari, J.; Lang, P.; Urrea, E.M.; Massari, P.; Mondragon-Ramirez, G.; Reyes-Acevedo, R.; Rice, K.; Rostaing, L.; Steinberg, S.; Xing, J.; Agarwal, M.; Harler, M.B.; Charpentier, B. *Am. J. Transplant.* **2012**, *12*, 210.
(58) Larsen, C.P.; Grinyó, J.; Medina-Pestana, J.; Vanrenterghem, Y.; Vincenti, F.; Breshahan, B.; Campistol, J.M.; Florman, S.; Rial Mdel, C.; Kamar, N.; Block, A.; Di Russo, G.; Lin, C.S.; Garg, P.; Charpentier, B. *Transplantation* **2010**, *90*, 1528.

(59) Vanrenterghem, Y.; Bresnahan, B.; Campistol, J.; Durrbach, A.; Grinyó, J.; Neumayer, H.H.; Lang, P.; Larsen, C.P.; Mancilla-Urrea, E.; Pestana, J.M.; Block, A.; Duan, T.; Glicklich, A.; Gujrathi, S.; Vincenti, F. *Transplantation* **2011**, *91*, 976.
(60) Durrbach, A.; Pestana, J.M.; Pearson, T.; Vincenti, F.; Garcia, V.D.; Campistol, J.; Rial Mdel, C.; Florman, S.; Block, A.; Di Russo, G.; Xing, J.; Garg, P.; Grinyó, J. *Am. J. Transplant.* **2010**, *10*, 547.
(61) Pestana, J.O.; Grinyo, J.M.; Vanrenterghem, Y.; Becker, T.; Campistol, J.M.; Florman, S.; Garcia, V.D.; Kamar, N.; Lang, P.; Manfro, R.C.; Massari, P.; Rial, M.D.; Schnitzler, M.A.; Vitko, S.; Duan, T.; Block, A.; Harler, M.B.; Durrbach, A. *Am. J. Transplant.* **2012**, *12*, 630.
(62) Centers for Disease Control, Prevention (CDC), *Morb. Mortal. Wkly. Rep.* **2002**, *51*, 371.
(63) Lo, M.S.; Tsokos, G.C. *Annu. N.Y. Acad. Sci.* **2012**, *1247*, 138.
(64) Stohl, W.; Hiepe, F.; Latinis, K.M.; Thomas, M.; Scheinberg, M.A.; Clarke, A.; Aranow, C.; Wellborne, F.R.; Abud-Mendoza, C.; Hough, D.R.; Pineda, L.; Migone, T.S.; Zhong, Z.J.; Freimuth, W.W.; Chatham, W.W.; on Behalf of the Bliss-52 and -76 Study Groups, *Arthritis Rheum.* **2012**, *64*, 2328.
(65) Baker, K.P.; Edwards, B.M.; Main, S.H.; Choi, G.H.; Wager, R.E.; Halpern, W.G.; Lappin, P.B.; Riccobene, T.; Abramian, D.; Sekut, L.; Sturm, B.; Poortman, C.; Minter, R.R.; Dobson, C.L.; Williams, E.; Carmen, S.; Smith, R.; Roschke, V.; Hilbert, D.M.; Vaughan, T.J.; Albert, V.R. *Arthritis Rheum.* **2003**, *48*, 3253.
(66) Furie, R.; Stohl, W.; Ginzler, E.M.; Becker, M.; Mishra, N.; Chatham, W.; Merrill, J.T.; Weinstein, A.; McCune, W.J.; Zhong, J.; Cai, W.; Freimuth, W.; the Belimumab Study Group, *Arthritis Res. Ther.* **2008**, *10*, R109.
(67) Navarra, S.V.; Guzmán, R.M.; Gallacher, A.E.; Hall, S.; Levy, R.A.; Jimenez, R.E.; Li, E.K.; Thomas, M.; Kim, H.Y.; León, M.G.; Tanasescu, C.; Nasonov, E.; Lan, J.L.; Pineda, L.; Zhong, Z.J.; Freimuth, W.; Petri, M.A.; the BLISS-52 Study Group, *Lancet* **2011**, *377*, 721.
(68) Furie, R.; Petri, M.; Zamani, O.; Cervera, R.; Wallace, D.J.; Tegzová, D.; Sanchez-Guerrero, J.; Schwarting, A.; Merrill, J.T.; Chatham, W.W.; Stohl, W.; Ginzler, E.M.; Hough, D.R.; Zhong, Z.J.; Freimuth, W.; van Vollenhoven, R.F.; the BLISS-76 Study Group, *Arthritis Rheum.* **2011**, *63*, 3918.
(69) Butt, A.A.; Kanwal, F. *Clin. Infect. Dis.* **2012**, *54*, 97.
(70) Asselah, T.; Marcellin, P. *Liver Int.* **2012**, *32*(s1), 88.
(71) World Health Organization, June 2011. http://www.who.int/mediacentre/factsheets/fs164/en.
(72) Venkatraman, S.; Bogen, S.L.; Arasappan, A.; Bennett, F.; Chen, K.; Jao, E.; Liu, Y.-T.; Lovey, R.; Hendrata, S.; Huang, Y.; Pan, W.; Parekh, T.; Pinto, P.; Popov, V.; Pike, R.; Ruan, S.; Santhanam, B.; Vibulbhan, B.; Wu, W.; Yang, W.; Kong, J.; Liang, X.; Wong, J.; Liu, R.; Butkiewicz, N.; Chase, R.; Hart, A.; Agrawal, S.; Ingravallo, P.; Pichardo, J.; Kong, R.; Baroudy, B.; Malcolm, B.; Guo, Z.; Prongay, A.; Madison, V.; Broske, L.; Cui, X.; Cheng, K.-C.; Hsieh, Y.; Brisson, J.-M.; Prelusky, D.; Korfmacher, W.; White, R.; Bogdanowich-Knipp, S.; Pavlovsky, A.; Bradley, P.; Saksena, A.K.; Ganguly, A.; Piwinski, J.; Girijavallabhan, V.; Njoroge, F.G. *J. Med. Chem.* **2006**, *49*, 6074.
(73) Prongay, A.J.; Guo, Z.; Yao, N.; Pichardo, J.; Fischmann, T.; Strickland, C.; Myers, J., Jr.; Weber, P.C.; Beyer, B.M.; Ingram, R.; Hong, Z.; Prosise, W.W.; Ramanathan, L.; Taremi, S.S.; Yarosh-Tomaine, T.; Zhang, R.; Senior, M.; Yang, R.-S.; Malcolm, B.; Arasappan, A.; Bennett, F.; Bogen, S.L.; Chen, K.; Jao, E.; Liu, Y.-T.; Lovey, R.G.; Saksena, A.K.; Venkatraman, S.; Girijavallabhan, V.; Njoroge, F.G.; Madison, V. *J. Med. Chem.* **2007**, *50*, 2310.
(74) Campas, C.; Pandian, R.; Bolos, J.; Castaner, R. *Drugs Fut.* **2009**, *34*, 697.
(75) Rizza, S.A.; Talwani, R.; Nehra, V.; Temesgen, Z. *Drugs Today* **2011**, *47*, 743.

(76) Poordad, F.; McCone, J., Jr.; Bacon, B.R.; Bruno, S.; Manns, M.P.; Sulkowski, M.S.; Jacobson, I.M.; Reddy, K.R.; Goodman, Z.D.; Boparai, N.; DiNubile, M.J.; Sniukiene, V.; Brass, C.A.; Albrecht, J.K.; J-P. Bronowicki for the SPRINT-2 Investigators, *N. Engl. J. Med.* **2011**, *364*, 1195.
(77) Bacon, B.R.; Gordon, S.C.; Lawitz, E.; Marcellin, P.; Vierling, J.M.; Zeuzem, S.; Poordad, F.; Goodman, Z.D.; Sings, H.L.; Boparai, N.; Burroughs, M.; Brass, C.A.; Albrecht, J.K.; Esteban, R.; for the RESPOND-2 Investigators, *N. Engl. J. Med.* **2011**, *364*, 1207.
(78) Siegel, R.; Naishadham, D.; Jemal, A. *CA Cancer J. Clin.* **2012**, *62*, 10.
(79) Younes, A.; Yasothan, U.; Kirkpatrick, P. *Nat. Rev. Drug Discov.* **2012**, *11*, 19.
(80) Francisco, J.A.; Cerveny, C.G.; Meyer, D.L.; Mixan, B.J.; Klussman, K.; Chace, D.F.; Rejniak, S.X.; Gordon, K.A.; DeBlanc, R.; Toki, B.E.; Law, C.L.; Doronina, S.O.; Siegall, C.B.; Senter, P.D.; Wahl, A.F. *Blood* **2003**, *102*, 1458.
(81) Younes, A.; Bartlett, N.L.; Leonard, J.P.; Kennedy, D.A.; Lynch, C.M.; Sievers, E.L.; Forero-Torres, A. *N. Engl. J. Med.* **2010**, *363*, 1812.
(82) Chen, R.; Gopal, A.K.; Smith, S.E.; Ansell, S.M.; Rosenblatt, J.D.; Klasa, R.; Connors, J.M.; Engert, A.; Larsen, E.K.; Kennedy, D.A.; Sievers, E.L.; Anas Younes, *Blood,* 2010, *116*, ASH Annual Meeting Abstracts # 283.
(83) Shustov, A.R.; Advani, R.; Brice, P.; Bartlett, N.L.; Rosenblatt, J.D.; Illidge, T.; Matous, J.; Ramchandren, R.; Fanale, M.A.; Connors, J.M.; Yang, Y.; Sievers, E.L.; Kennedy, D.A.; Pro, B. *Blood* **2010**, *116* ASH Annual Meeting Abstracts # 961.
(84) http://www.cancer.gov/cancertopics/types/lung.
(85) Cui, J.J.; Tran-Dube, M.; Shen, H.; Nambu, M.; Kung, P.-P.; Pairish, M.; Jia, L.; Meng, J.; Funk, L.; Botrous, I.; McTigue, M.; Grodsky, N.; Ryan, K.; Padrique, E.; Alton, G.; Timofeevski, S.; Yamazaki, S.; Li, Q.; Zou, H.; Christensen, J.; Mroczkowski, B.; Bender, S.; Kania, R.S.; Edwards, M.P. *J. Med. Chem.* **2011**, *54*, 6342.
(86) Christensen, J.G.; Zou, H.Y.; Arango, M.E.; Li, Q.; Lee, J.H.; McDonnell, S.R.; Yamazaki, S.; Alton, G.R.; Mroczkowski, B.; Los, G. *Mol. Cancer Ther.* **2007**, *6*, 3314.
(87) de Koning, P.D.; McAndrew, D.; Moore, R.; Moses, I.B.; Boyles, D.C.; Kissick, K.; Stanchina, C.L.; Cuthbertson, T.; Kamatani, A.; Rahman, L.; Rodriguez, R.; Urbina, A.; Sandoval, A.; Rose, P.R. *Org. Process Res. Dev.* **2011**, *15*, 1018.
(88) (a) Tan, W.; Wilner, K.D.; Bang, Y.; Kwak, E.L.; Maki, R.G.; Camidge, D.R.; Solomon, B.J.; Ou, S.I.; Salgia, R.; Clark, J.W. *J. Clin. Oncol.* **2010**, *28*(suppl; abstr 2596). (b) Li, C.; Alvey, C.; Bello, A.; Wilner, K.D.; Tan, W. *J. Clin. Oncol.* **2011**, *29* (suppl; abstr e13065).
(89) (a) Crinò, L.; Kim, D.; Riely, G.J.; Janne, P.A.; Blackhall, F.H.; Camidge, D.R.; Hirsh, V.; Mok, T.; Solomon, B.J.; Park, K.; Gadgeel, S.M.; Martins, R.; Han, J.; De Pas, T.M.; Bottomley, A.; Polli, A.; Petersen, J.; Tassell, V.R.; Shaw, A.T. *J. Clin. Oncol.* **2011**, *29*(suppl; abstr 7514). (b) Kwak, E.L.; Bang, Y.-J.; Camidge, D.R.; Shaw, A.T.; Solomon, B.; Maki, R.G.; Ou, S.-H.I.; Dezube, B.J.; Jänne, P.A.; Costa, D.B.; Varella-Garcia, M.; Kim, W.-H.; Lynch, T.J.; Fidias, P.; Stubbs, H.; Engelman, J.A.; Sequist, L.V.; Tan, W.; Gandhi, L.; Mino-Kenudson, M.; Wei, G.C.; Shreeve, S.M.; Ratain, M.J.; Settleman, J.; Christensen, J.G.; Haber, D.A.; Wilner, K.; Salgia, R.; Shapiro, G.I.; Clark, J.W.; Iafrate, A.J. *N. Engl. J. Med.* **2010**, *363*, 1693. (c) Camidge, D.R.; Bang, Y.; Kwak, E.L.; Shaw, A.T.; Iafrate, A.J.; Maki, R.G.; Solomon, B.J.; Ou, S.I.; Salgia, R.; Wilner, K.D.; Costa, D.B.; Shapiro, G.; LoRusso, P.; Stephenson, P.; Tang, Y.; Ruffner, K.; Clark, J.W. *J. Clin. Oncol.* **2011**, *29*(suppl; abstr 2501).
(90) http://www.daiichisankyo.com/news/20110422_305_E.pdf.

(91) Haginoya, N.; Kobayashi, S.; Komoriya, S.; Yoshino, T.; Suzuki, M.; Shimada, T.; Watanabe, K.; Hirokawa, Y.; Furugoori, T.; Nagahara, T. *J. Med. Chem.* **2004**, *47*, 5167.
(92) Yoshikawa, K.; Yokomizo, A.; Naito, H.; Haginoya, N.; Kobayashi, S.; Yoshino, T.; Nagata, T.; Mochizuki, A.; Osanai, K.; Watanabe, K.; Kanno, H.; Ohta, T. *Bioorg. Med. Chem.* **2009**, *17*, 8206.
(93) Yoshikawa, K.; Kobayashi, S.; Nakamoto, Y.; Haginoya, N.; Komoriya, S.; Yoshino, T.; Nagata, T.; Mochizuki, A.; Watanabe, K.; Suzuki, M.; Kanno, H.; Ohta, T. *Bioorg. Med. Chem.* **2009**, *17*, 8221.
(94) Furugohri, T.; Isobe, K.; Honda, Y.; Kamisato-Matsumoto, C.; Sugiyama, N.; Nagahara, T.; Morishima, Y.; Shibano, T. *J. Thromb. Haemost.* **2008**, *6*, 1542.
(95) Ogata, K.; Mendell-Harary, J.; Tachibana, M.; Masumoto, H.; Oguma, T.; Kojima, M.; Kunitada, S. *J. Clin. Pharm.* **2010**, *50*, 743.
(96) Camm, A.J.; Bounameaux, H. *Drugs* **2011**, *71*, 1503.
(97) Fujiwara, S. *J. Bone Miner. Metab.* **2005**, *23*, 83.
(98) Holick, M.F.; Siris, E.S.; Binkley, N.; Beard, M.K.; Khan, A.; Katzer, J.T.; Petruschke, R.A.; Chen, E.; de Papp, A.E. *J. Clin. Endocrinol. Metab.* **2005**, *90*, 3215.
(99) Avenell, A.; Gillespie, W.J.; Gillespie, L.D.; O'Connell, D. *Cochrane Database of Systematic Reviews*, **2009**, Issue 2. Art. No.: CD000227. doi: http://dx.doi.org/10.1002/14651858. CD000227.pub3.
(100) Hatakeyama, S.; Yoshino, M.; Eto, K.; Takahashi, K.; Ishihara, J.; Ono, Y.; Saito, H.; Kubodera, N. *J. Steroid Biochem. Mol. Biol.* **2010**, *121*, 25.
(101) Sanford, M.; McCormack, P.L. *Drugs* **2011**, *71*, 1755.
(102) Ito, M.; Nakamura, T.; Fukunaga, M.; Shiraki, M.; Matsumoto, T. *Bone* **2011**, *49*, 328.
(103) Hardesty, J.S.; Juang, P. *Pharmacotherapy* **2011**, *31*, 877.
(104) Lancaster, J.W.; Matthews, S.J. *Clin. Ther.* **2012**, *34*, 1.
(105) Coronelli, C.; Parenti, F.; White, R.; Pagani, H. US Patent 3978211, 1976.
(106) Shue, Y.-K.; Hwang, C.-K.; Chiu, Y.-H.; Romero, A.; Babakhani, F.; Sears, P.; Okumu, F. US Patent Application 2008/0269145, **2008**.
(107) Miller, M. *Expert Opin. Pharmacother.* **2010**, *11*, 1569.
(108) Dolgin, E. *Nat. Med.* **2011**, *17*, 10.
(109) Burke, R.A.; Faulkner, M.A. *Expert Opin. Pharmacother.* **2011**, *12*, 2905.
(110) Cundy, K.C.; Branch, R.; Chernov-Rogan, T.; Dias, T.; Estrada, T.; Hold, K.; Koller, K.; Liu, X.; Mann, A.; Panuwat, M.; Raillard, S.P.; Upadhyay, S.; Wu, Q.Q.; Xiang, J.-N.; Yan, H.; Zerangue, N.; Zhou, C.X.; Barrett, R.W.; Gallop, M.A. *J. Pharmacol. Exp. Ther.* **2004**, *311*, 315.
(111) Cundy, K.C.; Annamalai, T.; Bu, L.; De Vera, J.; Estrela, J.; Luo, W.; Shirsat, P.; Torneros, A.; Yao, F.; Zou, J.; Barrett, R.W.; Gallop, M.A. *J. Pharmacol. Exp. Ther.* **2004**, *311*, 324.
(112) Revill, P.; Bolos, J.; Serradell, N.; Bayes, M. *Drugs Fut.* **2006**, *31*, 771.
(113) Cundy, K.C.; Sastry, S.; Luo, W.; Zou, J.; Moors, T.L.; Canafax, D.M. *J. Clin. Pharmacol.* **2008**, *48*, 1378.
(114) Kushida, C.A.; Becker, P.M.; Ellenbogen, A.L.; Canafax, D.M.; Barrett, R.W.; the XP052 Study Group, *Neurology* **2009**, *72*, 439.
(115) Lee, D.O.; Ziman, R.B.; Perkins, A.T.; Poceta, J.S.; Walters, A.S.; Barrett, R.W. *the XP053 Study Group. J. Clin. Sleep Med.* **2011**, 7, 282.
(116) Bogan, R.K.; Cramer Bornemann, M.A.; Kushida, C.A.; Tran, P.V.; Barrett, R.W. *the XP060 Study Group. Mayo Clin. Proc.* **2010**, *85*, 512.
(117) http://www.accessdata.fda.gov/drugsatfda_docs/label/2011/022399s000lbl.pdf.
(118) Zeng, Q.Y.; Chen, R.; Darmawan, J.; Xiao, Z.Y.; Chen, S.B.; Wigley, R.; Chen, S.L.; Zhang, N.Z. *Arthritis Res. Ther.* **2008**, *10*, R17.

(119) Tanaka, K.; Kawasaki, H.; Kurata, K.; Aikawa, Y.; Tsukamoto, Y.; Inaba, T. *Jpn. J. Pharmacol.* **1995**, *67*, 305.
(120) Tanaka, K. *Rheumatol. Rep.* **2009**, *1*, e4.
(121) Inaba, T.; Tanaka, K.; Takeno, R.; Nagaki, H.; Yoshida, C.; Takano, S. *Chem. Pharm. Bull.* **2000**, *48*, 131.
(122) Hara, M.; Abe, T.; Sugawara, S.; Mizushima, Y.; Hoshi, K.; Irimajiri, S.; Hashimoto, H.; Yoshino, S.; Matsui, N.; Nobunaga, M.; Nakano, S. *Mod. Rheumatol.* **2007**, *17*, 1.
(123) Hara, M.; Abe, T.; Sugawara, S.; Mizushima, Y.; Hoshi, K.; Irimajiri, S.; Hashimoto, H.; Yoshino, S.; Matsui, N.; Nobunaga, M. *Mod. Rheumatol.* **2007**, *17*, 10.
(124) Lu, L.-J.; Teng, J.-L.; Bao, C.-D.; Han, X.-H.; Sun, L.-Y.; Xu, J.-H.; Li, X.-F.; Wu, H.-X. *Chin. Med. J.* **2008**, *121*, 615.
(125) Agarwala, S.S. *Expert Rev. Anticancer Ther.* **2009**, *9*, 587.
(126) Keler, T.; Halk, E.; Vitale, L.; O'Neill, T.; Blanset, D.; Lee, S.; Srinivasan, M.; Graziano, R.F.; Davis, T.; Lonberg, N.; Korman, A. *J. Immunol.* **2003**, *171*, 6251.
(127) Harmankaya, K.; Erasim, C.; Koelblinger, C.; Ibrahim, R.; Hoos, A.; Pehamberger, H.; Binder, M. *Med. Oncol.* **2011**, *28*, 1140.
(128) Hodi, F.S.; O'Day, S.J.; McDermott, D.F.; Weber, R.W.; Sosman, J.A.; Haanen, J.B.; Gonzalez, R.; Robert, C.; Schadendorf, D.; Hassel, J.C.; Akerley, W.; van den Eertwegh, A.J.; Lutzky, J.; Lorigan, P.; Vaubel, J.M.; Linette, G.P.; Hogg, D.; Ottensmeier, C.H.; Lebbé, C.; Peschel, C.; Quirt, I.; Clark, J.I.; Wolchok, J.D.; Weber, J.S.; Tian, J.; Yellin, M.J.; Nichol, G.M.; Hoos, A.; Urba, W.J. *N. Engl. J. Med.* **2010**, *363*, 711 Erratum in: *N. Engl. J. Med.*, 2010 Sep 23;b(13):1290.
(129) Wolchok, J.D.; Weber, J.S.; Hamid, O.; Lebbé, C.; Maio, M.; Schadendorf, D.; de Pril, V.; Heller, K.; Chen, T.T.; Ibrahim, R.; Hoos, A.; O'Day, S.J. *Cancer Immun.* **2010**, *10*, 9.
(130) Korn, E.L.; Liu, P.Y.; Lee, S.J.; Chapman, J.A.; Niedzwiecki, D.; Suman, V.J.; Moon, J.; Sondak, V.K.; Atkins, M.B.; Eisenhauer, E.A.; Parulekar, W.; Markovic, S.N.; Saxman, S.; Kirkwood, J.M. *J. Clin. Oncol.* **2008**, *26*, 527.
(131) Robert, C.; Thomas, L.; Bondarenko, I.; O'Day, S.; Garbe, C.; Lebbe, C.; Baurain, J.F.; Testori, A.; Grob, J.J.; Davidson, N.; Richards, J.; Maio, M.; Hauschild, A.; Miller, W.H.; Gascon, P.; Lotem, M.; Harmankaya, K.; Ibrahim, R.; Francis, S.; Chen, T.T.; Humphrey, R.; Hoos, A.; Wolchok, J.D. *N. Engl. J. Med.* **2011**, *364*, 2517.
(132) Gallwitz, B.; Haering, H.-U. *Diabetes Obes. Metab.* **2010**, *12*, 1.
(133) Neumiller, J.J.; Wood, L.; Campbell, R.K. *Pharmacotherapy* **2010**, *30*, 463.
(134) Gerish, J. *Diabetes Res. Clin. Pract.* **2010**, *90*, 131.
(135) Zettl, H.; Schubert-Zsilavecz, M.; Steinhilber, D. *Chem. Med. Chem.* **2010**, *5*, 179.
(136) Eckhardt, M.; Langkopf, E.; Mark, M.; Tadayyon, M.; Thomas, L.; Nar, H.; Pfrengle, W.; Guth, B.; Lotz, R.; Sieger, P.; Fuchs, H.; Himmelsbach, F. *J. Med. Chem.* **2007**, *50*, 6450.
(137) Thomas, L.; Eckhardt, M.; Langkopf, E.; Tadayyon, M.; Himmelsbach, F.; Mark, M. *J. Pharmacol. Exp. Ther.* **2008**, *325*, 175.
(138) Heise, T.; Graefe-Mody, E.U.; Huttner, S.; Ring, A.; Trommeshauser, D.; Dugi, K.A. *Diabetes Obes. Metab.* **2009**, *11*, 786.
(139) Blech, S.; Ludwig-Schwellinger, E.; Grafe-Mody, E.U.; Withopf, B.; Wagner, K. *Drug Metab. Dispos.* **2010**, *38*, 667.
(140) Singh-Franco, D.; McLaughlin-Middlekauff, J.; Elrod, S.; Harrington, C. *Diabetes Obes. Metab.* **2012**, *14*, 694.
(141) Homma, Y.; Yamaguchi, O.; Hayashi, K. *Br. J. Urol.* **2005**, *96*, 1314.
(142) Maruyama, T.; Suzuki, T.; Onda, K.; Hayakawa, M.; Moritomo, H.; Kimizuka, T.; Matsui, T. Patent Application WO 9920607, **1999**.

(143) Takasu, T.; Ukai, M.; Sato, S.; Matsui, T.; Nagase, I.; Maruyama, T.; Sasamata, M.; Miyata, K.; Uchida, H.; Yamaguchi, O. *J. Pharmacol. Exp. Ther.* **2007**, *321*, 646.
(144) Bhide, A.A.; Digesu, G.A.; Fernando, R.; Khullar, V. *Int. Urogynecol. J.* **2012**, (ahead of print). doi: 10.07/s00192-012-1724-0.
(145) Stafstrom, C.E.; Grippon, S.; Kirkpatrick, P. *Nat. Rev. Drug Discov.* **2011**, *10*, 729.
(146) Deeks, E.D. *CNS Drugs* **2011**, *25*, 887.
(147) Fattore, C.; Perucca, E. *Drugs* **2011**, *71*, 2151.
(148) Bialer, M.; White, H.S. *Nat, Rev. Drug Discov.* **2010**, *9*, 68.
(149) Gunthorpe, M.J.; Large, C.H.; Sankar, R. *Epilepsia* **2012**, *53*, 412.
(150) Rostock, A.; Tober, C.; Rundfeldt, C.; Bartsch, R.; Engel, J.; Polymeropoulos, E.E.; Kutscher, B.; Loscher, W.; Honack, D.; White, H.S.; Wolf, H.H. *Epilepsy Res.* **1996**, *23*, 211.
(151) Wickenden, A.D.; Yu, W.; Zou, A.; Jegla, T.; Wagoner, P.K. *Mol. Pharmacol.* **2000**, *58*, 591.
(152) Dieter, H.-R.; Engel, J.; Kutscher, B.; Polymeropoulos, E.; Szelenyi, S.; Nickel, B. US Patent 5384330, **1995**.
(153) http://www.accessdata.fda.gov/drugsatfda_docs/label/2012/022345s001lbl.pdf.
(154) Porter, R.J.; Partiot, A.; Sachdeo, R.; Nohria, V.; Alves, W.M. *Neurology* **2007**, *68*, 1197.
(155) French, J.A.; Abou-Khalil, B.W.; Leroy, R.F.; Yacubian, E.M.T.; Shin, P.; Hall, S.; Mansbach, H.; Nohria, V. *Neurology* **2011**, *76*, 1555.
(156) Brodie, M.J.; Lerche, H.; Gil-Nagel, A.; Elger, C.; Hall, S.; Shin, P.; Nohria, V.; Mansbach, H. *Neurology* **2010**, *75*, 1817.
(157) http://www.unaids.org/en/dataanalysis/epidemiology/; http://www.cdc.gov/hiv/topics/surveillance.
(158) Ghosh, R.K.; Ghosh, S.M.; Chawla, S. *Expert Opin. Pharmacother.* **2011**, *12*, 31.
(159) Guillemont, J.; Boven, K.; Crauwels, H.; de Bethune, M.-P. In *Antiviral Drugs: From Basic Discovery Through Clinical Trials*, 1st ed.; Kazmierski, W.M., Ed.; John Wiley & Sons, Inc.: Hoboken, NJ, USA, 2011; p 59.
(160) Janssen, P.A.J.; Lewi, P.J.; Arnold, E.; Daeyaert, F.; de Jonge, M.; Heeres, J.; Koymans, L.; Vinkers, M.; Guillemont, J.; Pasquier, E.; Kukla, M.; Ludovici, D.; Andries, K.; de Bethune, M.-P.; Pauwels, R.; Das, K.; Clark, A.D., Jr.; Frenkel, Y.V.; Hughes, S.H.; Medaer, B.; De Knaep, F.; Bohets, H.; De Clerck, F.; Lampo, A.; Williams, P.; Stoffels, P. *J. Med. Chem.* **2005**, *48*, 1901.
(161) Guillemont, J.; Pasquier, E.; Palandjian, P.; Vernier, D.; Gaurrand, S.; Lewi, P.J.; Heeres, J.; de Jonge, M.R.; Koymans, L.M.H.; Daeyaert, F.F.D.; Vinkers, M.H.; Arnold, E.; Das, K.; Pauwels, R.; Andries, K.; de Bethune, M.-P.; Bettens, E.; Hertogs, K.; Wiegerinck, P.; Timmerman, P.; Janssen, P.A.J. *J. Med. Chem.* **2005**, *48*, 2072.
(162) Garvey, L.; Winston, A. *Expert Opin. Invest. Drugs* **2009**, *18*, 1035.
(163) US FDA http://www.accessdata.fda.gov/drugsatfda_docs/label/2011/202022s000lbl.pdf.
(164) Molina, J.-M.; Cahn, P.; Grinsztejn, B.; Lazzarin, A.; Mills, A.; Saag, M.; Supparatpinyo, K.; Walmsley, S.; Crauwels, H.; Rimsky, L.T.; Vanveggel, S.; Boven, K. on behalf of the ECHO study group. *Lancet* **2011**, *378*, 238.
(165) Cohen, C.; Andrade-Villanueva, J.; Clotet, B.; Fourie, J.; Johnson, M.A.; Ruxrungtham, K.; Wu, H.; Zorrilla, C.; Crauwels, H.; Rimsky, L.T.; Vanveggel, S.; Boven, K.; on behalf of the THRIVE study group, *Lancet* **2011**, *378*, 229.
(166) http://rarediseases.info.nih.gov/GARD/Condition/8618/QnA/22942/Myelofibrosis.aspx.
(167) Quintás-Cardama, A.; Vaddi, K.; Liu, P.; Manshouri, T.; Li, J.; Scherle, P.A.; Caulder, E.; Wen, X.; Li, Y.; Waeltz, P.; Rupar, M.; Burn, T.; Lo, Y.; Kelley, J.; Covington, M.; Shepard, S.; Rodgers, J.D.; Haley, P.; Kantarjian, H.; Fridman, J.S.; Verstovsek, S. *Blood* **2010**, *115*, 3109.

(168) Lin, Q.; Meloni, D.; Pan, Y.; Xia, M.; Rodgers, J.; Shepard, S.; Li, M.; Galya, L.; Metcalf, B.; Yue, T.-Y.; Liu, P.; Zhou, J. *Org. Lett.* **2009**, *11*, 1999.
(169) Shi, J.G.; Chen, X.; McGee, R.F.; Landman, R.R.; Emm, T.; Lo, Y.; Scherle, P.A.; Punwani, N.G.; Williams, W.V.; Yeleswaram, S. *J. Clin. Pharmacol.* **2011**, *51*, 1644.
(170) ClinicalTrials.gov, NCT00952289 (2011) http://clinicaltrials.gov.
(171) ClinicalTrials.gov, NCT00934544 (2011) http://clinicaltrials.gov.
(172) Saraiva, M.J.M. *Expert Rev. Mol. Med.* **2002**, *4*, 1.
(173) Hammarström, P.; Wiseman, R.L.; Powers, E.T.; Kelly, J.W. *Science* **2003**, *299*, 713.
(174) Razavi, H.; Palaninathan, S.K.; Powers, E.T.; Wiseman, R.L.; Purkey, H.E.; Mohamedmohaideen, N.N.; Deechongkit, S.; Chiang, K.P.; Dendle, M.T.A.; Sacchettini, J.C.; Kelly, J.W. *Angew. Chem. Int. Ed.* **2003**, *42*, 2758.
(175) Yamamoto, T.; Muto, K.; Komiyama, M.; Canivet, J.; Yamaguchi, J.; Itami, K. *Chem. Eur. J.* **2011**, *17*, 10113.
(176) European Medical Agency assessment report, Procedure No.: EMEA/H/C/002294, **2011**.
(177) Kwong, A.D.; Kauffman, R.S.; Hurter, P.; Mueller, P. *Nat. Biotechnol.* **2011**, *29*, 993.
(178) Revill, P.; Serradell, N.; Bolós, J.; Rosa, E. *Drugs Fut.* **2007**, *32*, 788.
(179) Tanoury, G. Patent Application WO 2011/153423, 2011; Tanoury, G.; Chen, M.; Cochran, J.E.; Looker, A. Patent Application WO 20101/126881, **2010**.
(180) Reesink, H.W.; Zuezem, S.; Weegink, C.J.; Forestier, N.; Van Vliet, A.; Van de Wetering de Rooij, J.; McNair, L.; Purdy, S.; Kauffman, R.; Alam, J.; Jansen, P.L.M. *Gastroenterology* **2006**, *131*, 997.
(181) Jacobson, I.M.; McHutchison, J.G.; Dusheiko, G.; Di Bisceglie, A.M.; Reddy, K.R.; Bzowej, N.H.; Marcellin, P.; Muir, A.J.; Ferenci, P.; Flisiak, R.; George, J.; Rizzetto, M.; Shouval, D.; Sola, R.; Terg, R.A.; Yoshida, E.M.; Adda, N.; Bengtsson, L.; Sankoh, A.J.; Kieffer, T.L.; George, S.; Kauffman, R.S.; Zeuzem, S.; for the ADVANCE Study Team, *N. Engl. J. Med.* **2011**, *365*, 2405.
(182) Sherman, K.E.; Flamm, S.L.; Afdhal, N.H.; Nelson, D.R.; Sulkowski, M.S.; Everson, G.T.; Fried, M.W.; Adler, M.; Reesink, H.W.; Martin, M.; Sankoh, A.J.; Adda, N.; Kauffman, R.S.; George, S.; Wright, C.I.; Poordad, F.; for the ILLUMINATE Study Team, *N. Engl. J. Med.* **2011**, *365*, 1014.
(183) Zeuzem, S.; Andreone, P.; Pol, S.; Lawitz, E.; Diago, M.; Roberts, S.; Focaccia, R.; Younossi, Z.; Foster, G.R.; Horban, A.; Ferenci, P.; Nevens, F.; Müllhaupt, B.; Pockros, P.; Terg, R.; Shouval, D.; van Hoek, B.; Weiland, O.; Van Heeswijk, R.; De Meyer, S.; Luo, D.; Boogaerts, G.; Polo, R.; Picchio, G.; Beumont, M. *N. Engl. J. Med.* **2011**, *365*, 2417.
(184) http://www.endocrineweb.com/conditions/thyroid-cancer/thyroid-cancer.
(185) Wedge, S.R.; Ogilvie, D.J.; Dukes, M.; Kendrew, J.; Chester, R.; Jackson, J.A.; Boffey, S.J.; Valentine, P.J.; Curwen, J.O.; Musgrove, H.L.; Graham, G.A.; Hughes, G.D.; Thomas, A.P.; Stokes, E.S.E.; Curry, B.; Richmond, G.H.P.; Wadsworth, P.F.; Bigley, A.L.; Hennequin, L.F. *Cancer Res.* **2002**, *62*, 4645.
(186) Hennequin, L.F.; Stokes, E.S.E.; Thomas, A.P.; Johnstone, C.; Ple´, P.A.; Ogilvie, D.J.; Dukes, M.; Wedge, S.R.; Kendrew, J.; Curwen, J.O. *J. Med. Chem.* **2002**, *45*, 1300.
(187) Hennequin, L.F.; Stokes, E.S.E.; Thomas, A.P. Patent Application WO 0132651, **2001**.
(188) Langmuir, P.B.; Yver, A. *Clin. Pharm. Ther.* **2012**, *91*, 71.
(189) Wells, S.A.; Robinson, B.G.; Gagel, R.F.; Dralle, H.; Fagin, J.A.; Santoro, M.; Baudin, E.; Vasselli, J.R.; Read, J.; Schlumberger, M. *J. Clin. Oncol.* **2010**, *28 (15s),* abstr. 5503.
(190) http://www.cancer.gov/cancertopics/types/melanoma.

(191) Tsai, J.; Lee, J.T.; Wang, W.; Zhang, J.; Cho, H.; Mamo, S.; Bremer, R.; Gillette, S.; Kong, J.; Haass, N.K.; Sproesser, K.; Li, L.; Smalley, K.S.M.; Fong, D.; Zhu, Y.-L.; Marimuthu, A.; Nguyen, H.; Lam, B.; Liu, J.; Cheung, I.; Rice, J.; Suzuki, Y.; Luu, C.; Settachatgul, C.; Shellooe, R.; Cantwell, J.; Kim, S.-H.; Schlessinger, J.; Zhang, K.Y.J.; West, B.L.; Powell, B.; Habets, G.; Zhang, C.; Ibrahim, P.N.; Hirth, P.; Artis, D.R.; Herlyn, M.; Bollag, G. *Proc. Natl. Acad. Sci. U.S.A.* **2008**, *105*, 3041.

(192) Bollag, G.; Hirth, P.; Tsai, J.; Zhang, J.; Ibrahim, P.N.; Cho, H.; Spevak, W.; Zhang, C.; Zhang, Y.; Habets, G.; Burton, E.A.; Wong, B.; Tsang, G.; West, B.L.; Powell, B.; Shellooe, R.; Marimuthu, A.; Nguyen, H.; Zhang, K.Y.J.; Artis, D.R.; Schlessinger, J.; Su, F.; Higgins, B.; Iyer, R.; D'Andrea, K.; Koehler, A.; Stumm, M.; Lin, P.S.; Lee, R.J.; Grippo, J.; Puzanov, I.; Kim, K.B.; Ribas, A.; McArthur, G.A.; Sosman, J.A.; Chapman, P.B.; Flaherty, K.T.; Xu, X.; Nathanson, K.L.; Nolop, K. *Nature* **2010**, *467*, 596.

(193) Puzanov, I.; Flaherty, K.T.; Sosman, J.A.; Grippo, J.F.; Su, F.; Nolop, K.; Leed, R.J.; Bollag, G. *Drugs Fut.* **2011**, *36*, 191.

(194) Ibrahim, P.N.; Artis, D.R.; Bremer, R.; Mamo, S.; Nespi, M.; Zhang, C.; Zhang, J.; Zhu, Y.-L.; Tsai, J.; Hirth, K.-P.; Bollag, G.; Spevak, W.; Cho, H.; Gillette, S.; Wu, G.; Zhu, H.; Shi, S. Patent Application WO07002325, **2007**.

(195) United States Food and Drug Administration. Labeling information Zelboraf. FDA website (2011) http://www.accessdata.fda.gov/drugsatfda_docs/label/2011/202429s000lbl.pdf.

(196) Frampton, J.E. *CNS Drugs* **2011**, *25*, 615.

(197) Laughren, T.P.; Gobburu, J.; Temple, R.J.; Unger, E.F.; Bhattaram, A.; Dinh, P.V.; Fossom, L.; Hung, H.M.; Kilmek, V.; Lee, J.E.; Levin, R.L.; Lindberg, C.Y.; Mathis, M.; Rosloff, B.N.; Wang, S.-J.; Wang, Y.; Yang, P.; Yu, B.; Zhang, H.; Zhang, L.; Zineh, I. *J. Clin. Psychiatry* **2011**, *72*, 1166.

(198) (a) Citrome, L. *Int. J. Clin. Practice* **2012**, *66*, 356. (b) Berton, O.; Nestler, E.J. *Nat. Rev. Neurosci.* **2006**, *7*, 137.

(199) Page, M.E.; Cryan, J.F.; Sullivan, A.; Dalvi, A.; Saucy, B.; Manning, D.R.; Lucki, I. *J. Pharmacol. Exp. Ther.* **2002**, *302*, 1220.

(200) Heinrich, T.; Bottcher, H.; Gericke, R.; Bartoszyk, G.D.; Anzali, S.; Seyfried, C.A.; Greiner, H.E.; van Amsterdam, C. *J. Med. Chem.* **2004**, *47*, 4684.

(201) Hewitt, N.J.; Hewitt, P. *Xenobiotica* **2004**, *34*, 243.

(202) Hewitt, N.J.; Buhring, K.-U.; Dasenbrock, J.; Haunschild, J.; Ladstetter, B.; Utesch, D. *Drug Metab. Dispos.* **2001**, *29*, 1042.

(203) Reed, C.R.; Kajdasz, D.K.; Whalen, H.; Athanasiou, M.C.; Gallipoli, S.; Thase, M.E. *Curr. Med. Res. Opin.* **2012**, *28*, 27.

(204) Robinson, D.; Kajdasz, D.K.; Gallipoli, S.; Whalen, H.; Wamil, A.; Reed, C.R. *J. Clin. Psychopharm.* **2011**, *31*, 643.

KEYWORD INDEX, VOLUME 47

Note: Page numbers followed by "*f*" indicate figures, and "*t*" indicate tables.

A

D

E

F

G

J

K

L

M

CUMULATIVE CHAPTER TITLES KEYWORD INDEX, VOLUME 1 – 47

CUMULATIVE NCE INTRODUCTION INDEX, 1983–2011

GENERIC NAME	INDICATION	YEAR INTRODUCED	ARMC VOL., (PAGE)
abacavir sulfate	antiviral	1999	35 (333)
abarelix	anticancer	2004	40 (446)
abatacept	antiarthritic	2006	42 (509)
abiraterone acetate	anticancer	2011	47 (503)
acarbose	antidiabetic	1990	26 (297)
aceclofenac	antiinflammatory	1992	28 (325)
acemannan	wound healing agent	2001	37 (259)
acetohydroxamic acid	urinary tract/bladder disorders	1983	19 (313)
acetorphan	antidiarrheal	1993	29 (332)
acipimox	antihypercholesterolemic	1985	21 (323)
acitretin	antipsoriasis	1989	25 (309)
acrivastine	antiallergy	1988	24 (295)
actarit	antiinflammatory	1994	30 (296)
adalimumab	antiarthritic	2003	39 (267)
adamantanium bromide	antibacterial	1984	20 (315)
adefovir dipivoxil	antiviral	2002	38 (348)
adrafinil	sleep disorders	1986	22 (315)
AF-2259	antiinflammatory	1987	23 (325)
aflibercept	ophthalmologic (macular degeneration)	2011	47 (505)
afloqualone	muscle relaxant	1983	19 (313)
agalsidase alfa	Fabry's disease	2001	37 (259)
alacepril	antihypertensive	1988	24 (296)
alcaftadine	ophthalmologic (allergic conjunctivitis)	2010	46 (444)
alclometasone dipropionate	antiinflammatory	1985	21 (323)
alefacept	antipsoriasis	2003	39 (267)
alemtuzumab	anticancer	2001	37 (260)
alendronate sodium	osteoporosis	1993	29 (332)
alfentanil hydrochloride	analgesic	1983	19 (314)
alfuzosin hydrochloride	antihypertensive	1988	24 (296)
alglucerase	Gaucher's disease	1991	27 (321)
alglucosidase alfa	Pompe disease	2006	42 (511)
aliskiren	antihypertensive	2007	43 (461)
alitretinoin	anticancer	1999	35 (333)
alminoprofen	analgesic	1983	19 (314)
almotriptan	antimigraine	2000	36 (295)
alogliptin	antidiabetic	2010	46 (446)
alosetron hydrochloride	irritable bowel syndrome	2000	36 (295)
alpha-1 antitrypsin	emphysema	1988	24 (297)
alpidem	anxiolytic	1991	27 (322)
alpiropride	antimigraine	1988	24 (296)

GENERIC NAME	INDICATION	YEAR INTRODUCED	ARMC VOL., (PAGE)
alteplase	antithrombotic	1987	23 (326)
alvimopan	post-operative ileus	2008	44 (584)
ambrisentan	pulmonary hypertension	2007	43 (463)
amfenac sodium	antiinflammatory	1986	22 (315)
amifostine	cytoprotective	1995	31 (338)
aminoprofen	antiinflammatory	1990	26 (298)
amisulpride	antipsychotic	1986	22 (316)
amlexanox	antiasthma	1987	23 (327)
amlodipine besylate	antihypertensive	1990	26 (298)
amorolfine hydrochloride	antifungal	1991	27 (322)
amosulalol	antihypertensive	1988	24 (297)
ampiroxicam	antiinflammatory	1994	30 (296)
amprenavir	antiviral	1999	35 (334)
amrinone	congestive heart failure	1983	19 (314)
amrubicin hydrochloride	anticancer	2002	38 (349)
amsacrine	anticancer	1987	23 (327)
amtolmetin guacil	antiinflammatory	1993	29 (332)
anagrelide hydrochloride	antithrombotic	1997	33 (328)
anakinra	antiarthritic	2001	37 (261)
anastrozole	anticancer	1995	31 (338)
angiotensin II	anticancer adjuvant	1994	30 (296)
anidulafungin	antifungal	2006	42 (512)
aniracetam	cognition enhancer	1993	29 (333)
anti-digoxin polyclonal antibody	antidote, digoxin poisoning	2002	38 (350)
APD	osteoporosis	1987	23 (326)
apixaban	antithrombotic	2011	47 (507)
apraclonidine hydrochloride	antiglaucoma	1988	24 (297)
aprepitant	antiemetic	2003	39 (268)
APSAC	antithrombotic	1987	23 (326)
aranidipine	antihypertensive	1996	32 (306)
arbekacin	antibacterial	1990	26 (298)
arformoterol	antiasthma	2007	43 (465)
argatroban	antithrombotic	1990	26 (299)
arglabin	anticancer	1999	35 (335)
aripiprazole	antipsychotic	2002	38 (350)
armodafinil	sleep disorders	2009	45 (478)
arotinolol hydrochloride	antihypertensive	1986	22 (316)
arteether	antimalarial	2000	36 (296)
artemisinin	antimalarial	1987	23 (327)
asenapine	antipsychotic	2009	45 (479)
aspoxicillin	antibacterial	1987	23 (328)
astemizole	antiallergy	1983	19 (314)
astromycin sulfate	antibacterial	1985	21 (324)
atazanavir	antiviral	2003	39 (269)

GENERIC NAME	INDICATION	YEAR INTRODUCED	ARMC VOL., (PAGE)
atomoxetine	attention deficit hyperactivity disorder	2003	39 (270)
atorvastatin calcium	antihypercholesterolemic	1997	33 (328)
atosiban	premature labor	2000	36 (297)
atovaquone	antiparasitic	1992	28 (326)
auranofin	antiarthritic	1983	19 (314)
avanafil	male sexual dysfunction	2011	47 (510)
azacitidine	anticancer	2004	40 (447)
azelaic acid	acne	1989	25 (310)
azelastine hydrochloride	antiallergy	1986	22 (316)
azelnidipine	antihypertensive	2003	39 (270)
azilsartan	antihypertensive	2011	47 (511)
azithromycin	antibacterial	1988	24 (298)
azosemide	diuretic	1986	22 (316)
aztreonam	antibacterial	1984	20 (315)
balofloxacin	antibacterial	2002	38 (351)
balsalazide disodium	ulcerative colitis	1997	33 (329)
bambuterol	antiasthma	1990	26 (299)
barnidipine hydrochloride	antihypertensive	1992	28 (326)
beclobrate	antihypercholesterolemic	1986	22 (317)
befunolol hydrochloride	antiglaucoma	1983	19 (315)
belatacept	immunosuppressant	2011	47 (513)
belimumab	lupus	2011	47 (516)
belotecan	anticancer	2004	40 (449)
benazepril hydrochloride	antihypertensive	1990	26 (299)
benexate hydrochloride	antiulcer	1987	23 (328)
benidipine hydrochloride	antihypertensive	1991	27 (322)
beraprost sodium	antiplatelet	1992	28 (326)
besifloxacin	antibacterial	2009	45 (482)
betamethasone butyrate propionate	antiinflammatory	1994	30 (297)
betaxolol hydrochloride	antihypertensive	1983	19 (315
betotastine besilate	antiallergy	2000	36 (297)
bevacizumab	anticancer	2004	40 (450)
bevantolol hydrochloride	antihypertensive	1987	23 (328)
bexarotene	anticancer	2000	36 (298)
biapenem	antibacterial	2002	38 (351)
bicalutamide	anticancer	1995	31 (338)
bifemelane hydrochloride	nootropic	1987	23 (329)
bilastine	antiallergy	2010	46 (449)
bimatoprost	antiglaucoma	2001	37 (261)
binfonazole	sleep disorders	1983	19 (315)
binifibrate	antihypercholesterolemic	1986	22 (317)
biolimus drug-eluting stent	coronary artery disease, antirestenotic	2008	44 (586)

GENERIC NAME	INDICATION	YEAR INTRODUCED	ARMC VOL., (PAGE)
bisantrene hydrochloride	anticancer	1990	26 (300)
bisoprolol fumarate	antihypertensive	1986	22 (317)
bivalirudin	antithrombotic	2000	36 (298)
blonanserin	antipsychotic	2008	44 (587)
boceprevir	antiviral	2011	47 (518)
bopindolol	antihypertensive	1985	21 (324)
bortezomib	anticancer	2003	39 (271)
bosentan	antihypertensive	2001	37 (262)
brentuximab	anticancer	2011	47 (520)
brimonidine	antiglaucoma	1996	32 (306)
brinzolamide	antiglaucoma	1998	34 (318)
brodimoprin	antibacterial	1993	29 (333)
bromfenac sodium	antiinflammatory	1997	33 (329)
brotizolam	sleep disorders	1983	19 (315)
brovincamine fumarate	cerebral vasodilator	1986	22 (317)
bucillamine	immunomodulator	1987	23 (329)
bucladesine sodium	congestive heart failure	1984	20 (316)
budipine	Parkinson's disease	1997	33 (330)
budralazine	antihypertensive	1983	19 (315)
bulaquine	antimalarial	2000	36 (299)
bunazosin hydrochloride	antihypertensive	1985	21 (324)
bupropion hydrochloride	antidepressant	1989	25 (310)
buserelin acetate	hormone therapy	1984	20 (316)
buspirone hydrochloride	anxiolytic	1985	21 (324)
butenafine hydrochloride	antifungal	1992	28 (327)
butibufen	antiinflammatory	1992	28 (327)
butoconazole	antifungal	1986	22 (318)
butoctamide	sleep disorders	1984	20 (316)
butyl flufenamate	antiinflammatory	1983	19 (316)
cabazitaxel	anticancer	2010	46 (451)
cabergoline	antiprolactin	1993	29 (334)
cadexomer iodine	wound healing agent	1983	19 (316)
cadralazine	antihypertensive	1988	24 (298)
calcipotriol	antipsoriasis	1991	27 (323)
camostat mesylate	anticancer	1985	21 (325)
canakinumab	antiinflammatory	2009	45 (484)
candesartan cilexetil	antihypertensive	1997	33 (330)
capecitabine	anticancer	1998	34 (319)
captopril	antihypertensive	1982	13 (086)
carboplatin	antibacterial	1986	22 (318)
carperitide	congestive heart failure	1995	31 (339)
carumonam	antibacterial	1988	24 (298)
carvedilol	antihypertensive	1991	27 (323)
caspofungin acetate	antifungal	2001	37 (263)
catumaxomab	anticancer	2009	45 (486)

GENERIC NAME	INDICATION	YEAR INTRODUCED	ARMC VOL., (PAGE)
cefbuperazone sodium	antibacterial	1985	21 (325)
cefcapene pivoxil	antibacterial	1997	33 (330)
cefdinir	antibacterial	1991	27 (323)
cefditoren pivoxil	antibacterial	1994	30 (297)
cefepime	antibacterial	1993	29 (334)
cefetamet pivoxil hydrochloride	antibacterial	1992	28 (327)
cefixime	antibacterial	1987	23 (329)
cefmenoxime hydrochloride	antibacterial	1983	19 (316)
cefminox sodium	antibacterial	1987	23 (330)
cefodizime sodium	antibacterial	1990	26 (300)
cefonicid sodium	antibacterial	1984	20 (316)
ceforanide	antibacterial	1984	20 (317)
cefoselis	antibacterial	1998	34 (319)
cefotetan disodium	antibacterial	1984	20 (317)
cefotiam hexetil hydrochloride	antibacterial	1991	27 (324)
cefozopran hydrochloride	antibacterial	1995	31 (339)
cefpimizole	antibacterial	1987	23 (330)
cefpiramide sodium	antibacterial	1985	21 (325)
cefpirome sulfate	antibacterial	1992	28 (328)
cefpodoxime proxetil	antibacterial	1989	25 (310)
cefprozil	antibacterial	1992	28 (328)
ceftaroline fosamil	antibacterial	2010	46 (453)
ceftazidime	antibacterial	1983	19 (316)
cefteram pivoxil	antibacterial	1987	23 (330)
ceftibuten	antibacterial	1992	28 (329)
ceftobiprole medocaril	antibacterial	2008	44 (589)
cefuroxime axetil	antibacterial	1987	23 (331)
cefuzonam sodium	antibacterial	1987	23 (331)
celecoxib	antiarthritic	1999	35 (335)
celiprolol hydrochloride	antihypertensive	1983	19 (317)
centchroman	contraception	1991	27 (324)
centoxin	immunomodulator	1991	27 (325)
cerivastatin	antihypercholesterolemic	1997	33 (331)
certolizumab pegol	irritable bowel syndrome	2008	44 (592)
cetirizine hydrochloride	antiallergy	1987	23 (331)
cetrorelix	infertility	1999	35 (336)
cetuximab	anticancer	2003	39 (272)
cevimeline hydrochloride	antixerostomia	2000	36 (299)
chenodiol	gallstones	1983	19 (317)
CHF-1301	Parkinson's disease	1999	35 (336)
choline alfoscerate	cognition enhancer	1990	26 (300)
choline fenofibrate	antihypercholesterolemic	2008	44 (594)
cibenzoline	antiarrhythmic	1985	21 (325)

GENERIC NAME	INDICATION	YEAR INTRODUCED	ARMC VOL., (PAGE)
ciclesonide	antiasthma	2005	41 (443)
cicletanine	antihypertensive	1988	24 (299)
cidofovir	antiviral	1996	32 (306)
cilazapril	antihypertensive	1990	26 (301)
cilostazol	antithrombotic	1988	24 (299)
cimetropium bromide	antispasmodic	1985	21 (326)
cinacalcet	hyperparathyroidism	2004	40 (451)
cinildipine	antihypertensive	1995	31 (339)
cinitapride	gastroprokinetic	1990	26 (301)
cinolazepam	anxiolytic	1993	29 (334)
ciprofibrate	antihypercholesterolemic	1985	21 (326)
ciprofloxacin	antibacterial	1986	22 (318)
cisapride	gastroprokinetic	1988	24 (299)
cisatracurium besilate	muscle relaxant	1995	31 (340)
citalopram	antidepressant	1989	25 (311)
cladribine	anticancer	1993	29 (335)
clarithromycin	antibacterial	1990	26 (302)
clevidipine	antihypertensive	2008	44 (596)
clevudine	antiviral	2007	43 (466)
clobenoside	antiinflammatory	1988	24 (300)
cloconazole hydrochloride	antifungal	1986	22 (318)
clodronate disodium	calcium regulation	1986	22 (319)
clofarabine	anticancer	2005	41 (444)
clopidogrel hydrogensulfate	antithrombotic	1998	34 (320)
cloricromen	antithrombotic	1991	27 (325)
clospipramine hydrochloride	antipsychotic	1991	27 (325)
colesevelam hydrochloride	antihypercholesterolemic	2000	36 (300)
colestimide	antihypercholesterolemic	1999	35 (337)
colforsin daropate hydrochloride	congestive heart failure	1999	35 (337)
conivaptan	hyponatremia	2006	42 (514)
corifollitropin alfa	infertility	2010	46 (455)
crizotinib	anticancer	2011	47 (522)
crotelidae polyvalent immune fab	antidote, snake venom poisoning	2001	37 (263)
cyclosporine	immunosuppressant	1983	19 (317)
cytarabine ocfosfate	anticancer	1993	29 (335)
dabigatran etexilate	anticoagulant	2008	44 (598)
dalfampridine	multiple sclerosis	2010	46 (458)
dalfopristin	antibacterial	1999	35 (338)
dapiprazole hydrochloride	antiglaucoma	1987	23 (332)
dapoxetine	premature ejaculation	2009	45 (488)
daptomycin	antibacterial	2003	39 (272)
darifenacin	urinary tract/bladder disorders	2005	41 (445)
darunavir	antiviral	2006	42 (515)

GENERIC NAME	INDICATION	YEAR INTRODUCED	ARMC VOL., (PAGE)
dasatinib	anticancer	2006	42 (517)
decitabine	myelodysplastic syndromes	2006	42 (519)
defeiprone	iron chelation therapy	1995	31 (340)
deferasirox	iron chelation therapy	2005	41 (446)
defibrotide	antithrombotic	1986	22 (319)
deflazacort	antiinflammatory	1986	22 (319)
degarelix acetate	anticancer	2009	45 (490)
delapril	antihypertensive	1989	25 (311)
delavirdine mesylate	antiviral	1997	33 (331)
denileukin diftitox	anticancer	1999	35 (338)
denopamine	congestive heart failure	1988	24 (300)
denosumab	osteoporosis	2010	46 (459)
deprodone propionate	antiinflammatory	1992	28 (329)
desflurane	anesthetic	1992	28 (329)
desloratadine	antiallergy	2001	37 (264)
desvenlafaxine	antidepressant	2008	44 (600)
dexfenfluramine	antiobesity	1997	33 (332)
dexibuprofen	antiinflammatory	1994	30 (298)
dexlansoprazole	antiulcer	2009	45 (492)
dexmedetomidine hydrochloride	sleep disorders	2000	36 (301)
dexmethylphenidate hydrochloride	attention deficit hyperactivity disorder	2002	38 (352)
dexrazoxane	cardioprotective	1992	28 (330)
dezocine	analgesic	1991	27 (326)
diacerein	antiinflammatory	1985	21 (326)
didanosine	antiviral	1991	27 (326)
dilevalol	antihypertensive	1989	25 (311)
diquafosol tetrasodium	ophthalmologic (dry eye)	2010	46 (462)
dirithromycin	antibacterial	1993	29 (336)
disodium pamidronate	osteoporosis	1989	25 (312)
divistyramine	antihypercholesterolemic	1984	20 (317)
docarpamine	congestive heart failure	1994	30 (298)
docetaxel	anticancer	1995	31 (341)
dofetilide	antiarrhythmic	2000	36 (301)
dolasetron mesylate	antiemetic	1998	34 (321)
donepezil hydrochloride	Alzheimer's disease	1997	33 (332)
dopexamine	congestive heart failure	1989	25 (312)
doripenem	antibacterial	2005	41 (448)
dornase alfa	cystic fibrosis	1994	30 (298)
dorzolamide hydrochloride	antiglaucoma	1995	31 (341)
dosmalfate	antiulcer	2000	36 (302)
doxacurium chloride	muscle relaxant	1991	27 (326)
doxazosin mesylate	antihypertensive	1988	24 (300)
doxefazepam	anxiolytic	1985	21 (326)

GENERIC NAME	INDICATION	YEAR INTRODUCED	ARMC VOL., (PAGE)
doxercalciferol	hyperparathyroidism	1999	35 (339)
doxifluridine	anticancer	1987	23 (332)
doxofylline	antiasthma	1985	21 (327)
dronabinol	antiemetic	1986	22 (319)
dronedarone	antiarrhythmic	2009	45 (495)
drospirenone	contraception	2000	36 (302)
drotrecogin alfa	antisepsis	2001	37 (265)
droxicam	antiinflammatory	1990	26 (302)
droxidopa	Parkinson's disease	1989	25 (312)
duloxetine	antidepressant	2004	40 (452)
dutasteride	benign prostatic hyperplasia	2002	38 (353)
duteplase	anticoagulant	1995	31 (342)
ebastine	antiallergy	1990	26 (302)
eberconazole	antifungal	2005	41 (449)
ebrotidine	antiulcer	1997	33 (333)
ecabet sodium	antiulcer	1993	29 (336)
ecallantide	angioedema, hereditary	2009	46 (464)
eculizumab	hemoglobinuria	2007	43 (468)
edaravone	neuroprotective	2001	37 (265)
edoxaban	antithrombotic	2011	47 (524)
efalizumab	antipsoriasis	2003	39 (274)
efavirenz	antiviral	1998	34 (321)
efonidipine	antihypertensive	1994	30 (299)
egualen sodium	antiulcer	2000	36 (303)
eldecalcitol	osteoporosis	2011	47 (526)
eletriptan	antimigraine	2001	37 (266)
eltrombopag	antithrombocytopenic	2009	45 (497)
emedastine difumarate	antiallergy	1993	29 (336)
emorfazone	analgesic	1984	20 (317)
emtricitabine	antiviral	2003	39 (274)
enalapril maleate	antihypertensive	1984	20 (317)
enalaprilat	antihypertensive	1987	23 (332)
encainide hydrochloride	antiarrhythmic	1987	23 (333)
enfuvirtide	antiviral	2003	39 (275)
enocitabine	anticancer	1983	19 (318)
enoxacin	antibacterial	1986	22 (320)
enoxaparin	anticoagulant	1987	23 (333)
enoximone	congestive heart failure	1988	24 (301)
enprostil	antiulcer	1985	21 (327)
entacapone	Parkinson's disease	1998	34 (322)
entecavir	antiviral	2005	41 (450)
epalrestat	antidiabetic	1992	28 (330)
eperisone hydrochloride	muscle relaxant	1983	19 (318)
epidermal growth factor	wound healing agent	1987	23 (333)
epinastine	antiallergy	1994	30 (299)

GENERIC NAME	INDICATION	YEAR INTRODUCED	ARMC VOL., (PAGE)
epirubicin hydrochloride	anticancer	1984	20 (318)
eplerenone	antihypertensive	2003	39 (276)
epoprostenol sodium	antiplatelet	1983	19 (318)
eprosartan	antihypertensive	1997	33 (333)
eptazocine hydrobromide	analgesic	1987	23 (334)
eptilfibatide	antithrombotic	1999	35 (340)
erdosteine	expectorant	1995	31 (342)
eribulin mesylate	anticancer	2010	46 (465)
erlotinib	anticancer	2004	40 (454)
ertapenem sodium	antibacterial	2002	38 (353)
erythromycin acistrate	antibacterial	1988	24 (301)
erythropoietin	hematopoietic	1988	24 (301)
escitalopram oxolate	antidepressant	2002	38 (354)
eslicarbazepine acetate	anticonvulsant	2009	45 (498)
esmolol hydrochloride	antiarrhythmic	1987	23 (334)
esomeprazole magnesium	antiulcer	2000	36 (303)
eszopiclone	sleep disorders	2005	41 (451)
ethyl icosapentate	antithrombotic	1990	26 (303)
etizolam	anxiolytic	1984	20 (318)
etodolac	antiinflammatory	1985	21 (327)
etoricoxibe	antiarthritic	2002	38 (355)
etravirine	antiviral	2008	44 (602)
everolimus	immunosuppressant	2004	40 (455)
exemestane	anticancer	2000	36 (304)
exenatide	antidiabetic	2005	41 (452)
exifone	cognition enhancer	1988	24 (302)
ezetimibe	antihypercholesterolemic	2002	38 (355)
factor VIIa	haemophilia	1996	32 (307)
factor VIII	hemostatic	1992	28 (330)
fadrozole hydrochloride	anticancer	1995	31 (342)
falecalcitriol	hyperparathyroidism	2001	37 (266)
famciclovir	antiviral	1994	30 (300)
famotidine	antiulcer	1985	21 (327)
fasudil hydrochloride	amyotrophic lateral sclerosis	1995	31 (343)
febuxostat	gout	2009	45 (501)
felbamate	anticonvulsant	1993	29 (337)
felbinac	antiinflammatory	1986	22 (320)
felodipine	antihypertensive	1988	24 (302)
fenbuprol	biliary tract dysfunction	1983	19 (318)
fenoldopam mesylate	antihypertensive	1998	34 (322)
fenticonazole nitrate	antifungal	1987	23 (334)
fesoterodine	urinary tract/bladder disorders	2008	44 (604)
fexofenadine	antiallergy	1996	32 (307)
fidaxomicin	antibacterial	2011	47 (528)
filgrastim	immunostimulant	1991	27 (327)

GENERIC NAME	INDICATION	YEAR INTRODUCED	ARMC VOL., (PAGE)
finasteride	benign prostatic hyperplasia	1992	28 (331)
fingolimod	multiple sclerosis	2010	46 (468)
fisalamine	antiinflammatory	1984	20 (318)
fleroxacin	antibacterial	1992	28 (331)
flomoxef sodium	antibacterial	1988	24 (302)
flosequinan	congestive heart failure	1992	28 (331)
fluconazole	antifungal	1988	24 (303)
fludarabine phosphate	anticancer	1991	27 (327)
flumazenil	antidote, benzodiazepine overdose	1987	23 (335)
flunoxaprofen	antiinflammatory	1987	23 (335)
fluoxetine hydrochloride	antidepressant	1986	22 (320)
flupirtine maleate	analgesic	1985	21 (328)
flurithromycin ethylsuccinate	antibacterial	1997	33 (333)
flutamide	anticancer	1983	19 (318)
flutazolam	anxiolytic	1984	20 (318)
fluticasone furoate	antiallergy	2007	43 (469)
fluticasone propionate	antiinflammatory	1990	26 (303)
flutoprazepam	anxiolytic	1986	22 (320)
flutrimazole	antifungal	1995	31 (343)
flutropium bromide	antiasthma	1988	24 (303)
fluvastatin	antihypercholesterolemic	1994	30 (300)
fluvoxamine maleate	antidepressant	1983	19 (319)
follitropin alfa	infertility	1996	32 (307)
follitropin beta	infertility	1996	32 (308)
fomepizole	antidote, ethylene glycol poisoning	1998	34 (323)
fomivirsen sodium	antiviral	1998	34 (323)
fondaparinux sodium	antithrombotic	2002	38 (356)
formestane	anticancer	1993	29 (337)
formoterol fumarate	chronic obstructive pulmonary disease	1986	22 (321)
fosamprenavir	antiviral	2003	39 (277)
fosaprepitant dimeglumine	antiemetic	2008	44 (606)
foscarnet sodium	antiviral	1989	25 (313)
fosfluconazole	antifungal	2004	40 (457)
fosfosal	analgesic	1984	20 (319)
fosinopril sodium	antihypertensive	1991	27 (328)
fosphenytoin sodium	anticonvulsant	1996	32 (308)
fotemustine	anticancer	1989	25 (313)
fropenam	antibacterial	1997	33 (334)
frovatriptan	antimigraine	2002	38 (357)
fudosteine	expectorant	2001	37 (267)
fulveristrant	anticancer	2002	38 (357)
gabapentin	anticonvulsant	1993	29 (338)

GENERIC NAME	INDICATION	YEAR INTRODUCED	ARMC VOL., (PAGE)
gabapentin enacarbil	restless leg syndrome	2011	47 (530)
gadoversetamide	diagnostic	2000	36 (304)
gallium nitrate	calcium regulation	1991	27 (328)
gallopamil hydrochloride	antianginal	1983	19 (3190)
galsulfase	mucopolysaccharidosis VI	2005	41 (453)
ganciclovır	antiviral	1988	24 (303)
ganirelix acetate	infertility	2000	36 (305)
garenoxacin	antibacterial	2007	43 (471)
gatilfloxacin	antibacterial	1999	35 (340)
gefitinib	anticancer	2002	38 (358)
gemcitabine hydrochloride	anticancer	1995	31 (344)
gemeprost	abortifacient	1983	19 (319)
gemifloxacin	antibacterial	2004	40 (458)
gemtuzumab ozogamicin	anticancer	2000	36 (306)
gestodene	contraception	1987	23 (335)
gestrinone	contraception	1986	22 (321)
glatiramer acetate	multiple sclerosis	1997	33 (334)
glimepiride	antidiabetic	1995	31 (344)
glucagon, rDNA	antidiabetic	1993	29 (338)
GMDP	immunostimulant	1996	32 (308)
golimumab	antiinflammatory	2009	45 (503)
goserelin	hormone therapy	1987	23 (336)
granisetron hydrochloride	antiemetic	1991	27 (329)
guanadrel sulfate	antihypertensive	1983	19 (319)
gusperimus	immunosuppressant	1994	30 (300)
halobetasol propionate	antiinflammatory	1991	27 (329)
halofantrine	antimalarial	1988	24 (304)
halometasone	antiinflammatory	1983	19 (320)
histrelin	precocious puberty	1993	29 (338)
hydrocortisone aceponate	antiinflammatory	1988	24 (304)
hydrocortisone butyrate	antiinflammatory	1983	19 (320)
ibandronic acid	osteoporosis	1996	32 (309)
ibopamine hydrochloride	congestive heart failure	1984	20 (319)
ibritunomab tiuxetan	anticancer	2002	38 (359)
ibudilast	antiasthma	1989	25 (313)
ibutilide fumarate	antiarrhythmic	1996	32 (309)
icatibant	angioedema, hereditary	2008	44 (608)
idarubicin hydrochloride	anticancer	1990	26 (303)
idebenone	nootropic	1986	22 (321)
idursulfase	mucopolysaccharidosis II (Hunter syndrome)	2006	42 (520)
iguratimod	antiarthritic	2011	47 (532)
iloprost	antiplatelet	1992	28 (332)
imatinib mesylate	anticancer	2001	37 (267)
imidafenacin	urinary tract/bladder disorders	2007	43 (472)

GENERIC NAME	INDICATION	YEAR INTRODUCED	ARMC VOL., (PAGE)
imidapril hydrochloride	antihypertensive	1993	29 (339)
imiglucerase	Gaucher's disease	1994	30 (301)
imipenem/cilastatin	antibacterial	1985	21 (328)
imiquimod	antiviral	1997	33 (335)
incadronic acid	osteoporosis	1997	33 (335)
indacaterol	chronic obstructive pulmonary disease	2009	45 (505)
indalpine	antidepressant	1983	19 (320)
indeloxazine hydrochloride	nootropic	1988	24 (304)
indinavir sulfate	antiviral	1996	32 (310)
indisetron	antiemetic	2004	40 (459)
indobufen	antithrombotic	1984	20 (319)
influenza virus (live)	antiviral	2003	39 (277)
insulin lispro	antidiabetic	1996	32 (310)
interferon alfacon-1	antiviral	1997	33 (336)
interferon gamma-1b	immunostimulant	1991	27 (329)
interferon, b-1a	multiple sclerosis	1996	32 (311)
interferon, b-1b	multiple sclerosis	1993	29 (339)
interferon, gamma	antiinflammatory	1989	25 (314)
interferon, gamma-1	anticancer	1992	28 (332)
interleukin-2	anticancer	1989	25 (314)
ioflupane	diagnostic	2000	36 (306)
ipilimumab	anticancer	2011	47 (533)
ipriflavone	osteoporosis	1989	25 (314)
irbesartan	antihypertensive	1997	33 (336)
irinotecan	anticancer	1994	30 (301)
irsogladine	antiulcer	1989	25 (315)
isepamicin	antibacterial	1988	24 (305)
isofezolac	antiinflammatory	1984	20 (319)
isoxicam	antiinflammatory	1983	19 (320)
isradipine	antihypertensive	1989	25 (315)
itopride hydrochloride	gastroprokinetic	1995	31 (344)
itraconazole	antifungal	1988	24 (305)
ivabradine	antianginal	2006	42 (522)
ivermectin	antiparasitic	1987	23 (336)
ixabepilone	anticancer	2007	43 (473)
ketanserin	antihypertensive	1985	21 (328)
ketorolac tromethamine	analgesic	1990	26 (304)
kinetin	dermatologic, skin photodamage	1999	35 (341)
lacidipine	antihypertensive	1991	27 (330)
lacosamide	anticonvulsant	2008	44 (610)
lafutidine	antiulcer	2000	36 (307)
lamivudine	antiviral	1995	31 (345)
lamotrigine	anticonvulsant	1990	26 (304)

GENERIC NAME	INDICATION	YEAR INTRODUCED	ARMC VOL., (PAGE)
landiolol	antiarrhythmic	2002	38 (360)
laninamivir octanoate	antiviral	2010	46 (470)
lanoconazole	antifungal	1994	30 (302)
lanreotide acetate	growth disorders	1995	31 (345)
lansoprazole	antiulcer	1992	28 (332)
lapatinib	anticancer	2007	43 (475)
laronidase	mucopolysaccharidosis I	2003	39 (278)
latanoprost	antiglaucoma	1996	32 (311)
lefunomide	antiarthritic	1998	34 (324)
lenalidomide	myelodysplastic syndromes, multiple myeloma	2006	42 (523)
lenampicillin hydrochloride	antibacterial	1987	23 (336)
lentinan	immunostimulant	1986	22 (322)
lepirudin	anticoagulant	1997	33 (336)
lercanidipine	antihypertensive	1997	33 (337)
letrazole	anticancer	1996	32 (311)
leuprolide acetate	hormone therapy	1984	20 (319)
levacecarnine hydrochloride	cognition enhancer	1986	22 (322)
levalbuterol hydrochloride	antiasthma	1999	35 (341)
levetiracetam	anticonvulsant	2000	36 (307)
levobunolol hydrochloride	antiglaucoma	1985	21 (328)
levobupivacaine hydrochloride	anesthetic	2000	36 (308)
levocabastine hydrochloride	antiallergy	1991	27 (330)
levocetirizine	antiallergy	2001	37 (268)
levodropropizine	antitussive	1988	24 (305)
levofloxacin	antibacterial	1993	29 (340)
levosimendan	congestive heart failure	2000	36 (308)
lidamidine hydrochloride	antidiarrheal	1984	20 (320)
limaprost	antithrombotic	1988	24 (306)
linagliptin	antidiabetic	2011	47 (536)
linezolid	antibacterial	2000	36 (309)
liraglutide	antidiabetic	2009	45 (507)
liranaftate	antifungal	2000	36 (309)
lisdexamfetamine	attention deficit hyperactivity disorder	2007	43 (477)
lisinopril	antihypertensive	1987	23 (337)
lobenzarit sodium	antiinflammatory	1986	22 (322)
lodoxamide tromethamine	antiallergy	1992	28 (333)
lomefloxacin	antibacterial	1989	25 (315)
lomerizine hydrochloride	antimigraine	1999	35 (342)
lonidamine	anticancer	1987	23 (337)
lopinavir	antiviral	2000	36 (310)
loprazolam mesylate	sleep disorders	1983	19 (321)
loprinone hydrochloride	congestive heart failure	1996	32 (312)

GENERIC NAME	INDICATION	YEAR INTRODUCED	ARMC VOL., (PAGE)
loracarbef	antibacterial	1992	28 (333)
loratadine	antiallergy	1988	24 (306)
lornoxicam	antiinflammatory	1997	33 (337)
losartan	antihypertensive	1994	30 (302)
loteprednol etabonate	antiallergy	1998	34 (324)
lovastatin	antihypercholesterolemic	1987	23 (337)
loxoprofen sodium	antiinflammatory	1986	22 (322)
lulbiprostone	constipation	2006	42 (525)
luliconazole	antifungal	2005	41 (454)
lumiracoxib	antiinflammatory	2005	41 (455)
lurasidone hydrochloride	antipsychotic	2010	46 (473)
Lyme disease vaccine	Lyme disease	1999	35 (342)
mabuterol hydrochloride	antiasthma	1986	22 (323)
malotilate	hepatoprotective	1985	21 (329)
manidipine hydrochloride	antihypertensive	1990	26 (304)
maraviroc	antiviral	2007	43 (478)
masoprocol	anticancer	1992	28 (333)
maxacalcitol	hyperparathyroidism	2000	36 (310)
mebefradil hydrochloride	antihypertensive	1997	33 (338)
medifoxamine fumarate	antidepressant	1986	22 (323)
mefloquine hydrochloride	antimalarial	1985	21 (329)
meglutol	antihypercholesterolemic	1983	19 (321)
melinamide	antihypercholesterolemic	1984	20 (320)
meloxicam	antiarthritic	1996	32 (312)
mepixanox	respiratory stimulant	1984	20 (320)
meptazinol hydrochloride	analgesic	1983	19 (321)
meropenem	antibacterial	1994	30 (303)
metaclazepam	anxiolytic	1987	23 (338)
metapramine	antidepressant	1984	20 (320)
methylnaltrexone bromide	constipation	2008	44 (612)
mexazolam	anxiolytic	1984	20 (321)
micafungin	antifungal	2002	38 (360)
mifamurtide	anticancer	2009	46 (476)
mifepristone	abortifacient	1988	24 (306)
miglitol	antidiabetic	1998	34 (325)
miglustat	Gaucher's disease	2003	39 (279)
milnacipran	antidepressant	1997	33 (338)
milrinone	congestive heart failure	1989	25 (316)
miltefosine	anticancer	1993	29 (340)
minodronic acid	osteoporosis	2009	45 (509)
miokamycin	antibacterial	1985	21 (329)
mirabegron	urinary tract/bladder disorders	2011	47 (539)
mirtazapine	antidepressant	1994	30 (303)
misoprostol	antiulcer	1985	21 (329)
mitiglinide	antidiabetic	2004	40 (460)

GENERIC NAME	INDICATION	YEAR INTRODUCED	ARMC VOL., (PAGE)
mitoxantrone hydrochloride	anticancer	1984	20 (321)
mivacurium chloride	muscle relaxant	1992	28 (334)
mivotilate	hepatoprotective	1999	35 (343)
mizolastine	antiallergy	1998	34 (325)
mizoribine	immunosuppressant	1984	20 (321)
moclobemide	antidepressant	1990	26 (305)
modafinil	sleep disorders	1994	30 (303)
moexipril hydrochloride	antihypertensive	1995	31 (346)
mofezolac	analgesic	1994	30 (304)
mometasone furoate	antiinflammatory	1987	23 (338)
montelukast sodium	antiasthma	1998	34 (326)
moricizine hydrochloride	antiarrhythmic	1990	26 (305)
mosapride citrate	gastroprokinetic	1998	34 (326)
moxifloxacin hydrochloride	antibacterial	1999	35 (343)
moxonidine	antihypertensive	1991	27 (330)
mozavaptan	hyponatremia	2006	42 (527)
mupirocin	antibacterial	1985	21 (330)
muromonab-CD3	immunosuppressant	1986	22 (323)
muzolimine	diuretic	1983	19 (321)
mycophenolate mofetil	immunosuppressant	1995	31 (346)
mycophenolate sodium	immunosuppressant	2003	39 (279)
nabumetone	antiinflammatory	1985	21 (330)
nadifloxacin	antibacterial	1993	29 (340)
nafamostat mesylate	pancreatitis	1986	22 (323)
nafarelin acetate	hormone therapy	1990	26 (306)
naftifine hydrochloride	antifungal	1984	20 (321)
naftopidil	urinary tract/bladder disorders	1999	35 (344)
nalfurafine hydrochloride	pruritus	2009	45 (510)
nalmefene hydrochloride	addiction, opioids	1995	31 (347)
naltrexone hydrochloride	addiction, opioids	1984	20 (322)
naratriptan hydrochloride	antimigraine	1997	33 (339)
nartograstim	leukopenia	1994	30 (304)
natalizumab	multiple sclerosis	2004	40 (462)
nateglinide	antidiabetic	1999	35 (344)
nazasetron	antiemetic	1994	30 (305)
nebivolol	antihypertensive	1997	33 (339)
nedaplatin	anticancer	1995	31 (347)
nedocromil sodium	antiallergy	1986	22 (324)
nefazodone	antidepressant	1994	30 (305)
nelarabine	anticancer	2006	42 (528)
nelfinavir mesylate	antiviral	1997	33 (340)
neltenexine	cystic fibrosis	1993	29 (341)
nemonapride	antipsychotic	1991	27 (331)
nepafenac	antiinflammatory	2005	41 (456)
neridronic acide	calcium regulation	2002	38 (361)

GENERIC NAME	INDICATION	YEAR INTRODUCED	ARMC VOL., (PAGE)
nesiritide	congestive heart failure	2001	37 (269)
neticonazole hydrochloride	antifungal	1993	29 (341)
nevirapine	antiviral	1996	32 (313)
nicorandil	antianginal	1984	20 (322)
nif ekalant hydrochloride	antiarrhythmic	1999	35 (344)
nilotinib	anticancer	2007	43 (480)
nilutamide	anticancer	1987	23 (338)
nilvadipine	antihypertensive	1989	25 (316)
nimesulide	antiinflammatory	1985	21 (330)
nimodipine	cerebral vasodilator	1985	21 (330)
nimotuzumab	anticancer	2006	42 (529)
nipradilol	antihypertensive	1988	24 (307)
nisoldipine	antihypertensive	1990	26 (306)
nitisinone	antityrosinaemia	2002	38 (361)
nitrefazole	addiction, alcohol	1983	19 (322)
nitrendipine	antihypertensive	1985	21 (331)
nizatidine	antiulcer	1987	23 (339)
nizofenzone	nootropic	1988	24 (307)
nomegestrol acetate	contraception	1986	22 (324)
norelgestromin	contraception	2002	38 (362)
norfloxacin	antibacterial	1983	19 (322)
norgestimate	contraception	1986	22 (324)
OCT-43	anticancer	1999	35 (345)
octreotide	growth disorders	1988	24 (307)
ofatumumab	anticancer	2009	45 (512)
ofloxacin	antibacterial	1985	21 (331)
olanzapine	antipsychotic	1996	32 (313)
olimesartan Medoxomil	antihypertensive	2002	38 (363)
olopatadine hydrochloride	antiallergy	1997	33 (340)
omalizumab allergic	antiasthma	2003	39 (280)
omeprazole	antiulcer	1988	24 (308)
ondansetron hydrochloride	antiemetic	1990	26 (306)
OP-1	osteoinductor	2001	37 (269)
orlistat	antiobesity	1998	34 (327)
ornoprostil	antiulcer	1987	23 (339)
osalazine sodium	antiinflammatory	1986	22 (324)
oseltamivir phosphate	antiviral	1999	35 (346)
oxaliplatin	anticancer	1996	32 (313)
oxaprozin	antiinflammatory	1983	19 (322)
oxcarbazepine	anticonvulsant	1990	26 (307)
oxiconazole nitrate	antifungal	1983	19 (322)
oxiracetam	cognition enhancer	1987	23 (339)
oxitropium bromide	antiasthma	1983	19 (323)
ozagrel sodium	antithrombotic	1988	24 (308)
paclitaxal	anticancer	1993	29 (342)

GENERIC NAME	INDICATION	YEAR INTRODUCED	ARMC VOL., (PAGE)
palifermin	mucositis	2005	41 (461)
paliperidone	antipsychotic	2007	43 (482)
palonosetron	antiemetic	2003	39 (281)
panipenem/ betamipron carbapenem	antibacterial	1994	30 (305)
panitumumab	anticancer	2006	42 (531)
pantoprazole sodium	antiulcer	1995	30 (306)
parecoxib sodium	analgesic	2002	38 (364)
paricalcitol	hyperparathyroidism	1998	34 (327)
parnaparin sodium	anticoagulant	1993	29 (342)
paroxetine	antidepressant	1991	27 (331)
pazopanib	anticancer	2009	45 (514)
pazufloxacin	antibacterial	2002	38 (364)
pefloxacin mesylate	antibacterial	1985	21 (331)
pegademase bovine	immunostimulant	1990	26 (307)
pegaptanib	ophthalmologic (macular degeneration)	2005	41 (458)
pegaspargase	anticancer	1994	30 (306)
pegvisomant	growth disorders	2003	39 (281)
pemetrexed	anticancer	2004	40 (463)
pemirolast potassium	antiasthma	1991	27 (331)
penciclovir	antiviral	1996	32 (314)
pentostatin	anticancer	1992	28 (334)
peramivir	antiviral	2010	46 (477)
pergolide mesylate	Parkinson's disease	1988	24 (308)
perindopril	antihypertensive	1988	24 (309)
perospirone hydrochloride	antipsychotic	2001	37 (270)
picotamide	antithrombotic	1987	23 (340)
pidotimod	immunostimulant	1993	29 (343)
piketoprofen	antiinflammatory	1984	20 (322)
pilsicainide hydrochloride	antiarrhythmic	1991	27 (332)
pimaprofen	antiinflammatory	1984	20 (322)
pimecrolimus	immunosuppressant	2002	38 (365)
pimobendan	congestive heart failure	1994	30 (307)
pinacidil	antihypertensive	1987	23 (340)
pioglitazone hydrochloride	antidiabetic	1999	35 (346)
pirarubicin	anticancer	1988	24 (309)
pirfenidone	pulmonary fibrosis, idiopathic	2008	44 (614)
pirmenol	antiarrhythmic	1994	30 (307)
piroxicam cinnamate	antiinflammatory	1988	24 (309)
pitavastatin	antihypercholesterolemic	2003	39 (282)
pivagabine	antidepressant	1997	33 (341)
plaunotol	antiulcer	1987	23 (340)
plerixafor hydrochloride	stem cell mobilizer	2009	45 (515)
polaprezinc	antiulcer	1994	30 (307)

GENERIC NAME	INDICATION	YEAR INTRODUCED	ARMC VOL., (PAGE)
porfimer sodium	anticancer	1993	29 (343)
posaconazole	antifungal	2006	42 (532)
pralatrexate	anticancer	2009	45 (517)
pramipexole hydrochloride	Parkinson's disease	1997	33 (341)
pramiracetam sulfate	cognition enhancer	1993	29 (343)
pramlintide	antidiabetic	2005	41 (460)
pranlukast	antiasthma	1995	31 (347)
prasugrel	antiplatelet	2009	45 (519)
pravastatin	antihypercholesterolemic	1989	25 (316)
prednicarbate	antiinflammatory	1986	22 (325)
pregabalin	anticonvulsant	2004	40 (464)
prezatide copper acetate	wound healing agent	1996	32 (314)
progabide	anticonvulsant	1985	21 (331)
promegestrone	contraception	1983	19 (323)
propacetamol hydrochloride	analgesic	1986	22 (325)
propagermanium	antiviral	1994	30 (308)
propentofylline propionate	cerebral vasodilator	1988	24 (310)
propiverine hydrochloride	urinary tract/bladder disorders	1992	28 (335)
propofol	anesthetic	1986	22 (325)
prulifloxacin	antibacterial	2002	38 (366)
pumactant	respiratory distress syndrome	1994	30 (308)
quazepam	sleep disorders	1985	21 (332)
quetiapine fumarate	antipsychotic	1997	33 (341)
quinagolide	hyperprolactinemia	1994	30 (309)
quinapril	antihypertensive	1989	25 (317)
quinf amideamebicide	antiparasitic	1984	20 (322)
quinupristin	antibacterial	1999	35 (338)
rabeprazole sodium	antiulcer	1998	34 (328)
raloxifene hydrochloride	osteoporosis	1998	34 (328)
raltegravir	antiviral	2007	43 (484)
raltitrexed	anticancer	1996	32 (315)
ramatroban	antiallergy	2000	36 (311)
ramelteon	sleep disorders	2005	41 (462)
ramipril	antihypertensive	1989	25 (317)
ramosetron	antiemetic	1996	32 (315)
ranibizumab	ophthalmologic (macular degeneration)	2006	42 (534)
ranimustine	anticancer	1987	23 (341)
ranitidine bismuth citrate	antiulcer	1995	31 (348)
ranolazine	antianginal	2006	42 (535)
rapacuronium bromide	muscle relaxant	1999	35 (347)
rasagiline	Parkinson's disease	2005	41 (464)
rebamipide	antiulcer	1990	26 (308)
reboxetine	antidepressant	1997	33 (342)
remifentanil hydrochloride	analgesic	1996	32 (316)

GENERIC NAME	INDICATION	YEAR INTRODUCED	ARMC VOL., (PAGE)
remoxipride hydrochloride	antipsychotic	1990	26 (308)
repaglinide	antidiabetic	1998	34 (329)
repirinast	antiallergy	1987	23 (341)
retapamulin	antibacterial	2007	43 (486)
reteplase	antithrombotic	1996	32 (316)
retigabine	anticonvulsant	2011	47 (540)
reviparin sodium	anticoagulant	1993	29 (344)
rifabutin	antibacterial	1992	28 (335)
rifapentine	antibacterial	1988	24 (310)
rifaximin	antibacterial	1985	21 (332)
rifaximin	antibacterial	1987	23 (341)
rilmazafone	sleep disorders	1989	25 (317)
rilmenidine	antihypertensive	1988	24 (310)
rilonacept	genetic autoinflammatory syndromes	2008	44 (615)
rilpivirine	antiviral	2011	47 (542)
riluzole	amyotrophic lateral sclerosis	1996	32 (316)
rimantadine hydrochloride	antiviral	1987	23 (342)
rimexolone	antiinflammatory	1995	31 (348)
rimonabant	antiobesity	2006	42 (537)
risedronate sodium	osteoporosis	1998	34 (330)
risperidone	antipsychotic	1993	29 (344)
ritonavir	antiviral	1996	32 (317)
rivaroxaban	anticoagulant	2008	44 (617)
rivastigmin	Alzheimer's disease	1997	33 (342)
rizatriptan benzoate	antimigraine	1998	34 (330)
rocuronium bromide	muscle relaxant	1994	30 (309)
rofecoxib	antiarthritic	1999	35 (347)
roflumilast	chronic obstructive pulmonary disease	2010	46 (480)
rokitamycin	antibacterial	1986	22 (325)
romidepsin	anticancer	2009	46 (482)
romiplostim	antithrombocytopenic	2008	44 (619)
romurtide	immunostimulant	1991	27 (332)
ronafibrate	antihypercholesterolemic	1986	22 (326)
ropinirole hydrochloride	Parkinson's disease	1996	32 (317)
ropivacaine	anesthetic	1996	32 (318)
rosaprostol	antiulcer	1985	21 (332)
rosiglitazone maleate	antidiabetic	1999	35 (348)
rosuvastatin	antihypercholesterolemic	2003	39 (283)
rotigotine	Parkinson's disease	2006	42 (538)
roxatidine acetate hydrochloride	antiulcer	1986	22 (326)
roxithromycin	antiulcer	1987	23 (342)
rufinamide	anticonvulsant	2007	43 (488)

GENERIC NAME	INDICATION	YEAR INTRODUCED	ARMC VOL., (PAGE)
rufloxacin hydrochloride	antibacterial	1992	28 (335)
rupatadine fumarate	antiallergy	2003	39 (284)
ruxolitinib	anticancer	2011	47 (544)
RV-11	antibacterial	1989	25 (318)
salmeterol hydroxynaphthoate	antiasthma	1990	26 (308)
sapropterin hydrochloride	phenylketouria	1992	28 (336)
saquinavir mesvlate	antiviral	1995	31 (349)
sargramostim	immunostimulant	1991	27 (332)
sarpogrelate hydrochloride	antithrombotic	1993	29 (344)
saxagliptin	antidiabetic	2009	45 (521)
schizophyllan	immunostimulant	1985	22 (326)
seratrodast	antiasthma	1995	31 (349)
sertaconazole nitrate	antifungal	1992	28 (336)
sertindole	antipsychotic	1996	32 (318)
setastine hydrochloride	antiallergy	1987	23 (342)
setiptiline	antidepressant	1989	25 (318)
setraline hydrochloride	antidepressant	1990	26 (309)
sevoflurane	anesthetic	1990	26 (309)
sibutramine	antiobesity	1998	34 (331)
sildenafil citrate	male sexual dysfunction	1998	34 (331)
silodosin	urinary tract/bladder disorders	2006	42 (540)
simvastatin	antihypercholesterolemic	1988	24 (311)
sipuleucel-t	anticancer	2010	46 (484)
sitafloxacin hydrate	antibacterial	2008	44 (621)
sitagliptin	antidiabetic	2006	42 (541)
sitaxsentan	pulmonary hypertension	2006	42 (543)
sivelestat	antiinflammatory	2002	38 (366)
SKI-2053R	anticancer	1999	35 (348)
sobuzoxane	anticancer	1994	30 (310)
sodium cellulose phosphate	urinary tract/bladder disorders	1983	19 (323)
sofalcone	antiulcer	1984	20 (323)
solifenacin	urinary tract/bladder disorders	2004	40 (466)
somatomedin-1	growth disorders	1994	30 (310)
somatotropin	growth disorders	1994	30 (310)
somatropin	growth disorders	1987	23 (343)
sorafenib	anticancer	2005	41 (466)
sorivudine	antiviral	1993	29 (345)
sparfloxacin	antibacterial	1993	29 (345)
spirapril hydrochloride	antihypertensive	1995	31 (349)
spizofurone	antiulcer	1987	23 (343)
stavudine	antiviral	1994	30 (311)
strontium ranelate	osteoporosis	2004	40 (466)
succimer	antidote, lead poisoning	1991	27 (333)
sufentanil	analgesic	1983	19 (323)

GENERIC NAME	INDICATION	YEAR INTRODUCED	ARMC VOL., (PAGE)
sugammadex	neuromuscular blockade, reversal	2008	44 (623)
sulbactam sodium	antibacterial	1986	22 (326)
sulconizole nitrate	antifungal	1985	21 (332)
sultamycillin tosylate	antibacterial	1987	23 (343)
sumatriptan succinate	antimigraine	1991	27 (333)
sunitinib	anticancer	2006	42 (544)
suplatast tosilate	antiallergy	1995	31 (350)
suprofen	analgesic	1983	19 (324)
surfactant TA	respiratory surfactant	1987	23 (344)
tacalcitol	antipsoriasis	1993	29 (346)
tacrine hydrochloride	Alzheimer's disease	1993	29 (346)
tacrolimus	immunosuppressant	1993	29 (347)
tadalafil	male sexual dysfunction	2003	39 (284)
tafamidis	neurodegeneration	2011	47 (546)
tafluprost	antiglaucoma	2008	44 (625)
talaporfin sodium	anticancer	2004	40 (469)
talipexole	Parkinson's disease	1996	32 (318)
taltirelin	neurodegeneration	2000	36 (311)
tamibarotene	anticancer	2005	41 (467)
tamsulosin hydrochloride	benign prostatic hyperplasia	1993	29 (347)
tandospirone	anxiolytic	1996	32 (319)
tapentadol hydrochloride	analgesic	2009	45 (523)
tasonermin	anticancer	1999	35 (349)
tazanolast	antiallergy	1990	26 (309)
tazarotene	antipsoriasis	1997	33 (343)
tazobactam sodium	antibacterial	1992	28 (336)
tegaserod maleate	irritable bowel syndrome	2001	37 (270)
teicoplanin	antibacterial	1988	24 (311)
telaprevir	antiviral	2011	47 (548)
telavancin	antibacterial	2009	45 (525)
telbivudine	antiviral	2006	42 (546)
telithromycin	antibacterial	2001	37 (271)
telmesteine	expectorant	1992	28 (337)
telmisartan	antihypertensive	1999	35 (349)
temafloxacin hydrochloride	antibacterial	1991	27 (334)
temocapril	antihypertensive	1994	30 (311)
temocillin disodium	antibacterial	1984	20 (323)
temoporphin	anticancer	2002	38 (367)
temozolomide	anticancer	1999	35 (349)
temsirolimus	anticancer	2007	43 (490)
tenofovir disoproxil fumarate	antiviral	2001	37 (271)
tenoxicam	antiinflammatory	1987	23 (344)
teprenone	antiulcer	1984	20 (323)
terazosin hydrochloride	antihypertensive	1984	20 (323)

GENERIC NAME	INDICATION	YEAR INTRODUCED	ARMC VOL., (PAGE)
terbinafine hydrochloride	antifungal	1991	27 (334)
terconazole	antifungal	1983	19 (324)
tertatolol hydrochloride	antihypertensive	1987	23 (344)
tesamorelin acetate	lipodystrophy	2010	46 (486)
thrombin alfa	hemostatic	2008	44 (627)
thrombomodulin (recombinant)	anticoagulant	2008	44 (628)
thymopentin	immunomodulator	1985	21 (333)
tiagabine	anticonvulsant	1996	32 (319)
tiamenidine hydrochloride	antihypertensive	1988	24 (311)
tianeptine sodium	antidepressant	1983	19 (324)
tibolone	hormone therapy	1988	24 (312)
ticagrelor	antithrombotic	2010	46 (488)
tigecycline	antibacterial	2005	41 (468)
tilisolol hydrochloride	antihypertensive	1992	28 (337)
tiludronate disodium	Paget's disease	1995	31 (350)
timiperone	antipsychotic	1984	20 (323)
tinazoline	nasal decongestant	1988	24 (312)
tioconazole	antifungal	1983	19 (324)
tiopronin	urolithiasis	1989	25 (318)
tiotropium bromide	chronic obstructive pulmonary disease	2002	38 (368)
tipranavir	antiviral	2005	41 (470)
tiquizium bromide	antispasmodic	1984	20 (324)
tiracizine hydrochloride	antiarrhythmic	1990	26 (310)
tirilazad mesylate	subarachnoid hemorrhage	1995	31 (351)
tirofiban hydrochloride	antithrombotic	1998	34 (332)
tiropramide hydrochloride	muscle relaxant	1983	19 (324)
tizanidine	muscle relaxant	1984	20 (324)
tolcapone	Parkinson's disease	1997	33 (343)
toloxatone	antidepressant	1984	20 (324)
tolrestat	antidiabetic	1989	25 (319)
tolvaptan	hyponatremia	2009	45 (528)
topiramate	anticonvulsant	1995	31 (351)
topotecan hydrochloride	anticancer	1996	32 (320)
torasemide	diuretic	1993	29 (348)
toremifene	anticancer	1989	25 (319)
tositumomab	anticancer	2003	39 (285)
tosufloxacin tosylate	antibacterial	1990	26 (310)
trabectedin	anticancer	2007	43 (492)
trandolapril	antihypertensive	1993	29 (348)
travoprost	antiglaucoma	2001	37 (272)
treprostinil sodium	antihypertensive	2002	38 (368)
tretinoin tocoferil	antiulcer	1993	29 (348)
trientine hydrochloride	antidote, copper poisoning	1986	22 (327)

GENERIC NAME	INDICATION	YEAR INTRODUCED	ARMC VOL., (PAGE)
trimazosin hydrochloride	antihypertensive	1985	21 (333)
trimegestone	contraception	2001	37 (273)
trimetrexate glucuronate	antifungal	1994	30 (312)
troglitazone	antidiabetic	1997	33 (344)
tropisetron	antiemetic	1992	28 (337)
trovafloxacin mesylate	antibacterial	1998	34 (332)
troxipide	antiulcer	1986	22 (327)
ubenimex	immunostimulant	1987	23 (345)
udenafil	male sexual dysfunction	2005	41 (472)
ulipristal acetate	contraception	2009	45 (530)
unoprostone isopropyl ester	antiglaucoma	1994	30 (312)
ustekinumab	antipsoriasis	2009	45 (532)
vadecoxib	antiarthritic	2002	38 (369)
vaglancirclovir hydrochloride	antiviral	2001	37 (273)
valaciclovir hydrochloride	antiviral	1995	31 (352)
valrubicin	anticancer	1999	35 (350)
valsartan	antihypertensive	1996	32 (320)
vandetanib	anticancer	2011	47 (551)
vardenafil	male sexual dysfunction	2003	39 (286)
varenicline	addiction, nicotine	2006	42 (547)
vemurafenib	anticancer	2011	47 (552)
venlafaxine	antidepressant	1994	30 (312)
vernakalant	antiarrhythmic	2010	46 (491)
verteporfin	ophthalmologic (macular degeneration)	2000	36 (312)
vesnarinone	congestive heart failure	1990	26 (310)
vigabatrin	anticonvulsant	1989	25 (319)
vilazodone	antidepressant	2011	47 (554)
vildagliptin	antidiabetic	2007	43 (494)
vinflunine	anticancer	2009	46 (493)
vinorelbine	anticancer	1989	25 (320)
voglibose	antidiabetic	1994	30 (313)
voriconazole	antifungal	2002	38 (370)
vorinostat	anticancer	2006	42 (549)
xamoterol fumarate	congestive heart failure	1988	24 (312)
ximelagatran	anticoagulant	2004	40 (470)
zafirlukast	antiasthma	1996	32 (321)
zalcitabine	antiviral	1992	28 (338)
zaleplon	sleep disorders	1999	35 (351)
zaltoprofen	antiinflammatory	1993	29 (349)
zanamivir	antiviral	1999	35 (352)
ziconotide	analgesic	2005	41 (473)
zidovudine	antiviral	1987	23 (345)
zileuton	antiasthma	1997	33 (344)
zinostatin stimalamer	anticancer	1994	30 (313)

GENERIC NAME	INDICATION	YEAR INTRODUCED	ARMC VOL., (PAGE)
ziprasidone hydrochloride	antipsychotic	2000	36 (312)
zofenopril calcium	antihypertensive	2000	36 (313)
zoledronate disodium	osteoporosis	2000	36 (314)
zolpidem hemitartrate	sleep disorders	1988	24 (313)
zomitriptan	antimigraine	1997	33 (345)
zonisamide	anticonvulsant	1989	25 (320)
zopiclone	sleep disorders	1986	22 (327)
zucapsaicin	analgesic	2010	46 (495)
zuclopenthixol acetate	antipsychotic	1987	23 (345)

CUMULATIVE NCE INTRODUCTION INDEX, 1983–2011 (BY INDICATION)

GENERIC NAME	INDICATION	YEAR INTRODUCED	ARMC VOL., (PAGE)
gemeprost	abortifacient	1983	19 (319)
mifepristone	abortifacient	1988	24 (306)
azelaic acid	acne	1989	25 (310)
nitrefazole	addiction, alcohol	1983	19 (322)
varenicline	addiction, nicotine	2006	42 (547)
naltrexone hydrochloride	addiction, opioids	1984	20 (322)
nalmefene hydrochloride	addiction, opioids	1995	31 (347)
tacrine hydrochloride	Alzheimer's disease	1993	29 (346)
donepezil hydrochloride	Alzheimer's disease	1997	33 (332)
rivastigmin	Alzheimer's disease	1997	33 (342)
fasudil hydrochloride	amyotrophic lateral sclerosis	1995	31 (343)
riluzole	amyotrophic lateral sclerosis	1996	32 (316)
alfentanil hydrochloride	analgesic	1983	19 (314)
alminoprofen	analgesic	1983	19 (314)
meptazinol hydrochloride	analgesic	1983	19 (321)
sufentanil	analgesic	1983	19 (323)
suprofen	analgesic	1983	19 (324)
emorfazone	analgesic	1984	20 (317)
fosfosal	analgesic	1984	20 (319)
flupirtine maleate	analgesic	1985	21 (328)
propacetamol hydrochloride	analgesic	1986	22 (325)
eptazocine hydrobromide	analgesic	1987	23 (334)
ketorolac tromethamine	analgesic	1990	26 (304)
dezocine	analgesic	1991	27 (326)
mofezolac	analgesic	1994	30 (304)
remifentanil hydrochloride	analgesic	1996	32 (316)
parecoxib sodium	analgesic	2002	38 (364)
ziconotide	analgesic	2005	41 (473)
tapentadol hydrochloride	analgesic	2009	45 (523)
zucapsaicin	analgesic	2010	46 (495)
propofol	anesthetic	1986	22 (325)
sevoflurane	anesthetic	1990	26 (309)
desflurane	anesthetic	1992	28 (329)
ropivacaine	anesthetic	1996	32 (318)
levobupivacaine hydrochloride	anesthetic	2000	36 (308)
icatibant	angioedema, hereditary	2008	44 (608)
ecallantide	angioedema, hereditary	2009	46 (464)
astemizole	antiallergy	1983	19 (314)
azelastine hydrochloride	antiallergy	1986	22 (316)
nedocromil sodium	antiallergy	1986	22 (324)
cetirizine hydrochloride	antiallergy	1987	23 (331)
repirinast	antiallergy	1987	23 (341)

GENERIC NAME	INDICATION	YEAR INTRODUCED	ARMC VOL., (PAGE)
setastine hydrochloride	antiallergy	1987	23 (342)
acrivastine	antiallergy	1988	24 (295)
loratadine	antiallergy	1988	24 (306)
ebastine	antiallergy	1990	26 (302)
tazanolast	antiallergy	1990	26 (309)
levocabastine hydrochloride	antiallergy	1991	27 (330)
lodoxamide tromethamine	antiallergy	1992	28 (333)
emedastine difumarate	antiallergy	1993	29 (336)
epinastine	antiallergy	1994	30 (299)
suplatast tosilate	antiallergy	1995	31 (350)
fexofenadine	antiallergy	1996	32 (307)
olopatadine hydrochloride	antiallergy	1997	33 (340)
loteprednol etabonate	antiallergy	1998	34 (324)
mizolastine	antiallergy	1998	34 (325)
betotastine besilate	antiallergy	2000	36 (297)
ramatroban	antiallergy	2000	36 (311)
desloratadine	antiallergy	2001	37 (264)
levocetirizine	antiallergy	2001	37 (268)
rupatadine fumarate	antiallergy	2003	39 (284)
fluticasone furoate	antiallergy	2007	43 (469)
bilastine	antiallergy	2010	46 (449)
gallopamil hydrochloride	antianginal	1983	19 (3190)
nicorandil	antianginal	1984	20 (322)
ivabradine	antianginal	2006	42 (522)
ranolazine	antianginal	2006	42 (535)
cibenzoline	antiarrhythmic	1985	21 (325)
encainide hydrochloride	antiarrhythmic	1987	23 (333)
esmolol hydrochloride	antiarrhythmic	1987	23 (334)
moricizine hydrochloride	antiarrhythmic	1990	26 (305)
tiracizine hydrochloride	antiarrhythmic	1990	26 (310)
pilsicainide hydrochloride	antiarrhythmic	1991	27 (332)
pirmenol	antiarrhythmic	1994	30 (307)
ibutilide fumarate	antiarrhythmic	1996	32 (309)
nif ekalant hydrochloride	antiarrhythmic	1999	35 (344)
dofetilide	antiarrhythmic	2000	36 (301)
landiolol	antiarrhythmic	2002	38 (360)
dronedarone	antiarrhythmic	2009	45 (495)
vernakalant	antiarrhythmic	2010	46 (491)
auranofin	antiarthritic	1983	19 (314)
meloxicam	antiarthritic	1996	32 (312)
lefunomide	antiarthritic	1998	34 (324)
celecoxib	antiarthritic	1999	35 (335)
rofecoxib	antiarthritic	1999	35 (347)
anakinra	antiarthritic	2001	37 (261)
etoricoxibe	antiarthritic	2002	38 (355)

GENERIC NAME	INDICATION	YEAR INTRODUCED	ARMC VOL., (PAGE)
vadecoxib	antiarthritic	2002	38 (369)
adalimumab	antiarthritic	2003	39 (267)
abatacept	antiarthritic	2006	42 (509)
iguratimod	antiarthritic	2011	47 (532)
oxitropium bromide	antiasthma	1983	19 (323)
doxofylline	antiasthma	1985	21 (327)
mabuterol hydrochloride	antiasthma	1986	22 (323)
amlexanox	antiasthma	1987	23 (327)
flutropium bromide	antiasthma	1988	24 (303)
ibudilast	antiasthma	1989	25 (313)
bambuterol	antiasthma	1990	26 (299)
salmeterol hydroxynaphthoate	antiasthma	1990	26 (308)
pemirolast potassium	antiasthma	1991	27 (331)
pranlukast	antiasthma	1995	31 (347)
seratrodast	antiasthma	1995	31 (349)
zafirlukast	antiasthma	1996	32 (321)
zileuton	antiasthma	1997	33 (344)
montelukast sodium	antiasthma	1998	34 (326)
levalbuterol hydrochloride	antiasthma	1999	35 (341)
omalizumab allergic	antiasthma	2003	39 (280)
ciclesonide	antiasthma	2005	41 (443)
arformoterol	antiasthma	2007	43 (465)
cefmenoxime hydrochloride	antibacterial	1983	19 (316)
ceftazidime	antibacterial	1983	19 (316)
norfloxacin	antibacterial	1983	19 (322)
adamantanium bromide	antibacterial	1984	20 (315)
aztreonam	antibacterial	1984	20 (315)
cefonicid sodium	antibacterial	1984	20 (316)
ceforanide	antibacterial	1984	20 (317)
cefotetan disodium	antibacterial	1984	20 (317)
temocillin disodium	antibacterial	1984	20 (323)
astromycin sulfate	antibacterial	1985	21 (324)
cefbuperazone sodium	antibacterial	1985	21 (325)
cefpiramide sodium	antibacterial	1985	21 (325)
imipenem/cilastatin	antibacterial	1985	21 (328)
miokamycin	antibacterial	1985	21 (329)
mupirocin	antibacterial	1985	21 (330)
ofloxacin	antibacterial	1985	21 (331)
pefloxacin mesylate	antibacterial	1985	21 (331)
rifaximin	antibacterial	1985	21 (332)
carboplatin	antibacterial	1986	22 (318)
ciprofloxacin	antibacterial	1986	22 (318)
enoxacin	antibacterial	1986	22 (320)
rokitamycin	antibacterial	1986	22 (325)
sulbactam sodium	antibacterial	1986	22 (326)

GENERIC NAME	INDICATION	YEAR INTRODUCED	ARMC VOL., (PAGE)
aspoxicillin	antibacterial	1987	23 (328)
cefixime	antibacterial	1987	23 (329)
cefminox sodium	antibacterial	1987	23 (330)
cefpimizole	antibacterial	1987	23 (330)
cefteram pivoxil	antibacterial	1987	23 (330)
cefuroxime axetil	antibacterial	1987	23 (331)
cefuzonam sodium	antibacterial	1987	23 (331)
lenampicillin hydrochloride	antibacterial	1987	23 (336)
rifaximin	antibacterial	1987	23 (341)
sultamycillin tosylate	antibacterial	1987	23 (343)
azithromycin	antibacterial	1988	24 (298)
carumonam	antibacterial	1988	24 (298)
erythromycin acistrate	antibacterial	1988	24 (301)
flomoxef sodium	antibacterial	1988	24 (302)
isepamicin	antibacterial	1988	24 (305)
rifapentine	antibacterial	1988	24 (310)
teicoplanin	antibacterial	1988	24 (311)
cefpodoxime proxetil	antibacterial	1989	25 (310)
lomefloxacin	antibacterial	1989	25 (315)
RV-11	antibacterial	1989	25 (318)
arbekacin	antibacterial	1990	26 (298)
cefodizime sodium	antibacterial	1990	26 (300)
clarithromycin	antibacterial	1990	26 (302)
tosufloxacin tosylate	antibacterial	1990	26 (310)
cefdinir	antibacterial	1991	27 (323)
cefotiam hexetil hydrochloride	antibacterial	1991	27 (324)
temafloxacin hydrochloride	antibacterial	1991	27 (334)
cefetamet pivoxil hydrochloride	antibacterial	1992	28 (327)
cefpirome sulfate	antibacterial	1992	28 (328)
cefprozil	antibacterial	1992	28 (328)
ceftibuten	antibacterial	1992	28 (329)
fleroxacin	antibacterial	1992	28 (331)
loracarbef	antibacterial	1992	28 (333)
rifabutin	antibacterial	1992	28 (335)
rufloxacin hydrochloride	antibacterial	1992	28 (335)
tazobactam sodium	antibacterial	1992	28 (336)
brodimoprin	antibacterial	1993	29 (333)
cefepime	antibacterial	1993	29 (334)
dirithromycin	antibacterial	1993	29 (336)
levofloxacin	antibacterial	1993	29 (340)
nadifloxacin	antibacterial	1993	29 (340)
sparfloxacin	antibacterial	1993	29 (345)
cefditoren pivoxil	antibacterial	1994	30 (297)
meropenem	antibacterial	1994	30 (303)

GENERIC NAME	INDICATION	YEAR INTRODUCED	ARMC VOL., (PAGE)
panipenem/ betamipron carbapenem	antibacterial	1994	30 (305)
cefozopran hydrochloride	antibacterial	1995	31 (339)
cefcapene pivoxil	antibacterial	1997	33 (330)
flurithromycin ethylsuccinate	antibacterial	1997	33 (333)
fropenam	antibacterial	1997	33 (334)
cefoselis	antibacterial	1998	34 (319)
trovafloxacin mesylate	antibacterial	1998	34 (332)
dalfopristin	antibacterial	1999	35 (338)
gatilfloxacin	antibacterial	1999	35 (340)
moxifloxacin hydrochloride	antibacterial	1999	35 (343)
quinupristin	antibacterial	1999	35 (338)
linezolid	antibacterial	2000	36 (309)
telithromycin	antibacterial	2001	37 (271)
balofloxacin	antibacterial	2002	38 (351)
biapenem	antibacterial	2002	38 (351)
ertapenem sodium	antibacterial	2002	38 (353)
pazufloxacin	antibacterial	2002	38 (364)
prulifloxacin	antibacterial	2002	38 (366)
daptomycin	antibacterial	2003	39 (272)
gemifloxacin	antibacterial	2004	40 (458)
doripenem	antibacterial	2005	41 (448)
tigecycline	antibacterial	2005	41 (468)
garenoxacin	antibacterial	2007	43 (471)
retapamulin	antibacterial	2007	43 (486)
ceftobiprole medocaril	antibacterial	2008	44 (589)
sitafloxacin hydrate	antibacterial	2008	44 (621)
besifloxacin	antibacterial	2009	45 (482)
telavancin	antibacterial	2009	45 (525)
ceftaroline fosamil	antibacterial	2010	46 (453)
fidaxomicin	antibacterial	2011	47 (528)
enocitabine	anticancer	1983	19 (318)
flutamide	anticancer	1983	19 (318)
epirubicin hydrochloride	anticancer	1984	20 (318)
mitoxantrone hydrochloride	anticancer	1984	20 (321)
camostat mesylate	anticancer	1985	21 (325)
amsacrine	anticancer	1987	23 (327)
doxifluridine	anticancer	1987	23 (332)
lonidamine	anticancer	1987	23 (337)
nilutamide	anticancer	1987	23 (338)
ranimustine	anticancer	1987	23 (341)
pirarubicin	anticancer	1988	24 (309)
fotemustine	anticancer	1989	25 (313)
interleukin-2	anticancer	1989	25 (314)
toremifene	anticancer	1989	25 (319)

GENERIC NAME	INDICATION	YEAR INTRODUCED	ARMC VOL., (PAGE)
vinorelbine	anticancer	1989	25 (320)
bisantrene hydrochloride	anticancer	1990	26 (300)
idarubicin hydrochloride	anticancer	1990	26 (303)
fludarabine phosphate	anticancer	1991	27 (327)
interferon, gamma-1	anticancer	1992	28 (332)
masoprocol	anticancer	1992	28 (333)
pentostatin	anticancer	1992	28 (334)
cladribine	anticancer	1993	29 (335)
cytarabine ocfosfate	anticancer	1993	29 (335)
formestane	anticancer	1993	29 (337)
miltefosine	anticancer	1993	29 (340)
paclitaxal	anticancer	1993	29 (342)
porfimer sodium	anticancer	1993	29 (343)
irinotecan	anticancer	1994	30 (301)
pegaspargase	anticancer	1994	30 (306)
sobuzoxane	anticancer	1994	30 (310)
zinostatin stimalamer	anticancer	1994	30 (313)
anastrozole	anticancer	1995	31 (338)
bicalutamide	anticancer	1995	31 (338)
docetaxel	anticancer	1995	31 (341)
fadrozole hydrochloride	anticancer	1995	31 (342)
gemcitabine hydrochloride	anticancer	1995	31 (344)
nedaplatin	anticancer	1995	31 (347)
letrazole	anticancer	1996	32 (311)
oxaliplatin	anticancer	1996	32 (313)
raltitrexed	anticancer	1996	32 (315)
topotecan hydrochloride	anticancer	1996	32 (320)
capecitabine	anticancer	1998	34 (319)
OCT-43	anticancer	1999	35 (345)
alitretinoin	anticancer	1999	35 (333)
arglabin	anticancer	1999	35 (335)
denileukin diftitox	anticancer	1999	35 (338)
SKI-2053R	anticancer	1999	35 (348)
tasonermin	anticancer	1999	35 (349)
temozolomide	anticancer	1999	35 (349)
valrubicin	anticancer	1999	35 (350)
bexarotene	anticancer	2000	36 (298)
exemestane	anticancer	2000	36 (304)
gemtuzumab ozogamicin	anticancer	2000	36 (306)
alemtuzumab	anticancer	2001	37 (260)
imatinib mesylate	anticancer	2001	37 (267)
amrubicin hydrochloride	anticancer	2002	38 (349)
fulveristrant	anticancer	2002	38 (357)
gefitinib	anticancer	2002	38 (358)
ibritunomab tiuxetan	anticancer	2002	38 (359)

GENERIC NAME	INDICATION	YEAR INTRODUCED	ARMC VOL., (PAGE)
temoporphin	anticancer	2002	38 (367)
bortezomib	anticancer	2003	39 (271)
cetuximab	anticancer	2003	39 (272)
tositumomab	anticancer	2003	39 (285)
abarelix	anticancer	2004	40 (446)
azacitidine	anticancer	2004	40 (447)
belotecan	anticancer	2004	40 (449)
bevacizumab	anticancer	2004	40 (450)
erlotinib	anticancer	2004	40 (454)
pemetrexed	anticancer	2004	40 (463)
talaporfin sodium	anticancer	2004	40 (469)
clofarabine	anticancer	2005	41 (444)
sorafenib	anticancer	2005	41 (466)
tamibarotene	anticancer	2005	41 (467)
dasatinib	anticancer	2006	42 (517)
nelarabine	anticancer	2006	42 (528)
nimotuzumab	anticancer	2006	42 (529)
panitumumab	anticancer	2006	42 (531)
sunitinib	anticancer	2006	42 (544)
vorinostat	anticancer	2006	42 (549)
ixabepilone	anticancer	2007	43 (473)
lapatinib	anticancer	2007	43 (475)
nilotinib	anticancer	2007	43 (480)
temsirolimus	anticancer	2007	43 (490)
trabectedin	anticancer	2007	43 (492)
catumaxomab	anticancer	2009	45 (486)
degarelix acetate	anticancer	2009	45 (490)
mifamurtide	anticancer	2009	46 (476)
ofatumumab	anticancer	2009	45 (512)
pazopanib	anticancer	2009	45 (514)
pralatrexate	anticancer	2009	45 (517)
romidepsin	anticancer	2009	46 (482)
vinflunine	anticancer	2009	46 (493)
cabazitaxel	anticancer	2010	46 (451)
eribulin mesylate	anticancer	2010	46 (465)
sipuleucel-t	anticancer	2010	46 (484)
abiraterone acetate	anticancer	2011	47 (503)
brentuximab	anticancer	2011	47 (520)
crizotinib	anticancer	2011	47 (522)
ipilimumab	anticancer	2011	47 (533)
ruxolitinib	anticancer	2011	47 (544)
vandetanib	anticancer	2011	47 (551)
vemurafenib	anticancer	2011	47 (552)
angiotensin II	anticancer adjuvant	1994	30 (296)
enoxaparin	anticoagulant	1987	23 (333)

GENERIC NAME	INDICATION	YEAR INTRODUCED	ARMC VOL., (PAGE)
parnaparin sodium	anticoagulant	1993	29 (342)
reviparin sodium	anticoagulant	1993	29 (344)
duteplase	anticoagulant	1995	31 (342)
lepirudin	anticoagulant	1997	33 (336)
ximelagatran	anticoagulant	2004	40 (470)
dabigatran etexilate	anticoagulant	2008	44 (598)
rivaroxaban	anticoagulant	2008	44 (617)
thrombomodulin (recombinant)	anticoagulant	2008	44 (628)
progabide	anticonvulsant	1985	21 (331)
vigabatrin	anticonvulsant	1989	25 (319)
zonisamide	anticonvulsant	1989	25 (320)
lamotrigine	anticonvulsant	1990	26 (304)
oxcarbazepine	anticonvulsant	1990	26 (307)
felbamate	anticonvulsant	1993	29 (337)
gabapentin	anticonvulsant	1993	29 (338)
topiramate	anticonvulsant	1995	31 (351)
fosphenytoin sodium	anticonvulsant	1996	32 (308)
tiagabine	anticonvulsant	1996	32 (319)
levetiracetam	anticonvulsant	2000	36 (307)
pregabalin	anticonvulsant	2004	40 (464)
rufinamide	anticonvulsant	2007	43 (488)
lacosamide	anticonvulsant	2008	44 (610)
eslicarbazepine acetate	anticonvulsant	2009	45 (498)
retigabine	anticonvulsant	2011	47 (540)
fluvoxamine maleate	antidepressant	1983	19 (319)
indalpine	antidepressant	1983	19 (320)
tianeptine sodium	antidepressant	1983	19 (324)
metapramine	antidepressant	1984	20 (320)
toloxatone	antidepressant	1984	20 (324)
fluoxetine hydrochloride	antidepressant	1986	22 (320)
medifoxamine fumarate	antidepressant	1986	22 (323)
bupropion hydrochloride	antidepressant	1989	25 (310)
citalopram	antidepressant	1989	25 (311)
setiptiline	antidepressant	1989	25 (318)
moclobemide	antidepressant	1990	26 (305)
setraline hydrochloride	antidepressant	1990	26 (309)
paroxetine	antidepressant	1991	27 (331)
mirtazapine	antidepressant	1994	30 (303)
nefazodone	antidepressant	1994	30 (305)
venlafaxine	antidepressant	1994	30 (312)
milnacipran	antidepressant	1997	33 (338)
pivagabine	antidepressant	1997	33 (341)
reboxetine	antidepressant	1997	33 (342)
escitalopram oxolate	antidepressant	2002	38 (354)
duloxetine	antidepressant	2004	40 (452)

GENERIC NAME	INDICATION	YEAR INTRODUCED	ARMC VOL., (PAGE)
desvenlafaxine	antidepressant	2008	44 (600)
vilazodone	antidepressant	2011	47 (554)
tolrestat	antidiabetic	1989	25 (319)
acarbose	antidiabetic	1990	26 (297)
epalrestat	antidiabetic	1992	28 (330)
glucagon, rDNA	antidiabetic	1993	29 (338)
voglibose	antidiabetic	1994	30 (313)
glimepiride	antidiabetic	1995	31 (344)
insulin lispro	antidiabetic	1996	32 (310)
troglitazone	antidiabetic	1997	33 (344)
miglitol	antidiabetic	1998	34 (325)
repaglinide	antidiabetic	1998	34 (329)
nateglinide	antidiabetic	1999	35 (344)
pioglitazone hydrochloride	antidiabetic	1999	35 (346)
rosiglitazone maleate	antidiabetic	1999	35 (348)
mitiglinide	antidiabetic	2004	40 (460)
exenatide	antidiabetic	2005	41 (452)
pramlintide	antidiabetic	2005	41 (460)
sitagliptin	antidiabetic	2006	42 (541)
vildagliptin	antidiabetic	2007	43 (494)
liraglutide	antidiabetic	2009	45 (507)
saxagliptin	antidiabetic	2009	45 (521)
alogliptin	antidiabetic	2010	46 (446)
linagliptin	antidiabetic	2011	47 (536)
lidamidine hydrochloride	antidiarrheal	1984	20 (320)
acetorphan	antidiarrheal	1993	29 (332)
flumazenil	antidote, benzodiazepine overdose	1987	23 (335)
trientine hydrochloride	antidote, copper poisoning	1986	22 (327)
anti-digoxin polyclonal antibody	antidote, digoxin poisoning	2002	38 (350)
fomepizole	antidote, ethylene glycol poisoning	1998	34 (323)
succimer	antidote, lead poisoning	1991	27 (333)
crotelidae polyvalent immune fab	antidote, snake venom poisoning	2001	37 (263)
dronabinol	antiemetic	1986	22 (319)
ondansetron hydrochloride	antiemetic	1990	26 (306)
granisetron hydrochloride	antiemetic	1991	27 (329)
tropisetron	antiemetic	1992	28 (337)
nazasetron	antiemetic	1994	30 (305)
ramosetron	antiemetic	1996	32 (315)
dolasetron mesylate	antiemetic	1998	34 (321)
aprepitant	antiemetic	2003	39 (268)
palonosetron	antiemetic	2003	39 (281)
indisetron	antiemetic	2004	40 (459)

GENERIC NAME	INDICATION	YEAR INTRODUCED	ARMC VOL., (PAGE)
fosaprepitant dimeglumine	antiemetic	2008	44 (606)
oxiconazole nitrate	antifungal	1983	19 (322)
terconazole	antifungal	1983	19 (324)
tioconazole	antifungal	1983	19 (324)
naftifine hydrochloride	antifungal	1984	20 (321)
sulconizole nitrate	antifungal	1985	21 (332)
butoconazole	antifungal	1986	22 (318)
cloconazole hydrochloride	antifungal	1986	22 (318)
fenticonazole nitrate	antifungal	1987	23 (334)
fluconazole	antifungal	1988	24 (303)
itraconazole	antifungal	1988	24 (305)
amorolfine hydrochloride	antifungal	1991	27 (322)
terbinafine hydrochloride	antifungal	1991	27 (334)
butenafine hydrochloride	antifungal	1992	28 (327)
sertaconazole nitrate	antifungal	1992	28 (336)
neticonazole hydrochloride	antifungal	1993	29 (341)
lanoconazole	antifungal	1994	30 (302)
trimetrexate glucuronate	antifungal	1994	30 (312)
flutrimazole	antifungal	1995	31 (343)
liranaftate	antifungal	2000	36 (309)
caspofungin acetate	antifungal	2001	37 (263)
micafungin	antifungal	2002	38 (360)
voriconazole	antifungal	2002	38 (370)
fosfluconazole	antifungal	2004	40 (457)
eberconazole	antifungal	2005	41 (449)
luliconazole	antifungal	2005	41 (454)
anidulafungin	antifungal	2006	42 (512)
posaconazole	antifungal	2006	42 (532)
befunolol hydrochloride	antiglaucoma	1983	19 (315)
levobunolol hydrochloride	antiglaucoma	1985	21 (328)
dapiprazole hydrochloride	antiglaucoma	1987	23 (332)
apraclonidine hydrochloride	antiglaucoma	1988	24 (297)
unoprostone isopropyl ester	antiglaucoma	1994	30 (312)
dorzolamide hydrochloride	antiglaucoma	1995	31 (341)
brimonidine	antiglaucoma	1996	32 (306)
latanoprost	antiglaucoma	1996	32 (311)
brinzolamide	antiglaucoma	1998	34 (318)
bimatoprost	antiglaucoma	2001	37 (261)
travoprost	antiglaucoma	2001	37 (272)
tafluprost	antiglaucoma	2008	44 (625)
meglutol	antihypercholesterolemic	1983	19 (321)
divistyramine	antihypercholesterolemic	1984	20 (317)
melinamide	antihypercholesterolemic	1984	20 (320)
acipimox	antihypercholesterolemic	1985	21 (323)
ciprofibrate	antihypercholesterolemic	1985	21 (326)

GENERIC NAME	INDICATION	YEAR INTRODUCED	ARMC VOL., (PAGE)
beclobrate	antihypercholesterolemic	1986	22 (317)
binifibrate	antihypercholesterolemic	1986	22 (317)
ronafibrate	antihypercholesterolemic	1986	22 (326)
lovastatin	antihypercholesterolemic	1987	23 (337)
simvastatin	antihypercholesterolemic	1988	24 (311)
pravastatin	antihypercholesterolemic	1989	25 (316)
fluvastatin	antihypercholesterolemic	1994	30 (300)
atorvastatin calcium	antihypercholesterolemic	1997	33 (328)
cerivastatin	antihypercholesterolemic	1997	33 (331)
colestimide	antihypercholesterolemic	1999	35 (337)
colesevelam hydrochloride	antihypercholesterolemic	2000	36 (300)
ezetimibe	antihypercholesterolemic	2002	38 (355)
pitavastatin	antihypercholesterolemic	2003	39 (282)
rosuvastatin	antihypercholesterolemic	2003	39 (283)
choline fenofibrate	antihypercholesterolemic	2008	44 (594)
captopril	antihypertensive	1982	13 (086)
betaxolol hydrochloride	antihypertensive	1983	19 (315
budralazine	antihypertensive	1983	19 (315
celiprolol hydrochloride	antihypertensive	1983	19 (317
guanadrel sulfate	antihypertensive	1983	19 (319)
enalapril maleate	antihypertensive	1984	20 (317)
terazosin hydrochloride	antihypertensive	1984	20 (323)
bopindolol	antihypertensive	1985	21 (324)
bunazosin hydrochloride	antihypertensive	1985	21 (324)
ketanserin	antihypertensive	1985	21 (328)
nitrendipine	antihypertensive	1985	21 (331)
trimazosin hydrochloride	antihypertensive	1985	21 (333)
arotinolol hydrochloride	antihypertensive	1986	22 (316)
bisoprolol fumarate	antihypertensive	1986	22 (317)
bevantolol hydrochloride	antihypertensive	1987	23 (328)
enalaprilat	antihypertensive	1987	23 (332)
lisinopril	antihypertensive	1987	23 (337)
pinacidil	antihypertensive	1987	23 (340)
tertatolol hydrochloride	antihypertensive	1987	23 (344)
alacepril	antihypertensive	1988	24 (296)
alfuzosin hydrochloride	antihypertensive	1988	24 (296)
amosulalol	antihypertensive	1988	24 (297)
cadralazine	antihypertensive	1988	24 (298)
cicletanine	antihypertensive	1988	24 (299)
doxazosin mesylate	antihypertensive	1988	24 (300)
felodipine	antihypertensive	1988	24 (302)
nipradilol	antihypertensive	1988	24 (307)
perindopril	antihypertensive	1988	24 (309)
rilmenidine	antihypertensive	1988	24 (310)
tiamenidine hydrochloride	antihypertensive	1988	24 (311)

GENERIC NAME	INDICATION	YEAR INTRODUCED	ARMC VOL., (PAGE)
delapril	antihypertensive	1989	25 (311)
dilevalol	antihypertensive	1989	25 (311)
isradipine	antihypertensive	1989	25 (315)
nilvadipine	antihypertensive	1989	25 (316)
quinapril	antihypertensive	1989	25 (317)
ramipril	antihypertensive	1989	25 (317)
amlodipine besylate	antihypertensive	1990	26 (298)
benazepril hydrochloride	antihypertensive	1990	26 (299)
cilazapril	antihypertensive	1990	26 (301)
manidipine hydrochloride	antihypertensive	1990	26 (304)
nisoldipine	antihypertensive	1990	26 (306)
benidipine hydrochloride	antihypertensive	1991	27 (322)
carvedilol	antihypertensive	1991	27 (323)
fosinopril sodium	antihypertensive	1991	27 (328)
lacidipine	antihypertensive	1991	27 (330)
moxonidine	antihypertensive	1991	27 (330)
barnidipine hydrochloride	antihypertensive	1992	28 (326)
tilisolol hydrochloride	antihypertensive	1992	28 (337)
imidapril hydrochloride	antihypertensive	1993	29 (339)
trandolapril	antihypertensive	1993	29 (348)
efonidipine	antihypertensive	1994	30 (299)
losartan	antihypertensive	1994	30 (302)
temocapril	antihypertensive	1994	30 (311)
cinildipine	antihypertensive	1995	31 (339)
moexipril hydrochloride	antihypertensive	1995	31 (346)
spirapril hydrochloride	antihypertensive	1995	31 (349)
aranidipine	antihypertensive	1996	32 (306)
valsartan	antihypertensive	1996	32 (320)
candesartan cilexetil	antihypertensive	1997	33 (330)
eprosartan	antihypertensive	1997	33 (333)
irbesartan	antihypertensive	1997	33 (336)
lercanidipine	antihypertensive	1997	33 (337)
mebefradil hydrochloride	antihypertensive	1997	33 (338)
nebivolol	antihypertensive	1997	33 (339)
fenoldopam mesylate	antihypertensive	1998	34 (322)
telmisartan	antihypertensive	1999	35 (349)
zofenopril calcium	antihypertensive	2000	36 (313)
bosentan	antihypertensive	2001	37 (262)
olimesartan Medoxomil	antihypertensive	2002	38 (363)
treprostinil sodium	antihypertensive	2002	38 (368)
azelnidipine	antihypertensive	2003	39 (270)
eplerenone	antihypertensive	2003	39 (276)
aliskiren	antihypertensive	2007	43 (461)
clevidipine	antihypertensive	2008	44 (596)
azilsartan	antihypertensive	2011	47 (511)

GENERIC NAME	INDICATION	YEAR INTRODUCED	ARMC VOL., (PAGE)
butyl flufenamate	antiinflammatory	1983	19 (316)
halometasone	antiinflammatory	1983	19 (320)
hydrocortisone butyrate	antiinflammatory	1983	19 (320)
isoxicam	antiinflammatory	1983	19 (320)
oxaprozin	antiinflammatory	1983	19 (322)
fisalamine	antiinflammatory	1984	20 (318)
isofezolac	antiinflammatory	1984	20 (319)
piketoprofen	antiinflammatory	1984	20 (322)
pimaprofen	antiinflammatory	1984	20 (322)
alclometasone dipropionate	antiinflammatory	1985	21 (323)
diacerein	antiinflammatory	1985	21 (326)
etodolac	antiinflammatory	1985	21 (327)
nabumetone	antiinflammatory	1985	21 (330)
nimesulide	antiinflammatory	1985	21 (330)
amfenac sodium	antiinflammatory	1986	22 (315)
deflazacort	antiinflammatory	1986	22 (319)
felbinac	antiinflammatory	1986	22 (320)
lobenzarit sodium	antiinflammatory	1986	22 (322)
loxoprofen sodium	antiinflammatory	1986	22 (322)
osalazine sodium	antiinflammatory	1986	22 (324)
prednicarbate	antiinflammatory	1986	22 (325)
AF-2259	antiinflammatory	1987	23 (325)
flunoxaprofen	antiinflammatory	1987	23 (335)
mometasone furoate	antiinflammatory	1987	23 (338)
tenoxicam	antiinflammatory	1987	23 (344)
clobenoside	antiinflammatory	1988	24 (300)
hydrocortisone aceponate	antiinflammatory	1988	24 (304)
piroxicam cinnamate	antiinflammatory	1988	24 (309)
interferon, gamma	antiinflammatory	1989	25 (314)
aminoprofen	antiinflammatory	1990	26 (298)
droxicam	antiinflammatory	1990	26 (302)
fluticasone propionate	antiinflammatory	1990	26 (303)
halobetasol propionate	antiinflammatory	1991	27 (329)
aceclofenac	antiinflammatory	1992	28 (325)
butibufen	antiinflammatory	1992	28 (327)
deprodone propionate	antiinflammatory	1992	28 (329)
amtolmetin guacil	antiinflammatory	1993	29 (332)
zaltoprofen	antiinflammatory	1993	29 (349)
actarit	antiinflammatory	1994	30 (296)
ampiroxicam	antiinflammatory	1994	30 (296)
betamethasone butyrate propionate	antiinflammatory	1994	30 (297)
dexibuprofen	antiinflammatory	1994	30 (298)
rimexolone	antiinflammatory	1995	31 (348)
bromfenac sodium	antiinflammatory	1997	33 (329)

GENERIC NAME	INDICATION	YEAR INTRODUCED	ARMC VOL., (PAGE)
lornoxicam	antiinflammatory	1997	33 (337)
sivelestat	antiinflammatory	2002	38 (366)
lumiracoxib	antiinflammatory	2005	41 (455)
nepafenac	antiinflammatory	2005	41 (456)
canakinumab	antiinflammatory	2009	45 (484)
golimumab	antiinflammatory	2009	45 (503)
mefloquine hydrochloride	antimalarial	1985	21 (329)
artemisinin	antimalarial	1987	23 (327)
halofantrine	antimalarial	1988	24 (304)
arteether	antimalarial	2000	36 (296)
bulaquine	antimalarial	2000	36 (299)
alpiropride	antimigraine	1988	24 (296)
sumatriptan succinate	antimigraine	1991	27 (333)
naratriptan hydrochloride	antimigraine	1997	33 (339)
zomitriptan	antimigraine	1997	33 (345)
rizatriptan benzoate	antimigraine	1998	34 (330)
lomerizine hydrochloride	antimigraine	1999	35 (342)
almotriptan	antimigraine	2000	36 (295)
eletriptan	antimigraine	2001	37 (266)
frovatriptan	antimigraine	2002	38 (357)
dexfenfluramine	antiobesity	1997	33 (332)
orlistat	antiobesity	1998	34 (327)
sibutramine	antiobesity	1998	34 (331)
rimonabant	antiobesity	2006	42 (537)
quinf amideamebicide	antiparasitic	1984	20 (322)
ivermectin	antiparasitic	1987	23 (336)
atovaquone	antiparasitic	1992	28 (326)
epoprostenol sodium	antiplatelet	1983	19 (318)
beraprost sodium	antiplatelet	1992	28 (326)
iloprost	antiplatelet	1992	28 (332)
prasugrel	antiplatelet	2009	45 (519)
cabergoline	antiprolactin	1993	29 (334)
acitretin	antipsoriasis	1989	25 (309)
calcipotriol	antipsoriasis	1991	27 (323)
tacalcitol	antipsoriasis	1993	29 (346)
tazarotene	antipsoriasis	1997	33 (343)
alefacept	antipsoriasis	2003	39 (267)
efalizumab	antipsoriasis	2003	39 (274)
ustekinumab	antipsoriasis	2009	45 (532)
timiperone	antipsychotic	1984	20 (323)
amisulpride	antipsychotic	1986	22 (316)
zuclopenthixol acetate	antipsychotic	1987	23 (345)
remoxipride hydrochloride	antipsychotic	1990	26 (308)
clospipramine hydrochloride	antipsychotic	1991	27 (325)
nemonapride	antipsychotic	1991	27 (331)

GENERIC NAME	INDICATION	YEAR INTRODUCED	ARMC VOL., (PAGE)
risperidone	antipsychotic	1993	29 (344)
olanzapine	antipsychotic	1996	32 (313)
sertindole	antipsychotic	1996	32 (318)
quetiapine fumarate	antipsychotic	1997	33 (341)
ziprasidone hydrochloride	antipsychotic	2000	36 (312)
perospirone hydrochloride	antipsychotic	2001	37 (270)
aripiprazole	antipsychotic	2002	38 (350)
paliperidone	antipsychotic	2007	43 (482)
blonanserin	antipsychotic	2008	44 (587)
asenapine	antipsychotic	2009	45 (479)
lurasidone hydrochloride	antipsychotic	2010	46 (473)
drotrecogin alfa	antisepsis	2001	37 (265)
tiquizium bromide	antispasmodic	1984	20 (324)
cimetropium bromide	antispasmodic	1985	21 (326)
romiplostim	antithrombocytopenic	2008	44 (619)
eltrombopag	antithrombocytopenic	2009	45 (497)
indobufen	antithrombotic	1984	20 (319)
defibrotide	antithrombotic	1986	22 (319)
alteplase	antithrombotic	1987	23 (326)
APSAC	antithrombotic	1987	23 (326)
picotamide	antithrombotic	1987	23 (340)
cilostazol	antithrombotic	1988	24 (299)
limaprost	antithrombotic	1988	24 (306)
ozagrel sodium	antithrombotic	1988	24 (308)
argatroban	antithrombotic	1990	26 (299)
ethyl icosapentate	antithrombotic	1990	26 (303)
cloricromen	antithrombotic	1991	27 (325)
sarpogrelate hydrochloride	antithrombotic	1993	29 (344)
reteplase	antithrombotic	1996	32 (316)
anagrelide hydrochloride	antithrombotic	1997	33 (328)
clopidogrel hydrogensulfate	antithrombotic	1998	34 (320)
tirofiban hydrochloride	antithrombotic	1998	34 (332)
eptilfibatide	antithrombotic	1999	35 (340)
bivalirudin	antithrombotic	2000	36 (298)
fondaparinux sodium	antithrombotic	2002	38 (356)
ticagrelor	antithrombotic	2010	46 (488)
apixaban	antithrombotic	2011	47 (507)
edoxaban	antithrombotic	2011	47 (524)
levodropropizine	antitussive	1988	24 (305)
nitisinone	antityrosinaemia	2002	38 (361)
sofalcone	antiulcer	1984	20 (323)
teprenone	antiulcer	1984	20 (323)
enprostil	antiulcer	1985	21 (327)
famotidine	antiulcer	1985	21 (327)
misoprostol	antiulcer	1985	21 (329)

GENERIC NAME	INDICATION	YEAR INTRODUCED	ARMC VOL., (PAGE)
rosaprostol	antiulcer	1985	21 (332)
roxatidine acetate hydrochloride	antiulcer	1986	22 (326)
troxipide	antiulcer	1986	22 (327)
benexate hydrochloride	antiulcer	1987	23 (328)
nizatidine	antiulcer	1987	23 (339)
ornoprostil	antiulcer	1987	23 (339)
plaunotol	antiulcer	1987	23 (340)
roxithromycin	antiulcer	1987	23 (342)
spizofurone	antiulcer	1987	23 (343)
omeprazole	antiulcer	1988	24 (308)
irsogladine	antiulcer	1989	25 (315)
rebamipide	antiulcer	1990	26 (308)
lansoprazole	antiulcer	1992	28 (332)
ecabet sodium	antiulcer	1993	29 (336)
tretinoin tocoferil	antiulcer	1993	29 (348)
polaprezinc	antiulcer	1994	30 (307)
pantoprazole sodium	antiulcer	1995	30 (306)
ranitidine bismuth citrate	antiulcer	1995	31 (348)
ebrotidine	antiulcer	1997	33 (333)
rabeprazole sodium	antiulcer	1998	34 (328)
dosmalfate	antiulcer	2000	36 (302)
egualen sodium	antiulcer	2000	36 (303)
esomeprazole magnesium	antiulcer	2000	36 (303)
lafutidine	antiulcer	2000	36 (307)
dexlansoprazole	antiulcer	2009	45 (492)
rimantadine hydrochloride	antiviral	1987	23 (342)
zidovudine	antiviral	1987	23 (345)
ganciclovir	antiviral	1988	24 (303)
foscarnet sodium	antiviral	1989	25 (313)
didanosine	antiviral	1991	27 (326)
zalcitabine	antiviral	1992	28 (338)
sorivudine	antiviral	1993	29 (345)
famciclovir	antiviral	1994	30 (300)
propagermanium	antiviral	1994	30 (308)
stavudine	antiviral	1994	30 (311)
lamivudine	antiviral	1995	31 (345)
saquinavir mesvlate	antiviral	1995	31 (349)
valaciclovir hydrochloride	antiviral	1995	31 (352)
cidofovir	antiviral	1996	32 (306)
indinavir sulfate	antiviral	1996	32 (310)
nevirapine	antiviral	1996	32 (313)
penciclovir	antiviral	1996	32 (314)
ritonavir	antiviral	1996	32 (317)
delavirdine mesylate	antiviral	1997	33 (331)
imiquimod	antiviral	1997	33 (335)

GENERIC NAME	INDICATION	YEAR INTRODUCED	ARMC VOL., (PAGE)
interferon alfacon-1	antiviral	1997	33 (336)
nelfinavir mesylate	antiviral	1997	33 (340)
efavirenz	antiviral	1998	34 (321)
fomivirsen sodium	antiviral	1998	34 (323)
abacavir sulfate	antiviral	1999	35 (333)
amprenavir	antiviral	1999	35 (334)
oseltamivir phosphate	antiviral	1999	35 (346)
zanamivir	antiviral	1999	35 (352)
lopinavir	antiviral	2000	36 (310)
tenofovir disoproxil fumarate	antiviral	2001	37 (271)
vaglancirclovir hydrochloride	antiviral	2001	37 (273)
adefovir dipivoxil	antiviral	2002	38 (348)
atazanavir	antiviral	2003	39 (269)
emtricitabine	antiviral	2003	39 (274)
enfuvirtide	antiviral	2003	39 (275)
fosamprenavir	antiviral	2003	39 (277)
influenza virus (live)	antiviral	2003	39 (277)
entecavir	antiviral	2005	41 (450)
tipranavir	antiviral	2005	41 (470)
darunavir	antiviral	2006	42 (515)
telbivudine	antiviral	2006	42 (546)
clevudine	antiviral	2007	43 (466)
maraviroc	antiviral	2007	43 (478)
raltegravir	antiviral	2007	43 (484)
etravirine	antiviral	2008	44 (602)
laninamivir octanoate	antiviral	2010	46 (470)
peramivir	antiviral	2010	46 (477)
boceprevir	antiviral	2011	47 (518)
rilpivirine	antiviral	2011	47 (542)
telaprevir	antiviral	2011	47 (548)
cevimeline hydrochloride	antixerostomia	2000	36 (299)
etizolam	anxiolytic	1984	20 (318)
flutazolam	anxiolytic	1984	20 (318)
mexazolam	anxiolytic	1984	20 (321)
buspirone hydrochloride	anxiolytic	1985	21 (324)
doxefazepam	anxiolytic	1985	21 (326)
flutoprazepam	anxiolytic	1986	22 (320)
metaclazepam	anxiolytic	1987	23 (338)
alpidem	anxiolytic	1991	27 (322)
cinolazepam	anxiolytic	1993	29 (334)
tandospirone	anxiolytic	1996	32 (319)
dexmethylphenidate hydrochloride	attention deficit hyperactivity disorder	2002	38 (352)
atomoxetine	attention deficit hyperactivity disorder	2003	39 (270)

GENERIC NAME	INDICATION	YEAR INTRODUCED	ARMC VOL., (PAGE)
lisdexamfetamine	attention deficit hyperactivity disorder	2007	43 (477)
finasteride	benign prostatic hyperplasia	1992	28 (331)
tamsulosin hydrochloride	benign prostatic hyperplasia	1993	29 (347)
dutasteride	benign prostatic hyperplasia	2002	38 (353)
fenbuprol	biliary tract dysfunction	1983	19 (318)
clodronate disodium	calcium regulation	1986	22 (319)
gallium nitrate	calcium regulation	1991	27 (328)
neridronic acide	calcium regulation	2002	38 (361)
dexrazoxane	cardioprotective	1992	28 (330)
nimodipine	cerebral vasodilator	1985	21 (330)
brovincamine fumarate	cerebral vasodilator	1986	22 (317)
propentofylline propionate	cerebral vasodilator	1988	24 (310)
formoterol fumarate	chronic obstructive pulmonary disease	1986	22 (321)
tiotropium bromide	chronic obstructive pulmonary disease	2002	38 (368)
indacaterol	chronic obstructive pulmonary disease	2009	45 (505)
roflumilast	chronic obstructive pulmonary disease	2010	46 (480)
levacecarnine hydrochloride	cognition enhancer	1986	22 (322)
oxiracetam	cognition enhancer	1987	23 (339)
exifone	cognition enhancer	1988	24 (302)
choline alfoscerate	cognition enhancer	1990	26 (300)
aniracetam	cognition enhancer	1993	29 (333)
pramiracetam sulfate	cognition enhancer	1993	29 (343)
amrinone	congestive heart failure	1983	19 (314)
bucladesine sodium	congestive heart failure	1984	20 (316)
ibopamine hydrochloride	congestive heart failure	1984	20 (319)
denopamine	congestive heart failure	1988	24 (300)
enoximone	congestive heart failure	1988	24 (301)
xamoterol fumarate	congestive heart failure	1988	24 (312)
dopexamine	congestive heart failure	1989	25 (312)
milrinone	congestive heart failure	1989	25 (316)
vesnarinone	congestive heart failure	1990	26 (310)
flosequinan	congestive heart failure	1992	28 (331)
docarpamine	congestive heart failure	1994	30 (298)
pimobendan	congestive heart failure	1994	30 (307)
carperitide	congestive heart failure	1995	31 (339)
loprinone hydrochloride	congestive heart failure	1996	32 (312)
colforsin daropate hydrochloride	congestive heart failure	1999	35 (337)
levosimendan	congestive heart failure	2000	36 (308)
nesiritide	congestive heart failure	2001	37 (269)
lulbiprostone	constipation	2006	42 (525)

GENERIC NAME	INDICATION	YEAR INTRODUCED	ARMC VOL., (PAGE)
methylnaltrexone bromide	constipation	2008	44 (612)
promegestrone	contraception	1983	19 (323)
gestrinone	contraception	1986	22 (321)
nomegestrol acetate	contraception	1986	22 (324)
norgestimate	contraception	1986	22 (324)
gestodene	contraception	1987	23 (335)
centchroman	contraception	1991	27 (324)
drospirenone	contraception	2000	36 (302)
trimegestone	contraception	2001	37 (273)
norelgestromin	contraception	2002	38 (362)
ulipristal acetate	contraception	2009	45 (530)
biolimus drug-eluting stent	coronary artery disease, antirestenotic	2008	44 (586)
neltenexine	cystic fibrosis	1993	29 (341)
dornase alfa	cystic fibrosis	1994	30 (298)
amifostine	cytoprotective	1995	31 (338)
kinetin	dermatologic, skin photodamage	1999	35 (341)
gadoversetamide	diagnostic	2000	36 (304)
ioflupane	diagnostic	2000	36 (306)
muzolimine	diuretic	1983	19 (321)
azosemide	diuretic	1986	22 (316)
torasemide	diuretic	1993	29 (348)
alpha-1 antitrypsin	emphysema	1988	24 (297)
telmesteine	expectorant	1992	28 (337)
erdosteine	expectorant	1995	31 (342)
fudosteine	expectorant	2001	37 (267)
agalsidase alfa	Fabry's disease	2001	37 (259)
chenodiol	gallstones	1983	19 (317)
cisapride	gastroprokinetic	1988	24 (299)
cinitapride	gastroprokinetic	1990	26 (301)
itopride hydrochloride	gastroprokinetic	1995	31 (344)
mosapride citrate	gastroprokinetic	1998	34 (326)
alglucerase	Gaucher's disease	1991	27 (321)
imiglucerase	Gaucher's disease	1994	30 (301)
miglustat	Gaucher's disease	2003	39 (279)
rilonacept	genetic autoinflammatory syndromes	2008	44 (615)
febuxostat	gout	2009	45 (501)
somatropin	growth disorders	1987	23 (343)
octreotide	growth disorders	1988	24 (307)
somatomedin-1	growth disorders	1994	30 (310)
somatotropin	growth disorders	1994	30 (310)
lanreotide acetate	growth disorders	1995	31 (345)
pegvisomant	growth disorders	2003	39 (281)

GENERIC NAME	INDICATION	YEAR INTRODUCED	ARMC VOL., (PAGE)
factor VIIa	haemophilia	1996	32 (307)
erythropoietin	hematopoietic	1988	24 (301)
eculizumab	hemoglobinuria	2007	43 (468)
factor VIII	hemostatic	1992	28 (330)
thrombin alfa	hemostatic	2008	44 (627)
malotilate	hepatoprotective	1985	21 (329)
mivotilate	hepatoprotective	1999	35 (343)
buserelin acetate	hormone therapy	1984	20 (316)
leuprolide acetate	hormone therapy	1984	20 (319)
goserelin	hormone therapy	1987	23 (336)
tibolone	hormone therapy	1988	24 (312)
nafarelin acetate	hormone therapy	1990	26 (306)
paricalcitol	hyperparathyroidism	1998	34 (327)
doxercalciferol	hyperparathyroidism	1999	35 (339)
maxacalcitol	hyperparathyroidism	2000	36 (310)
falecalcitriol	hyperparathyroidism	2001	37 (266)
cinacalcet	hyperparathyroidism	2004	40 (451)
quinagolide	hyperprolactinemia	1994	30 (309)
conivaptan	hyponatremia	2006	42 (514)
mozavaptan	hyponatremia	2006	42 (527)
tolvaptan	hyponatremia	2009	45 (528)
thymopentin	immunomodulator	1985	21 (333)
bucillamine	immunomodulator	1987	23 (329)
centoxin	immunomodulator	1991	27 (325)
schizophyllan	immunostimulant	1985	22 (326)
lentinan	immunostimulant	1986	22 (322)
ubenimex	immunostimulant	1987	23 (345)
pegademase bovine	immunostimulant	1990	26 (307)
filgrastim	immunostimulant	1991	27 (327)
interferon gamma-1b	immunostimulant	1991	27 (329)
romurtide	immunostimulant	1991	27 (332)
sargramostim	immunostimulant	1991	27 (332)
pidotimod	immunostimulant	1993	29 (343)
GMDP	immunostimulant	1996	32 (308)
cyclosporine	immunosuppressant	1983	19 (317)
mizoribine	immunosuppressant	1984	20 (321)
muromonab-CD3	immunosuppressant	1986	22 (323)
tacrolimus	immunosuppressant	1993	29 (347)
gusperimus	immunosuppressant	1994	30 (300)
mycophenolate mofetil	immunosuppressant	1995	31 (346)
pimecrolimus	immunosuppressant	2002	38 (365)
mycophenolate sodium	immunosuppressant	2003	39 (279)
everolimus	immunosuppressant	2004	40 (455)
belatacept	immunosuppressant	2011	47 (513)
follitropin alfa	infertility	1996	32 (307)

GENERIC NAME	INDICATION	YEAR INTRODUCED	ARMC VOL., (PAGE)
follitropin beta	infertility	1996	32 (308)
cetrorelix	infertility	1999	35 (336)
ganirelix acetate	infertility	2000	36 (305)
corifollitropin alfa	infertility	2010	46 (455)
defeiprone	iron chelation therapy	1995	31 (340)
deferasirox	iron chelation therapy	2005	41 (446)
alosetron hydrochloride	irritable bowel syndrome	2000	36 (295)
tegaserod maleate	irritable bowel syndrome	2001	37 (270)
certolizumab pegol	irritable bowel syndrome	2008	44 (592)
nartograstim	leukopenia	1994	30 (304)
tesamorelin acetate	lipodystrophy	2010	46 (486)
belimumab	lupus	2011	47 (516)
Lyme disease vaccine	Lyme disease	1999	35 (342)
sildenafil citrate	male sexual dysfunction	1998	34 (331)
tadalafil	male sexual dysfunction	2003	39 (284)
vardenafil	male sexual dysfunction	2003	39 (286)
udenafil	male sexual dysfunction	2005	41 (472)
avanafil	male sexual dysfunction	2011	47 (510)
laronidase	mucopolysaccharidosis I	2003	39 (278)
idursulfase	mucopolysaccharidosis II (Hunter syndrome)	2006	42 (520)
galsulfase	mucopolysaccharidosis VI	2005	41 (453)
palifermin	mucositis	2005	41 (461)
interferon, b-1b	multiple sclerosis	1993	29 (339)
interferon, b-1a	multiple sclerosis	1996	32 (311)
glatiramer acetate	multiple sclerosis	1997	33 (334)
natalizumab	multiple sclerosis	2004	40 (462)
dalfampridine	multiple sclerosis	2010	46 (458)
fingolimod	multiple sclerosis	2010	46 (468)
afloqualone	muscle relaxant	1983	19 (313)
eperisone hydrochloride	muscle relaxant	1983	19 (318)
tiropramide hydrochloride	muscle relaxant	1983	19 (324)
tizanidine	muscle relaxant	1984	20 (324)
doxacurium chloride	muscle relaxant	1991	27 (326)
mivacurium chloride	muscle relaxant	1992	28 (334)
rocuronium bromide	muscle relaxant	1994	30 (309)
cisatracurium besilate	muscle relaxant	1995	31 (340)
rapacuronium bromide	muscle relaxant	1999	35 (347)
decitabine	myelodysplastic syndromes	2006	42 (519)
lenalidomide	myelodysplastic syndromes, multiple myeloma	2006	42 (523)
tinazoline	nasal decongestant	1988	24 (312)
taltirelin	neurodegeneration	2000	36 (311)
tafamidis	neurodegeneration	2011	47 (546)

GENERIC NAME	INDICATION	YEAR INTRODUCED	ARMC VOL., (PAGE)
sugammadex	neuromuscular blockade, reversal	2008	44 (623)
edaravone	neuroprotective	2001	37 (265)
idebenone	nootropic	1986	22 (321)
bifemelane hydrochloride	nootropic	1987	23 (329)
indeloxazine hydrochloride	nootropic	1988	24 (304)
nizofenzone	nootropic	1988	24 (307)
alcaftadine	ophthalmologic (allergic conjunctivitis)	2010	46 (444)
diquafosol tetrasodium	ophthalmologic (dry eye)	2010	46 (462)
verteporfin	ophthalmologic (macular degeneration)	2000	36 (312)
pegaptanib	ophthalmologic (macular degeneration)	2005	41 (458)
ranibizumab	ophthalmologic (macular degeneration)	2006	42 (534)
aflibercept	ophthalmologic (macular degeneration)	2011	47 (505)
OP-1	osteoinductor	2001	37 (269)
APD	osteoporosis	1987	23 (326)
disodium pamidronate	osteoporosis	1989	25 (312)
ipriflavone	osteoporosis	1989	25 (314)
alendronate sodium	osteoporosis	1993	29 (332)
ibandronic acid	osteoporosis	1996	32 (309)
incadronic acid	osteoporosis	1997	33 (335)
raloxifene hydrochloride	osteoporosis	1998	34 (328)
risedronate sodium	osteoporosis	1998	34 (330)
zoledronate disodium	osteoporosis	2000	36 (314)
strontium ranelate	osteoporosis	2004	40 (466)
minodronic acid	osteoporosis	2009	45 (509)
denosumab	osteoporosis	2010	46 (459)
eldecalcitol	osteoporosis	2011	47 (526)
tiludronate disodium	Paget's disease	1995	31 (350)
nafamostat mesylate	pancreatitis	1986	22 (323)
pergolide mesylate	Parkinson's disease	1988	24 (308)
droxidopa	Parkinson's disease	1989	25 (312)
ropinirole hydrochloride	Parkinson's disease	1996	32 (317)
talipexole	Parkinson's disease	1996	32 (318)
budipine	Parkinson's disease	1997	33 (330)
pramipexole hydrochloride	Parkinson's disease	1997	33 (341)
tolcapone	Parkinson's disease	1997	33 (343)
entacapone	Parkinson's disease	1998	34 (322)
CHF-1301	Parkinson's disease	1999	35 (336)
rasagiline	Parkinson's disease	2005	41 (464)
rotigotine	Parkinson's disease	2006	42 (538)

GENERIC NAME	INDICATION	YEAR INTRODUCED	ARMC VOL., (PAGE)
sapropterin hydrochloride	phenylketouria	1992	28 (336)
alglucosidase alfa	Pompe disease	2006	42 (511)
alvimopan	post-operative ileus	2008	44 (584)
histrelin	precocious puberty	1993	29 (338)
dapoxetine	premature ejaculation	2009	45 (488)
atosiban	premature labor	2000	36 (297)
nalfurafine hydrochloride	pruritus	2009	45 (510)
pirfenidone	pulmonary fibrosis, idiopathic	2008	44 (614)
sitaxsentan	pulmonary hypertension	2006	42 (543)
ambrisentan	pulmonary hypertension	2007	43 (463)
pumactant	respiratory distress syndrome	1994	30 (308)
mepixanox	respiratory stimulant	1984	20 (320)
surfactant TA	respiratory surfactant	1987	23 (344)
gabapentin enacarbil	restless leg syndrome	2011	47 (530)
binfonazole	sleep disorders	1983	19 (315)
brotizolam	sleep disorders	1983	19 (315)
loprazolam mesylate	sleep disorders	1983	19 (321)
butoctamide	sleep disorders	1984	20 (316)
quazepam	sleep disorders	1985	21 (332)
adrafinil	sleep disorders	1986	22 (315)
zopiclone	sleep disorders	1986	22 (327)
zolpidem hemitartrate	sleep disorders	1988	24 (313)
rilmazafone	sleep disorders	1989	25 (317)
modafinil	sleep disorders	1994	30 (303)
zaleplon	sleep disorders	1999	35 (351)
dexmedetomidine hydrochloride	sleep disorders	2000	36 (301)
eszopiclone	sleep disorders	2005	41 (451)
ramelteon	sleep disorders	2005	41 (462)
armodafinil	sleep disorders	2009	45 (478)
plerixafor hydrochloride	stem cell mobilizer	2009	45 (515)
tirilazad mesylate	subarachnoid hemorrhage	1995	31 (351)
balsalazide disodium	ulcerative colitis	1997	33 (329)
acetohydroxamic acid	urinary tract/bladder disorders	1983	19 (313)
sodium cellulose phosphate	urinary tract/bladder disorders	1983	19 (323)
propiverine hydrochloride	urinary tract/bladder disorders	1992	28 (335)
naftopidil	urinary tract/bladder disorders	1999	35 (344)
solifenacin	urinary tract/bladder disorders	2004	40 (466)

GENERIC NAME	INDICATION	YEAR INTRODUCED	ARMC VOL., (PAGE)
darifenacin	urinary tract/bladder disorders	2005	41 (445)
silodosin	urinary tract/bladder disorders	2006	42 (540)
imidafenacin	urinary tract/bladder disorders	2007	43 (472)
fesoterodine	urinary tract/bladder disorders	2008	44 (604)
mirabegron	urinary tract/bladder disorders	2011	47 (539)
tiopronin	urolithiasis	1989	25 (318)
cadexomer iodine	wound healing agent	1983	19 (316)
epidermal growth factor	wound healing agent	1987	23 (333)
prezatide copper acetate	wound healing agent	1996	32 (314)
acemannan	wound healing agent	2001	37 (259)

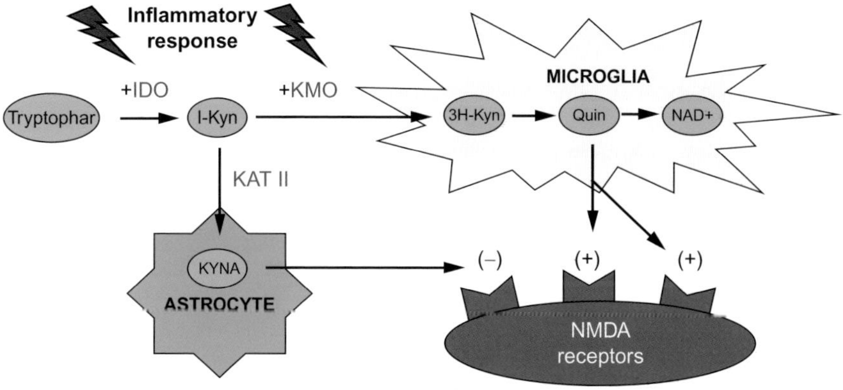

Plate 4.1 Kynurenine pathway. Enzymes regulating kynurenine metabolism in the CNS are reportedly upregulated in response to inflammation. Selective inhibition at points within the kynurenine pathway may be beneficial in treating NI-related conditions. Indolamine 2,3-dioxygenase (IDO), kynurenine 3-monooxygenase, kynurenine aminotransferase II (KAT II).

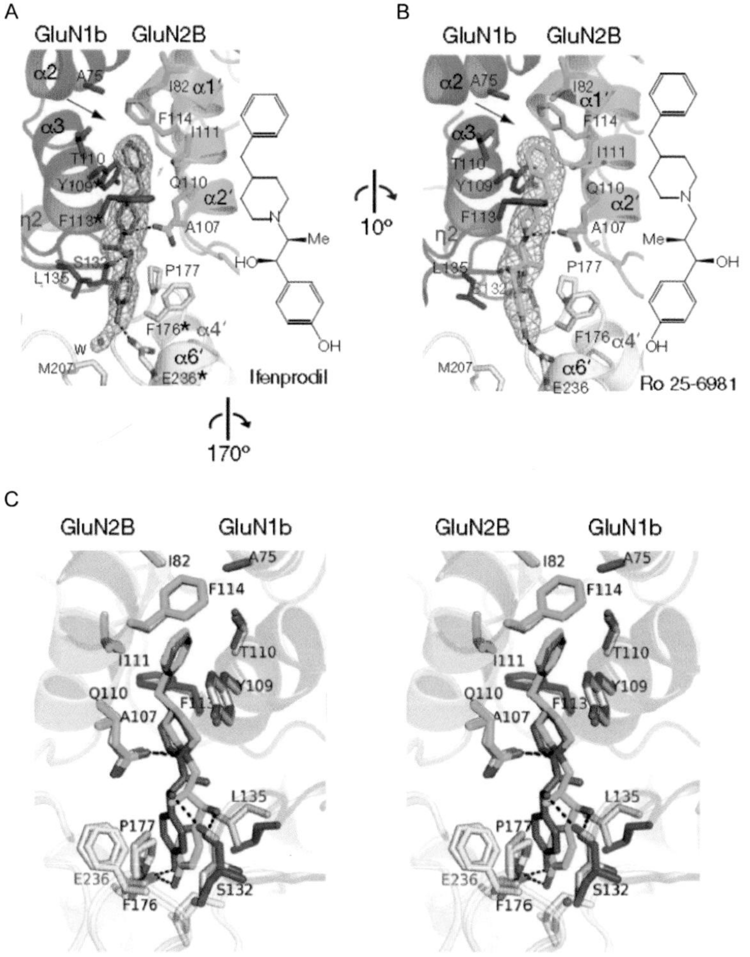

Plate 7.1 Binding of ifenprodil (A) and Ro 25-6981 (B) at the binding pocket; (C) comparison of binding patterns of ifenprodil (gray) and Ro 25-6981 (lime) in stereoview (reproduced with permission from Ref. 17 Nature Publishing Group).

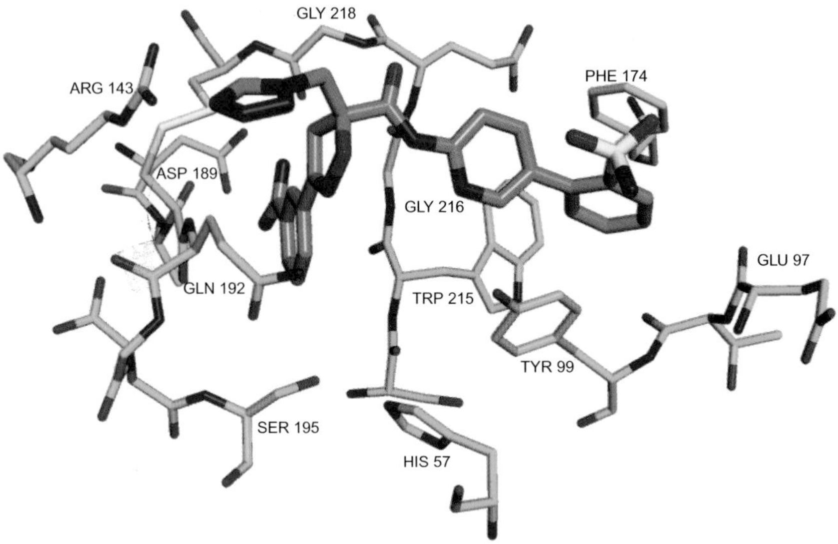

Plate 9.5 Model of isoxazoline **12** in the active site of FXa.

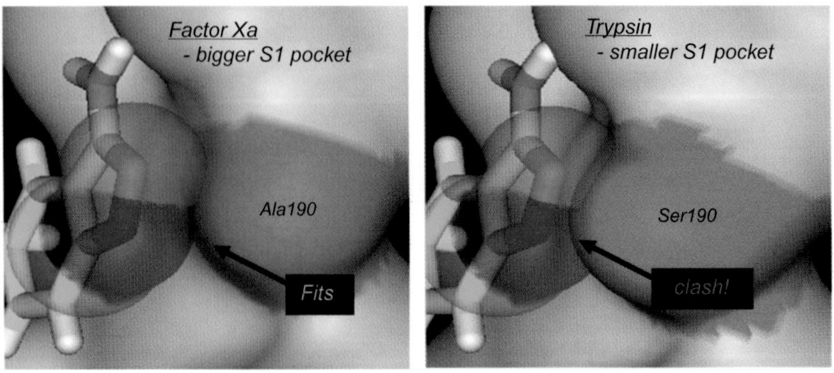

Plate 9.10 Aminobenzisoxazole **23** in the S1 specificity pocket of FXa and trypsin.

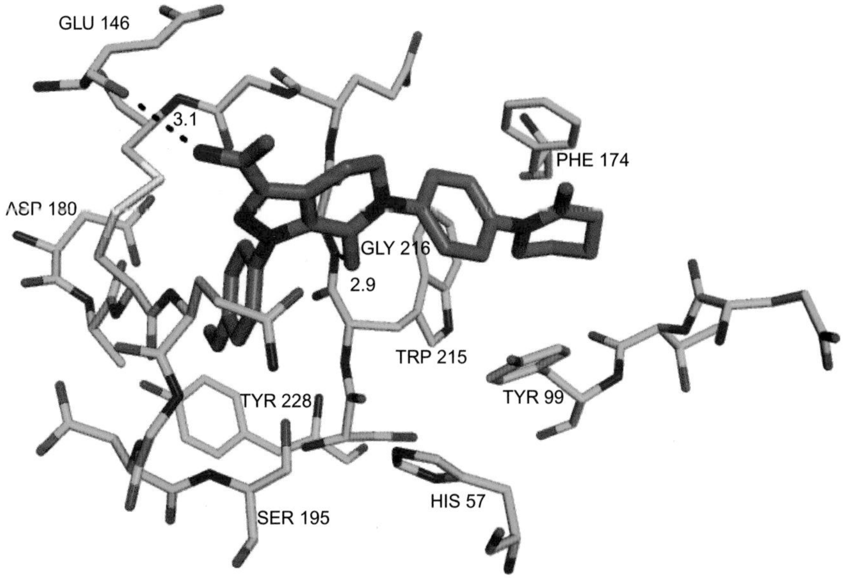

Plate 9.13 FXa-bound X-ray structure of apixaban **1**.

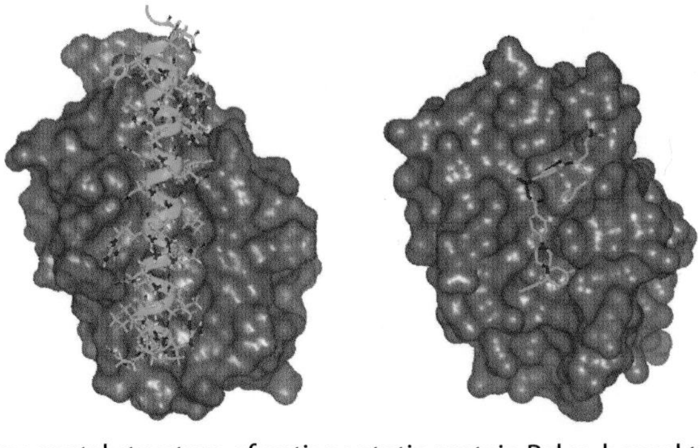

Plate 17.2 X-ray crystal structure of antiapoptotic protein Bcl-x_L bound to the Bim BH3-domain peptide (left), and the small-molecule BH3 mimetic ABT-737 (right).

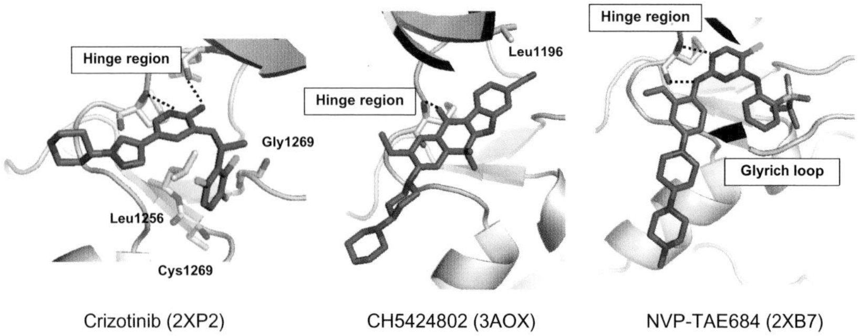

Plate 19.1 Binding modes of representative ALK inhibitors.

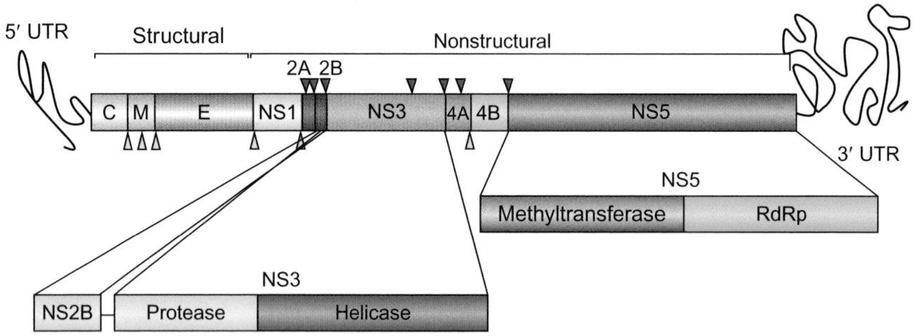

Plate 20.1 Schematic of the DENV genome.

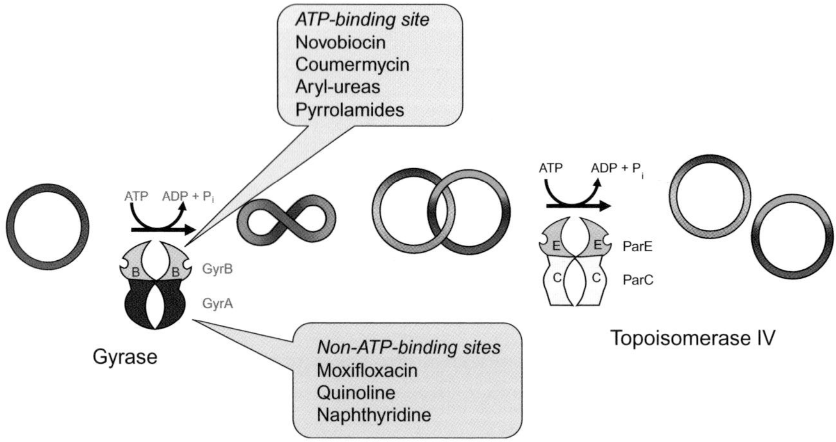

Plate 21.1 Schematic diagram of bacterial type II topoisomerases.

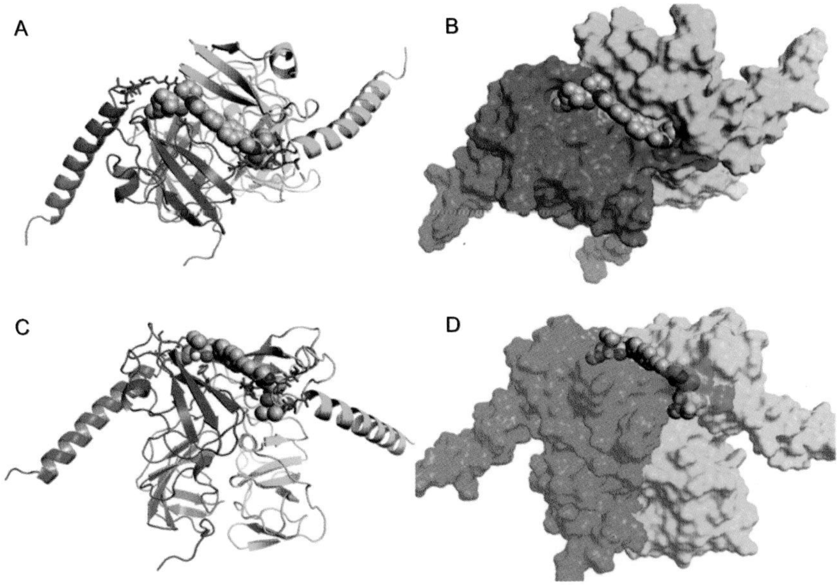

Plate 22.4 Putative binding mode of BMS-790052 with NS5A gt1b domain I, based on the crystal structure of the ZnF domain[34] and homology modeling of the AH and ZnF-AH loop units (residues 1–36).[10] Monomer units are color-coded red and cyan. BMS-790052-resistant mutant sites (gt1a: Met28, Gln30, Leu31, Tyr93) are color-coded blue. (A) Dimeric interaction mode of BMS-790052 (CPK spheres) looking down the C2-symmetry axis of the NS5A homodimer. The bis-phenyl imidazole core of BMS-790052 spans the NS5A dimer interface, positioning the Pro-Val-carbamate caps at the hinge loops, between the ZnF and AH units. (B) Transparent surface rendering of (A). (C) View of the dimer interface, perpendicular to the C2 axis. (D) Transparent surface rendering of (C).

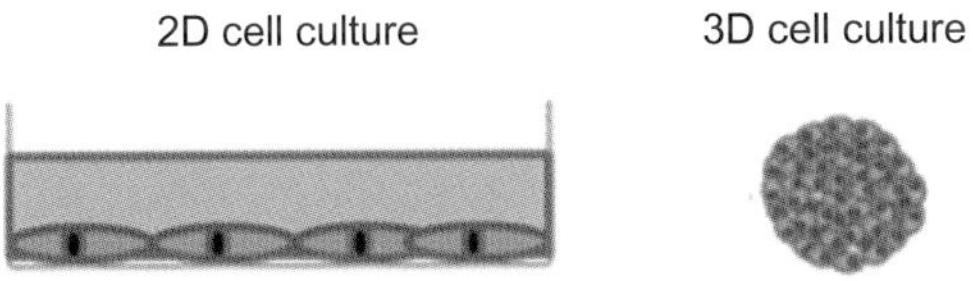

Plate 24.1 Culturing cells in 2D monolayers versus 3D. In 2D monolayer cultures, cells are grown on a flat, impermeable surface such as the plastic surface of a culture dish. These cells flatten out on the plastic, with exposure to plastic and medium and very little cell–cell contact. In 3D cell cultures, cells are completely exposed to other cells, or matrix.

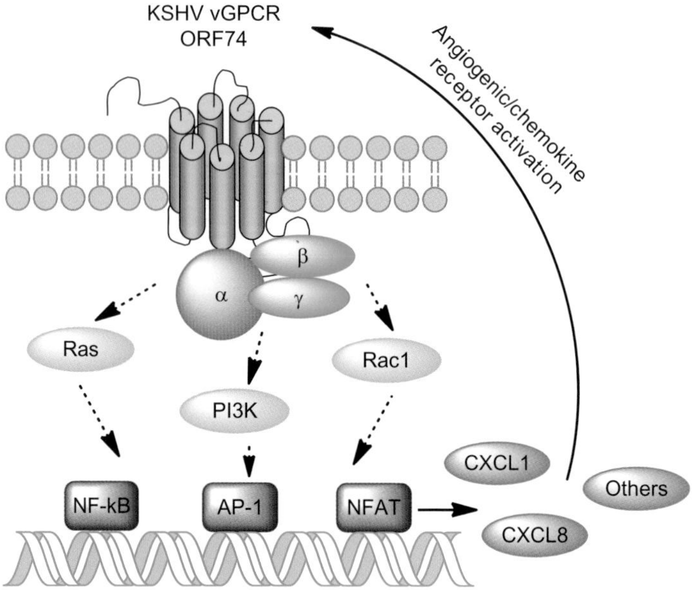

Plate 25.1 Interaction of vGPCRs with the host's signaling machinery, leading to reprogramming of cell signaling, activation of several transcription factors, and subsequent production of growth factors and chemokines.

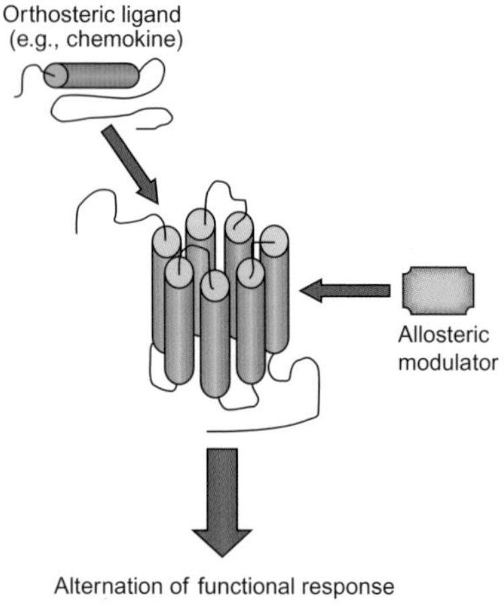

Plate 25.2 Allosteric modulation of vGPCRs. Allosteric modulators bind to a site topographically distinct from the orthosteric site on the receptor to alter either the affinity or efficacy of an endogenous ligand (chemokine) and thus shift the functional response. Allosteric modulators can also reduce the basal activity of the receptor in the absence of an endogenous ligand.

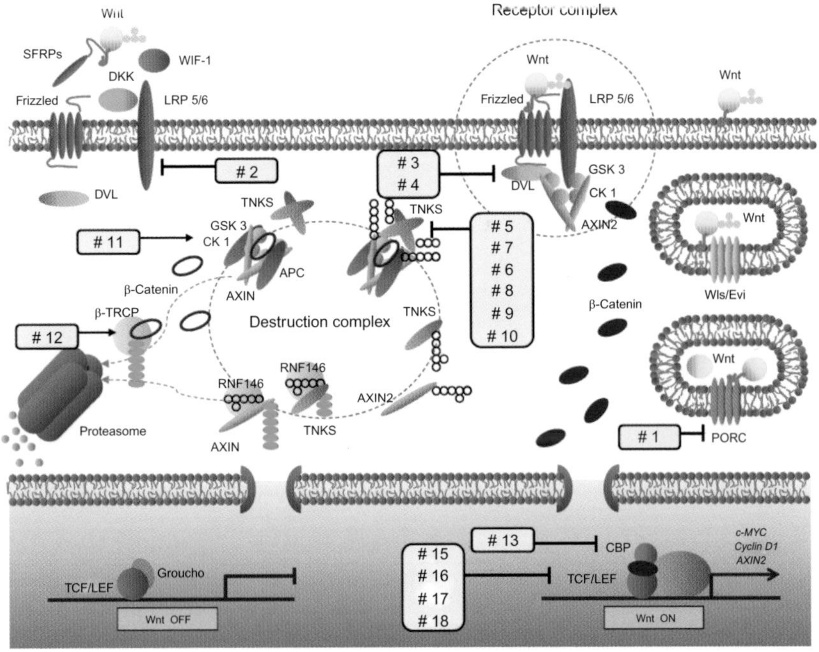

Plate 26.1 Model for Wnt signaling. Newly synthesized Wnt is palmitoylated by PORC and N-glycosylated (gray dots). Upon Wnt binding to the Frizzled/LRP5/6 receptor complex, β-catenin (blue circles) ceases to be degraded and enters the nucleus where it binds TCF/LEF. In the absence of active Wnt, β-catenin becomes phosphorylated (blue rings) in the destruction complex. Subsequently, it is ubiquitinated (green circles) and degraded in the proteasome. Destruction complex proteins are gradually poly-ADP-ribosylated (black rings) by TNKS. Poly-ADP-ribosylation of AXIN and TNKS leads to a destabilization of the destruction complex, followed by ubiquitination and degradation of AXIN and TNKS in the proteasome. Inhibitory substances are marked by a bar, while activating substances are marked by an arrow. The numbers (#) refer to the compounds in the text.

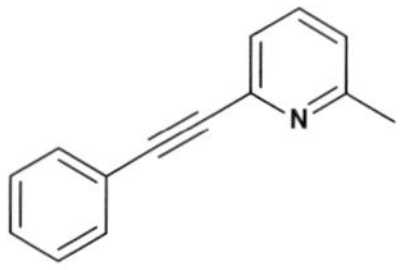

4

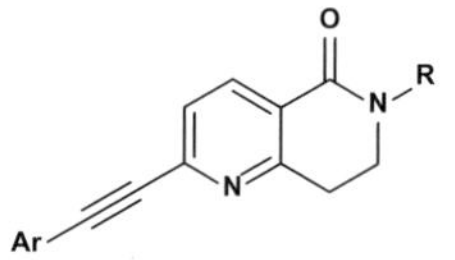

5 (Ar=Ph)

6 (Ar=3-F-Ph)

Compound	R	mGlu$_5$ EC$_{50}$ or IC$_{50}$ (nM)	% Glu Max	Functionality
6a	H	290	72	PAM
6b	CH_3	170	34	NAM
6c	*t*-Bu	54	40	PAM
6d	*n*-Bu	56	46	PAM
6e	$CH_2(CH_2)_3$	660	34	NAM
6f	$CH_2(CH_2)_4$	130	53	PAM

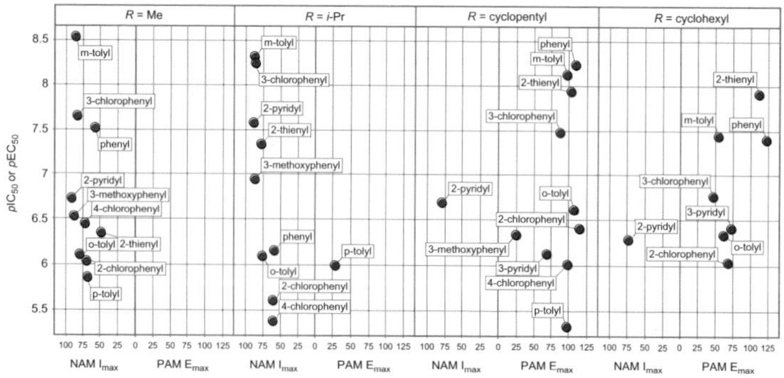

Plate 28.1 Switches in functional potency and efficacy for a series of *N*-substituted 2-(arylethynyl)-7,8-dihydro-1,6-naphthyridin-5(6H)-one analogs (**5**), Ar as specified in the table.

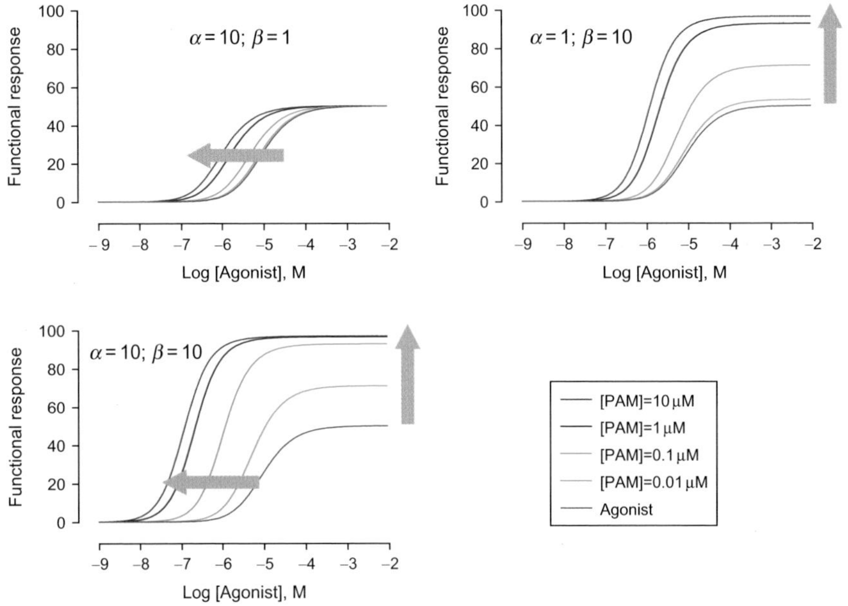

Plate 28.2 Computer simulation of curves showing the impact of allosteric modulation on an agonist functional response for different combinations of α and β. Calculations based on Leach equation [6], using the following parameters: $pK_A = 5$; $pK_B = 7$; $E_m = 100$; $\tau_A = 1$; $\tau_B = 0$; $n_H = 1.5$.

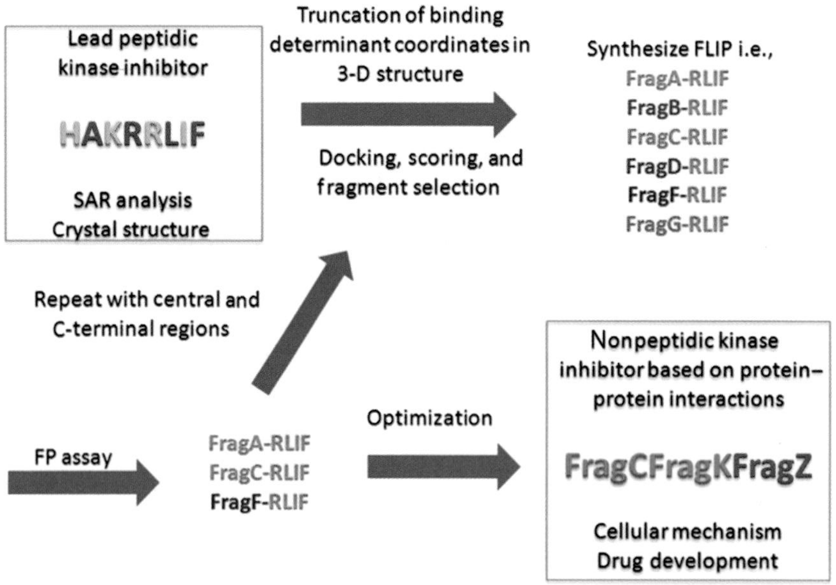

Plate 29.1 Overview of the REPLACE strategy for generation of more drug-like protein–protein interaction inhibitors.

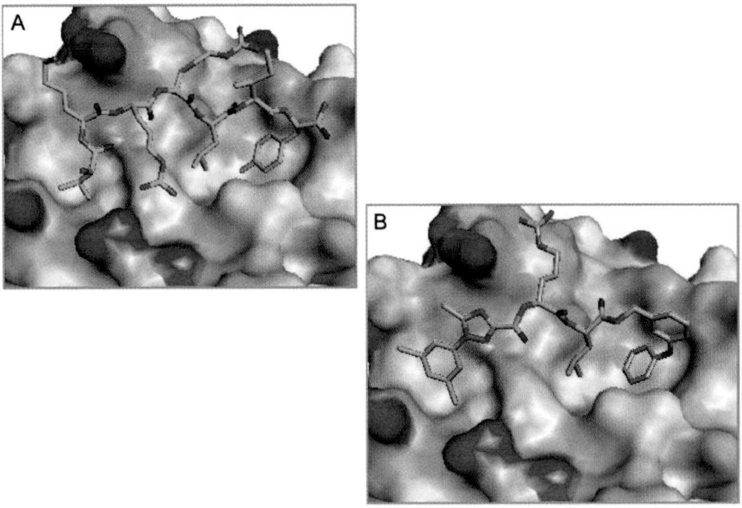

Plate 29.2 REPLACE-mediated conversion of the cyclin groove inhibitor HAKRRLIF (A) into an N- and C-terminally capped dipeptide (B).

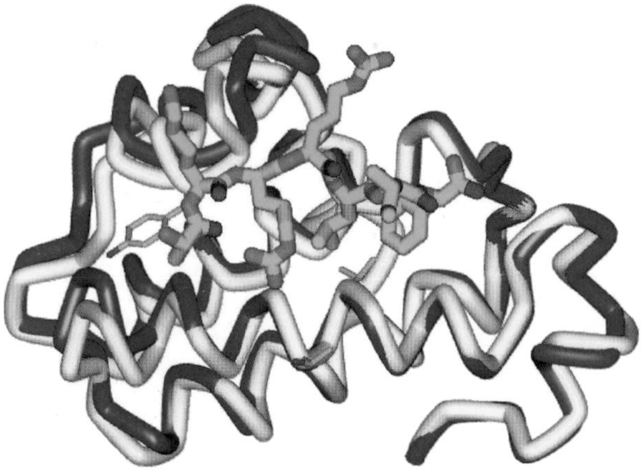

Plate 29.4 The cyclin-binding grooves of cyclin A (blue ribbon) and cyclin D1 (yellow ribbon) are compared through overlay. Unique residues of cyclin D1 shown in orange are Tyr127, Thr62, and Val60 (left to right).

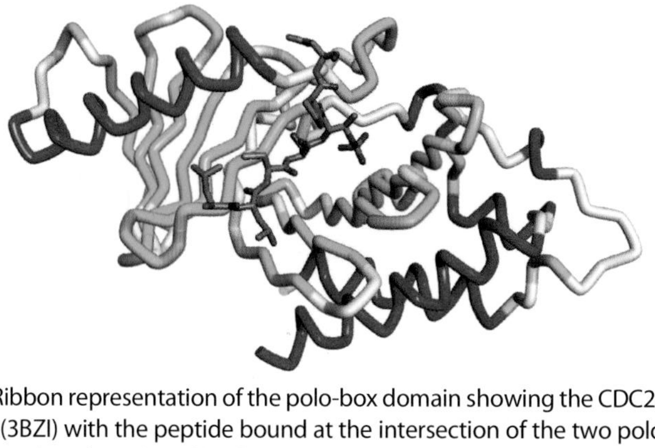

Plate 29.5 Ribbon representation of the polo-box domain showing the CDC25c PBD crystal structure (3BZI) with the peptide bound at the intersection of the two polo boxes. The Leu-Leu-Cys tripeptide replaced by fragment alternatives is on the left side of the figure.